Effects of acid deposition
and tropospheric ozone
on forest ecosystems in Sweden

Ecological Bulletins No. 44

Effects of acid deposition and tropospheric ozone on forest ecosystems in Sweden

Edited by
H. Staaf and G. Tyler

Ecological Bulletins

ECOLOGICAL BULLETINS are published in cooperation with the ecological journals Ecography and Oikos. Ecological Bulletins consists of monographs, reports and symposia proceedings on topics of international interest, published on a non-profit making basis. Orders for volumes should be placed with the publisher. Discounts are available for standing orders.

Editor-in-Chief and Editorial Office:
Pehr H. Enckell
Joint Editorial Office
Ecology Building
S-223 62 Lund
Sweden
Fax: +46-46 222 37 90

Technical Editors:
Linus Svensson and Gunilla Andersson

Editorial Board:
Björn E. Berglund, Lund.
Tom Fenchel, Helsingør.
Erkki Leppäkoski, Turku.
Ulrik Lohm, Linköping.
Nils Malmer (Chairman), Lund.
Hans M. Seip, Oslo.

Published and distributed by:
MUNKSGAARD International Publishers Ltd.
P.O. Box 2148
DK-1016 Copenhagen K, Denmark
Fax: +45-33 12 93 87

and

MUNKSGAARD International Publishers Ltd.
238 Main Street
Cambridge, MA 02142-9740
USA
Fax: +1-617 547 7489

Suggested citation:
Author's name. 1995. Title of paper. – In: Staaf, H. and Tyler, G. (eds), Effects of acid deposition and tropospheric ozone on forest ecosystems in Sweden. – Ecol. Bull. (Copenhagen) 44: 000–000.

Cover photo: Björn Uhr/N ©

Ecological Bulletins still available

Prices excl. VAT and postage.

11. *Ecology in Semi-arid East Africa* (1971). S. Ulfstrand. Price DKK 69.60 (US $ 11.60, £ 7.33, DEM 18.08).
12. *Natural Resources Research in East Africa* (1971). M. Zumer. Price DKK 69.60 (US $ 11.60, £ 7.33, DEM 18.08).
18. *Scandinavian Aerobiology* (1973). Editor S. Nilsson. Price DKK 69.60 (US $ 11.60, £ 7.33, DEM 18.08).
19. *Biocontrol of Rodents* (1975). Editors L. Hansson and B. Nilsson. Price DKK 149.60 (US $ 24.93, £ 15.75, DEM 38.86).
22. *Nitrogen, Phosphorus and Sulphur – Global Cycles.* SCOPE Report 7 (1979, 2nd reprinted edition). Editors B. H. Svensson and R. Söderlund. Price DKK 149.60 (US $ 24.93, £ 15.75, DEM 38.86).
23. *Energetical Significance of the Annelids and Arthropods in a Swedish Grassland Soil* (1977). T. Persson and U. Lohm. Price DKK 138.40 (US $ 23.07, £ 14.57, DEM 35.95).
24. *Peut-on-Arrêter l'Extension des Deserts?* (1976). Rédacteurs A. Rapp, H. N. le Houérou et B. Lundholm. Price DKK 196.00 (US $ 32.67, £ 20.63, DEM 50.91).
25. *Soil Organisms as Components of Ecosystems* (1977). Editors U. Lohm and T. Persson. Price DKK 288.00 (US $ 48.00, £ 30.32, DEM 74.81).
26. *Environmental Role of Nitrogen-fixing Blue-green Algae and Asymbiotic Bacteria* (1978). Editor U. Granhall. Price DKK 218.40 (US $ 36.40, £ 22.99, DEM 56.73).
27. *Chlorinated Phenoxy Acids and Their Dioxins. Mode of Action, Health Risks and Environmental Effects* (1978). Editor C. Ramel. Price DKK 196.00 (US $ 32.67, £ 20.63, DEM 50.91).
31. *Environmental Protection and Biological Forms of Control of Pest organisms* (1980). Editors B. Lundholm and M. Stackerud. Price DKK 172.80 (US $ 28.80, £ 18.19, DEM 44.88).

34. *Fish Gene Pools. Preservation of Genetic Resources in Relation to Wild Fish Stocks* (1981). Editor N. Ryman. Price DKK 127.20 (US $ 21.20, £ 13.39, DEM 33.04).
35. *Environmental Biogeochemistry* (1983). Editor R. Hallberg. Price DKK 344.80 (US $ 57.42, £ 36.29, DEM 89.56).
36. *Ecotoxicology* (1984). Editor L. Rasmussen. DKK 344.80 (US $ 57.47, £ 36.29, DEM 89.56).
37. *Lake Gårdsjön. An acid forest lake and its catchment* (1985). Editors F. Andersson and B. Olsson. Price DKK 425.60 (US $ 70.93, £ 44.80, DEM 110.55).
38. *Research in Arctic life and earth sciences: present knowledge and future perspectives.* Proceedings of a symposium held 4–6 September, 1985 at Abisko, Sweden (1987). Editor M. Sonesson. Price DKK 184.80 (US $ 30.80, £ 19.45, DEM 48.00).
39. *Ecological implications of contemporary agriculture.* Proceedings of a symposium held 7–12 September, 1986 at Wageningen (1988). Editors H. Eijsackers and A. Quespel. Price DKK 288.00 (US $ 48.00, £ 30.32, DEM 74.81).
40. *Ecology of arable land – organisms, carbon and nitrogen cycling* (1990). Editors O. Andrén, T. Lindberg, K. Paustian and T. Rosswall. Price DKK 331.20 (US $ 55.20, £ 34.86, DEM 86.03).
41. *The cultural landscape during 6000 years in southern Sweden – the Ystad Project* (1991). Editor B.E. Berglund. Price DKK 714.00 (US $ 119.07, £ 75.20, DEM 185.56).
42. *Trace gas exchange in a global perspective* (1992). Editors D. S. Ojima and B. H. Svensson. Price DKK 250.00 (US $ 41.67, £ 64.94, DEM 26.32).
43. *Environmental constraints on the structure and productivity of pine forest ecosystems: a comparative analysis* (1994). Editors H. L. Gholtz, S. Linder and R. E. McMurtie. Price DKK 300.00 (US $ 53.00, £ 33.30, DEM 75.00).

Contents

Integrated studies and modelling

Preface

This volume presents results from an integrated research programme "Effects of Air Pollutants and Acidification" funded by the Swedish Environmental Protection Agency (SNV) between 1988/89 and 1992/93. The programme has focussed on the fluxes and environmental effects of acidifying substances and tropospheric ozone in terrestrial, natural ecosystems. During the programme period, 71.5 million SEK were allocated to about 60 projects at different Swedish universities and research institutes.

Most of the contributions in this book are original articles based on results from some of these projects. In addition, there are some papers reviewing earlier and recent research. When considered as a whole, the published papers represent all major areas of research addressed within the programme. A publication list from the programme is given at the end of the volume.

Comprehensive research on the ecological effects of acidifying substances has been carried out in many countries in Europe and North America since the 1970s. In Sweden, the SNV initiated such research a few years after Svante Odén presented his "discovery" of the Acid Rain problem. Since 1977, Swedish acidification research has been organized as integrated programmes, initially concentrated on water and soil acidification but gradually including projects on forest damage. From 1988/89 and onwards the research has been divided into different programmes with participation from several funding agencies. In 1988, the Swedish Environmental Protection Agency started two programmes for studying the effects of acidification and other air pollution problems in natural ecosystems; one on surface waters and one dealing with the terrestrial environment. Later, a third programme dealing with effects on human health was initiated. Other agencies have administered research on the impact of air pollution on crops, technical materials, historical buildings and monuments, below-ground constructions, and archaeological sites.

Much of the research concerning the terrestrial environment has been performed in forest ecosystems, the reason being that almost 60% of Sweden is covered by forests, and that forests play a very important role in Sweden's economy and as a source of recreation. Furthermore, studies of soils and how they react to acid deposition has been a major research theme. Today, most scientists consider soil changes as the major pollution-mediated threat to forest organisms in southern Sweden, while tropospheric ozone is considered to be the most important gaseous pollutant.

We are greatly indebted to the funding committee of the research programme for encouraging us to produce this book. During the programme period the following persons have been members of the committee: Erik Eriksson, Ursula Falkengren-Grerup, Sune Linder, Jan Mulder, Lennart Rasmussen, Lennart Schotte, Carl Olof Tamm and Tore Österaas. Ulla Bertills, Ulf von Brömssen, Lars-Erik Liljelund, Jan Nilsson and Håkan Staaf have represented the Swedish Environmental Protection Agency. We are grateful to all reviewers, who proposed valuable criticism and improvements to the submitted manuscripts. Special thanks to Margaret Jarvis, who corrected the English in most of the articles, and to Linus Svensson, technical editor of Ecological Bulletins. Finally, we would like to extend our thanks to all authors, other colleagues, and technical personnel who made it possible to finalize this volume.

Solna and Lund, January 1995

Håkan Staaf   Germund Tyler

Ecological Bulletins 44: 11–16. Copenhagen 1995

Acidification research in Sweden – national and international perspectives

Carl Olof Tamm

Tamm, C.O. 1995. Acidification research in Sweden – national and international perspectives. – Ecol. Bull. (Copenhagen) 44: 11–16.

Acidification research since 1968, especially in terrestrial ecosystems in Sweden, is briefly described, as reflected in the Swedish contributions to five important meetings and the international discussion during them. These meetings were held in Stockholm, June 1972, Columbus, Ohio, May 1975, Telemark, Norway, June 1976, Sandefjord, Norway, March 1980 and Muskoka, Ontario, Canada, September 1985. For the period after 1985 some important research themes are highlighted: paleoecology, soil resistance to acidification, processes at the soil/surface water interface, and last but not least, deposition of nitrogen compounds both as contributing to acidification and as causing other problems, described as nitrogen saturation. It is concluded that the scientific community has become much more alert to alarm signals concerning environmental risks during the last three decades, and that this has also been reflected to a certain extent in political decisions. Concerning the acidic precipitation and its environmental consequences there have been changes in scientific focussing, in research funding, and in general attitude in many countries during this period. Still, it is believed that the Swedish part of the story merits a special interest, both because of stimulating hypotheses put forward at an early stage, and because these hypotheses, as well as later ones, have been followed up in a fairly consistent way.

C.O. Tamm, Dept of Ecology and Environmental Sciences, Swedish Univ. of Agricultural Sciences, P.O. Box 7072, S-750 07 Uppsala, Sweden.

Introduction

There is a long tradition in the Nordic countries of research on the functioning of forests and other ecosystems under acid conditions. More than 70 years ago, C. Olsen in Denmark made an extensive study of the distribution of wild plants in relation to soil acidity, using the recently introduced pH scale. At the same time, O. Tamm in Sweden studied the process of podzolisation and found that there might be considerable acidification of new soils in a 100-yr-period, although it took much longer for a mature podzol profile to develop. In all Scandinavian countries, attempts were made to find relationships between forest types and forest production on the one hand and site factors on the other, work begun by Cajander in Finland. In peatlands, a postglacial succession was described from shallow lakes or circumneutral fens via nutrient-poor fen types to acid raised bogs. However, limnologists had found that lakes, even in regions with predominantly acid soils under the coniferous canopy, often had some alkalinity, as long as most of the water flowing into the lake originated from groundwater. Exceptions were many "dystrophic" lakes, with their water coloured brown by acid humic substances.

Internationally, a debate was in full progress during much of the first half of the 20th century on so-called soil-forming factors: climate, topography, parent material, organisms, and time. Most soil scientists believed that climate was the decisive factor for the development of geological material into different soil types. However, the parent material (texture and mineralogical type), together with topography, might modify the climatic influence, resulting in "azonal" soils. It was recognized that vegetational changes, often caused by human land use, could change the productivity of the land, usually for the worse, but major changes in soil profile development were considered unlikely or at least very slow. There were few critics of this view, although already in the 19th century P. E. Müller had observed mull formation beneath scattered oak shrubs on an otherwise podzolised Danish heathland with a humus layer of the mor type. In Germany, a debate begun in the 1930's as to whether or not soil degradation was induced by planting Norway spruce on former hardwood land, but a majority of soil scientists did not accept the idea of rapid changes.

The idea that soil profiles had developed slowly over centuries or even millennia was of course correct on the whole, but it does not necessarily lead to the conclusion that soil profiles are stable by nature. However, this belief came to be a commonly accepted "paradigm". It might be of some value to remember this as a background to the last decades' debate of the acid deposition effects and related problems in Sweden, as well as in other countries.

Early research in Sweden

The history of acid deposition research has been well described by Cowling (1982) and Gorham (1989). Some of the landmarks have clear links to the Swedish scientific community and also to the way in which Swedish politicians responded to signals from scientists and acted both nationally and internationally. This is one of the reasons for the publication of this book, which is an attempt to present in a condensed form the Swedish research on effects of acid precipitation on the terrestrial environment.

The first of these "landmarks" was constituted by Odén's alarming newspaper article in 1967 and his "Bulletin" (Odén 1968). The "Bulletin" was followed by Sweden's "case study" to the United Nations' Conference on the Human Environment in Stockholm 1972: "Air pollution across national boundaries. The impact on the environment of sulfur in air and precipitation" (Anonymous 1971). However, the scientific argumentation by Odén and in the Swedish case study was not unanimously accepted. Even in Sweden and Norway, the two countries where fishery decline linked to fresh water acidification had first been observed, many scientists were critical of Odén's argument that this phenomenon was caused by the increasing acidity of the precipitation. One reason for scepticism was no doubt that measured data from the atmospheric chemistry network only covered the period from 1950 onwards. The attitude among Swedish politicians and the general public was more positive. The use of high sulphur fuel oil (>2.5% S) was banned in the late 1960's, but the decision was primarily based on concern for health problems in cities. Further restrictions followed later, in which the regional acidification problems played a dominant role.

The value of a hypothesis depends not only on its correctness. The testability is equally important. Odéns comprehensive "Bulletin" from 1968 contained a number of testable hypotheses, and they were tested, mainly against already existing data, both in the case study and in

a publication by Malmer (1974). Malmer added new information on surface water acidification by comparing extensive data sets for lake water chemistry from different occasions during the period 1930–1970 in central south Sweden. More extensive testing of Odén's hypotheses in specially designed experiments and other investigations required active cooperation from funding agencies. In Sweden, only small-scale pilot tests could be started during the early 1970's, mainly within the framework of existing activities. A comprehensive project, the "Effects of acid rain on fish and forest" (SNSF) was initiated in Norway as early as 1972 and extensive experiments and other field work were set up. The Swedish Environmental Protection Board organized lake inventories and other studies, partly by their own staff. The agency soon started funding individual projects dealing with acidification effects, from 1977 onwards within integrated research programmes. The establishment of a number of research groups interested in acid deposition effects in both Sweden and Norway in the mid 1970's formed a platform for good scientific cooperation. At this time it was not so much a question of joint projects – the financing of such was far too uncertain, at least in Sweden – but there were a number of informal meetings where experiences were exchanged, and experts from one country were often asked for advice from the other.

"Acid rain" becomes an international issue
The first international meeting in Columbus, Ohio

The status of the acid deposition research, both internationally and in Scandinavia, can be discussed in relation to some important international meetings on the subject. The main emphasis here, as in the rest of this book, will be on effects in terrestrial ecosystems, with only brief reference to problems with air pollutant transport and with aquatic systems. The first of these meetings, the "First International Symposium on Acid Precipitation and the Forest Ecosystem" (held in Columbus, Ohio, in 1975), marked the international recognition of "acid rain" as a possibly serious threat, at least in some regions, and therefore worthy of more study than had so far been devoted to local emission effects. There were important contributions to this symposium from Sweden, Norway, Canada, United States, and others (see Dochinger and Seliga 1976). Some of the overview papers came from Sweden, the most important one a key-note speech by Odén. Other Swedish contributions dealt with methods to measure strong acids in precipitation and in throughfall beneath conifers, and with the effect of experimental acidification on soil nitrogen turnover. On the aquatic side, there were Swedish papers on acidification-induced vegetational changes and on the interaction between acidification and thermal stratification in lakes. Norway contributed with SNSF results and with reports from the first extensive international study of long-range transport of air pollutants, LRTAP, organized by OECD 1972–1975, but centred in Oslo. Contributions from countries other than Scandinavia, Canada and parts of the United States mostly concerned effects in strongly polluted areas, but a paper by Ulrich and Mayer stated that the forest ecosystems in Solling, Germany, deviated markedly from balance between incoming and outgoing protons. Incoming protons were to a large extent exchanged for calcium, magnesium and aluminium ions, and it was said that this in the future might lead to nutritional disturbances.

Meetings in Telemark and Sandefjord, Norway

The next meeting to be discussed here was held in Telemark, Norway, one year later (Anonymous 1976), on the invitation of the Norwegian Government, who wanted an international discussion of the "mid-term" results from the SNSF project together with those from the LRTAP study. Both the Norwegian and the Swedish Governments were interested in international conventions on "transboundary pollution", but before negotiations in that direction could take place, some sort of scientific concensus was required. Naturally the SNSF material dominated the scientific presentations. There were also some overview papers from Sweden, as well as more specific reports on other environmental problems, related to long-range air pollutants, such as lead enrichment in marine bivalves and PCB (polychlorinated biphenyls) in Baltic seals. The question of climate change caused by anthropogenic emissions also came up during the meeting, which ended with a number of conclusions and recommendations (Anonymous 1976).

In 1980, a large international meeting was held at Sandefjord, Norway (Drabløs and Tollan 1980). The meeting was organized by the SNSF project, which published its final report in the same year. The number of participants from countries in central and western Europe had increased in comparison with earlier meetings, though the interest still appeared to be strongest from Scandinavia, Canada, and eastern United States. Many important contributions came of course from Norway and the SNSF project, but it is interesting to note that several soil scientists earlier uninterested in effects of acid deposition had now changed their attitude. The conference summary, written by F. T. Last, UK, G. E. Likens, USA, B. Ulrich, Germany, and L. Walløe, Norway, accepted the view that acid precipitation was the main cause of observed ecological effects in aquatic systems, but pointed out a number of quantification problems concerning dry deposition, several soil processes including weathering, land-use changes, responses of organisms to aluminium, and effects on forest growth. Swedish contributions to the conference dealt with, among other things, the new problem of ground water acidification, the possibility of using the soil samples collected by the Forest

Survey for acidification studies, and effects of acid application on forest soils and their organisms, including the trees.

The Muskoka Conference (Ontario, Canada)

The next big international meeting on effects of acid precipitation was held at Muskoka, in 1985 (Martin 1987). The size of that meeting reflected the strong change in attitude (and funding) which had started about the time of the Sandefjord meeting five years earlier. Much more evidence of regional acidification of freshwater systems had become available from both Europe and North America. After some dry summers in the early 1980's extensive forest damage was observed in Germany, which greatly increased the awareness of air pollution problems in general but more particularly of the soil changes earlier described by Ulrich. The discoloration of needles, followed by premature needle fall and later death of branches and entire trees was often called "neuartigen Waldschäden" or in English "novel forest damage". Several German scientists ascribed the damage to other causes than soil acidification, such as specific nutrient disturbances, ozone, pathogens, etc. Much of this discussion also came up at Muskoka, without reaching consensus. On the aquatic side, an increasing interest in paleoecology was demonstrated, with studies of diatoms and other microfossils in lake sediments made in order to find out about acidity conditions in the past.

There were many Swedish contributions to the Muskoka Conference, reflecting the increasing research in the field, described in the rest of this volume. One of the Swedish papers, by Grennfelt and Hultberg, dealt in a comprehensive way with the increasing proportion of nitrogen compounds in acidifying deposition and warned about "nitrogen saturation". Other Swedish contributions dealt with results from the integrated project "Lake Gårdsjön. An acid forest lake and its catchment" (see also Andersson and Olsson 1985). Effects of liming fresh water systems on both water chemistry and aquatic organisms were discussed. Comparisons between recent and historical measurements of soil acidity had shown that considerable soil acidification had taken place during 50 years in SW Sweden (Tamm and Hallbäcken, in Martin 1986). This investigation had been performed in an area with relatively few local emissions but exposed to air masses transported with the predominantly south-westerly winds from industrialized regions in Europe. The study, which was soon confirmed and extended by others, also offered the possibility of separating, at least to some extent, the influence of deposition from that of the vegetation (e.g. stand age and tree species).

Some milestones 1972–1985

Employing the simplification of using international meetings as "milestones" in acid deposition research, we may summarize what has been said in the following way:

Stockholm 1972: Sweden's case study – a challenge to the scientific community.

Columbus 1975: Acid deposition recognized as a problem for certain sensitive regions, mainly in aquatic systems.

Telemark 1976: Long-range transport of air pollutants confirmed, establishing a basis for political discussion. Existence of sensitive freshwater systems confirmed, but serious effects on terrestrial ecosystems not proven.

Sandefjord 1980: The extention of regions sensitive to acid deposition greatly enlarged. Research beginning on biological mechanisms in acidified lakes and rivers. Soil changes observed, but effects on tree growth absent or in an unexpected direction.

Muskoka 1985: The severity of the acidification problem recognized in most industrialized countries. Regional soil acidification well established, even if its connection to observed "novel forest decline" remained far from clear. Liming as a possible countermeasure was discussed. A growing insight into the role of nitrogen.

Some trends during the last decade

The series of meetings begun at Columbus, Ohio, in 1975 has been continued with a meeting in Glasgow, Scotland, in 1990 (Last and Watling 1991). A further meeting is planned for 1995 in Gothenburg, Sweden. However, there have been so many activities dealing with acid deposition and related matters since 1985 that it is difficult to use any particular meeting as a "milestone". A number of international research programmes have been initiated, several of them with Swedish participation. One of these activities, the "Surface Waters Acidification Programme" (SWAP) 1984–1989, was sponsored by the British energy industry and managed by a group selected by the Royal Society of London together with the Norwegian and Swedish Academies of Sciences. The primary goal was to clarify and quantify the various causes, including long-distance-transport of air pollutants, of fresh water acidification with associated biological effects. Although SWAP was focussed on freshwater acidification, research funded in this way stimulated research on resistance to acidification and processes at land/water interfaces generally (see Mason 1990).

With the rising flood of information it is becoming increasingly more difficult to judge which results are most important. Nor should it be necessary to do so in the introduction to a book reporting Swedish acidification research during the last decade. That task might be left to the reader's own judgment and to the future. However, it may be permissible to highlight some of the mainstreams in research on effects of acidic deposition, where we believe that Swedish activities have been important, even in an international perspective. Detailed references are not given, as they will be available elsewhere in the book.

The first topic concerns paleoecology, where the acidification history of a number of lakes in different parts of Sweden has been elucidated by Renberg and coworkers. There is a general pattern of slow and more or less continuous acidification since the deglaciation. Particularly in southwest Sweden, this slow process has changed to a rapid loss of alkalinity, accompanied by low pH-values. The rapid acidification phase coincides with increasing occurrence in the sediments of soot particles from combustion of fossil fuels. Recently Renberg et al. (1993) have found evidence for the occurrence of alkalization periods beginning c. 2000 years ago and coinciding with signs of deforestation, e.g. charcoal fragments and pollen from grass or weed species. However, during earlier periods of higher acidity in lakes the pH values stayed well above 5, even when the landscape was dominated by forest, unlike in many recently acidified waters.

Another important theme has been the acidification sensitivity of soils and eventually of groundwater and surface waters. It soon became clear that silicate weathering was one of the main counteracting forces to acid deposition in humid climates, and particularly so in regions with relatively coarse-textured soils, formed from rocks dominated by acid silicates, such as in most of Sweden. In all soils, there is a store of exchangeable "base cations", which is depleted first, when protons associated with anions of strong mineral acids (sulphate, nitrate and chloride ions) are added. The long-term resilience of the system depends on the rate of release of new base cations: in a glacial till or outwash sand typical for much of Sweden this is mainly by weathering of primary minerals. Much effort has been devoted to determinations of weathering rates, and of the roles of climate, mineralogy, texture, pH, and presence or absence of organic chelating substances. Swedish research has resulted, for example, in simulation models for weathering, maps of soil sensitivity to acidification, and estimates of critical loads for acid deposition, which if exceeded over long periods will cause permanent damage to the environment.

There is also increasing interest in processes at the soil/water interfaces. Groundwater acidification has already been mentioned, as well as the observation that pristine clear water lakes may have alkalinity, even if they are more or less surrounded by acid peat deposits. The source of alkalinity might then be groundwater inflow at the lake bottom. However, processes active at the soil/water interface are complex, including purely chemical reactions and processes mediated by organisms. Examples are precipitation of iron hydroxides "lake ore", denitrification and other nitrogen transformations, and release of free aluminium ions from complexes with chelating organic substances.

The last important theme to be mentioned here has concerned the role of nitrogen compounds for acidification but also the new problem of "nitrogen saturation". Research on nitrogen nutrition and nitrogen cycling is an old tradition in Swedish forest ecology: early in this century nitrogen supply was recognized as a main limiting factor for forest growth. In 1968, Odén looked for evidence that forest growth had increased as a result of the increase in nitrogen deposition between 1950 and 1965. However, the growth data available were not accurate enough at that time.

The results of early fertilization experiments led to a widespread use of nitrogen fertilizers on forest land from the 1960's onwards in order to increase forest yield.

Parallel with this practical fertilization, forest research organisations laid out various types of experiments, some of which were designed to find "optimum nitrogen levels" (which of course also required supraoptimum treatments to fulfil this aim). Around 1970, it was already clear from the first of these experiments that forest plots could be nitrogen saturated in the sense that the trees were growing less than on control plots, and that much of the added nitrogen was lost from the ecosystem. However, the heavily fertilized spruce trees showed a considerable resilience, in the sense that symptoms of reduced vitality were rather unspecific (mainly unusually large litterfall). The design of these early experiments was not suitable for long-term follow-up, but new and better experiments were laid out in 1967–1974.

The discussion of possible negative effects of further application of nitrogen to forest soils led to a recommendation issued by the Swedish Board of Forestry and the Environmental Protection Board in 1984, that forest owners in south Sweden should refrain from the use of nitrogeneous fertilizers. This recommendation is still in force, and was extended in 1991 to warn against high accumulated additions of nitrogen in middle and northern Sweden, the limit set to 300 and 600 kg nitrogen ha^{-1}, respectively. The recommendation has been well accepted by those responsible for forest fertilization in Sweden (mainly forest companies and the managers of the state forests), though it was based on informed judgment rather than on hard facts. However, the need for more facts stimulated further research on various aspects of the nitrogen cycle in Sweden, e.g. various possibilities of decreasing nitrate leaching from both forests and agricultural land.

Internationally, interest in adverse environmental effects of high nitrogen supply by fertilizers or deposition also increased. In Germany, Ulrich (1984) stressed the decoupling in space and time of processes in the nitrogen cycle, especially nitrification and uptake by roots, as a cause of "acid surges" in the soil, causing root damage and possibly the forest damage observed after dry summers. The role of nitrogen deposition caused even more concern in the Netherlands, where the ammonia emission from agricultural activities had become exceptionally high and started to cause obvious changes in the environment. Later, other countries with intensive animal production have encountered similar problems, and a number of international conferences have been devoted to problems related to nitrogen saturation.

It may be taken as an indication of a change in the intellectual climate that the issue of nitrogen saturation became internationally recognized as an important problem in less than five years. It should be remembered that it took more than three times as long to get the international scientific community to accept acid deposition effects as an ecological problem affecting more of the natural environment than some salmon rivers in Norway and lake fishery in parts of Sweden.

Concluding remarks

Scarcity of natural resources has always occurred and has often led to social and political problems. Environmental degradation has also been observed for a long time, as a consequence of deforestation or mining, but until the middle of this century both decision-makers and most scientists looked upon environmental problems as phenomena of a local character or at most affecting limited regions. Even the serious health consequences of the London smog in the 1950's could be overcome by changes in local heating systems – and taller stacks. Nature was believed to have self-healing properties and the supply of non-renewable resources was to a large extent considered as an economic problem. Scarcity increased price and made it possible to use low-grade raw material or expensive substitutes.

A forceful attack on this attitude was delivered in 1962 by Rachel Carson in her famous book "Silent Spring", but Svante Odén's "Bulletin" of 1968 strongly contributed to the new thinking, that many of the problems created by modern technology affect large regions or even the global environment and have to be dealt with in a very positive way. Wherever possible, pollution problems should be stopped at the source. As the Swedish contributions to this "shift of paradigm" have been far from unimportant, it is hoped that this book, describing the research on acidification effects in our country, but in continuous contact with the larger international scientific community, can be read as a new "Case Study" of scientific progress in an important sector of environmental research.

References

Andersson, F. and Olsson, B. (eds) 1985. Lake Gårdsjön. An acid forest lake and its catchment. – Ecol. Bull. (Stockholm) 37.
Anonymous 1971. Air pollution across national boundaries. The impact on the environment of sulfur in air and precipitation. Sweden's case study for the United Nations conference on the human environment. – Roy. Min. Foreign Affairs and Roy. Min. Agricult. Stockholm.

Anonymous 1976. Report from the International Conference on the Effects of Acid Precipitation in Telemark, Norway, June 14–19, 1976. – Ambio 5: 200–263.

Carson, R. 1962. Silent spring. – Boston.

Cowling, E. B. 1982. A historical resumé of progress in scientific and public understanding of acid precipitation and its consequences. – Environ. Sci. Technol. 16: 110A–123A.

Dochinger, L. S. and Seliga T. A. (eds) 1976. Proceedings of the First International Symposium on Acid Precipitation and the Forest Ecosystem. – USDA For. Serv. Gen. Tech. Rep. NE-23.

Drabløs, D. and Tollan, A. (eds) 1980. Ecological impact of acid precipitation. – Oslo-Ås.

Gorham, E. 1989. Scientific understanding of ecosystem acidification: A historical review. – Ambio 18: 150–154.

Last, F. T. and Watling, R. 1991. Acidic Deposition. Its nature and impacts. – Proc. Roy. Soc. Edinburgh 97B.

Malmer N. 1974. On the effects on water, soil and vegetation of an increasing atmospheric supply of sulphur. – Swed. Environmental Protection Board SNV PM 402 E, Solna.

Martin, H. C. 1986. Acidic Precipitation. – Proc. Int. Symp. on Acidic Precipitation, Muskoka, Ontario, September 15–20, 1985. Reprinted from Water Air and Soil Pollution, Vol. 30 and Vol. 31.

Mason, B. J. 1990. Surface waters acidification programme. – Cambridge Univ. Press, Cambridge.

Odén, S. 1968. The acidification of air and precipitation and its consequences for the natural environment. – Bull. Ecol. Comm. Stockholm No. 1: 1–86, (in Swedish, English translation, Arlington Va).

Renberg, I., Korsman, T. and Anderson N. J. 1993. A temporal perspective of lake acidification in Sweden. – Ambio 22: 264–271.

Ulrich, B. 1984. Ion cycle and forest ecosystem stability. – In: Ågren. G. I. (ed.), State and change of forest ecosystems – indicators in current research. Swed. Univ. Agric. Sci. Dept Ecology and Environm. Res. Rep. No. 13: 207–233, Uppsala.

Ecological Bulletins 44: 17–34. Copenhagen 1995

Deposition of acidifying substances in Sweden

Gun Lövblad, Karin Kindbom, Peringe Grennfelt, Hans Hultberg and Olle Westling

Lövblad, G., Kindbom, K., Grennfelt, P., Hultberg, H. and Westling, O. 1995. Deposition of acidifying substances in Sweden. – Ecol. Bull. (Copenhagen) 44: 17–34.

The present deposition of sulphur and nitrogen in Sweden is described from the aspect of spatial and temporal variations, based mainly on data from national, regional and local environmental monitoring programmes but also on model calculations. Monitoring data are available to provide nationwide patterns for wet deposition and air quality of sulphur and nitrogen compounds, and also to provide data on deposition on an ecosystem scale as a base for effect assessments. Monitoring data are lacking for dry deposition, except for net throughfall of sulphur. Development of a national model, as a complement to the environmental monitoring programme, will need validation data for dry deposition, especially for nitrogen. Large variations in deposition are apparent, attributable to the gradient from south and southwest to the north, with the largest deposition in the south and on the Swedish west coast. The gradient is caused by the influence of European emissions, in combination with the differences caused by local variations in land use and local pollution. The long-term variations observed in pollution loads are mainly caused by emission changes but to some extent also by weather variations. The sulphur deposition has clearly decreased in the last decade, in terms of wet deposition, throughfall deposition and the occurrence of gaseous as well as particulate sulphur compounds in the atmosphere, with considerable decreases mainly in the last six years. There are up to 40–50% lower amounts than in 1980 in large parts of the country, because of decreases in European sulphur emissions. The decrease indicated by the model agrees with the measured decrease. For nitrogen deposition no trends are apparent. The emissions of nitrogen oxides over Europe are also fairly constant. Comparisons are made between models and measurement data. A good agreement is seen from many aspects, such as the large scale gradients over Sweden and the deposition trends. The differences between models and measurements in their ability to describe the spatial variations in particular, are also indicated. The gap in our knowledge about nitrogen deposition is pointed out. Prognoses for the future sulphur deposition are made based on the new sulphur protocol. Further decreases in sulphur deposition are expected. However, in the most sensitive ecosystems in many parts of Sweden the reductions planned up to 2010 will not be sufficient to get down to or below the critical loads.

G. Lövblad, K. Kindbom, P. Grennfelt, H. Hultberg and O. Westling, IVL, P.O. Box 47086, S-402 58 Göteborg, Sweden.

Introduction

Deposition of acidifying air pollutants to sensitive ecosystems, will eventually lead to the risk of acidification of the ecosystem. Acidification is mainly caused by deposition of sulphur but deposition of nitrogen compounds, oxidized as well as reduced, may contribute. The first effect of elevated nitrogen loads is a fertilization of the ecosystems and increased productivity. However, in the long-term, higher deposition will lead to more profound ecosystem changes and nitrogen will start to leach out of the system. In areas with a high nitrogen deposition, deposition of ammonium may lead to nitrification and thus to further acidification. Alkaline particles, e.g. oxides of calcium, magnesium and potassium, neutralize acid deposition. The impact of alkaline compounds on the ecosystems is, however, small in most areas in Sweden, except in the vicinity of emission sources such as forest industries, steel works, etc. (Lövblad 1987).

To understand the effects on the ecosystem due to deposition of sulphur and nitrogen, it is essential to have access to reliable data on deposition, on a relevant scale. To develop relevant abatement strategies, it is also important to link the pollutants deposited with their source. For this, modelling, as well as monitoring, is a necessary tool.

The aim of this paper is to present the sulphur and

nitrogen deposition to the ecosystems in Sweden, using available monitoring and modelling data. The information may serve as basic information for the overall risk assessment for effects on Swedish ecosystems caused by the deposition of sulphur and nitrogen compounds.

Deposition processes and methods of measurement

Deposition of airborne pollutants to the ecosystems take place via three processes: wet deposition (deposition of pollutants with precipitation, rain, snow or hail), dry deposition (deposition or absorption of gases and/or particles directly from the atmosphere to the receptor) and fog and cloud water deposition. Wet deposition is relatively easy to quantify using measurements, while dry deposition and fog and cloud water deposition are more complicated to monitor and to model. Nevertheless, important contributions are made by all these processes and must be considered in all areas.

The present state of knowledge, including how the different processes contribute to the deposition over large areas and how deposition can be estimated by monitoring and modelling, is summarized in a number of articles. In some recent theses from the Netherlands, processes as well as methods to measure and estimate deposition are summarized (Ivens 1990, Erisman 1992, Draaijers 1993). There are further reviews, e.g. by Hicks et al. (1989), Murphy and Sigmon (1990) and Davidson and Wu (1990). In national review reports (the United Kingdom Review Group on Acid Rain 1990, Heij and Schneider 1991, Kenttämies 1991, Ulrich and Wiliot 1993, NADP/ NTN 1994), deposition processes as well as their specific importance in relation to the total deposition in European and North American regions are described. European deposition and air quality data are compiled in EMEP reports (Davies et al. 1992, Sandnes 1993, Schaugh et al. 1994). In addition, the present knowledge on deposition processes was summarized at a workshop in Göteborg 1992 by a large number of European and North American scientists (Lövblad et al. 1993). The aim of the workshop was to state the knowledge of deposition process mechanisms with the objective of evaluating the possibilities of improving model simulation of deposition and monitoring procedures and, in turn of improving the estimates of deposition to ecosystems with respect to critical loads exceedances. The state of knowledge was described in a number of background documents and discussed and summarized by the different working group reports.

Wet deposition

The wet deposition processes include incorporation of gases and particles into water drops and snowflakes in the cloud droplets (rain-out), and into the falling hydro-mete-

ors (wash-out) in the lower atmosphere. Wet deposition processes have been extensively studied. The working group report (Fowler et al. 1993a), as well as the background document on wet deposition (Fowler et al. 1993b) from the deposition workshop in Göteborg states that the wet deposition can be estimated well by monitoring and modelling over large areas and long integrating times. However, there are areas with deviations between models and monitoring results; for example coastal areas in north western Europe, where models seem to underpredict the enhanced wet deposition, and high altitude areas, where the wet deposition collectors seem to undercapture precipitation, mainly snow.

Differences in wet deposition depend on variations in precipitation amount and concentrations of pollutants in precipitation. Local variations in wet deposition in Sweden is mainly due to differences in precipitation amount (Granat 1991). In areas where there are large variations in local precipitation amount, such as the area from the Swedish west coast with an increasing altitude towards inland, considerable wet deposition gradients are seen. Within a distance of < 50 km, the yearly wet deposition is more than doubled.

Measurements of wet deposition should be made using wet-only collectors, i.e. collectors which are open only during precipitation periods. In areas where the contribution from dry deposition of particles and gases into the collectors is small, bulk collectors open all the time may be used. However, to find out the magnitude of the dry deposition contributions, parallel sampling is necessary during some period.

Dry deposition

The dry deposition includes deposition to all parts of the earth surface, such as different types of forests and low vegetation, and to lake surfaces and to snow. The dry deposition is by nature more complicated and may be influenced by many more factors than the wet deposition. Atmospheric turbulence, which is one of the main factors controlling the mass flow of pollutants from the atmosphere to the surface of receptors, is greatly influenced by the physical structure of the receptor. Physical-chemical characteristics of the pollutant and of the receptor surface increase or decrease the total resistance to mass flow from the atmosphere to the ecosystems. Dry deposition is less studied than wet deposition, to a large extent because of the lack of methods for routine monitoring. The present state of knowledge of dry deposition of sulphur dioxide, nitrogen oxides, ammonia, ammonium and particles, is summarized in background documents and working group reports from the workshop in Göteborg. (Erisman and Baldocchi 1993, Duyzer and Fowler 1993, Sutton et al. 1993, Ruijgrok et al. 1993, Erisman et al. 1993b, Braun et al. 1993, Andersen et al. 1993, Hummelshøj et al. 1993). Among the most serious lack of knowledge is the effect of wet surfaces and the understanding of parti-

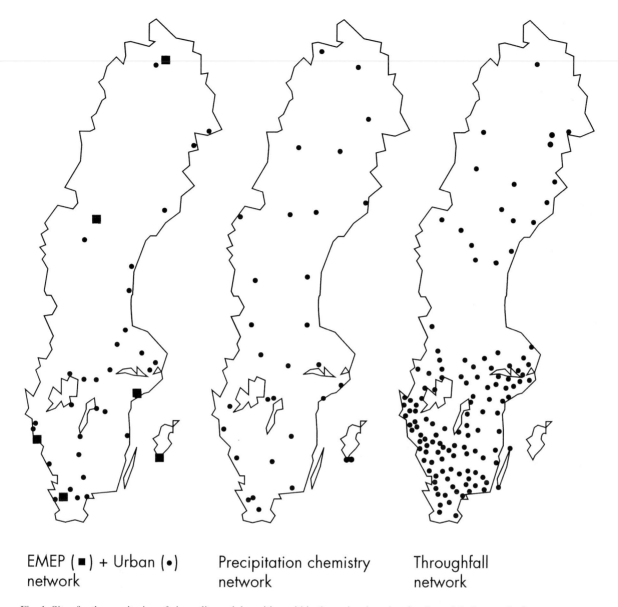

EMEP (■) + Urban (•)
network

Precipitation chemistry
network

Throughfall
network

Fig. 1. Sites for the monitoring of air quality and deposition within the national, regional and municipal networks for measurements in rural air (see also Table 1).

cle (including fog droplets) deposition. The methods available today for monitoring dry deposition, are very intensive and are restricted to limited areas. Measurements of dry deposition have mainly been made within research projects. However, measurement results from one site cannot easily be extrapolated to larger areas. The models used today can only to some extent explain monitoring results. Large deviations between modelled and measured dry deposition are sometimes found (Lövblad and Erisman 1992, Erisman et al. 1993a).

One way of estimating total deposition of some components to forests and to low vegetation is to measure throughfall, i. e. the precipitation collected under the

canopies (see e.g. Ivens 1990, Schaefer and Reiners 1990, Draaijers 1993). A conclusion from the deposition workshop (Erisman et al. 1993b, Beier et al. 1993) was that both field and experimental studies have shown evidence that throughfall provide estimates of the total deposition of sulphate, sodium and possibly also chloride. These are limited to ions which do not take part in the uptake and leaching processes in the canopy, unless the uptake and leaching can be quantified. Sulphur in the form of sulphate is not taken up or leached from the canopy in amounts which are of importance in relation to the amounts deposited in south to middle Sweden, at least not as a net process over a period of time (Erisman et al.

Table 1. Monitoring data available in Swedish background areas.

Network	Number of stations	Sampling frequency	Collector	Parameters
Precipitation				
EMEP	4	Daily	wet-only	SO_4^{2-}, NO_3^-, Cl, NH_4^+, Na, K, Mg, Ca, pH, $\varkappa$
Precipitation chemistry	35	Monthly	bulk	SO_4^{2-}, NO_3^-, Cl, NH_4^+, Na, K, Mg, Ca, pH, $\varkappa$
Regional throughfall	125*	Monthly	bulk	SO_4^{2-}, NO_3^-, Cl, NH_4^+, pH, $\varkappa$ (Na, K, Mg, Ca, Mn)**
Air quality				
EMEP	6	Daily		SO_2, NO_2, SO_4^{2-}
EMEP	4	Daily		total NO_3, total NH_4
EMEP	5	Hourly		Ozone
Urban***	20–40	Monthly	diffusion samplers	SO_2, NO_2

*open field + throughfall, **analyzed for some stations, ***measurements are made at 2 sites, c. 20 km outside each urban area.

1993a). If the wet deposition is subtracted from the measured throughfall flux, the remaining "net" throughfall can be used to estimate the amount of pollutants deposited during dry periods to the canopy. Net throughfall includes not only the dry part of the deposition, but also the possible contribution from fog and cloud water deposition.

Another way of estimating the total deposition of sulphur is to measure the sulphur run-off from a defined catchment, provided that sulphur is not being retained in any way in the soil. In defined forested catchments, throughfall and run-off measurements have been found to give comparable results (Hultberg 1985, Likens et al. 1990, Hultberg and Grennfelt 1992).

Fog and cloud water deposition

Fog and cloud water deposition is generally studied to a much lower degree than the other deposition processes. Routine fog/cloud water monitoring is carried out in only a few areas in Europe. The studies have indicated that the physical processes of the cloud droplet deposition can be relatively well described in a simplified way (Fowler et al. 1993a). However, the concentrations of ions in fog/cloud water are high, often much higher than in precipitation. These chemical properties cannot easily be ex-

plained. Data are still insufficient for estimating the contribution of these processes to the total deposition and for including simulation of fog and cloud water deposition in model estimates over larger areas (Fowler et al. 1993a).

Deposition data in Sweden
Available monitoring data

Monitoring networks providing nationwide data on air quality and wet deposition in Sweden are the EMEP network and the national precipitation chemistry network (see Fig. 1 and Table 1). The EMEP programme (UN ECE Environmental Monitoring and Evaluation Programme) includes measurements of air quality and precipitation in background air at c. 100 stations over Europe. Six of them are Swedish. More details can be found in Schaug et al. (1994), Sandnes (1993) and Kindbom et al. (1993b). The national precipitation network is run at 35 stations in Sweden. Further details are presented by Persson et al. (1993). In addition, throughfall measurements are made by so many regional authorities, that there is a nationwide network, using the same equipment and procedure (Hallgren Larsson and Westling 1994 a,b,c). See Fig. 1 and Table 1.

Wet deposition data

Most of the precipitation sampling in Sweden is made using bulk collectors, except for the EMEP-network using wet-only collectors. Within the national precipitation chemistry network, some sites are run continuously with parallel bulk and wet-only sampling. The results indicate small differences for sulphate and nitrate, and somewhat larger deviations for ammonia (Table 2a). The differences in Table 2a represent possible contribution by dry deposition and influence on deposition processes by the sam-

Table 2a. Mean ratio between bulk and wet-only precipitation sampled in parallel during five years at three stations. Bulk sampling is made within the precipitation chemistry network (monthly means) and wet-only sampling is made within the EMEP network (daily sampling).

Station	SO_4^{2-}-S	NO_3^--N	NH_4^+-N
Vavihill	1.05	1.04	0.86
Aspvreten	1.05	1.00	1.14
Bredkälen	0.91	0.95	1.02

Table 2b. Results from parallel bulk and wet-only sampling within the precipitation chemistry network during several years based on monthly sampling (Granat 1989). Results are presented for different regions and given in µeq l^{-1} as the difference between the sampled concentrations, mean, and difference in percent of the mean concentration.

Area of Sweden		SO_4^{2-}-S	NO_3^--N	NH_4^+-N
Southern and	Difference	2.5	2.5	2.9
southwestern	Mean	80	45	52
Sweden	Relative difference (%)	3	6	6
Central Sweden	Difference mid S	1.4	0.7	1.2
and northern coast	Difference N coast	1.1	0.2	1.0
	Mean	53	21	24
	Relative difference (%)	2	2	5
Inland of	Difference	0	0	0.2
northern Sweden	Mean	31	11	10
	Relative difference (%)	<3	<10	≈2

plers which are of different size and aerodynamic design. The wet-only sampling is done daily and the bulk sampling is monthly. A comparison made by Granat (1989) based on parallel monthly bulk and wet-only sampling for several years (Table 2b), showed relative differences around or < 5% for sulphate, nitrate and ammonium in most parts of Sweden.

There are no monitoring data on total deposition of sulphur to other types of vegetation in Sweden. The dry deposition part is assumed to be low and wet deposition, monitored using bulk samplers, is assumed to be realistic estimates of the sulphur deposition to non-forested areas. This assumption is based mainly on measurement results from the Lake Gårdsjö area, but also on the fact that the air concentrations of sulphur compounds are low in many parts of the country. At Lake Gårdsjön (Fig. 2, Hultberg 1985) sulphur run-off from many "mini catchments" was continuously measured, along with bulk deposition in the area. The run-off flux of sulphur had been found to be equal to the throughfall flux. In one of the catchments, all the trees were cut down. Within short, the sulphur run-off decreased within a short time to the level of bulk deposition, while it remained the same in the other catchments where the forest was intact.

Dry deposition data

No routine measurements of dry deposition are carried out within the environmental monitoring set-up in Sweden. Except for the results from some research studies the main data, available to estimate the magnitude of dry deposition, are net throughfall results for sulphate. At sites where the fog and cloud water deposition is important, this contribution will be included in the net throughfall (see below). Total deposition of sulphur measured with throughfall has been found to agree well with data on sulphur run-off from catchments on the Swedish west coast (Hultberg and Grennfelt 1992). Throughfall data for Sweden are available from the regional environmental

monitoring (Hallgren Larsson and Westling 1994a,b,c). Data from the Nordic countries are available from a joint Nordic throughfall monitoring programme run during one year 1991/92 (Lövblad et al. 1994). In the evaluation of these data, results from the regional monitoring were included. For dry deposition of nitrogen no monitoring data whatsoever are available.

Fog and cloud water deposition

No measurement data are available yet to estimate the general contribution from fog and cloud water deposition in Sweden. Up to now there have been only few measurements made of the concentrations of ions in the fog/cloud water solution. There are still problems with quantifying the amount transferred to the ecosystems. Fog and cloud water deposition is assumed to be of minor importance in many areas, but is suspected to contribute significantly in the northern mountaineous regions. So far, research has been initiated, but no data are yet available.

Model data used to estimate deposition in Sweden

In addition to monitored data, model-calculated data add to knowledge of pollution load. Presently, a nationwide model is being developed in Sweden with the aim of being a support to the national environmental monitoring. The model is an enlargement of the west coast "MATCH" model (Persson et al. 1994). It has been developed for the description of the mesoscale pattern of air quality and deposition of sulphur and nitrogen compounds and is able to describe the wet deposition field well because a good agreement is obtained between modelled and monitoring data, from the EMEP and from the national precipitation chemistry network. Dry deposition is also modelled and is presented as mean dry deposition over 20 × 20 km grid squares. Comparisons of the calculated deposition with

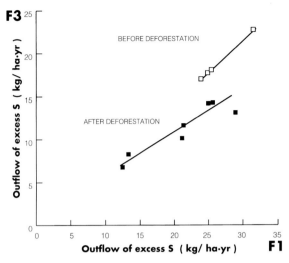

Fig. 2a. Output of excess sulphur (kg S ha^{-1} yr^{-1}) before and after deforestation of a catchment (F3) in the Lake Gårdsjön area, in relation to the reference catchment (F1) (From Hultberg 1985).

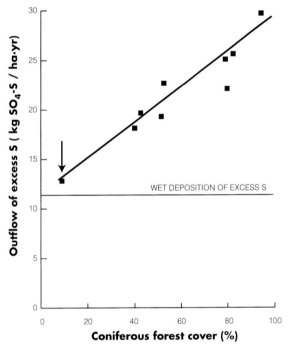

Fig. 2b. Linear regression between output of excess sulphur in the different catchments in the Lake Gårdsjön area in relation to the cover of conifers (% of area). The deforested catchment is marked with an arrow (From Hultberg 1985).

"point" throughfall measurements in forests are difficult since there are large variations within the grid square. Reasonable agreement is found, but the uncertainties of the dry deposition calculations are larger than for the wet deposition. There is still not sufficient knowledge to include fog and cloud water deposition in the model simulation. In most areas the fog and cloud water deposi-

tion is also assumed to make only minor contributions to the total deposition.

Besides the monitoring network, the EMEP programme includes model calculations of transport and deposition of sulphur and nitrogen emitted from sources in Europe. Further details are found in Sandnes (1993). On a large scale, the EMEP model provides sulphur and nitrogen deposition patterns over Sweden on a 150 × 150 km scale. The EMEP model calculations also generate source receptor matrices, from which the origin of the pollutants can be derived. Swedish ecosystems are influenced not only by pollution emitted via Swedish sources but to a large degree also by emissions in other European countries (Sandnes 1993). The EMEP model results show the magnitude of contributions to the deposition of sulphur and oxidized nitrogen received from sources in different countries (Table 3).

Another approach was used to estimate deposition for calculating critical loads exceedances. Because of lack of relevant data, deposition with a spatial resolution sufficient to cover variations on an ecosystem scale, had to be estimated (Lövblad et al. 1992). Deposition of sulphur to forests was mapped using monitored wet deposition and monitored air concentration data. Dry deposition factors were derived from "net" throughfall deposition at the monitoring sites and extrapolated to all parts of the country. Deposition was estimated separately for spruce forests and for pine and deciduous forests. Deposition to open field and lake surfaces was estimated to be equal to the bulk deposition on the basis of results from Lake Gårdsjön (Hultberg 1985 and Hultberg pers. comm.).

For nitrogen compounds, total deposition was estimated based on monitored wet deposition and estimated dry deposition (Lövblad et al. 1992). Dry deposition was estimated using measured air concentrations and deposition velocities from the international literature (including a review by Voldner et al. 1986).

The estimates for sulphur are believed to be realistic since they to a large extent are based on monitoring data. For nitrogen, relatively large uncertainties are involved. Besides the uncertainties in deposition velocities, assumptions had to be made about the ratio between gaseous and particulate nitrate and ammonium, based on few monitoring results, since only total nitrate and total ammonium is measured continuously within the EMEP network.

Deposition of sulphur and nitrogen compounds

Measured deposition of sulphur

Data from spruce forest measurements in Sweden and the other Nordic countries (except for sites in northern Finland where pine forests are the most common forest type), show the mesoscale pattern of sulphur deposition. Figure 3 shows the results for throughfall, mainly to

Table 3. Contribution to the total deposition in Sweden from emission sources in different countries in Europe, 1992 (Sandnes 1993).

Contribution from sources in	Deposition of sulphur (%)	Deposition of oxidized nitrogen (%)	Deposition of reduced nitrogen (%)
Sweden	12	12	28
Denmark	6	7	11
Germany	20	18	18
Poland	7	5	5
Great Britain	11	16	4
Undetermined origin	21	14	15

spruce forest, and open field (bulk) deposition of non-marine sulphur 1991/92 (Lövblad et al. 1994). Measurement sites in grown-up forests of similar characteristics were chosen. Coarse isolines are drawn to show the pattern for deposition to mature spruce forest, which is the dominating forest type in many areas. Clear large scale gradients are seen from south and south-west to the north, and from the coast to the inner northern parts of the

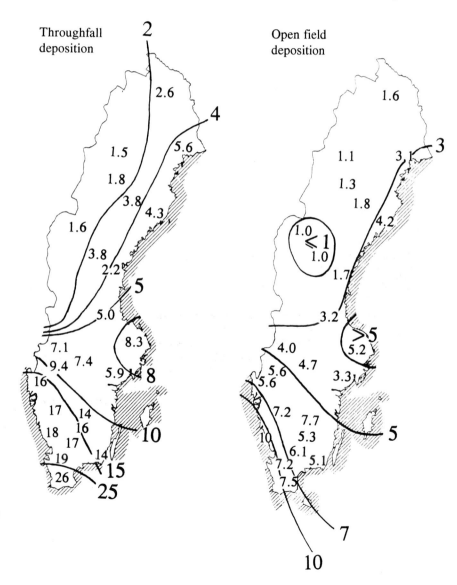

Fig. 3. Throughfall and open field deposition of non-marine sulphur to spruce forest during 1991/92 (kg S ha⁻¹ yr⁻¹). (Lövblad et al. 1994).

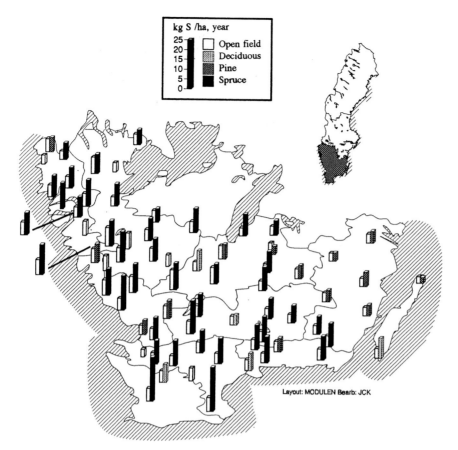

Fig. 4. Deposition of excess sulphur to open field (bulk deposition) and to different types of forests (throughfall) in southern Sweden during the hydrological year 1992/93. (Hallgren Larsson and Westling 1994a).

Layout: MODULEN Bearb: JCK

Nordic countries. In addition, in some areas, e.g. the Stockholm area, an influence due to deposition of locally emitted air pollution is observed. To some extent the throughfall deposition pattern is similar to that of wet deposition, with the similar large scale gradients, but there is an additional gradient effect attributable to the dry deposition.

In the inland parts of central and northern Sweden the sulphur deposition is < 3 kg S ha^{-1} yr^{-1}. The dry deposition contribution is 0.5–1 kg S ha^{-1} yr^{-1}. In these low-polluted areas, throughfall data should be validated against other types of measurements, to assess that there is no significant internal cycling or canopy uptake of sulphur compared to the small amounts deposited. The measurements are made at low altitudes, where most of the forested areas are found. The results are therefore not representative of deposition to high altitude areas. Further south, and along the coasts, the sulphur deposition is higher. In southern to south-western Sweden it reaches 15–25 kg S ha^{-1} yr^{-1}. The relative increase from north to south in net throughfall – i.e. the difference between throughfall and wet deposition, which for sulphur is assumed to be equal to the dry deposition – is larger than the increase in bulk deposition, indicating an increasing contribution of dry deposition towards the south. Wet deposition is between

7 and 10 kg S ha^{-1} yr^{-1} in south western Sweden and 1–3 kg S ha^{-1} yr^{-1} in the north.

Monitoring data from southern Sweden, where measurements are made in different types of forests (Fig. 4, Hallgren Larsson and Westling 1994a) show differences in the magnitude of the deposition between forest types. In southern Scandinavia, the deposition of sulphur to mature spruce forests is c. 50% higher than the deposition to pine and deciduous forests (Westling et al. 1992) at near-by sites with the same pollution load. The difference between spruce and pine were shown to be to some extent related to the difference in biomass (as needles and branches, kg dry weight ha^{-1}, Ivens et al. 1990). Because of this, it is important to distinguish between different land use categories, and also different types of forests, when describing the variations in deposition.

Deposition to open fields is less than deposition to forested ecosystems (see Fig. 2b). Data on wet deposition measured using bulk samplers are assumed to give reasonable estimates. For some types of open field to which the dry deposition, at least during part of the year, may be important, for example agricultural areas, bulk deposition may underestimate the total deposition.

The estimated sulphur deposition on an ecosystem scale has been compared to deposition calculated by the

24

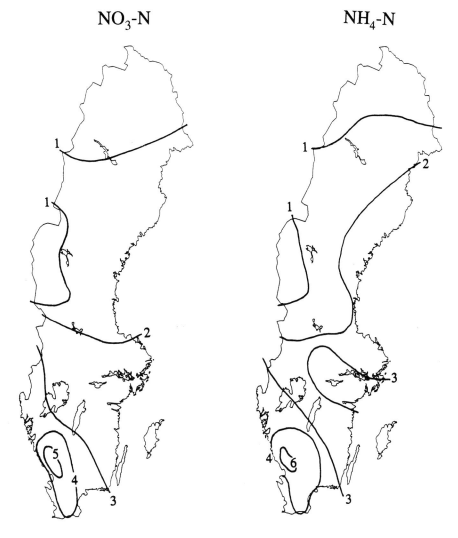

Fig. 5. Bulk deposition of nitrate and ammonium 1993 (kg N ha⁻¹ yr⁻¹). (Kindbom et al. 1994b).

EMEP model (Erisman 1993c, Lövblad 1994). Land-use weighted deposition to the EMEP grid squares, calculated from measured deposition to various land use categories combined with land use data for the grid square, was compared to the EMEP model results. The comparison, made for 1989, indicates systematic discrepancies between the various grid squares. To some extent the discrepancies seem to be related to differences in landuse and local emissions within the squares. Local emissions are not taken into account by the deposition estimation approach, which is based on rural monitoring results. However, local emissions will not contribute more than marginally, < 10% according to the EMEP calculations except in some squares with large emissions (Tuovinen et al. 1994). On the other hand, land-use within the different grid squares is not taken into account by the EMEP model. Deviations may also be caused by uncertainties in the assumptions made for ecosystems other than forests.

The EMEP calculated dry deposition is more than the estimated dry deposition from monitoring results, as a total for Sweden, and in the coastal, more densely populated grid squares. The calculated dry deposition for the grid squares, which are largely forested, is less than extrapolated measurements. Summing up the total deposition over Sweden, and comparing the two approaches, the total "measured" deposition is 24% more than the calculated. The "measured" dry deposition is in relatively good agreement, 94%, with the model calculated. There are large deviations between measured and modelled data for the wet deposition.

Measured deposition of nitrogen compounds

Nitrogen deposition to forests in Sweden cannot at present be quantified by monitoring data. Wet (bulk) deposition data of nitrate and ammonium may be assumed to represent the deposition to open fields (Fig. 5), at least in

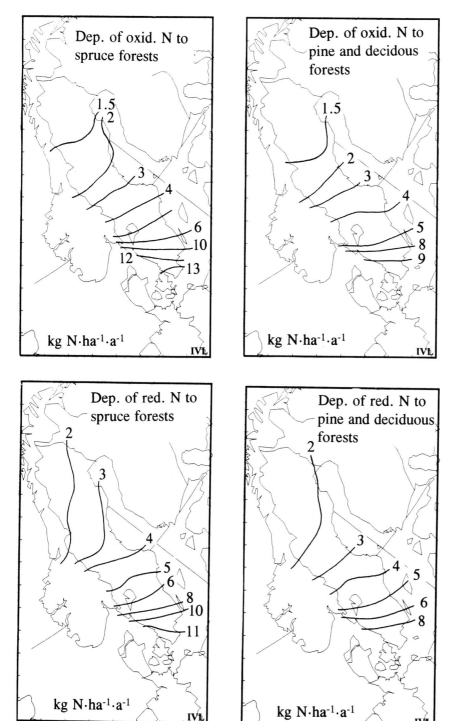

Fig. 6. Estimated deposition of oxidized and reduced nitrogen to Swedish forests (kg N ha⁻¹ yr⁻¹), (Lövblad et al. 1992).

areas without large ammonia emissions. Coarse isolines are drawn to indicate the wet deposition pattern. A comparison between different parts of the country shows that the open field deposition of nitrogen (sum of oxidized and reduced) in the north is < 2 kg N ha⁻¹ yr⁻¹ c. 6 times less than in the areas with most wet deposition in the south, which is 10–15 kg N ha⁻¹ yr⁻¹.

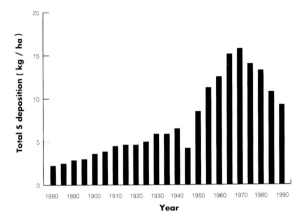

Fig. 7. Calculated historical deposition (mg S m⁻² yr⁻¹) to an area in southern Sweden – EMEP grid square 20,21 (Mylona 1993).

Estimated deposition of nitrogen

As mentioned, it is not possible to estimate total deposition of nitrogen from throughfall data. No dry deposition monitoring is carried out in Sweden. The only remaining possibility is theoretical estimates. Deposition estimates of nitrogen compounds to Swedish forests for 1985 – 1989 are shown in Fig. 6 (Lövblad et al. 1992). The estimates contain uncertainties in the assumptions made, caused by deposition velocities used and by the lack of some air concentration data. For example, only data for total (i.e. the sum of gaseous and particulate) nitrate and ammonium are available. The gaseous part of the total ammonium and nitrate, with a higher deposition velocity, has to be estimated from only a few monitoring data.

The estimated deposition of nitrogen compounds shows the same geographical pattern as the deposition of sulphur, with a gradient from the southwest to the north. The data also indicate more deposition along the coast of the Bothnian Sea and Bothnian Bay than inland. The total deposition to spruce forests of nitrogen (sum of nitrate and ammonium) is estimated to be > 20 kg N ha⁻¹ yr⁻¹ in the south of Sweden, and 3–6 kg N ha⁻¹ yr⁻¹ in the north.

Deposition of nitrogen in Sweden will in the future be modelled with the recently developed model (Persson et al. 1994). Monitoring data on dry deposition will be a necessary complement to the model.

Nitrogen deposition modelling is done within the EMEP project with respect to the large scale deposition and results are presented for grid squares of 150 × 150 km and without consideration to ecosystem type (Tuovinen 1994). The data indicate a nitrogen deposition of 2–4 kg N ha⁻¹ yr⁻¹ in the north of Sweden and 10–12 kg N ha⁻¹ yr⁻¹ in the south.

Trends over time of air quality and deposition

The occurrence of air pollution, as air quality and deposition, is subject to variation not only on a geographical scale but also on a time scale. Temporal variations can be divided into long-term changes, trends over centuries and decades, medium term changes between years, or seasons and short-term variations over days (episodes). The long-term changes are mainly caused by changes in emission, while short-term variations can also be attributed to changes in weather.

Data to study long-term trends are available as emission estimates, over > 100 yr, which in turn can be the base for deposition estimates. Measurement data available for trend studies in Sweden are monitoring data for precipitation chemistry since 1955 and EMEP monitoring results from the end of 1970's. The total deposition trend can be estimated from trends in wet deposition and trends in air pollution concentrations. During the last decade throughfall monitoring provide data on sulphur deposition to forests.

Long-term changes in sulphur and nitrogen emission and its influences on deposition

Since the middle of the 19th century up to the second world war, there has been a more or less constant increase in emission of sulphur and of nitrogen oxides (Kindbom et al. 1993a, Mylona 1993). The dramatic increase during the 1950's and 1960's reversed during the 1970's. Since then emissions of sulphur have begun to decrease in Sweden and western Europe. Emissions of nitrogen oxides in Sweden have been relatively constant from 1980.

A sulphur deposition trend for the period 1870–1990 was calculated by Mylona (1993) using the EMEP model and estimated historic SO_2 emissions. The sulphur deposition in the beginning of the period, to the inner parts of southern Sweden (Fig. 7) is calculated to be of the magnitude 0.2–0.3 g m⁻² yr⁻¹, i.e. about a third of the present load and similar to the estimated 95 percentile of critical loads of sulphur in that area, 0.21 g S m⁻² yr⁻¹ (Sandnes 1993). The calculated deposition of sulphur passed through a maximum of 1.5 g m⁻² yr⁻¹ around 1970 and then decreased during the last 20 years to be around 0.75 g m⁻² yr⁻¹ in 1990.

The calculated trends should be verified with measurements. No monitoring data are available for the whole period. Indirect monitoring data, pH values and deposition of oil soot particles, obtained from sediment studies in lakes on the Swedish west coast, support the magnitude of the calculated sulphur trends (Renberg and Wik 1985).

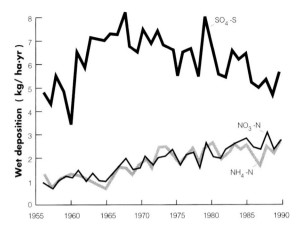

Fig. 8. Trends for concentration (µeq/l) of sulphur and nitrogen in precipitation and wet deposition of sulphur and nitrogen (kg ha^{-1} yr^{-1}) in central Sweden 1955 to 1990 (kg ha^{-1} yr^{-1}), (Bernes 1992).

Trends of sulphur and nitrogen wet deposition during the last decades

Precipitation chemistry data are available from 1955 at some stations in Sweden (Granat unpubl.). The wet deposition of sulphur as a mean over the stations in central Sweden (Fig. 8, Bernes 1992), increased during the 1960's. From 1970, as can be expected, a decrease is observed, but the decrease is somewhat less than the sulphur emissions decrease in Sweden and Europe. Based on the EMEP monitoring data, the wet deposition of sulphur has, in large parts of Sweden, decreased by 30–50% or more from 1980 to 1992 (Kindbom et al. 1994a).

For wet deposition of oxidized and reduced nitrogen in central Sweden, there has been a slow but steady increase since measurements begun in 1955 (Bernes 1992). The EMEP results on wet deposition of nitrogen between 1980 and 1992, show large variations between years, and no significant changes (Kindbom 1994a).

Trends for concentrations in air of sulphur and nitrogen compounds

To study the trend for total deposition of sulphur and nitrogen, the trend for dry deposition or at least for the air concentrations of relevant compounds is needed in addition to wet deposition data. Trends for air concentrations of sulphur and nitrogen compounds have been studied based on EMEP measurements from 1980 – 1992 (Fig. 8, Kindbom et al. 1994a). Statistically significant decreases are observed for sulphur dioxide at all Swedish stations in operation during the whole period. A decrease between 40% and up to > 50% is found from 1980 to 1992 at all stations. The magnitude of the decrease is the same as for the wet deposition.

The trends observed for wet deposition of sulphur and air concentrations are similar to the trends in Norway and Finland (Finnish Meteorological Inst. 1993, Statens Forurensningstilsyn 1992).

There is, however, a risk for an overestimation of the trend for SO$_2$, arising from the mild winters 1990–1992. The average concentrations of SO$_2$ during the year have been found to depend largely on the frequency of episodes with transport of polluted air masses from central and eastern Europe to Scandinavia. The episodes observed during the most recent years with relatively mild winters, have been characterized rather by high concentrations of nitrogen dioxide – higher than previously observed – than by high concentrations of sulphur dioxide. The air masses have passed more frequently over western Europe, where the air is more influenced by traffic exhausts than by combustion of high sulphur fossil fuels. There have been very few easterly episodes with air masses passing over eastern Europe during the last winters (Table 4, Kindbom et al. 1993b). Whether the "sulphur dioxide episode" is a phenomenon which is becoming unimportant in Sweden, as a result of decreasing emission in central Europe or whether it has just not occurred because of the prevailing weather conditions in recent years, cannot be determined at present.

Table 4. Frequency of increased daily mean concentrations and annual means (µg m^{-3}) of SO$_2$ and NO$_2$ in air in southern Sweden.

Year	Number of days SO$_2$-S >15	Annual mean	Number of days NO$_2$-N >10	Annual mean
1986	16		0	
1987	20		3	
1988	14		0	
1989	8		9	
1990	9		7	
1991	9		6	
1992	5	2.1	0	1.7

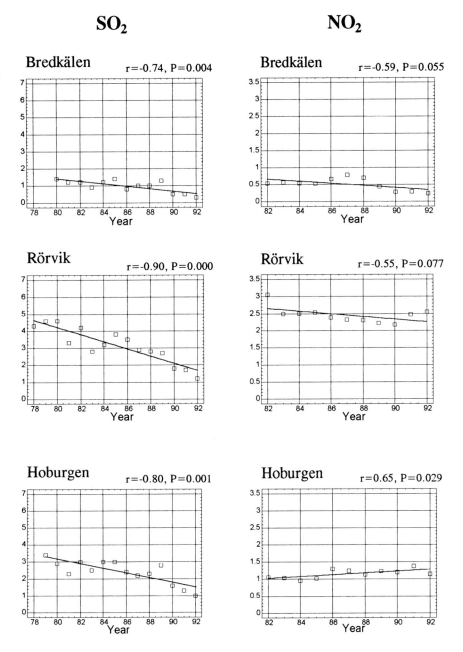

Fig. 9. Annual average concentrations of sulphur dioxide and nitrogen dioxides at the Swedish EMEP stations 1978–1992 (μg m⁻³ S or N) (Kindbom et al. 1994a).

Trends for total deposition of sulphur

The EMEP results all point in the same direction, towards a 30–40% decrease for sulphur compounds in the air as well as in precipitation. Thus a similar decrease can be expected for the total deposition of sulphur. For sulphur, throughfall data from the time period may give an indication of trends for total sulphur deposition in Sweden. Today there is an extensive monitoring programme including throughfall monitoring in forests at 125 stations all over Sweden (Hallgren Larsson and Westling 1994 a,b,c). Most of these stations have been in operation for

only a few years. The longest time series available are for Blekinge county (since 1986) and for Kronoberg county (since 1987) in southeastern Sweden (Westling et al. 1992). Even if the period is short, the trend for sulphur is clearly decreasing (Fig. 9). The measurements at lake Gårdsjön research site on the west coast, make up one of the longest Swedish time series for throughfall. The ten year trend reported in Hultberg and Grennfelt (1992) shows no change for sulphur between 1980 and 1988. However, if the years 1990–1993 are included a decreasing trend is also seen here (Fig. 10), (Hultberg unpubl.). The decrease in sulphur concentration at Lake Gårdsjön

Open field   Spruce forest   Pine forest

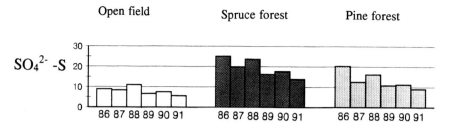

SO_4^{2-} -S

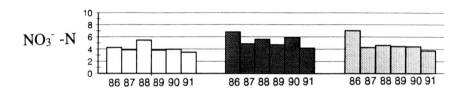

NO_3^- -N

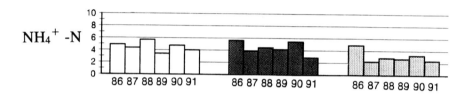

NH_4^+ -N

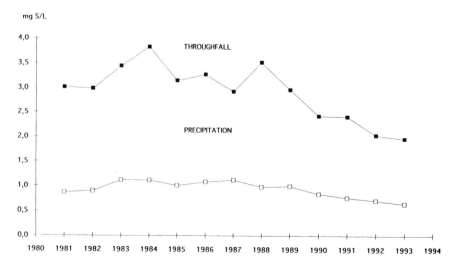

Fig. 10. Throughfall and open field (bulk) deposition of sulphur and nitrogen in spruce and pine forests from 1986 to 1991. Results from the regional forest monitoring programme in Blekinge county in south-eastern Sweden (kg S or N ha^{-1} yr^{-1}), (Westling et. al. 1992).

is c. 40% in precipitation as well as in throughfall, for the period 1980–1993.

Trends for all measurements of sulphur point in the same direction; the sulphur pollution load has decreased significantly in most parts of Sweden since 1980. The greatest changes have been during the last five to six years. Since the emissions outside Sweden have an important effect on total deposition loads in Sweden (Table 3), the large scale model calculations are essential. A calculation for sulphur (Fig. 12, Berge pers. comm.), indicates that the decrease in sulphur deposition between

1980 and 1990 was c. 40% in southern Sweden. In the eastern parts of the northern coast the calculated decrease was somewhat larger, at around 50%. This is in agreement with the results of the measurements. There is no obvious trend in the deposition of nitrogen compounds, as would be expected from emissions inventories and from air and precipitation measurements.

mg S/L

Fig. 11. Annual weighted concentration (mg S l^{-1}) of non-marine sulphur in throughfall and precipitation at Lake Gårdsjön in south-western Sweden during the period 1980–1993 (Hultberg et al. unpubl.).

30

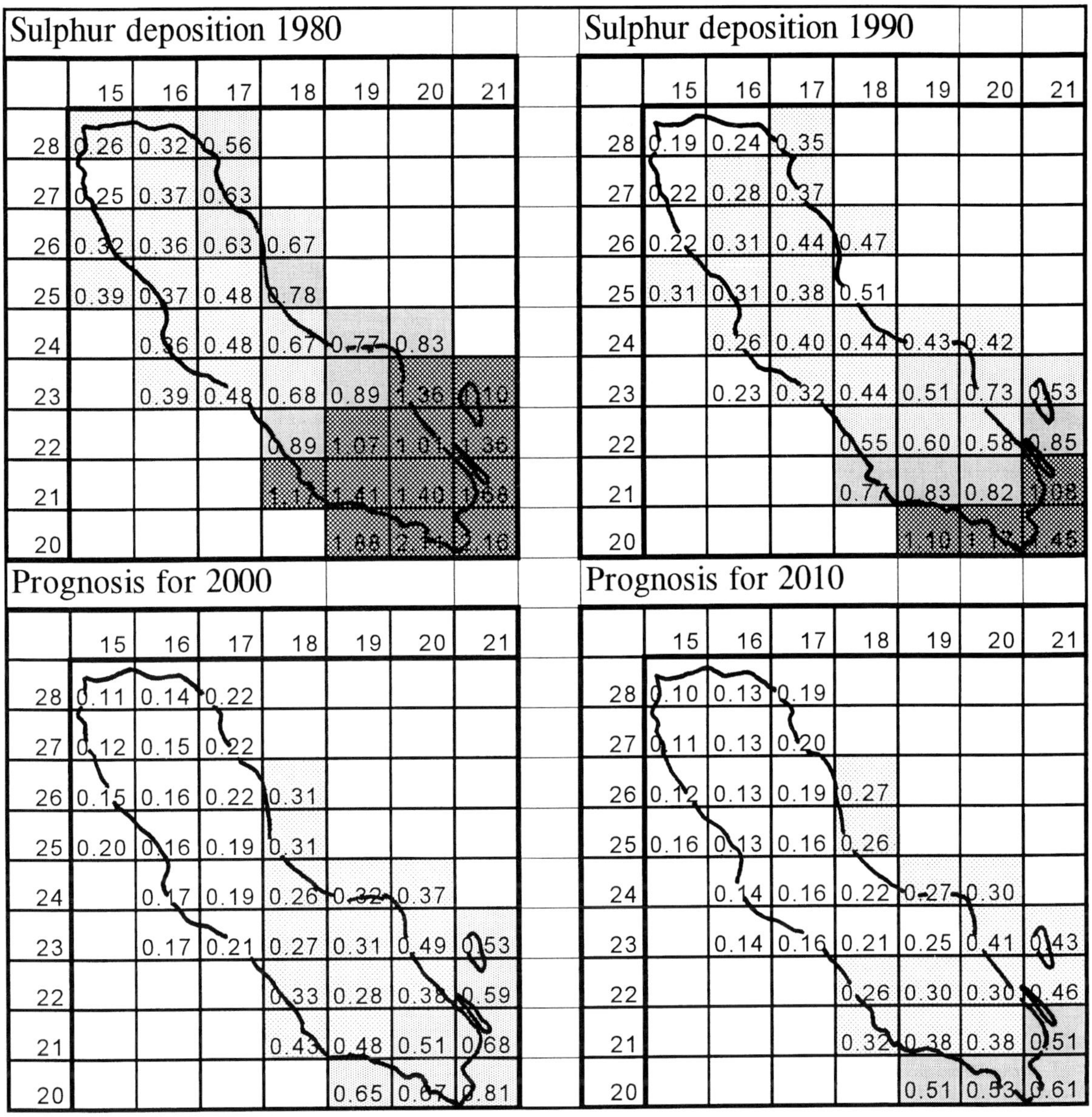

Sulphur deposition 1980

	15	16	17	18	19	20	21
28	0.26	0.32	0.56				
27	0.25	0.37	0.63				
26	0.32	0.36	0.63	0.67			
25	0.39	0.37	0.48	0.78			
24		0.36	0.48	0.67	0.77	0.83	
23		0.39	0.48	0.68	0.89	1.36	
22				0.89	1.07	1.0	
21				1.1	1.41	1.40	
20					1.88	2.	1.

Sulphur deposition 1990

	15	16	17	18	19	20	21
28	0.19	0.24	0.35				
27	0.22	0.28	0.37				
26	0.22	0.31	0.44	0.47			
25	0.31	0.31	0.38	0.51			
24		0.26	0.40	0.44	0.43	0.42	
23		0.23	0.32	0.44	0.51	0.73	0.53
22				0.55	0.60	0.58	0.85
21				0.77	0.83	0.82	1.08
20					1.10	1.	1.45

Prognosis for 2000

	15	16	17	18	19	20	21
28	0.11	0.14	0.22				
27	0.12	0.15	0.22				
26	0.15	0.16	0.22	0.31			
25	0.20	0.16	0.19	0.31			
24		0.17	0.19	0.26	0.32	0.37	
23		0.17	0.21	0.27	0.31	0.49	0.53
22				0.33	0.28	0.38	0.59
21				0.43	0.48	0.51	0.68
20					0.65	0.67	0.81

Prognosis for 2010

	15	16	17	18	19	20	21
28	0.10	0.13	0.19				
27	0.11	0.13	0.20				
26	0.12	0.13	0.19	0.27			
25	0.16	0.13	0.16	0.26			
24		0.14	0.16	0.22	0.27	0.30	
23		0.14	0.16	0.21	0.25	0.41	0.43
22				0.26	0.30	0.30	0.46
21				0.32	0.38	0.38	0.51
20					0.51	0.53	0.61

Fig. 12. Sulphur deposition (mg S m^{-2} yr^{-1}) to the EMEP grid squares calculated for the year 1980 (base year for the agreements under the protocol) and 1990. Prognoses for the sulphur deposition are made based on emissions according to the new sulphur protocol, (Berge pers. comm.).

Future deposition amounts

EMEP model calculations are complementary tools in trend studies, and, more important, are a tool for the prognoses of future deposition amounts. To predict future amounts, data on future emissions are needed. The recently signed sulphur protocol to the Convention on Long Range Transboundary Air Pollution (UN ECE) provides the prognoses for emissions in 2000 and 2010.

The results (Fig. 17) indicate decreases in Sweden of between 70 and 80% of the 1980 year sulphur deposition by 2010. Even though the decrease is considerable, it is not enough to achieve the level of the 95 percentile critical loads (Sandnes 1993). Especially in the western parts of Sweden, the most sensitive ecosystems will still have a deposition exceeding this load (Amann et al. 1993).

For nitrogen, the present NO$_x$ protocol will only result in minor decreases in Scandinavia (Saltbones 1989). For most countries the protocol involves only a freezing of emissions at the 1985 level; for other countries, including Sweden, it involves a 30% decrease by 1997. For reduced nitrogen emissions there is at present no protocol. Some countries have national or regional plans for the reduction of ammonia emissions. The work for a new NO_x protocol is just beginning. Until the guidelines for this protocol are drawn it is not possible to predict future amounts of nitrogen deposition.

Conclusions

Sulphur and nitrogen are deposited from the atmosphere to the ecosystems via wet, fog/cloud water and dry deposition processes. Dry deposition is of importance mainly in the southern to central parts of Sweden. There are data available for estimating the wet deposition from measurements and from models. For dry deposition few monitoring data are available, except for net throughfall of sulphur. The validation of models will need more dry deposition measurements, especially for nitrogen.

There are large variations in deposition over Sweden. A gradient is seen from the south and southwest towards the north, resulting from the large influence of emissions from Europe on depositions to Swedish ecosystems. In addition to the large-scale gradient, there are local variations caused by differences in land use and by differences in local pollution. Forested areas for example, receive larger amounts of acidifying input because of the large uptake of pollution during dry periods by forests as compared to open fields in Sweden, which are assumed to receive mainly wet deposition. Critical loads are exceeded in many areas, especially in southern and central Sweden. This aspect will be dealt with in detail in a number of other articles within this publication.

The more long term temporal variations are mainly caused by variations in emissions but partly also by weather variations. The trend in sulphur deposition is clearly decreasing for wet deposition and throughfall of sulphur and for the occurrence of gaseous as well as particulate sulphur compounds in the atmosphere. Considerable decreases have been observed mainly during the last six years. Models as well as measurements agree and indicate 40 to 50% lower amounts in most parts of the country than in 1980 year. No trends are apparent for nitrogen deposition. The emissions of nitrogen oxides over Europe are also fairly constant.

Based on the new sulphur protocol, further sulphur deposition decrease is expected. However, in the most sensitive ecosystems, especially in western parts of Sweden, critical loads for sulphur will still be exceeded, even after the year 2010.

Available results show that the sulphur pollution situation for Sweden as an average, at present and historically is relatively well known. The future situation for Sweden as a whole can also probably be predicted relatively well. For nitrogen, however, there is a lack of data and much work needs still to be done in order to obtain a reliable and complete picture of present and future deposition.

References

Amann, M., Klaassen, G. and Schöpp, W. 1993. Closing the gap between the 1990 deposition and the critical sulfur deposition values. – The UN/ECE Task Force on Integrated Assessment Modelling. Int. Inst. Applied Systems Analysis, Laxenburg, Austria.

Andersen, H. V., ApSimon, H., Asman, W., Barrett, K., Kulweit, J., Schjørring, J. and Sutton, M. 1993. Dry deposition of reduced nitrogen. – In: Lövblad, G., Erisman, W. and Fowler, D. (eds), Models and methods for the quantification of atmospheric input to ecosystems. Nordic Council of Ministers Report 1993:573, Copenhagen, pp. 25–32.

Beier, C., Braun, S., Brook, J., Campbell, G., Draaijers, G., Hansen, K., Kallweit, D., Lenz, R., Lindberg, S., Staaf, H. and Stary, R. 1993. Deposition monitoring. – In: Lövblad, G., Erisman, W. and Fowler, D. (eds), Models and methods for the quantification of atmospheric input to ecosystems. Nordic Council of Ministers Report 1993:573, Copenhagen, pp. 37–41.

Bernes, C. 1992. Försurning och kalkning av svenska vatten. – Monitor 12. Swedish Environmental Protection Agency, Solna, (in Swedish).

Braun, S., Duyzer, J., Grennfelt, P., Meixner, F. X., Neftel, A., Pilegaard, K., van Pul, A. and San José, R. 1993. Dry deposition of nitrogen oxides. – In: Lövblad, G., Erisman, W. and Fowler, D. (eds), Models and methods for the quantification of atmospheric input to eosystems. Nordic Council of Ministers Report 1993:573, Copenhagen, pp. 17–23.

Davidson, C. I. and Wu, Y.-L. 1990. Dry deposition of particles and vapors. – In: Lindberg, S. E., Page, A. L. and Norton, S. A. (eds), Acidic precipitation. Springer, New York, pp. 103–216.

Davies, T. D., Barthelmie, R. J., Glynn, S. and Schaug, J. 1992. European precipitation chemistry atlas. Vol. 3. – EMEP/CCC-Report 6/92. School of Environ. Sci., Univ. of East Anglia, UK.

Draaijers, G. 1993. The variability of atmospheric deposition to forests: the effects of canopy structure and forest edges. – Ph.D. thesis, Univ. of Utrecht, The Netherlands.

Duyzer, J. and Fowler, D. 1993. Dry deposition of nitrogen oxides. – In: Lövblad, G., Erisman, J. W. and Fowler, D. (eds), Models and methods for the quantification of atmospheric input to ecosystems. Nordic Council of Ministers Report 1993:573, Copenhagen, pp. 95–123.

Erisman, J. W. 1992. Atmospheric deposition of acidifying compounds in the Netherlands. – Ph.D. thesis, Univ. of Utrecht, The Netherlands.

– and Baldocchi, D. 1993. Dry deposition of sulphur dioxide. – In: Lövblad, G., Erisman, W. and Fowler, D. (eds), Models and methods for the quantification of atmospheric input to ecosystems. Nordic Council of Ministers Report 1993:573, Copenhagen, pp. 75–93.

– , Beier, C., Draaijers, G. and Lindberg, S. 1993a. Deposition monitoring. – In: Lövblad, G., Erisman, W. and Fowler, D. (eds), Models and methods for the quantification of atmospheric input to ecosystems. Nordic Council of Ministers Report 1993:573, Copenhagen, pp. 163–183.

– , Ferm, M., Fowler, D., Granat, L., Hultberg, H., Kesselmeier, J., Köble, R., Lovett, G., Moldan, F., Padro, J., Runge, E., Simmons, C., Slanina, S. and Zapletal, M. 1993b. Dry deposition of sulphur dioxide. – In: Lövblad, G., Erisman, W. and Fowler, D. (eds), Models and methods for the quantification of atmospheric input to ecosystems. Nordic Council of Ministers Report 1993:573, Copenhagen, pp. 13–15.

– , van Jaarsveld, H., van Pul, A., Fowler, D., Smith, R. and Lövblad, G. 1993c. Comparison between small scale and long-range transport modelling. – In: Lövblad, G., Erisman, W. and Fowler, D. (eds), Models and methods for the quantification of atmospheric input to ecosystems. Nordic Council of Ministers Report 1993:573, Copenhagen, pp. 187–198.

Finnish Meteorological Institute 1993. Air quality measurements 1992. – Helsinki.

Fowler, D., Granat, L., Koble, R., Lovett, G., Moldan, F., Simmons, C., Slanina, S. and Zapletal, M. 1993a. Working group report 2. Wet, cloud water and fog deposition. – In: Lövblad, G., Erisman, W. and Fowler, D. (eds), Models and methods for the quantification of atmospheric input to ecosystems. Nordic Council of Ministers Report 1993:573, Copenhagen, pp. 9–12.

– , Gallagher, M. W. and Lovett, G. M. 1993b. Fog, cloudwater and wet deposition. – In: Lövblad, G., Erisman, W. and Fowler, D. (eds), Models and methods for the quantification of atmospheric input to ecosystems. Nordic Council of Ministers Report 1993:573, Copenhagen, pp. 51–73.

Granat, L. 1989. Luft- och nederbördskemiska nätet. Rapport från verksamheten 1988. – Swedish Environ. Protec. Agency, Report 3649, Solna, (in Swedish, English summary).

– , 1991. Luft- och nederbördskemiska stationsnätet inom PMK. Rapport från verksamheten 1990. – Swedish Environ. Protec. Agency, Report 3942, Solna, (in Swedish, English summary).

Hallgren Larsson, E. and Westling, O. 1994a. Luftföroreningar i södra Sverige; Nedfall och effekter oktober 1992 till september 1993. – IVL Report B1108. Swedish Environ. Res. Inst., IVL-Aneboda Lammhult, Sweden, (in Swedish, English summary).

– and Westling, O. 1994b, Luftföroreningar i norra Sverige; Nedfall och effekter oktober 1992 till september 1993. IVL Report B1150. Swedish Environ. Res. Inst., IVL-Aneboda Lammhult, Sweden, (in Swedish, English summary).

– and Westling, O. 1994c, Luftföroreningar i mellersta Sverige; Nedfall och effekter oktober 1992 till september 1993. – IVL Report B1139. Swedish Environ. Res. Inst., IVL-Aneboda Lammhult, Sweden, (in Swedish, English summary).

Heij, G. J. and Schneider, T. (eds) 1991. Dutch Priority Programme on Acidification. Final report. – Report no. 200–09, National Inst. Public Health Environ. Protection, Bilthoven, The Netherlands.

Hicks, B. B., Albritton, D. L., Davidson, C. I., Dodge, M., Draxler, R. R., Fehsenfeld, F. C., Hales, J. M., Lindberg, S. E., Meyers, T. P., Schwartz, S. E., Tanner, R. L., Wesely, M. L. and Vong, R. L. 1989. Atmospheric processes research and processes model development. – State of science/technology Report no. 2. National acid precipitation assessment program.

Hultberg, H. 1985. Budgets of base cations, chloride, nitrogen and sulphur in the acid Lake Gårdsjön catchment, SW Sweden. – Ecol. Bull. (Copenhagen) 37: 133–157.

– and Grennfelt, P. 1992. Sulphur and seasalt deposition as reflected by throughfall and run-off chemistry in forested catchments. – Environ. Poll. 75: 215–222.

Hummelshøj, P., Ruijgrok, W., Semb, A., Westling, O. and Wyers, P. 1993. Dry deposition of particles. – In: Lövblad, G., Erisman, W. and Fowler, D. (eds), Models and methods for the quantification of atmospheric input to ecosystems. Nordic Council of Ministers Report 1993:573, Copenhagen, pp. 33–36.

Ivens, W. 1990. Atmospheric deposition onto forests: an analysis of the deposition variability by means of throughfall measurements. – Ph.D. thesis, Univ. of Utrecht, The Netherlands.

– , Lövblad, G., Westling O. and Kauppi, P. 1990. Throughfall monitoring as a means of monitoring deposition to forest ecosystems – evaluation of European data. – Nord 1990:120, The Nordic Council of Ministers, Copenhagen.

Kenttämies, K. (ed.) 1991. Acidification research in Finland. Review of the results of the Finnish acidification research programme (HAPRO) 1985–1990. – Government Printing Centre, Helsinki.

Kindbom, K., Lövblad, G. and Sjöberg, K. 1993a. Beräkning av ackumulerad syrabelastning från atmosfären till de svenska ekosystemen. Delrapport 1: Emissioner av svavel, kväve och alkaliskt stoft i Sverige 1900–1990. – IVL Report B1109, (in Swedish, English summary).

– , Lövblad, G., Persson, K. och Sjöberg, K. 1993b. Atmosfärkemisk övervakning vid IVL's stationer inom EMEP. – Swedish Environ. Protect. Agency Report 4211, Solna, (in Swedish, English summary).

– , Lövblad, G. och Sjöberg, K. 1994a. Sulphur and nitrogen compounds in air and precipitation in Sweden 1980 to 1992. – IVL Report B1144.

– , Lövblad, G., Munthe, J. och Sjöberg, K. 1994b. Luft- och Nederbördskemiska stationsnätet inom PMK. Övervakning av svavel- och kväveföreningar, baskatjoner och tungmetaller. Rapport från verksamheten 1993. – Swedish Environ. Protect. Agency, Report, Solna, (in Swedish, English summary).

Likens, G. E., Bormann, F. H., Hedin, L. O., Driscoll, C. T. and Eaton, J. S. 1990. Dry deposition of sulphur: a 23 year record for the Hubbard Brook Forest ecosystem. – Tellus, 42B: 319–329.

Lövblad, G. 1987. Utsläpp till luft av alkali. IVL Report B858. – Swedish Environ. Res. Inst., Göteborg, Sweden, (in Swedish, English summary).

– 1993. Comparison of estimated land-use specific sulphur deposition with EMEP deposition data for Sweden. – In: Lövblad, G., Erisman, N. and Fowler, D. (eds), Models and methods for the quantification of atmospheric imput to ecosystems. Nordic Council of Ministers Report 1993:573, Copenhagen pp. 199–216.

– 1994. Comparison between measured and modelled deposition in Sweden. – In: Berg, T. and Shavy, J. (eds), EMEP/CCC Report 2/94, Norwegian Inst. for Air Res., Kjeller, Norway, pp. 267–277.

– and Erisman, J. W. 1992. Deposition of nitrogen in Europe. – In: Grennfelt, P. and Thörnelöf, E. (eds), Critical loads for nitrogen – a workshop report. Nord 1992:41, The Nordic Council of Ministers, Copenhagen, pp. 239–286.

– , Andersen, B., Hovmand, M., Joffre, S., Pedersen, U., and Reissell, A., 1992. Mapping deposition of sulphur, nitrogen and base cations in the Nordic Countries. – IVL Report B1055. Swedish Environ. Res. Inst., Göteborg, Sweden.

– , Erisman, J. W. and Fowler, D. (eds) 1993. Models and methods for the quantification of atmospheric input to ecosystems. – Nordiske Seminar- og Arbejdsrapporter 1993:573. The Nordic Council of Ministers, Copenhagen.

– , Aamlid, D., Hovmand, M., Hyvärinen, A., Reissell, A., Schaug, J. and Westling O. 1994. Throughfall monitoring in the Nordic countries. – IVL Report B1132. Swedish Environ. Res. Inst., Göteborg, Sweden.

Murphy, C. E. Jr. and Sigmon, J. T. 1990. Dry deposition of sulfur and nitrogen oxide gases to forest vegetation. – In: Lindberg, S. E., Page, A. L. and Norton, S. A. (eds), Acidic precipitation. Springer, New York, pp. 217–240.

Mylona, S. 1993. Trends of sulphur dioxide emissions, air concentrations and depositions of sulphur in Europe since 1880. – EMEP/MSC-W Report 2/93. Norwegian Meteorological Inst., Blindern, Norway.

NADP/NTN – Annual data summary. 1994. Precipitation chemistry in the United States, 1993. – Natural Resource Ecol. Lab., Colorado State Univ., Fort Collins, CO.

Persson, K., Lövblad, G. and Munthe, J. 1993. Luft- och nederbördskemiska stationsnätet inom PMK samt tungmetaller inklusive kvicksilver. Rapport från verksamheten 1992. –

Swedish Environ. Protect. Agency, Report 4213, Solna, (in Swedish, English summary).

Persson, C., Langner, J. and Robertsson, L. 1994. Regional spridningsmodell för Göteborgs och Bohus, Hallands samt Älvsborgs län. Regional luftmiljöanalys för år 1991. – SMHI RMK Report No 65, (in Swedish, English summary).

Renberg, I. and Wik, M. 1985. Carbonaceous particles in lake sediments-pollutants from fossil fuel combustion. – Ambio 14: 161–163.

Ruijgrok, W., Davidson, C. I. and Nicholson, K.W. 1993. Dry deposition of particles. – In: Lövblad, G., Erisman, W. and Fowler, D. (eds), Models and methods for the quantification of atmospheric input to ecosystems. Nordic Council of Ministers Report 1993:573, Copenhagen, pp. 145–161.

Saltbones, J., Eliassen, A. and Sandnes, H. 1989. Estimated reductions of deposition of sulphur and oxides of nitrogen in Europe due to planned emission reductions. – EMEP/MSC-W Note 3/89, Norwegian Meteorological Inst., Blindern, Norway.

Sandnes, H. 1993. Calculated budgets for airborne acdifying components in Europe 1985, 1987, 1988, 1990, 1991 and 1992. – EMEP/MSC-W Report 1/93. Norwegian Meteorological Inst., Blindern, Norway.

Schaefer, D. A. and Reiners, W. A. 1990. Throughfall chemistry and canopy processing mechanisms. – In: Lindberg, S. E., Page, A. L. and Norton, S. A. (eds), Acidic precipitation. Springer, New York, pp. 241–284.

Schaug, J., Arnesen, K., Bartonova, A., Pederson, U. and Skjelmoen, J. E. 1994. Data report 1992. – EMEP/CCC-Report 4/94.

Statens Forurensningstilsyn 1992. Overvåking av langtransportert forurenset luft og nedbør. Årsrapport 1991. – Rapport 506/92, (in Norwegian, English summary).

Sutton, M. A., Asman, W. A. H. and Schjørring, J. K. 1993. Dry deposition of reduced nitrogen. – In: Lövblad, G., Erisman, W. and Fowler, D. (eds), Models and methods for the quantification of atmospheric input to ecosystems. Nordic Council of Ministers Report 1993:573, Copenhagen, pp. 125–143.

Tuovinen, J.-P., Barrett, K. and Styve, H. 1994. Transboundary Acidifying Pollution in Europe: Calculated fields and budgets 1985–93. – EMEP/MSC-W Report 1/94, Norwegian Meteorological Inst., Blindern, Norway.

Ulrich, E. and Williot, B. 1993. Les dépots atmosphériques en France de 1850 à 1990. – Office National des Forets, Fontainebleau, France.

United Kingdom Review Group on Acid Rain. 1990. Acid deposition in the United Kingdom 1986–1988. Third Report. – Warren Spring Lab., UK.

Voldner, E. C., Barrie, L. A. and Sirois, A. 1986. A literature review of dry deposition of oxides of sulphur and nitrogen with emphasis on long-range transport modelling in north America. – Atmospher. Environ. 20: 2101–2123.

Westling O., Hallgren Larsson, E., Lövblad, G. and Sjöberg, K. 1992. Deposition och effekter av luftföroreningar i södra och mellersta Sverige. – IVL Report B1079. Swedish Environ. Res. Inst., IVL-Aneboda Lammhult, Sweden, (in Swedish, English summary).

Ecological Bulletins 44: 35–42. Copenhagen 1995

Concentrations of tropospheric ozone in Sweden

Karin Kindbom, Gun Lövblad, Kjell Peterson and Peringe Grennfelt

Kindbom, K., Lövblad, G., Peterson, K. and Grennfelt, P. 1995. Concentrations of tropospheric ozone in Sweden. – Ecol. Bull. (Copenhagen) 44: 35–42.

Results of measurements of the concentrations of ozone in the lower layers of the atmosphere are presented, with respect to variations in time and space. Exceedance of critical exposure doses is calculated from ozone concentrations measured at the Swedish background monitoring sites. The critical exposure doses used in the calculations of exceedance are those discussed and proposed for forest trees and for agricultural crops at the UN ECE Workshop on Critical Levels of Ozone, 1994. The variations of ozone concentrations have been investigated as yearly means, seasonal variations, short term variations (episodes) and diurnal variations. No great differences in yearly means between the southern and northern part of the country were found. Seasonal differences are evident; there are higher monthly mean values in the north in the winter and an earlier rise of ozone concentration in early spring, while monthly means during summer are higher in the south. Episodes with high concentrations of ozone, mainly in the summer, are more frequent in the southern part of the country, but the numbers and magnitude of the episodes vary greatly between years. Diurnal variations follow a general pattern with the lowest ozone concentrations measured during early morning hours and the highest in the afternoons. Calculations of AOT40, accumulated exposure over the threshold concentration of 40 ppb, show that the critical dose proposed for forest trees was exceeded regularly in the south while it was not exceeded at the two stations in the north during 1990–1993. The critical exposure doses suggested for crops, estimated to cause production losses of 5 or 10%, were both exceeded. The calculations indicate that there are risks of ozone damage to forest trees primarily in the south, as well as of production losses for crops in the whole country, although more extensively in the south.

K. Kindbom, G. Lövblad, K. Peterson and P. Grennfelt, IVL, Box 47086, S-402 58 Göteborg, Sweden.

Introduction

Tropospheric ozone and other photochemical oxidants are formed by photochemical oxidation processes as a result of the anthropogenic emissions of nitrogen oxides (NO_x) and volatile organic compounds (VOC). The ozone formation occurs during anticyclonic conditions, warm and sunny weather with low wind speeds. Episodes with high concentrations of ozone may damage living organisms, including human health. Increased long-term concentrations of ozone will damage crops, forests and natural vegetation, but also materials such as polymers.

The occurrence of ground-level ozone in Europe is mainly a large scale pollution problem. Ozone is formed in polluted air masses moving slowly over the continent. Local emissions, adding precursors, are mixed into the air masses and ozone is formed under the influence of sunlight. Ozone reacts, very fast, with nitrogen monoxide

emitted to the air, causing a destruction of ozone. The ozone concentrations found in urban areas are therefore generally lower than in rural areas with cleaner air.

The background concentrations of ozone over Europe have increased significantly during the last 50–100 years. Measurements during the last decades of the 19th century in France and from the 1950's until the present in Germany, indicate earlier long term mean concentrations of 10–15 ppb (Volz and Kley 1988). The yearly mean concentrations of ozone in Sweden are presently c. 30 ppb.

Emissions in Europe as a whole contribute to the ozone episodes observed in Sweden. It is not possible to calculate budgets of the ozone formation over Europe in the same way as for sulphur and nitrogen within EMEP (European Monitoring and Evaluation Programme). This is mainly because ozone formation is caused by both nitrogen oxides and volatile organic compounds. Sources for these compounds cannot be considered separately since their role in the ozone formation is interrelated. Moreover, there are still large uncertainties with respect to emissions and chemistry. Preliminary results indicate, however, that emissions occurring at some distance from Sweden contribute more than the Swedish emissions to the episodic ozone (Simpson 1991, 1992). A reason for this is that the emission density relevant to ozone formation is generally much less in Scandinavia than on the continent. In Germany, for example, the annual NO_x emission averaged over the country is 39 kg ha^{-1}, given as nitrogen, while it is 7 kg ha^{-1} in the southern parts of Sweden (up to c. 59°N) and much less further north (Sandnes 1993). The Swedish emissions will in turn contribute to the ozone concentrations in other countries.

In order to reduce, primarily, episodic ozone in Europe a protocol on a 30% reduction of VOC by 1999 was adopted within the United Nations Economic Commission for Europe (UN ECE) Convention on Long-Range Transboundary Air Pollution, in 1991. The effect on ozone concentrations by reducing the emissions of precursors is difficult to estimate, depending on the combined effect of nitrogen oxides and volatile organic compounds. Reductions according to the protocol are expected to result in the largest reductions of ozone concentrations in central and western Europe. However, the commitments made so far by the European countries will not result in large reductions. In Sweden, the main result of the 30% reduction in the emissions of volatile organic compounds in Europe is expected to be reduced peak concentrations during episodes, and probably fewer episodes. The average concentrations are not expected to change by more than a few percent (Simpson and Styve 1992).

This paper presents the concentrations of ozone in Sweden, with temporal and spatial variations. The exceedances of the critical exposure doses are calculated from measured ozone concentrations at the Swedish monitoring stations.

Critical levels of ozone

Within the UN ECE Convention, critical levels have been defined for ozone, based on the present scientific knowledge of effects on crops and forests. These critical levels have been modified gradually and the concept of critical levels for ozone is presently based on a cumulative exposure value, given as ppb-hours, exceeding a given threshold level of ozone during a specified time period.

At the UN ECE workshop on Critical Levels for Ozone in Bern in November 1993, the meeting agreed upon a critical level concept with respect to agricultural crops as well as to forest trees (UN ECE 1994). The proposed long-term critical level for ozone was expressed as cumulative exposure over the threshold concentration of 40 ppb for both agricultural crops and forest trees. The exposure index is referred to as AOT40, accumulated

exposure over a threshold of 40 ppb. The way the critical levels for ozone are to be formulated is still under consideration. There is however a general agreement that the exposure above a threshold level is the most relevant effect measure. It is however not clear that 40 ppb is the best threshold value. It may be lower or higher.

The way the critical levels are set is not only important for the effects, it may also be very important for the control strategies. Since many of the monitored one-hour values are close to 40 ppb in northern Scandinavia, the exceedances of critical levels will increase substantially if, for example, 30 ppb is chosen as the threshold value instead of 40 ppb. Calculations of AOT30, AOT40 and AOT50 for Scandinavia and central Europe show that the exceedances in ppb-h are more sensitive to the choice of base concentration in the outskirts of Europe than in central Europe (Grennfelt and Beck 1994). In the following, the critical level will be referred to as critical exposure dose.

At the UN ECE workshop, a provisional AOT40 value of 10 000 ppb-h during six months and 24 h a day was proposed for forest trees. In view of the small data-base available, the critical level proposed is provisional in character and is based on observed decreases in whole plant biomass in relation to a control. It is expected to be re-evaluated at later workshops as more data become available (UN ECE 1994).

For crops, data from four countries in Europe concerning production losses for spring wheat caused by ozone exposure were used as a basis for the recommendations of critical levels. Regression calculations indicated that a dose exceeding 5 300 ppb-h, above 40 ppb, during May–July, daylight hours, would result in a production loss of 10%. Accordingly an exceedance of 2 600 ppb-h would result in a 5% production loss. Recommendations were made at the UN ECE workshop not to use AOT40 values for yield reductions below 10%. The uncertainty in experimental determination of yield losses between 0% and 10% was considered to be too great (UN ECE 1994).

An estimate of physical and economic production losses in Sweden, caused by anthropogenic ozone, suggests an average yield loss of c. 9% for all crops. Because of uncertainty in the basic material, mainly in the dose-response functions, a probable range of 6–21% for yield losses was given by Hasund et. al. (1990). These authors plotted the concentrations against yield to produce linear regression lines, so the concept of critical dose was not used. When evaluated on the basis of domestic prices, the yield losses in Sweden were estimated to 1.4 billion SEK in 1988 prices, with a probable range of 0.97–3.3 billion SEK. As an average for the total area of arable land utilized, the yield loss amounted to 520 SEK ha^{-1}.

A source of uncertainty when estimating damage to crops is that the concentration of ozone, which is used as the scientific base for the critical exposure levels, is determined at a height just over the plants (10 cm). There is, however, a well developed height gradient over a crop field. The ozone concentrations from the Swedish mon-

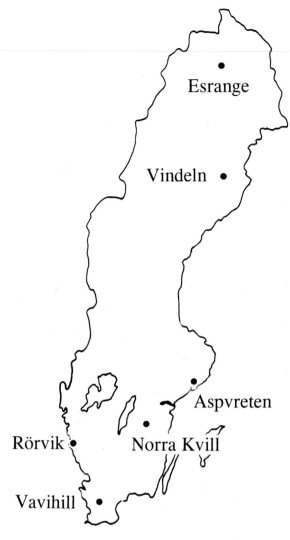

Fig. 1. Location of the Swedish monitoring sites for ozone.

of a one-hour guideline in the range of 75–100 ppb. In order to reduce the potential for adverse acute and chronic effects, and to provide an additional margin for protection, an eight-hour guideline for exposure to ozone of 50–60 ppb is recommended (WHO 1987). Within the European Community, a system of exchange of information and warnings about high ozone concentrations has been suggested. The present level for warning is set at 90 ppb as a one-hour mean value. This warning system will be introduced in Sweden in 1995.

Swedish ozone data

The monitoring data from six sampling sites for surface ozone in Sweden, five of which are included in the EMEP network, have been evaluated. The location and description of the sites are given in Fig. 1 and in Table 1. All measurements are made in an open landscape, but at different altitudes above sea level and in different surroundings and climates. The EMEP measurement sites are selected, as far as possible, not to be influenced by night inversions and by local vegetation uptake. Open places, preferably situated at altitudes higher than the local surroundings, have been selected. These sites are normally exposed to higher ozone concentrations than areas close to forests and at low altitudes. The effect of local topography on measured ozone concentrations was clearly shown in measurements made during three summers in an area on the Swedish west coast, Stenungsund (Sjöberg and Grennfelt 1990). However, it cannot be determined if the frequency of exceedances of critical exposure dose, calculated from available monitoring data, are significantly influenced by local topography.

The ozone concentrations, measured as one hour averages all year around, have been monitored since 1985 at all of the stations except Esrange where measurements began in late 1990.

itoring network, used in the present calculations, are measured at a height of between 3–5 m. These data are used directly to estimate the exposure dose, without compensation for the gradient, which will lead to an overestimation of the damage.

According to WHO guidelines, existing data on the health effects of ozone, have led to the recommendation

Variations in ground-level ozone concentrations in Sweden

The variations in ozone concentrations can be considerable, on a daily, a seasonal or a yearly basis. Most of

Table 1. Locations of the Swedish monitoring stations for ozone.

Station	Latitude/longitude	Altitude m a.s.l.	Site
Esrange	67°53'N 21°04'E	475	Northern inland, north of the arctic circle
Vindeln	64°14'N 19°46'E	230	In the north, 50 km from the east coast
Aspvreten	58°48'N 17°23'E	20	Close to the Baltic Sea, 100 km south of Stockholm
Norra Kvill	57°43'N 15°43'E	255	On the south Swedish highlands
Rörvik	57°25'N 11°56'E	10	Close to the sea, 50 km south of Gothenburg
Vavihill	56°01'N 13°09'E	172	In the extreme south, on the ridge of Söderåsen

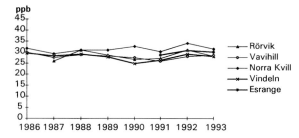

Fig. 2. Yearly mean values of ground-level ozone (ppb) at Swedish EMEP stations 1986–1993.

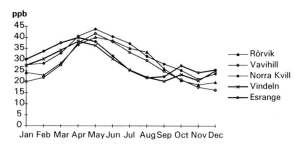

Fig. 3. Monthly averages of ground-level ozone 1991–1993 (ppb) at Swedish EMEP stations.

these temporal variations can be explained by the weather conditions. However, variations on a regional or local scale are also attributable to the influence of local pollution and to local deposition climate.

Annual means

The annual mean values of ground-level ozone at all the Swedish monitoring sites are of the order of 30 ppb. There is no clear distinction between the yearly average in the southern and the northern part of Sweden (Fig. 2). However, the annual mean values are not of particular interest when evaluating the threat to vegetation and health. In order to make a risk analysis, a more detailed examination of the data is necessary.

Seasonal variations

The seasonal variations of ozone concentrations depend mainly on the occurrence of favourable conditions for chemical formation processes in the atmosphere, which prevail on the light and warm part of the year, spring and summer. The seasonal variations of ozone, presented as averages for specified months during 1991, 1992 and 1993, are shown in Fig. 3. In the north, at Esrange and Vindeln, the rise in ozone concentrations occurs earlier in spring, March–May, than at the stations further south where the highest concentrations are from April to June. In winter the monthly mean values are higher in the north than in the south, while in summer they are higher in the south. The early spring peak, especially pronounced at the northern sites, has not been fully evaluated. These higher concentrations in early spring might partly be attributable to increased transport of stratospheric ozone during the winter and early spring, and partly to the photochemical degradation of VOC and NO_x accumulated in the polar troposphere during the winter months. The intensive solar radiation caused by the snow reflec-

tion in the north during this period of the year may also contribute to the formation of ozone.

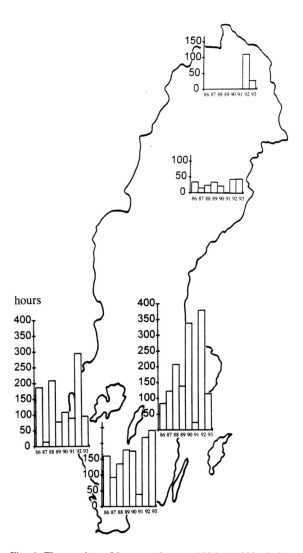

Fig. 4. The number of hours each year, 1986 to 1993, during which the hourly mean value of ozone exceeded 60 ppb. (Measurements at Esrange began in late 1990).

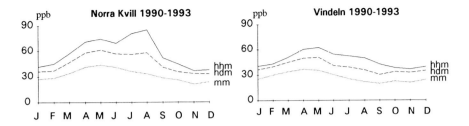

Fig. 5. Highest hourly mean (hhm), highest daily mean (hdm) and monthly mean (mm) each month, averaged over the years 1990–1993 at the EMEP stations Norra Kvill and Vindeln.

Short term variations

Episodes with high concentrations of ozone occur mainly during the summer season, April to August. Hourly means >60 ppb are more frequent in the southern part of the country than in the north but the numbers of exceedances vary greatly between years. In Fig. 4 the numbers of hourly mean values >60 ppb in 1986 to 1993 at the different monitoring sites are given.

At the northern stations, Vindeln and Esrange, the gap between monthly mean concentrations and the highest daily and hourly means is not as pronounced as it is during the summer season further south. In Fig. 5 the averages of the highest hourly and daily means each month, as well as the monthly mean, are given for Vindeln and Norra Kvill, as an example of the situation at a northern and at a southern site. The averages are computed for four years, 1990–1993. The differences between the stations can be explained by the fact that episodes with high ozone concentrations, caused by European emissions, influence the southern parts of the country much more frequently than the northern parts.

Diurnal variations

The pattern of variation in diurnal ozone concentrations is mainly regulated by sunlight. A calculation of the hourly means during the growing season, April to September, 1992 and 1993, was made to study the diurnal variation of ozone concentrations at the different sites (Fig. 6). The highest concentrations are measured during the after-

noons and the lowest in the early morning hours. The low concentrations are partly attributable to night time inversions and subsequent deposition of ozone to the ground or to the vegetation. The frequency of inversions at night is in turn dependent on local meteorology at the location of the monitoring site, influenced mainly by

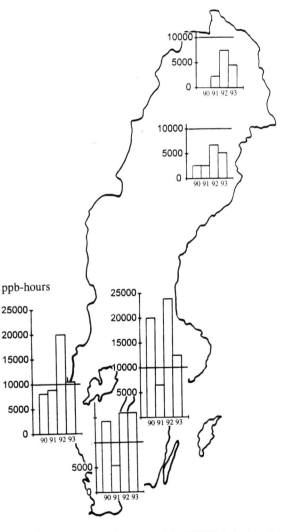

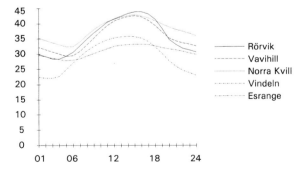

Fig. 6. Hourly mean values of ozone (ppb) averaged during April–September, 1992 and 1993.

Fig. 7. Ozone exposure doses as ppb-h AOT40 during April–September, 1990–1993. Lines are at 10 000 ppb-h AOT40, the critical exposure dose suggested for forest trees. (Measurements at Esrange began in late 1990).

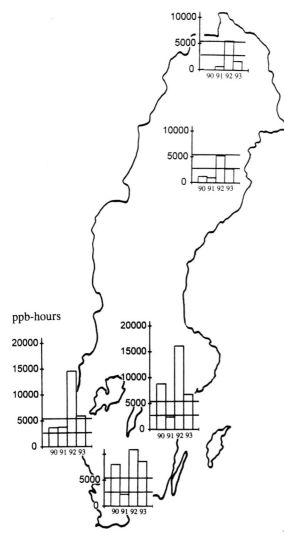

ppb-hours

Fig. 8. Ozone exposure doses as ppb-h AOT40 during daylight hours in May–July, 1990–1993. Lines are at 5 300 ppb-h and 2 600 ppb-h AOT40, the critical exposure doses suggested for crops. (Measurements at Esrange began in late 1990).

altitude. A site located in a valley is more likely to have inversions at night than a site on a local hill, but the degree of night time destruction of ozone is also dependent on the local surrounding vegetation. In this respect there are differences between the stations. In Esrange, for example (Fig. 6), the diurnal variation, as an average over the growing season in 1992, is smaller than at the other stations.

Ozone concentrations in urban areas

Ozone concentrations in urban areas generally are lower than in rural, cleaner air. In urban areas, with more air pollution, reactions between ozone and nitrogen monox-

ide cause a destruction of ozone. Measurements in 1992 and 1993 in the urban area of Gothenburg indicate that the monthly mean values are 20–30% lower than at Rörvik, in a rural area 50 km south of the city (Brandberg 1993 and 1994, Kindbom et al. 1993 and unpubl. data). However, in the plume from Gothenburg, ozone precursors contribute to the rural ozone concentrations down-wind (Sjöberg and Grennfelt 1990).

Critical exposure doses

Below, the risk of damage to Swedish forest trees caused by ozone is evaluated, based on exceedance of the six months critical exposure dose of 10 000 ppb-h >40 ppb. Furthermore, the production losses for crops are evaluated from the exceedance of 40 ppb during daylight hours in May, June and July. Only measurement data from the six Swedish EMEP monitoring sites, at 3 to 5 m height, are used.

Critical exposure doses, calculated according to the proposals at the UN ECE workshop, are presented in Fig. 7 and 8 for the years 1990 to 1993. In the south the critical exposure dose of 10 000 ppb-h >40 ppb, suggested for forest trees, was exceeded in two or three of the four calculation periods (Fig. 7). There were no exceedances at the two northern stations, Esrange and Vindeln.

Table 2. Ozone exposure doses >40 ppb during April–September and during daylight hours in May–July. The availability of data is given as a percentage of the total number of hours for each calculation period.

Station	Year	ppb-h >40 ppb Apr–Sep	% available	ppb-h >40 ppb May–July, daylight	% available
Esrange	1991	2243	97	649	99
	1992	7411	96	5581	96
	1993	4505	87	1675	73
Vindeln	1990	2591	100	1213	100
	1991	2558	96	974	82
	1992	6714	98	5222	82
	1993	5123	98	2684	83
Aspvreten	1990	9781	85	3869	72
	1991	5716	87	2455	77
	1992	12732	92	8548	89
	1993	7646	87	4979	94
N. Kvill	1990	20010	97	8827	97
	1991	6476	97	2393	94
	1992	23853	97	16127	100
	1993	12429	96	6738	95
Rörvik	1990	8075	96	3696	95
	1991	8847	100	3749	100
	1992	19953	98	14567	100
	1993	10450	100	5895	99
Vavihill	1990	14204	97	7986	99
	1991	5360	90	2257	88
	1992	15911	99	10877	98
	1993	15918	98	8613	98

Table 3. Exceedances (ppb-h) >30, 40 and 50 ppb at TOR (Tropospheric Ozone Research) and EMEP sites below 1000 m a.s.l., April–September 1989. N denotes number of monitoring stations (from Grennfelt and Beck 1994).

Area	N	Number of ppb-h exceeding		
		30 ppb	40 ppb	50 ppb
N Scandinavia, >62°N	6	4000–16400	1200–5200	80–1100
S Scandinavia, <62°N	17	10600–24300	2800–10500	230–4000
European Continent, coastal	4	25000–49000	8700–27800	3200–13500
European Continent, inland	18	9800–60000	5200–41500	2300–26500

The critical exposure dose suggested for crops, 5 300 ppb-h >40 ppb during daylight hours in May, June and July, was exceeded at the three stations in the south of Sweden in three of the four years included in the calculations. The exception was 1991, which was a very cold and rainy summer with exceptionally low concentrations of ozone throughout the country (Fig. 8 and Table 2). In the north, at Vindeln and Esrange, 5 300 ppb-h AOT40 was exceeded, or very nearly exceeded, in 1992. The alternative critical exposure dose for crops, 2 600 ppb-h >40 ppb during daylight hours in May to July, was exceeded at the four southern stations in 1990 and at all monitoring sites in 1992 and 1993, with the exception of Esrange (Fig. 8 and Table 2). In 1991, the AOT40 of 2 600 ppb-h was only exceeded at one site, Rörvik.

In the evaluation of critical exposure doses, which is a cumulative measure, the availability of data is an important factor. In Table 2, monitoring data available for the calculations are given as a percentage of the total number of hours for each calculation period. With few exceptions the availability of data is >90%. Calculation periods in which data cover is <90% of the time are the May–July period 1991 at Vavihill, 1991–93 at Vindeln and 1993 at Esrange. In 1993 at Esrange, the availability of data was <90% also in the April–September calculation period. At Aspvreten the data cover is >90% only in May–July 1993 and during April–September in 1992.

As for the AOT40 calculations, the availability of data in the north, Vindeln and Esrange, was lower in 1992 and 1993 than at the southern stations, with the exception of Aspvreten. The cumulative exposure in the north may thus have been underestimated for these years, compared to the southern stations. In order to estimate the possible contribution from missing values, corrective calculations have been made for Vindeln and Esrange. The calculations are based on the assumption that the frequency distribution of ozone concentrations for the missing data is equal to the measured data. At Vindeln, the exposure dose for crops in 1992 and 1993 increases by 22% and 20%, respectively, when correcting for missing values. In 1992, the exposure increases from 5 222 ppb-h AOT40 to 6 352, clearly above the critical level of 5 300 ppb-h for crops. In 1993, the corrected exposure results in an increase from 2 684 to 3 227 ppb-h AOT40. At Esrange, corrections with respect to missing values in 1993 result in an increase in exposure for crops of 36%, from 1 675 to

2 284 ppb-h AOT40, while the exposure dose to forest trees increases 14%, from 4 505 to 5 148 ppb-h. None of the calculations implies any important changes in exceedance of the proposed critical levels, even if the exposure doses increased when corrected with respect to missing values.

The consequence of using AOT40, as proposed in the UN ECE Workshop, to calculate exceedance of critical doses in Scandinavia can be discussed from several aspects. This includes factors such as time periods for calculation, climatic factors, important crops in different regions and the choice of base level. In northern Scandinavia, for example, the growing season is shorter than further south and does not correspond to the suggested time periods. Additionally, the AOT40 concept does not take into account modifying factors, such as increased sensitivity of forest trees to ozone in a harsher climate. Furthermore, the critical level for crops is based on experimental results for spring wheat. In the north, no wheat is grown while other crops, such as pasture are important. Pasture is only slightly less sensitive to ozone than wheat (Pleijel pers. comm.).

The magnitude of exceedance of critical levels in different parts of Europe is influenced by the choice of base levels. In Table 3 the changes in exposure dose when using 30, 40 or 50 ppb as base level are illustrated. In northern Scandinavia the exposure above the base level increases by a factor of more than three if AOT30 is used instead of AOT40. In southern Scandinavia (<62°N) the increase is slightly less while the increase at inland locations in continental Europe is by a factor of only one and a half to two (Grennfelt and Beck 1994). When using AOT50 instead of AOT40, the differences are even larger; the decrease in exposure from AOT40 to AOT50 in northern Scandinavia is between 5 and 15 times while in continental Europe the decrease is only a factor of two.

Conclusion

Ozone concentrations have increased significantly during recent decades as a consequence of emissions of nitrogen oxides and volatile organic compounds, in Sweden as well as in the rest of Europe (Volz and Kley 1988, Koechler et al. 1988, Areskoug 1991). Every spring and summer, crops and forests are exposed to ozone concentrations which may have negative effects. Damage to agricultural crops has been observed and damage to forest trees is suspected to be caused at least partly by ozone. Ozone concentrations which may threaten human health have only been observed on a few occasions during the last five years in Sweden.

A conclusion of the calculations of AOT40 for forest trees and crops, proposed at the UN ECE workshop and applied to ozone measurement data in Sweden, is that exceedances occur regularly in the southern part of Sweden and to some extent in the north. There is risk of damage to forest trees, mainly in the south, as well as production losses for crops in the whole country, although possibly more extensively in the south.

The availability of data during the May–July calculation periods was somewhat less at the two northern stations. This loss of data might influence the results on exceedance, calculated as ozone exposure dose. Corrections of the exposure doses with respect to missing values were made. The calculations showed an increase in exposure dose but did not change the overall picture significantly.

In the future no great changes in ozone exposure are expected. The reduction of emissions of VOC and NO_x, both precursors of ozone, are subject to international regulations in two protocols within the UN ECE Convention on long-range transboundary air pollution. The NO_x Protocol involves a 30% reduction of emissions in only a few of the European countries, while most have agreed on a freezing of emissions at the levels of the mid-1980's. In the VOC Protocol a 30% reduction by 1999, compared to an annual level between 1984 and 1990, is agreed upon. The result of implementing the protocols is probably a faster reduction in VOC-concentrations than in NO_x-concentrations in Europe. Since the formation of ozone in Europe is limited by the availability of VOC, rather than by NO_x, this will probably result in fewer episodes with high ozone concentrations in the future (Simpson and Styve 1992).

In Swedish rural air the concentrations of ozone are, however, not expected to decrease in the immediate future. The ozone formation in these areas is primarily limited by NO_x, rather than VOC. The emissions of NO_x in Sweden are expected to decrease by c. 20% from 1991 to 2000 (Swedish Environmental Protection Agency 1994). These emission reductions, primarily from road traffic, are expected mainly to influence urban air. On the other hand, NO_x emissions from sources influencing rural air, such as emissions from air traffic and shipping are expected to increase in Sweden during the next five years. It is difficult to predict how these expected changes in emissions of VOC and NO_x in Sweden, in connection with long range transport, will influence future ozone concentrations.

References

Areskoug, H. 1991. Övervakning av marknära ozon vid Aspvreten. – Swedish Environ. Protect. Agency, Solna, Report 3941, (in Swedish).

Brandberg, J. 1993. Årsrapport luftföroreningar i Göteborg 1992. – Miljö- och Hälsoskydd, Göteborg, 1993–11, (in Swedish).

– 1994. Årsrapport luftföroreningar i Göteborg 1993. – Miljö- och Hälsoskydd Göteborg, 1994–1, (in Swedish).

Grennfelt, P. and Beck, J. 1994. Ozone concentrations in Europe in relation to different concepts of the critical levels. – In: Fuhrer, J. and Achermann, B. (eds), Critical levels for ozone – a UN ECE workshop report. Schriftenreihe der FAC Liebefeld-Bern, No. 16.

Hasund, K. P., Hedevåg, L. and Pleijel, H. 1990. Ekonomiska konsekvenser av det marknära ozonets påverkan på jordbruksgrödor. – Swedish Environ. Protect. Agency, Solna, Report 3862, (in Swedish with English summary).

Kindbom, K., Persson, K., Sjöberg, K. and Lövblad, G. 1993. Atmosfärkemisk övervakning vid IVL's stationer inom EMEP. – Swedish Environ. Protect. Agency, Solna, Report 4211, (in Swedish with English summary).

Koechler, U., Wege, K., Hartmannsgruber, R. and Claude, H. 1988. Vergleich und Bewertung von verschiedene Gersets zur Messing des atmosphaerischen Ozons zur Absicherung von Trendaussagen. – BPT-Bericht 1/88, GFS, München.

Sandnes, H. 1993. Calculated budgets for airborne acidifying components in Europe, 1985, 1987, 1988, 1989, 1990 and 1991. – EMEP/MSC-W Report 1/92, Norwegian Meteorol. Inst., Oslo.

Simpson, D. 1991. Long period modelling of photochemical oxidants in Europe, some properties of targeted VOC emission reductions. – EMEP MSC-W Note 1/91.

– 1992. Long period modelling of photochemical oxidants in Europe, calculations for April–September 1985, April–October 1989. – EMEP/MSC-W Report 2/1991.

– and Styve, H. 1992. The Effects of the VOC protocol on ozone concentrations in Europe. – EMEP MSC-W Note 4/92.

Sjöberg, K. and Grennfelt, P. 1990. Ozonmätningar kring Stenungsund 1989. – IVL Report L89/377, (in Swedish).

Swedish Environmental Protection Agency 1994. Strategy for sustainable development. Proposals for a Swedish programme – Enviro '93. – Swedish Environ. Protect. Agency, Solna, Report 4266.

UN ECE 1994. Critical levels for ozone – a UN-ECE Workshop report. – Schriftenreihe der FAC Liebefeld, No. 16. Swiss Federal Res. Stat. Agricult. Chemistry Environ. Hygiene, Liebefeld-Bern, Switzerland.

WHO 1987. Air quality guidelines for Europe. – WHO Regional Publ., European Series No. 23, Copenhagen.

Volz, A. and Kley, D. 1988. Evaluation of the Montsouris series of ozone measurements made in the nineteenth century. – Nature 332: 240–242.

Ecological Bulletins 44: 43–53. Copenhagen 1995

Acid deposition and soil acidification at a southwest facing edge of Norway spruce and European beech in south Sweden

Anna-Maj Balsberg Påhlsson and Bo Bergkvist[1]

Balsberg Påhlsson, A.-M. and Bergkvist, B. 1995. Acid deposition and soil acidification at a southwest facing edge of Norway spruce and European beech in south Sweden. – Ecol. Bull. (Copenhagen) 44: 43–53.

The aim of the study was to use the deposition gradient across a forest edge to study the influence of acidifying substances on soil changes as well as on soil acidification rate and processes. Element fluxes in throughfall and soil solution from the B horizon were measured during one year, at different distances from the forest edges of a spruce and a beech forest in Scania, southern Sweden. The exchangeable amounts of elements in the soil were also determined. Bulk deposition was collected in a non-forested area adjacent to the forests.

In the spruce stand, element fluxes were considerably influenced by the forest edge. Throughfall fluxes (as a substitute for total deposition) at the spruce forest edge were 48 kg SO_4-S ha^{-1} yr^{-1} and 76 kg N_{inorg} ha^{-1} yr^{-1} (55% NH_4-N and 45% NO_3-N). In the interior of the spruce stand, the deposition was only 30 kg SO_4-S ha^{-1} yr^{-1} and 32 kg N_{inorg} ha^{-1} yr^{-1}.

The throughfall fluxes of base cations were also substantially increased at the spruce forest edge. Nitrogen deposited at the edge was quantitatively leached through the B horizon as NO_3, while almost all N deposited in the interior of the stand was taken up by tree roots or immobilised in the soil. The throughfall fluxes of SO_4-S and base cations were only slightly higher at the beech forest edge than in the interior of the beech stand, while there was no difference in N flux.

Proton budgets showed the calculated total proton load (TPL) to be c. 7 times higher at the spruce forest edge, than in the interior of the stand. TPL to the beech stand was not influenced by distance to the forest edge and was similar to that in the interior of the spruce stand. In the spruce forest, just behind the edge, 60% of the TPL was attributable to N transformations, while inside the stand direct H^+ deposition was the main H^+ source and N transformations were of minor importance. Protons were buffered mainly by base cation exchange and silicate mineral weathering. Weathering rates may be too low to keep pace with base cation losses, particularly at the spruce forest edge. As a consequence, exchangeable base cations and pH-H_2O were lower at the spruce forest edge and soil acidity was increased in comparison to the interior of the stand.

A.-M. Balsberg Påhlsson and B. Bergkvist, Dept of Ecology, Plant Ecology, Univ. of Lund, Ecology Building, S-223 62 Lund, Sweden.

Introduction

During recent decades an extensive acidification of forest soils of southern Sweden has become evident. Decreased pH has been measured both in the topsoil and in the lower mineral horizons (Falkengren-Grerup 1986, Hallbäcken and Tamm 1986). The exchangeable amounts of base cations in different soil horizons have decreased by c. 50% in 35 years, while exchangeable Al has doubled

(Falkengren-Grerup et al. 1987, Falkengren-Grerup and Tyler 1992). In southwest Sweden, where the acid deposition rate is highest, soil acidification has been detected at a depth well below 2 m, but in the northeast, where acid deposition rate is lower, soil acidification has not penetrated as deeply (Eriksson et al. 1992).

A deposition gradient of much smaller scale may be found in an exposed forest edge and a few meters into the stand; atmospheric deposition is highest at the forest

---

[1]Corresponding author

edge, and decreases steeply toward the interior of a stand (Hasselrot and Grennfelt 1987, Draaijers et al. 1988, Beier and Gundersen 1989). Deposition of coarse fraction particles is strongly influenced by the forest edge, while deposition of fine particles and gases is less influenced by distance to the edge (Wiman and Lannefors 1985, Wiman and Ågren 1985, Beier 1991).

Floristic changes caused by atmospheric deposition have occurred at forest edges in northeastern France (Thimonier et al. 1992). Soil changes are also likely to be found along the deposition gradient of an exposed forest edge, which may be regarded as a long term acidification "experiment". Studies at a forest edge may therefore give indications of future soil changes and acidification rates and processes in conditions of constant or increasing amounts of acidifying substances in the atmosphere.

The aims of the study were to quantify, as a function of distance from the forest edge into the spruce and the beech stands: i) the fluxes of H^+ and acidifying substances, base cations and Al in throughfall water and in the soil solution; ii) the H^+ load to the soil; iii) the influence of different H^+-producing and H^+-consuming processes; iv) soil acidity and exchangeable element soil pools.

Site description

The study was carried out in western Scania, the southernmost province of Sweden. This part of Sweden is densely populated (c. 100 inhabitants km^{-2}) and includes large areas of intensely managed agricultural land. Adjacent to large cities and to continental Europe, the province is subjected to large emissions of atmospheric pollutants, including S and N compounds. The recent bulk deposition (collected with permanently open funnels) to the region is 10 kg S ha^{-1} yr^{-1} and 13 kg N ha^{-1} yr^{-1} (IVL 1991).

The study area is situated at Stubbaröd on the southwestern slopes of the hill Söderåsen (58°02'N, 13°04'E', altitude c. 115–120 m a.s.l.), 37 km north of Lund and 24 km east of the strait Öresund. The climate is temperate with a mean annual temperature of 7.2°C and a precipitation of 795 mm, according to a meteorological station 10 km NE of Stubbaröd. Usually, the precipitation is fairly evenly distributed over the year, though somewhat lower during the spring months. From December to March part of the precipitation may fall as snow, but snow cover is seldom continuous throughout the winter. As the prevailing wind and precipitation directions are southwesterly the study area is exposed to a high degree of atmospheric pollution which makes it suitable for these studies. To the SW of the area the land is open farmland mainly dominated by pastures. Except locally, where some bushes are present, there is a sharp transition between the spruce and beech forest stands and the pastures.

The Norway spruce *Picea abies* L. forest is an even-aged stand and the first generation on a former tilled field, planted c. 30 yr ago. The canopy is quite closed, c. 3000 trees ha^{-1} and a forest floor vegetation is absent. With the exception of some trees at the edge, which show symptoms of damage (crown thinning etc.) the stand shows no clear symptoms of forest decline.

The European beech *Fagus sylvatica* L. forest is composed of mainly even-aged trees, c. 100 yr old. The canopy is closed and the number of trees is c. 400 trees ha^{-1}. The shrub layer is sparse (young beech plants) and the forest floor cover is patchy (mainly *Lamium galeobdelon, Maianthemum bifolium, Oxalis acetosella* and *Deschampsia flexuosa*).

The bed-rock is gneiss, covered by moraine originating from siliceous rocks. The soil (a sandy loam) has a high boulder content, the fine earth contains c. 15% clay. The site is well drained with a Dystric Cambisol (FAO). In the interior of the spruce stand the top soil (0 to 5 cm) has a pH-H_2O of c. 5.0, and pH-H_2O of the mineral soil (20–30 cm) is c. 5.4. Corresponding pH-H_2O values for the beech stand varied between 4.5 and 5.3 in the top soil and in the mineral soil.

Materials and methods
Installation and sampling

Bulk deposition was collected in four samplers (shaded polyethylene, PE, funnels, diam. 20 cm, with PE sieves, connected to 10 l PE bottles) placed 3 m above ground in the pasture area, 50 m apart from the front of the study area.

At each of five distances from the south-west forest edge into the spruce and the beech stands, samples were taken of throughfall, soil solution and soil. The sampling stations were situated 1, 5, 15, 25 and 50 m from the forest edge into the spruce stand, and 2, 10, 25, 50 and 75 m into the beech stand. At each station, throughfall was collected in five shaded samplers (funnels as above, 5 l bottles) placed c. 0.5 m above ground. Bulk deposition and throughfall were weighed in the field. Bulk deposition samples were analysed individually but the five throughfall samples at each station were pooled before analysis. A 1.3 l subsample was brought to the laboratory. Soil solution from the B horizon (at 30–40 cm soil depth) was collected from three ceramic cup lysimeters (P 80, continuous vacuum 60 to 80 kPa) per sampling station. Collecting bottles (Duran glass) were placed in a PVC container below ground to keep samples dark and cool. The three lysimeter samples were pooled before analysis. To allow the lysimeters to equilibrate with the soil solution, the first few lysimeter solutions, obtained within 2 months of installation, were discarded. Afterwards, soil solution, bulk deposition and throughfall were sampled each month from May 1991 to May 1992.

Soil samples were taken in spring 1991 from the A (0 to 5 cm) and the B (20 to 30 cm) horizons. At each distance six pits were sampled separately using a steel cylinder, area 38 cm^2; four samples from each pit and soil depth were bulked together. All samples were stored in PE bags at c. 4°C prior to analysis.

Sample pretreatment and analysis

Unfiltered water subsamples were taken within 14 h for pH determination in the laboratory. Soil solution pH was determined at 6 to 8°C (annual mean soil temperature of the region) in view of the highly temperature sensitive Al chemistry (Lydersen 1990). As Al concentrations of bulk deposition and throughfall waters are almost negligible, pH was determined at room temperature.

Throughfall samples were filtered (OOR, Munktell, STORA) whereas bulk deposition and soil-solution samples were not. A subsample was taken for determination of NH_4, NO_3, SO_4, Cl and dissolved organic carbon (DOC). Soil solution and throughfall samples were acidified (HNO_3, anal. gr.) prior to metal analysis. Subsamples of bulk deposition (500 to 1000 ml) for metal analysis were transferred to a 1 l Erlenmeyer flask which was covered with a hood of filter paper and evaporated at 105°C. The residue was treated with 10 ml conc. HNO_3 (anal. gr.) to destruct organic matter and redissolve the metals. The volume was reduced by evaporation to c. 2 ml and diluted with deionized water to 25 ml prior to analysis.

Fresh soil samples were sieved (nylon net, mesh size 2 mm). Dry weight was determined at 105°C and loss of ignition at 600°C. Soil pH was determined after shaking fresh soil (15 g of A horizon and 25 g of B horizon samples) with 50 ml deionized water. Exchangeable base cations, Al, and H$^+$ were extracted with 1 M NH_4Cl (10 g fresh soil of the A horizon and 25 g of the B horizon, 100 ml extractant); the extracts were evaporated and digested in Erlenmeyer flasks as described above.

The following analytical methods were used: pH in water and NH_4Cl extracts were determined electrometrically; NH_4 was determined colorimetrically (flow injection analysis, FIA); NO_3, SO_4 and Cl by ion chromatography. Dissolved organic carbon (DOC) was analysed by infrared technique. In the NH_4Cl extracts, Na, K, Ca, Mg, and Al, were analysed using Inductively Coupled Plasma Emission spectroscopy (ICP). In aqueous samples, metals (total concentrations) were analysed by acetylene-air flame AAS; K and Na with 1000 ppm Cs (as CsCl) in sample and standard solutions; Ca and Mg with 10 000 ppm La (as $LaCl_3$) in solutions; Al: acetylene – N_2O flame.

Calculations and data handling

Element fluxes were calculated as follows: For bulk precipitation and throughfall, the element fluxes (concentration times water volume) were calculated for each of the

sampling occasions. The soil-solution fluxes were calculated for the period using measured lysimeter soil-solution concentrations and calculated soil-water fluxes, assuming the flux of Cl in throughfall to be equal to the flux of Cl in the soil solution (Mulder et al. 1989).

As Cl is often considered to pass through ecosystems with little tendency for storage or reaction, the Cl budget calculations should give appropriate water fluxes through the soil. The inertness of Cl was verified in a long-term (10 yr) study on the Swedish west coast (Hultberg and Grennfelt 1992), but in an integrated forest study in USA, Canada and Norway (Johnson and Lindberg 1992) it was shown that individual years, Cl may temporarily be retained or released within the soil. Though calculation of soil water fluxes using the Cl budget may be uncertain for an one-year period, it is probably the best method to use in the present study. In exposed forest edges a hydrological model would probably give more uncertain soil water fluxes because of the temperature and evapo-transpiration gradient along the forest edges and into the stands.

Total atmospheric input of H$^+$ to the soil was calculated by summing H$^+$ in bulk deposition and in dry deposition. Dry-deposited H$^+$ was estimated according to Mulder et al. (1987) as

$$H^+_{dry} = (SO_4^{2-})_{TF} - (SO_4^{2-})_{BD} + (NO_3^-)_{TF} - (NO_3^-)_{BD} + (NH_4^+)_{BD} - (NH_4^+)_{TF}$$

where TF denotes throughfall and BD bulk deposition.

Proton load calculations for a soil reveal the intensity of the current soil acidification rate, and enable quantification of the importance of different proton-producing and -consuming processes (van Breemen et al. 1983, 1984). The annual H$^+$ load to the soil caused by the acid production of the transport or transformation of H$^+$, NH_4^+, NO_3^- and SO_4^{2-}, at different distances from the forest edge, was calculated according to Mulder et. al. (1987) (expressed in $mmol_c$ m^{-2} yr^{-1}):

$$H^+ \text{ load} = (H^+_{BD+dry} - H^+_{SO}) + (SO_4^{2-}_{SO} - SO_4^{2-}_{TF}) + (NH_4^+_{TF} - NH_4^+_{SO}) + (NO_3^-_{SO} - NO_3^-_{TF})$$

BD denotes bulk deposition, TF throughfall and SO soil output from the B horizon.

In the NH_4Cl extracts, the H$^+$ concentration was calculated from pH according to Meiwes et al. (1984). Differences in NH_4Cl extractable soil content of elements between soil samples taken at different distances from the south-west exposed forest edge of the spruce stand were evaluated using analysis of variance and the Tukey HSD test (with Tukey-Kramer adjustment for unequal n) for pairwise comparison of means (SYSTAT 1989).

As the beech stand had never been ploughed and had a much longer forest history than the spruce forest, the soil conditions were less uniform than expected initially, making soil measurement values and calculations diffi-

Table 1. pH and fluxes of water (l m^{-2} yr^{-1}) and major solutes (mg m^{-2} yr^{-1}) in bulk deposition, throughfall and the soil solutions (from the B horizon at 30 to 40 cm soil depth) at different distances from the forest edge at the Stubbaröd site. Soil water fluxes were estimated assuming that Cl$^-$ input equals Cl$^-$ output.

		H_2O	pH	NH_4^+-N	NO_3^--N	SO_4^{2-}-S	Al	Ca	K	Mg	Na	DOC
Bulk deposition (BD)		1080	4.3	1300	990	1400	43	400	170	210	2200	1000
Spruce												
Throughfall (TF)												
Distance (m)	1	601	4.3	4200	3400	4800	250	2900	3100	1800	13000	7700
	5	562	4.5	2800	2200	3200	120	1700	2500	1000	8000	5400
	15	641	4.5	2000	1600	2800	79	1200	2100	570	4100	4900
	25	596	4.6	2000	1400	2700	79	1300	2500	540	3400	6200
	50	726	4.4	1800	1400	3000	88	1200	2700	520	3300	6000
Soil output (SO)												
Distance	1	181	4.5	220	9000	9100	1400	9200	2000	4200	23000	4000
	5	194	5.8	74	1400	7800	68	8800	180	1200	10000	2400
	15	302	6.1	210	410	5000	48	4300	180	670	4700	2100
	25	226	5.6	110	680	4100	63	5500	240	800	5000	1500
	50	354	4.6	67	230	2900	350	1700	340	530	3400	1000
Beech												
Throughfall (TF)												
Distance	2	676	5.1	801	930	2200	67	1300	3800	660	5000	3500
	10	712	5.0	1400	1000	1900	59	1100	3000	570	4600	4800
	25	723	4.9	1200	1000	1700	50	1100	2600	580	4100	4400
	50	672	5.0	1200	880	1700	41	980	2600	510	3600	4700
	75	761	4.8	1400	1000	1700	57	940	2500	470	3500	4300
Soil output (SO)												
Distance	2	400	4.5	75	81	2800	1100	780	250	620	4500	3400
	10	415	4.4	98	320	2800	1400	1200	380	650	4900	5700
	25	435	6.3	190	640	2600	120	3800	1300	1300	4500	3800
	50	411	5.0	89	1700	2500	520	3200	950	760	4400	1600
	75	480	4.5	210	310	2700	1300	740	490	660	4100	6100

cult to interpret. Comparison with the spruce stand must therefore be done with care.

Acid deposition and soil acidification rate

Bulk deposition and throughfall fluxes

Annual mean pH of bulk deposition was 4.3 (Table 1). In passing through the beech crowns the throughfall water was partially neutralized and pH increased by 0.5 to 0.8 pH units. The pH of spruce forest throughfall water was only slightly higher than that of the bulk deposition, with annual pH averaging 4.3 to 4.6. In both forests the pH increase in throughfall water was greatest during the growing season.

In either forest, no clear effects of the forest edge could be detected in the pH values. The same was true for the fluxes of atmospheric potentially acidifying agents into the beech forest. In the spruce forest, however, the edge effect was evident in the fluxes of SO_4^{2-}-S, NO_3^--N and NH_4^+-N (Fig. 1). On an annual basis, the bulk deposition supplied 14 kg ha^{-1} of SO_4^{2-}-S and 23 kg ha^{-1} of inorganic N to the open field. In the spruce forest front (1 m from the forest edge) the supply of S was increased by a factor of 3.2, and that of N by 3.4, in comparison to the bulk

deposition in the open field. With distance from the edge the fluxes of S and N decreased exponentially and 25 m into the forest they were 1.9 (S) and 1.5 (N) times higher than those of bulk deposition.

The pattern especially of SO_4^{2-} in throughfall water was evidently seasonal (Fig. 1). From the middle of October to May the supply of S to the spruce forest soil was on average twice that from May to the middle of October. The throughfall water, especially that sampled during November and December, accounted for the transfer of large amounts of SO_4^{2-}-S to the soil, averaging 45 to 50% of the total supply during the winter half of the year.

In both the spruce and the beech forests, the fluxes of base cations in throughfall water were several times higher than the bulk deposition (Table 1, Fig. 2). The gradient from the forest front to inside the forest was especially steep for the sea salts Na and Mg. In comparison to the open field, the supply to the edge of the spruce forest was increased by a factor of 5.9 for Na, 8.6 for Mg and 7.2 for Ca. The corresponding factors inside the forest were 1.5 for Na, 2.5 for Mg and 3 for Ca. The edge effect was less pronounced for the beech forest. Compared to the open field, the supply of Na, Mg and Ca to the beech forest edge was increased by 2.3 to 3.2 times and inside the forest, 75 m from the front, by 1.6 to 2.4 times. The fluxes of K also seemed to be affected by the forest edge. Because atmospheric deposition of K is

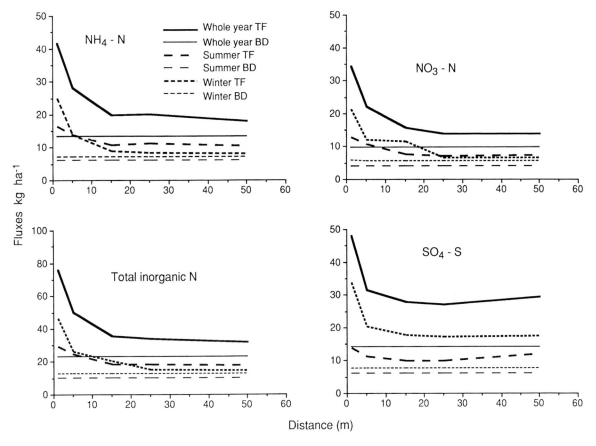

Fig. 1. Mean bulk deposition (BD) and throughfall (TF) fluxes of SO_4-S, NH_4-N, NO_3-N and total-N_{inorg} at the spruce stand during the summer half year (May 1991 to October 1991), the winter half year (October 1991 to May 1992), and the whole year, at the Stubbaröd site. (Note the differing scale of the total-nitrogen illustration.)

small, this indicates that not only dry and wet deposition may be responsible for the observed edge effects on base cations, but also leaching of base cations from the foliage, partly in exchange for H^+ from the atmosphere (H^+ buffering).

In a beech forest, stemflow may be an important pathway to the soil for many elements. Results from a number of beech forest localities in the same region as the present study show that the stemflow contribution to the total amount reaching the ground on a stand-area basis (as throughfall + stemflow) is c. 5% for NH_4^+ and for NO_3^-. For Al, Ca and Mg the corresponding percentage is c. 10%, and for SO_4^{2-}, Cl, K and Na c. 15% (Bergkvist and Folkeson 1995, Falkengren-Grerup and Bergkvist unpubl. data, Bergkvist unpubl. data). Stemflow, creating steep flux gradients near trunks (Falkengren-Grerup 1989), is thus important for the element transport to the ground only within a radius of 1 m from the trunks. These gradients were beyond the aim of the present study, however, as was quantification of fluxes on a stand-area basis. The present study used the steep deposition gradient at an exposed forest edge to examine the influence

of different amounts of acid deposition on soil solute fluxes, soil acidification and H^+-producing and -consuming processes. Inclusion of element fluxes with stemflow to the area close to trunks would have led to an overestimation of element fluxes to the ground at the sampling stations. Being based on throughfall measurements and lysimeters placed well away from trunks, our fluxes should be valid for the different sampling stations along the forest edge transect with the exception of the areas within 1 m of the trunks, which were calculated to constitute < 13% of the stand area.

Soil solute fluxes

The increased deposition rate of elements to the soil at the spruce forest edge was reflected by higher element fluxes through the soil profile (Fig. 2, Table 1). The fluxes through the soil deceased gradually with distance into the stand. Total annual fluxes of elements in the front (1 m from the forest edge) were six times as high as those in the interior of the stand (at 50 m). At 5 m into the stand,

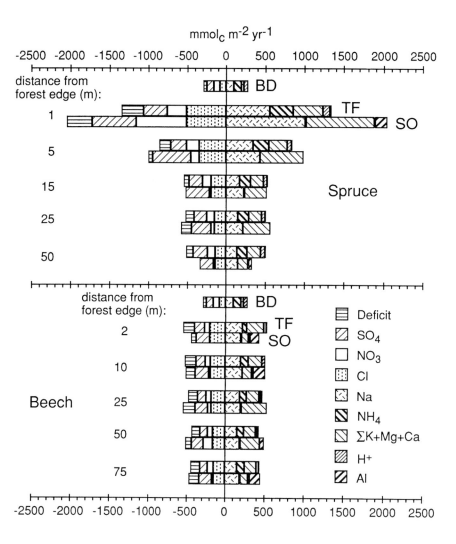

Fig. 2. Solution charge balance for bulk deposition (BD), throughfall (TF) and soil output (SO) for the spruce and beech stands at different distances from the forest edge. The anion deficit was calculated as the difference between the added fluxes of measured inorganic cations and anions.

Legend:
- Deficit
- SO_4
- NO_3
- Cl
- Na
- NH_4
- $\Sigma K+Mg+Ca$
- H^+
- Al

the fluxes were still 3 times greater than those at 50 m. As a result of the limited "edge effect" in deposition rates of elements and the heterogeneous soil conditions (see "calculations and data handling") in the beech forest, the element fluxes through the soil profile were not correlated with distance from the forest edge.

All the inorganic N (55% as NH_4-N, 45% as NO_3-N) supplied by throughfall water to the soil of the spruce forest edge was quantitatively leached from the soil profile as NO_3-N. Leaching of NH_4-N through the soil was small (<220 mg m^{-2} yr^{-1}) at all distances from the forest edge. The NO_3-N leaching from the soil at 1 m from the spruce forest edge even exceeded the total input of inorganic N with throughfall, indicating significant amounts of organic N in throughfall and/or mineralisation of soil organic N. The amount of inorganic N, expressed as a percentage of throughfall N, leached from the B horizon decreased steeply with distance from the edge; at 5 m only 30% was leached and at 50 m < 10%, indicating that most of the N may either be taken up by the roots or immobilized in the soil.

In the interior (at 50 m) of the spruce stand, input and output of SO_4 were balanced but the closer the forest edge the more SO_4 was leached from the soil, resulting in a net release.

The soil solution had an anion deficit, which was larger at the spruce forest edge than in the interior of the stand (Fig. 2). At the prevailing pH of the soil solution at the edge, the anion deficit mainly consists of dissociated organic acids; the DOC leaching was also higher at the edge. In the interior of the stand bicarbonate may also have contributed to the anion deficit, as pH was > 5.5.

Base cations contributed most to the total amount of cationic charge leached from the soil; H^+ and Al were of minor importance. The leaching of base cations was greatest at the edge, and decreased with distance into the stand. Aluminium was of some importance, making up c. 8%, only at the edge. Further from the edge, negligible amounts of H^+ and Al were leached from the soil.

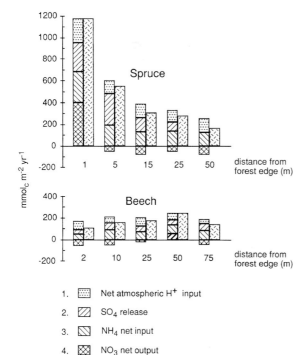

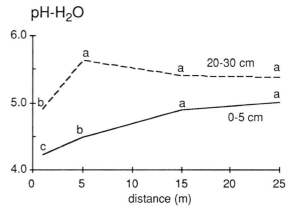

Fig. 4. pH-H₂O of the spruce forest soil in the upper A horizon (0 to 5 cm soil depth) and in the B horizon (20 to 30 cm soil depth) at different distances from the forest edge. Means lacking common letters differ significantly ($p < 0.05$) between distances (Tukey test) ($n = 6$).

1. ▦ Net atmospheric H⁺ input
2. ▨ SO₄ release
3. ▨ NH₄ net input
4. ▩ NO₃ net output
5. ▨ Total H⁺ load (Σ1+2+3+4)

Fig. 3. Total proton load (TPL) to the soil at different distances from the forest edge of the spruce and the beech stands at the Stubbaröd site.

Proton load and buffering

An evident "edge effect" was also observed in total proton load (TPL). The TPL at the edge of the spruce stand was 7 times higher than at 50 m into the stand (Fig. 3). In the beech stand no "edge effect" was seen and at all distances the TPL was close to that of the interior of the spruce stand. Direct net H⁺ input from the atmosphere to the stand was about twice as high at the spruce forest front than in the interior. Sulphate release contributed H⁺ to the soil at all plots, except at the innermost one (50 m in). Nitrogen transformations made up more than half the TPL in the spruce forest front. It is apparent that nitrification and NO₃ leaching was a consequence of the high NH₄ input to the forest front, making N transformations the dominating H⁺ source. At greater distances from the forest edge, no net NO₃-N was leached; rather NO₃-N retention in the interior of the stand acted as a H⁺ sink.

The H⁺ produced within the soil or deposited from the atmosphere to the spruce stand was mainly buffered by base cation exchange and silicate mineral weathering. This is evident from Fig. 2, if cation mobilization in the soil is estimated (as output with the soil solution minus input to the soil with throughfall). In the spruce stand Al mobilization was almost negligible, but at the forest edge Al leaching was increased because of the more acidic conditions. The greater H⁺ load at the edge also resulted

in a greater net loss of base cations. Here, substantially more Σ (K + Mg + Ca) was leached from the soil than was supplied by throughfall. Only in the innermost plot was Σ (K + Mg + Ca) leaching from the soil profile balanced by deposition to the soil.

Because of the acid soil conditions, in some plots in the beech stand, Al mobilization was as important a H⁺ buffering mechanism as it was at the edge of the spruce stand.

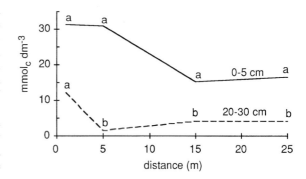

Fig. 5. Exchangeable acidity of the spruce forest soil in the upper A horizon (0 to 5 cm soil depth) and in the B horizon (20 to 30 cm soil depth) at different distances from the forest edge. Means lacking common letters differ significantly ($p < 0.05$) between distances (Tukey test) ($n = 6$).

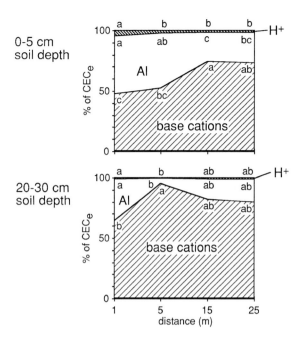

Fig. 6. Effective base saturation (V_e) (ΣNa + K + Ca + Mg as percent of CEC_e [see Table 2]) and Al and H^+ as percent of CEC_e of the spruce forest soil in the upper A horizon (0 to 5 cm soil depth) and in the B horizon (20 to 30 cm soil depth) at different distances from the forest edge. Means lacking common letters differ significantly (p < 0.05) between distances (Tukey test) (n = 6).

Table 2. Effective cation exchange capacity (CEC_e) of the spruce forest soil at different distances from the forest edge at the Stubbaröd site. Mean, and in brackets standard deviation (n = 6). Means lacking common letters differ significantly (p < 0.05) between distances (Tukey test).

Distance from forest edge (m)	CEC_e (mmol$_c$ dm^{-3})
0–5 cm soil depth	
1	59.1 (29.6)a
5	65.2 (3.9)a
15	63.3 (11.1)a
25	53.5 (26.4)a
20–30 cm soil depth	
1	39.5 (8.4)a
5	34.3 (3.0)a
15	23.9 (3.7)b
25	20.7 (3.3)b

kvist 1987, Bergkvist and Folkeson 1992). As expected for this part of Sweden the supply of potentially acidifying agents by deposition was comparatively large. The bulk deposition of SO_4-S was 14 kg ha^{-1} yr^{-1} and that of inorganic N compounds was 23 kg ha^{-1} yr^{-1}, of which the contribution of NH_4-N was almost 60%. The deposition of N, in particular, was higher than reported from nearby study areas (within 100 km) in southern Sweden (Grennfelt 1987, IVL 1991, Bergkvist and Folkeson 1992, 1994) but lower than in an area in Denmark 50 km north of Copenhagen and c. 40 km west of Stubbaröd (Beier and Gundersen 1989). The bulk deposition of both N and S

Trends in soil chemistry along the deposition gradient in the spruce stand

Only the soil samples from the spruce stand were chemically analysed, as element deposition to the beech stand showed no significant trend with distance from the forest edge (Fig. 2, Table 1). The samples from the innermost plot (50 m) were also omitted, as the element deposition to that plot did not differ significantly from the plots at 25 and 15 m.

Soil acidity was highest in the uppermost soil layer and in the B horizon near the forest edge (Fig. 4 and 5). Effective base saturation (V_e) (as % of CEC_e) also varied along the deposition gradient (Fig. 6, Table 2) being c. 50% in the uppermost 5 cm closest to the edge and c. 75%; at 15 m. An evident change in V_e was also observed in the B horizon (at 20 to 30 cm) varying from c. 65% at 1 m from the edge to c. 95% at 5 m. The main exchangeable base cation was Ca (Fig. 7) but closest to the forest edge Na and Mg contributed substantially to V_e.

Discussion

The acidity of bulk deposition was in the same range as reported during the latest decade for south Sweden (Berg-

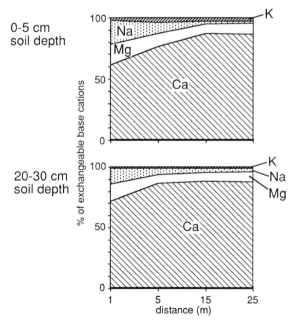

Fig. 7. Calcium, Mg, Na and K as percent of total amount of exchangeable base cations of the spruce forest soil in the upper A horizon (0 to 5 cm soil depth) and in the B horizon (20 to 30 cm soil depth) at different distances from the forest edge.

were almost the same as at Solling, Germany (Matzner and Meiwes 1994).

Although no error analysis of throughfall and soil solution has been conducted, we assume that uncertainties in the major solute fluxes of the present study will be similar to those reported for our earlier studies (Bergkvist and Folkeson 1992, 1994). Errors, as the standard deviation of the mean of the major solute fluxes of the throughfall waters $(n = 5)$, were $\pm < 5$ to 15%, and for soil solutions $(n = 3)$ ± 50 to 90% for NH_4 and NO_3, $\pm$ c. 10% for SO_4 and Cl, and ± 25 to 50% for Al, Ca, K, Mg, Na and DOC. The uncertain error estimation and the lack of statistical confidence testing must be taken into consideration in interpreting the results.

The input of all measured elements to the soil by throughfall water was generally considerably larger than the input by bulk deposition reaching the non-forested soil. The filtering effect by the tree canopies was obvious in both forests, but the forest "edge effect" was pronounced only in the spruce forest. The absence of a strong edge effect in the beech forest may be the result of differences in forest structure and aerodynamics. These factors govern aerosol deposition to a forest, in a complex interplay with particle size and concentration (Wiman and Lannefors 1985, Wiman and Ågren 1985). Different species have different "filter capacity", with spruce being more effective in capturing atmospheric particles than beech. Determinant factors would be total leaf area (particularly different when beech is defoliated during winter) and leaf surface characteristics, together with the general structure of the trees.

The annual flux of S and inorganic N-compounds beneath the spruce-forest front trees was increased by a factor of 3.2 for S and 3.4 for N, in relation to the fluxes in the adjacent non-forested area. This resulted in an annual supply to the soil at the forest edge of 48 kg SO_4-S, 34 kg NH_4-N and 42 kg NO_3-N ha^{-1} yr^{-1}. These values are in accordance with those for the most polluted areas in western continental Europe (van Breemen et al. 1987, Mulder et al. 1987, Rasmussen 1988, Matzner and Meiwes 1994). Total N inputs may be even higher than throughfall inputs of $NH_4 + NO_3$, because of direct assimilation by foliage (Eilers et al. 1992, Brumme et al. 1992). The edge effect occurred < 25 m from the front. Inside the forest, 25–50 m from the forest edge, the corresponding values were 30 kg ha^{-1} yr^{-1} SO_4-S , 14 kg ha^{-1} yr^{-1} NH_4-N and 18 kg ha^{-1} yr^{-1} NO_3-N. Similar values have been measured recently in spruce forests in Scania (IVL 1991, Bergkvist and Folkeson 1994) and are representative of current deposition rates for spruce forests in this part of Sweden.

Soil acidification may be considered as a decrease of the acid neutralizing capacity (ANC) of the soil, and changes of ANC (ΔANC) may be quantified by developing a H$^+$ budget (van Breemen et al. 1983). ANC decreased by nearly 12 kmol$_c$ ha^{-1} yr^{-1} just inside the forest edge. The soil acidification rate further decreased exponentially along the deposition gradient. In the inte-

rior of the stand, ΔANC was c. -2 kmol$_c$ ha^{-1} yr^{-1}. The higher soil acidification rate close to the forest edge of the spruce stand was a result of the much higher deposition rate of H$^+$ and acidifying substances to the forest edge. In a European perspective, soil acidification rate of the plots closest to the forest edge must be considered as high, while the soil of the interior stand is more slowly acidified than European forest soils in general (van Breemen et al. 1984).

The change in nitrogen dynamics induced by the high N input at the edge (c. 75 kg N ha^{-1} yr^{-1}) was the main reason for the large difference between soil acidification rate at the forest edge and that in the interior of the stand. The N transformations were of little importance to the H$^+$ load of the stand interior, but close to the forest edge more than half the H$^+$ load was attributable to N transformations. In the Netherlands, van Breemen et al. (1986, 1987) reported that N transformations generated amounts of H$^+$ similar to those measured at our spruce forest edge. Similar amounts of H$^+$ attributable to N transformations were produced when a beech forest in the same region as this study was treated with repeated additions of NH_4NO_3 for five years (Bergkvist and Folkeson 1992).

The margin of the spruce forest stand must be considered to be N saturated (Nihlgård 1985, Aber et al. 1989). No net retention of N was observed at the forest edge; c. 90 kg N ha^{-1} yr^{-1} was leached as NO_3-N. The N deposition and leaching at this edge was comparable to that in forests of western Continental Europe (van Breemen et al. 1987, Mulder et al. 1987, Gundersen and Rasmussen 1990). A general trend in central and northern Europe is an increase of the NO_3 and NH_4 deposition during recent decades, and a simultaneous decrease of H$^+$ and SO_4 deposition (Christophersen et al. 1990, Matzner and Meiwes 1994). Many forest areas may therefore gradually become N-saturated, resulting in leaching of NO_3 and N transformations which will accelerate soil acidification.

Rather more inorganic N was leached through the soil during the studied period than was deposited to the soil with throughfall. The reason for this is unknown. The soil flux calculations may be uncertain, but other reasons may exist. Nitrogen addition to a soil may inhibit the mineralisation of organic material (particularly lignin compounds), resulting in increased leaching of N-containing water-soluble organic substances from the O horizon (Fog 1988). The increased leaching of DOC (and the increased anion deficit) through the B horizon at the forest edge may be a result of this N effect. In the N-treated plots in the experiment with repeated additions of NH_4NO_3 to a beech forest (Bergkvist and Folkeson 1992), a comparable increase of the DOC leaching was also observed. We can only speculate that if these N containing organic substances in the soil solution were partly mineralized in passing the mineral soil, an increase of the NO_3 content in the collected soil solution of the B horizon would have been observed.

Because atmospheric SO_4 deposition has decreased

during the last decade (Matzner and Meiwes 1994), previously adsorbed SO_4 may desorb and leach from the soil. A decrease in SO_4 deposition to the stands of the present study would have been most pronounced at the forest edge (Beier 1991). The observed large net release of SO_4 at the forest edge may be a result of the decreased SO_4 deposition. This will temporarily prevent reversibility of soil acidification (Alewell and Matzner 1993) as net release of SO_4 contributes H^+ to the soil (Reuss and Johnsson 1986), particularly at the forest edge.

As was mentioned above, the DOC leaching was greater closer to the spruce forest edge, and so was also the "anion deficit" of the soil solution. The anion deficit mainly consists of dissociated organic acids, which function as mobile anions. If base cations are transported out of the soil profile with these mobile organic anions, the ANC of the soil will decrease and soil acidification will increase (van Breemen et al. 1984). Thus N deposition may acidify the soil not only by N transformations, but also indirectly by increasing the solubility and leaching of organic substances, partly functioning as mobile anions.

The fact that the forest edge and the interior of the spruce stand have received different acid loads for several years may explain the differences in soil chemical properties observed along the transect. The soil was clearly more acidified closer to the forest edge; in the upper A horizon the V_e was decreased by c. 25%, pH-H_2O was 0.8 units lower and exchangeable acidity doubled in comparison to the interior soil; in the B horizon the changes were somewhat smaller. The studied soil was mainly buffering H^+, entering or produced within the soil, by base cation exchange and silicate mineral weathering. Al solubilization was a minor buffering mechanism in this soil (see Ulrich 1983). Dissolution of silicate minerals consumes H^+ and releases the equivalent amount of base cations. In the long term perspective, the relation of total H^+ load and maximum of H^+ consumption due to silicate mineral weathering in a soil is crucial for soil development. If silicate mineral weathering rate is too slow to keep pace with the rate of H^+ addition, exchangeable bases will decrease and H^+, and later also Al, will increase on the exchange complex.

Studies of maximum silicate mineral weathering rates (and H^+ consumption) have been reviewed by Fölster (1985). Regardless of the methods used in those studies, weathering rates varied in the range of c. 0.2 to 2 $kmol_c$ ha^{-1} yr^{-1}, mostly depending on content of weatherable silicates in the soil and soil hydrology. Warfvinge et al. (1993 and pers. comm.) report weathering rate of Dystric Cambisols in the same region as this study and with similar properties, to be c. 0.4 $kmol_c$ ha^{-1} yr^{-1}. The H^+ load to the soil at the forest edge of the present study was 30 times the weathering rate reported by Warfvinge et al. (1993 and pers. comm.) and 6 times the weathering rate in the upper range reported by Fölster (1985). The H^+ load decreased gradually into the stand and in the interior part the H^+ addition to the soil was 4 and 0.8 times the

reported weathering rates. Thus, at least close to the forest edge, the soil chemical status will continue to change with time, and the soil chemical status of the interior of the stand will most probably also change if the acidifying deposition is not radically reduced.

While the deposition of acidity and acidifying substances was greater near the forest edge, so was the deposition of base cations, which may partly counteract soil acidification. Deposition of Na and Mg increased more at the edge than did deposition of Ca and K. This difference in deposition rate probably explains the higher percentage of Na and Mg of the V_e closer to the forest edge. Consequently, without the increased base cation deposition, the soil would have changed even more close to the forest edge.

Thus, the leached airborne substances deposited on the foliage cannot only be seen as negative to the soil. From an ecological view, the filtering effect and wash out of airborne base cations, especially Mg and Ca, may be regarded as an important input to the forest ecosystem, somewhat counteracting soil losses. However, the trend in northern Europe and North America is towards steep declines in the atmospheric concentrations of base cations during the past decades (Hedin et al. 1994). To varying degrees, these base-cation trends have offset recent reductions in acidity and SO_4 deposition, and may have contributed to increased acidity sensitivity of poorly buffered ecosystems.

Acknowledgments – We are greatful to A. Jonshagen for field assistance, A. Balogh and S. Billberg for laboratory work, A. Persson for typing and to the landowners for land-use permission. The study was financed by the National Swedish Environment Protection Agency.

References

Aber, J.D., Nadelhoffer, K.J., Stender, P. and Melillo, J.M. 1989. Nitrogen saturation in northern forest ecosystems. – BioScience 39: 378–386.

Alewell, C. and Matzner, E. 1993. Reversibility of soil solution acidity and sulphate retention in acid forest soils. – Water Air Soil Poll. 71: 155–165.

Beier, C. 1991. Separation of gaseous and particulate dry deposition of sulphur at a forest edge in Denmark. – J. Environ. Qual. 20: 460–466.

– and Gundersen, P. 1989. Atmospheric deposition to the edge of a spruce forest in Denmark. – Environ. Poll. 60: 257–272.

Bergkvist, B. 1987. Soil solution chemistry and metal budgets of spruce forest ecosystems in S. Sweden. – Water Air Soil Poll. 33: 131–154.

– and Folkeson, L. 1992. Soil acidification and element fluxes of a *Fagus sylvatica* forest as influenced by simulated nitrogen deposition. – Water Air Soil Poll. 65: 111–133.

– and Folkeson, L. 1995. The influence of tree species on acid deposition, proton budgets and element fluxes in south Swedish forest ecosystems. – Ecol. Bull. (Copenhagen) 44: 90–99.

Breemen, N. van, Mulder, J. and Driscoll, C.T. 1983. Acidification and alkalinization of soils. – Plant Soil 75: 283–308.

– , Driscoll C.T. and Mulder, J. 1984. Acidic deposition and internal proton sources in acidification of soils and waters. – Nature 307: 599–604.

–, de Visser, P. H. B. and van Grinsven, J. J. M. 1986. Nutrient and proton budgets in four soil-vegetation systems underlain by *Pleistocene alluvial* deposits. – J. Geol. Soc. (Lond.) 143: 659–666.

–, Mulder, J. and van Grinsven, J. J. M. 1987. Impacts of acid atmospheric deposition on woodland soils in the Netherlands: II. Nitrogen transformations. – Soil Sci. Soc. Am. J. 51: 1634–1640.

Brumme, R., Leimcke, U. and Matzner, E. 1992. Interception and uptake of NH_4 and NO_3 from wet deposition by above ground parts of young beech (*Fagus sylvatica* L.) trees. – Plant Soil 142: 273–279.

Christophersen, N., Robson, A., Neal, C., Whitehead, P. G., Vigerust, B. and Henriksen, A. 1990. Evidence for long-term deterioration of stream water chemistry and soil acidification at the Birkenses catchment, southern Norway. – J. Hydrol. 116: 63–76.

Draaijers, G. P. J., Ivens, W. P. M. F. and Bleuten, W. 1988. Atmospheric deposition in forest edges measured by monitoring canopy throughfall. – Water Air Soil Poll. 42: 1209–136.

Eilers, G., Brumme, R. and Matzner, E. 1992. Above ground N-uptake from wet deposition by Norway spruce (*Picea abies* Karst.) – For. Ecol. Manage. 51: 239–249.

Eriksson, E., Karltun, E. and Lundmark, J.-E. 1992. Acidification of forest soils in Sweden. – Ambio 21: 150–154.

Falkengren-Grerup, U. 1986. Soil acidification and vegetation changes in deciduous forests in southern Sweden. – Oecologia 70: 339–347.

– 1989. Effect of stemflow on beech forest soils and vegetation in southern Sweden. – J. Appl. Ecol. 26: 341–352.

– and Tyler, G. 1992. Changes since 1950 of mineral pools in the C-horizon of Swedish deciduous forest soils. – Water Air Soil Poll. 64: 495–501.

–, Linnermark, N. and Tyler, G. 1987. Changes in soil acidity and cation pools of south Sweden soils between 1949 and 1985. – Chemosphere 16: 2239–2248.

Fölster, H. 1985. Proton consumption rates in holocene and present day weathering of acid forest soils. – In: Drever, J. I. (ed.), The chemistry of weathering. D. Reidel, Hingham, MA, pp. 197–209.

Fog, K. 1988. The effect of added nitrogen on the rate of decomposition of organic matter. – Biol. Rev. 63: 433–462.

Grennfelt, P. 1987. Deposition processes for acidifying compounds. – Environ. Technol. Lett. 8: 515–527.

Gundersen, P. and Rasmussen, L. 1990. Nitrification in forest soils: effects from nitrogen deposition on soil acidification and aluminium release. – Rev. Environ. Contam. Toxicol. 113: 1–45.

Hallbäcken, L. and Tamm, C. O. 1986. Changes in soil acidity from 1927 to 1982–1984 in a forest area of south-west Sweden. – Scand. J. For. Res. 1: 219–232.

Hasselrot, B. and Grennfelt, P. 1987. Deposition of air pollutants in a wind-exposed forest edge. – Water Air Soil Poll. 34: 135–143.

Hedin, L. O., Granat, L., Likens, G. E., Buishand, T. A., Galloway, S. N., Butler, T. J. and Rodhe, H. 1994. Steep declines in atmospheric base cations in regions of Europe and North America. – Nature 367: 351–354.

Hultberg, H. and Grennfelt, P. I. 1992. Sulphur and seasalt deposition as relfected by throughfall and runoff chemistry in forested catchments. – Environ. Poll. 75: 215–222.

IVL 1991. Miljöatlas. Resultat från IVLs undersökningar i miljön 1991. – Inst. Vatten- och Luftvårdsforskning, Stockholm.

Johnson, D. W. and Lindberg, S. E. 1992. Atmospheric deposition and forest nutrient cycling. A synthesis of the Integrated Forest Study. – Ecol. Stud. 91. Springer, New York.

Lydersen, E. 1990. The solubility and hydrolysis of aqueous aluminium hydroxides in dilute fresh waters at different temperatures. – Nord. Hydrol. 21: 195–204.

Matzner, E. and Meiwes, K. S. 1994. Long-term development of element fluxes with bulk precipitation and throughfall in two German forests. – J. Environ. Qual. 23: 162–166.

Meiwes, K.-J., König, N., Kharma, P. K., Prenzel, J. and Ulrich, B. 1984. Chemische Untersuchungsverfahren für Mineralboden, Auflagehumus und Wurzeln zur Charakterisierung und Bewertung der Versauerung in Waldböden. – Berichte des Forschungszentrums Waldökosysteme/Waldsterben 7: 14–16.

Mulder, J., van Grinsven, J. J. M. and van Breemen, N. 1987. Impacts of acid atmospheric deposition on woodland soils in the Netherlands: III. Aluminium chemistry. – Soil Sci. Soc. Am. J. 51: 1640–1646.

–, Breemen, N. van, Rasmussen, L. and Driscoll, C. T. 1989. Aluminium chemistry of acid sandy soils with various inputs of acidic deposition in the Netherlands and in Denmark. – In: Lewis, T. E. (ed.), Environmental chemistry and toxicity of aluminium. 194th Annual Meeting of the American Chemical Society, New Orleans, August 30–September 4, Lewis Publ., Chelsea, Michigan, pp. 171–194.

Nihlgård, B. 1985. The ammonium hypothesis. An additional explanation to the forest dieback in Europe. – Ambio 14: 2–8.

Rasmussen, L. 1988. Report from laboratory of environmental science and ecology. – Technical Univ. of Denmark, Lyngby.

Reuss, J. O. and Johnson D. W. 1986. Acid deposition and the acidification of soils and waters. – Ecol. Stud. 59. Springer, New York.

SYSTAT, Inc. 1989. Wilkinson, Leland. SYSTAT: The System for Statistics. – Evanston, IL., USA.

Thimonier, A., Dupory, J. L. and Timbal, J. 1992. Floristic changes in the herb-layer vegetation of a deciduous forest in the Lorraine plain under the influence of atmospheric deposition. – For. Ecol. Manage. 55: 149–167.

Ulrich, B. 1983. Soil acidity and its relations to acid deposition. – In: Ulrich, B. and Pankrath, J. (eds), Effects of accumulation of air pollutants in forest ecosystems. D. Reidel Publishing Company, Dordrecht, pp. 127–146.

Warfvinge, P., Falkengren-Grerup, U., Sverdrup, H. and Andersen B. 1993. Modelling long-term cation supply in acidified forest stands. – Environ. Poll. 80: 209–221.

Wiman, B. L. B. and Lannefors, H. O. 1985. Aerosol characteristics in a mature coniferous forest. Methodology, composition, sources and spatial concentrtion variations. – Atmos. Environ. 19: 349–362.

– and Ågren, G. I. 1985. Aerosol depletion and deposition in forests. A model analysis. – Atmos. Environ. 19: 335–348.

Ecological Bulletins 44: 54–64. Copenhagen 1995

Acidification-induced chemical changes of forest soils during recent decades – a review

S. Ingvar Nilsson and Germund Tyler

Nilsson, S.I. and Tyler, G. 1995. Acidification-induced chemical changes of forest soils during recent decades – a review. – Ecol. Bull. (Copenhagen) 44: 54–64.

Reports on acidification-induced changes in chemical properties of forest soils in northern Europe, particularly Sweden, are reviewed and evaluated. Most studies have been performed retrospectively by comparing either old data or old preserved samples with recently sampled soils. Almost without exception, decreases in soil pH have been reported for the last 2–5 decades from all horizons of podsols and related soil types. The pH decreases have been accompanied by decreases in exchangeable base cations and increases in exchangeable Al pools. The relative importance of acid deposition and other H^+ sources to the soil chemical changes is discussed.

S.I. Nilsson, Dept of Soil Sciences, Swedish Univ. of Agricultural Sciences, P.O. Box 7014, S-750 07 Uppsala, Sweden. – G. Tyler, Dept of Ecology, Soil-Plant Research, Univ. of Lund, Ecology Building, S-223 62 Lund, Sweden.

Introduction

Acidification-induced changes in chemical properties of forest soils have become a matter of great concern in central and northern Europe during the last one or two decades. According to an appreciable number of reports, to be discussed in this review, base saturation and pH have decreased considerably, both in upper and lower soil horizons. Concentrations of exchangeable and soil solution Al have risen. Increased fluxes of strong mineral acids originating from atmospheric deposition are usually considered to be one of the main reasons for these changes. Large areas of forest soils are currently at a level of soil acidity at which any input of acid to the ecosystem increases the release rates of base cations, Al and some heavy metals (Tyler 1981).

External sources as well as internal processes may bring about changes in soil chemistry. Internal H^+ sources include soil respiration (formation of carbonic acid) plant and microbial uptake of base cations and NH_4^+ with an equivalent release of H^+, nitrification of NH_4^+, oxidation of sulphides and other reduced sulphur compounds, and the production of organic acids. External sources include the deposition of acidic or potentially acidifying N and S compounds and the use of acid or potentially acidifying fertilizers in forestry and agriculture.

A net increase in the input of H^+ is consequently possible not only because of changes in deposition rates of acid compounds but also as the result of increased primary productivity and harvest rates, as well as of changes in land use. Particularly important is the afforestation of open land and the replacement of deciduous forest by conifers which are characterized by a greater capacity to capture acidic aerosols (dry deposition) and, possibly, by a higher growth rate.

Internal processes and reactions which neutralize H^+ include the complete mineralization of organic matter, whereby base cations and NH_4^+ are also released, the chemical weathering of minerals, and plant and microbial uptake of anions (e.g. NO_3^-) resulting in the release of OH^-. A decreased rate of these processes may thus accelerate the net input of H^+ to the ecosystem.

Acidification-induced changes in chemical properties of soils may be termed "soil acidification". It is essentially the question of a decrease in the acid-neutralizing capacity of the soil. An apparent symptom of soil acidification is a decrease in soil pH. However, major pH changes as a result of increased input of H^+ only occur when the capacity of soil buffering systems is insufficient. Many pristine soils contain $CaCO_3$ and only marginal pH decreases are possible as long as any $CaCO_3$ remains. When $CaCO_3$ is exhausted, there is a sudden drop in pH and a base cation buffering stage is attained (Ulrich 1981, Ulrich and Pankrath 1983). The gradual

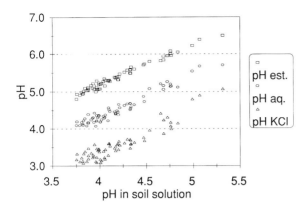

Fig. 1. The pH of soil solutions, obtained by high-speed centrifugation of beech forest topsoil samples, and the pH (0.2 M KCl) and pH (H_2O) of the same samples. An estimated value (pH_{est}) of pH (H_2O) was calculated as the H^+ concentration of the soil solution diluted by the water volume used in the pH (H_2O) measurement. These data can be used for converting a pH (KCl) or a pH (H_2O) value to the pH in the soil solution. (From Falkengren-Grerup and Tyler 1993).

exhaustion of $CaCO_3$, as well as other mineral depletions resulting from increased H^+ input, and changes in amounts and properties of soil organic matter, are processes of soil acidification. Further acidification includes the appearance of exchangeable and soil solution Al and a gradual decline of the exchangeable base cation pools. Far-reaching pH-decreases of the soil solution are counteracted by base cation exchange and, gradually, by the release of Al^{3+} from mineral and organic Al pools. Most forest topsoils in southern Sweden seem to have passed from a stage of base cation buffering to a stage of Al buffering during recent decades.

The aims of this paper are to review reports on acidification-induced changes in chemical properties of mainly Swedish forest soils during recent decades and to evaluate other possible explanations of the chemical changes observed. We do not consider possible biological effects of these changes. Soil names are according to FAO (1988), if not stated otherwise.

Evidence for soil chemical changes

Reports on decreasing soil pH

pH (negative 10 logarithm of the H^+ activity) in soils is usually measured in extracts or suspensions of the soil, though the soil solution would be the ideal medium for analysis. The pH of acid soils measured in water suspensions, pH (H_2O), is usually higher than in soil solutions expelled by high speed centrifugation, but still lower than expected from the dilution introduced by the distilled water added (Falkengren-Grerup and Tyler 1993). This is caused by dissociation of organic acids and Al hydrolysis, for example. pH may also be measured in soil extracts

with neutral salt solutions, e.g. 0.2 M KCl, yielding lower pH-values than for soil solution, because of displacement of exchangeable H^+ and Al^{3+} by the excess K^+ introduced. Typically, pH (0.2 M KCl) of forest topsoils is c. 1 unit lower than pH (H_2O), with soil solution pH intermediate (Fig. 1).

Since the early 1980's, long-term chemical changes of forest soils, in particular pH decreases, have been widely reported from continental and northern Europe. In a comprehensive study by von Butzke (1981), from Nordrhein-Westfalen in Germany, comprising mainly pseudogleys and pseudogley podsols, according to the German soil classification system, consistent declines in soil pH (1 M KCl) between 1959/61 and 1981 were measured. Most decreases were in the range 0.2–0.5 pH-units, though larger decreases were recorded for less acid soils, partly as a consequence of the logarithmic nature of the pH-scale. Topsoil pH (0.1 M KCl) of mor and moder podsols of *Picea*, *Quercus* and *Pinus* stands in the Hamburg area, with a S deposition of 25–40 kg ha^{-1} yr^{-1}, decreased on average by 0.4 pH-units between 1950 and 1981 (von Buch 1982).

In contrast to these reports, an appreciable number of Bavarian soils, originally sampled in 1953/70 and re-sampled in 1981, exhibited only marginal pH decreases, usually <0.2 units (Wittmann and Fetzer 1982). However, changes in management of many sites might have influenced the result. Cessation of litter raking, which was formerly a common practice in this area, might be an important factor, though it is not always evident how this has affected soil pH. Less consistent pH-changes (0.1 M KCl) of the topsoil were also measured after three decades in spruce stands on loess in Hessen (Riebling and Schaefer 1984). Much greater decreases, on average c. 1 pH (KCl) unit, were reported for forest cambisols in the Berlin area between 1950 and 1980/81 (von Grenzius 1984). These data are consistent with those of von Butzke (1981) from Nordrhein-Westfalen, as the original pH (KCl) was usually >4.5.

Decreasing pH of forest soils during recent decades is also reported from Austria by Stöhr (1984) and Glatzel et al. (1985). In the Orlickeé Mountains, Czech Republic, with spruce forest soils of high acidity exposed to considerable deposition of air pollutants, pH (H_2O) in the topsoil decreased by c. 0.7 units, and in the B horizon by 0.5 units, between 1953 and 1981 (Pelisek 1984).

The first reports from Scandinavia of long-term pH decreases of forest soils, based on comprehensive documentation, were published by Hallbäcken and Tamm (1986) and Falkengren-Grerup (1986). It is true that the gradual pH decreases resulting from changes in land-use, forest management and succession/stand age have long been known. In the cited studies, however, sites were selected in such a way that influence from these and related conditions/processes were minimized.

Hallbäcken and Tamm (1986) reinvestigated 90 soil profiles in 1982–84 at the Tönnersjöheden Experimental Park, southwest Sweden, originally studied in 1927. The

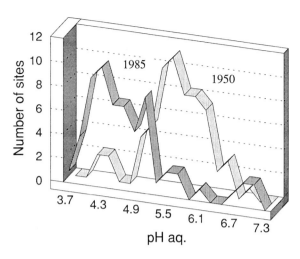

Fig. 2. Changes in soil pH (H_2O) between 1950 and 1985 in south Sweden; forest topsoils. (From Berggren et al. 1992).

geographical location and a comparatively high mean annual precipitation makes this area particularly exposed to deposition of long-range transported air pollutants. Samples at field moisture were used for pH (H_2O) determination on both occasions, using a quinhydrone electrode method.

In all soil horizons in this study, the pH values measured were considerably lower than in 1927. The average pH decrease for different stand types (including both continuous and discontinuous forest cover) varied between 0.3 and 0.9 units in the O horizon and between 0.3 and 0.7 units in the mineral soil layers. The median pH (H_2O) of the humus layer in stands with a continuous beech canopy had decreased from 4.5 to 3.8. In spruce stands developed from other ecosystem types (heathland or deciduous forests), decreases varied to some extent with stand age and original pH. There was also a tendency towards smaller pH-decreases in sites with originally lower pH values. However, no clear stand age trend in pH was measured in the C horizons and it was concluded that increasing acidity, even in the deeper soil layers, was difficult to explain without assuming an influence of acid deposition.

Retrospective studies by Falkengren-Grerup (1986, 1987) concerned soil pH and vegetational changes, mainly in deciduous forest, at a large number of sites in Scania, south Sweden (Fig. 2). They were originally studied by different researchers in 1949–1970 and re-sampled in 1984–1985. Only mature stands with a continuous known forest history were included, and a wide range of soil acidity conditions was represented, though most soils could tentatively be classified as dystric cambisols or in a few cases as podsols. All pH determinations were carried out electrometrically using glass/calomel electrodes and identical soil:solution proportions on the two measuring occasions.

As in the study by von Butzke (1981), there was a

distinct relationship between original pH and the observed pH decrease. At an original pH (H_2O) of 5.5–6, the mean pH decrease in the humus layer was c. 1.0, compared to a decrease of <0.5 at pH (H_2O) 4–5. There was also less pH decrease in sites with shorter sampling intervals (15–20 yr) than in sites with longer intervals (30–35 yr). However, this difference was probably accentuated by some over-representation of very acid sites with shorter sampling intervals.

There was also a considerable pH decrease measured in deeper horizons of Scanian deciduous forest soils, particularly in those originally sampled as early as 1949–50 (Falkengren-Grerup 1987). pH (H_2O) changes were usually larger than pH (KCl) changes and averaged 0.5–1.0 pH-unit, often largest in the upper C horizon (50–100 cm depth). It is also evident that the mineral horizons of most soils studied in a few decades had shifted from a pH range with only small amounts of exchangeable or soil solution Al to a pH range with a far-reaching transformation of solid Al compounds into exchangeable and soluble species.

The Tönnersjöheden and the Scanian sites discussed so far are all situated in the extreme south of Sweden, where the forest is currently exposed to a total (wet + dry) deposition of N + S which approaches or even slightly exceeds c. 40 kg ha^{-1} yr^{-1}. In a study by Tamm and Hallbäcken (1988), the situation at Tönnersjöheden was compared with a similar revision of a study from 1926–27 in north-central Sweden (64°12'N, 19°37'E) in an area with an estimated deposition not exceeding 15 kg N + S ha^{-1} yr^{-1}. The lower soil horizons of coniferous forests already had a lower pH (H_2O) in the southern than in the northern area in 1926–27. In contrast to the southern site, no measurable pH (H_2O) changes had occurred in the C horizon of the northern site during almost six decades. Changes were also smaller in the B horizon, while pH decreased mainly with stand age in the upper horizons of the conifer stands of the northern site. The authors concluded that the B and C horizons were much less affected by biological acidification than the topsoil, and that the most probable explanation of the differences between the southern and the northern site in the pH time trends of these horizons was the difference in the acid deposition load.

Comparative retrospective studies of long-term soil pH changes have also been conducted in areas in south, south-central and northern Sweden (Jacks et al. 1989, Jacks 1992, Gustafsson et al. 1993). This work differed from the studies reported above in that it was based on pH-measurements of preserved vs freshly taken air-dried samples. In one of the southern areas (Hallandsåsen), quite considerable pH (H_2O) decreases (1.5–2 pH units) between 1951 and 1987 were measured in different horizons of soils with originally high pH (6–7), whereas pH-changes in originally more acid soils (pH <5) were smaller (<1 pH unit). Both open land and forest sites were sampled; the pH (H_2O) decline was less pronounced in soils from open land.

pH changes of the central and north Swedish sites, originally sampled in 1954–1957, were smaller, sometimes insignificant. However, there was always some pH decrease in the uppermost 10 cm of the coniferous forest sites. The pH of originally quite acid soil actually increased in the south-central sites, which might be the result of the sites changing from forest to open land (Jacks 1992). No pH difference (p>0.05) between old and new samples was observed in the B and C horizons of the central and north Swedish sites, whereas pH decreases were highly significant (p<0.001) in the B and C horizons of the investigated south Swedish sites (Gustafsson et al. 1993). The increased soil acidity in the south was primarily explained by acid deposition. It was also concluded that, at most sites, the total pH range was smaller in the new than in the old samples and that pH seemed to have approached a steady state. This steady state pH displayed a south–north gradient with the lowest values in the south. An approaching steady state in pH of acid topsoils has recently been reported for south Swedish beech forests (Falkengren-Grerup and Tyler 1991).

Evidence for acidification of soils influenced by stemflow water

The ground in the close vicinity of tree trunks, especially those of large beeches, often receives much more precipitation water than average. This is caused by the structural properties of trees, which favour a channelling of throughfall water along certain branches and stem tracks. The acidity of the stemflow water may sometimes be higher than that of rainfall (Bergkvist et al. 1987), but most important to the soil is the greater flux of acids, including organic acids, mineral nutrients, etc. to the ground adjacent to the stem.

Effects of stemflow water on soil chemical properties were described for continental Europe by e.g. Jochheim (1985) and Wittig and Neite (1985) and for Sweden by Nihlgård (1970). That this phenomenon is highly valid in south Sweden was also demonstrated by Falkengren-Grerup (1989). Soil pH (KCl) was, on average, 0.4–0.5 units lower along the flow tracks close to beech trunks than between trunks, both in the A and B horizons of cambisols in Scania. In podsols there was only a slight variability of pH (KCl) along the flow tracks and there was even a tendency for higher pH values close to the stems. This is also an indication that a steady state of soil acidity may have been attained under ambient deposition conditions in south Sweden, though only in the most acid soils. In a later study (Falkengren-Grerup and Björk 1991), the problem of reversibility of soil acidification was approached by measuring soil pH along transects from dated beech stumps. The study showed that soil acidity (exchangeable H^+) was about halved in soil close to the stem within fifteen years after the beech was felled and only small further changes occurred over the next few years.

Reports on changes in soil pools of mineral elements

Changes in soil pH are always accompanied by, or are partly the effect of, other changes in soil chemical properties. As shown in several studies (e.g. Hallbäcken and Popovic 1985, Berdén et al. 1987) there is a close positive relationship between pH and base saturation of soils. Base saturation is defined as the equivalent sum of the exchangeable pool of base cations (Ca, Mg, K, Na in $mmol_c$ kg^{-1} dry soil), calculated as the percentage of the cation exchange capacity (CEC) of the soil. The CEC is the equivalent sum, in $mmol_c$ kg^{-1} dry soil, of mainly exchangeable base cations, and exchangeable or titratable acidity. A large fraction of the acidity consists of Al. Soils with a soil solution pH >5 have typically quite a small amount of exchangeable Al, whereas there is a rapid increase in exchangeable Al below pH 4.5.

As expected, decreasing soil pH is accompanied by, and partly the result of, a replacement of base cations for H^+ and Al^{3+} on the exchange complex. As the effective CEC (i.e. CEC at ambient pH) decreases with decreasing soil pH, to the extent that variable charge sites are protonated, one would also expect diminishing exchangeable cation pools in the soil, but not necessarily decreasing concentrations of base cations in the soil solution, as long as the acidification process continues (see Tyler 1981, Falkengren-Grerup and Tyler 1993).

Several reports on long-term trends in exchangeable cation concentrations of forest soils, mainly explained as an effect of acid deposition, have been published since the early 1980's. Exchangeable Al pools of the lower horizons in a 120-year-old beech forest on loess soil (dystric cambisol), in Germany increased considerably between 1966 and 1973/79, apparently as a consequence of a shift from a base cation to an Al buffering stage (Ulrich et al. 1980). Large decreases in exchangeable Ca and Mg (though not K) pools were reported from spruce stands on podsols and dystric cambisols in the Harz Mountains (Hauhs 1985).

Troedsson (1985) studied the sensitivity of Swedish forest soils to acidification using c. 2 500 O horizon samples from 40–60-year-old forest stands in central Sweden. Exchangeable $H^+ + Al$ tended to increase in most soil types between 1961/63 and 1971/73, though differences were usually not significant. There was some indication of diminishing exchangeable Ca and Mg pools and a significant decrease of exchangeable K. The differences observed were explained, at least partly, as a result of increased primary productivity of the forests, as only the topsoil was considered and the main element showing a decrease was K.

Falkengren-Grerup et al. (1987) studied the exchangeable cation pools in ten profiles (usually 100 cm deep) in different ecosystems in Scania, south Sweden. Soil samples preserved in an air-dried condition since 1949/54 and comparable samples from the same sites in 1984 were analyzed simultaneously in 1986, using identical methods (extraction with ammonium acetate of pH 7.0 for topsoils and pH 4.8 for samples of the mineral soil horizons). On average, the base cation pools of the whole profiles had decreased by c. 50% during these 35 yr. There was no consistent trend of relative change with soil depth, nor was the base cation loss clearly related to the original acidity state of the soils. The decrease was equally apparent in an open heathland developed on a non-calcareous sandstone moraine as in the deciduous and coniferous forest sites. A comparable decrease was measured in exchangeable Zn, an element behaving almost like base cations with respect to soil acidity, whereas exchangeable H^+ and Al had usually increased their pools.

These results from Scania were largely confirmed by Falkengren-Grerup and Eriksson (1990), who studied the soil chemistry of the upper C horizon of 19 beech and oak stands in south Sweden, originally sampled in 1947 and resampled in 1988. Exchangeable cation pools had decreased in the order Na > Mn > Zn > Ca, Mg > K, from 20% to c. 70% remaining in the 1988 samples. The change was greatest for soils with originally high concentrations and the deficit was mainly compensated for by an increase in exchangeable Al. The same and other C horizon samples from south Sweden were treated further by Falkengren-Grerup and Tyler (1992), including a strong acid (0.1 M HNO_3) extraction. The HNO_3 and the NH_4Ac extractions showed similar time trends, though the relative differences between the 1947/52 and 1988 samples were smaller with HNO_3, as a usually much larger fraction of the total pools was obtained with this extractant. As the acid-soluble pools of base cations had also decreased significantly during the time interval studied, it was concluded that there had been mineralogical changes, which could possibly counteract a complete reversibility of the acidification-induced changes in pH and base saturation.

That a depletion of the exchangeable cation pool of Scanian forest topsoils is still in progress, in spite of an approaching pH steady state, was indicated by Falkengren-Grerup and Tyler (1991). They compared measurements of samples at field moisture performed with identical methods in 1979 and 1989, representing c. 100 mature beech forest stands. It was concluded that the rate of decrease of the exchangeable pools of K, Ca, and Mg estimated from different time series had been almost constant in relative terms since c. 1950, at 1–2% annually. Extrapolation of the data may indicate that, on average, only 20% of the pools of 1950 would remain in the topsoil in the year 2050, at the current depletion rate. However, changes in cation selectivity and other feedback mechanisms may invalidate such prognoses.

Acidification and chemical properties of coniferous forest soils in recharge areas were investigated along three SW–NE transects across south, south-central and northern Sweden (Eriksson et al. 1992). The pools of base cations decreased between 1947/63 and 1988/90 at 80–100 cm depth in most soils of the southern and south-central transects. No clear direction of change was ob-

served in the northern transect. Changes in the deep horizons of the two southern transects were mainly attributed to percolation of strong mineral acids, mainly sulphuric acid, and it was concluded that the acidification pattern was similar to the general acid deposition pattern in Sweden.

Sjöström and Qvarfort (1992) followed a different approach in assessing changes of soil chemical properties with time. They compared two podsolic soil profiles developed from a sandy glacial till in south-central Sweden. One of the profiles had been excluded from the influence of percolating water being protected by a barn built in 1890 on a ventilated stone foundation. This soil profile was compared with a similar profile from outside the barn. The exposed soil had a higher exchangeable acidity and a lower base saturation in several horizons than the soil which had been covered. The concentrations of base cations, Mn and Fe not extractable with 1 M NH_4Ac, pH 4.8, were lower in the exposed site, whereas the extractable concentrations were higher. This difference was interpreted by the authors as an effect of a higher weathering rate, mostly as a result of infiltration of acidic precipitation. However, other differences in hydrology imposed by the inhibited percolation below the barn might have influenced the result, as well as the complete absence of fresh organic matter. The higher acidity, the lower base saturation and the higher concentrations of extractable cations outside the barn may to a large extent be explained simply by a higher organic matter content.

Long-term changes in soil anion pools, mainly sulphate, have also been considered in a few Swedish studies. By model calculations of sulphate adsorption on goethite and gibbsite, using data on soil pH and sulphate, Eriksson et al. (1992) concluded that sulphate saturation had been attained in southwest Sweden whereas little adsorption had taken place in the north. In a study by Gustafsson and Jacks (1993) it was demonstrated that, in southwest Sweden, neither organic S nor extractable sulphate concentrations increased significantly between 1951 and 1989 in podsolic B horizons. These findings were explained by high concentrations of organic carbon compared to those measured in a site in northern Sweden, where experimental application of H_2SO_4 during 5 yr caused a 40% adsorption of the S applied.

A steady state for sulphate has been indicated in forest sites in Scania. This conclusion is based on budget studies, particularly of spruce forest soils (Bergkvist and Folkesson 1995). Essentially all sulphate deposited in the stands was found to be transported in the percolation water through the soil profiles, giving no net retention of sulphate.

This was also the case in small untreated forest catchments around Gårdsjön in southwestern Sweden (Hultberg 1985) and in an experimental catchment treated with Na_2SO_4 (Hultberg et al. 1990). Similar results have been obtained with podsolized soils (entisols and spodosols) in United States and Canada, whereas old, strongly weath-

Table 1. The south–north (SN) and west–east (WE) distribution of some soil chemical variables expressed as % of the effective CEC, shown by linear regressions. Veff is base saturation. Also shown are mean values and standard deviations (SD). n.s. = not significant.

Variable	Regression	Mean	SD
Veff O horizon	Veff = 69.5 + 1.03 SN $R^2 = 0.32$; $p < 0.0001$; WE n.s.	83	12
Veff "C" horizon	Veff = 10.8 + 0.56 SN + 2.52 WE $R^2 = 0.38$; $p < 0.0001$	36	17
Mg^{2+} saturation O horizon	Mgsat = 15.7 + 0.37 SN − 0.84 WE $R^2 = 0.21$; $p < 0.0001$	16	4
Na^+ saturation O horizon	Nasat = 3.56 − 0.18 WE $R^2 = 0.20$; $p < 0.0001$; SN n.s.	3	1
K^+ saturation O horizon	Ksat = 10.0 + 0.54 SN − 0.65 WE $R^2 = 0.39$; $p < 0.0001$	14	5
Na^+ saturation "C" horizon	Nasat = 2.82 + 0.12 SN $R^2 = 0.17$; $p < 0.0001$; WE n.s.	6	3
Ca^{2+} saturation "C" horizon	Casat = 0.28 SN + 1.55 WE $R^2 = 0.37$; $p < 0.0001$, intercept n.s.	13	10
Mg^{2+} saturation "C" horizon	Mgsat = 3.0 + 0.33 WE $R^2 = 0.18$; $p < 0.0001$; SN n.s.	6	3

ered soils rich in clay (ultisols) had a considerable sulphate adsorption capacity (Rochelle et al. 1987).

The sulphate adsorption capacity of oxide surfaces is pH dependent, increasing with decreasing pH. In acid soils rich in N, nitrification can be a potent internal soil acidification process, as 2 mol H^+ are formed for each mol ammonium which is converted to nitrate. In such soils, several researchers have observed an inverse relationship between the concentrations of sulphate and nitrate in the soil solution. This phenomenon has been interpreted as an increase in the sulphate adsorption capacity of the oxides resulting from H^+ adsorption, or possibly as formation of basic Al sulphates, such as jurbanite, resulting from an increased Al dissolution (Nilsson 1985, 1991, Nodvin et al. 1988). Nitrogen "saturation", resulting in nitrification and a high concentration and leaching rate of nitrate (Tamm 1991, Johnson 1992), may therefore retard the sulphate leaching, at least temporarily.

It was demonstrated in a laboratory experiment that a vigorous field vegetation of Deschampsia flexuosa, with a high N uptake capacity, stimulated sulphate leaching from a strongly nitrifying soil. The sulphate leaching from reconstructed podsol profiles (O, E and upper B horizon) with Deschampsia, compared to that of bare podsol profiles, did not decrease to the extent predicted from the increase in evapotranspiration. This was reflected in the sulphate:chloride ratio of the soil leachates, which was significantly higher in the profiles with Deschampsia. The chloride ion is not adsorbed in podsolized soils and can therefore be used as a hydrological tracer. It was suggested that the extra sulphate leaching was caused

by desorption initiated by an increase in pH related to root uptake of nitrate (Berdén and Nilsson 1995).

Current chemical state of Swedish forest soils

Regional differences in base saturation of coniferous forest soils

The concentration of exchangeable base cations, the titratable acidity at pH 7 (TA), the base saturation and the concentration of exchangeable Al were recently determined at Swedish coniferous forest sites dominated by Norway spruce Picea abies or Scots pine Pinus sylvestris on mostly podsolized soils (Nilsson 1988, 1993, Johnson et al. 1991). The extractants were 1 M NH_4OAc buffered at pH 7 (exchangeable base cations and TA) and 1 M KCl (exchangeable Al). Base saturation increased in the S–N and W–E directions in all investigated horizons (O and E horizon, upper 5 cm of the B horizon and the "C" horizon at 50 cm depth).

A similar soil chemistry pattern was found by Eriksson et al. (1992) for a smaller number of sites. The regional gradients of soil chemistry coincided with S–N or W–E gradients in soil age, humidity, annual tree growth and atmospheric deposition of S and N (Nilsson 1990, Eriksson et al. 1992, Nilsson 1993). The sampling sites formed a subset of some 20 000 permanent plots covering the whole forested part of the country and used in an extensive forest monitoring programme which began in 1983 (Ranneby et al. 1987).

We have reanalyzed the data and included only soils with a base saturation (at pH 7) of <35% in the O horizon or <20% in the mineral soil horizons. These restrictions were designed to include mainly podsolized soils; they had only a minor influence on the number of soils investigated. For logistical reasons the necessary data for a complete soil classification were not available. The choice of base saturation restrictions was entirely based on data collected in control plots of old Swedish or Finnish forest liming experiments reported by Hallbäcken and Popovic (1985) and Derome et al. (1986), respectively.

Base saturation was defined as Σ base cations × 100:CEC, where CEC is calculated either as Σ base cations + TA, or as Σ base cations + Al (effective CEC). The unit is $mmol_c$ kg^{-1} dry soil. The base saturation (effective base saturation, Veff) and metal ion saturations reported here are based on the CEC expression containing Al, as this will give a more true absolute value for both, at least in the mineral soil horizons, although relative trends would be affected to a much smaller extent by the choice of CEC expression (see Ulrich 1966, Nilsson 1995). Metal ion saturation is defined as for Ca^{2+}: $(Ca^{2+}) \times 100$:CEC, where (Ca^{2+}) is the concentration of exchangeable Ca^{2+} in $mmol_c$ kg^{-1} dry soil.

The base saturation and metal ion saturation values

Table 2. Properties of beech forest soils in Scania (95 sites, sampled in 1981–85; dystric cambisols and podsols). Data given are P_{10}, P_{50} and P_{90} values of the ranges measured. Sampling depth: 0–5 cm (calculated from the lower boundary of the O_i horizon) and 30–40 cm (B horizon). Exchangeable Ca and Al measured in 1 M KCl (0–5 cm samples) and 1 M NH_4OAc, pH 4.8 (30–40 cm samples). See also Tyler (1987).

Sampling depth	0–5 cm			30–40 cm		
percentile (n = 95)	P_{10}	P_{50}	P_{90}	P_{10}	P_{50}	P_{90}
pH (0.2 M KCl)	2.97	3.43	3.99	3.73	3.92	4.19
pH (H_2O)	–	–	–	4.20	4.52	5.05
Ca:Al, mol. ratio	0.68	3.2	29	0.02	0.06	1.3
clay %	1.7	4.1	15.0	2.0	5.0	16.5
organic matter %	9.6	14.7	47.0	1.1	3.5	4.7

were used as dependent variables in separate linear regressions with the S–N and W–E numbers of the topographical maps of Sweden (scale 1:100 000) as independent variables. The values were averaged by horizon for each map before the regression calculations were made. Regressions, means and standard deviations are shown in Table 1.

In the O horizon there was a pronounced base saturation increase towards the north. In the "C" horizon an increase could be demonstrated both in the S–N and in the W–E direction. The other horizons showed similar patterns. The Mg^{2+} saturation of the O horizon increased towards the north, as already indicated by the base saturation gradient. There was a decrease towards the east, however, and this was also the case for Na^+. This pattern strongly suggested an influence of deposition of Mg^{2+} and Na^+ salts brought by prevailing westerly winds from the Atlantic. The pattern was obscured in the mineral soil horizons by Mg^{2+} and Na^+ contributions from the weathering of soil minerals.

Exchangeable K^+ in the O horizon also increased towards the north. A principal component analysis (Nilsson 1995) clearly demonstrated, however, that the K^+ saturation displayed a pattern of its own that could not simply be related to the base saturation.

We suggest that the K^+ saturation pattern, at least partly, was related to decrease in tree growth towards the north (Nilsson 1990). Among the base cations, K^+ is in most demand as a plant nutrient, though K^+ is not the main limiting nutrient for forest production (see Tamm 1991). Thus an increase in exchangeable K^+ in the O horizon from S to N should be detectable as a result of the decrease in tree growth.

In the "C" horizon, exchangeable Ca^{2+} and Na^+ both increased from·S to N, while exchangeable Mg^{2+} and K^+ (not shown) did not. Ca^{2+} and Mg^{2+} increased significantly towards the east. This is in accordance with the fact that Ca^{2+} and Na^+ have high concentrations in the parent material in the northeastern part of the country, indicating the presence of plagioclases, while other minerals rich in

Ca^{2+} and Mg^{2+} increase towards the east (Melkerud et al. 1992).

Beech forest soils

Deciduous forest, usually dominated by beech *Fagus sylvatica* or oak *Quercus* spp., constitutes a main vegetation type of non-arable land in the extreme south of Sweden. Though considerably reduced by man, beech forest is the potential ecosystem of most well-drained land in Scania. Beech soils in this area may be podsols or dystric cambisols, eutric cambisols occurring only locally and mainly in sites with calcareous parent material. Gleyic subtypes of both podsols and cambisols are also widespread in conditions of impeded soil drainage and aeration. Some of these soils may be true gleysols according to the FAO classification.

A study comprising 95 beech forests in Scania, selected to cover most of the edaphic variability encountered, was performed during the 1980's. It revealed that almost no soils still contain $CaCO_3$ in the A or B horizons (Table 2). pH (0.2 M KCl) mostly ranged from 3.0–4.0 in the topsoil and 3.7–4.2 in the B horizon. The molar ratio of exchangeable Ca:Al (extractant 1 M NH_4OAc pH 4.8) averaged 0.06 in the B horizon with the 10 percentile value as low as 0.02. In the top 5 cm, characterized by appreciably higher organic matter contents, the Ca:Al ratio was usually above 1, though values as low as 0.1 were measured in a few sites.

One hundred and fifty oak and hornbeam (*Carpinus betulus*) topsoils in Scania had acidity and Ca:Al ranges similar to the beech forest soils (Tyler 1989). The pH (KCl) range was smaller (no values below 3.2) and the 50 percentile value was as high as 3.9. Chemical ranges for oak soils differed only a little from those of beech soils.

Discussion

Retrospective soil investigations almost consistently show that pH and/or exchangeable base cation stores have decreased, particularly in south Sweden, during recent decades. We will now discuss possible sources of error which may bias these findings, and to what extent factors other than acid deposition have contributed to the increased soil acidity.

Reliability of data – evaluation of methods

Inconsistent reporting of studies performed on, e.g. temporal changes in environmental conditions may bias any conclusions on the general validity of the findings. For researchers and scientific journals it is certainly not very attractive to publish reports if no changes have been detected or statistically ascertained. However, from our survey of the materials available for retrospective work on forest soil chemistry in Sweden we are almost able to exclude sources of error resulting from missing "aberrant" information. There are few reported divergences from the general pattern and these are usually explained by particular local conditions, e.g. far-reaching mechanical disturbance of the soil, forest liming, or extensive thinning of the forest stand.

Another problem in repetitive work is how to localize the exact spot on the ground where soil was sampled several decades earlier. It is certainly not recommended to sample this spot again, because of the inevitable disturbance and mixing of materials from different soil horizons at the first sampling. At least in beech forest, it is more important to know how the soil was sampled with respect to distance from major tree trunks, because of the variable influence of stemflow. This is sometimes even more important than to know which part of the forest stand was sampled. Several of the retrospective studies cited in this review, however, have utilized permanent plots or detailed field notes from the first sampling. In some cases the person who took the samples in the original study has participated in the repeated work. As most studies are based on resampling of many different sites, it seems unlikely that bias arising from differences in the location of sampling points has distorted the results, though it has probably introduced some additional variability of the data.

Retrospective studies on soil chemistry have to rely either on old data or on the re-analysis of preserved soil samples, sources of error being possible in both cases. There is some evidence in support of the view that the pH of air-dried soil samples may change, mainly decrease, over years of storage (Falkengren-Grerup 1995). A real ambient decrease in soil pH may thus be underestimated or completely concealed, if a study is based on simultaneous analysis of new and old samples. However, most retrospective work on soil pH has relied on old pH measurements, which have been compared with new ones. The reliability of old pH data may be argued but comparisons of methods, made by, e.g. Hallbäcken and Tamm (1986) and Falkengren-Grerup (1986, 1987), indicate that systematic differences between old and new pH measurements are of minor importance.

Repetitions of old studies on exchangeable base cations, however, have usually to be performed on preserved samples, as sensitivity or precision of old analyses (if they exist) is frequently insufficient. Little is known about changes in cation pools during long-term storage of air-dried samples. A minor study over 6 yr was not conclusive (Falkengren-Grerup 1995). Exchangeable Ca seemed to increase and Na to decrease, leaving K and Mg more or less unchanged. Part of the differences measured in Na and Ca might have resulted from analytical variability or inhomogeneity of the small sample volumes possible to extract for analysis. More work on this problem is definitely needed, as it is of fundamental importance to any use of dried soil samples for determination of labile fractions of mineral elements.

Relative importance of acid deposition and other sources of soil acidity

For several reasons, it is difficult to distinguish one single factor as wholly controlling the decline in soil pH or base saturation. In Sweden, and particularly in the southern part of the country, extensive changes in land-use and increases in S and N deposition took place almost simultaneously during the first half of the 20th century (Nilsson 1993, Renberg et al. 1993). Large areas in south Sweden were almost totally devoid of trees during the 18th and 19th centuries. The open areas were used for haymaking and cattle grazing and included large *Calluna* heaths which were regularly burnt to improve the forage quality. By the end of the 19th century, tree plantation began on a regular basis (Malmström 1937, Nilsson 1990).

However, most of the data reported from south Sweden on soil chemical changes during recent decades are for mature or aged forest stands, some of them even characterized by forest continuity for centuries (Brunet 1993). Effects of changes in land-use will then be of minor importance. Instead, soil chemical changes reported might underestimate the changes on a general scale, which have taken place in this region over areas where land-use has shifted from extensive grazing (ground with few trees) to forest, or from deciduous to coniferous forest, during the last century. Conifers, particularly spruce plantations, which are nowadays widespread even south of the natural border of *Picea abies*, are characterized both by a usually high primary productivity and by a high capturing capacity for acidic particles from the atmosphere, because of the larger total surface area of the foliage (Nihlgård 1970, Bergkvist 1987a,b). Greater acid strength of the organic matter produced in spruce than in deciduous forest may also be of importance (Binkley and Valentine 1991). The soils below spruce stands on formerly open ground have, therefore, probably changed significantly more than indicated by the retrospective studies.

More important for the interpretation and explanation of the reports on higher soil acidity is the increased growth and primary productivity of Swedish forests during recent decades, evidenced in both coniferous and deciduous forest (Falkengren-Grerup and Eriksson 1990, Eriksson and Johansson 1993). It is often assumed that the elevated N deposition rate during recent decades has favoured primary productivity of forest ecosystems which are traditionally considered to be N-limited. Biological soil acidification as a result of increased cation accumulation in perennial biomass, or the nitrification of deposited ammonium, may thus have contributed to the observed changes in soil chemistry.

However, soil acidification at a particular site cannot be fully understood unless all quantitatively important H^+ sources of the ecosystem are accounted for (Ulrich et al. 1979, van Breemen et al. 1983, Nilsson 1985, Bredemeier et al. 1990). Many H^+ budgets for entire ecosystems based on the mass balance and charge balance concepts have been published (e.g. Ulrich et al. 1979, van Breemen et al. 1984, cf. Berdén et al. 1987). In western Europe, including the southern part of Scandinavia, atmospheric deposition of strong mineral acids and acid precursors (SO_2, NO_x and NH_4^+) may account for 30–70% of the acid input to the soil. Much of the acid input from the atmosphere is through particle capture by the tree canopies. The other main effect of trees on soil acidification, root uptake of excess cations from the soil, is of importance in all aggrading and managed forests because of the non-reversible export of cations through tree harvesting. Soil acidification, particularly in connection with whole-tree harvesting could be similar to or even greater than acidification caused by deposition (Berdén et al. 1987, Nilsson 1993).

A dynamic soil chemistry model (considering leaching, cation exchange, mineral weathering, nutrient uptake and solution equilibrium reactions) was applied to retrospective data on forest soil chemical changes in south Sweden (Warfvinge et al. 1993). From this study it might be concluded that the observed decrease in exchangeable base cation pools is largely determined by acidity input (including differences in dry deposition among tree species and stand types) and that leaching losses at the site are much more important than accumulation in tree biomass for the cation depletion of the soil. This conclusion may have been influenced by the model structure, however, which as in many other soil acidification models strongly emphasizes the external acid input compared to the internal acidification processes (van Oene 1994). pH effects attributable to long-term changes in the quality of soil organic matter may also be of importance.

The retrospective study at Tönnersjöheden in south Sweden (Hallbäcken and Tamm 1986) showed a pH decline, presumably indicating a decline in exchangeable base cations, in all soil horizons of the podsols studied. Declines influenced by increasing stand age, related to base cation accumulation in biomass and organic matter, were clearly demonstrated for the O and E horizons. It was concluded that internal acidification processes were not important for the pH decrease in the B and C horizons. However, the measured pH values suggest that carbonic acid dissociation through the formation of dissolved H^+ and HCO_3^- could have been important, at least during the first part of the time considered (1927–1984). Several lysimeter studies also show that soluble organic compounds (determined as dissolved organic carbon) amount to several mg C l^{-1} in soil solutions collected from the lower part of the B horizon or from the C horizon in podsolized soils (Nilsson and Bergkvist 1983, Bergkvist and Folkeson 1992, Berggren 1992). The dissolved organic compounds in these soil horizons mainly consist of fairly strong dissolved organic acids (Cronan and Aiken 1985), which may influence soil pH and base cation leaching. However, as there was no consistent variability of pH related to stand age in the B and C horizons of the study by Hallbäcken and Tamm (1986), a

pH influence by organic acids transported from upper horizons is not a likely main cause of the temporal pH decrease. Furthermore, the regional study by Eriksson et al. (1992) neatly demonstrated the covariation between a high soil storage of sulphate and extensive temporal declines in exchangeable base cations in podsolized well-drained Swedish forest soils.

Concluding remarks

Although there are several factors behind observed declines in soil pH and in base cation pools, there is strong evidence that the atmospheric deposition of acid and potentially acidifying S and N compounds has significantly influenced the chemical properties of forest soils in south Sweden. However, it is not justified to state that the H^+ deposition input is the only factor of importance.

Indications of significant soil changes in forest lake catchments had already been found in the 1960's and 1970's, long before any retrospective soil investigations had been made. These indications consisted of observations of declining pH and alkalinity and increasing Al concentrations in lakewater and streamwater (Bernes 1991). It could be shown that the input of acidity from catchments into streams and lakes was strongly dependent on the concentration of the relatively mobile sulphate ion, which mainly originated from atmospheric deposition (Seip 1980).

The S cycling in forest ecosystems is influenced by both chemical processes, such as sulphate adsorption/ desorption, and biological mineralization and immobilization processes. The latter, in particular, are still not well understood. However, they seem to be important for the leaching of sulphate and dissolved organic S after forest liming (Valeur and Nilsson 1993, Andersson et al. 1994).

Because the N retention in forest ecosystems is usually much greater than the S retention, effects of S and N deposition on soil acidification are not additive and should be treated separately. Nitrogen is the most important limiting nutrient in boreal forest ecosystems, and there are many feed-back processes between vegetation, soil microflora and soil humic compounds which will influence the soil acidifying effect of a given amount of N input. Important processes in the N cycle, such as nitrification and denitrification, are still poorly understood, and it is therefore premature to include them in mechanistic soil models in order to forecast the nitrate leaching.

It is evident, in a situation with declining S deposition and constant or increasing N deposition, that the S and N cycling must be better understood, if more reliable estimates of the critical loads of S and N, and of the potential for soil recovery, are to be made.

References

Andersson, S, Valeur, I. and Nilsson, I. 1994. Influence of lime on soil respiration, leaching of DOC, and C/S relationships in the mor humus of a Haplic Podsol. – Environ. Int. 20: 81–88.

Berdén, M. and Nilsson, S.I. 1995. Influence of added ammonium sulphate on the leaching of aluminium, nitrate and sulphate – a laboratory experiment. – Water Air Soil Pollut., in press.

– , Nilsson, S.I., Rosén, K. and Tyler, G. 1987. Soil acidification – Extent, causes and consequences. An evaluation of literature information and current research. – Nat. Swedish Environ. Protect. Board, Report 3292, Solna, Sweden.

Berggren, D. 1992. Speciation and mobilization of aluminium and cadmium in podzols and cambisols of S. Sweden. – Water Air Soil Pollut. 62: 125–156.

– , Bergkvist, B., Falkengren-Grerup, U. and Tyler, G. 1992. Influence of acid deposition on metal solubility and fluxes in natural soils. – In: Låg, J. (ed.), Chemical climatology and geomedical problems. Norwegian Acad. Sci. Letters, Otta, Norway, pp. 141–154.

Bergkvist, B. 1987a. Leaching of metals from forest soils as influenced by tree species and management. – For. Ecol. Manage 22: 29–56.

– 1987b. Soil solution chemistry and metal budgets of spruce forest ecosystems in S. Sweden. – Water Air Soil Pollut. 32: 131–154.

– and Folkesson, L. 1992. Soil acidification and element fluxes of a *Fagus sylvatica* forest as influenced by simulated nitrogen deposition. – Water Air Soil Pollut. 61: 111–133.

– and Folkesson, L. 1995. The influence of tree species on acid deposition, proton budgets and element fluxes in some Swedish forest ecosystems. – Ecol. Bull. (Copenhagen) 44: 90–99.

– , Folkesson, L. and Olsson, K. 1987. Metal fluxes in *Picea abies*, *Fagus sylvatica* and *Betula pendula* forest ecosystems. – In: Lindberg, S.E. and Hutchinson, T.C. (eds), Int. Conf. heavy metals in the environment, New Orleans, September 1987. CEP Consultants, Edinburgh 2: 407–409.

Bernes, C. 1991. Försurning och kalkning av svenska vatten. – Monitor 12. Nat. Swedish Environ. Protect. Board, Solna, Sweden.

Binkley, D. and Valentine, D. 1991. Fifty-year biogeochemical effects of green ash, white pine, and Norway spruce in a replicated experiment. – Forest Ecol. Manage. 40: 13–25.

Bredemeier, M., Matzner, E. and Ulrich, B. 1990. Internal and external proton load to forest soils in northern Germany. – J. Environ. Qual. 19: 469–477.

Breemen, N. van, Mulder, J. and Driscoll, C.T. 1983. Acidification and alkalinization of soils. – Plant Soil 75: 283–308.

– , Driscoll, C.T. and Mulder, J. 1984. Acidic deposition and internal proton sources in acidification of soils and waters. – Nature 307: 599–604.

Brunet, J. 1993. Environmental and historical factors limiting the distribution of rare forest grasses in south Sweden. – For. Ecol. Manage 61: 263–275.

Buch, M.W. von 1982. Humusformen umweltbelasteter Bestände Hamburger Waldungen. – Forstarchiv 53: 46–51.

Butzke, H. von 1981. Versauern unsere Wälder? Erste Ergebnisse der Überprüfung 20 Jahre alten pH-Wert-Messungen in Waldböden Nordrhein-Westfalens. – Forst Holzwirt 21: 542–548.

Cronan, C.S. and Aiken, G.R. 1985. Chemistry and transport of soluble humic substances in forested watersheds of the Adirondack Park, New York. – Geochim. Cosmochim. Acta 49: 1697–1705.

Derome, J., Kukkola, M. and Mälkönen, E. 1986. Forest liming on mineral soils. Result of Finnish experiments. – Nat. Swedish Environ. Protect. Board, Report 3084, Solna, Sweden.

Eriksson, E., Karltun, E. and Lundmark, J.-E. 1992. Acidification of forest soils in Sweden. – Ambio 21: 150–154.

Eriksson, H. and Johansson, U. 1993. Yields of Norway spruce (*Picea abies* (L.) Karst.) in two consecutive rotations in southwestern Sweden. – Plant Soil 154: 239–247.

Falkengren-Grerup, U. 1986. Soil acidification and vegetation changes in deciduous forest in southern Sweden. – Oecologia 70: 339–347.

– 1987. Long-term changes in pH of forest soils in southern Sweden. – Environ. Pollut. 43: 79–90.

– 1989. Effects of stemflow on beech forest soils and vegetation in southern Sweden. – J. Appl. Ecol. 26: 341–352.

– 1995. Effects of long-term storage on chemical properties of soil samples. – Ecol Bull. (Copenhagen) 44: 129–132.

– and Eriksson, H. 1990. Changes in soil, vegetation and forest yield between 1947 and 1988 in beech and oak sites of southern Sweden. – For. Ecol. Manage 38: 37–53.

– and Björk, L. 1991. Reversibility of stemflow induced soil acidification in Swedish beech forest. – Environ. Pollut. 74: 31–37.

– and Tyler, G. 1991. Changes of cation pools of the topsoil in south Swedish beech forests between 1979 and 1989. – Scand. J. For. Res. 6: 145–152.

– and Tyler, G. 1992. Changes since 1950 of mineral pools in the upper C-horizon of Swedish deciduous forest soils – Water Air Soil Pollut. 64: 495–501.

– and Tyler, G. 1993. The importance of soil acidity, moisture, exchangeable cation pools and organic matter solubility for the cationic composition of beech forest (*Fagus sylvatica* L.) soil solution. – Z. Pflanzenernähr. Bodenkd. 156: 365–370.

– , Linnermark, N. and Tyler, G. 1987. Changes in acidity and cation pools of south Swedish soils between 1949 and 1985. – Chemosphere 16: 2239–2248.

FAO. 1988. Soil map of the world. Revised legend. – World soil resources, report 60. FAO-Unesco, Rome.

Glatzel, G., Kilian, W., Sterba, H. and Stöhr, C. 1985. Waldbodenversauerung in Österreich: Ursachen – Auswirkungen. – Allgem. Forstz. Febr., pp. 35–36.

Grenzius, R. von 1984. Starke Versauerung der Waldböden Berlins. – Forstwiss. Centralbl. 103: 131–139

Gustafsson, J.P. and Jacks, G. 1993. Sulphur status in some Swedish podzols as influenced by acidic deposition and extractable organic carbon. – Environ. Pollut. 81: 185–191.

– , Jacks, G., Stegmann, B. and Ross, H.B. 1993. Soil acidity and adsorbed anions in Swedish forest soils – long-term changes. – Agric. Ecosyst. Environm. 47: 103–115.

Hallbäcken, L. and Popovic, B. 1985. Effects of forest liming on soil chemistry – Investigation of Swedish liming experiments. – Nat. Swedish Environ. Protect. Board, Report 1880, Solna, Sweden, (in Swedish with English summary).

– and Tamm, C.O. 1986. Changes in soil acidity from 1927 to 1982–1984 in a forest area of south-west Sweden. – Scand. J. For. Res. 1: 219–232.

Hauhs, M. 1985. Wasser- und Stoffhaushalt im Einzugsgebiet der Langen Bramke (Harz). – Ber. Forschungsz. Waldökosysteme/Waldsterben (Göttingen) 17: 1–206.

Hultberg, H. 1985. Budgets of base cations, chloride, nitrogen and sulfur in the acid lake Gårdsjön catchment. – Ecol. Bull. (Copenhagen) 37: 133–157.

– , Lee, Y.-H., Nyström, U. and Nilsson, S.I. 1990. Chemical effects on surface-, ground- and soilwater of adding acid and neutral sulphate to catchments in southwest Sweden. – In: Mason, B.J. (ed.), The surface waters acidification programme. Cambridge Univ. Press, pp. 167–182.

Jacks, G. 1992. Acidification of soil and water below the highest holocene shoreline on Södertörn, central eastern Sweden. – Sv. Geol. Undersökn., Ser. Ca 81: 145–148.

– , Andersson, S. and Stegmann, B. 1989. pH-changes over 30–40 years along a depositional gradient. – Proc. Symp. Environm. threats For. Nat. Ecosyst., Oulu, Finland, Nov. 1–4, 1988, pp. 61–68.

Jochheim, H. 1985. Der Einfluss des Stammablaufwassers auf

dem chemischen Bodenzustand und die Vegetationsdecke in Altbuchenbeständen verschiedenen Waldgesellschaften. – Ber. Forschungszentrums Waldökosysteme/Waldsterben (Göttingen) 13: 1–226.

Johnsson, D. W. 1992. Nitrogen retention in forest soils. – J. Environ. Qual. 21: 1–12.

– , Cresser, M. S., Nilsson, S. I., Turner, J., Ulrich, B., Binkley, D. and Cole, D. W. 1991. Soil changes in forest ecosystems: Evidence for and probable causes. – Proc. Royal Soc. Edinburgh, Sect. B. Biol. Sci. 97: 81–116.

Malmström, C. 1937. Der Versuchsrevir Tönnersjöheden in Halland. Ein Beitrag zur Kenntnis der südwestschwedischen Wälder, Heiden und Torfmoore. – Medd. Statens Skogsförsöksanst. 30: 323–358, (in Swedish with German summary).

Melkerud, P. A., Olsson, M. T. and Rosén, K. 1992. Geochemical atlas of Swedish forest soils. – Reports in forest ecology and forest soils 65. – Dept Forest Soils, Swedish Univ. Agricult. Sci., Uppsala, Sweden.

Nihlgård, B. 1970. Precipitation, its chemical composition and effect on soil water in a beech and a spruce forest in south Sweden. – Oikos 21: 208–217.

Nilsson, N. E. (ed.) 1990. Skogen. – Sveriges Nationalatlas. Bra Böcker, Höganäs, Sweden, (in Swedish).

Nilsson, S. I. 1985. Why is lake Gårdsjön acid? – An evaluation of processes contributing to soil and water acidification. – Ecol. Bull. (Copenhagen) 37: 311–318.

– 1988. Acidity properties in Swedish forest soils. Regional patterns and implications for forest liming. – Scand. J. For. Res. 3: 417–424.

– 1991. The interaction between mobile anions and aluminium in a shallow podzol. – In: Robitaille, G. (ed.), Proc. acid rain forest resour. conf. Environment Canada, Canadian Forestry Service, U.S. Dept Agric. For. Serv. Quebec City, pp. 532–538.

– 1993. Acidification of oligotrophic forest lakes. Interactions between deposition, forest growth and effects on lake-water quality. – Ambio 22: 272–276.

– 1995. Current chemical state of Swedish forest soils investigated by principal component analysis (PCA) and linear regression. – Ecol. Bull. (Copenhagen) 44: 65–74.

– and Bergkvist, B. 1983. Aluminium chemistry and acidification processes in a shallow podzol on the Swedish westcoast. – Water Air Soil Pollut. 20: 311–329.

Nodvin, S. C., Driscoll, C. T. and Likens, G. E. 1988. Soil processes and sulphate loss at the Hubbard Brook Experimental Forest. – Biogeochemistry 5: 185–199.

Oene, H. van 1994. Modelling the impact of sulphur and nitrogen deposition. – Ph.D. thesis. Report 68. Dept Ecol. Environ. Res., Swedish Univ. Agricult. Sci., Uppsala.

Pelisek, J. 1984. Changes in the acidity of forest soils of the Orlickee Mountains, caused by acid rain. – Lesnctvi 30: 955–962, (in Czech with English summary).

Ranneby, B., Cruse, T., Hägglund, B., Jonasson, H. and Swärd, J. 1987. Designing a new national forest survey for Sweden. – Stud. For. Suec. 177: 1–29.

Renberg, I., Korsman, T. and Andersson, N. J. 1993. A temporal perspective of lake acidification in Sweden. – Ambio 22: 264–271.

Riebling, R. and Schaefer, C. 1984. Jahres- und Langzeitentwicklung der pH-Werte von Waldböden in hessischen Fichtenbeständen. – Forst Holzwirt 7: 177–182.

Rochelle, B. P., Church, R. and David, M. B. 1987. Sulfur retention at intensively studied sites in the U.S. and Canada. – Water Air Soil Pollut. 33: 73–84.

Seip, H. M. 1980. Acidification of freshwater – sources and mechanisms. – In: Drabløs, D. and Tollan, A. (eds), Ecological impact of acid precipitation. SNSF Project Oslo-Ås, pp. 358–366.

Sjöström, J. and Qvarfort, U. 1992. Long-term changes of soil chemistry in central Sweden. – Soil Sci. 154: 450–457.

Stöhr, D. (ed.) 1984. Waldbodenversauerung in Österreich. – Österr. Forstverein Forschungsinitiative gegen das Waldsterben. Inst. f. Forstökologie, Wien.

Tamm, C. O. 1991. Nitrogen in terrestrial ecosystems. Questions of productivity, vegetational changes and ecosystem stability. – Ecological Studies 81, Springer, New York.

– and Hallbäcken, L. 1988. Changes in soil acidity from the 1920's to the 1980's in two forest areas with different acid deposition. – Ambio 17: 56–61.

Troedsson, T. 1985. Sensitivity of Swedish forest soils to acidification related to site characteristics. – Nat. Swedish Environ. Protect. Board, Report 3001, Solna, Sweden.

Tyler, G. 1981. Leaching of metals from the A-horizon of a spruce forest soil. – Water Air Soil Pollut. 15: 353–369.

– 1987. Acidification and chemical properties of *Fagus sylvatica* L. forest soils. – Scand. J. Fors. Res. 2: 263–271.

– 1989. Edaphical distribution patterns of macrofungal species in deciduous forest of south Sweden. – Acta Oecol. Oecol. Gen. 10: 309–326.

Ulrich, B. 1966. Kationenaustausch – Gleichgewichte in Böden. – Z. Pflanzenernähr. Düng. Bodenkd. 113: 141–159.

– 1981. Ökologische Gruppierung von Böden nach ihrem chemischen Bodenzustande. – Z. Pflanzenernähr. Düng. Bodenkd. 144: 289–305.

– and Pankrath, J. (eds) 1983. Effects of accumulation of air pollutants in forest ecosystems. – D. Reidel Publ. Company, Dordrecht.

– , Mayer, R. and Khanna, P. K. 1979. Deposition von Luftverunreinigungen und ihre Auswirkungen in Waldökosystemen im Solling. – Schriften Forstlichen Fak. Univ. Göttingen 58.

– , Mayer, R. and Khanna, P. K. 1980. Chemical changes due to acid precipitation in a loess-derived soil in central Europe. – Soil Sci. 130: 193–199.

Valeur, I. and Nilsson, S. I. 1993. Effects of lime and two incubation techniques on sulfur mineralization in a forest soil. – Soil Biol. Biochem. 25: 1343–1350.

Warfvinge, P., Falkengren-Grerup, U., Sverdrup, H. and Andersen, B. 1993. Modelling long-term cation supply in acidified forest stands. – Environ. Pollut. 80: 209–221.

Wittig, R. and Neite, H. 1985. Acid indicators around the trunk base of *Fagus sylvatica* in limestone and loess beechwoods: distribution patterns and phytosociological problems. – Vegetatio 64: 113–119.

Wittmann, O. and Fetzer, K. D. 1982. Aktuelle Bodenversauerung in Bayern. – Bayerisches Staatsministerium für Landesentwicklung und Umweltfragen. – Stockert, München.

Ecological Bulletins 44: 65–74. Copenhagen 1995

Current chemical state of Swedish forest soils investigated by principal component (PCA) and regression analysis

S. Ingvar Nilsson

Nilsson, S.I. 1995. Current chemical state of Swedish forest soils investigated by principal component (PCA) and regression analysis. – Ecol. Bull. (Copenhagen) 44: 65–74.

The acid/base soil chemistry at 400–700 Swedish forest sites (the number of sites depending on soil horizon), was characterized by a PCA analysis for each soil horizon (O, E, B and "C" horizon). The "C" horizon represented a soil depth of 50 cm. The variables used in the PCA model were the percentage saturation values of exchangeable Al^{3+}, Ca^{2+}, Mg^{2+}, K^+ and Na^+, defined as e.g. $[Ca^{2+}] \cdot 100/CEC$, where $[Ca^{2+}]$ is exchangeable Ca^{2+} and CEC is the cation exchange capacity in $mmol_c \cdot kg^{-1}$ defined as (Σ exchangeable base cations + exchangeable Al). Exchangeable base cations were obtained after extraction in 1 M NH_4OAc at pH 7 and exchangeable Al^{3+} after extraction in 1 M KCl. The first three principal components accounted for >90% of the variation. The principal components were used as dependent variables with the south-north (SN) and west-east (WE) topographic map numbers (scale 1:100 000) as the independent variables. Component 1 described the base saturation and was positively correlated with the S–N and/or W–E directions. Component 2 described the increase in exchangeable Mg^{2+} and Na^+ towards the west caused by atmospheric deposition of marine Mg^{2+} and Na^+ salts. This trend was most evident in the O horizon. In the mineral soil horizons, component 2 also reflected the distribution of soil minerals rich in Ca^{2+} and Mg^{2+}. In the O horizon, the second and particularly the third principal component showed a strong relationship with increasing amounts of exchangeable K^+ towards the north. The K^+ gradient was interpreted as an indicator of declining tree growth and ensuing declining K^+ uptake towards the north. In the "C" horizon, the third principal component was correlated with the regional distribution of plagioclase and to a minor extent with that of mica and K^+ feldspar. Linear regressions indicated that $CuCl_2$-extractable exchangeable plus complexed Al accounted for most of the titratable acidity at pH 7 in the mineral soil horizons, while acid humic compounds seemed to be more important in the O horizon. There were indications of an extensive Al hydrolysis in the B and "C" horizons. The importance of organically bound Al as a pH and Al buffer is discussed.

S.I. Nilsson, Dept of Soil Sciences, Swedish Univ. of Agricultural Sciences, P.O. Box 7014, S-750 07 Uppsala, Sweden.

Introduction

The regional variation in acid-base soil chemistry depends on the soil-forming factors as defined by Jenny (Birkeland 1984), i.e. parent material, landscape topography, vegetation and climate. Excess atmospheric deposition of acid or potentially acidifying sulphur and nitrogen compounds is an additional factor which will affect the chemical properties of the soil.

Previous studies of a limited number of Swedish forest soils showed quite distinct south-north (S–N) and west-east (W–E) gradients in pH and base saturation in podsolized forest soils, in that these variables tended to increase towards the north and towards the east (Hallbäcken and Popovic 1985, Eriksson et al. 1992, Karltun 1994).

Concentrations of exchangeable base cations, titratable acidity at pH 7, base saturation and concentrations of exchangeable and complexed Al were recently determined for a larger number of Swedish coniferous forest sites (c. 400–700 depending on soil horizon), usually dominated by either Norway spruce *Picea abies* (L.) Karst. or Scots pine *Pinus sylvestris* L. on mostly podsolized soils. Base saturation in all investigated soil layers, i.e. the O and E horizons, the upper 5 cm of the B horizon and the soil layer at a depth of 50 cm (the "C" horizon), increased as expected in the S–N and W–E directions (Nilsson 1988, Johnson et al. 1991, Nilsson 1993).

The regional gradients in soil chemistry found by these authors coincide more or less with declining S–N or W–E gradients in soil age, humidity, annual tree growth and atmospheric deposition of S and N (see Nilsson 1990, Bernes 1991, Nilsson 1993). This paper describes a further analysis of the regional soil chemistry data previously reported by Nilsson (1988), Johnson et al. (1991) and Nilsson (1993), consisting of a principal component analysis (PCA), (Gauch 1982) for each soil horizon. I also discuss the relationship between titratable acidity and exchangeable and complexed Al in different soil horizons.

Materials and methods
Sampling

Samples were collected between 1983 and 1986. The forest sites represented three main field vegetation types, dominated by blueberry *Vaccinium myrtillus* L., by narrow-leafed grasses, or by broad-leafed grasses, according to the terminology given by Lundmark and Odell (1983). After collection and air-drying, the samples were stored at ambient temperature in an unheated building close to the laboratory.

The sampling sites were a subset of some 20 000 permanent sites which were chosen to represent the entire forested part of the country, and which are presently being used in an extensive forest site monitoring programme which began in 1983 (Ranneby et al. 1987).

Analytical methods

Exchangeable base cations and titratable acidity at pH 7 (TA) were determined by AAS (C_2H_2-air flame) and by titration with 0.1 M NaOH, respectively, using 1 M NH_4OAc, buffered at pH 7, as extractant. The AAS analyses were run without any addition of lanthanum, which may give erroneous Ca^{2+} and Mg^{2+} values, for example, because of interference by aluminium or phosphorus (Allen 1989). However, linear regressions showed only minor deviations from a 1:1 relationship between lanthanum addition and no addition, except for the Ca^{2+} values in E horizon samples. In this case the regression equation was used to correct the Ca^{2+} values. Exchangeable Al was determined by AAS (N_2O–C_2H_2 flame), using 1 M KCl as extractant (Nilsson 1988). Exchangeable plus complexed Al was also determined by AAS, on a smaller number of samples, using 0.5 M $CuCl_2$ as extractant (see Nilsson and Lundmark 1986). The number of sites included in the latter determination was c. 300–500, depending on the soil horizon.

Data treatment

Only soils which had a base saturation (at pH 7) < 35% in the O horizon or < 20% in the mineral soil horizons were included in the calculations. The number of sites was only slightly reduced by this restriction (2–7% depending on horizon). The rationale was that only acid and more or less podsolized soils should be included in the dataset. The base saturation restrictions were based on data obtained in control plots of old Swedish and Finnish forest liming experiments, reported by Hallbäcken and Popovic (1985) and Derome et al. (1986), respectively. For logistical reasons, the data for a complete soil classification were not available.

PCA analyses were run using the SAS PCA procedure (SAS 1989). The variables used were the % saturation values of exchangeable Al (or, alternatively, titratable acidity at pH 7, TA) and of exchangeable Ca^{2+}, Mg^{2+}, K^+ and Na^+ as defined for Ca^{2+}:

$$[Ca^{2+}] \cdot 100/CEC$$

$[Ca^{2+}]$ was the concentration of exchangeable Ca^{2+}, and CEC was the cation exchange capacity defined as either (Σ exchangeable base cations + exchangeable Al), or (Σ exchangeable base cations + TA). Σ base cations, TA, Al, CEC and the individual base cation concentrations were in $mmol_c \cdot kg^{-1}$ dry soil. Exchangeable Al was assumed to be trivalent, Al^{3+} (see Bloom et al. 1979, Thomas and Hargrove 1984). Base saturation (V%) was calculated according to the equation V = Σ base cations $\cdot$ 100/CEC. Acidity relationships (TA in 1 M NH_4OAc extracts versus Al in 0.5 M $CuCl_2$ extracts) were analyzed by linear regressions, and the degree of Al hydrolysis was estimated from the regression slopes.

Results and discussion
PCA analysis

Only the PCA analyses and base saturation values based on CEC = (Σ exchangeable base cations + exchangeable Al), will be discussed. In the mineral soil horizons this CEC estimate corresponds closely to the effective CEC as defined by Ulrich (1966), for example, i.e. CEC at the actual soil pH. This is not entirely true for the O horizon, in which exchangeable H^+ is also important.

However, the two sets of PCA analyses yielded essentially the same qualitative information about each soil horizon, including the O horizon, both in terms of axis loadings and regional variation (see below). For regional variation, this was also true of the two sets of base saturation values.

The first three principal components (PRIN1, PRIN2 and PRIN3) accounted for approximately 60, 21 and 12% of the variation of the data set in the E, B and "C"

Table 1. Eigenvectors of the first three principal components PRIN1, PRIN2 and PRIN3, of the O, E, B and "C" horizons (*a* equations), and linear regressions between each of the three principal components and the south-north (SN) and west-east (WE) topographical map numbers (*b* equations). N.S. means not significant. For further explanations, see Figs 1 and 2.

O horizon
a. PRIN1 = -0.70Al $+ 0.59$Ca $+ 0.23$Mg $+ 0.30$K -0.11Na
b. PRIN1 = $-0.76 + 0.06$SN; WE N.S.; $R^2 = 0.36$; $p < 0.0001$

a. PRIN2 = -0.02Al -0.35Ca $+ 0.59$Mg $+ 0.41$K $+ 0.60$Na
b. PRIN2 = $0.35 + 0.07$SN -0.19WE; $R^2 = 0.28$; $p < 0.0001$

a. PRIN3 = -0.10Al $+ 0.26$Ca $+ 0.27$Mg -0.81K $+ 0.44$Na
b. PRIN3 = $0.63 -0.05$SN; WE N.S.; $R^2 = 0.23$; $p < 0.0001$

E horizon
a. PRIN1 = -0.55Al $+ 0.39$Ca $+ 0.49$Mg $+ 0.44$K $+ 0.33$Na
b. PRIN1 N.S.

a. PRIN2 = 0.15Al -0.66Ca -0.07Mg $+ 0.40$K $+ 0.62$Na
b. PRIN2 = $0.55 -0.10$WE; SN N.S.; $R^2 = 0.07$; $p < 0.003$

a. PRIN3 = -0.02Al $+ 0.34$Ca -0.26Mg -0.56K $+ 0.70$Na
b. PRIN3 N.S.

B horizon
a. PRIN1 = -0.56Al $+ 0.32$Ca $+ 0.49$ Mg $+ 0.44$K $+ 0.39$Na
b. PRIN1 = $-0.94 + 0.11$WE; SN N.S.; $R^2 = 0.30$; $p < 0.0001$

a. PRIN2 = 0.17Al -0.75Ca -0.07Mg $+ 0.43$K $+ 0.46$Na
b. PRIN2 = $0.39 -0.11$WE; SN N.S.; $R^2 = 0.08$; $p < 0.0005$

a. PRIN3 = 0.07Al $+ 0.19$Ca -0.17Mg -0.56K $+ 0.78$Na
b. PRIN3 N.S.

"C" horizon
a. PRIN1 = -0.58Al $+ 0.39$Ca $+ 0.45$Mg $+ 0.43$K $+ 0.35$Na
b. PRIN1 = $-1.06 + 0.03$SN $+ 0.11$WE; $R^2 = 0.35$; $p < 0.0001$

a. PRIN2 = 0.06Al -0.73Ca -0.10Mg $+ 0.55$K $+ 0.38$Na
b. PRIN2 = $0.45 -0.08$WE; SN N.S.; $R^2 = 0.08$; $p < 0.0008$

a. PRIN3 = 0.002Al $+ 0.21$Ca -0.46Mg -0.35K $+ 0.79$Na
b. PRIN3 = $-0.44 + 0.02$SN; WE N.S.; $R^2 = 0.08$; $p < 0.0007$

---

horizons. The corresponding figures for the O horizon were 40, 34 and 16%.

These three principal components were used as dependent variables in separate linear regressions with the south-north (SN) and west-east (WE) numbers of the topographical maps of Sweden (scale 1:100 000) as independent variables. The component values were averaged by horizon for all sites included on each topographical map before the regression calculations were made. The combined axis loadings (eigenvectors) of PRIN1, PRIN2 and PRIN3 were used in the interpretation of the linear regressions. Eigenvectors and linear SN/WE regressions are shown in Table 1. The linear regressions of the O and "C" horizons are also illustrated in Figs 1 and 2. The first principal component (PRIN1) described the regional base saturation pattern: the eigenvectors had negative Al^{3+} loadings and positive base cation loadings (except for Na^+ in the O horizon). Figures 3 and 4 illustrate the regional base saturation gradients in the O and "C" horizons. PRIN1 was positively correlated with the W–E and/or S–N direction in all horizons except for the E horizon, whether or not the Ca^{2+} values of the E horizon were corrected. R^2 was in the range 0.30–0.36 in the O, B

and "C" horizons. The eigenvector of the second principal component (PRIN2) had Ca^{2+} and Mg^{2+} loadings of opposite sign (− and +, respectively) in the O horizon, but of the same sign (−) in the E, B, and "C" horizons. The Mg^{2+} loading varied from a high positive value in the O horizon (+0.59) to slightly negative values in the mineral soil (−0.07– −0.10), while the Na^+ loading was positive in all horizons. PRIN2 was negatively correlated with the W–E direction in all horizons, but the coefficient of determination was much higher in the O horizon ($R^2 = 0.28$) than in the mineral soil horizons ($R^2 = 0.07$– 0.08). The high positive Mg^{2+} and Na^+ loadings in the O horizon suggested a soil chemical gradient influenced by a declining W–E gradient in dry and wet deposition of Mg^{2+} and Na^+ attributable to prevailing westerly winds from the Atlantic.

In the mineral soil horizons, the chemical pattern was likely to be influenced by the weathering of soil minerals. Mg^{2+}- and Ca^{2+}-rich minerals in the parent material increase towards the east and northeast (Melkerud et al. 1992). This pattern was partly reflected by the linear regressions. The regional distribution of K^+ and Na^+ in the parent material does not explain their loadings, nor does that for Al^{3+}. The loadings of these three exchangeable cations may be partly a consequence of competition with Ca^{2+} and Mg^{2+} for exchange sites. Furthermore, the Na^+ saturation in the mineral soil may to some extent be influenced by the atmospheric deposition of Na^+ salts. In the O horizon, PRIN2 was also positively correlated with the S–N direction. This was not simply a reflection of the S–N base saturation gradient, because that would have required a positive Ca^{2+} loading together with the positive loadings of Na^+, K^+ and Mg^{2+} (see Table 1). The third principal component (PRIN3) of the O horizon was negatively correlated with the S–N direction. In this case K^+ had a high negative loading, while the loadings of the other base cations were positive. As these significant S–N correlations with PRIN2 and PRIN3 were only found in the O horizon, where the density of fine roots is usually high (e.g. Majdi 1994), I suggest that they could be related to a difference in the K^+ requirement of the trees between the south and the north caused by a declining S–N gradient in annual tree growth (Nilsson 1990, Nilsson 1993). Among the base cations, K^+ is the element in most demand as a plant nutrient, although nitrogen is normally the main limiting nutrient (see Tamm 1991). Slower tree growth towards the north, caused by either nitrogen limitation or some other growth-limiting factor, implies a lower K^+ uptake by the roots. This should be possible to trace in the O horizon as an increasing S–N gradient in exchangeable K^+, as inferred by the PRIN2 regression and also particularly, the PRIN3 regression for this horizon.

In the mineral soil horizons, the Ca^{2+} and Na^+ loadings of PRIN3 were positive, while the Mg^{2+} and K^+ loadings were negative. In the E and B horizons, no geographical gradients could be shown, while in the "C" horizon PRIN3 was positively correlated with the S–N direction.

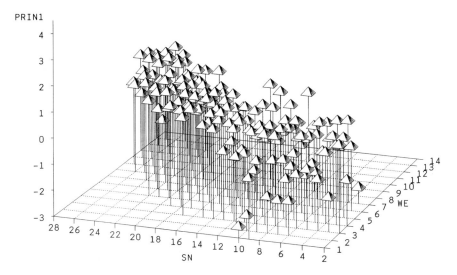

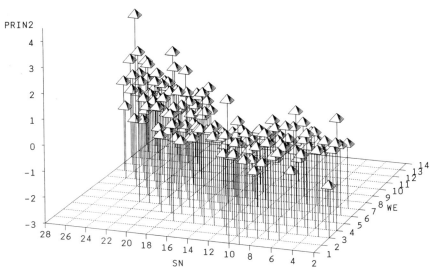

Fig. 1. The first three principal components PRIN1, PRIN2 and PRIN3 of the O horizon versus the map number in the south-north (SN) and west-east (WE) directions. Corresponding linear regressions are found in Table 1. Map numbers refer to the topographical map of Sweden (scale 1:100 000). The graphs only include those parts of Sweden where samples were collected. Map numbers 2 and 28 on the SN axis correspond approximately to 55°N and 67°N. Map numbers 1 and 14 on the WE axis correspond approximately to 13°E and 23°E.

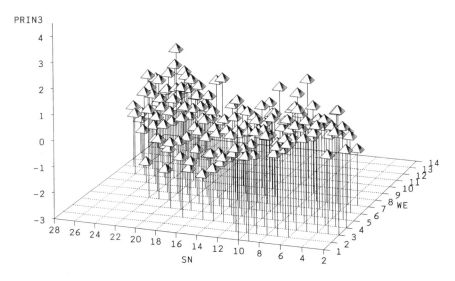

Fig. 2. The first three principal components PRIN1, PRIN2 and PRIN3 of the "C" horizon versus map number in the south-north (SN) and west-east (WE) directions. The corresponding linear regressions are found in Table 1. The "C" horizon corresponds to a defined soil depth, 50 cm. For further explanation, see Fig. 1.

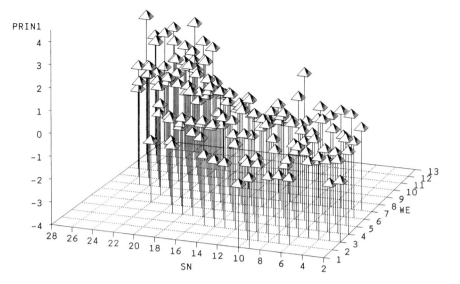

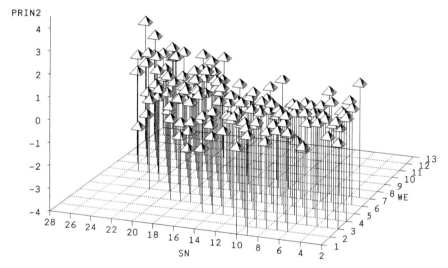

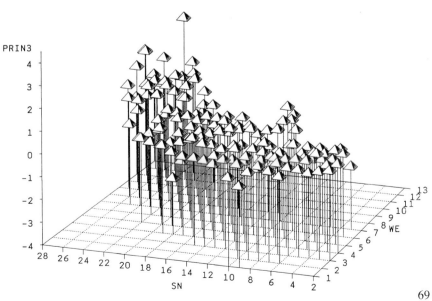

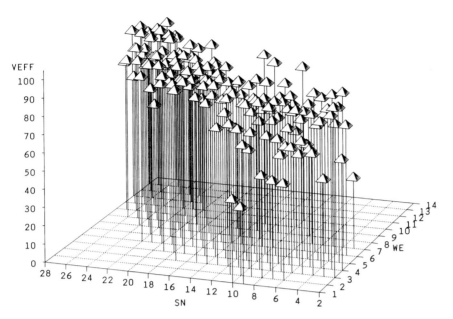

Fig. 3. The effective base saturation Veff in the O horizon versus map number in the south-north (SN) and west-east (WE) directions. Veff = 69.5 + 1.03 SN; WE N.S.; $R^2 = 0.32$; $p < 0.0001$. For further explanation, see the text and Fig. 1.

The coefficient of variation was quite low ($R^2 = 0.08$). However, a tentative explanation for the significant correlation could be that PRIN3 was positively related to the Ca^{2+} and Na^+ content in the parent material. Both cations have a high concentration in the northeastern part of the country, a pattern which coincides with the distribution of plagioclase and oligoclase (Melkerud et al. 1992). The negative loadings of Mg^{2+} and K^+ are more difficult to explain. The K^+ loading was in accordance with the distribution of mica and potassium feldspar in the sedimentary forest soils, but not with that of till soils (Melkerud et al. 1992).

Soil acidity relationships

Based on the sites where the $CuCl_2$-extractable aluminium was determined, the average Al_{KCl}/Al_{CuCl_2} molar ratios were 0.30, 0.58, 0.25 and 0.16 in the O, E, B and "C" horizons, respectively. The amount of adsorbed exchangeable plus complexed Al obtained in the $CuCl_2$ extracts was thus considerably higher than the KCl extractable (exchangeable) fraction. This is in accordance with findings for other soils, e.g. ultisols (Juo and Kamprath 1979, Hargrove and Thomas 1981). Al which is not exchangeable may consist of interlayer polymer

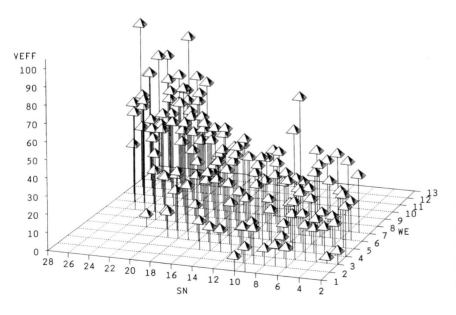

Fig. 4. The effective base saturation Veff in the "C" horizon versus map number in the south-north (SN) and west-east (WE) directions. Veff = 10.8 + 0.56 SN + 2.52 WE; $R^2 = 0.38$; $p < 0.0001$. For further explanation, see the text and Fig. 1.

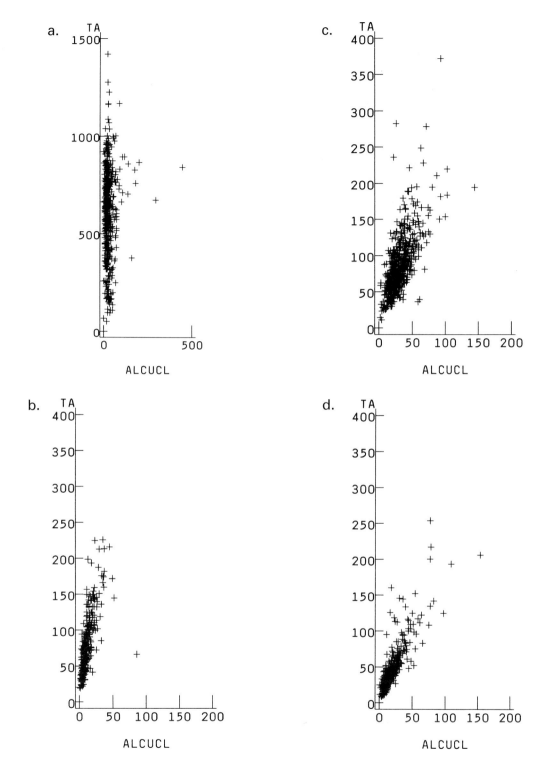

Fig. 5. Scatter plots of titratable acidity at pH 7 (TA) versus CuCl$_2$-extractable (exchangeable plus complexed) Al. The unit is mmol·kg^{-1} dry soil. Note the difference in scale between the O horizon plot and the three plots of the mineral soil.
a. O horizon. − N.S.
b. E horizon. TA = 3.11 Al + 42.3; R^2 = 0.48, p < 0.0001
c. B horizon. TA = 1.67 Al + 35.7; R^2 = 0.46, p < 0.0001
d. "C horizon". TA = 1.69 Al + 13.6; R^2 = 0.70, p < 0.0001

complexes in 2:1 clay minerals such as vermiculite, or of Al dissolved from amorphous oxides by the acid $CuCl_2$ solution (Juo and Kamprath 1979, Oates and Kamprath 1983). It may also consist of stable organic Al complexes (Bloom et al. 1979, Juo and Kamprath 1979, Hargrove and Thomas 1981, Oates and Kamprath 1983, Walker et al. 1990) and in sandy podzolic soils such as the Swedish forest soils referred to in this paper, this fraction would be expected to be the most important (Walker et al. 1990).

In order to estimate the contribution of $CuCl_2$-extractable Al to the titratable acidity linear regressions with titratable acidity (TA) at pH 7 (1 M NH_4OAc extracts) and Al in 0.5 M $CuCl_2$ extracts (both in $mmol \cdot kg^{-1}$ dry soil), as dependent and independent variable, respectively, were calculated for each soil horizon. Scatter plots are shown in Fig. 5a–d and the linear regressions are given in the figure legends. In the O horizon there was no significant relationship, probably because much of the acidity in this horizon could be accounted for by dissociation of organic acid groups rather than by Al hydrolysis. In the E horizon the regression had a slope of 3.1, indicating that the titratable acidity at pH 7 may be largely accounted for by exchangeable plus complexed Al with a charge of +3 according to the formula

$$Al^{3+} + 3OH^- <=> Al(OH)_3 \text{ (s)}$$

which ideally should give a regression slope of exactly 3.0. In the B and "C" horizons the regression slopes were 1.67 and 1.69, respectively. There may thus have been an extensive previous Al hydrolysis in these horizons, partly because of a higher pH than in the horizons above (Kissel et al. 1971, Bloom et al. 1977, Hargrove and Thomas 1982, Thomas and Hargrove 1984). If the hypothesis concerning Al hydrolysis is correct, there should be a mixture of exchangeable Al^{3+} and organically complexed Al^{3+} and $Al(OH)_n^{(3-n)+}$ surface complexes, where the surface complexes would dominate. According to the regression slopes, the average OH–Al stoichiometry would be $Al(OH)_{1.33}^{1.67+}$ and $Al(OH)_{1.31}^{1.69+}$ in the B and "C" horizons, respectively (Fig. 5c,d).

General discussion

PCA analysis

PCA analysis turned out to be an efficient tool for distinguishing between factors affecting the forest soil chemistry in different parts of Sweden. The first principal component (PRIN1), which accounted for 40–60% of the variability, was closely related to the S–N or W–E base saturation gradients in individual soil horizons, (Table 1, Figs 1a, 2a, 3 and 4). Because of the close connection between base saturation and PRIN1, the determining factors should have been the same, namely declining S–N and/or W–E gradients with respect to soil age, humidity, annual tree growth and atmospheric deposition of S and

N (Nilsson 1990, Bernes 1991, Eriksson et al. 1992, Nilsson 1993).

Recently, Karltun (1994) made a PCA analysis using data from entire soil profiles at 25 coniferous forest sites, distributed along three SW–NE transects, representing southern, central and northern Sweden. These are the same sites as those referred to by Eriksson et al. (1992) and Eriksson and Karltun (1995). The profiles extended to a maximum soil depth of 260 cm. Besides base cation saturation and Al saturation values, the PCA analysis included pH_{H_2O} and the concentration of H_2O-extractable SO_4^{2-}. Karltun made the interesting observation that the first principal component was significantly correlated with the local S deposition as well as with the base mineral index (negative and positive correlation, respectively). The base mineral index gives an indication of the amounts of easily weatherable soil minerals (Tamm 1934). A high base mineral index implies a high base saturation.

In my study the second principal component (PRIN2), accounting for 21–34% of the variability, was partly correlated with declining W–E gradients in the atmospheric deposition of Na^+ and Mg^{2+} salts of a marine origin, and partly with the regional distribution of Ca^{2+} and Mg^{2+} in the parent material. The marine influence was particularly evident in the O horizon, but the Na^+ loadings suggested that it may have extended to the mineral soil horizons as well.

The third principal component (PRIN3) accounted for 12–16% of the variability. In the "C" horizon PRIN3 seemed to be partly correlated with the contents of Ca^{2+} and Na^+ in the parent material (plagioclase or oligoclase) and also, although to a minor extent, with the contents of K^+ in mica and feldspars.

One of the most striking features revealed by the PCA analysis was the increasing S–N gradient with respect to exchangeable K^+ in the O horizon, which coincided with a declining S–N gradient in tree growth (Nilsson 1990). This gradient, and the W–E gradient attributed to the deposition of marine Na^+ and Mg^{2+}, emphasized that the chemical properties of the O horizon (forest floor) were to a great extent influenced by climatic and biotic factors, rather than by the properties of the parent material. The influence of climate and biota was less obvious in the mineral soil horizons.

The distinctive S–N and W–E increasing gradients in base saturation, seen in most soil layers, would have developed over time-periods of several thousand years (cf. Renberg et al. 1993). It is therefore rather remarkable that extensive temporal declines in exchangeable base cation storage have occurred in forest soils in southern Sweden over the relatively short time period of the last 30–60 yr (Bernén et al. 1987, Falkengren-Grerup et al. 1987, Falkengren-Grerup and Eriksson 1990, Johnson et al. 1991, Eriksson et al. 1992). Several factors have probably contributed to the decline of base cations. One of these may have been the change in landuse, particularly the increasing afforestation of formerly open land, which began by the end of the 19th century. The significance of the afforestation for the observed changes in soil chemistry is poorly known, however, as practically all published retrospective soil studies were of sites which were covered by forest during the whole time-period considered. However, soil acidification after a change of tree species, e.g. after planting Norway spruce on former beech forest soils, is well documented (Nihlgård 1971, Nilsson and Tyler 1995).

Both total tree production and annual growth rate have increased considerably during the 20th century, particularly in southern Sweden (Nilsson 1990). This implies an increased uptake of base cations by the trees, and a corresponding increase in the export of these base cations at the time of tree harvest. There has also been a considerable increase in the atmospheric deposition of acid S and N compounds (the S deposition is now decreasing again). A high soil leaching of relatively mobile SO_4^{2-}, and in some areas also of NO_3^- ions, has caused an equally high cation leaching i.e. soil acidification.

Interactions between deposition and tree growth have probably also been important. For instance, increasing N deposition at initially N deficient forest sites is one of the factors expected to contribute to an increase in tree growth, and thereby also to an increase in the base cation depletion of the soil (Tamm 1991). Eriksson and Johansson (1993) showed that the volumetric growth of Norway spruce was significantly greater during the second rotation than during the first, on some permanent plots in southwestern Sweden, studied from 1880 to 1989. The authors assumed that the increasing deposition of N was one of the main explanations for this increase, in growth.

Soil acidity relationships

The $CuCl_2$-extractable Al pool in sandy podzolic soils consists of exchangeable Al^{3+} plus organic Al complexes (Walker et al. 1990). Much of this Al is not immediately available for leaching or plant uptake, but contributes significantly to the titratable soil acidity in the mineral soil horizons. It has therefore been used for simple estimates of the lime requirement, based on soil acidity considerations (Oates and Kamprath 1983, Nilsson and Lundmark 1986). Recent research has shown that the Al activity in the soil solution may be determined by solid Al-organic matter complexes rather than by gibbsite or other Al minerals (Mulder et al. 1989, Walker et al. 1990, Berggren and Mulder 1995). $CuCl_2$-extractable aluminium could therefore be a most important pH and Al buffer. Mulder et al. (1989) estimated that the organically bound Al in sandy nutrient-poor Dutch forest soils may be totally depleted within a time-period of a few decades at the present rate of Al leaching. Such a depletion will cause a sudden drop in pH when the soil eventually changes from an "aluminium buffer stage" to an "iron buffer stage." A similar calculation based on my data is of interest. Let us assume that the average dry bulk

density of the O, E and B horizons was 0.1, 1.0 and 1.2 kg dm^{-3}, respectively, and that the corresponding horizon thicknesses were 5, 5 and 40 cm. The average amount of $CuCl_2$ extractable Al down to 50 cm would then be 162 $kmol \cdot ha^{-1}$. (The average Al_{CuCl_2} concentrations calculated for the whole data set were 28.6, 12.4 and 32.3 mmol· kg^{-1} dw in the O, E and B horizons, respectively). If we further assume an annual percolation of 400 mm H_2O, and that the Al concentration in the water leaving the B horizon varies between 1 and 5 mg Al dm^{-3} (0.037–0.18 mmol Al dm^{-3}), the Al pool would be theoretically depleted within the next 220–1090 years. Although this is likely to be an underestimate, as the Swedish forest soils are much richer in weatherable primary minerals than the sandy Dutch ones, one should observe that Al concentrations of the order of 5 mg dm^{-3} are not uncommon in the podsolized B horizons in southern Sweden (Anonymous 1991). The dynamics of the $CuCl_2$-extractable Al pool therefore definitely requires further investigation.

Acknowledgements – This work was financed by the Swedish Environmental Protection Agency and the Swedish University of Agricultural Sciences. The soil sampling was performed within the framework of the Swedish Forest Site Monitoring Programme. Soil samples and laboratory facilities were put at my disposal by J.-E. Lundmark and K. Rosén, former and present managers respectively, of the Forest Site Monitoring Programme. E. Carlsson helped me with the sample selection. My sincere thanks to B. Kindingstad and B. Kranz, who performed the lion's share of the chemical analyses. I also thank G. Tyler who encouraged me to write this paper, and the gals and pals on the "nightshift" at "Marklära" for their good company.

References

Allen, S.E. (ed.) 1989. Chemical analysis of ecological materials (2nd ed.). – Blackwell Scientific Publications, Oxford, England.

Anonymous 1991. Miljöatlas. Resultat från IVLs undersökningar i miljön. – Göteborg, in Swedish.

Berdén, M., Nilsson, S.I., Rosén, K. and Tyler, G. 1987. Soil acidification – Extent, causes and consequences. An evaluation of literature information and current research. – Swedish Environmental Protection Agency, Solna, Sweden, Report 3292.

Berggren, D. and Mulder, J. 1995. The role of organic matter in controlling aluminium solubility in acidic mineral soil horizons. – Geochimica et Cocmochimica Acta, (in press).

Bernes, C. 1991. Försurning och kalkning av svenska vatten. – Monitor 12. Swedish Environmental Protection Agency, Solna, Sweden, in Swedish.

Birkeland, P.W. 1984. Soils and Geomorphology. – Oxford Univ. Press, Oxford.

Bloom, P.R., McBride, M.B. and Chadbourne, B. 1977. Adsorption of aluminum by a smectite: I. Surface hydrolysis during Ca^{2+}–Al^{3+} exchange. – Soil Sci. Soc. Am. J. 41: 1068–1073.

–, McBride, M.B. and Weaver, R.M. 1979. Aluminum organic matter in acid soils: Buffering and solution aluminum activity. – Soil Sci. Soc. Am. J. 43: 488–493.

Derome, J., Kukkola, M. and Mälkönen, E. 1986. Forest liming on mineral soils. Results of Finnish experiments. – Swedish Environmental Protection Agency, Solna, Sweden, Report 3084.

Eriksson, E., Karltun, E. and Lundmark, J.-E. 1992. Acidification of forest soils in Sweden. – Ambio 21: 150–154.

Eriksson, H. and Johansson, U. 1993. Yields of Norway spruce (*Picea abies* [L.] Karst.) in two consecutive rotations in southwestern Sweden. – Plant and Soil 154: 239–247.

Falkengren-Grerup, U. and Eriksson, H. 1990. Changes in soil, vegetation and forest yield between 1947 and 1988 in beech and oak sites of southern Sweden. – For. Ecol. Manage. 38: 37–53.

– , Linnermark, N. and Tyler, G. 1987. Changes in acidity and cation pools of south Swedish soils between 1949 and 1985. – Chemosphere 16: 2239–2248.

Gauch Jr., H. 1982. Multivariate analysis in community ecology. – Cambridge Univ. Press, Cambridge, MA.

Hallbäcken, L. and Popovic, B. 1985. Effects of forest liming on soil chemistry. Investigation of Swedish liming experiments. – Swedish Environmental Protection Agency, Solna, Sweden, Report 1880, in Swedish with English summary.

Hargrove, W. L. and Thomas, G. W. 1981. Extraction of aluminum from aluminum-organic matter complexes. – Soil Sci. Soc. Am. J. 45: 151–153.

– and Thomas, G. W. 1982. Titration properties of Al-organic matter. – Soil Sci. 134: 216–225.

Johnson, D. W., Cresser, M. S., Nilsson, S. I., Turner, J., Ulrich, B., Binkley, D. and Cole, D. W. 1991. Soil changes in forest ecosystems: Evidence for and probable causes. – Proc. Royal Soc. Edinb. Sec. B. Biol. Sci. 97: 81–116.

Juo, A. S. R. and Kamprath, E. J. 1979. Copper chloride as an extractant for estimating the potentially reactive aluminum pool in acid soils. – Soil Sci. Soc. Am. J. 43: 35–38.

Karltun, E. 1994. Principal geographic variation in the acidification of Swedish forest soils. – Water Air Soil Poll. 76: 353–362.

Kissel, D. E., Gentzsch, E. P. and Thomas, G. W. 1971. Hydrolysis of nonexchangeable acidity in soils during salt extractions of exchangeable acidity. – Soil Sci. 111: 293–297.

Lundmark, J. E. and Odell, G. 1983. Fältinstruktion för ståndortskartering av permanenta provytor vid riksskogstaxeringen. – Report, Dept Forest Soils, Swedish Univ. of Agricultural Sci., Uppsala, in Swedish.

Majdi, H. 1994. Effects of nutrient applications on fine-root dynamics and root/rhizosphere chemistry in a Norway spruce stand. – Ph.D. thesis, Dept Ecology and Environmental Research, SUAS, Uppsala.

Melkerud, P.-A., Olsson, M. T. and Rosen, K. 1992. Geochemical atlas of Swedish forest soils. – Reports in Forest Ecology and Forest Soils 65. Dept Forest Soils, Swedish Univ. of Agricultural Sciences, Uppsala.

Mulder, J., van Breemen, N. and Eijck, H. C. 1989. Depletion of soil aluminium by acid deposition and implications for acid neutralization. – Nature 337: 347–349.

Nihlgård, B. 1971. Pedological influence of spruce planted on former beech forest soils in Scania, south Sweden. – Oikos 22: 302–314.

Nilsson, N. E. (ed.) 1990. Sveriges Nationalatlas. Skogen. – Bra Böcker, Höganäs, in Swedish.

Nilsson, S. I. 1988. Acidity properties in Swedish forest soils. – Regional patterns and implications for forest liming – Scand. J. For. Res. 3: 417–424.

– 1993. Acidification of oligotrophic forest lakes. Interactions between deposition, forest growth and effects on lake-water quality. – Ambio 22: 272–276.

– and Lundmark, J. E. 1986. Exchangeable aluminium as a basis for lime requirement determinations in acid forest soils. – Acta Agricult. Scand. 36: 173–185.

– and Tyler, G. 1995. Acidification induced chemical changes of forest soils during recent decades. A review. – Ecol. Bull. (Copenhagen) 44: 54–64.

Oates, K. M. and Kamprath, E. J. 1983. Soil acidity and liming: II. Evaluation of using aluminum extracted by various chloride salts for determining lime requirements. – Soil Sci. Soc. Am. J. 47: 690–692.

Ranneby, B., Cruse, T., Hägglund, B., Jonasson, H. and Swärd, J. 1987. Designing a new national forest survey for Sweden. – Stud. For. Suec. 177.

Renberg, I., Korsman, T. and Anderson, N. J. 1993. A temporal perspective of lake acidification in Sweden. – Ambio 22: 264–271.

SAS Institute Incorporated 1989. SAS/STAT User's Guide. Ver. 6, 4th ed. – Cary, N.C.

Tamm, C. O. 1991. Nitrogen in terrestrial ecosystems. Questions of productivity, vegetational changes and ecosystem stability. – Ecological Studies 81, Springer, New York, NY.

Tamm, O. 1934. Eine Schnellmethode für mineralogische Bodenuntersuchung – Svenska Skogsvårdsföreningens Tidskrift 1–2: 231–250, in Swedish with German summary.

Thomas, G. W. and Hargrove, W. L. 1984. The chemistry of soil acidity. – In: Soil Acidity and Liming, Agronomy Monograph no. 12, ASA-CSSA-SSSA Madison, WI, pp. 3–56.

Ulrich, B. 1966. Kationenaustausch-Gleichgewichte in Böden. – Z. Pflanzenern. Düng. Bodenk. 113: 141–159.

Walker, W. J., Cronan, C. S. and Bloom, P. R. 1990. Aluminum solubility in organic soil horizons from northern and southern forested watersheds. – Soil Sci. Soc. Am. J. 54: 369–374.

Ecological Bulletins 44: 75–89. Copenhagen 1995

Critical loads of acidity for Swedish forest ecosystems

Harald Sverdrup and Per Warfvinge

Sverdrup, H. and Warfvinge, P. 1995. Critical loads of acidity for Swedish forest ecosystems. – Ecol. Bull. (Copenhagen) 44: 75–89.

The critical loads of acid deposition have been mapped for Swedish forest soils using PROFILE at 1804 within the Swedish Forest Inventory. The results of the calculations show that Swedish forest soils are sensitive to acid deposition. A deposition reduction of 85% for sulphur deposition and 40% for nitrogen deposition in relation to the 1980 deposition level is required in order to protect 95% of the forest resource in Sweden. This corresponds to a target load of >6 kg S ha^{-1} yr^{-1} in southern Sweden, 4 kg S ha^{-1} yr^{-1} in center Sweden and 3 kg ha^{-1} yr^{-1} in northern Sweden. Under 1992 deposition level, still 75% of the forested area receive higher acid deposition than the critical load. The critical loads will be exceeded in 30% of the forest area under the deposition levels to be reached in 2010 according to current emission reduction plans. Forest soils in Sweden are at present still undergoing further acidification.

H. Sverdrup and P. Warfvinge, Dept of Chemical Engineering II, Inst of Technology, Box 124, Chemical Center of Lund Univ., S-221 00 Lund, Sweden.

The critical load concept

The critical load for sulphur and nitrogen was defined by the International workshop on critical loads held at Skokloster, Sweden, in 1988 as: The maximum amount of sulphur and nitrogen deposition that will not cause long-term damage to ecosystem structure and function.

During the course of 1989, all nations in Europe agreed to map critical loads in their countries as a basis for negotiations of sulphur and nitrogen emission reductions. The work is coordinated by a Task Force on Mapping (TFM) through a coordination centre located at the Federal Dutch Inst. for Health and Environment (RIVM) in Bilthoven, Netherland, and on behalf of the United Nations Economic Commission for Europe (UNECE). The critical load maps will be used for optimizing the abatement strategies for the whole of Europe (Hettelingh et al. 1991). A manual has been developed to provide guidelines for the mapping work (Sverdrup et al. 1990, Hettelingh et al. 1991). Critical loads are being determined for a number of different receptors, such as forest, grassland, lakes, streams or groundwater. The critical load is not calculated for a lake or forest as a physical entity as such, but for the lake or forest as an ecosystem.

The importance of weathering

The chemical state of a soil results from the interaction of many processes. The most important soil ecosystem processes for the control of base cation fluxes are chemical weathering, base cation deposition, vegetation base cation uptake, hydrological transport and ion exchange. The ion exchange can be ignored if the system is at steady state. In addition, acidic deposition, nitrogen uptake and nitrification influence other soil processes by modifying or partially controlling the chemical conditions in the soil. Temporal imbalances may make it necessary to consider biomass decomposition and vegetation recycling of nutrients. It is important to realise that it is not possible to assess the impact of any single soil process, without taking all other significant processes quantitatively into account.

In the calculation of critical loads, all sources of acidity in the system are balanced against all sources of alkalinity in the ecosystem. A steady state, when no acidification of the system is permitted, the exchange alkalinity term must be zero, and we can write the sources of acidity input to the ecosystem as balanced by the only long term net source of alkalinity, the weathering.

The rate of neutralization is dependent on the geochemical and physical properties of the soil, as well as the chemical conditions. Weathering plus base cation deposition provide the major supply of base cations to replace

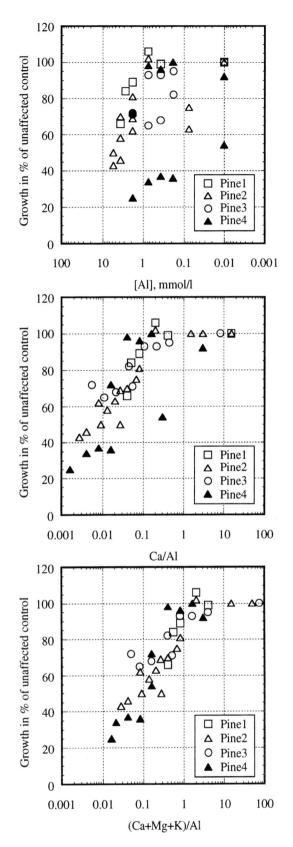

base cations removed by leaching and uptake and are important factors determining the chemical status of the soil. Whereas natural leaching processes take 15–30 000 yr to act, anthropogenic acidification has caused greater change in a mere 120 yr (Sverdrup et al. 1995).

Effects of acid deposition on trees

Experiments and field observations have shown that high soil solution Al concentration and low pH may cause adverse growth effects on trees (Ulrich 1983, 1985a,b, Schulze 1987, Moseholm et al. 1988, Bonneau 1991, Sverdrup and Warfvinge 1993b). Acid deposition above the critical load will deplete the base saturation of the soil and drive the soil solution towards a stready state where soil pH and Al concentrations are above limits at which adverse effects occur.

The relation between plant growth and the soil solution concentration of Al is best expressed through the BC/Al ratio (Sverdrup and Warfvinge 1993b). Results from an experiment with Scots pine by Arovaara and Ilvesniemi (1990) are shown in Fig. 1. Plant growth has been plotted against Al concentration, Ca/Al ratio and BC/Al ratio. It is evident that the BC/Al-ratio gives the best relationship. A similar pattern can be seen for a large number of trees and plants (Sverdrup and Warfvinge 1993b). The results for Norway spruce are expressed by the BC/Al ratio (Fig. 2). The latter data were extracted from a large number of laboratory bioassays (Schulze 1987, Sverdrup and Warfvinge 1993b) and a small number of field studies in Germany and France (Ulrich 1985a,b, Bonneau 1991).

Coniferous trees in Scandinavia are adapted to nitrogen deficiency and increased nitrogen deposition will greatly enhance the growth potential. During the last 150 yr, nitrogen deposition in southern Sweden has increased from 1–2 kg N ha^{-1} yr^{-1} to 15–30 kg ha^{-1} yr^{-1}. During the same period, forestry management methods and techniques have improved the utilization of available soil nutrients by trees significantly, increasing forest growth (Tamm 1991). Acid deposition causing base cation leaching from the ion exchange complex, has made extra base cations temporarily available in the soil solution for increased growth. However, as a result of acid deposition, the accelerated base cation leaching will rapidly deplete the forest reservoir of base cations at sites where the weathering rate is insufficient to resupply the lost amount.

Several criteria have been proposed as diagnostic effect parameters. For example: in the E-horizon ≥ pH 4.0 and in the B-horizon ≥ pH 4.4, total Al ≤ 4.0 mg l^{-1} or

Fig. 1. The three diagrams show growth response for seedlings Scots pine in comparison with the Al concentrations, Ca/Al ratio and the BC/Al ratio in laboratory experiments from Finland (data from Arovaara and Ilvesniemi 1990).

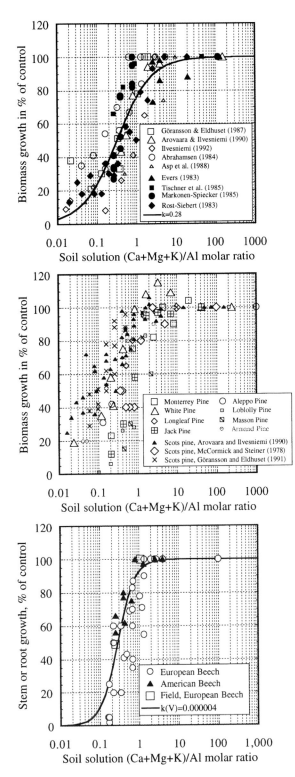

Fig. 2. Relationship between shoot and/or root growth and soil solution BC/Al ratio of various tree species in laboratory experiments. The sources of data can be found in the review by Sverdrup and Warfvinge (1993b). The upper diagram shows Norway spruce from Swedish and German experiments, the center diagram shows Scots pine from American, Finnish and Swedish experiments and the lower diagram shows data for beech.

inorganic labile Al ≤ 2.0 mg l⁻¹ (Ulrich 1983, Meiwes et al. 1986, Schulze 1987, Sverdrup et al. 1990). BC/Al = 1.0 was used as a limit in the critical load calculation, since this appears to be best supported by different types of experiments and observations (Sverdrup et al. 1990, Sverdrup and Warfvinge 1993b).

Methods

The critical loads have been calculated using two methods: 1) the Simple Mass Balance model (SMB) (Sverdrup and de Vries 1994) and 2) the PROFILE model (Warfvinge and Sverdrup 1992, Sverdrup et al. 1994). The SMB model has been employed by all European countries. PROFILE is a four-layer integrated soil chemistry model for stready state calculations. The PROFILE model is at present used for UN/ECE convention work by Sweden, Denmark, Norway, Switzerland, Great Britain, Germany, Russia and Poland. PROFILE is being tested and used for research purposes in universities and research institutes in more than 27 countries worldwide. The overlap in methodology allows for a test of consistency between the different methods.

Receptors mapped and limits used

The present forest critical loads map applies to Sweden's forest of 267 000 km². For the forest assessment each calculation point was area weighted in relation to the absolute local density of sites in relation to receptor density. The forest vegetation includes the southern Swedish deciduous forest, the coniferous forest and the northern birch forest. In the critical loads calculation, the linking of ecological response to deposition level is a central principle (Sverdrup et al. 1990, Hettelingh et al. 1991, Sverdrup et al. 1992). An indicator organism is used to assess the ecosystem response. A chemical limit is determined which marks the level below which there is no adverse response. The limit is entered into the chemical mass balance equation, which also includes acid deposition. Thus, the ecosystem becomes connected to deposition. BC/Al = 1.0 was used as a chemical limit in the critical load calculation.

The mass balance approach

The critical load of actual acidity is derived from a mass balance, in which all inputs of acidity to the system are to be balanced by all sources of acid neutralizing capacity; ANC, with no net depletion of the base saturation:

$$D + BC_U + BC_{IM} + ANC_L = N_U + N_{IM} + N_{DE} + W \quad (1)$$

where D is acidity deposition, BC_U and BC_{IM} are acidity from net base cation uptake and net immobilization. N_U is the corresponding ANC production from net N uptake and N_{IM} from net N immobilization in the soil, N_{DE} is the ANC produced by denitrification. Here immobilization is a permanent uptake in a solid phase in the ecosystem. W is weathering. It is important to realize that it is necessary to work with ANC, a conservative entity in ecosystems due to a number of biological responses and solution, buffer equilibria, proton budgets usually run into difficulties. The critical load is found by balancing deposition (D = CL) against the terms for net neutralization through weathering or by leaching of acidity transfer out of the system:

$$CL = W - ANC_L \qquad (2)$$

where ANC_L is the leached ANC from the system under critical load deposition. The ANC leaching is calculated by applying the chemical limit to the soil solution. This implies that the uptake and the immobilization terms have been considered as a part of the load into the system, and they are included in the calculation of exceedance. In the concept of sustainability, harvest is a burden on the capacity of the system.

Exceedance is derived by subtracting the critical load from today's total input of acidity:

$$EX = D + SA - CL =$$
$$D + (BC_U + BC_{IM} - N_U - N_{IM} - N_{DE}) - W - ANC_L. \quad (3)$$

We define the acidity input from internal soil processes as SA:

$$SA = (BC_U + BC_{IM} - N_U - N_{IM} - N_{DE}). \qquad (4)$$

The only assumption behind Eqs 3–4 is that there is no net leaching of ammonium from the soil. It is equivalent for the ANC balance whether ammonium is first nitrified and then taken up, or ammonium taken up and the residual nitrified. Thus, all N is counted as acidity and the amount of N taken up as alkalinity (Sverdrup and de Vries 1994). The Simple Mass Balance model (SMB) is the standard European model for calculating critical loads (Sverdrup et al. 1990, Sverdrup and de Vries 1994). We can derive the formula for calculation, by starting with the definition in Eq. 2:

$$CL = W - ANC_L. \qquad (2)$$

The expression for exceedance becomes dependent on all inputs of acidity, both from atmospheric deposition (D) and from soil processes.

In forest ecosystems, applying a BC/Al ratio of 1, ANC of the leachate from the root zone will be negative, and the carbonate species in the ANC expression will be very small. We chan thus write:

$$ANC_L = -H_L^+ - Al_L^{3+} \qquad (5)$$

where Al_L^+ is Al leaching and H_L^+ is H^+ leaching. The limiting ANC leaching is determined by the maximum permitted leaching of H and Al, by applying a simplified expression of ANC. The limiting Al-flux is determined by the molar BC/Al-ratio used as critical chemical limit. This leads to:

$$Al_L^{3+} = \frac{BC_L}{(BC/Al)_{crit}} \qquad (6)$$

where is BC_L base cation amount available for leaching, calculated from a mass balance:

$$BC_L = W(CaMgK) + BC_D - BC_U \qquad (7)$$

where BC_D is the base cation deposition and W(CaMgK) the weathering rate of Ca+Mg+K.

In the mass balance equation for base cations, c. 30% of released base cations from weathering are Na, providing no protection against Al and H^+ for plants. The production of Ca, Mg and K from weathering is:

$$W(CaMgK) = x_{CaMgK} \cdot W \qquad (8)$$

where $x_{Ca+Mg+K}$ is the fraction of weathering as Ca+Mg+K (0.7).

Some base cations will escape uptake because plants have difficulty in taking up base cations from a solution with concentration <2 meq m^{-3} (Sverdrup and Warfvinge 1993b). Additional amounts may escape due to the inability of plants to take up nutrients during the winter. This minimum leaching cannot however, be larger than that available from weathering and deposition:

$$BC_{min} = Q \cdot [BC]_{min}. \qquad (9)$$

Q is percolation, $[BC]_{min}$ is the limiting concentration for uptake, provided there is enough, if:

$$BC_{min} > x_{CaMgK} \cdot W + BC_D \qquad (10)$$

then:

$$BC_{min} = x_{CaMgK} \cdot W + BC_D. \qquad (11)$$

Uptake cannot be larger than the amount available for uptake, if:

$$BC_U > x_{CaMgK} \cdot W + BC_D - BC_{min} \qquad (12)$$

then:

$$BC_U = x_{CaMgK} \cdot W + BC_D - BC_{min}. \qquad (13)$$

This Al leaching can be written as:

$$Al_L^{3+} = 1.5 \cdot \frac{x_{CaMgK} \cdot W + BC_D - BC_U}{(BC/Al)_{crit}}. \qquad (14)$$

78

Operationally, the H⁺ concentration can be calculated using the gibbsite equation:

$$[H^+] = \left(\frac{[Al]^{3+}}{K_{gibb}}\right)^{1/3} \qquad (15)$$

where K_{gibb} is the gibbsite coefficient (A value of 300 m⁶ eq⁻² or $-pK_{gibb} = 8.5$ was used).

Accordingly, the limiting H⁺-concentration corresponding to a certain Al concentration in the soil is calculated from the Al³⁺-flux calculated above, dividing by the flow and the gibbsite coefficient:

$$[H^+]_{limit} = \left(\frac{Al_L^{3+}}{Q \cdot K_{gibb}}\right)^{1/3}. \qquad (16)$$

By inserting the expression for the Al-limiting flux in the expression and multiplying by flow Q to get from H⁺-concentration to flow, we get:

$$H_L^+ = \left(1.5 \cdot \frac{x_{CaMgK} \cdot W + BC_D - BC_U}{(BC/Al)_{crit} \cdot Q \cdot K_{gibb}}\right)^{1/3}. \qquad (17)$$

The SMB equation for critical load of acidity in keq ha⁻¹ yr⁻¹ thus becomes:

$$CL = ANC_W + \left(1.5 \cdot \frac{x_{CaMgK} \cdot W + BC_D - BC_U}{(BC/Al)_{crit} \cdot K_{gibb}}\right)^{1/3}$$
$$\cdot Q^{2/3} + 1.5 \cdot \frac{x_{CaMgK} \cdot W + BC_D - BC_U}{(BC/Al)_{crit}}. \qquad (18)$$

The factor 1.5 derives from the conversion of critical loads and base cation concentrations in equivalents to molar ratio.

The equation is based on the following assumptions: 1) The soil profile is assumed to be one stirred tank. a) The same gibbsite coefficient is assumed to apply through the soil profile. b) The weathering rate is evenly distributed over the soil profile. c) Uptake is evenly distributed over the soil profile. 2) The weathering rate is independent of chemical conditions. 3) The BC/Al ratio is assumed to have a value such that the value of ANC_L always is negative. 4) Complexing of Al by organic acids is ignored in the soil below the organic horizon.

Strictly, none of these assumptions are valid. In order to work with a more realistic representation of the forest soil and the chemical reactions in it, as well as to address the uncertainty introduced by the above-mentioned assumptions, the PROFILE model was constructed for calculation of critical loads for Sweden.

Soil stability

Net soil aluminium depletion may cause structural changes in soils. For many soils, secondary aluminium phases and complexes are important structural components in the soil. Stability of these soils depend on the stability of the reservoir of these substances. In high precipitation areas, acid deposition may potentially lead to aluminium leaching in excess of aluminium produced by weathering:

$$Al_L > W(Al) \qquad (19)$$

where Al_W is the production of Al from weathering of minerals. The production of Al from weathering of minerals is related to the production of base cations from weathering through the stoichiometry of the minerals. An approximation, using typical mineralogy of north European soils, would imply:

$$W(Al) = 2 \cdot W. \qquad (20)$$

The Al criterion (leaching not greater than weathering of aluminium) leads to the equation for critical load of acidity.

$$CL(Al) = 3 \cdot W + \left(\frac{2 \cdot W}{K_{Gibb}}\right)^{1/3} \cdot Q^{2/3} \qquad (21)$$

where Cl(Al) is the critical load of acidity based on soil stability.

The effective critical load will be the minimum of the ecosystem critical load of acidity calculated from plant tolerance of aluminium (BC/Al-ratio) and the ecosystem critical load of acidity calculated from soil stability criteria.

The PROFILE model

Critical loads for forest soils in Sweden were calculated using the steady state mass balance model, implemented as the MacIntosh version of the PROFILE model. Units used are keq ha⁻¹ yr⁻¹ unless otherwise specified. In PROFILE, the soil profile is divided into four layers, using input data for the thickness of each soil layer (O, A/E, B, C). The chemical limits were applies to the soil layers contained within upper 0.5 m of the soil for coniferous trees, on the assumption that this is the typical rooting depth for this type of forest in Sweden.

Processes included

PROFILE is based on a conceptual model of a forest soil. It contains the following chemical subsystems: 1) Deposition, leaching and accumulation of dissolved chemical components. 2) Chemical weathering reactions of soil minerals with the soil solution. 3) The net result of soil reactions with N-compounds; nitrification, denitrification. 4) Internal cycling of N and base cations in the canopy, such as uptake, canopy exchange, litterfall and mineralization. 5) Net uptake of base cations and nitro-

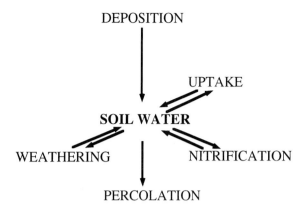

DEPOSITION

UPTAKE

SOIL WATER

WEATHERING   NITRIFICATION

PERCOLATION

Fig. 3. The processes in PROFILE are linked through the soil solution. Uptake and immobilization are primarily derived from input data, but modified depending on soil chemistry conditions.

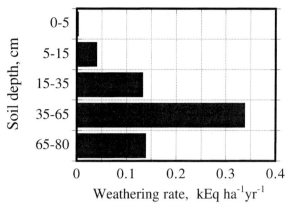

Fig. 4. The weathering rate at different soil depths in the soil profile at Lake Gårdsjön. The difference in rate between layers makes it important to consider each soil horizon separately in the calculation of critical loads.

gen removed by tree harvest. 6) Solution equilibrium reactions involving the carbonate system, CO_2, speciation and complexing involving Al and organic acids.

Naturally, these processes only represent a selection of the chemical reactions in the soil. Among processes that have not been included are sulphate adsorption, a series of reactions that may change the CEC of the soil matrix, store sulphur irreversibly or affect the ANC balance in certain soils. All processes included in the model have been subject to some degree of simplification. In the PROFILE model, all soil processes are linked via the soil solution (Fig. 3). Temperature dependence is considered for nitrification, denitrification, immobilization, weathering and solution equilibria. The PROFILE model has been set in an automated framework reading input data from a file containing the regional database.

The weathering rate submodel

The outstanding difference between PROFILE and other comparable soil chemistry models is that the weathering rate is calculated from independent geophysical properties of the soil system. This reduces the degrees of freedom in the model and reduces the need for information that makes the model mathematically constrained.

Several chemical reactions between soil minerals and constituents in the soil solution contribute to the base cation release rate, and the total chemical weathering rate will be the sum of weathering rates of all individual elementary chemical reactions with each mineral. For most minerals, several reactions have been identified experimentally, the present model formulations includes: 1) The reaction with H_2O, Al and cations of the parent mineral. 2) The reaction with H^+, Al and cations of the parent mineral. 3) Reactions with dissociated organic acid ligands. 4) The reaction with dissolved CO_2.

The reaction orders and the rection rate coefficients were determined experimentally for most common pri-

mary and secondary minerals (Sverdrup 1990, Sverdrup and Warfvinge 1993a). The rate coefficients themselves are functions of the base cation and Al concentrations, as discussed in Sverdrup (1990).

The rate of dissolution is inhibited by the reaction products due to complexing on active dissolution sites. Thus Al, and the base cations Ca, Mg, Na and K are important weathering rate inhibitors. The presence of organic acids affects weathering in two ways:

1) The reaction between the mineral and organic acids is of little importance for most soil minerals, especially

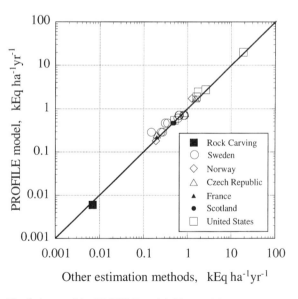

Fig. 5. A test of the PROFILE model. The model was sent out to researchers in Europe and America with a request to test the model on field measurements. The returned calculated values and corresponding independent field measurements were used to construct the diagram. The correlation coefficient is r = 0.99 (Sverdrup and Warfvinge 1993a).

those with a high Madelung site energy for the base cation in T4 and S4 sites, and important for low energy minerals such as nesosilicates, pyroxenes and amphiboles (Sverdrup 1990).

2) Organic acids may enhance weathering of minerals with a strong Al product inhibition, by complexing Al and base cations. This effect is sometimes cancelled by the fact that organic acids occur in the top layers of the soil, where the mineral content is low (Sverdrup 1990).

The activity of exposed mineral surfaces is less than unity in unsaturated soils where the availability of moisture is limited. All surfaces that participate in chemical reactions must be wetted, but wetness alone is insufficient; there must also be sufficient soil solution present for the weathering process to interact with other soil processes. Thus mineral surface reactivity is assumed to be proportional to soil moisture saturation.

It is important to test model performance in calculating the weathering rate from geochemical soil properties, since the critical load is so closely connected with the weathering rate. The PROFILE model was sent out to researchers in Europe and America with a request to test the model on field measurements. The returned calculations and independent measurements were used to construct Fig. 5 using determinations at 29 sites. The test shows that the weathering rate can be estimated with an accuracy better than ±8% (Sverdrup and Warfvinge 1988, 1993a, Jönsson et al. 1995). Further comparisons can be found in Lång (1995), and similar studies for Norway and Scotland are in progress. The calculated weathering rate values include the specific reaction coefficients for the reaction with organic acids as determined in laboratory experiments.

The weathering rate increases down the soil profile (Fig. 4), a fact that also affects the critical load. The weathering rate is low in the upper soil layers, since the content of weatherable minerals is low there.

The Al growth response submodel

The PROFILE model may also be used to calculate the steady state chemistry in response to a certain deposition input. To translate the steady state soil chemistry to ecological effects, a response expression fitted to data for spruce, as shown in Fig. 2, was used (Sverdrup and Warfvinge 1993b):

$$f = \frac{[BC]}{[BC] + K \cdot [Al]} \qquad (22)$$

where $k = 0.35$, the value is derived from laboratory experiments (Sverdrup and Warfwinge 1993b). [Al] was defined as:

$$[Al] = (3 \cdot [Al^{3+}] + 2 \cdot [AlOH^{2+}] + [Al(OH)_2^+])/3 \qquad (23)$$

and "BC" is defined as:

$$[BC] = [Ca^{2+}] + [Mg^{2+}] + [K^+]. \qquad (24)$$

Al and BC complexed with organic ligands are excluded from the response expression. In the PROFILE model, complexation is calculated using a simplified version of the Tipping model, exploiting the Oliver equation (Warfvinge and Sverdrup 1992). Thus, high DOC concentrations in the upper soil layers reduce the response to a high Al concentration, since a large part of the available Al is complexed. This is considered in the PROFILE model. The field data used in Fig. 2 are based on growth versus the BC/Al ratio in the B-layer. In PROFILE, growth reduction is calculated for each soil horizon, and the plant is allowed to reallocate uptake from one layer to another, if supply is insufficient in one particular horizon. In this way, the plant is allowed to optimize uptake in the PROFILE model.

The BC/Al-response can be derived as a purely empirical expression from laboratory data. But it may equally well be interpreted as a result of competition between base cations and Al at root uptake receptor sites (Ericsson et al. 1995), and the equation given above can be theoretically derived from such a principle (Sverdrup and Warfvinge 1993b).

Input data

All data in this study are based on the framework of the Swedish National Forest Survey (Riksskogstaxeringen/ ståndortskarteringen, Kempe et al. 1992). It consists of a network of 27 000 stations evenly spread over the complete forest area of Sweden (267 000 km^2, SOU 1992:76). Soil samples down to the C-layer at c. 60 cm soil depth were collected from 1804 sites. The basic soil parameters for this study were analysed on these samples. The list below represents the input data required for the PROFILE model for each site in the calculations:

Site identity of the Swedish Forest Inventory, precipitation in mm, runoff in mm, total values for deposition of S, NO$_3$, Cl, NH$_4$, Na, Ca, K, Mg, net uptake of base cations and N, all in keq ha^{-1} yr^{-1}, temperature, soil texture in m^2 m^{-3}, weight fraction in the soil of the following minerals; K-feldspar, plagioclase, pure albite, hornblende, pyroxene, epidote, calcite, biotite, muscovite, chlorite, vermiculite and apatite, soil type, moisture class (1–5), thickness of the organic horizon and the A/E-layer, a soil field texture class in case the soil texture is not available, longitude, latitude, soil rooting depth, area weight of each site, and nutrient limitation ratios for Mg/N, K/N, Ca/N, and a code for the Swedish province.

Field estimates of growth in the Forest Inventory were used as input to the Swedish forest yield model HUGIN to estimate the average yield over one rotation period (Kempe et al. 1992). This was combined with stem, branch and needle analysis data from the Swedish Forest Inventory to yield uptake of N and base cations. The inventory data were collected during 1983–1985.

Table 1. The mineralogy analysis scheme required each soil sample to be divided into 4 parts, based on density and particle size.

	Light fraction ≤ 2700 kg m^{-3}	Heavy fraction ≥ 2700 kg m^{-3}
Sand, silt	I	II
≥ 2 μm	Feldspars Quartz Muscovite Biotite	Pyroxene Amphiboles Epidotes Biotite Nesosilicates
Clay	III	IV
≤ 2 μm	Vermiculite Montmorillonite Quartz	Chlorite

Several other parameters, such as CO_2 pressure, soil solution dissolved organic carbon, distribution of uptake and evapotranspiration and gibbsite coefficients, have less influence and are entered as standard values taken from the literature and generally remain constant between runs and sites. Annual average air temperature, precipitation and runoff were taken from the official statistics of the Swedish Meteorological and Hydrological Inst. for each NILU grid (50 × 50 km). Other input data were derived strictly in accordance with the "Mapping critical loads" by Sverdrup et al. (1990).

Uptake

Forest growth, soil type and moisture class were measured at 27 000 stations, soil properties required by the model and certain chemical parameters at the 1804 soil sampling sites. For each site the net long term uptake is specified, based on measurements in the survey of base cation and nitrogen contents of stem and branch, combined with site specific estimates of net long-term forest growth. Each tree species is considered separately, and the total uptake is weighted together for each calculation point. In the model, base cation uptake will occur from each layer, but uptake is stopped if the base cation soil solution concentration falls below 15 μeq l^{-1}. Residual uptake is then moved to the next layer.

Mineralogy

Field data

Soil mineralogy for 124 sites was measured by the Swedish Geological Survey at Uppsala. Fifteen samples were also analysed at the Czech Geological Survey, Praha. The subset of samples for mineralogy analyses, was derived by placing 5 samples within a series of geological provinces, subjectively chosen using the bedrock geology map and a map of till deposits in Sweden. Total elemental

content of the soil after complete dissolution was determined for all 1804 soil sampling sites. Total analysis implies that the minerals in a soil sample are totally dissolved in hot hydrofluoric acid or a lithium borate-alkali melt, diluted in water and analysed using standard methods. They were analysed for Ca, Mg, Na, K, Al, Si, Fe, Ti and trace metals.

The mineralogy was determined by separation of the sample into two density classes and two particle size classes (Table 1). Each subsample is analysed for total elemental composition using wet chemistry methods, and subsequently analysed for elements and minerals using XRF and XRD, respectively, calibrated on standards of mixtures of pure mineral phases. The fast weathering minerals will tend to concentrate in the heavier fraction; the analyses can then be carried out on this enriched fraction. Separation of the mineral sample in a liquid with density in the range 2.680 g cm^{-3}, will generally concentrate easily weathered minerals 30–70 times. The mineralogical determination is generally carried out on the silt and clay fraction of the soil, since all particles > 250 μm dissolve too slowly to be significant for the weathering rate. These data are used to estimate the mineralogy of the whole soil sample.

Normative mineralogy

The model UPPSALA is a normative back-calculation model for reconstructing the mineralogy from the total analysis in order to provide input to models like PROFILE from simple survey data.

For each of the geological province the observed mineralogy was correlated to total analysis. In retrospect it appeared that the optimal strategy should have been first to run the total analysis, before selecting the sampling sites for mineralogy. Originally, it was believed that the application of seven geological provinces, would correspond to seven total analysis-mineralogy correlations. In this first phase, one normative correlation was used, as the seven a priori determined provinces were not optimal for the purpose. For the next phase, the procedure has been revised, by defining the geological provinces based on total analysis results.

The mineralogy model is based on a priori knowledge of the stoichiometric composition of the minerals of the particular soil. The minerals have been grouped into assemblies of minerals with similar composition and dissolution rate. Accordingly, for soils of granitic origin or soils derived from rock of secondary origin (sandstones), as well as originating from different types of schists and shales, it is assumed that the following mineralogical groupings are valid. Muscovite is assumed to comprise muscovite, secondary di-octahedral illite, di-octahedral chlorite and vermiculite of secondary weathered type. Chlorite comprises tri-octahedral chlorite, primary illite and tri-octahedral vermiculite of primary type as well as biotite and phogopite and glauconite. Hornblende implies all amphiboles, such as hornblende, riebeckite, arfvedsonite, glauconite and tremolite. Epidote comprises all epidotes, zoisites and pyroxenes. Plagioclase has been assumed to be oligoclase with 80% albite feldspar component, for Sweden. K-feldspar is assumed to be 10% albite feldspar component. All phosphorus has been attributed to apatite. The equations used in the UPPSALA model for calculation of the % weight content of the individual minerals are: 1) K-Feldspar = 5.88 K_2O – 0.588 Na_2O. 2) Plagioclase = 11.1 Na_2O – 0.22 K-Feldspar. 3) Apatite = 2.24 P_2O_5. 4) Hornblende = 6.67 CaO – 3.67 Apatite – 0.2 Plagioclase. 5) Muscovite = 2.08 K_2O – 0.208 Na_2O. 6) Chlorite = 3.85 MgO – 0.39 Hornblende – 0.39 Muscovite. 7) Epidote = 0.1 Hornblende + 0.03 Plagioclase – 0.3. 8) Quartz = SiO_2 – 0.63 Plagioclase – 0.68 K-Feldspar – 0.38 Muscovite – 0.33 Chlorite – 0.45 Hornblende – 0.42 Epidote. 9) Al-residual = Al_2O_3 – 0.1 Plagioclase – 0.1 K-Feldspar – 0.26 Muscovite – 0.09 Chlorite – 0.01 Hornblende – 0.025 Epidote.

Residual Al is used to form clay minerals in a group called vermiculite if they can be matched by a small amount of residual base cations, often K, whereas other residual Al is considered inert and added to quartz. The calculation is checked by calculating the amount of quartz, and only such sites which lie within the range 95–105% are accepted. When no more adequate ions are available for the formation of a mineral, then the content of that mineral is set to zero. For 1804 out of 1913 sites calculated, the mineralogy was accepted, and of the rejected sites c. 30 were estimated to be calcareous soils. These calcareous sites were separated out, and a modified UPPSALA model was applied to them. All Ca not occupied by feldspar and apatite is allocated to calcite. Mafic minerals are omitted, since weathering of calcite will dominate: 1) K-Feldspar = 5.88 K_2O – 0.588 Na_2O. 2) Plagioclase = 11.1 Na_2O – 0.22 K-Feldspar. 3) Apatite = 2.24 P_2O_5. 4) Calcite = 1.79 CaO – 3.67 Apatite – 0.2 Plagioclase. 5) Quartz = SiO_2 – 0.63 Plagioclase – 0.68 K-Feldspar. 6) Al-residual = Al_2O_3 – 0.1 Plagioclase – 0.1 K-Feldspar – 0.26 Muscovite.

Quartz is calculated to check whether the mineralogy obtained adds up to a number close to 100%. The weak point in the UPPSALA model is the expression for epidote. Epidote occurs in small amounts only, in most soils of granitic origin. For Sweden the expression for hornblende is better since it is omnipresent in Scandinavian soils. The expressions do not correspond to normative calculations based on absolute stoichiometries for pure minerals, but rather represent a set of empirical field stoichiometries.

Texture

Texture was measured by granulometry and BET-adsorption analysis on the 124 mineralogical analysis samples, and correlated against field texture classification. The textures for all 1804 sites were read from the correla-

tion using the field classification. The simplest way to calculate the soil texture is from a particle size distribution. The conversion from a standard sieve curve to surface area was carried out using:

$$A_{tot} = (x_{clay} \cdot 8.0 + x_{silt} \cdot 2.2 + x_{sand} \cdot 0.3) \cdot \frac{\varrho_{oil}}{1000} \qquad (25)$$

where A has the dimension 10^6 m^3 m^{-3} soil, and where we have the condition:

$$x_{clay} + x_{silt} + x_{sand} + x_{coarse} = 1 \qquad (26)$$

and x_{clay} is the fraction of soil particles <2 μm, x_{silt} the fraction of 2–60 μm, x_{sand} the fraction of 60–200 μm, and x_{coarse} the fraction of soil matrix coarser than 200 μm. ϱ_{soil} is the bulk density of the soil in kg m^{-2}. The granulometric approach is used, due to difficulties of interference from secondary precipitates and organic matter with the BET measurements on whole soil samples. BET measurement is good for soil samples not affected by these factors. The granulometric relation was determined by measuring BET surface on a series of soil samples divided up into sand, silt and clay fractions. Untreated fractions and such where the precipitates and organic matter had been removed were used. A relation between the field classification of soil texture and laboratory measurements of exposed surface was found:

$$A_{tot} = 0.093 \cdot e^{0.51 \cdot TX} \qquad (27)$$

where TX is the texture class, A_{tot} is the total surface in 10^6 m^2 m^{-3} soil. The relation is based on determinations of surface area and texture class on 100 soil samples. The texture class is determined in the field by placing a small soil sample in the palm of the hand, spitting on it and rolling a small as possible string. The texture class is judged from the thickness of the string.

Deposition

Deposition data were prepared by the Swedish Environmental Res. Inst. (IVL), from the Swedish deposition monitoring network and EMEP data (Lövblad et al. 1992, Mylona 1993), modified with filtering factors for different vegetation types. The total deposition is calculated for each calculation point using filtering factors and the vegetation mixture for that point.

Results

The input data were kept separate for the 1804 individual points in the calculation with PROFILE. The results for 1804 points were distributed to the grid cells, by means of the longitude-latitude coordinates for each point. The

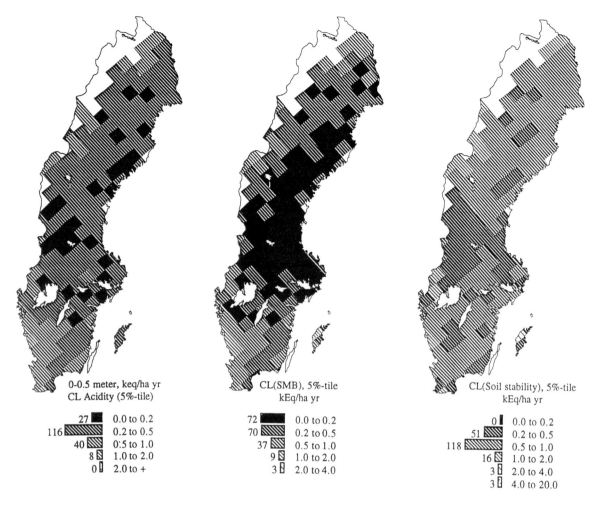

Fig. 6. The critical loads for Swedish soils if the objective is to protect 95% of the area. Calculated using 1) the PROFILE model, 2) the SMB model, 3) the soil stability criterium.

value for each grid cell is calculated from the results of 3–20 sites. All together there are 192 NILU grid cells, with an average of 9 points in each grid.

Critical loads of acidity

Critical loads for forest ecosystems are illustrated in Fig. 6. For Sweden, both the SMB and the PROFILE model were employed, the two models show good consistency. The weathering rates in both calculations were obtained from PROFILE calculations. Calculations of critical loads were made using both the soil stability (SMB) and the BC/Al criteria (SMB, PROFILE). The 5 percentile of the minimum critical load from all three approaches are shown in Fig. 8.

Areas where critical loads have been exceeded are illustrated according to the PROFILE model, SMB and the soil stability criterion (Fig. 7). Even the soil stability criterion is violated in Sweden under the present load; the

consequence of a long-term violation of the soil stability criterion is unknown. The critical loads in Sweden are low, and hence the area with exceedance is large. For a target of protecing 95% of conifer ecosystems in a grid cell, nearly 85% of the country has exceedance with respect to 1988 deposition level. Today, deposition levels are 30% lower for S but unchanged for N, as compared with 1988, and the area exceeded is somewhat smaller. The maps express the critical load in terms of acidity.

The difference between the PROFILE map and the SMB map may arise mainly from two factors. First, in the SMB model complexing of Al with organic acids is ignored, whereas in PROFILE both speciation of Al and complexing with organic acids have been included. Secondly, in PROFILE, the soil is stratified into several layers, each considering the appropriate processes and properties of each horizon. Thus in PROFILE weathering will vary between layers, as will the presence of organic acids.

Calculation of the 95 percentile of the maximum crit-

84

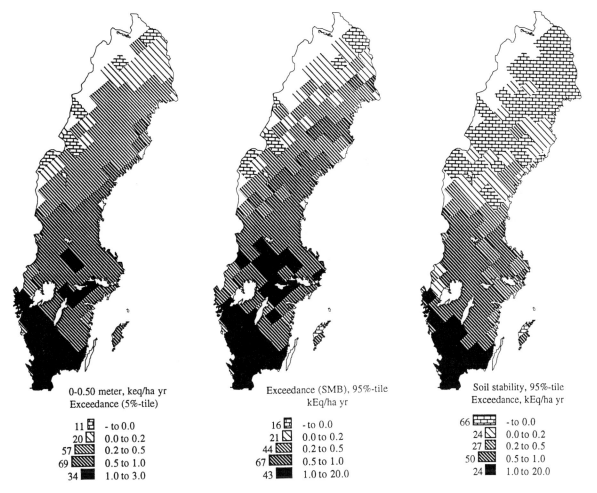

Fig. 7. Areas where the critical loads have been exceeded for Swedish soils if the objective is to protect 95% of the area. 82% of Sweden's forested area receives more acidic deposition than the critical load. Calculating using 1) the PROFILE model, 2) the SMB model, 3) the soil stability criterium.

ical loads from all three approaches shows that 88% of Sweden's forest area has deposition in excess of the critical loads (Fig. 8).

The required reduction in S deposition (Fig. 9) shows that both S and N deposition must be reduced to have no exceedance in 95% of the area.

The role of forestry and weathering

The major effect of forestry on exceedance is not through a large acidity input, but rather by 1) lowering of BC in the BC/Al response parameter, 2) by increasing deposition through scavenging. This makes the impact of forestry on soil acidification difficult to quantify in a unique way. The effect can be seen by comparing Fig. 8 with Fig. 7, where net uptake in Fig. 8 is set to zero, assuming virgin forest everywhere (no harvesting at all).

At present, very approximately, the portion of excee-

dance caused by forestry and landuse is 10–20%, that by N deposition is 30–40%, and the portion caused by S deposition is 50–60%. The proportions vary within Sweden. Approximately 60% of the N deposition comes from agricultural activities and 40% from emissions of oxidized N, mainly from transport.

Weathering in Swedish forest soils can support forest harvests up to 80–85 million m^3 yr^{-1} without any adverse effects, provided there is no acid deposition. But there is not enough weathering to support both the amount of wood harvested by present forestry and the neutralization of a total acidity deposition equivalent to 400 000 tons of sulphuric acid annually.

In political terms, the issue of transboundary acid deposition, concerns who has the right of ownership to the weathering rate of a country. Since the weathering rate is an expression of the mineralogy, it could be considered under the same legal issues as rights to national mineral deposits. This might be a way of establishing a legally

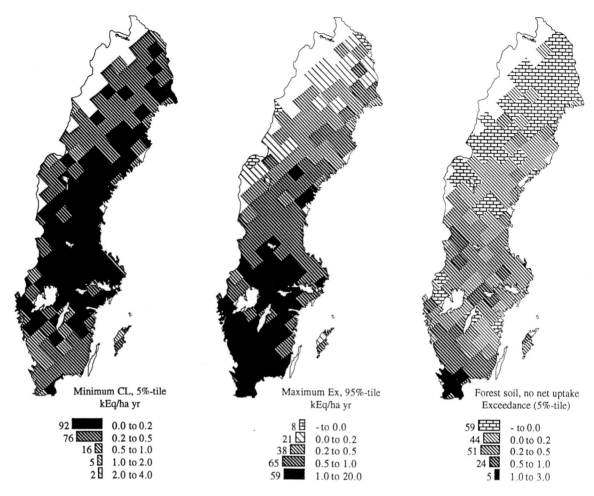

Fig. 8. The minimum critical loads for the 5 percentile and maximum exceedance for the 95 percentile of the area, using all three modes of calculation. 88% of Sweden's forested area receives more acidic deposition than the critical load. The third map shows the exceedance assuming no net N of BC uptake (virgin forest) in the PROFILE calculation.

binding liability with the polluting states, according to the "polluter pays" principle.

Uncertainties

It is stressed that the PROFILE model consists of tested and well-established subcomponents. The BC/Al ratio used as limit can also be considered to be well-established from laboratory experiments.

A sensitivity analysis of the PROFILE model shows that the model can calculate the weathering rate for a given soil within a maximum error range of ±40%, the ANC leaching within ±25%, and the BC/Al ratio within ±30% of the best estimate. The analysis was performed by Monte Carlo simulation, with an assumed error range in each parameter of ±10–100% depending on parameter type. The error range apply to the 10th and 90th percen-

tiles, that is 80% of the values fall in this range. The most important factors for weathering rate calculations are soil physical parameters, that is soil density, soil specific area and soil moisture content. The second most important parameters are soil solution equilibrium constants, especially soil P_{CO_2}, while climatic parameters, deposition and uptake parameters and soil stratification are of less importance for the weathering rate calculations. For the calculation of the BC/Al ratio, deposition and uptake parameters are important, as well as soil physical constants and soil solution equilibrium constants. As for the calculation of ANC leaching, the most important factors are percolation, deposition, uptake and mineralogy.

The uncertainties in the model calculations will naturally induce some uncertainty in the calculation of critical loads. This uncertainty may alter the exact numbers presented here but not the magnitude of the numbers. When taking all parameter uncertainties into consideration according to the sensitivity analysis, the resulting critical

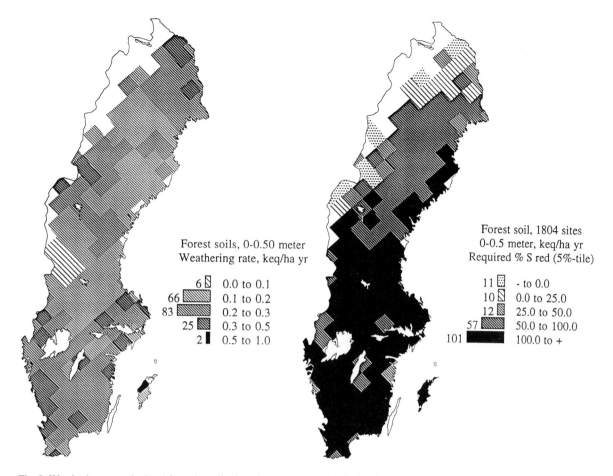

Fig. 9. Weathering rate calculated from the soil mineralogy, texture and soil chemistry using the PROFILE model. Tests indicate that uncertainty in the weathering rates are c. ±5% on the absolute value. Necessary reduction in S deposition if the 5 percentile critical load is applied. The map illustrates the fact that both S and N deposition must be reduced to achieve no exceedance in 95% of the sites.

loads varies within a maximum interval of ±40%. For an examination of the critical load calculation sensitivity, 20 field data points within a 50 × 50 km area in southern Sweden were used. The cumulative frequency distributions of the calculated critical load for each of the 20 points are shown in Fig. 10. The figure shows that variations between different sites in the same geographical area are more important than variations caused by uncertainties in the model calculations. Furthermore, when comparing the 10/90 and 25/75 percentiles with the medians of the critical load distributions for the 20 points, the uncertainty interval for different predictions can be studied.

The calculation of critical loads have followed a best available approach principle. This implies that the values will change, as knowledge of the involved processors develop and as the values can get revised and updated. The observant reader may then notice that the maps sometimes change a little due to such improvements. It is felt, however, that the major features are already incorporated in the present maps, and that no drastic changes should be expected.

Maintaining a BC/Al near or >1 will probably protect coniferous trees, but there are many other plants that are much more sensitive to Al and pH. Many ground vegetation species will be subject to significant chemical stress if presently calculated critical loads are applied as the maximum deposition level.

Coniferous trees are not the most sensitive species in many forest ecosystems, and other plants may be used for setting the limit. Several terrestrial ecosystems of interest for natural conservation, do not have trees as an important element. Taking such systems into account, will require new calculations and the application of different BC/Al limits and response functions, but the same models and regional database can be used. The final critical load would in such a case be obtained by some type of weighted minimum value approach.

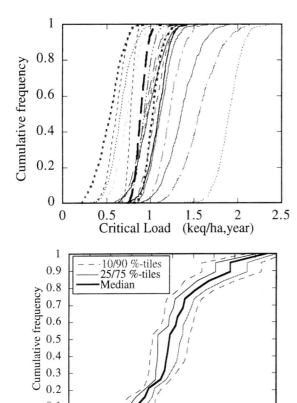

Fig. 10. The upper diagram shows the cumulative frequency distributions of critical load values at each of the 20 sites within NILU grid 58–61. It can be seen that the variation between sites is larger than the uncertainty for the individual site. The lower diagram shows the distribution of critical load values within NILU grid 58–61, showing the cumulative distribution of the median values, the upper and lower 25/75 percentile and the 10/90 percentile.

Conclusions

Sweden is receiving acid deposition in excess of critical load for >80% of its forest resources. We can conclude that a minimum deposition reduction over Sweden of 80% in sulphur deposition and 40–50% in nitrogen deposition is required for 95% of the forest resource. The UN/ECE Oslo protocol, results in an average reduction in S-deposition over Sweden of 60%. This is a great step forward, but is still far from what is required to meet the critical load (Sverdrup et al. 1992). Suggested target loads for Sand N deposition are given in Table 2.

It must be concluded that exceedance of the critical load of acidity arises mainly from deposition of S and N, and that the role of forestry is modest, but not insignificant. The role of forestry is mostly passive, it causes acid deposition on a land surface to become more efficient, and to a lesser degree acid input from forest growth.

Table 2. Suggested target loads based on regional critical loads for Sweden, based on the present calculations.

Target load	Southern Sweden	Central Sweden	Northern Sweden
kg S ha^{-1} yr^{-1}	6.0	4.0	3.0
kg N ha^{-1} yr^{-1}	6.0	3.0	1.0

References

Arovaara, H. and Ilvesniemi, H. 1990. The effect of soluble inorganic aluminium and nutrient imbalances on *Pinus sylvestris* and *Picea abies* seedlings. – In: Kaupi, P. (ed.), Acidification in Finland. Springer, Berlin, pp. 715–733.

Bonneau, M. 1991. Effects of atmospheric pollution via the soil. – In: Landmann, G. (ed.), French research into forest decline – DEFORPA programme. 2nd report. – Ecole Nationale du Génie Rural, des Eaux et des Forets, pp. 87–100.

Ericsson, T., Göransson, A., van Oene, H. and Gobran, G. 1995. Interaction between aluminium, calcium and magnesium – impacts on nutrition and growth of trees. – Ecol. Bull. (Copenhagen) 44: 191–196.

Hettelingh, J., Downing, R. and de Smet, P. 1991. Mapping critical loads for Europe. – CCE Techn. Report No. 1, RIVM Report No. 259101001, Coordination Center for Effects, RIVM.

Jönsson, C., Warfvinge, P. and Sverdrup, H. 1995. Uncertainty in prediction of weathering rate and environmental stress factors with the PROFILE model. – Water Air Soil Pollut. 78: 37–55.

Kempe, G., Toet, H., Magnusson, P.H. and Bergstedt, N.N. 1992. The Swedish national forest inventory 1983–1987. – Dept of Forest Survey, Swedish Univ. of Agricult. Sci., Report 51, Umeå.

Lång, L.O. 1995. Mineral weathering rates and mineral depletion in forest soils, SW Sweden. – Ecol. Bull. (Copenhagen) 44: 100–113.

Lövblad, G., Amann, M., Andersen, B., Hovmand, M., Joffre, S. and Pedersen, U. 1992. Deposition of sulfur and nitrogen in the Nordic countries: Present and future. – Ambio 21: 339–347.

Meiwes, K.J., Khanna, P.K. and Ulrich, B. 1986. Parameters for describing soil acidification and their relevance to the destability of forest ecosystems. – For. Ecol. Manage. 15: 161–179.

Moseholm, L., Andersen, B. and Johnsen, I. 1988. Acid deposition and novel forest decline in central and northern Europe. – Miljørapport 1988: 9, Nordic Council of Ministers, Copenhagen.

Mylona, S. 1993. Trends of sulphur dioxide emissions, air concentrations and depositions of sulphur in Europe since 1880. – Meteorological Synthesizing Centre-West, Report 2/92, Norwegian Meteorological Inst., Oslo, Norway, EMEP.

Schulze, E.-D. 1987. Tree response to acid deposition into the soil, a summary of the COST workshop at Juelich 1985. – In: Mathy, P. (ed.), Air pollution and ecosystems. D. Reidel Publ. Co., pp. 225–241.

SOU 1992. Skogspolitiken inför 2000-talet. – Statens offentliga utredningar, 1992: 76, Vol. 1, Vol. 2, Vol. 3. Allmänna Förlaget, Stockholm, (in Swedish).

Sverdrup, H. 1990. The kinetics of base cation release due to chemical weathering. – Lund University Press, Lund.

– and Warfvinge, P. 1988. Weathering of primary silicate minerals in the natural soil environment in relation to a chemical weathering model. – Water Air Soil Pollut. 38: 387–408.

– and Warfvinge, P. 1993a. Calculating field weathering rates using a mechanistic geochemical model – PROFILE. – J. Appl. Geochem. 8: 273–283.

– and Warfvinge, P. 1993b. Effect of soil acidification on the growth of trees and plants as expressed by the (Ca+Mg+K)/Al ratio. – Lund University, Rep. Environ. Engineering Ecol. 2: 1–123.

– and de Vries, W. 1994. Calculating critical loads for acidity with the simple mass balance method. – Water Air Soil Pollut. 72: 143–162.

– , de Vries, W. and Henriksen, A. 1980. Mapping critical loads. – Miljørapport 1990: 14. Nordic Council of Ministers, Copenhagen.

– , Warfvinge, P., Frogner, T., Håöya, A.O., Johansson, M. and Andersen, B. 1992. Critical loads for forest soils in the Nordic countries. – Ambio 21: 348–355.

– , Warfvinge, P. and Jönsson, C. 1994. Critical loads of acidity for forest soils, groundwater and first order streams in Sweden. – In: Hornung, M., Delve, J. and Skeffington, R. (eds), Critical loads; concepts and applications. Her Majesty's stationary office publications, pp. 54–67.

– , Warfvinge, P., Blake, L. and Goulding, K. 1995. Modeling recent and historic soil data from the Rothamsted Experimental Station, England, using SAFE. – Agricult. Ecosyst. Environ. 53: 161–177.

Tamm, C.O. 1991. Nitrogen in terrestrial ecosystems. – Ecol. Stud. 81, Springer, Berlin.

Ulrich, B. 1983. An ecosystem oriented hypothesis on the effect of air pollution on forest ecosystems. – In: Persson, G. and Jernelöv, A. (eds), Ecological effects of acid deposition. Swedish Environmental Protection Board, Solna, PM 1636, pp. 221–231.

– 1985a. Interaction of indirect and direct effects of air pollutants in forests. – In: Troyanowsky, C. (ed.), Air pollution and plants. Gesell. Deutsche Chemiker, VCH Verlagsgesellschaft, Weinheim, pp. 149–181.

– 1985b. Stability and destabilization of central European forest ecosystems, a theoretical data based approach. – In: Colley, J. and Galley, F. (eds), Trends in ecological research for the 1980's. Nato Conference Series, pp. 217–237.

Warfvinge, P. and Sverdrup, H. 1992. Calculating critical loads of acid deposition with PROFILE – a steady-state soil chemistry model. – Water Air Soil Pollut. 63: 119–143.

Ecological Bulletins 44: 90–99. Copenhagen 1995

The influence of tree species on acid deposition, proton budgets and element fluxes in south Swedish forest ecosystems

Bo Bergkvist and Lennart Folkeson

Bergkvist, B. and Folkeson, L. 1995. The influence of tree species on acid deposition, proton budgets and element fluxes in south Swedish forest ecosystems. – Ecol. Bull. (Copenhagen) 44: 90–99.

Bulk deposition, throughfall, stemflow, litterfall, soil solution (B horizon), above-ground biomass increment and soil were sampled and the annual fluxes of water, H^+, Na, K, Ca, Mg, Al, Mn, NH_4, NO_3, SO_4 and Cl were quantified in adjoining stands of *Picea abies*, *Fagus sylvatica* and *Betula pendula* at two sites in southernmost Sweden. The total atmospheric deposition of H^+ to the spruce canopies was two to eight times the deposition to beech or birch canopies. The corresponding figures for NH_4, NO_3 and SO_4 deposition were 1.5 to 3.0. Soil budgets showed a net loss of base cations from all soils, especially from the spruce soils. Proton budgets showed the calculated total proton load to the spruce stands to be two to five times the load to the beech or birch stands. The total proton load was mainly attributable to atmospheric H^+ input (50 to 83%) and base-cation incorporation into the above-ground biomass (26 to 43%).

Much of the acidity was exported in the the form of Al ions to deeper soil layers and Al buffering was the major buffer mechanism. Compared to beech and birch, spruce greatly promoted soil acidification.

The results indicate that unless acid and nitrogen deposition is much reduced, soil acidification and mineral nutrient loss are likely to continue in a majority of south Swedish forest soils with low weathering rates, particularly in the spruce forest soils, because of the much greater acidifying influence of spruce than of deciduous trees.

B. Bergkvist, Dept of Ecology, Plant Ecology, Ecology Building, S-223 62 Lund, Sweden. – L. Folkeson, Swedish Road and Transport Res. Inst., S-581 95 Linköping, Sweden.

Introduction

Considerable acidification of forest soils as a result of air pollution has been documented in areas well away from major air-pollution sources. The pool of exchangeable base cations in the upper 100 cm of forest soils in south Sweden has been reduced by 60 to 70% since 1930 (Hallbäcken 1992) and by c. 50% since 1950 (Falkengren-Grerup et al. 1987), whereas the exchangeable pool of Al has doubled since 1950. In Sweden, the soil acidification closely follows the regional gradient of acid deposition if the tree species is the same (Eriksson et al. 1992). Soil acidification has been traced well below 2 m in coniferous forest soils of southwest Sweden where acid deposition is high, whereas soils on the east coast, where acid deposition is less, are less affected.

Land use also has a decisive influence on the chemical state of soils. Compared with a deciduous forest, a coniferous forest usually receives a greater load of dry-de-posited protons (and most other elements) as a result of its evergreen canopy and greater aerosol-trapping capacity (larger leaf area, smaller collector dimensions, etc.) (see e.g. Wiman et al. 1990). This, together with differences in foliage chemistry, root-uptake features and biomass production (nutrient accumulation), results in substantial variation in the properties of soils in stands of different tree species (Nihlgård 1971). Biogeochemical cycling of elements also varies considerably between stands of different tree species (Heinrichs and Mayer 1977).

To examine the influence of tree species on current soil-acidifying processes and intensity in south Sweden, solute fluxes were studied in a spruce, a beech and a birch forest ecosystem, at each of two sites. The three stands of each site were similar in soil parent material and exposure to atmospheric deposition. Bulk deposition, throughfall, stemflow, litterfall, above-ground biomass increment and soil solution were analysed. The aims were: i) to quantify

Table 1. Stand properties of the spruce, beech and birch forest ecosystems at the Munkarp and Nyhem sites in the province of Skåne, southernmost Sweden.

Species	Age (yr)	Density (trees ha⁻¹)	Basal area (m² ha⁻¹)	Canopy cover (%)	Field + ground layer cover (%)
Munkarp					
Spruce	48	1280	53	70	< 1
Beech	80–120	395	35	100	20
Birch	20–40	1450	29	70	100
Nyhem					
Spruce	55	1100	42	80	10
Beech	70–110	800	42	100	30
Birch	30–50	710	27	70	100

the influence of the tree species on the current soil acidification rate through studies of the fluxes of protons, metals and anions and ii) to evaluate the importance of different proton-producing and proton-consuming processes.

Site description

The investigation was carried out at two sites in the province of Skåne, southernmost Sweden: Munkarp (55°57'N, 13°29'E; alt. 95 m) and Nyhem (56°13'N, 13°27'E; alt. 130 m).

At each site, adjoining pure stands (area 0.2 to 0.8 ha) of Norway spruce *Picea abies* L., European beech *Fagus sylvatica* L. and silver birch *Betula pendula* Roth were studied (Table 1).

All stands are devoid of shrubs. The spruce stands are almost devoid of a field layer; there are scattered mosses (*Dicranum* and *Hypnum* spp.) at Nyhem but not at Munkarp. In the beech stands, the sparse field layer is dominated by *Deschampsia flexuosa* L. and *Vaccinium myrtillus* L. at Munkarp and by *D. flexuosa* at Nyhem. In the birch stands, the ground is totally covered by grasses (mainly *D. flexuosa* and *Poa pratensis* L. and occasional

Agrostis capillaris L.) with other species (e.g. *Melampyrum pratense* L., *Oxalis acetosella* and *V. myrtillius*) occasionally intermingled.

The average annual mean temperature of the region is 7.2°C. The coldest months are January and February (−1.2°C) and the warmest month is July (16°C). The average annual precipitation is 795 mm.

The soil profiles are Haplic Podzols (FAO) developed on sandy moraines (sandy loam; clay content 5.5 to 7.5% in the upper 30 cm). The top-soil is most acidic in the spruce stands (pH-H₂O 3.8 to 3.9) and least acidic in the birch stands (4.3 to 4.5). The pH increases down the profile (Table 2). The base saturation of the mineral soil is low in all stands, 3 to 13% of the effective CEC (cation exchange capacity). Aluminium occupies 85 to 95% of the CEC.

The sites studied were utilized during the 19th and the early 20th century for cultivation, grazing and haymaking. Most of the ground was cleared of stones probably already prior to the 19th century, today evidenced by scattered stone walls and moss-covered mounds of stones. The spruce stands are planted and form the first spruce generation. The beech stands evolved from grazed sparse beech woods, adjoining the present-day spruce stands. At Munkarp, the birch stand evolved from parts of a grazed sparse wood which had a higher interspersion of

Table 2. Selected chemical properties of the spruce, beech and birch forest soils of the Munkarp and Nyhem sites. Effective CEC (cation exchange capacity), base cations (Na, K, Ca, Mg), Al and H⁺ in 1 M NH₄Cl.

Species	pH–H₂O		Effective CEC (µmol꜀ g⁻¹)		% of CEC					
					Base cations		Al		H⁺	
	0–5 cm	40–50 cm	0–5 cm	40–50 cm	0–5 cm	40–50 cm	0–5 cm	40–50 cm	0–5 cm	40–50 cm
Munkarp										
Spruce	3.90	4.29	156	14	62	13	32	85	7	2
Beech	4.05	4.49	44	13	35	3	43	95	23	2
Birch	4.46	4.63	104	10	69	9	18	89	13	2
Nyhem										
Spruce	3.82	4.20	239	58	72	7	11	90	17	4
Beech	4.03	4.42	69	12	39	3	47	94	14	2
Birch	4.28	4.62	89	7	62	9	24	86	14	5

birch trees. At Nyhem, the birch stand evolved in a wooded meadow.

No point sources of pollutants of any importance to the deposition of acid or metals occur within 50 km of the sites.

Materials and methods

Installation and sampling

At each site, bulk deposition was sampled using four continuously open polyethylene (PE) funnels (20 cm diam.) with PE sieves in the bottom and shaded PE bottles placed in c. 2.5 m high towers in young tree plantations adjoining the forest stands.

In each stand, five collectors of throughfall (and litterfall) were distributed along each of three lines below the canopies but > 1 m away from trunks. The collectors were similar to those used for bulk deposition, but with the funnel edge c. 0.5 m above the ground. The litter from the five collectors was pooled.

Four trees in each stand were chosen for stemflow measurement. The trees were representative of the stand with regard to vigour, height, canopy structure, branch insertion, etc, but were not necessarily close to the median or mean D^2H (see below). Stemflow was sampled using c. 2 cm broad silicon-glue coated polyurethane-foam collectors applied around the trunks at c. 1.5 m height. The stemflow was led through a funnel with a sieve and a silicon tube to a PE bottle (for chemical sampling) placed over a large shaded plastic container in which overflow was collected for volume measurement. The very voluminous stemflow of some beech trees (all four at Munkarp and one at Nyhem) necessitated volume estimation with a mechanical tipping counter.

Soil solution from the lower part of the B horizon (at 40 to 50 cm soil depth) was collected with three ceramic cup lysimeters (P80, vacuum 60 to 80 kPa) in each stand. Collecting bottles (Duran glass) were placed in a PVC container below ground to keep samples dark and cool. The three lysimeter samples were pooled before analysis. To allow the lysimeters to equilibrate with the soil solution, the first few lysimeter solutions, obtained within 2 months of installation, were discarded.

During the spring of 1984, soil was sampled in five 50 cm deep pits in each stand, using a steel cylinder, 38 cm² in area. Samples were taken from 0 to 5, 5 to 10, 10 to 20, 20 to 30, 30 to 40 and 40 to 50 cm depth; five samples from each level were bulked together. All samples were stored in PE bags at c. 4°C prior to analysis.

Diameter at breast height (D) and height (H) were determined for each tree of the six stands. The trees were ranked according to D^2H. In November 1986, three trees for destructive sampling were selected in each stand: the tree with median D^2H, the tree next smaller and the tree next larger than the median. Some departures from this had to be made because of location of sampling vessels,

marked deviations in tree-growth habit etc. The trees were felled, leaving stumps as short as possible, cut into specified fractions and weighed in the field. Representative parts of logs and twigs were brought to the laboratory for chemical analyses. No roots were sampled.

Equipment for collection of bulk deposition, throughfall, stemflow and litterfall was installed during the period October 1983 to April 1984; continuous sampling began in June 1984 and ended in June 1987. Period length usually varied between 6 and 11 weeks depending on the amount of precipitation, snow depth, frost, etc. Sampling continued during the winters. The ceramic cup lysimeters were installed in the autumn of 1990 and sampling was performed on eleven occasions, usually once a month, between December 1990 and December 1991.

Sample pretreatment

Water volume (weight), pH and conductivity were determined in the laboratory immediately after each collection. The throughfall samples were pooled five and five to form three bulk samples for each stand. Samples which were judged as contaminated by bird droppings, etc, according to the pH and conductivity results, were excluded from pooling.

Throughfall and stemflow samples were filtered (OOR, Munktell, STORA) whereas bulk deposition and soil-solution samples were not. A subsample was taken for immediate determination of NH_4^+, NO_3^-, SO_4^{2-} and Cl^-. Soil-solution samples were acidified (HNO_3, anal. gr.) prior to metal analysis. Subsamples of bulk deposition, throughfall and stemflow water were taken out for base-cation analysis. For Al and Mn analysis, a 700 to 1000 ml subsample was transferred to a 1 l Erlenmeyer flask which was sealed with a hood of fine-grained filter paper and evaporated at 105°C until dry. The residue was treated with 10 ml conc. HNO_3 (anal. gr.) to destroy organic matter and convert the metals into a chemically uniform and soluble state. The sample was diluted to 25 ml prior to analysis.

Fresh soil samples were sieved (nylon net, mesh 2 mm). Dry weight was determined at 105°C. Extracts for determination of the exchangeable soil store of base cations were obtained using 1 M acidic ammonium acetate (pH 4.8); these extracts were evaporated and digested in Erlenmeyer flasks as described above. Extracts for the analysis of effective CEC (base cations + Al + H⁺) were obtained using 1 M NH_4Cl.

Litterfall samples were dried in paper bags at 40°C to constant weight. Dry weight was determined and a 2.5 g subsample was digested in 30 ml conc. HNO_3 (anal. gr.) in an Erlenmeyer flask as described above for chemical analysis.

The biomass fractions were treated to provide information on tree age, above-ground production during the last three years, dry/fresh weight relations and representative samples for chemical analysis of twigs, branches and wood of logs produced during the last three years. No rinsing was performed prior to analysis. For dry-weight determinations, representative subsamples were dried in paper bags at c. 40°C to constant weight. For chemical analysis, dried subsamples (3 to 10 g; not ground) were digested in 30 ml conc. HNO_3 (anal. gr.) in Erlenmeyer flasks for six days as described above.

Sample analysis

Soil texture, including clay content using the pipette method, was determined. All glassware and plastic material was acid cleaned before use. In the laboratory, at least one blank in each series of six samples was included throughout sample treatment and analysis.

The following analytical methods were used: pH and conductivity were determined electrometrically at c. 20°C. The NH_4Cl extracts and the digestion residues of litterfall samples were analysed for Na, K, Ca, Mg and Al, for litterfall also for Mn and S, using Inductively Coupled Plasma spectrometry. Aqueous samples and digestion residues of biomass samples and acid ammonium acetate extracts were analysed for metals (total concentrations) using acetylene-air flame AAS; K and Na with 1000 ppm Cs (as CsCl) and Ca and Mg with 10 000 ppm La (as $LaCl_3$) in sample and standard solutions; Al: acetylene-N_2O flame. Ammonium was determined by Flow Injection Analysis. The anions NO_3^-, SO_4^{2-} and Cl^- were determined by ion chromatography. In biomass samples, total N was analysed by Kjeldahl analysis.

Calculations

Aluminium was calculated as bivalent Al where the yearly mean pH of the soil solution was ≥ 5.0. Such high pH values were only found in the beech and birch stands at Nyhem. In the other stands, Al was calculated as Al^{3+}.

Dry-deposited acidity was estimated according to Mulder et al. (1987) as

$$H^+_{dry} = (SO_4^{2-})_{TF+SF} - (SO_4^{2-})_{BD} + (NO_3^-)_{TF+SF} - (NO_3^-)_{BD} + (NH_4^+)_{BD} - (NH_4^+)_{TF+SF}$$

where TF denotes throughfall, SF stemflow and BD bulk deposition.

This equation was developed in a region with high N deposition (the Netherlands) but seems to be applicable also to the conditions of lower N deposition in south Sweden. This is evident from a comparison with the result obtained using the calculation procedure of Bredemeier et al. (1990) where the net H^+ input with throughfall and stemflow water and the calculated canopy buffering of H^+ were added together. When the median net atmospheric H^+ input to the stands was calculated using the two methods, the difference was <5%.

Quantitative model-generated values for the dry deposition of K, Ca, Mg and Mn to the coniferous stands were calculated from Wiman (1984 and pers. comm). The steady-state model uses a system of partial differential equations from the concept of forests as volume sinks. The model incorporates submodels of forest structure (leaf area index, vertical distribution of foliage), forest aerodynamics (windspeed, eddy diffusivity) and aerosol characteristics (collection efficiency, particle size distribution) and analyses the interplay between these submodels.

Of the model-generated quantitative results for aerosol dry deposition to coniferous forest stands, only half the annually deposited amounts were assumed to be deposited to the deciduous stands, as beech and birch are defoliated during the winter half-year.

Annual element fluxes were calculated as follows: i) For bulk deposition and throughfall, the element fluxes (concentration × water volume) were calculated for each sampling occasion. The summed fluxes for the 3-yr period were divided by 3 to give annual fluxes. ii) Element fluxes in stemflow were calculated in the same way using water volumes and stand tree density. iii) The soil-solution fluxes were calculated by summing the fluxes for each of the eleven sampling occasions, using element concentrations and simulated soil-water fluxes. Soil-water fluxes were simulated using a numerical model (SOIL; Jansson 1991), considering plant and soil properties. The driving variables of the model, precipitation, air temperature, vapour pressure, windspeed and cloudiness, were obtained from national meteorological statistics (SMHI). The daily precipitation amounts used in the SOIL model, obtained from the Ljungbyhed meteorological station (within 21 km of the sites) for the period from December 1990 to December 1991 (annual precipitation sum: 909 mm), were reduced by a certain percentage to fit the mean annual precipitation sum measured at each site during the period from June 1984 to June 1987 (701 mm at Munkarp, 848 at Nyhem). iv) Annual element fluxes in litterfall were calculated by multiplying the concentrations in the collected needle-litter and leaf-litter fractions by the dry weight of the pooled 3-yr total litterfall, and dividing the product by 3. v) For each of the six stands, the amounts of elements in the annual aboveground biomass increment were calculated as follows: for each biomass fraction representing the last three years, the mean concentration in the three sample trees in the stand was calculated for each element, the biomass distribution on fractions was averaged for each stand; dry weight and element content of the different biomass fractions were averaged for each stand, and then divided by 3.

Using element fluxes, solute budgets (SB) were calculated in two ways: i) SB = BD + DD − BI − SO, for elements with important internal canopy fluxes (H^+, K, Ca, Mg, Al and Mn) and ii) SB = TF + SF − BI − SO, for elements for which internal canopy fluxes were negligible (Na, NH_4-N, NO_3-N, SO_4-S and Cl). (BD denotes bulk deposition, DD dry deposition, TF throughfall, SF

Table 3. Fluxes and calculated solute budgets of water and elements at the forest sites. BD = bulk deposition, DD = dry deposition, TF = throughfall, SF stemflow, LF = litterfall, BI = above-ground biomass increment, SO = soil output, SB = solute budget. DD of H+ (acidity) estimated according to Mulder et al. (1987), DD of Na, NH4-N, NO3-N, SO4-S and Cl considered to be part of TF+SF (Mulder et al. 1987). DD of K, Ca, Mg and Mn from Wiman (1984 and pers. comm.), DD of Al considered negligible (Mulder et al. 1987). Means of TF, SF and SO, respectively, lacking common letters differ significantly (p < 0.05) between tree species (Tukey test, SO statistics calculated on concentrations). Solute budget calculated as SB = BD + DD − BI − SO (for H+, K, Ca, Mg, Al and Mn) or as SB = TF + SF − BI − SO (for Na, NH4-N, NO3-N, SO4-S and Cl).

	H_2O (l m^{-2})	H$^+$	Na	K	Ca	Mg	Al	Mn (mg m^{-2} yr^{-1})	NH$_4$-N	NO$_3$-N	SO$_4$-S	Cl
MUNKARP												
BD	701	36.6	859	198	523	152	33.2	7.2	1050	599	1460	2400
s.d. n=3	114	8.9	302	18	322	38	11.0	4.6	135	43	296	868
Spruce												
DD	–	166.0	–	230	130	70	–	7.0	–	–	–	–
TF	496b	51.7a	2890a	2090b	1930a	563a	89.9a	516.0a	935a	918a	3070a	5370a
s.d. n=9	79	12.8	621	485	785	124	26.1	156.0	184	138	605	1130
SF	18b	11.2a	262ab	241a	419a	72a	12.1a	44.0a	53a	83a	523a	599ab
s.d. n=12	8	4.9	133	183	355	47	6.2	40.0	22	29	301	311
LF	–	–	90	527	2040	201	101.0	711.0	–	–	452	–
BI	–	–	44	1710	907	263	37.0	285.0	2860	–	–	–
SO	335	15.0a	4840a	543a	1100a	659a	1760.0a	885.0a	312a	886a	4110a	6330a
SB	–	188.0	-1730	-1830	-1350	-700	-1760.0	-1160.0	-2180	115	-517	-361
Beech												
DD	–	69.0	–	115	65	35	–	3.5	–	–	–	–
TF	526b	13.7b	2160b	2800a	1590a	448a	33.8b	151.0b	798c	663b	1930b	4490ab
s.d. n=9	77	3.4	337	567	898	99	16.6	30.0	60	76	265	694
SF	77a	4.5b	188b	283a	101b	25b	3.3b	7.4b	55a	44b	296a	422b
s.d. n=12	35	3.0	96	95	73	14	2.3	4.3	29	28	157	194
LF	–	–	57	434	1080	158	31.6	203.0	–	–	275	–
BI	–	–	20	860	570	120	5.2	81.0	1650	–	–	–
SO	402	21.4a	3650b	203b	262b	282b	1100.0b	25.0b	80a	36b	2400b	3930b
SB	–	84.2	-1320	-750	-244	-215	-1070.0	-95.3	-877	671	-174	982
Birch												
DD	–	79.0	–	115	65	35	–	3.5	–	–	–	–
TF	643a	13.5b	1430c	1850b	1470a	452a	29.9b	222.0b	541c	593b	1730b	3600b
s.d. n=9	104	1.3	197	513	798	87	8.2	63.0	68	93	695	764
SF	71a	9.5a	319a	215a	283ab	102a	10.6a	48.0a	50a	62ab	433a	727a
s.d. n=12	14	4.6	104	139	187	54	7.1	25.0	31	42	224	230
LF	–	–	28	389	1030	229	34.1	284.0	–	–	274	–
BI	–	–	19	613	545	117	5.8	113.0	1600	–	–	–
SO	411	20.4a	2120c	207b	1030a	278b	769.0b	105.0b	164a	726a	1830b	2220b
SB	–	95.2	-390	-507	-987	-208	-742.0	-207.0	-1170	-71	333	2110
NYHEM												
BD	848	45.4	1410	131	402	208	38.1	5.7	687	602	1430	3570
s.d. n=3	126	4.5	592	26	207	42	11.6	1.9	97	57	169	1500
Spruce												
DD	–	309.0	–	230	130	70	–	7.0	–	–	–	–
TF	534b	126.0a	3700a	3080a	3460a	842a	100.0a	279.0a	1320a	1430a	5770a	7750a
s.d. n=9	98	11.6	1190	568	1520	195	17.5	70.0	214	260	453	1710
SF	12c	7.5a	122b	141b	186a	34b	6.5a	10.3b	39a	61a	348a	270a
s.d. n=12	5	3.1	49	72	112	19	3.3	5.5	21	50	188	131
LF	–	–	65	460	1300	165	83.4	145.0	–	–	385	–
BI	–	–	29	2090	683	212	24.8	103.0	2020	–	–	–
SO	430a	22.8a	10200a	556a	905a	978a	3030.0a	39.0a	593a	417a	6320a	14000a
SB	–	332.0	-6410	-2290	-1060	-912	-3020.0	-129.0	-1250	1070	-202	-5980
Beech												
DD	–	41.0	–	115	65	35	–	3.5	–	–	–	–
TF	619ab	14.6c	1500b	1950b	1500b	369c	26.4b	220.0ab	527b	509b	1750b	4290b
s.d. n=9	108	3.7	309	136	709	113	16.8	56.0	93	96	285	519
SF	58a	3.9a	319a	227a	194a	55ab	3.4b	21.1a	28a	15b	280a	666a
s.d. n=12	16	2.4	171	63	118	21	1.8	8.3	23	7	100	306
LF	–	–	113	535	1630	212	35.3	392.0	–	–	296	–
BI	–	–	13	460	490	95	2.1	103.0	1350	–	–	–
SO	519	12.4b	3340b	442ab	370b	340b	715.0b	49.0a	206b	65b	1910b	2390b
SB	–	74.0	-1530	-656	-393	-192	-679.0	-143.0	-1000	459	120	2570
Birch												
DD	–	71.0	–	115	65	35	–	3.5	–	–	–	–
TF	731a	44.3b	2060b	1330c	1640b	611b	48.3b	168.0b	464b	602b	2000b	5600b
s.d. n=9	120	13.6	370	227	710	159	25.3	39.0	99	140	475	1150
SF	40b	8.0a	329a	106b	144a	61a	6.6a	18.9a	21a	42ab	264a	734a
s.d. n=12	17	4.1	238	50	94	29	3.6	9.8	21	25	196	485
LF	–	–	51	215	677	215	16.3	108.0	–	–	183	–
BI	–	–	9	213	199	64	1.7	26.0	710	–	–	–
SO	510	2.4b	3060b	211b	565b	288b	129.0c	47.0a	264b	80b	1670b	3230b
SB	–	114.0	-680	-178	-297	-109	-92.6	-63.8	-489	564	594	3100

stemflow, BI above-ground biomass increment and SO soil output). Calculation ii) may possibly somewhat over-estimate the input of Na, SO_4-S and Cl. Further, the use of throughfall and stemflow fluxes might underestimate the input of N, since a substantial proportion of the total N deposition might be taken up directly by the foliage (Matzner and Meiwes 1994).

In the NH_4Cl extracts, the H^+ concentration was calculated from pH according to Meiwes et al. (1984).

Differences in soil-solution concentrations, stemflow and throughfall fluxes between the tree species were evaluated using analysis of variance and the Tukey HSD test (with the Tukey-Kramer adjustment for unequal n) for pairwise comparison of means.

Results and discussion

Atmospheric deposition and canopy interactions

The dry deposition of K, Ca, Mg and Mn, as calculated from Wiman (1984 and pers. comm.), made an important contribution to the total deposition (bulk + dry) to the canopies (Table 3). Dry deposition of Al was considered negligible (Mulder et al. 1987). In contrast, dry deposition (included in TF + SF) made an important contribution to the total fluxes of SO_4-S, Na and Cl. For these ions and for NH_4-N and NO_3-N, Σ(throughfall + stemflow) fluxes were taken as input (Mulder et al. 1987). The total deposition (throughfall + stemflow) of NH_4-N, of NO_3-N and of SO_4-S to the spruce canopy was 1.5 to 3 times that to the deciduous canopies (Table 3). Deposited amounts to the beech and birch stands were quite similar.

The annually deposited (throughfall + stemflow) amounts of NH_4-N, NO_3-N and SO_4-S to the stands were large, despite the great distance from pollution sources, and were similar to those found in another study (Berg-kvist and Folkeson 1992) and in monitoring studies of the same region (IVL 1991). Total inorganic N deposition to our stands was 1 to 2.8 g m^{-2} yr^{-1}. Considerably larger amounts of N deposition, 3.0 to 6.0 g m^{-2} yr^{-1}, have been reported from deciduous and coniferous forests in Denmark and western continental Europe (Breemen et al. 1987, Mulder et al. 1987, Rasmussen 1988).

Dry-deposited acidity (H^+), estimated according to Mulder et al. (1987), dominated over bulk deposition, and was 2 to 8 times as high to the spruce canopies as to the deciduous canopies. By far the major part of the deposited H^+ was neutralized by the foliage, as revealed by much lower amounts of free acidity (H^+) in through-fall + stemflow than in bulk + dry deposition (Table 3). The buffering capacity of the canopy may mainly be achieved by ion exchange with base cations in the foliage, with subsequent restoration of foliar buffer capacity and excretion of H^+ to the soil (Matzner and Ulrich 1984, Johnson and Lindberg 1992). In part, strong acidity deposited from the atmosphere may also be buffered by

protonation of organic-acid anions released from the canopy (Johnson and Lindberg 1992).

As calculated by subtracting bulk + dry deposition (Table 3) from throughfall + stemflow, net foliage leaching of K, Ca and Mg was found to be substantial from the canopies of all tree species. The quantities leached this way amounted to 5 to 9 times the total (bulk + dry) deposition for K, to 2 to 6 times for Ca and to 2 to 4 times for Mg. Manganese leaching from the canopy by far exceeded deposition; leaching being 20 to 40 times deposition for spruce, and 10 to 20 times for beech and birch. In all stands, the contribution of stemflow to the element fluxes to the soil was usually < 10 to 15%.

An estimate of the total element uptake by the roots into the above-ground part of the trees may be calculated from Table 3 by summing: i) net foliage leaching; ii), litterfall (part of the internal cycling); and iii) above-ground biomass increment. The net foliage leaching (calculated as throughfall + stemflow − bulk deposition − dry deposition) is part of the internal cycling. For Na, NH_4, NO_3, SO_4 and Cl, net foliage leaching was set to zero as throughfall + stemflow fluxes were considered as total input. Only a small part of the elements annually taken up from the soil was stored in the above-ground tree biomass. At both sites this incorporation was largest in spruce. Elements taken up by the roots were returned to the soil mainly by litterfall for Ca, Al and Mn and by canopy leaching for K and Mg.

Soil solute losses

The amounts of elements leaving the B horizon with the soil solution (at 50 cm soil depth) were often largest in the spruce stands, differences between beech and birch being rather small (Table 3).

Generally, N leaching from the soil profile seems to mirror the N deposition to the stands. Amounts of both deposition and leaching of N were generally largest in spruce stands. An exception was that the NO_3 leaching was similar in the birch and spruce soils of Munkarp.

An output of NO_3-N in the range of 1.1 to 8.7 g N m^{-2} yr^{-1} has been reported from deciduous and coniferous forests in Denmark and western continental Europe where the N input was in the range of 3.0 to 6.0 g N m^{-2} yr^{-1} (Breemen et al. 1987, Mulder et al. 1987, Rasmussen 1988). Compared to this, the output from the spruce, beech and birch soils of the present study, < 0.1 to 1.2 g N m^{-2} yr^{-1}, must be designated as rather low, but they are of the same magnitude as previously reported for the region (IVL 1991, Bergkvist and Folkeson 1992).

In the spruce and beech stands of both sites, practically all SO_4 deposited to the soil was leached through the soil profile, indicating SO_4 saturation (the solution is close to equilibrium with the solid phase). The birch soils of these sites seemed to be less SO_4 saturated and obviously have some ability to retain SO_4.

In the soil solution, the anions SO_4 and NO_3 were

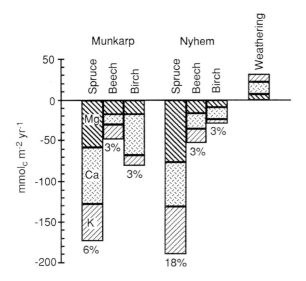

Fig. 1. Current annual soil base-cation (Mg, Ca and K) net losses of the upper 50 cm soil layer in the spruce, beech and birch stands studied, and the calculated maximum weathering rate according to Sverdrup and Warfvinge (pers. comm.). Figures are annual percentage losses of the ammonium acetate-extractable (pH 4.8) soil store of the base cations (assuming zero weathering). Solute budget: see Table 3.

net release or accumulation of Na and Cl are still far from clear.

The deviation from zero of the Na and Cl balances may also result from inaccuracy in the estimate of the soil fluxes resulting from spatial variability and from errors in the hydrologic simulations.

Generally, the calculations showed net losses of most elements from the soils studied (Table 3). All stands were losing base cations, Al and Mn. As a rule, the solute budget was most negative (output greater than input) in the spruce stands. The spruce stands showed the largest amounts of leaching of Na, K, Ca, Mg and Al.

The annual net export of base cations from the soil, resulting from leaching and biomass uptake, has important implications for the long-term nutrition and acidity status of a forest ecosystem. The replenishment of the soil store of base cations ultimately depends on the weathering rate of the soil, though quantities supplied in deposition are not negligible. The great uncertainty in the estimations of weathering rate makes it difficult to evaluate the potential change in the soil store of base cations.

In Fig. 1, annual soil base-cation budgets are compared with simulated weathering rates (Sverdrup and Warfvinge pers. comm.) in a soil (from the same region) with mineralogical properties similar to those of the present study. The percentage losses of the exchangeable base-cation $(K + Ca + Mg)$ pool (assuming zero weathering) are also illustrated.

In all stands, leaching + biomass incorporation were greater than the total atmospheric input of base cations. The spruce stands showed the largest net export of base cations and the largest percentage loss of exchangeable base cations. The differences between the beech and birch stands were small. The spruce stand of Nyhem had by far the largest export of base cations relative to the exchangeable store. Magnesium, Ca and K (mol_c m^{-2} yr^{-1}) were lost in similar amounts from that stand.

Biomass increment was the major cause of the removal of $K + Ca + Mg$ from the soil in the beech stand at Munkarp, making up c. 60% of the total soil export (Table 3). In all the other stands, biomass increment accounted for between 25 and 45% of the soil export, and leaching was the principal component.

A comparison of measured depletion rates with simulated weathering rates strongly suggests that the present weathering cannot keep pace with the intense base-cation depletion from the soils under study, particularly in the spruce stands (Fig. 1). Only in the birch stand of Nyhem will the chemical state of the soil remain virtually unchanged, as demonstrated by the nearly balanced base-cation export and weathering. In the other stands, the soil chemistry will change and, given the current deposition rate, the exchangeable stores of base cations will decrease. This has been shown to have occurred during the past decades in a number of forest soils in south Sweden (Falkengren-Grerup et al. 1987, Hallbäcken 1992). Thus, the trend towards decreasing exchangeable pools of base cations seems to continue.

mainly balanced by Ca, Mg and Al (Table 3). The more acidic the soil, the more important was Al in the solution. The largest amounts of Al were leached from the soil of the spruce stands. Aluminium mobilization is known to be strongly pH sensitive (Berggren 1992), and as pH of the soil solutions of the mineral soil decreases beneath 4.5, Al solubilization will increase drastically. The differences in Al, Ca and Mg mobilization between tree species reveal that buffering processes as well as proton-buffering rate differ between species. These differences can be attributed to differences in soil acidity. It is apparent that Al solubilization was an important proton-buffering process in the soils studied. An exception was the less acidic soil of the birch stand of Nyhem, where the most important proton-buffering process was cation exchange.

Sodium and Cl are often considered to pass through ecosystems with little tendency to storage or reaction. Thus, their ecosystem balances may indicate the degree of accuracy of the calculated fluxes. The Na and Cl balances were here found to deviate substantially from zero for most stands (Table 3). However, similar results have been reported for a number of forest stands in an integrated forest study in USA, Canada and Norway (Johnson and Lindberg 1992). In a monitoring study on the Swedish west coast (Hultberg and Grennfelt 1992), the long-term (10 yr) mean ecosystem balances of Na and Cl were close to zero, but for individual years the balance could substantially deviate from zero, indicating temporary retention or release. Mechanisms responsible for

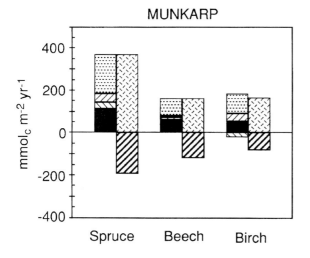

MUNKARP

1. ▦ Net atmospheric H⁺ input

2. ▨ N transformations

3. ▧ SO₄ release or retention

4. ■ Cation incorporation into biomass

5. ▨ Total H⁺ load (Σ1+2+3+4)

6. ▨ "Cation acid" (Al) output

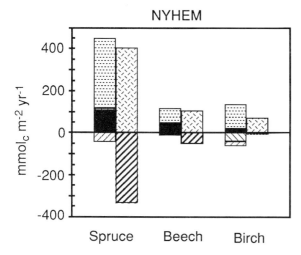

NYHEM

Fig. 2. Total acidity budgets of the spruce, beech and birch forest ecosystems of the two sites.

Large decreases in exchangeable Ca and Mg (though not K) pools have been reported from spruce stands on podsols and dystric cambisols in the Harz Mountains, Germany (Haus 1985). In North America, forest soils have also been shown to have lost base cations in excess of additions during the past decades (pine, yellow-poplar

and chestnut oak in the Walker Branch watershed; Johnson et al. 1988). For Ca, biomass accumulation exceeded leaching at all their sites, but for Mg, leaching exceeded biomass accumulation in the stands with the highest atmospheric input of SO₄ (Johnson and Todd 1990). There was a low soil net export of Mg in a Douglas-fir forest in western USA with an acid deposition slightly less than that of the deciduous stands of the present study (Homann et al. 1992). The net export of Ca from the Douglas-fir soil was comparable to that of our spruce stands, as the dominating acid buffering mechanism in the less acidic American soil was cation exchange/silicate mineral weathering. For the base cations altogether, the American study showed biomass accumulation and soil leaching to be equally important.

Element fluxes in Norway spruce and European beech forest ecosystems have been studied for more than two decades in the German Solling area, and the strong influence of the tree species has been convincingly demonstrated (Bredemeier et al. 1990, Matzner and Meiwes 1994). The magnitude of the element fluxes was also similar to that of the present study.

In a French study (Lelong et al. 1990), input and output fluxes of K, Mg and Ca in small catchments with Norway spruce or European beech were also comparable to those found for spruce and beech in the present study.

Proton budgets

To develop H⁺ budgets for a soil, processes that generate H⁺ can be tracked by accounting for the major processes that produce and consume H⁺ (Breemen et al. 1983, Bredemeier et al. 1990). H⁺-producing or H⁺-consuming processes which contribute significantly to the total H⁺ load of the soils studied are the transport or transformation of H⁺, NH₄, NO₃, SO₄ and the biomass incorporation of base cations.

Because uptake of base cations far exceeds that of H₂PO₄ and SO₄, the ionic uptake results in a net proton flux to the soil (nitrogen uptake is treated separately). Therefore, only base-cation uptake was considered in the calculations of the H⁺ load generated by ionic biomass incorporation, though somewhat overestimating this H⁺ load.

A weak acid such as H₂CO₃ is an important proton source only at pH >5.5 and pH values as high as that were never recorded in the acidic soils of the present study.

The annual total proton load (TPL) to the soil was calculated by summing four terms (expressed in mmol$_c$ m⁻² yr⁻¹): i) the net atmospheric input of H⁺ to the soil,

$$(H^+)_{BD+DD} - (H^+)_{SO}$$

ii) NH₄ in excess of NO₃ uptake from deposition + net nitrification of deposited NH₄ and organic N,

$$(NH_4 - NO_3)_{TF+SF} + (NO_3 - NH_4)_{SO}$$

iii) SO_4 release from or retention in the soil,

$$(SO_4)_{SO} - (SO_4)_{TF+SF}$$

iv) the incorporation of base cations into the above-ground biomass increment (see Table 3 for subscripts).

The total proton load (TPL) to the spruce stands, 370 to 410 $mmol_c$ m^{-2} yr^{-1}, was generally considerably higher than that to the adjoining deciduous stands, 70 to 165 $mmol_c$ m^{-2} yr^{-1} (Fig. 2). The net H^+ loading from the atmospheric deposition made up 50 to 83% of the TPL to the stands. Base-cation incorporation into the above-ground biomass made up 26 to 43% of the TPL to the stands. Transformations of NH_4, NO_3 and SO_4 were of minor importance for TPL. In the birch stands the SO_4 retention had a partly H^+ neutralizing (H^+ consuming) influence, as had N transformations in all stands at Nyhem.

Acidity was exported in considerable quantities from the ecosystems (negative axis in Fig. 2), predominantly as dissolved Al ("cation acid"). Half or more of the TPL was exported by Al leaching in all stands, except for the birch stand at Nyhem, where the acidity output was negligible. In this stand, Al was less soluble, as a result of the much less acidic soil conditions.

A comparison between TPL and acidity output reveals the extent to which the soil acts as an acid neutralizer. Leaching of Al from the soil profile means transfer of acidity from the ecosystem into its surroundings. Upon transport to environments with a higher pH, e.g. deeper soil layers, ground water or aquatic ecosystems, the Al^{3+}, acting as a "cation acid", will release protons and possibly precipitate as the hydroxide. The more acidity is leached from the ecosystem, the less does the soil act as an acid neutralizer between atmospheric (or internal) sources and aquatic environments.

Our stands, of varying age, represent common types of forest on soils typical of SW Sweden. With respect to TPL (Fig. 2), our birch/beech stands and spruce stands seem to represent a range from low to intermediate rates of forest-soil acidification in a European perspective (Breemen et al. 1986, Bredemeier et al. 1990). The Nyhem spruce stand is slightly less acidified than the Solling spruce stand in central Germany (Bredemeier et al. 1990), the atmospheric deposition rates of H^+, NH_4, NO_3 and SO_4 of the two sites being almost the same.

Conclusions

In representative forest stands in southernmost Sweden, the total atmospheric deposition of H^+ to spruce canopies was 2 to 8 times the deposition to adjoining beech or birch canopies, while 1.5 to 3.0 times more NH_4, NO_3 and SO_4 were deposited to the spruce canopies.

In most of the soils, Al buffering was of equal or greater importance than base-cation buffering.

Solute budgets revealed a net loss of base cations from all soils, the spruce stands having the greatest losses. Simulated weathering rates were much smaller than the measured base-cation losses, particularly in the spruce stands. These soils will be further acidified and a further reduction in the soil store of exchangeable base cations is to be expected in the years to come, unless acid deposition is greatly diminished.

Proton-budget calculations showed that the total proton load (TPL) to the spruce stands was many times higher than the TPL to the adjoining deciduous stands. In all stands, the predominant part of TPL was attributable to acid deposition. About one-third of the TPL was internally generated as a result of base-cation incorporation into the above-ground biomass increment. Much acidity was exported, in the form of Al ("cation acid"), from the ecosystems to deeper soil layers.

Acknowledgments – We are grateful to D. Berggren and G. Tyler for scientific discussions, U. Emanuelsson for information on stand history, P.-E. Jansson for assistance with the SOIL model, F. Larsson, A. Jonshagen, L.-G. Olsén and I. Persson for field assistance, K. Olsson for field assistance and data handling, A. Balogh, S. Billberg, M.-B. Larsson, T. Olsson, A. Persson and E. Sjöström for laboratory work, A. Persson for typing and the landowners for land-use permission. The study was financed by the National Swedish Environment Protection Agency and the Swedish Council for Forestry and Agricultural Research.

References

Berggren, D. 1992. Speciation and mobilization of aluminium and cadmium in Podzols and Cambisols of S. Sweden. – Water Air Soil Poll. 62: 125–156.

Bergkvist, B. and Folkeson, L. 1992. Soil acidification and element fluxes of a *Fagus sylvatica* forest as influenced by simulated nitrogen deposition. – Water Air Soil Poll. 65: 111–133.

Bredemeier, M., Matzner, E. and Ulrich, B. 1990. Internal and external proton load to forest soils in northern Germany. – J. Environ. Qual. 19: 469–477.

Breemen, N. van, Mulder, J. and Driscoll, C. T. 1983. Acidification and alkalinization of soils. – Plant Soil 75: 283–308.

– , de Visser, P. H. B. and van Grinsven, J. J. M. 1986. Nutrient and proton budgets in four soil-vegetation systems underlain by Pleistocene alluvial deposits. – J. Geol. Soc. (Lond.) 143: 659–666.

– , Mulder, J. and van Grinsven, J. J. M. 1987. Impacts of acid atmospheric deposition on woodland soils in the Netherlands: II. Nitrogen transformations. – Soil Sci. Soc. Am. J. 51: 1634–1640.

Eriksson, E., Karltun, E. and Lundmark, J.-E. 1992. Acidification of forest soils in Sweden. – Ambio 21: 150–154.

Falkengren-Grerup, U., Linnermark, N. and Tyler, G. 1987. Changes in soil acidity and cation pools of south Sweden soils between 1949 and 1985. – Chemosphere 16: 2239–2248.

Hallbäcken, L. 1992. Long term changes of base cation pools in soil and biomass in a beech and a spruce forest of southern Sweden. – Z. Pflanzenernaehr. Bodenkd. 155: 51–60.

Haus, H. 1985. Wasser- und stoffhaushalt im Einzugsgebiet der Langen Bramke (Harz). – Ber. Forschungszentrums Wald-ökosysteme/Waldsterben (Göttingen) 17: 1–206.

Heinrichs, H. and Mayer, R. 1977. Distribution and cycling of major and trace elements in two central European forest ecosystems. – J. Environ. Qual. 6: 402–407.

Homann, P. S., van Miegroet, H., Cole, D. W. and Wolfe, G. V. 1992. Cation distribution, cycling and removal from mineral soil in Douglas-fir and red alder forests. – Biogeochemistry Dordr. 16: 124–150.

Hultberg, H. and Grennfelt, P. I. 1992. Sulphur and seasalt deposition as reflected by throughfall and runoff chemistry in forested catchments. – Environ. Poll. 75: 215–222.

IVL 1991. Miljöatlas. Resultat från IVLs undersökningar i miljön 1991. – Inst. Vatten- och Luftvårdsforskning. Stockholm.

Jansson, P.-E. 1991. Simulation model for soil water and heat conditions. Description of the SOIL model. – Swedish Univ. of Agricultural Sciences, Uppsala, Sweden, report 165.

Johnson, D. W. and Todd, D. E. 1990. Nutrient cycling in forests of Walker Branch watershed, Tennessee, USA. Roles of uptake and leaching in causing soil changes. – J. Environ. Qual. 19: 97–104.

– and Lindberg, S. E. 1992. Atmospheric deposition and forest nutrient cycling. A synthesis of the Integrated Forest Study. – Ecol. Stud. 91. Springer, New York.

– , Henderson, G. E. and Todd, D. E. 1988. Changes in nutrient distribution in forests and soils of Walker Branch watershed, Tennessee, USA, over an eleven-year period. – Biogeochemistry Dordr. 5: 275–294.

Lelong, F., Dupraz, C., Durand, P. and Didon-Lescot, J. F. 1990. Effects of vegetation type on the biogeochemistry of small catchments (Mont Lozere, France). – J. Hydrol. 116: 125–145.

Matzner, E. and Ulrich, B. 1984. Raten der Deposition, der internen Produktion und des Umsatzes von Protonen in Waldökosystemen. – Z. Pflanzenernaehr. Bodenkd. 147: 290–308.

– and Meiwes, K. J. 1994. Long-term development of element fluxes with bulk precipitation and throughfall in two German forests. – J. Environ. Qual. 23: 162–166.

Meiwes, K.-J., König, N., Kharma, P. K., Prenzel, J. and Ulrich, B. 1984. Chemische Untersuchungsverfahren für Mineralboden, Auflagehumus und Wurzeln zur Charakterisierung und Bewertung der Versauerung in Waldböden. – Berichte des Forschungszentrums Waldökosysteme/Waldsterben 7: 14–16.

Mulder, J., van Grinsven, J. J. M. and van Breemen, N. 1987. Impacts of acid atmospheric deposition on woodland soils in the Netherlands: III. Aluminium chemistry. – Soil Sci. Soc. Am. J. 51: 1640–1646.

Nihlgård, B. 1971. Pedological influence of spruce planted on former beech forest soils in Scania, south Sweden. – Oikos 22: 302–314.

Rasmussen, L. 1988. Report from laboratory of environmental science and ecology. – Technical Univ. of Denmark, Lyngby.

Wiman, B. 1984. Aerosol dry deposition of heavy metals and acids to forest ecosystems. – Swedish Environ. Protection Board, Solna, report 1908.

Wiman, B. L. B., Unsworth, M. H., Lindberg, S. E., Bergkvist, B., Jaenicke, R. and Hansson, H.-C. 1990. Perspectives on aerosol deposition to natural surfaces: interactions between aerosol residence times, removal processes, the biosphere and global environmental change. – J. Aerosol Sci. 21: 313–338.

Ecological Bulletins 44: 100–113. Copenhagen 1995

Mineral weathering rates and primary mineral depletion in forest soils, SW Sweden

Lars-Ove Lång

Lång, L.-O. 1995. Mineral weathering rates and primary mineral depletion in forest soils, SW Sweden. – Ecol. Bull. (Copenhagen) 44: 100–113.

Weathering rates for the upper 0.5 m soil at 22 coniferous forested sites in SW Sweden have been calculated by two different methods. The significance of sediment mineralogy and of grain-size distributions for these calculations with existing models is discussed. Estimations of long-term weathering rates are based upon depletion of < 2 mm material. Weathering rates of individual minerals are used by the PROFILE model for calculating total current weathering rates (here using coarse-silt mineralogy). Calcium and Mg minerals are reliably determined for the coarse-silt fraction, as shown by a good correlation between ICP-analyses and calculations from petrographic microscopy for both CaO and MgO contents. Similarities in the geochemistry of coarse silt and < 2 mm material for the till sites suggest reasonably well defined geochemical conditions for comparing the total weathering rates calculated by the long-term and by the PROFILE methods. The rates calculated by both methods are c. 50–90 meq m^{-2} yr^{-1} in tills. For most sites, the long-term rates are lower than the PROFILE rates. A weak correlation exists between the rates calculated by the two methods. Grain-size distribution and the heavy-mineral size sorting have a considerable influence on the PROFILE total weathering rates for reworked and glaciofluvial sediments. Epidote content significantly enhances Ca weathering in the PROFILE calculations. Estimates of long-term Ca weathering emphasise the importance of hornblende and apatite depletion.

L.-O. Lång, Dept of Geology, Earth Science Centre, Göteborg Univ., S-413 81 Göteborg, Sweden.

Introduction

In Sweden, remedial and preventive programmes to combat acidification are in progress at different levels of decision making. Local authorities incorporate the idea of critical load in their environmental protection agendas. This requires detailed information about the current release of cations (Ca, Mg, K and Na) to soilwater and groundwater. Hence, knowledge of the rate of chemical weathering is one of the most essential requirements for critical load assessment.

The need for estimating field weathering rates in acidification research has focussed considerable attention upon biogeochemical models. Thus, the inherent problem of combining the perspective of geological long-term and large-scale weathering processes with the current cation field release rates is being addressed (Bricker et al. 1994).

The interaction between the kinetic mineral weathering and the soil chemical environment involves several concurrent processes. Detailed study of these processes requires well defined laboratory conditions. Weathering mechanisms, have been described for feldspars (Holdren and Berner 1979, Fung and Sanipelli 1982), for pyroxene and amphiboles (Berner et al. 1980, Velbel 1989) and for biotite (Turpault and Trotignon 1994). The pH dependence of the kinetic dissolution of primary minerals has been studied by Chou and Wollast (1984), Murphy and Helgeson (1987) and Acker and Bricker (1992). Sjöberg (1989) included documentation of the effects of temperature. Welch and Ullman (1993) reported increasing dissolution rates attributable to organic substances. From combined laboratory and field studies, Lundström (1990) suggested that weathering rates increased 2–3.5 times in the presence of natural dissolved organics.

Field weathering rates are lower than laboratory determined rates for specific minerals or for uniform bulk mineralogy (e.g. Paces 1983, Velbel 1985, Schnoor 1990, Swoboda-Colberg and Drever 1992). The most common methods for calculating field weathering rates are compiled in Jacks (1990). The method of interpreting dif-

Fig 1. Investigation sites.

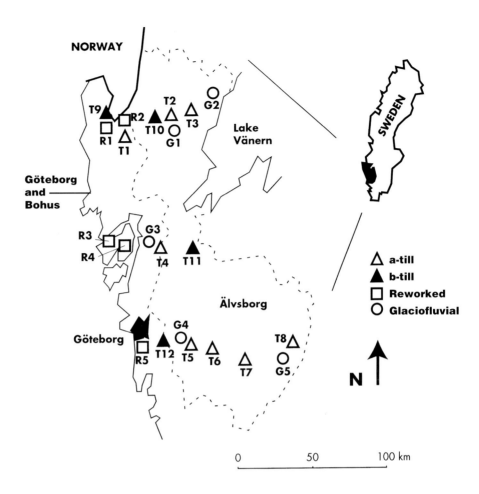

ferent $^{87}Sr/^{86}Sr$ isotope ratios in precipitation and bedrock has been used by Åberg et al. (1989), Miller et al. (1993) and other authors. Comparisons of weathering rates calculated from different starting-points include mass balance and long-term rates (Cronan 1985, Frogner 1990) and also biogeochemically modelled rates (Bain and Langan 1994).

In addition to detailed field and laboratory studies for model development, the weathering rate methods need to be tested against other criteria. Sediment mineralogy and grain-size distributions determine the potential cation source and availability for release to soil water, thus considerably delimiting weathering-rate estimates. For example, several studies describe the tendency for individual primary minerals to be dependent upon crystal size and to be concentrated in various grain-sizes, for example in tills within the Precambrian bedrock areas of Scandinavia (Lindén 1975, Nevalainen 1989, Melkerud 1991). Consideration of the geological properties of the soils is therefore essential. Geologically derived weathering rates are uncertain, but their evaluation will help to define where improvements in methods are required.

The objectives of this paper are: 1) to compare weathering rates in forest soils calculated by a long-term and a biogeochemical model, 2) to consider the rates with respect to the different coarse-grained deposits and 3) to evaluate the importance of individual minerals for the cation release. The aim is not to develop weathering-rate models or kinetic equations to describe weathering reaction rates of individual minerals. Instead, the evaluation is centred on detailed mineralogical data, which is not normally involved in weathering-rate assessments.

Applied models

In Sweden, weathering rates in forest soils have been mapped by Olsson et al. (1993), using a long-term weathering approach, and by Warfvinge and Sverdrup (1992), using the integrated soil-chemistry model PROFILE. Both methods are employed in this study.

The long-term method extrapolates the element loss within soil profiles to the time of sediment exposure (Teveldal et al. 1990, Olsson and Melkerud 1991, Bain et al. 1993). Zirconium, associated with the mineral zircon (ZrSiO$_4$), is taken as a conservative element with regard to pedogenic weathering. Initially, Marshall and Hase-

Table 1A. Assumed percentage contents of CaO and MgO in heavy minerals and phyllosilicates, from literature.

CaO	%	MgO	%
Apatite	56	Chlorite	15
Titanite	27	Hornblende	13
Epidote	22	Biotite	10
Unidentified	17	Pyroxene	10
Hornblende	12	Unidentified	7
Garnet	10	Garnet	1
Pyroxene	10		

man (1942) used zircon as a tool for a quantitative study of losses and gains of materials within a soil profile.

The PROFILE model (Sverdrup and Warfvinge 1992) calculates the final soil-water chemistry at specific acid deposition rates, as well as total soil-weathering rates, calculated on the basis of the amounts of individual minerals in the sampled soil. The model takes account of several simultaneous reactions (involving soil-water solution content of hydrogen, aluminium and base cations, CO_2 and organic acids) that occur in the soil and affect weathering rates. Since weathering reactions take place on wetted surfaces, the exposed surface area and the soil moisture content are both required for the calculations. Information on the kinetic dissolution of individual minerals at different pH was obtained from laboratory analysis and from evaluation of published data (Sverdrup 1990).

These two methods have different starting-points for calculating mineral weathering rates in forest soils. PROFILE calculates weathering rate from the amounts of

minerals left in the soils, while the long-term method is based on the amounts lost since deglaciation and the initiation of weathering.

Methods

The 22 forest soils sampled in two counties, Göteborg and Bohus, and Älvsborg (Fig. 1), are developed on glacial sediments deposited between 13 000 and 10 000 BP. Twelve sites are from tills, five from glaciofluvial deposits and five from "reworked sediments" (poorly sorted deposits below the former marine limit). Of the tills, eight sites represent the a-type (compact, possibly a basal till) and the remaining four are of b-type (less compact, possibly a supraglacial till). Samples are from three layers: the most leached horizon (E or Ao/E), a lower B horizon at 0.5 m depth, and the uppermost C horizon.

Although the limited number of soil layers sampled precludes the determination of weathering depths at each site, it is reasonable to assume that most of the biological and geochemical interactions and changes occur in the top 0.5 m. This depth has previously been used in investigations at Gårdsjön, 50 km north of Göteborg (Olsson and Melkerud 1991), for long-term weathering calculations (Olsson et al. 1993) and in soil acidification modelling in forests (DeVries et al. 1989).

Long-term weathering rates are calculated from total element concentrations, whereas PROFILE rates depend

Table 1B. Calculated amounts of CaO and MgO in the main minerals (mineral weight∗percentage content), for site T6. Total analyses by ICP from the same subsamples are included for comparison.

	Sample E		Sample B		Sample C	
	mineral weight (%)	calculated CaO/MgO (%)	mineral weight (%)	calculated CaO/MgO (%)	mineral weight (%)	calculated CaO/MgO (%)
CaO						
Epidote	2.18	0.48	1.73	0.38	2.27	0.50
Hornblende	4.69	0.56	7.88	0.95	8.56	1.03
Titanite	0.11	0.03	0.27	0.07	0.50	0.14
Garnet	0.05	0.01	0.30	0.03	0.11	0.01
Apatite	0	0	0.20	0.11	0.24	0.13
Unidentified	0.37	0.06	0.47	0.08	0.61	0.10
Feldspar $Na_2O*0.30$	2.89	0.87	2.98	0.89	3.10	0.93
Sum		2.04		2.62		2.98
ICP		1.88		2.56		2.92
MgO						
Hornblende	4.69	0.61	7.88	1.02	8.56	1.11
Garnet	0.05	0	0.30	0	0.11	0
Biotite	0.01	0	0.50	0.05	1.02	0.10
Unidentified	0.37	0.03	0.47	0.03	0.61	0.04
Sum		0.64		1.10		1.25
ICP		0.49		0.98		1.12

Table 2. The five-layer model to 0.5 m soil depth, used in PROFILE calculations.

Layer	Soil horizons	Thickness (m)	
		till, reworked	glaciofluvial
1	Oi, Oe, Ao	0.06–0.12	0.07–0.11
2	E, Ao/E	0.01–0.05	0.01–0.03
3	Bh, E/B, A/B	0.01–0.11	0.02–0.26
4	Bs1	0.07–0.30	0.10–0.30
5	Bs2, Bs3	0.04–0.29	0.04–0.22

on individual mineral contents. Three mineralogical and geochemical analyses have been selected for comparison:

A. The heavy mineral and phyllosilicate contents in the coarse-silt fraction (grain-size of 45–63 μm) were determined by petrographic microscopy and the contents reported in terms of number of grains. This fraction is suitable for optical identification and it includes almost only monomineralic grains. It was selected because weathering is expected to be most efficient in fine-grained fractions, because of the large exposed surface area. For the a-tills studied, c. 1/3 (by weight) of the < 2

Table 3. Parameters required for PROFILE calculations. A. Measured at each site. B. Information from regional maps. C. Information from the literature (specified in Table 4).

		Range	
		layer 2	layer 5
A	Coarse-silt mineralogy (%):		
	Hornblende	0.4–9.2	1.1–7.0
	Epidote	1.2–15.0	0.9–14.0
	Garnet	0 –0.7	0 –0.6
	Pyroxene	0 –0.3	0 –0.4
	Apatite	0 –0.2	0 –0.9
	Biotite	0 –5.9	0.4–31.8
	Muscovite	0 –1.9	0 –1.4
	Chlorite	0	0 –0.2
	Surface area $*10^6$ ($m^2 m^{-3}$)	0.9–2.2	0.6–3.3
B	Temperature (°C)	5.5–7.3	Tessler (1972)
	Precipitation (m yr^{-1})	0.80–1.08	Eriksson (1980)
	Runoff (m yr^{-1})	0.32–0.53	Eriksson (1980)
	Deposition of SO_4 (meq $m^{-2} yr^{-1}$)	70–116	Lövblad et al. (1992)
	Deposition of NO_3, NH_4 (meq $m^{-2} yr^{-1}$)	32–64	Lövblad et al. (1992)
	Deposition of non marine Mg+Ca+K (meq $m^{-2} yr^{-1}$)	10–35	Lövblad et al. (1992)
C	Base cation uptake	Nitrification rates	
	Nitrogen uptake	Soil moisture	
	Layer percolation	Soil bulk density	
	CO_2	DOC	
	log K	pK H/Al	
	pK H/Ca		

mm material is finer grained than the 45–63 μm fraction while 2/3 is coarser. For the glaciofluvial deposits, this fraction represents the finest portion of the sediment distribution.

The grains were density-separated at 2.68 g cm^{-3}, using heavy liquids (tetrabromethane or polytungstate), and mounted on glass slides. Counts of the heavy-mineral component included 300–320 non-opaque heavy-mineral grains and accompanying phyllosilicates, opaque and light minerals. Also, > 2000 grains were counted in each light-mineral component, in order to obtain the total sample contents. The contribution from the light-mineral component was calculated using published individual mineral densities and the weight-ratios between the light and the heavy components. Guidelines for optical identification of grains are available in Krumbein and Pettijohn (1938) and Mange and Maurer (1992).

B. Element analyses of the coarse-silt (45–63 μm) and the < 2 mm fractions by ICP (plasma-emission spectrometry) were done for major elements (SiO_2, TiO_2, Al_2O_3, Fe_2O_3, MnO, MgO, CaO, Na_2O, K_2O, P_2O_5) and some trace elements (Ba, Sc, Sr, Y and Zr). The results were recalculated to 100% dry-matter compensated for loss on ignition (LOI) (Melkerud et al. 1992). All element contents related to the ICP analyses and comparisons to this data are given in the oxide form.

C. Calciumoxide and MgO contents calculated from the coarse-silt mineralogy are based upon published values (Table 1A). Since there have been no comprehensive studies of heavy mineral and phyllosilicate geochemical composition of the study area, the mineral compositions of varieties of granodiorite (Deer et al. 1992) are used to approximate the overall bedrock source conditions. Ahlin (1976) and Eliasson (1993) are additional sources for the mineral geochemical compositions.

The mineral counts were transformed to weight percentages, using the individual mineral densities, and subsequently calculated to amount of oxides (Table 1B). All amphiboles are assigned to hornblende. No corrections for grain shape have been made. About 5% of the heavy minerals were not satisfactorily identified, mainly because of precipitate cover. However, there are indications that grains from the epidote and amphibole groups strongly dominate and the CaO and MgO contents of unidentified grains were estimated from the average composition of these two minerals. Unidentified grains were of very limited importance in the CaO and MgO calculations, as shown by the example in Table 1B.

Magnesiumoxide is associated only with heavy minerals and phyllosilicates. Plagioclase is an important additional source for CaO. Assuming all feldspar Na_2O to be within plagioclase feldspar, and excluding up to 2% of Na_2O in hornblende, the feldspar CaO content was estimated for each sample from the ICP-analyses by multiplying the Na_2O content by 0.30 (using the Na_2O/CaO ratio for oligoclase). The total CaO is then the amount calculated from the heavy mineral and phyllosilicate geochemistry plus that from the plagioclase CaO estimate.

Table 4. Specified values for the layers for each site (not measured in this study). *Values for glaciofluvial sites. Applied moisture content for all layers is 0.2.

Layer	soil bulk density (kg m^{-3})	CO_2 (atm)	% water entering-leaving	uptake % of max (Valid for Mg+Ca+K, and for N)	DOC (mg l^{-1})	log K gibbsite	pK H/Al exchange	pK H/Ca exchange
1	400	3	100–90, 100–94*	20	24	6.5	3.0	2.0
2	1000	5	90–82, 94–90*	15	20	7.5	3.5	3.0
3	1200	8	82–72, 90–84*	20	16	8.0	4.0	3.0
4	1200	10	72–60, 84–76*	25	12	8.2	4.5	4.0
5	1400	15	60–50, 76–70*	20	8	8.5	5.0	4.0

Long-term weathering

We make use of the routine of Olsson et al. (1993) for calculating long-term weathering rates for base cations, using Zr as the reference element. Regression equations are developed that depend on correlations between the element losses and site properties, such as geochemical composition in < 2 mm material, and climate (expressed as a temperature sum in Odin et al. 1983). The annual mean weathering rates (W), (in mg m^{-2} yr^{-1}), are calculated for Ca, Mg, K and Na according to Olsson et al. (1993, Na unpubl.) as:

$$W_{Ca} = -111.16 + 0.260\ X_{Ca} \quad W_{Mg} = -29.28 + 0.285\ X_{Mg}$$
$$W_K = -311.89 + 0.208\ X_K \quad W_{Na} = -113.29 + 0.189\ X_{Na}$$

where X = temperature sum*concentration of the element (%) in < 2 mm material at 0.5 m depth.

Calculations with PROFILE

PROFILE total weathering rates were calculated by the 2.3 Macintosh-version. Podsol development at all study sites motivated the use of a five-layer soil model for the calculations (Table 2). Tables 3 and 4 summarise parameters, information sources and applied values.

Mineralogical analyses were done for soil layers 2 (E horizon) and 5 (lower B horizon at 0.5 m depth). The amounts of heavy minerals and phyllosilicates are derived from the numbers of identified grains. Feldspar grains could not be properly identified, so approximate feldspar contents were calculated from ICP-analyses of the light components of the coarse-silt and fine-sand fractions by using the IGPET programme (Carr 1990). These samples are from 10 sites distributed over the area in different sediment types. The limited variation of K-feldspar and plagioclase percentages between sites allows the use of fixed ratios between quartz, K-feldspar and plagioclase. Within the soil profiles, the mean amounts of K-feldspar are fairly similar, whereas the plagioclase increases slightly with depth. The average K-feldspar percentage of the light minerals is 16%, a value adopted for all layers. Plagioclase (oligoclase composition as-

sumed) was set to 28% and 31% for layers 2 and 5, respectively. These amounts of K-feldspar, plagioclase and quartz were subsequently multiplied by the percentage of light minerals in the individual sample to obtain the sample content.

The mineral contents of layers 1, 3 and 4 are estimated. Layer 1 mineralogy for PROFILE was taken as half the content of layer 2 (except for K-feldspar, with the same content) because long-term weathering was probably more intense in the organically rich layer 1. Largely because of the small sediment surface area in layer 1, cation contributions are quite limited within the PRO-FILE total weathering rates of the complete 0.5 m column. The mineral contents of layer 3 and 4 were calculated using ratios of the contents (L) in layer 2 and 5, as: $L_3 = 2/3 L_2 + 1/3 L_5$ and $L_4 = 1/3 L_2 + 2/3\ L_5$.

Surface areas were estimated from grain-size distributions in layers 2 and 5 and correlation factors for the sand, silt and clay fractions, according to Sverdrup et al. (1990). A layer 1 value of $5*10^5$ m^2 m^{-3} was taken for all sites and values for layers 3 and 4 were again obtained by interpolation.

The mean annual air temperature of 1931–60 is taken to approximate temperature. The total atmospheric deposition of sulfur and nitrogen to forest and non-marine base cations refers to the $50*50$ km EMEP-system grid values. Regional throughfall data (Hallgren Larsson and Westling 1992) suggest equal NO_3 and NH_4 deposition. Deposition in spruce forests is different from that in mixed pine and deciduous forests (Lövblad et al. 1992). The deposition map for Norway spruce was used when spruce comprises > 70% of the forest. At a few sites the spruce content is 30–70% and the remainder is Scots pine and deciduous. We then used an average value for deposition.

Values for maximum tree uptake of base cations and nitrogen are data compiled in Sverdrup et al. (1990), where 0.5 for base cations corresponds to Norway spruce in coastal sites in southern Sweden (0.4 is used for eight higher altitude sites, 150–300 m a.s.l.). If Scots pine makes up > 40% of the forest, the reduction in maximum uptake is 15%. The values for nitrogen maximum uptake equal the base cation uptake. Nitrification rates of 0.01 kmol m^{-3} yr^{-1} represent a medium rate in the model.

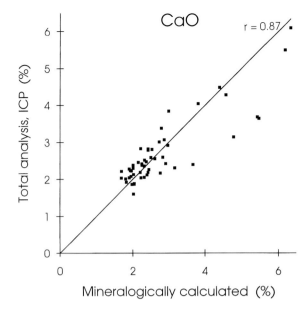

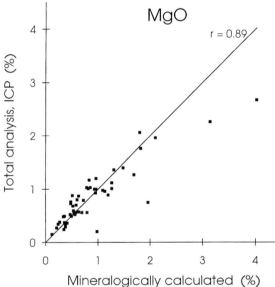

Fig. 2. Relationships for MgO and CaO between total analyses (ICP) and mineralogically calculated in the coarse-silt fraction. Number of analyses are 51 for MgO and 52 for CaO.

Table 4 shows standardised input data for the remaining parameters for all sites. Moisture content, percolation between layers, CO_2 pressure and DOC are factors that need to be extensively monitored to obtain reliable site values.

After calculation of total weathering rates by PRO-FILE, the proportion of Ca, Mg, K and Na depletion was estimated. This procedure utilises average composition of principal minerals and assumes a similarity in cation release conditions as well as dissolution with no imme-

diate clay mineral formation. Contributions (in equivalent units) from mineral weathering are set to:

$$Ca_{tot} = \text{apatite} + \text{epidote} + 0.4*\text{hornblende} + 0.4*\text{oligoclase}$$
$$Mg_{tot} = 0.6*\text{hornblende} + 0.8*\text{biotite}$$
$$K_{tot} = \text{K-feldspar} + 0.2*\text{biotite}$$
$$Na_{tot} = 0.6*\text{oligoclase}$$

Results and discussion

Mineralogical and geochemical comparisons

Calculated MgO and CaO contents compare well with those from ICP analyses (Fig. 2). Hornblende usually accounts for $> 80\%$ of the MgO content. Opaque minerals (ilmenite and magnetite) may contain minor amounts ($< 0.05\%$) of MgO not accounted for. The plagioclase CaO (estimated from the Na_2O content) makes up c. 36% (s.d. = 10%) of the calculated CaO content.

The few samples that clearly deviate, with higher calculated values, are for MgO in samples with biotite contents of 15% or more and for CaO in four glaciofluvial samples with high heavy-mineral (especially epidote) contents. Grain-shape and mineral-size variations are likely to influence MgO and CaO within the analysed coarse-silt fraction. Estimates of the degree of importance require grain measurements. Thin biotite grains probably exaggerate the calculated MgO, and epidote grains smaller than the average grain-size fraction may overestimate the calculated CaO in samples with large epidote contents.

In tills, there are good relationships between the cation element contents of the coarse-silt fraction and the < 2 mm material. There is slightly more CaO in the coarse-

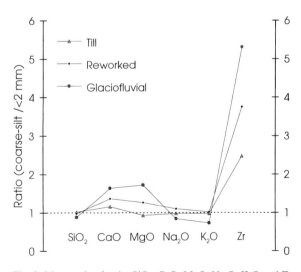

Fig. 3. Mean ratios for the SiO_2, CaO, MgO, Na_2O, K_2O and Zr contents for the coarse-silt and < 2 mm fractions. Analyses include 44 till, 10 reworked and 8 glaciofluvial samples.

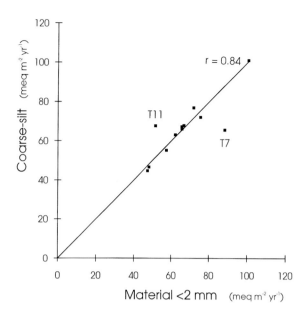

Fig. 4. Relationship of long-term total weathering rates derived from both the coarse-silt and the < 2 mm material.

silt (Fig. 3). In the reworked samples the contents are also fairly similar, although less so than for the tills. In glaciofluvial samples, the coarse-silt CaO and MgO contents are considerably higher than in the <2 mm material, while SiO_2, Na_2O and K_2O contents are lower.

The enrichment of Zr is substantial in the coarse-silt fraction in all three sediment types. This trend is most certainly caused by the original crystal size of zircon in the bedrock sources. Mange and Maurer (1992) point out that zircon tends to occur as small grains. Relative Zr concentration consistency between different size fractions in soil profiles must be taken into account in detailed consideration of long-term weathering. The median values for the element ratios are almost identical to the mean values in Fig 3. Zirconium content is lower for reworked and glaciofluvial sediments.

The mineralogical and geochemical relationships in Figs 2 and 3 suggest that: 1) The mineral composition of the coarse-silt fraction is verified by the geochemical data and is applicable for PROFILE calculations. 2) In till soils the average cation contents are quite similar in the coarse-silt and the <2 mm material.

The last observation does not imply that the two fractions have the same mineral assemblages. However, the calculated long-term total weathering rates (the sum of Ca, Mg, K and Na release) for the two fractions at twelve till sites (Fig. 4), correlate well (r = 0.84; r = 0.99, omitting sites T7 and T11). This relationship suggests that for 10 till sites the <2 mm material can be used as an alternative to coarse-silt estimates for long-term calculations.

Total weathering rates

Total weathering rates calculated by the long-term method and the PROFILE model are within the same order of magnitude, as were calculated of both long-term and PROFILE rates for Scottish catchments (Bain and Langan 1994). The weathering rates for tills are comparable to those presented in regional maps for the area (Ca, Mg and K) by Olsson et al. (1993), using the long-term weathering method, but are higher than those calculated by Warfvinge and Sverdrup (1992) with the PROFILE model.

On average, the PROFILE calculated rates are 15 meq m^{-2} yr^{-1} higher than the long-term rates for the 10 selected till sites (Fig. 5). Sverdrup and Warfvinge (1991) demonstrate an average difference of 12 meq m^{-2} yr^{-1} between the two methods (from the sum of Ca and Mg contents). Figure 5 shows that the long-term rates fall within a narrower range than do the PROFILE rates and that the correlation between the weathering rates is weak. PROFILE rates are considerably more sensitive to the specific site conditions, which can only partly be illustrated with the present data.

PROFILE calculates total weathering rates between 45 and 150 meq m^{-2} yr^{-1} (Fig. 6). The maximum rates are for R3, mainly because of its fine-grained composition, and for T10, with a high heavy-mineral content. The a-till rates are 64–82 meq m^{-2} yr^{-1}, with the exception of site T3 (45 meq m^{-2} yr^{-1}) which has a low heavy-mineral content. The a-till rates are similar throughout the study area and can provide a reference for considering rate variations between the sediment types. The tendency for higher rates in b-tills than in adjacent a-tills may be

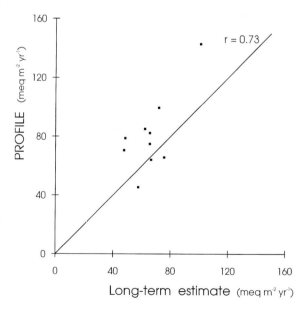

Fig. 5. Relationship between total weathering rates for 10 till sites, calculated by PROFILE and by the long-term weathering method.

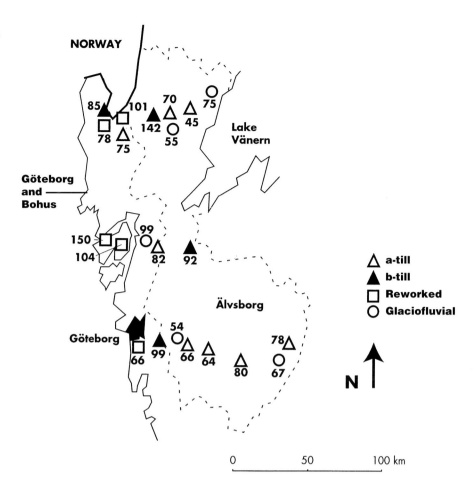

Fig. 6. Regional distribution of total weathering rates (in meq m^{-2} yr^{-1}) calculated by PROFILE.

related to the deposition conditions e.g. b-tills have presumably been more influenced by meltwaters as they have a less compact structure.

The reworked sites R2, R3 and R4 have higher rates than the a-tills and the sandy R1 and R5 sites have similar rates. Large variations in the fine fraction content influence the calculated PROFILE weathering rates for the reworked sediments. Thus, in areas where poorly sorted sandy-silty reworked sediments prevail, the PROFILE rates are presumed to be of the same order as those for tills within the same mineral environment.

The soils on glaciofluvial sediments have comparable or only slightly lower total weathering rates than the a-tills. However, there are two main differences in the geological conditions for the glaciofluvial deposits: 1) the substantially smaller surface area lowers the rates, although pedogenic formation of clay and fine-silt in the studied profiles implies a greater surface area than in the original deposit, 2) more Ca and Mg in heavy minerals in the coarse-silt fraction give higher calculated PROFILE weathering rates.

Unpublished data for G2 and G5 sites also demonstrate comparable Ca and Mg enrichment in the fine sand in all samples, independent of soil layer. Mineral sorting effects during sediment deposition are a plausible explanation for this.

Sensitivity analyses of PROFILE for mineralogy and surface area

Of the individual minerals, epidote and oligoclase together account for 80–83% of the PROFILE total weathering rates (Fig. 7). Hornblende, K-feldspar and biotite weathering each account for 3–7%, apatite for 2–3% and garnet for 1%. The contribution from pyroxene, muscovite and chlorite only occasionally exceeds 0.1%. Till and reworked sites have quite similar mineral distributions. Glaciofluvial sites have more epidote and biotite and less feldspars.

Figure 8 illustrates how the total weathering rates vary when mineralogy and surface area changes, for till sites T2 and T6. Although these till locations differ in geography and in heavy-mineral assemblages (epidote dominates in T2 and hornblende in T6), the total weathering rates are similar. Values up to 50% higher or lower than in the original calculations are set for a specific mineral or surface area, keeping all other parameters constant. An

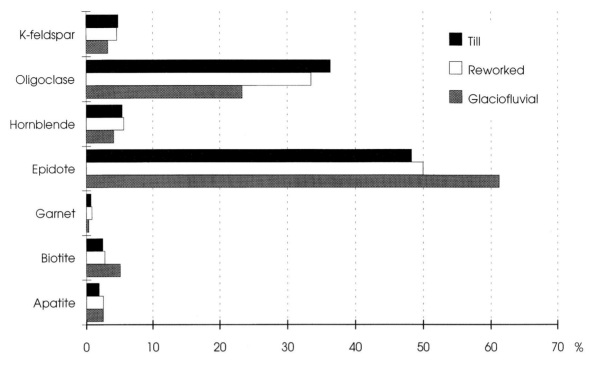

Fig. 7. Individual mineral contribution to total weathering rates (PROFILE calculations, including Ca, Mg, K and Na). Based on mean values.

increase in the amount of a mineral is balanced by a corresponding decrease in quartz content.

Changing epidote or oligoclase contents by 50% alters total weathering rates by 10–15 meq m^{-2} yr^{-1}. For K-feldspar and hornblende, the analogous response is <5 meq m^{-2} yr^{-1}. Epidote is possibly the most dependent of the main heavy minerals on grain-size, in the mineralogical environment of SW Sweden. It often occurs as small grains in the bedrock. However, no studies of epidote grain-size distribution in bedrock or quaternary deposits have been made. For example, in 12 samples the epidote content was c. 45% higher in the analysed coarse-silt fraction than in the fine-sand of 74–125 μm (which corresponds to 1% higher in terms of total mineral amounts). If the epidote content is lowered by 1%, the PROFILE total weathering rate will decrease c. 10 meq m^{-2} yr^{-1}.

Apatite and biotite affect the total weathering rates simularly (2.1 and 2.2 meq m^{-2} yr^{-1}, respectively). Apatite content is 0.2% or less in the coarse-silt fraction of the two test sites. A higher concentration of apatite would increase total weathering rates considerably. For example, 1% apatite throughout the T2 profile would result in a total weathering rate of 13 meq m^{-2} yr^{-1}.

For biotite, the effect upon total weathering rates varies much more (0.4–14.7 meq m^{-2} yr^{-1}), and exceptionally high biotite values may be expected to exaggerate the total weathering rates. Soil contents of phyllosilicates can be quite difficult to determine, because of their form, grain-size dependence, variable composition and density.

Table 5 shows the significance of large errors in biotite contents for PROFILE total weathering rates. The rates for sites T1, T10 and G2 (with the highest biotite contents), are compared to rates derived by setting a maximum 5% biotite throughout the three profiles. This operation includes recalculation of (and consequently increase in) all other mineral contents requested in PROFILE. The results show that a slight decrease in the total weathering rates occurs for the two till sites with the lower-biotite alternative. In contrast, the G2 rate increases, because 5% biotite gives higher epidote contents, which contribute relatively more to the total weathering rate than the decreasing biotite content. The three examples show that considerable variations in biotite have only a limited influence on the PROFILE calculations.

The PROFILE-model includes vermiculite as a weatherable mineral. Vermiculite is almost always present in the clay fraction of forest podsol soils (e.g. April et al. 1986, Bain et al. 1990) and is recorded within the study area at Gårdsjön (Melkerud 1984, Schweda et al. 1991). Adding a vermiculite content of 50% of the clay fraction of layer 2 and 20% of layer 5 in site T6, which has a clay content of 11%, causes the total calculated weathering rate to increase by 1.0 meq m^{-2} yr^{-1}, or by 1.6%.

Titanite (sphene) occurs in all analysed samples (up to 0.9%). Titanite is the only mineral present with a potential of weatherable cations (Ca) that has not been included in the model calculations.

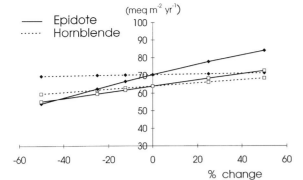

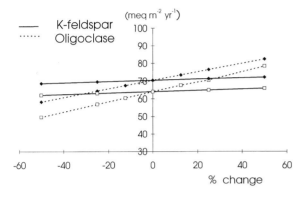

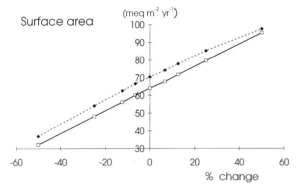

Fig. 8. Sensitivity analysis of total weathering rates at sites T2 and T6 calculated by PROFILE for main minerals and surface area. X values are the assessed increase and decrease in percentage of the initial values (O). ◆ = T2 and □ = T6.

The PROFILE total weathering rates vary greatly with values of the exposed surface area (Fig. 8). The grain-size distributions of the till sites included in this study are representative for the fine-sandy tills which predominate in SW Sweden. Clay and silt occur more frequently in the upper soil horizons. The pedogenic influence of grain size is exemplified by the 22% decrease in average total weathering rates for tills, when calculated with a grain size distribution similar to the C horizon for the whole soil profile. Silty-sandy aeolian deposits may in places contribute to the increased fine-grain content in superficial layers.

Table 5. Comparison between PROFILE total weathering rates calculated with original (A) and lower (B) biotite contents.

	SITE		
	T1	T10	G2
Biotite content in layer 2 (%)	1.9	3.2	5.9
Biotite content in layer 5 (%)	15.6	12.2	31.8
A. Total weathering rates (meq m^{-2} yr^{-1})	74.7	142.5	74.6
B. Total weathering rates with max 5% biotite (meq m^{-2} yr^{-1})	71.8	141.5	79.5
Difference (B–A) (%)	–3.9	–0.7	+6.6

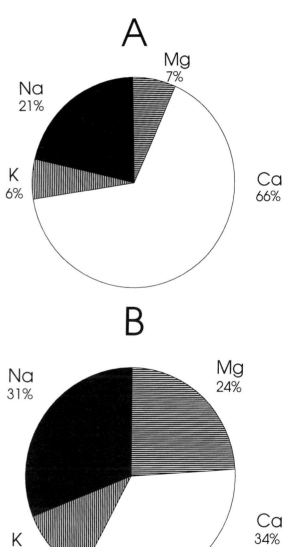

Fig. 9. The average Ca, Mg, K and Na contribution to the total weathering rates for 10 till sites. A. PROFILE. B. Long-term weathering.

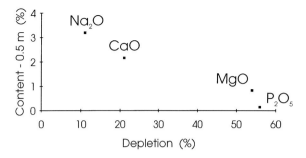

Fig. 10. Average depletion percentages for P_2O_5, MgO, CaO and Na$_2$O at six till sites. Element depletion is calculated from the contents of the two levels, using the ratio: (0.50 m - E)/ 0.50 m, for individual sites.

Available data restricts the evaluation of PROFILE parameters to mineralogical composition and surface area. Other main parameters in the model may affect the total weathering rates to the same or greater extent. For example, the total weathering rates increase 10% with an assumed mean temperature addition for T2 and T6 of 25%. Moisture content and soil bulk density are also crucial, although approximated with standard values in this paper, and an increase or a reduction of 25% in either parameter corresponds to a change of 20-25% in the total weathering rate. Warfvinge and Sandén (1992) also stress the importance of moisture content, soil bulk density and exposed surface area in PROFILE calculations.

Depletion of base cations and minerals

In addition to the weathering rate calculations, the data can also be used to illustrate the long-term mineral depletion by: A) comparing the percentage contents of specific cations calculated with PROFILE and the long-term method, and B) estimating Ca-mineral depletion from geochemical data. Calculations were made for the previously selected 10 till sites.

Figure 9 shows the percentage contents of Mg, Ca, K and Na in long-term and PROFILE total weathering rate calculations. The evaluation for PROFILE includes only feldspars, apatite, epidote, hornblende and biotite, since the other minerals contributed insignificant amounts of cations. Release of Ca strongly dominates in the PRO-FILE calculations (9A) whereas Mg and, to a lesser extent K release, is more important for the long-term rates (9B). The PROFILE model computed on average slightly higher total rates than the long-term method for the 10 till sites. Hence the calculated weathering rates for Na and K are fairly comparable for the two methods. The main difference is in the Ca and Mg release.

The long-term depletion of Ca minerals can be roughly estimated by geochemical analyses of <2 mm material. The Ca minerals of importance (except for epidote) also contain any of the three elements Mg, Na and P. The evaluation relies on simultaneous relative increase in

CaO, MgO, Na$_2$O and P_2O_5 and decrease in SiO_2 and Zr analysed contents between the leached E horizon and 0.5 m soil depth. Six sites (T1, T2, T4, T5, T6 and T12) of the 10 till sites fulfil all the geochemical requirements and are considered suitable for assessment of mineral decomposition. The differences in MgO, Na$_2$O and P_2O_5 sample contents are used to calculate CaO depletion for individual sites from chemical compositions of the predominating minerals. The Ca analyses are used to check the depletion estimates.

A rough approximation of the biotite and hornblende contribution to the Mg weathering is made from the content in the coarse-silt fraction. The hornblende MgO decrease is converted to CaO depletion using the CaO/ MgO composition ratio of 12/13. The difference in P_2O_5 multiplied by 1.3 gives the apatite CaO depletion. Finally, the plagioclase CaO portion is derived from the Na-depletion by Na$_2$O*0.30, assuming Na and Ca release in proportion to the amounts in plagioclase. No correction for Na-hornblende depletion is included.

The average degree of depletion is not proportional to the element concentrations (Fig. 10), reflecting unequal mineral-weathering rates. The most pronounced depletion is of P_2O_5 and MgO. The amount of P_2O_5 present in the leached E horizons of the six till soils averages only 0.08%.

Hornblende accounts for 61–96% of the MgO depletion. The mean calculated CaO depletion is c. 20% higher than the analysed CaO (only hornblende, apatite and plagioclase taken into account), implying a fairly good consistency between the two CaO values. Figure 11 shows the contribution to the total Ca depletion from the different minerals. Hornblende is found to be the dominant mineral of the Ca depletion.

The calculation ignores Na depletion in hornblende. Sodiumoxide is c. 15–20% of CaO in hornblende. Assuming weathering of Na and Ca from hornblende to be proportional to their concentrations permits utilisation of depletion data (Fig. 10). Figure 11 indicates that hornblende accounts for about half of the Ca depletion, corresponding to 0.22% CaO or 0.03% Na$_2$O (10% of the Na$_2$O depletion). The overestimation of the plagioclase in the Ca depletion would be c. 2–4%.

In spite of the number of uncertain data and necessary

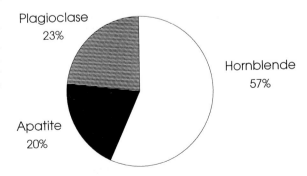

Fig. 11. Summary of long-term Ca mineral weathering in tills.

approximations, reasonable agreement between the calculated and the analysed values of Ca depletion suggests that the relative distribution of the main Ca minerals in the long-term depletion trends is fairly reliable. The relative mineral weathering can be estimated by dividing the mineral sources for Ca depletion by the average amounts in the C horizons of the till sites (apatite 0.25%, hornblende 4% and Ca-plagioclase 7%). An approximate relationship derived for the long-term weathering contributions from apatite, hornblende and plagioclase is 100:10:5.

Hornblende, epidote and saussurite

Hornblende is a major long-term Ca source in the area. This is consistent with the considerable Mg depletion in Fig. 10, as well as results from the Kalix river watershed in northern Sweden (Öhlander et al. 1991). For the present calculations hornblende is the predominant depleted heavy-mineral whereas epidote is the most important mineral in the PROFILE calculations. The influence of epidote on the long-term weathering cannot be evaluated, but the estimate above suggests that hornblende, apatite and plagioclase alone can account for most of the Ca depletion in the tills. Warfvinge and Sverdrup (1992) calculated average epidote contents of 1–2% from geochemical analyses for the study area. Considerably higher epidote contents are applied in the present calculations and different epidote values are a probable reason for the obtained differences in PROFILE total weathering rates.

An important example of mineral transformation that must be considered is the formation of very fine-grained saussurite (epidote composition) from plagioclase, affecting the Ca release. Saussurite occurs frequently in SW Sweden as a result of the repeated metamorphic activity (e.g. Samuelsson 1985). Snäll (1992) notes the problems arising in the laboratory when density-separating plagioclase from epidote with pronounced saussurite formation. This source of error is accounted for here by analyses of both heavy and light components. In the coarse-silt fraction, saussurite is quite frequently found on the plagioclase grain surfaces as very small individual grains or in aggregations. Detached grains also occur and were classified as epidote. Saussurite formation presumably increases the surface area, important during plagioclase/epidote weathering and for the Ca release rates. Saussurite is an illustrative example of the interdependence between mineral composition, grain-sizes and active surface area, all factors of significance for weathering rate assessments.

Conclusions

For the long-term calculations of total weathering rates in tills, the coarse-silt fraction used in this study gave results that were consistent with the bulk geochemistry. However, the control of the mineralogical grain-size dependence is insufficient for detailed applications of the PROFILE model.

Calculated long-term and PROFILE total weathering rates for the upper 0.5 m of till soil were of the same order of magnitude. Since PROFILE calculates cumulative total weathering rates with soil depth, the selected depth will determine the PROFILE rates. Both methods are under continuous revision and the calculated rates must be taken as current best estimates.

The PROFILE model calculates similar weathering rates for the well sorted glaciofluvial deposits and for the tills. The total weathering rates may be overestimates, particularly in the glaciofluvial deposits, because of the use of a size fraction likely to be enriched in heavy minerals. However, the high content of heavy minerals in the silt and sand fractions of permeable material and the successive increase of weatherable mineral surface resulting from grain degradation promotes mineral weathering. This study suggests that in areas where epidote (and saussurite) is common, such as SW Sweden, the mineral size sorting and the selection of fractions in which weathering is most efficient are significant for the PROFILE total weathering rates.

The most important mineralogical feature is the extent of depletion of hornblende, apatite and, most probably, biotite in the uppermost soil layers by long-term decomposition. In addition to hornblende and apatite, the abundance of epidote, saussurite and plagioclase determine the Ca release. For estimating individual cation release, it is crucial to consider whether the model relates weathering rates to the loss of minerals and cations since deglaciation or to what is still left in the soil. Using present mineralogy (as in the PROFILE model) would give lower rates of Mg and K weathering compared to the mean long-term weathering.

To increase the accuracy of critical-load estimates, it is essential in sensitivity analyses of current field weathering rates to consider the interdependence of mineralogy and grain size.

Acknowledgements – This work has been supported by the Swedish Environmental Protection Agency, and by a research scholarship at the Göteborg Univ., Sweden. I wish to thank L. Börjesson, L. Karlsson and B. Svenhede-Lång for laboratory assistance, L. Lundqvist for information on mineral chemistry and R. Stevens and E.-L. Tullborg for critically reviewing the manuscript. H. Sverdrup and P. Warfvinge made the utilisation of the PROFILE model possible.

References

Åberg, G. A., Jacks, G. and Hamilton, P. J. 1989. Weathering rates and [87]Sr/[86]Sr ratios: An isotopic approach. – J. Hydrol. 109: 65–78.

Acker, J. G. and Bricker, O. P. 1992. The influence of pH on biotite dissolution and alteration kinetics at low temperature. – Geochim. Cosmochim. Acta 56: 3073–3092.

Ahlin, S. 1976. The compositional relationships of biotite and garnet in the Göteborg area of south-western Sweden and their thermometric implications. – Geol. Fören. Stockh. Förh. 98: 337–342.

April, R. H., Hluchy, M. M. and Newton, R. M. 1986. The nature of vermiculite in Adirondack soils and till. – Clays Clay Min. 34: 549–556.

Bain, D. C. and Langan, S. J. 1994. Comparison of methods of calculation of weathering rates in soils as applied to catchments. – In: Helios Rybicka, E. and Sikora, W. S. (eds), Abstracts, 3rd Int. Symp. Environ. Geochem., Univ. Mining and Metallurgy Krakow, Fac. Geol., Geophysics Environ. Protec., Krakow, pp. 27–28.

–, Mellor, A. and Wilson, M. J. 1990. Nature and origin of an aluminous vermiculitic weathering product in acid soils from upland catchments in Scotland. – Clay Min. 25: 467–475.

–, Mellor, A., Robertson-Rintoul, M. S. E. and Buckland, S. T. 1993. Variations in weathering processes and rates with time in a chronosequence of soils from Glen Feshie, Scotland. – Geoderma 57: 275–293.

Berner, R. A., Sjöberg, E. L., Velbel, M. A. and Krom, M. D. 1980. Dissolution of pyroxenes and amphiboles during weathering. – Science 207: 1205–1206.

Bricker, O. P., Paces, T., Johnson, C. and Sverdrup, H. 1994. Weathering and erosion aspects of small catchment research. – In: Moldan, B. and Cerny, J. (eds), Biogeochemistry of small catchments: A tool for environmental research. John Wiley and Sons, Chichester, pp. 85–105.

Carr, M. J. 1990. IGPET 3.0. Unpublished manual. – Terra Softa, Somerset, New Jersey.

Chou, L. and Wollast, R. 1984. Study of the weathering of albite at room temperature and pressure with a fluidized bed reactor. – Geochim. Cosmochim. Acta 48: 2205–2217.

Cronan, C. S. 1985. Chemical weathering and solution chemistry in acid forest soils: Differential influence of soil type, biotic processes, and H[+] deposition. – In: Drever, J. I. (ed.), The chemistry of weathering. D. Reidel, Holland, pp. 175–195.

De Vries, W., Posch, M. and Kämäri, J. 1989. Simulation of the long-term soil response to acid deposition in various buffer ranges. – Water Air Soil Pollut. 48: 349–390.

Deer, W. A., Howie, R. A. and Zussman, J. 1992. An introduction to the rock-forming minerals. 2nd ed. – Longman Scientific and Technical, Harlow.

Eliasson, T. 1993. Mineralogy, geochemistry and petrophysics of the red coloured granite adjacent to fractures. – SKB Technical Report 93–06, Swedish Nuclear Fuel and Waste Management, Stockholm.

Eriksson, B. 1980. The water balance of Sweden. Annual mean values (1931–60) of precipitation, evaporation and run-off. – Reports RMK nr 18, RHO nr 21, SMHI, Norrköping, (in Swedish, English summary).

Frogner, T. 1990. The effect of acid deposition on cation fluxes in artificially acidified catchments in western Norway. – Geochim. Cosmochim. Acta 54: 769–780.

Fung, P. C. and Sanipelli, G. G. 1982. Surface studies of feldspar dissolution using surface replication combined with electron microscopic and spectroscopic techniques. – Geochim. Cosmochim. Acta 46: 503–512.

Hallgren Larsson, E. and Westling, O. 1992. Nedfall av luftföroreningar i Älvsborgs och Göteborgs och Bohus län. Årsrapport 1990–1991. – Miljörapport 1992:4, Länsstyrelsen i Göteborgs and Bohus län, Göteborg, (in Swedish).

Holdren Jr, G. R. and Berner, R. A. 1979. Mechanism of feldspar weathering I. Experimental studies. – Geochim. Cosmochim. Acta 43: 1161–1171.

Jacks, G. 1990. Mineral weathering studies in Scandinavia. – In: Mason, B. J. (ed.), The surface waters acidification programme. Cambridge Univ. Press, Cambridge, pp. 215–222.

Krumbein, W. C. and Pettijohn, F. J. 1938. Manual of sedimentary petrography. – Appleton-Century-Crofts, New York.

Lindén, A. 1975. Till petrographical studies in an archean bedrock area in south central Sweden. – Striae 1, Uppsala.

Lövblad, G., Andersen, B., Joffre, S., Pedersen, U., Hovmand, M. and Reissell, A. 1992. Mapping deposition of sulphur, nitrogen and base cations in the Nordic countries. – Swedish Environ. Res. Inst., Stockholm, Report B 1055.

Lundström, U. 1990. Laboratory and lysimeter studies of chemical weathering. – In: Mason, B. J. (ed.), The surface waters acidification programme. Cambridge Univ. Press, Cambridge, pp. 267–274.

Mange, M. A. and Maurer, H. F. W. 1992. Heavy minerals in colour. – Chapman and Hall, London.

Marshall, C. E. and Haseman, J. F. 1942. The quantitative evaluation of soil formation and development by heavy mineral studies: A grundy silt loam profile. – Soil Sci. Soc. Am. Proc. 7: 448–453.

Melkerud, P.-A. 1984. Distribution of clay minerals in soil profiles – A tool in chronostratigraphical and lithostratigraphical investigations of till. – Striae 20: 31–37.

– 1991. The importance of fine-textured material for chemical weathering, soil formation and aluminum mobilization. – In: Rosén, K. (ed.), Chemical weathering under field conditions. Rep. Forest Ecol. Forest Soils 63. Dept Forest Soils, Swedish Univ. Agricult. Sci., Uppsala, pp. 119–138.

–, Olsson, M. T. and Rosén, K. 1992. Geochemical atlas of Swedish forest soils. – Rep. Forest Ecol. Forest Soils 65. Dept Forest Soils, Swedish Univ. Agricult. Sci., Uppsala.

Miller, E. K., Blum, J. D. and Friedland, A. J. 1993. Determination of soil exchangeable cation loss and weathering rates using Sr isotopes. – Nature 362: 438–441.

Murphy, W. M. and Helgeson, H. C. 1987. Thermodynamic and kinetic constraints on reaction rates among minerals and aqueous solutions. III. Activated complexes and the pH-dependence of the rates of feldspar, pyroxene, wolastonite and olivine hydro lysis. – Geochim. Cosmochim. Acta 51: 3117–3153.

Nevalainen, R. 1989. Lithology of fine till fractions in the Kuhmo greenstone belt area, eastern Finland. – In: Perttunen, M. (ed.), Transport of glacial drift in Finland. Geological Survey of Finland, Special Paper 7, Espoo, pp. 59–65.

Odin, H., Eriksson, B. and Perttu, K. 1983. Temperature climate maps for Swedish forestry. – Rep. Forest Ecol. Forest Soils 45. Dept Forest Soils, Swedish Univ. Agricult. Sci., Uppsala, (in Swedish, English summary).

Öhlander, B., Ingri, J. and Pontér, C. 1991. Geochemistry of till weathering in the Kalix river watershed, northern Sweden. – In: Rosén, K. (ed.), Chemical weathering under field conditions. Rep. Forest Ecol. Forest Soils 63. Dept Forest Soils, Swedish Univ. Agricult. Sci., Uppsala, pp. 1–18.

Olsson, M. and Melkerud, P.-A. 1991. Determination of weathering rates based on geochemical properties in soil. – In: Pulkkinen, E. (ed.), Environmental geochemistry in northern Europe, Geological Survey of Finland, Special Paper 9, Espoo, pp. 69–78.

–, Rosén, K. and Melkerud, P.-A. 1993. Regional modelling of base cation losses from Swedish forest soils due to whole-tree harvesting. – Appl. Geochem., Suppl. Issue 2: 189–194.

Paces, T. 1983. Rate constants in dissolution derived from the measurements of mass balances in catchments. – Geochim. Cosmochim. Acta 47: !855–1863.

Samuelsson, L. 1985. Description to the map of solid rocks Göteborg NO. – Geological Survey of Sweden, Af 136, Uppsala, (in Swedish, English summary).

Schnoor, J. L. 1990. Kinetics of chemical weathering: A comparison of laboratory and field weathering rates. – In: Stumm, W. (ed.), Aquatic chemical kinetics: Reaction rates of processes in natural waters. John Wiley and Sons, New York, pp. 475–504.

Schweda, P., da Rocha Araujo, P. and Sjöberg, L. 1991. Soil chemistry, clay mineralogy and non-crystalline phases in

112

soil profiles from southern Sweden and Gårdsjön. – In: Rosén, K. (ed.), Chemical weathering under field conditions. Rep. Forest Ecol. Forest Soils 63. Dept Forest Soils, Swedish Univ. Agricult. Sci., Uppsala, pp. 49–62.

Sjöberg, L. 1989. Kinetics and non-stoichiometry of labradorite dissolution. – In: Miles, D. L. (ed.), Water-rock interaction, WRI-6, Malvern, A. A. Balkema, Rotterdam, pp. 639–642.

Snäll, S. 1992. Mineralogisk analys av moräner. – Geological Survey of Sweden, BRAP 92005, Uppsala, (in Swedish).

Sverdrup, H. U. 1990. The kinetics of base cation release due to chemical weathering. – Lund Univ. Press, Lund.

– and Warfvinge, P. 1991. On the geochemistry of chemical weathering. – In: Rosén, K. (ed.), Chemical weathering under field conditions. Rep. Forest Ecol. Forest Soils 63. Dept Forest Soils, Swedish Univ. Agricult. Sci., Uppsala, pp. 79–118.

– and Warfvinge, P. 1992. PROFILE – A mechanistic geochemical model for calculation of field weathering rates. – In: Kharaka, V. K. and Maest, A. S. (eds), Proc. 7th Int. Symp. water-rock interaction, A. A. Balkema, Rotterdam, pp. 585–590.

– , De Vries, W. and Henriksen, A. 1990. Mapping critical loads. – Nordic Council of Ministers, Nord 1990:98.

Swoboda-Colberg, N. G. and Drever, J. I. 1992. Mineral dissolution rates: A comparison of laboratory and field studies. – In: Kharaka, V. K. and Maest, A. S. (eds), Proc. 7th Int. Symp. water-rock interaction, A. A. Balkema, Rotterdam, pp. 115–118.

Tessler, R. 1972. Klimatdata för Sverige. – Bäckmans tryckerier, (in Swedish).

Teveldal, S., Jörgensen, P. and Stuanes, A. O. 1990. Long-term weathering of silicates in a sandy soil at Nordmoen, southern Norway. – Clay Min. 25: 447–465.

Turpault, M.-P. and Trotignon, L. 1994. The dissolution kinetics of biotite single crystals in dilute HNO_3 at 24°C: Evidence of an anisotropic corrosion process of micas in acidic solutions. – Geochim. Cosmochim. Acta 58: 2761–2775.

Velbel, M. A. 1985. Geochemical mass balances and weathering rates in forested watersheds of the southern Blue Ridge. – Amer. J. Sci. 285: 904–930.

– 1989. Weathering of hornblende to ferruginous products by a dissolution-reprecipitation mechanism: petrography and stoichiometry. – Clays Clay Min. 37: 515–524.

Warfvinge, P. and Sandén, P. 1992. Sensitivity analysis (chapter 5). – In: Sandén, P. and Warfvinge P. (eds), Modelling groundwater response to acidification. RH No. 5, SMHI, Norrköping, pp. 119–146.

– and Sverdrup, H. 1992. Modelling regional soil mineralogy and weathering rates. – In: Kharaka, V. K. and Maest, A. S. (eds), Proc. 7th Int. Symp. water-rock interaction, A. A. Balkema, Rotterdam, pp. 585–590.

Welch, S. A. and Ullman. W. J. 1993. The effect of organic acids on plagioclase dissolution rates and stoichiometry. – Geochim. Cosmoshim. Acta 57: 2725–2736.

Ecological Bulletins 44: 114–122. Copenhagen 1995

Use of the aluminium species composition in soil solution as an indicator of acidification

Ulla S. Lundström and Reiner Giesler

Lundström, U. S. and Giesler, R. 1995. Use of the aluminium species composition in soil solution as an indicator of acidification. – Ecol. Bull. (Copenhagen) 44: 114–122.

Soil solution chemistry was investigated for 3–4 yr at two Swedish sites, Svartberget (N Sweden) and Gårdsjön (SW Sweden), with different acid loads and also at Tisa in the north of the Czech Republic. At Svartberget the aluminium concentrations were highest in the eluvial horizon (35 µM) and 75% of the aluminium was organically bound. The concentrations of aluminium were low below the illuvial horizon. At Gårdsjön, the concentration of inorganic aluminium was high in the illuvial horizon, and constituted 80% of the total aluminium (70 µM), but decreased with depth. At Tisa, inorganic aluminium in high concentrations (1000 µM) leached throughout the whole profile. When the process of podzolization is occurring, organic acids form complexes with aluminium, whereby the weathering rate is enhanced and the eluvial horizon forms. In the illuvial horizon aluminium and iron are immobilized. High concentrations of aluminium, of which inorganic aluminium constitutes the largest proportion, indicate that the process of podzolization is perturbed. The equilibria of Al^{3+} with solid phases were evaluated, of which none of the tested phases seemed possible. It was concluded that the concentration and distribution of aluminium species, sulphate and pH in a soil solution profile indicate the degree of acidification and could be used in monitoring. Free drainage centrifugation and aluminium speciation were found to be useful tools.

U. S. Lundström and R. Giesler, Dept of Forest Ecology, Swedish Univ. of Agricultural Sciences, S-901 83 Umeå, Sweden.

Introduction

The anthropogenic pollutants SO_2 and NO_x have caused severe damage to forests in central Europe. In the northern part of the Czech Republic, forest trees have died over vast areas, and afforestation is difficult. Deposition in these areas is very high (e.g. sulphur 300–450 mmol m^{-2} a^{-1}, Moldan and Schnoor 1992). In Sweden, the maximum deposition of sulphur in the southern part is c. 85 mmol m^{-2} a^{-1} (Lövblad et al. 1992) but the soils are sensitive to acidification because they are mainly formed from igneous rocks (granite and gneiss).

In Sweden, the soil pH is lower in areas with high deposition than in areas with a lower pollution load (Lundmark 1989). Furthermore, the exchangeable pool of base cations is smaller in areas with acidic deposition (Nilsson 1988).

In the south of Sweden, the region most severely affected by acidic deposition, the pH in the upper soil layers has decreased by up to 1.5 pH units during the last few decades (Falkengren-Grerup 1987), and the soil acid-

ity has progressed into the C-horizon (Tamm and Hallbäcken 1988). In addition, the pool of exchangeable base cations has decreased (Falkengren-Grerup et al. 1987). In southwestern Sweden the base cation pool down to 70 cm depth decreased by about two thirds between 1927 and 1984 (Hallbäcken 1992).

Nevertheless, the annual forest production per unit area in Sweden has risen strongly since the 1920's, especially in the southern part (Nihlgård 1990). This increase probably results from the introduction of improved silvicultural techniques, but could also be the result of elevated nitrogen deposition (Eriksson and Johansson 1993).

Despite the increased soil acidification, it has been difficult to find significant evidence of acidification damage in trees. Needle loss (Wijk et al. 1993), nutrient balance in needles (Nohrstedt 1990) and root growth (Majdi and Persson 1993) have been studied in this respect. Soil chemistry may be a useful tool for monitoring acidification of the ecosystem and may reflect conditions for trees.

In general, podzolized soils develop in regions with

humid climates and medium-textured soils supporting coniferous forest. Podzolization is the major soil-forming process in Scandinavian forests. The mor layer releases organic acids, which form strong complexes with aluminium and thereby promote the weathering of primary minerals (Lundström 1993). In the upper soil layers, where these acids are present, the enhanced weathering rate results in the formation of an eluvial horizon. In this horizon, high concentrations of organically bound aluminium indicate that weathering is occurring. Deeper in the soil profile, aluminium and iron will be immobilized. Organic acids might be microbiologically degraded, and resulting aluminium and iron ions can precipitate in an aluminium-iron-silicate phase in the illuvial horizon (Lundström 1994). Complexes might also bind additional aluminium and iron, thereby becoming electrically neutral and, consequently, precipitating in the illuvial horizon (Petersen 1976).

Anthropogenic soil acidification probably perturbs the weathering process because organic acids are less dissociated at lower pHs. This, in turn, will result in a reduced complex-forming ability. The leaching of inorganic aluminium from the illuvial horizon is indicative of a disturbance of the podzolization process. Thus the distribution of Al between different horizons and the partitioning of the different aluminium species in soil solution in a profile reveal the state of the soil-forming process. Altered soil chemical processes might influence the growth of trees in the long run, e.g. laboratory experiments indicate that inorganic aluminium in soil solution interferes with the growth of tree roots (Rost-Siebert 1983, 1985, Hutchinson et al. 1986).

The soil solution chemistry of two soil profiles in Sweden, with different acidic pollution loads, were monitored for 3–4 yr. Two sites in the Czech Republic in a highly acidified area were also investigated. The soil solution chemistry was used to evaluate the degree of perturbance of the process of podzoliztion. Special attention was directed to the distribution of different aluminium species in soil solution in the profiles in order to evaluate the stage of acidification. A centrifugation drainage technique has been developed (Giesler and Lundström 1990, 1993), by which samples can be easily taken in well defined horizons and analysed without having to store them. This technique facilitates the sampling of soil solution for monitoring work. Possible solid phase equilibria were evaluated in order to explain the measured Al concentrations. The molar quotient (Ca + Mg)/(monomeric inorganic Al) in the soil solution was calculated, as it has been used to predict favourable or adverse conditions for forest growth.

Experimental

Site description and sampling

At Svartberget Forest Research Station, 70 km NW of Umeå in the north of Sweden (64°14'N, 19°46'E), three sets of percolation lysimeters were installed at ten depths in a Scots pine stand, *Pinus sylvestris*, with a mean age of 70 yr. Cores of mineral soils were taken to the different depths and each core was placed in a polyethene tube (18.7 cm diameter) with a filter at its perforated base. The tubes were installed from the side of a pit and positioned under an undisturbed mor layer. The percolated water could be pumped to the ground surface (see Lundström 1993). The soil is a haplic podzol (FAO 1988), with a 5 cm O horizon, a 1–3 cm thick eluvial horizon and an illuvial horizon, c. 30 cm thick, which has been described elsewhere (Lundström 1993). Mean annual precipitation is c. 720 mm. Samples were taken from 1987 to 1990 in May, June, August and October. Aluminium was speciated in a profile on 22 occasions for a total of 100 samples. To collect throughfall, five polyethene bottles (10 dm³) with funnels on their tops (20 cm diameter) were placed randomly around the stand. During winter the bottles with funnels were replaced with buckets to collect snowfall.

At Gårdsjön, situated close to the Swedish west coast 40 km north of Gothenburg (58°04'N, 12°03'E), samples were taken in a Norway spruce stand, *Picea abies*, 80–100 yr old, with a small component of Scots pine, *Pinus sylvestris*. The soil is a haplic podzol (FAO 1988) with O, E and B horizons 10, 5 and 40 cm thick respectively; details of this profile are in Giesler et al. (unpubl.). From the face of a pit (1 × 1 m) ten soil cores (100 g) were taken from each horizon and were then combined to form composite samples. Eight horizons down to 65 cm depth were sampled. The pit was refilled after each sampling and re-excavated at the next sampling occasion, when it was also lengthened by 20 cm. The free drainage centrifugation technique was used to obtain the soil solution (Giesler and Lundström 1990, 1993). Mean annual precipitation is c. 1000 mm. Samples were taken 3–4 times each year from 1989 to 1991. Aluminium speciation was carried out on five sampling occasions.

The sites in the Czech Republic were both situated at the northern border with Germany: Tisa (50°47'N, 14°04'E), 16 km N of Usti n. Labem, and Dl. Louka (50°39'N, 13°39'E), 30 km W of Usti n. Labem. At Tisa, the soil is a dystric cambisol (FAO 1988) formed from sandstone, and at Dl. Louka it is a haplic podzol (FAO 1988) formed from granite. Norway spruce, *Picea abies*, was the dominant tree species at both sites. At Dl. Louka the soil had been limed for several years, but at Tisa the soil was not limed. Samples from these sites were taken in the same way as at Gårdsjön, at 5–6 depths, except that the soil cores from each horizon were centrifuged separately and the soil solutions bulked for each horizon. Samples were taken once in October, 1993.

Table 1. Mineral equilibria and logK values at 25°C.

Equilibrium	Solid phase	logK	Reference
$Al(OH)_3 + 3H^+ =$	synthetic gibbsite	8.11	(May et al. 1979)
$Al^{3+} + 3H_2O$	natural gibbsite	8.77	(May et al. 1979)
	microcrystalline gibbsite	9.35	(Hem et al. 1973),
	amorph. Al-trihydroxide	10.80	(Stumm and Morgan 1981)
$1/2\ Al_2Si_2O_5(OH)_4 + 3H^+ =$	kaolinite	3.30	(Stumm and Morgan 1981)
$Al^{3+} + H_4SiO_4 + 1/2\ H_2O$	halloysite	5.64	(Stumm and Morgan 1981)
$Al(OH)SO_4 + H^+ =$	jurbanite	−3.80	(Nordström 1982).
$Al^{3+} + SO_4^{2-} + H_2O$			

Centrifugation

The centrifugation of the Gårdsjön samples was performed with a Beckman J21 high speed centrifuge and a JA 14 rotor. In the Czech Republic a portable centrifuge, Beckman GS-15R, and a S4180 rotor was used. A free drainage technique was used, in which the soil sample was placed in a soil-holding bucket with perforated bottom on top of a collecting cup (Giesler and Lundström 1993). The Gårdsjön samples were centrifuged at a speed of 14 000 rpm for 80 min and the Czech samples at a speed of 4500 rpm for 30 min at a constant temprature of 5°C.

Analyses and evaluation

Soil solution pH was measured immediately. Total concentrations of cations were determined by ICP-AES or ICP-MS (Perkin-Elmer, plasma II, Emission Spectrometer; Perkin-Elmer, Sciex Elan, 500) after acidification of the samples; anion concentrations were measured using IC (Dionex 4000i). Dissolved organic carbon (DOC) was determined by ICP using the method of Emteryd et al. (1991). At Svartberget and Gårdsjön, aluminium was speciated according to the ion-exchange method of Driscoll (1984) and determined using the method of Barnes (1975). The determined aluminium passing the ion-exchanger is referred to here as "organically bound monomeric aluminium", and "inorganic monomeric aluminium" is calculated as the difference between determined "monomeric aluminium" and organically bound monomeric aluminium. "Polymeric aluminium" is referred to the acid-digestible part. At Tisa and Dl. Louka, quickly reacting aluminium (inorganic aluminium species without Al-F complexes) was determined using the method described by Clarke et al. (1992), and total aluminium was determined by ICP. Thus the polymeric part of the total aluminium cannot be determined.

The chemical equilibrium model AlCHEMI (Schecher and Driscoll 1987, 1988) was used to speciate the inorganic aluminium species. The speciation of inorganic aluminium is temperature-dependent, because of the equilibrium constants. The influence of temperature is greatest for the aluminium hydroxy complex, the concentration of which will decline as the temperature decreases. The measured soil temperatures were used. However, the determination to distinguish between organic and inorganic aluminium was carried out at room temperature.

In the ALCHEMI model we investigated the conditions for the equilibria for synthetic gibbsite, natural gibbsite, microcrystalline gibbsite, kaolinite, amorphous aluminium trihydroxide, halloysite and for jurbanite, being the most probable sources for inorganic aluminium in soil solution. The solubility product constants are given in Table 1.

The saturation indices (SI) for the different mineral phases were calculated as the logarithm of the measured ion activity product (IAP) divided by the temperature-corrected equilibrium constant (K_{eq}), $SI = \log(IAP/K_{eq})$.

Results and discussion
Distribution of soil solution components in the soil profiles

Forest soils in northern Sweden have only been slightly affected by acid deposition. In the Svartberget research area the bulk deposition rate of sulphur is c. 13 mmol m^{-2} a^{-1}, and that of nitrogen is c. 22 mmol m^{-2} a^{-1} (Anon. 1988, 1989). The throughfall deposition of sulphur at the site at Svartberget was about the same as the bulk deposition. At Gårdsjön, on the west coast of southern Sweden, the total deposition of sulphur and nitrogen is c. 85 and 125 mmol m^{-2} a^{-1}, respectively. (Hultberg et al. 1992). In the north of the Czech Republic in Bohemia, the total deposition of sulphur is 300–450 mmol m^{-2} a^{-1} (Moldan and Schnoor 1992).

Various techniques have been used to collect the soil solution. In another study at Svartberget, samples from percolation lysimeters were compared to soil solution samples obtained by centrifugation (14 000 rpm) during two years of sampling (Giesler and Lundström unpubl.). For the parameters discussed in this article there was no significant discrepancy between the methods. Generally concentrations in soil solutions obtained by centrifugation were more consistent and varied less than those in soil solution from lysimeters. A comparison has also been made between centrifugation at different centrifugation

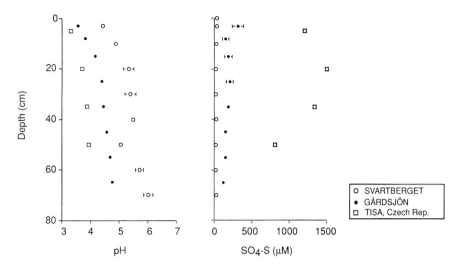

Fig. 1. Means of pH and concentrations of sulphate (μM) in soil solutions at different soil depths at Svartberget, Gårdsjön (Sweden) and Tisa in the (Czech Republic). Standard error is given by bars. At Svartberget n = 22 (fewer at some depths), at Gårdsjön n = 5–7 and at Tisa n = 1.

speeds and times, showing that the differences in soil solution concentrations were too small to affect the comparison made in this study (Giesler and Lundström 1990).

The average pH at Svartberget was 4.4–5.9 in the mineral soil throughout the profile (Fig.1). At Gårdsjön, the pH was almost 1.0 pH unit lower than that at Svartberget, 3.6–4.8 throughout the mineral soil profile. At Dl. Louka and Tisa, the sites in the Czech Republic, the pH was even lower, 3.3–4.3 throughout the mineral soil; i.e. generally c. 0.5 pH unit lower than at Gårdsjön. At Svartberget, the pH in the upper soil was probably determined largely by the organic acids leached from the mor layer.

Titrations of organic acids were made for 19 samples, and the pK_a was calculated to be 4.0–4.7 (Lundström 1993), using a monoprotic acid model. Concentrations of organic acids in the upper soil were 40–290 μM. Thus the pH exceeded the pK_a, and as a result most of the organic acids were dissociated. The low pHs in the Czech Republic and probably also at Gårdsjön were a result of the deposition of acidic, anthropogenic pollutants. At these low pHs in the upper soil, the portion of the organic acids in dissociated form will be smaller. Assuming that the pK_a at these sites is similar to that at Svartberget, only 5–20% will be dissociated in the upper soil. Thus at the

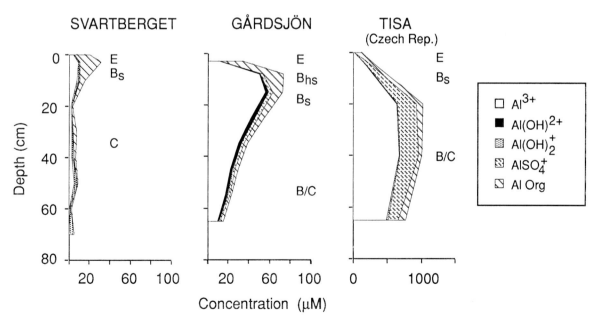

Fig. 2. The mean distribution with soil depth of different aluminium species at Svartberget (100 samples), Gårdsjön (28 samples) and Tisa (5 samples). In Tisa ▨ deenotes the sum of organically bound and polymeric aluminium. Soil horizons are indicated in the figure.

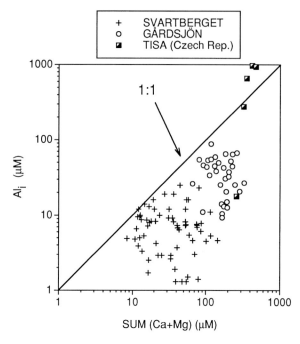

Fig. 3. Concentrations of monomeric inorganic aluminium related to the molar sum of Ca and Mg. Scales are logarithmic.

polluted sites the ability of the organic acids to form complexes in the soil solution would be reduced, provided that the concentration of the acids remain unchanged. Hereby the weathering attributable to complex formation will decrease and at sufficiently low pH, inorganic aluminium would be leached from the illuvial horizon.

The anthropogenic contribution was reflected in the sulphate concentration in soil solution, which was very high (up to 1500 µM) at both Dl. Louka and Tisa. By contrast, at Gårdsjön the concentration of sulphate was c. 200 µM below the eluvial horizon. At Svartberget, the sulphate concentration was <40 µM, while the proportion of marine origin was only a few µM. The hydrogen ion input attributable to the acid deposition at Svartberget seemed to be neutralized before the soil solution entered the mineral soil. The buffering capacity of the organic acids in mineral soil solution, which was fairly limited, was not utilized. At Tisa the concentrations of nitrate were high (1 mM) in the bottom of the profile, which indicated that the load of nitrogen was in excess of that which the vegetation could assimilate. At Svartberget and Gårdsjön the concentrations of nitrate were mostly not detectable.

The distribution of aluminium species in the different horizons differed, depending on the acidic load. At Svartberget, the concentrations of organically bound aluminium were high in the upper soil equilibrated by low concentrations of inorganic aluminium (Fig. 2). Monomeric inorganic aluminium constituted c. 25% of the total aluminium (35 µM) in the eluvial horizon and c. 40% of the

total aluminium (5–22 µM) at 10–20 cm depth in the B horizon. This horizon-related variation in the relative proportions of the various species was similar on all 22 profile-sampling occasions. The total concentration of aluminium in this unpolluted area changed with season, the concentrations being higher in autumn than during the rest of the year (Lundström 1993). The concentration of polymeric aluminium was on average <5 µM. An apparent equilibrium constant for Al+Org = AlOrg was calculated to be log K_{AlOrg} = 5.4±0.3 in summer and spring and log K_{AlOrg} = 4.9±0.1 in autumn, which showed that the AlOrg complex was strong (Lundström 1993). In Washington, USA, in a podzolized soil which was minimally affected by anthropogenic pollution, the portion of organically bound aluminium was dominant, as at Svartberget (Dahlgren and Ugolini 1989a).

At Gårdsjön, the concentration of monomeric inorganic aluminium was high in the illuvial horizon. Inorganic aluminium constituted 80% of the total aluminium (70 µM) in the upper illuvial horizon but only 40% of the total aluminium (20 µM) at 65 cm depth. The standard error was 4–30% for all species at all depths, with the highest values at 7 cm and 65 cm depths. The concentration of polymeric aluminium was <10 µM throughout the profile.

At Tisa, 80–90% of the total aluminium (280–970 µM) was in monomeric, inorganic form, most of which was Al^{3+} and $AlSO_4^+$. At Dl. Louka, 60–85% of the total aluminium was inorganic, monomeric aluminium. The concentrations of total aluminium were lower (50–145 µM) than at Tisa. The sampled site at Dl. Louka had been limed several times during the few last years, which may have influenced the soil solution chemistry. The soil at Tisa was formed from sandstone, whereas that at Dl. Louka was formed from granite, which also may have contributed to the differences. Concentrations of inorganic aluminium in soil solution as high as those found at Tisa were also reported from Nacetin, c. 45 km W of Usti n. Laben (Dambrine et al. 1993). In the Netherlands, high concentrations of inorganic aluminium similar to that in the Czech Republic have been found (Mulder et al. 1987).

At Svartberget and Gårdsjön the horizon-related variation in relative proportions of the various aluminium species, together with the pH and sulphate concentration in soil solution represents an almost unaffected and an affected coniferous-forest soil by acid deposition. The temporal variations of the concentrations at each site were low compared to the differences caused by acidic load. The situation exampled by the sampling at Tisa, in the Czech Republic, show a soil strongly affected by acid deposition.

The concentration of inorganic aluminium and its relation to base cations is known to influence the growth of roots in greenhouse experiments (Rost-Siebert 1983, 1985, Hutchinson et al. 1986). At the sites with high concentrations of monomeric inorganic aluminium, the concentrations of calcium and magnesium were also high (Fig. 3). There was no correlation between monomeric

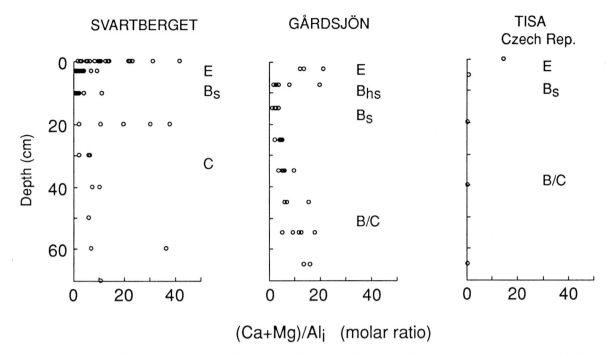

Fig. 4. The molar quotient (Ca+Mg)/(monomeric inorganic Al) in soil solution at different depths at Svartberget and Gårdsjön (Sweden) and Tisa (Czech Republic). Soil horizons are indicated in the figure.

inorganic aluminium and the molar sum of calcium and magnesium in the three investigated profiles, and the quotient was almost always >1.

At Svartberget, this quotient was 16 ± 13 in soil solution from the mor layer, c. 3.0 ± 2.5 in the eluvial and upper illuvial horizon, and 5–40 in the deeper soil (Fig. 4). At Gårdsjön, the quotient was 18 ± 5 in the eluvial horizon, 4 ± 3 in the B_{hs} horizon and 2.7 ± 1.0 in the B_s horizon, which was the lowest. From 25 cm depth to 65 cm the quotient increased from 4 to 15. At Gårdsjön the marine influence is important and consequently Mg makes up a large proportion of the base cations. Even if the marine Mg is excluded, the molar quotient will be more than one unit. At Tisa the molar quotient was 15 in soil solution in the mor layer, 1.2 in the eluvial horizon and c. 0.5 deeper down in the profile. At Nacetin, in the Czech Republic, this quotient was about the same (0.4–4.9) (Dambrine et al. 1993). At Tisa the concentration of calcium was 100–300 μM and at Nacetin 100–175 μM (Dambrine et al. 1993); both sites were unlimed.

The index Ca/Al in soil solution was proposed by Meiwes et al. (1986) to evaluate the degree of acidification. The molar ratio (Ca+Mg)/Al has been used for the top 50 cm of soil to calculate the critical loads for forest soils (Sverdrup et al. 1992). The soil solutions from our three sites all show that these quotients had a large variation and a horizon-related distribution. At both Swedish sites the quotient (Ca+Mg)/(monomeric inorganic Al) was >1 in all horizons. The horizon-related distribution of aluminium species at Gårdsjön indicates that the soil is

acidified, which was not indicated by the quotient criterion. In Tisa the quotient was >1 in the upper soil and <1 deeper down, despite the leaching of inorganic aluminium throughout the profile. Thus the molar quotient (Ca+Mg)/Al seems difficult to use as a criterion without considering the horizon distribution and the fact that only monomeric inorganic Al is relevant in the calculation of the quotient. For monitoring soil acidification, horizon-related distribution of aluminium species concentrations, pH, sulphate and nitrate concentrations seem to be more relevant criteria.

Equilibria with solid phases.

A wide range of pH values and Al^{3+} concentrations were available in this investigation and were used to evaluate the equilibria of soil solution with solid phases. In the eluvial horizon at all sites, the soil solution was undersaturated with respect to gibbsite (Fig. 5), and deeper in the profile the linear relationship between pAl^{3+} and pH had a slope <3. These findings suggest that the Al^{3+} concentration is not primarily influenced by any of the gibbsite forms. An exception was the deep soil at Svartberget, where an equilibrium of microcrystalline gibbsite was plausible. Throughout the soil profile, acidic sites were more undersaturated with respect to gibbsite than were the less acidic sites. With respect to kaolinite, the soil solutions at all sites were undersaturated in the upper soil and oversaturated in the deeper soil. Oversaturation in

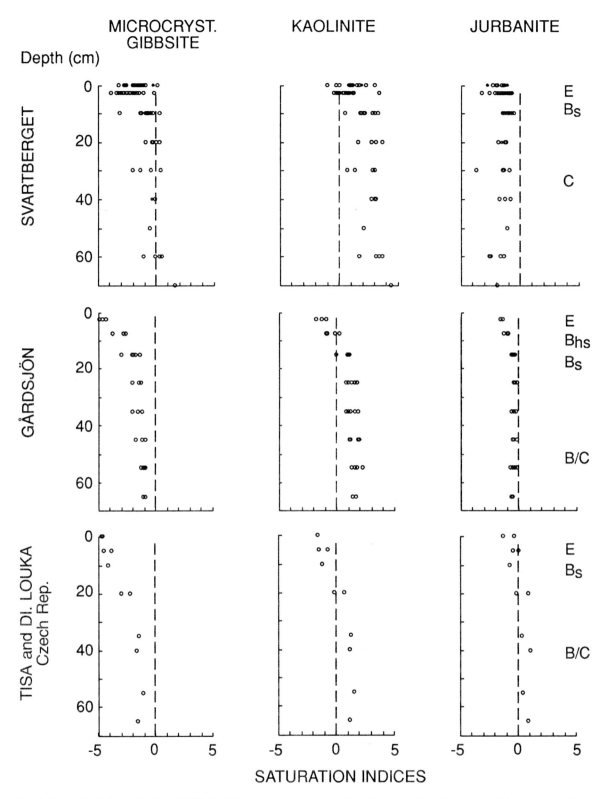

Fig. 5. Saturation indices, SI= log (IAP/K$_{eq}$) (IAP ion activity product), for microcrystalline gibbsite, kaolinite and jurbanite at different depths at Svartberget and Gårdsjön (Sweden) Tisa and Dl. Louka (Czech Republic). Soil horizons are indicated in the figure.

deep soil was less pronounced in the acidic sites. With respect to jurbanite, the soil solution was undersaturated at the Svartberget and Gårdsjön site, whereas at the more acidic Czech sites it was oversaturated. Comparing a varity of forest stands and soils in southern Sweden, Berggren (1990) found lower saturation of the soil solution with respect to gibbsite at low pH values than at higher ones. For jurbanite, however, the opposite was true. Thus our findings are in accordance with those of Berggren. In a spodosol investigated in Washington State, USA, (Dahlgren and Ugolini 1989b), the soil solution was undersaturated in the eluvial horizon with respect to gibbsite, imogolite and kaolinite, and the linear relationship between pAl^{3+} and pH had a slope < 3. In the Bs and C horizons there appeared to be an equilibrium between interlayer – Al and imogolite. In the Netherlands, soil solution was also highly undersaturated with respect to gibbsite and jurbanite in the upper soil and less undersaturated deeper down, where solutions were close to equilibrium with jurbanite (Mulder et al. 1987). Soil solutions investigated in the Netherlands were all oversaturated with respect to kaolinite.

At Svartberget, a monoprotic acid model was used by titration of organic acids (pK_a 4.0–4.7) (Lundström 1993). The apparent complex formation constants (K_{AlOrg}) for Al^{3+} + Org = AlOrg were calculated to be $logK_{AlOrg}$ = 5.4 ± 0.3 in spring and summer and $logK_{Alorg}$ = 4.9 ± 0.1 in autumn (Lundström 1993). The Al^{3+} concentrations occurring in Svartberget were thus better explained by this equilibrium model with organics than by any of the solid-phase models considered. The importance of solid alumino-organics as a major source of dissolved aluminium by acidification has also been shown by Mulder et al. (1989). The role of organics in aluminium soil chemistry seems to be important.

Conclusions

Soil solution chemistry, with special attention directed towards aluminium speciation, pH and sulphate concentration, could be used in monitoring acidification. The temporal variation at the two Swedish sites was small compared to the differences between the sites caused by acidic load. This enables monitoring surveys to be made with few sampling occasions. By using free drainage centrifugation, which enables well-defined horizons to be sampled and analysed promptly without storing the samples, soil solution was easily obtained. With the aluminium speciation method of Clarke et al. (1992), we were able to speciate aluminium with a small sampling volume.

High concentrations of organically bound aluminium in the eluvial horizon in a podzolized soil indicate an ongoing process of podzolization, with dissociated organic acids forming complexes with aluminium and thereby enhancing the weathering rate (Lundström and Öhman 1990, Lundström 1990, Ugolini et al. 1991). This process is probably not extensive at Gårdsjön and is almost completely inhibited at Tisa. High concentrations of inorganic aluminium leaching from the illuvial horizon indicate that the podzolization process is perturbed. At Tisa, the most damaged of the studied areas, leaching of inorganic Al occurred throughout the profile. At all investigated sites, the quotient (Ca+Mg)/(monomeric inorganic Al) > 1 in the upper soil, where most of the root uptake will take place. The quotient criterion (Ca+Mg)/Al for soil solution seemed to be nonrelevant for acidification, especially if it is not closely related to horizons.

None of the equilibrium evaluations with solid phases could explain the variation in soil solution aluminium concentrations. When a monoprotic acid model that considered aluminium complex formation was applied in Svartberget, the aluminium concentration was better explained. As long as the aluminium chemistry in forest soil remains poorly understood and there are no known equilibria to explain the aluminium concentrations, it is difficult to model soil solution chemistry. Therefore it will be important to monitor the soil. Because the soil solution transports the products of all soil formation processes, its properties quickly change in response to altered conditions in the soil. Free drainage centrifugation (Giesler and Lundström 1990, 1993) and aluminium speciation are useful tools, which enable surveys to be easily made.

Acknowledgement – The authors thank H. Hultberg, I. Andersson, E. Lakomaa, J. Tichý and M. Drechsler for making the sampling at Gårdsjön and in the Czech Republic possible. These projects were financially supported by the Swedish Environmental Protection Agency.

References

Anonymous, 1988 and 1989. Climate and chemistry of water at Svartberget. Reference measurements 1987 and 1988. – Swedish Univ. Agricult. Sci., Vindeln Experimental Forest Station.

Barnes, R. B. 1975. The determination of special forms of aluminium in natural water. – Chem. Geol. 15: 177–191.

Berggren, D. 1992. Speciation and mobilization of aluminium and cadmium in acid forest soils of S. Sweden. – Water Air Soil Pollut. 62: 125–156.

Clarke, N., Danielsson, L.-G. and Sparén, A. 1992. The determination of quickly reacting aluminium in natural waters by kinetic discrimination in a flow system. – Int. J. Anal. Chem. 48: 77–100.

Dahlgren, R. A. and Ugolini, F. C. 1989 a. Aluminium fractionation of soil solutions from unperturbed and tephra-treated spodosols, Cascade Range, Washington, USA. – Soil Sci. Soc. Am. J. 53: 559–566.

– and Ugolini, F. C. 1989 b. Formation and stability of imogolite in a tephritic spodosol, Cascade Range, Washington, U.S.A. – Geochim. Cosmochim. Acta 53: 1897–1904.

Dambrine, E., Kinkor, V., Jehlicka, J. and Gelhaye, D. 1993. Fluxes of dissolved mineral elements through a forest ecosystem submitted to extremely high atmospheric pollution inputs (Czech Republic). – Ann. Sci. For. 50: 147–157.

Driscoll, C. T. 1984. A procedure for the fractionation of aqueous aluminium in dilute acid waters. – Int. J. of Environ. Anal. Chem. 16: 267–283.

Emteryd, O., Andersson, B. and Wallmark, H. G. 1991. Determination of organic carbon in natural water with inductively coupled plasma/ atomic emission spectrometry after evaporation of inorganic carbon. – Microchem. J. 43: 87–93.

Eriksson, H. and Johansson, U. 1993. Yields of Norway spruce (*Picea abies* (L.) Karst.) in two consecutive rotations in southwestern Sweden. – Plant Soil 154: 239–247.

Falkengren-Grerup, U., Linnermark, N. S. and Tyler, G. 1987. Changes in acidity and cations pools of south Swedish soils between 1949 and 1985. – Chemospere 16: 2239–2248.

FAO-Unesco 1988. Soil map of the world. Revised legend. – World Soil Resources Report 60. Rome.

Giesler, R. and Lundström, U. 1990. Centrifugation method for soil and its influence on the soil solution.– Report 15. Dept of Forest Site Research. Swedish Univ. Agricult. Sci., Umeå.

– and Lundström, U. S. 1993. Soil solution chemistry – The effects of bulking soil samples and spatial variation. – Soil Sci. Soc. Am. J. 57: 1283–1288.

Hallbäcken, L. 1992. Long-term changes of base cation pools in soil and biomass in a beech and spruce forest of southern Sweden. – Zeitschr. Pflanzenernähr. Bodenk. 155: 51–60.

Hem, J. D., Roberson, C. F., Lind, C. J. and Polzer, W. L. 1973. Chemical interactions of aluminium with aqueous silica at 25°C. – U.S. Geol. Surv. Water Supply 1827-E.

Hultberg, H., Andersson, B. I. and Moldan, F. 1992. The covered catchment. An experimental approach to reversal of acidification in a forest ecosystem. – In: Rasmussen, L., Brydges, T. and Mathy, P. (eds), Experimental manipulations of biota and biogeochemical cycling in ecosystems. Ecosystem Res., Rep. No. 4, Comm. European Communities, pp. 46–54.

Hutchinson, T. C., Bozic, L. and Munoz-Vega, G. 1986. Responses of five species of conifer seedlings to aluminium stress. – Water Air Soil Poll. 31: 283–294.

Lundmark, J. E. 1989. Marktillstånd och markförändringar – Swedish Univ. Agricult. Sci., Uppsala, Skogsfakta Konferens 12: 11–19, (in Swedish).

Lundström, U. 1990. Laboratory and lysimeter studies of chemical weathering. – In: Mason, J. (ed.), The surface waters acidification programme. Cambridge Univ. Press, Cambridge, pp. 267–274.

– 1993. The role of dissolved organics in soil solution chemistry in a podzolized soil. – J. Soil Sci. 44: 121–133.

– 1994. Significance of organic acids for weathering and the podzolization process. – Environm. Int. 20:1: 21–30.

– and Öhman, L.-O. 1990. Dissolution of feldspars in the presence of natural, organic solutes. – J. Soil Sci. 41: 359–369.

Lövblad, G., Amann, M., Andersen, B., Hovmand, M., Joffre, S. and Pedersen, U. 1992. Deposition of sulfur and nitrogen in the Nordic countries: Present and future. – Ambio 21: 339–347.

Majdi, H. and Persson, H. 1993. Spatial distribution of fine roots, rhizosphere and bulk-soil chemistry in an acidified *Picea abies* stand. – Scand. J. For. Res. 8: 147–155.

May, H. M., Helmke, P. A. and Jackson, M. L. 1979. Gibbsite solubility and thermodynamic properties of hydroxyaluminium ions in aqueous solutions at 25°C. – Geochim. Cosmochim. Acta 43: 861–868.

Meiwes, K. J., Khanna, P. K. and Ulrich, B. 1986. Parameters for describing soil acidification and their relevance to the stability of forest ecosystems. – For. Ecol. Manage. 15: 161–179.

Moldan, B. and Schnoor, J. 1992. Czechoslovakia examining a critically ill environment. – Environ. Sci. Technol. 26: 14–21.

Mulder, J., van Grinsven, J.J.M. and van Breemen, N. 1987. Impacts of acid atmospheric deposition on woodland soils in the Netherlands: III. Aluminium chemistry. – Soil Sci. Soc. Am. J. 51: 1640–1646.

– , van Breemen, N. and Eijck, H.C. 1989. Depletion of soil aluminium by acid deposition and implications for acid neutralization. – Nature 337: 247–249.

Nihlgård, B. 1990. Skog. – In: Brodin, Y.-W. (ed.), Effekter av svavel och kvävebelastning på skogsmark, yt- och grundvatten. Swedish Environ. Protect. Agency, Solna, Report 3762 pp. 65–83, (in Swedish).

Nilsson, S.I. 1988. Acidity properties in Swedish forest soils. Regional patterns and implications for forest liming. – Scand. J. For. Res. 3: 417–424.

Nohrstedt, H.-Ö. 1990. Halter och kvoter av makronäringsämnen i årsbarr hos medelålders till äldre gran och tall. – In: Liljelund, L.-E., Lundmark, J.-E., Nihlgård, B., Nohrstedt, H.-Ö. and Rosén, K. (eds), Vitality fertilization-present knowledge and research needs. – Swedish Environ. Protect. Agency, Solna, Report 3813, pp. 85–102, (in Swedish with English summary).

Nordström, D.K. 1982. Effects of sulfate on aluminium concentrations in natural waters. – Geochim. Cosmochim. Acta 46: 681–682.

Petersen, L. 1976. Podzols and podzolization. – DSR-Copenhagen.

Rost-Siebert, K. 1983. Aluminium-Toxizität und – Toleranz an Keimpflanzen von Fichte (*Picea abies* Karst.) und Buche (*Fagus sylvatica* L.). – Allg. Forstz. 38: 686–689.

– 1985. Untersuchungen zur H- und Al-ionen-toxicotät an Keimpflanzen von Fichte (*Picea abies* I.) und Buche (*Fagus sylvatica* L.) in Lösungskultur. – Ber. Forschungsz. Waldökosyst./Waldsterb, Univ. Göttingen, Band 12.

Schecher, W. D. and Driscoll, C. 1987. An evaluation of uncertainty associated with aluminium equilibrium calculations. – Water Resour. Res. 24:4: 525–534.

– and Driscoll, C. 1988. An evalation of the equilibrium calculations within acidification models: The effect of uncertainty in measure chemical components. – Water Resour. Res. 24:4: 533–540.

Stumm, W. and Morgan, J. J. 1981. Aquatic chemistry. – Wiley, New York.

Sverdrup, H., Warfvinge, H., Frogner, T., Håoya, A. O. and Johansson, M. 1992. Critical loads for forest soils in the Nordic countries. – Ambio 21: 348–355.

Tamm, C. O. and Hallbäcken, L. 1988. Changes in soil acidity in two forest areas with different acid deposition. – Ambio 17: 56–61.

Ugolini, F.C., Dahlgren, R., LaManna, J., Nuhn, W. and Zachare, J. 1991. Mineralogy and weathering processes in recent and holocene tephra deposits of the Pacific Northwest, USA. – Geoderma 51: 277–299.

Wijk, S., Berghäll, S., Wulff, S. and Söderberg, U. 1993. Skogsskador i Sverige, 1992. Resultat av skogsskadebevakningen från rikskogstaxeringen och skogsvårdsorganisationens observationsytor. – The National Board of Forestry, Jönköping, Report 1993: 3, (in Swedish with English summary).

Ecological Bulletins 44: 123–128. Copenhagen 1995

Effect of acidification on retention, leaching and loss of selenium in a podzol profile – a laboratory study

Lars Johnsson

Johnsson, L. 1995. Effect of acidification on retention, leaching and loss of selenium in a podzol profile – a laboratory study. – Ecol. Bull. (Copenhagen) 44: 123–128.

Selenite in synthetic throughfall solution adjusted to pH 3.5 or 4.5 was added to columns containing soil horizons from a reconstructed podzol profile in order to study acid rain effects on the Se dynamics. Most of the added selenite was retained in the upper part of the O horizon. Below the Bs1 horizon the concentrations were <0.1 µg dm^{-3}. Selenium was transported as organic complexes or fixed to colloidal Fe compounds. No inorganic Se could be detected in the leachates. Budget calculations showed that 21% and 74% of added Se was lost from the columns in the pH 3.5 and pH 4.5 treatments, respectively, probably through volatilization. Results from a batch experiment suggest that selenite retention in the organic soil horizon is a combined microbial and chemical process and that selenite in the mineral soil was retained exclusively by chemical adsorption. Soils from all horizon retained between 98 and 99% of added selenite and there was little or no effect of pH on Se retention. The strong retention of selenite in the organic horizon means that the vast majority of the deposited atmospheric Se will be fixed in the mor layer unless lost through volatilization. This study shows that there is a need for further research on processes governing Se dynamics in forest soils before large scale liming of forest soils can be recommended.

L. Johnsson, Dept of Soil Sciences, Swedish Univ. of Agricultural Sciences, P.O. Box 7014, S-750 07 Uppsala, Sweden.

Introduction

Selenium is it self an essential element and may also be able to counteract the toxicity caused by other elements, for example Cd and Hg (Parizek et al. 1974, Groth 1976). Selenium seems to reduce Hg bioavailability in lakes by forming HgSe (Håkanson et al. 1990), which is insoluble in water. It has been suggested by several authors that acidification causes reduced availability of selenium (e.g. Cary et al. 1967, Frost and Griffin 1977, Hamdy and Gissel-Nielsen 1977). However, these investigations were made on agricultural soils or with purified minerals. In a previous study it has been shown that the influence of pH on Se retention in the mor layer of a forest soil was small (Gustafsson and Johnsson 1994). Most of the native Se was preferentially incorporated in low-molecular-weight fractions of the organic carbon pool (e.g. humates and fulvates), while rather small amounts were recovered in the solid-phase humin (Gustafsson and Johnsson 1992). The prevailing mechanism for selenium retention in the mor layer involves a microbially mediated reduction to lower oxidation states (lower than Se(IV)), and a subsequent incorporation into humic substances. In these soil horizons, complexation of inorganic selenite to humic substances is unimportant except as an initial (and transitory) step whereby dissolved selenite is retained in the soil matrix (Gustafsson and Johnsson 1994).

In the forest ecosystem, deposition of atmospheric Se plays an important role in the cycling of Se. The Se content in the mor layers in Swedish forest soils shows that a large proportion of the Se originates from marine or anthropogenic sources (Johnsson 1989). The effect of acidification on retention and transport of deposited Se in the podzol profile is not well understood. The objective of this work was to investigate the influence of acidification on the retention of deposited selenite in a podzol profile and the way in which Se is transported within the profile.

Materials and methods

Soil material from different horizons was collected from a haplic podzol (FAO 1988) situated at Stråsan in central Sweden (60°92'N, 16°02'E). The soil was stored at +2°C

Table 1. Background data for the different horizons of the Stråsan soil; Particle size distribution (Cl = clay; Si = silt, Sa = sand) in per cent of the mineral substance (min. sub.), organic carbon (org. C) in % of dry weight (DW), total Se-concentration (µg kg⁻¹), pH in water, exchangeable cations in 0.5 M BaCl₂, cation exchange capacity (CEC) in 0.5 M BaCl₂, base saturation (V) in % of CEC in 0.5 M BaCl₂ and dithionite (Dit), pyrophosphate (Pyr) and oxalate (Ox) extractable Fe and Al.

Horizon	Particle size distribution, % of min. sub.			Org. C, % DW	Total Se conc. µg kg⁻¹	pH in H₂O	Exchangeable cations, mmol$_c$ kg⁻¹						CEC BaCl₂ mmol$_c$ kg⁻¹	V, %	Fe, g kg⁻¹			Al, g kg⁻¹		
	Cl	Si	Sa				Ca	Mg	K	Na	Al	H			Dit	Pyr	Ox	Dit	Pyr	Ox
O	19	39	42	27	576	4.23	77.5	12.6	9.14	3.13	62.5	61.8	227	45	1.24	0.10	0.91	0.74	0.41	0.74
A	16	32	52	9	425	3.92	13.6	3.18	2.35	0.43	32.4	85.9	138	14	1.98	0.14	1.00	0.67	0.39	0.66
E	6	40	54	1	9.1	4.28	0.94	0.40	0.51	0.13	13.3	15.8	31.5	6.3	0.66	0.03	0.13	0.19	0.13	0.23
Bs1	19	39	42	5	814	4.87	13.9	1.20	0.56	0.28	75.1	40.0	131	12	38.1	1.66	22.5	11.1	6.15	9.98
Bs2	16	42	42	4	1054	5.15	8.25	0.75	0.32	0.14	29.9	16.5	55.8	17	29.3	0.66	16.4	14.2	6.53	19.2
Bs3	8	40	52	2	420	5.12	1.46	0.11	0.14	0.04	8.20	3.51	13.5	13	14.3	0.16	9.72	9.62	3.48	19.7
C1	8	34	58	–	200	5.20	0.29	0.05	0.29	0.64	1.63	1.85	4.75	26	5.23	0.34	3.24	3.68	1.49	9.31
C2	8	41	51	1	181	5.21	0.17	0.03	0.15	0.24	2.11	1.45	4.10	14	3.31	0.23	1.56	2.84	1.34	5.97

Table 2. Horizon combinations used in the columns.

Column type	Horizon used	Length of column (cm)
1	O (6 cm) + A (2 cm)	8
2	as 1 + E (8 cm)	16
3	as 2 + Bs1 (1 cm)	17
4	as 3 + Bs2 (7 cm)	24
5	as 4 + Bs3 (6 cm)	30
6	as 5 + C2 (20 cm)	50

before analysis of chemical and physical parameters (Table 1). The investigation included one column and one batch experiment.

Column experiment

The column experiment, performed in duplicate consisted of two treatments, each comprising a set of six different types of columns filled with homogenized material from the soil horizons, according to Table 2. The soils from the organic and mineral horizons were passed through a 4 mm and a 2 mm sieve, respectively. A high density polyethene filter (average pore size 80 µm) was placed at the bottom of each column to retain the fine material. The column outlet was connected to a peristaltic pump adjusted to a flow of 30 ml h⁻¹ to prevent reducing conditions at the bottom of the columns. Solutions were prepared to simulate the throughfall at Stråsan (Table 3). One set of columns was leached with leaching solution adjusted to pH 3.5 and the other with a similar solution adjusted to pH 4.5. The pH was adjusted by adding a mixture of H_2SO_4 and HNO_3 in a molar ratio of 2:1. The soil columns were leached nine times per week with 6 ml solution. For the first 35 d of the experiment, the soil columns were leached with solution without added Se. After this period, selenium in the form of selenite (SeO_3^{2-}) was added to the solutions (100 µg Se dm⁻³).

Table 3. Added ion species and concentrations (µmol$_c$ dm⁻³) of leaching solution used in the column experiment after pH adjustment to pH 3.5 and pH 4.5.

Ion species	pH 3.5 Concentration, µmol$_c$ dm⁻³	pH 4.5 Concentration, µmol$_c$ dm⁻³
Ca²⁺	9	9
K⁺	3	3
Mg²⁺	5	5
Na⁺	10	10
NH₄⁺	42	42
H⁺	316	31.6
Cl⁻	13	13
NO₃⁻	112	56
SO₄²⁻	304	27
SeO₃²⁻	2	2

Table 4. Se budget for the soil columns containing the O+A-horizons and leached with precipitation adjusted to pH 3.5 and 4.5. Mean values and range for column duplicates.

	pH 3.5			pH 4.5		
	Se, μg	Range, μg	% of input	Se, μg	Range, μg	% of input
Input (I)	86	3.8	100	85	0.18	100
Leaching from O+A horizon (L)	1.6	0.16	1.9	0.53	0.13	0.5
Net storage in O+A horizon (S)	66	14	76	21	0.33	25
Se not accounted for (I-S-L)	18	10	21	63	0.67	74

The soil columns were then leached for 112 d. The experiment was performed at 10°C in the dark. The total amount of leaching solution added during 147 d was c. 1.2 dm^3 and the amount of selenium added was c. 85 μg.

Batch experiment

In the batch experiment the effect of pH and different anions on Se retention in the O, Bs1, Bs2 and C2 horizons was studied. The following acids were used: a mixture of H_2SO_4 and HNO_3 (molar ratio 2:1), H_2SO_4, HNO_3, HCl and $ClCH_2COOH$ (monochloroacetic acid). Two grams of fresh soil were first shaken for 1 h with 20 ml of the corresponding acid in 0.005 M $CaCl_2$, to equilibrate the soil at the desired pH. Thereafter, Se in the form of SeO_3^{2-} was added (0.25 μmol). After equilibration (64 h for the organic soil and 2 h for the mineral soils (Gustavsson and Johnsson 1994)), the samples were centrifuged (10 000 rpm; 10 min) and the selenite concentrations in the supernatants were determined.

Sampling and chemical analyses

In the column experiment, leachates were collected each week and analysed for total Se concentration, pH, colour (420 nm) and conductivity. In a few leachates inorganic selenite and selenate concentrations were determined. Most leachates were analysed for total organic carbon (TOC), Fe and Al. At the end of the experiment the columns containing the whole profile (type 6, Table 2) were destructively sampled, each horizon being divided into two or more segments and total Se-, C-, N-, S-concentrations and pH(H_2O) determined. In the batch experiment, the equilibrium solutions were analysed for inorganic selenite. Total Se concentrations in leachates and soils were determined by AAS (atomic absorption spectrophotometry) using the hydride method after digestion in a mixture of concentrated acid (HNO_3, $HClO_4$ and H_2SO_4 in the ratio 7:2:1) and reduction in 5 M HCl (Johnsson 1989). Determination of selenite and selenate in leachates and extracts and TOC in leachates is described by Gustafsson and Johnsson (1992). Fe and Al in leachates and soils were determined by AAS. Total C-, N-

and S-concentrations in the soil were determined by elemental analysis (Leco, CHN-932, elemental analyser, model 600–800–332).

The results were analysed statistically using simple, stepwise regression and analysis of variance in the SYSTAT-MGLH-program. Figures were produced by using the SYSTAT-SYGRAPH-program (Wilkinson 1990a,b).

Result and discussion

A Se-budget for the soil columns containing the O+A-horizons showed a large difference between the two pHs in the amounts of Se unaccounted for (Table 4): 21% of added Se for pH 3.5 and 74% for pH 4.5. The losses were of the same order of magnitude as in an earlier column experiment (Gustafsson and Johnsson 1992). The losses were considered to have been caused by volatilization. The higher variation between duplicates at pH 3.5 than at pH 4.5 was probably because of a larger degree of flow along the wall in one of the columns at pH 3.5.

Volatilization of Se in the soil is a microbial process that involves methylation of organic and inorganic Se compounds, a process influenced by amounts of available organic matter, temperature, water content (Doran 1982) and pH (Hamdy and Gissel-Nielsen 1977). Earlier studies on Se volatilization were generally carried out on agricultural mineral soils. The amounts of Se lost by volatilization when added as selenite have generally been small, 0–8% of the amount added (Haygarth et al. 1991). Additions of organic Se compounds resulted in larger amounts of volatilized Se, between 7 and 87% of total soil Se depending on the compound used (Doran and Alexander 1977). Possible explanations of the large amounts of Se being volatilized from the mor layer could be large amounts of microbially available organic compounds and a low chemical adsorption of inorganic Se to the mineral fraction. An additional reason for the high Se volatilization from the mor layer may be the fast, microbially mediated reductive incorporation of added inorganic selenite to humic substances (Gustavsson and Johnsson 1994). Thus, Se added as inorganic compounds will change to organic Se compounds that are more readily available for methylation. The results from this study

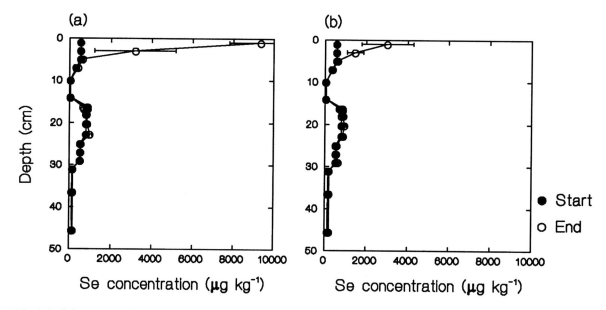

Fig 1. Soil Se concentrations at different soil depths in soil columns containing an artificial podzol profile used in a leaching experiment. The Se concentration in leaching solutions was 100 μg dm⁻³ in the form of selenite. Filled circles show the initial Se concentrations, and unfilled circles show the Se concentrations after the experiment. (a) Leaching solution adjusted to pH 3.5 and (b) pH 4.5. Error bars show standard deviation for column duplicates.

suggest that Se volatilization could be a very important part of the Se cycle in forest soils and that this process is strongly pH dependent.

In the column experiment, most of the added Se was retained in the upper part of the O horizon (0–2 cm) unless volatilized (Fig. 1). The total Se concentrations in the leachates from all types of columns were low (Table 5). In leachates below the Bs1 horizon the concentrations were at or below the detection limit (0.1 μg Se dm⁻³). No inorganic Se could be detected in the leachates. Significantly more Se and total organic carbon (TOC), Fe and Al were leached from columns at pH 3.5 than at pH

4.5 and the concentration increased over time. At the low pH, a correlation between total Se and TOC concentrations was found in leachates from the columns containing the organic horizons, O and A ($Se_{leachates}$ = −0.82 + 0.02$TOC_{leachates}$; R^2 = 0.55; p < 0.000; n = 24). The total Se concentration in leachates from the Bs1 horizon was correlated with the Fe concentration ($Se_{leachates}$ = −0.18 + 1.90$Fe_{leachates}$; R^2 = 0.59; p < 0.000; n = 24). The results indicate that Se was transported associated to, or incorporated in, dissolved organic matter or fixed to colloidal Fe compounds. The high retention of Se to iron compounds is described by several authors (Plotnikov

Table 5. Mean values and standard deviations (SD) for concentrations of Se (n = 30), total organic carbon (TOC; n = 24; *n = 18), Fe (n = 24), Al (n = 24) and pH (n = 30) in leachates from the column types described in Table 2, leached with artificial precipitation adjusted to pH 3.5 and 4.5. Figures do not include the first 35 d of the experiment, when no Se was added.

Column type	pH	Se, μg dm⁻³	SD, μg dm⁻³	TOC, mg dm⁻³	SD, mg dm⁻³	Fe, mg dm⁻³	SD, mg dm⁻³	Al, mg dm⁻³	SD, mg dm⁻³	pH	SD
1	3.5	1.70	0.90	122	22.5	2.03	0.41	1.92	0.78	4.10	0.22
1	4.5	0.63	0.57	72.3	12.1	1.13	0.16	0.77	0.34	4.37	0.12
2	3.5	1.99	0.77	140	26.5	6.88	1.29	4.01	1.14	4.39	0.13
2	4.5	0.63	0.19	73.6	7.83	2.51	0.59	1.23	0.40	4.50	0.08
3	3.5	0.22	0.14	10.6*	1.69	0.23	0.05	7.59	3.46	4.12	0.10
3	4.5	0.11	0.09	8.54*	1.57	0.14	0.02	1.64	0.56	4.45	0.08
4	3.5	<0.1	–	6.77*	1.83	0.03	0.01	1.11	0.52	4.68	0.16
4	4.5	<0.1	–	6.03*	1.34	0.04	0.02	0.36	0.18	4.95	0.07
5	3.5	<0.1	–	3.83*	1.43	0	0	0.52	0.27	4.85	0.07
5	4.5	<0.1	–	3.25*	1.02	0	0	0.41	0.16	4.85	0.09
6	3.5	<0.1	–	2.52*	0.67	0	0	0.31	0.17	4.92	0.09
6	4.5	<0.1	–	3.63*	1.04	0	0	1.57	0.37	4.61	0.07

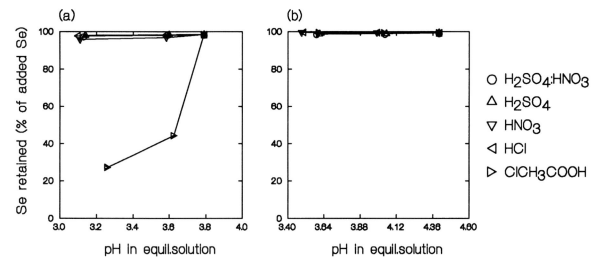

Fig. 2. Sorption of added Se (0.25 μmol) as a function of pH and added amounts of different anions. (a) Soil from the O horizon and (b) from the Bs1 horizon. Error bars show standard deviation for sample duplicates.

1960, Hamdy and Gissel-Nielssen 1977, Balistrieri and Chao 1990). The variations in the concentrations of Se and the other elements analysed, TOC, Fe and Al, were larger at pH 3.5 than at pH 4.5 (Table 5). This differences was largest for the column types 1 to 3 and was a result of the increase in concentration of these elements over time.

In the batch experiment, in which the influence of different anions on the retention of added selenite was studied, addition of monochloroacetic acid (ClCH₂COOH) to the organic soil caused a decrease of the amount of selenite retained of c. 70% compared to the other acid tretments (Fig. 2a). This effect was probably caused by its toxicity, that may have brought about total or partial reduction of the microbial activity in the soil. In another experiment, addition of an antimicrobial agent, sodium azide, to soil from the O horizon had a similar effect on Se retention (Gustafsson and Johnsson 1994). An earlier experiment showed that the kinetics for selenite retention differed greatly between the organic and the mineral horizons. The time needed for sorption of the same amount of added selenite (99.4% of 0.25 μmol Se added) was 64 h for the organic soil and <1 h for the mineral soil (Gustavsson and Johnsson 1994). The fast retention in the mineral soil indicates that chemical adsorption was the dominating mechanism. According to Balistrieri and Chao (1990), the amount of selenite adsorbed in mineral soils depends on the amounts of active substances such as Fe and Al-oxides. Both experiments clearly indicate the microbial participation in the retention mechanism of selenite to soil organic matter. A comparison of the amounts of selenite retained after one hour in the kinetic experiment, 36%, with the amounts retained when monochloric acetic acid was added, between 27% and 43%, shows that they are similar and might represent the amount of added selenite that was abiotically retained by the organic matter.

In order to compare the sorption capacity for selenite of the O, Bs1, Bs2 and C2 horizons at different pHs, 0.25 μmol Se was added to fresh soil samples adjusted to suitable pH by the addition of a mixture of H₂SO₄ and HNO₃ (molar ratio 2:1). Between 98 and 99% of the added Se was retained by all soil horizons, although the Se addition was between 10 to 70 times higher than the native Se content. There was little or no effect of pH on the Se retention in any of the soils. The small influence of pH as well as of different anions was also evident when the soils were treated with different acids, except for monochloroacetic acid added to the organic soil (Fig. 2). The low influence of pH on selenite retention in the mor layer was in accordance with another investigation (Gustafsson and Johnsson 1994). The low Se concentration in precipitation, <0.1 nmol dm⁻³ (Ross pers. comm.), mostly as selenite (Ross 1990), and the very strong retention of selenite in the organic horizon, means that by far the largest part of the deposited atmospheric Se will be fixed in the mor layer unless lost through volatilization.

Conclusions

The Se dynamics in podsol soils are closely related to the turnover of the soil organic matter. The strong retention of Se in the organic soil horizon means that by far the major proportion of deposited atmospheric Se is retained in the organic matter, mostly by microbial incorporation into humic compounds. Leaching of Se from the mor layer occurs as soluble complexes with organic matter and no or very small amounts of inorganic Se were leached to deeper horizons. The largest losses of selenium from the soil in the column experiment were through volatilization, a process that was strongly pH dependent. An increase in pH will increase the losses of selenium, probably because of greater microbial availability of the organic matter (Persson and Wirén 1989).

Acknowledgements – The author would like to thank J. P. Gustafsson for stimulating cooperation and D. Berggren for critically reviewing the manuscript. The National Swedish Environmental Protection Agency provided financial support for this study.

References

Balistrieri, L. S. and Chao, T. T. 1990. Adsorption of selenium by amorphous iron oxyhydroxide and manganese dioxide. – Geochim. Cosmochim. Acta 54: 739–751.

Cary, E. E., Weiczorek, G. A. and Allaway, W. H. 1967. Reactions of selenite selenium added to soils that produce low-selenium forage. – Soil Sci. Soc. Am. Proc. 31: 21–26.

Doran, J. W. 1982. Microorganisms and the biological cycling of selenium. – In: Marshall, K. C. (ed.) Advances in microbial ecology. Vol. 6, Plenum Press, New York, pp. 1–32.

– and Alexander, M. 1977. Microbial formation of volatile Se compounds in soil. – Soil Sci. Soc. Am. J. 40: 687–690.

FAO 1988. FAO/Unesco soil map of the world, revised legend. World Res. Report 60. – FAO, Rome. Reprinted as Technical Paper 20, Isric, Wageningen, 1989.

Frost, R. R. and Griffin, R. A. 1977. Effect of pH on adsorption of arsenic and selenium from landfill leachate by clay minerals. – Soil Sci. Soc. Am. J. 41: 53–56.

Gustafsson, J. P. and Johnsson, L. 1992. Selenium retention in organic matter of Swedish forest soils. – J. Soil Science 43: 461–472.

– and Johnsson, L. 1994. The association between selenium and humic substances in forested ecosystems – laboratory evidence. – Appl. Org. Met. Chem. 8: 141–147.

Groth, D. H. 1976. Interactions of mercury, cadmium, selenium, tellurium arsenic and beryllium. – In: Nordberg, G. F. (ed.), Effects and dose-response relationship of toxic metals. Elsevier Scientific, Amsterdam, pp. 527–543.

Håkanson, L., Andersson, P., Andersson, T., Bengtsson, Å., Grahn, P., Johansson, J.-Å., Jönsson, C.-P., Kvarnäs, H., Lindgren, G. and Nilsson, Å. 1990. Åtgärder mot höga kvicksilverhalter i insjöfisk. – Swedish Environ. Protect. Agency, Report 3918, Solna, (in Swedish).

Hamdy, A. A. and Gissel-Nielsen, G. 1977. Fixation of selenium by clay minerals and iron oxides. – Z. Pflanzenernaerhr. Bodenkd. 140: 63–70.

Haygarth, P. M., Jones, K. C. and Harrison, A. F. 1991. Selenium cycling through agricultural grassland in UK: Budgeting the role of the atmosphere. – Sci. Total Environ. 103: 89–111.

Johnsson, L. 1989. Se-levels in the mor layer of Swedish forest soils. – Swedish J. Agric. Res. 19: 21–28.

Parizek, J., Kalouskova, J., Babicky, A., Benes, J. and Paulik, L. 1974. Interaction of selenium with mercury, cadmium and other toxic metals. – In: Hoekstra, W.G., Suttic, J.W., Gantu, H.E. and Mertz, W. (eds), Trace elements metabolism in animals 2. Butterworths, London, pp. 119–131.

Persson, T. and Wirén, A. 1989. Microbial activity in forest soils in relation to acid/base and carbon/nitrogen status. – Medd. Nor. Inst. Skogsforsk. 42: 83–94.

Plotnikov, V. I. 1960. Coprecipitation of selenium and tellurium with metal hydroxides. – Russ. J. Inorg. Chem. 5: 351–354.

Ross, H. B. 1990. Biogeochemical cycling of atmospheric selenium. – In: Gücer, S., Adams, F., Klockow, D. and Izdar, E. (eds), Metal speciation in the environment. NATO ASI 23, Springer, Berlin, pp. 523–544.

Wilkinson, L. 1990a. SYSTAT: the system for statistics. – SYSTAT Evanston, IL. USA.

– 1990b. SYGRAPH: the system for graphics. – SYSTAT, Evanston, IL. USA.

Ecological Bulletins 44: 129–132. Copenhagen 1995

Effects of long-term storage on some chemical properties of forest soil samples

Ursula Falkengren-Grerup

Falkengren-Grerup, U. 1995. Effects of long-term storage on some chemical properties of forest soil samples. – Ecol. Bull. (Copenhagen) 44: 129–132.

A common method to preserve soil samples for future analyses has been to store them air-dried in paper bags in storage rooms at ambient temperature, but there is little knowledge of effects of storage conditions and storage time longer than a few weeks on the chemical properties of soils. Effects on pH and exchangeable and easily extractable Al, Na, K, Sr, Mg and Ca were studied in two extractants (1 M NH_4Ac, pH 4.8 and 0.2 M HNO_3) over 2–40 yr (pH) and 2–6 yr (cations) of storage in unheated rooms. The first analysis was two months after sampling. All analyses were repeated using the same methods and equipment. pH decreased gradually over time. The decrease was insignificant during the first two years (changes caused by drying or storage during the first two months were not measured) but was 0.8 pH-unit, as an average, over a few decades. Largest pH-changes were found in the originally least acid soils.

Exchangeable base cations changed little over two years for soils sampled in the 1950's and analyzed in 1988 and 1990. This also applied to soils sampled in 1988 using the strongest extractant (0.2 M HNO_3). In extractions with NH_4Ac, Mg and Sr were unaffected while smaller amounts of Al, Na and, partly, K were found after storage. Calcium was the only element that increased in concentration, but only after six years of storage. Some changes may be attributable to inaccuracy of analysis, others to storage. The present study demonstrates some effects of storage but further studies must be made to assess the general validity of the findings.

U. Falkengren-Grerup, Soil-Plant Res., Ecology Building, Lund University, S-223 62 Lund, Sweden.

Introduction

A common method to preserve soil samples for future analyses has been to store them air-dried in paper bags in unheated storage rooms. This will probably induce some fluctuations in soil moisture during storage and temperature may vary substantially. There is little knowledge of effects of time and storage conditions on the chemical properties of the soil as most studies have been made on drying effects over short periods, from a few days to some weeks, while the actual storage time may be several years.

There are indications that changes occur both during the drying process and on rewetting prior to analysis. The mineralization of nitrogen and carbon increases (Birch 1958, 1960), possibly because of the breaking of hydrogen bonds within the soil organic matter and increased microbial activity. Hydrolysis of water associated with exchangeable cations increases the adsorption capacity and proton-donating ability of the soil (Haynes and Swift 1985).

A higher drying temperature seems to accentuate these effects. Extractable (Mehlich-1) P, Fe and Mn increased considerably with temperature, also in the air-dried soil as compared to the moist soil. The changes in base cation concentrations (Ca, Mg, K, Zn) were not consistent and were probably dependent on initial soil properties, such as organic matter content and clay mineralogy (Payne and Rechcigl 1989). In some peat soils, on the other hand, extractable Fe (NH_4Ac-EDTA) decreased with higher temperature while P (NH_4Ac) increased and changes in other nutrient concentrations (K, Ca, Mg, Mn, Cu, Zn, Mo) were negligible (Saarinen 1989).

The effect of storage time, studied over 16 wk, was a continuous change towards an increase in surface acidity, exchangeable Mn, solubility and oxidizability of organic matter (Bartlett and James 1980). pH usually decreased as an immediate effect of drying and then continued to

Table 1. Datasets used in analyses of long-term storage effects. The first analysis was made within two months (datasets I–IV) or c. 40 yr (dataset V) after soil sampling.

	Years of sampling	Soil horizon	Forest type	Years between 1st and repeated analysis	Analysis	Reference
I	1949–54	A, B, C	deciduous and spruce	35, 40*	pH H$_2$O pH 1 M KCl	(a) Linnermark 1960
II	1984–85	A, C	deciduous and spruce	2, 6	pH H$_2$O pH 1 M KCl	(b) Falkengren-Grerup et al. 1987
III	1984	A	beech	6	pH H$_2$O	(c) Falkengren-Grerup 1986
IV	1988	C	beech and oak	2	pH H$_2$O pH 0.2 M KCl Cations: HNO$_3$; NH$_4$Ac, pH 4.8	(d) Falkengren-Grerup and Eriksson 1990
				2, 6		
V	1947–52	C	beech and oak	2	Cations: HNO$_3$; NH$_4$Ac, pH 4.8	(d) Falkengren-Grerup and Eriksson 1990 (e) Falkengren-Grerup and Tyler 1992

*The first analysis according to literature data.

decrease over a period of 4–12 months. The decrease was most pronounced during the first two months for pH (H$_2$O) while changes in pH(CaCl$_2$) were small (Davey and Conyers 1988).

The aim of this paper is to present results of repeated chemical analyses of pH, exchangeable and easily extractable cations in soils which had been stored air-dried for 2–40 yr. The results are from various studies of forest soils and can therefore not be as specific as if derived from an experiment particularly designed for this purpose. The aim is to show that changes may occur for some elements and in some soils and to encourage further studies on this problem.

Materials and methods

Four data sets, mostly for deciduous forests in southern Sweden, were used in the study (Table 1). The soils were sampled from the A, B or C-horizons in the 1950's and

1980's and were kept in paper bags (datasets II–V) or in glass tubes with stoppers (dataset I) in unheated storage rooms in south Sweden. Soil samples of data set V were moved from central to south Sweden in 1988. Site characteristics and details of chemical analyses are given in the primary publications (see Table 1).

The soil was air-dried immediately after sampling and sieved (mesh size 0.6 or 2 mm) before storage (dataset IV stored unsieved). pH was measured electrometrically in H$_2$O, 0.2 or 1 M KCl extracts. Exchangeable and easily extractable cations were extracted in 1 M NH$_4$Ac, pH 4.8, and 0.2 M HNO$_3$ and analyzed by plasma emission spectroscopy. The first analysis was made within two months of sampling. Determination of pH was repeated two or six years later and of cations two years later, using the same methods, soil:solution proportions and equipment as in the first analysis. pH-comparisons over longer periods were made using literature data for the first analysis (Linnermark 1960, Falkengren-Grerup et al. 1987). Soils sampled in 1988 were also analyzed six years later for NH$_4$Ac-exchangeable base cations.

Table 2. Mean pH changes in soil samples stored in unheated storage rooms for 2–40 yr after first analysis. Mean pH at first analysis and pH changes were calculated from the H-ion concentration, giving the soils with high pH and large pH decreases less weight. The first analysis was made on air-dried soil within two months after sampling. References as in Table 1.

Years after 1st analysis	Mean pH change	Mean pH at 1st analysis	pH extraction	Sampling depth cm	No. of samples	Reference
2	0.00	5.0	H$_2$O	70–80	18	(d)
2	−0.06	4.3	0.2 M KCl	70–80	18	(d)
2	+0.05	3.3	1 M KCl	0–20	14	(b)
6	−0.09	4.9	H$_2$O	60–85	15	(b)
6	−0.02	4.4	1 M KCl	60–85	15	(b)
6	−0.12	4.5	H$_2$O	0–10	52	(c)
35	−0.42	3.8*	1 M KCl	0–20	16	(b)
40	−0.80	5.2*	H$_2$O	0–100	101	(b)

*First analysis according to literature data.

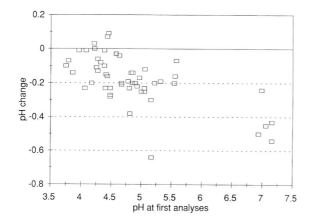

Fig. 1. pH(H$_2$O) changes as related to the first analysis in 52 topsoil samples (0–10 cm) after six years of storage. Linear correlation coefficient is –0.65 (p < 0.001).

Results and discussion

pH decreased gradually over time, the decrease being insignificant during the first years but amounting to an average of 0.8 unit over a few decades (Table 2). An increased acidity with time was also found by Davey and Conyers (1988), though studied for only 4–12 months. The changes were smaller when measured in KCl than in H$_2$O. Potassium cloride, being a stronger extractant, exchanges Al and a larger amount of H ions from the solid phase, which gives a lower pH. The greatest H ion difference was found in the most acid soils, but an analysis in pH units is not precise enough to measure small changes in H ion concentration in such soils.

However, the average pH decrease differs considerably between soils, with only small changes in the most acid soils (Fig. 1). Over a six year period, the most acid soils decreased by < 0.2, while the least acid soils decreased by up to 0.6 pH-unit. The average pH decrease was only 0.12. A linear negative correlation between initial pH and pH decrease was significant (p < 0.05) for all soils stored

for six years or more. In a study of dried soils stored for 50 yr, pH decreased by up to 0.6 unit over the whole pH range (pH (H$_2$O) 3.0–5.0; Östling et al. unpubl. data). The most acid soils varied in organic matter content (O.M., loss on ignition) from 20% to > 80% and the pH decreases were to some degree positively related to O.M. No such relationship was found in the present study for the C horizon samples stored for two years only, which were lower in O.M. (< 5%). The same applied to the topsoils with 5–10% O.M. and six years of storage (r = –0.14, p > 0.10).

Differences in the exchangeable ion concentrations between years may result from inaccuracy of the analyses or from de facto changes during storage. By comparing soils with a long storage time, where no further storage effects are expected, with soils with a short storage time, the errors in analysis are excluded. Differences in concentration ratios (Student's t-test) between soils sampled in the 1950's and soils sampled in the 1980's were therefore tested (Table 3).

Cations were analyzed twice within two years for soils stored for two or for 42 yr. Extractants were 1 M NH$_4$Ac, pH 4.8, and 0.2 M HNO$_3$, the latter extracting 2–4 times more Na, K and Sr and 15–20 times more Al, Mg and Ca (Falkengren-Grerup and Tyler 1992). The concentration ratios between the 1988 and 1990 concentrations for soils stored since the 1950's were usually 0.95–1.05. The same also applies to HNO$_3$-extracted cations for soils stored since 1988. Only K had significantly different ratios for soils stored for two or for 42 yr. The additional amounts extracted by 0.2 M HNO$_3$, as compared to 1 M NH$_4$Ac, should be less affected, as this fraction is more strongly adsorbed to the soil particles. A small change of a relatively large HNO$_3$ pool is also more difficult to detect. However, for Na and K, two elements with a comparatively small additional pool extracted by the HNO$_3$, storage effects were found only for K.

Soils sampled and analyzed for the first time in 1988 were extracted by NH$_4$Ac in 1990 and again in 1994 to see whether effects of storage were persistent or continuous. Aluminium and Na decreased by 20–30% during

Table 3. Base cations in the C-horizon of 18 beech and oak forest soils, given as the geometric mean of the ratio between the first (1988) and repeated analysis of soils stored 2, 6 or 42 yr. The extraction was made in 1 M NH$_4$Ac (pH 4.8) and in 0.2 M HNO$_3$. Differences between the logarithmic ratios were tested by Student's t-test, probabilities given as p-values.

Analysis		Sampling year	Base for ratio (Year of first to repeated analysis)	Al	Na	K	Sr	Mg	Ca
NH$_4$Ac	i	1947–52	1988:1990	1.17	1.04	1.05	1.07	1.01	1.04
	ii	1988	1988:1990	1.40	1.28	1.34	1.19	1.14	0.96
	iii	1988	1988:1994	1.49	1.25	0.98	1.03	0.91	0.69
t-test (p)		(i against ii)		<0.001	0.002	0.002	0.09	0.21	0.35
		(i against iii)		<0.001	0.003	0.13	0.35	0.28	<0.001
HNO$_3$	iv	1947–52	1988:1990	1.02	0.93	0.90	1.04	1.06	0.95
	v	1988	1988:1990	1.03	0.91	1.04	1.09	1.00	0.96
t-test (p)		(iv against v)		0.83	0.33	0.03	0.28	0.50	0.88

the first two years with no further changes in the following four years. The decrease for K during the first two years was not repeated in the second analysis. Strontium and Mg were not significantly affected by storage. Calcium was affected only in the repeated analysis and was the only element that increased. The repeated analysis after six years indicated that measurable effects, if any, occurred within the first two years, possibly soon after the analysis made after two months (decrease in Al, Na; no change in Sr, Mg). The changes in K and Ca cannot be explained by this study.

A stronger retention of ions was not commonly found as a direct effect of air-drying or of higher drying temperatures (Payne and Rechcigl 1989, Saarinen 1989) nor as an effect of storage up to four months (Bartlett and James 1980). Both drying and storage seemed to increase decomposition of organic matter content and increase acidity in these studies, while extractable amounts of different ions in some cases increased and in others were unaffected, probably depending on the initial soil characteristics.

Care must be taken in the interpretation of results derived from analyses of soils stored for a long time before analysis. The mechanisms behind the changes in soil chemistry must be studied to increase the understanding and to enable assessment of the general validity of the findings.

References

Bartlett, R. and James, B. 1980. Studying dried, stored soil samples – some pitfalls. – Soil Sci. Soc. Am. J. 44: 721–724.

Birch, H. F. 1958. The effect of soil drying on humus decomposition and nitrogen availability. – Plant Soil 10: 9–31.

– 1960. Nitrification in soils after different periods of dryness. – Plant Soil 12: 81–96.

Davey, B. G. and Conyers, M. K. 1988. Determining the pH of acid soils. – Soil Sci. 146: 141–150.

Falkengren-Grerup, U. 1986. Soil acidification and vegetation changes in deciduous forest in southern Sweden. – Oecologia 70: 339–347.

– and Eriksson, H. 1990. Changes in soil, vegetation and forest yield between 1947 and 1988 in beech and oak sites of southern Sweden. – For. Ecol. Manage. 38: 37–53.

– and Tyler, G. 1992. Changes since 1950 of mineral pools in the upper C-horizon of Swedish deciduous forest soils. – Water Air Soil Pollut. 64: 495–501.

–, Linnermark, N. and Tyler, G. 1987. Changes in acidity and cation pools of South Swedish soils between 1949 and 1985. – Chemosphere 16: 2239–2248.

Haynes, R. J. and Swift, R. S. 1985. Effects of air-drying on the adsorption and desorption of phosphate and levels of extractable phosphate in a group of acid soils. – N. Zeal. Geoderma 35: 145–157.

Linnermark, N. 1960. Podsol och brunjord (Podsols and brown soils) I–II. – Inst. Mineral Paleontol. Quart. Geol., Univ. Lund, Sweden, 75: 1–233, (English summary).

Payne, G. G. and Rechcigl, J. E. 1989. Influence of various drying techniques on the extractability of plant nutrients from selected Florida soils. – Soil Sci. 148: 275–283.

Saarinen, J. 1989. Effect of drying temperature on the extractable macro- and micronutrients and pH of different peat types. – Suo 40: 149–153.

Ecological Bulletins 44: 133–146. Copenhagen 1995

Tropospheric ozone – a stress factor for Norway spruce in Sweden

Lena Skärby, Göran Wallin, Gun Selldén, Per Erik Karlsson, Susanne Ottosson, Sirkka Sutinen and Peringe Grennfelt

Skärby, L., Wallin, G., Selldén, G., Karlsson, P.E., Ottosson, S., Sutinen, S. and Grennfelt, P. 1995. Tropospheric ozone – a stress factor for Norway spruce in Sweden. – Ecol. Bull. (Copenhagen) 44: 133–146.

Nutrient and chlorophyll content, ultrastructural changes, photosynthesis, stem growth and biomass were studied in Norway spruce, *Picea abies*, exposed to three different concentrations of ozone for five seasons (1985–89) in open-top chambers. The treatments were charcoal-filtered air (CF), non-filtered, ambient air (NF) and non-filtered air enriched with ozone (NF+). The results show that ozone can change the ultrastructure of needles, decrease the content of chlorophyll and decrease the rate of photosynthesis, and that the extent of these changes depends on the length of exposure and the age of the needles and shoots. No significant effects of ozone on above-ground biomass were observed. Based on the results on photosynthesis, an attempt was made to model the canopy photosynthesis of 8 and 89 yr old trees. The results indicate that the ambient ozone concentrations in Sweden can reduce the canopy photosynthesis, and that the effect is small in young trees and becomes greater in old trees, mainly because of the different proportions of needle age classes. This paper is a summary document of the results, some of which have already been published elsewhere.

L. Skärby, P.E. Karlsson and P. Grennfelt, Swedish Environmental Res. Inst., Box 47086, S-402 58 Göteborg, Sweden. – G. Wallin, G. Selldén and S. Ottoson, Dept of Plant Physiology, Botanical Inst., Univ. of Göteborg, Carl Skottbergs gata 22, S-413 19 Göteborg, Sweden. – S. Sutinen, The Finnish Forest Res. Inst., Suonenjoki Res. Stat., FIN-776 00 Suonenjoki, Finland.

Introduction

Boreal forests in the northern hemisphere are an important resource for production of timber, paper and other cellulose products as well as for recreation. During recent years there have been several signs that forests are threatened by human activities and air pollution is considered to be one serious stress factor.

During the last century the atmosphere over Europe has changed in the composition of many trace components. Since the middle of the 19th century it has been known that ozone is a natural constituent of the atmosphere. Although all early measurements suffer from lack of accuracy, because of interferences (Lefohn et al. 1992), recent analyses of historical ozone data, such as the data from Montsouris outside Paris, from 1876 to 1910, and from Montcalieri, in northern Italy, from 1868 to 1893, indicate that the concentrations of tropospheric ozone in Europe have at least doubled over the last 100 yr (Volz and Kley 1988, Anfossi et al. 1991). Thus, the average background concentrations of ozone in parts of Europe may have been as low as 20 μg m^{-3}, as compared to current annual means of 50–60 μg m^{-3} (Volz and Kley 1988, Anfossi et al. 1991). Furthermore, measurements started in the early fifties in the former GDR show that the annual mean concentration of ozone has increased by 1–3% yr^{-1} (Feister and Warmbt 1987). Superimposed on the long-term increase in ozone background concentrations, short-term episodes of ozone occur every summer in Europe, especially in the central and southern parts. During such episodes the ozone concentrations may rise to 200–300 μg m^{-3} over large areas (Beck and Grennfelt 1994).

Plants are the main sink for ozone in the lower troposphere. Uptake of ozone is always associated with a destruction of ozone, in which ozone oxidizes compounds on the surface and/or within the plant. The oxidation may lead to irreparable damage to the plant or the need for repair reactions. Thus, uptake of ozone tends to have negative effects on plants. The extent to which the

Table 1. Soil pH and total organic carbon in soil (TOC; % d.wt.) in 1986, 1987 and 1989 (means ± standard deviation). CF = charcoal-filtered air; NF = non-filtered air; NF+ = non-filtered air enriched with ozone.

Treatment	1986			1987			1989		
	pH_{H_2O}	pH_{CaCl_2}	TOC	pH_{H_2O}	pH_{CaCl_2}	TOC	pH_{H_2O}	pH_{CaCl_2}	TOC
CF	5.4±0.1	4.8±0.1	3.6±0.1	5.6±0.4	4.8±0.2	3.2±0.5	5.2±0.1	4.7±0	3.2±0.4
NF	5.3±0.1	4.7±0.1	4.0±0.8	5.6±0.1	4.8±0.1	3.7±0.5	5.1±0	4.6±0	4.0±0.3
NF+	5.2±0	4.6±0	4.1±0.2	5.7±0	4.7±0	4.0±0.5	5.2±0.1	4.6±0	4.0±0.3

increased concentrations of ozone damage plants is therefore an important issue.

Early short-term laboratory exposures of young trees to high concentrations of ozone clearly indicated that the trees were sensitive to ozone (Botkin et al. 1971, Wilkinson and Barnes 1973). However, a central problem in air pollution research is the extrapolation from the test situation to the field situation. Increasing awareness of this difficulty has caused air pollution research to move from test systems with a very low ecological realism to test systems with a higher ecological realism, i.e. long-term exposures of trees using low concentrations of ozone in open-top field chambers (Seufert and Arndt 1985, Kats et al. 1985, Krause and Höckel 1988). To some extent, of course, the use of open-top chambers also affects temperature, light, wind-speed and humidity in relation to the real field situation. Although any form of enclosure within a chamber will affect the microclimate of the plants, the open-top chamber at present represents the most realistic exposure technique when the aim is to conduct a dose-response study using ozone concentrations below and above ambient.

In 1985, an open-top chamber experiment was set up in Sweden in order to study effects of long-term exposure to ozone on Norway spruce. Soil-grown trees were exposed to charcoal-filtered air, non-filtered ambient air and moderately elevated concentrations of ozone for five years. The aim was to identify effects of ozone on gas exchange, chlorophyll content and needle ultrastructure of shoots of different ages. Some results on growth and biomass are included, although it was recognized early that the number of replicates (n = 2) was too small to be able to detect statistically significant effects. Several of these results have been published previously. This paper is a final summary document of the results obtained, as well as an attempt to quantify the effects of ozone on canopy photosynthesis. A model based on gas exchange parameters in different shoot age classes and distribution of biomass was used.

Experimental facility and plant material

The experimental site was situated at Rörvik, on the Swedish west coast, 35 km south of Göteborg (57°40'N, 11°90'E), 0.5 km from the coastline. In 1985, two trees of one clone of Norway spruce, *Picea abies* (L.), a Polish provenance, Rycerka S21K74266, were planted in the ground inside each of six chambers. The 4 yr old pot-grown spruce cuttings were obtained from 30 yr old trees.

The design of the chambers was based on that of Heagle et al. (1973). Each chamber was 2.5 m high (excluding frustum) and 3.0 m in diameter. The frustum reduced the diameter of the top of the chamber to 1.75 m. The walls of the chambers were made from clear corrugated PVC, and these remained in place all year around. The flow through the chambers corresponded to 3–3.5 air changes per minute. The fans were not turned off at night, but were turned off in the winter periods.

The soil was a sandy clay loam. Soil samples were taken from each chamber plot in December 1986, 1987 and 1989. The samples were sent to the Swedish Univ. of Agricultural Sci. for determination of pH_{H_2O}, pH_{CaCl_2} and total organic carbon (TOC). These parameters varied little between treatments as well as between years (Table 1). The effective cation exchange capacity, determined once in 1986, varied between 54 and 56 me/100 g d.wt. The base saturation varied between 86% and 93%. The soil content of total and available N as well as of total contents of P, K, Mg, Ca did not vary significantly between treatments (data not shown). For further information on experimental facilities and plant maintenance see Wallin et al. (1990).

Experimental design

Two chambers were used in each treatment. Two spruce trees per chamber were used in the study and all results are based on the chamber mean (i.e. four individual trees where n = 2).

Climate

Air temperature was recorded all year round, while the relative humidity of the air was recorded between May and November each year, in one chamber and at an ambient plot, using shielded and aspirated psychrometers with wet and dry bulb Pt-100 sensors (In Situ Instruments, Ockelbo, Sweden). On a monthly basis (June-September) the temperature inside the chamber was between 0.5 and 1.6°C higher than outside. During the rest of the year, the temperature difference was less. Monthly mean temperatures and monthly precipitation at Rörvik between 1985 and 1990 are presented in Table 2. The winters of 85/86 and 86/87 were cold while the following winters were mild.

Pollution climate

The monthly means of O_3, SO_2 and NO_2 for 1985 to 1990 are presented in Table 3. The pollutants were measured at a nearby EMEP-station, 5 m above the ground. The monthly mean ozone concentrations always exceeded 50 $\mu g\ m^{-3}$ between April and September. Throughout the experiment the ambient air concentrations of SO_2 and NO_2 were low, especially during April–September.

Ozone generation and potential formation of dinitrogen pentoxide

Ozone was generated using ambient air (not dried) in an electric discharge generator (Sonozaire 630A, Howe-Baker Engineers Inc. Texas). The ozone-enriched air from the generators was passed through 2 m long aluminium tubes (0.06 m in diam.) and introduced into the air stream at the fan inlet, situated 2.5 m in front of the chamber inlet.

When using air-operated electric discharge generators and ambient air, there is a potential risk that by-products such as N_2O_5 can be formed (Harris et al. 1982). Consequently, it has been debated if artificially generated ozone may produce artefacts, such as an increased nitrogen deposition (Brown and Roberts 1988, Krause 1988, Olszyk et al. 1990, Taylor et al. 1993). N_2O_5 is rapidly converted to NO_3^- and the rate of this conversion is strongly dependent on air humidity. A relative humidity of 8% is enough rapidly to convert virtually all N_2O_5 to NO_3^- (Harris et al. 1982), which in turn is deposited to all surfaces. In the present experiment the relative humidity of the ambient air was always above 8%.

In order to test the amounts of N_2O_5, gaseous and particle-bound nitrate were measured with a combination of Na_2CO_3 coated denuders and Na_2CO_3 impregnated filters (Ferm 1986, Febo et al. 1987). Sampling took place between 11.00 and 16.00 h local time, on three different days with clear, sunny weather during July–September 1989. Air was sampled at four different points (with double sampling at each point): 1) in ambient air, about 1 m in front of the fan to one NF chamber, 2) in one NF chamber, 3) in one NF+ chamber, and 4) in one NF++ chamber (this chamber was used in a different experiment and the addition was c. double that of NF+). In the chamber measurements, the openings of the denuders were positioned in the middle of the manifold where the

air entered the chamber from the fan (see Wallin (1990) for details of the chamber and manifold). After sampling, the denuders and filters were separately leached with water and the nitrate content determined by ion chromatography (Ferm 1986).

The concentrations of nitrate were generally low. The gas phase nitrate concentrations were reduced by 50% in the NF treatment compared to ambient air (Table 4), even though the residence time for the air in the fan and air duct was only about one second. Furthermore, the concentrations of gas phase nitrate were between 7% and 40% higher in the NF+ treatment and between 52% and 155% higher in the NF++ treatment, compared to NF (Table 4), indicating that some N_2O_5 was generated by the ozonizers. However, the nitrate concentrations in the NF+ treatment were always below that of the ambient air.

Ozone exposure

Three concentrations of ozone were used: charcoal-filtered air (CF), non-filtered air (NF) and non-filtered air enriched with ozone (NF+). A constant amount of ozone was added daily between 11.00–18.00 h, local time. The periods of filtration and ozone exposure are presented in Table 5. The aim was to maintain the exposure periods from 1 April to 15 November. However, because of variations in climate and technical problems, especially in the first year, the exposure periods varied between the years (Table 5). Exposure period means (24 and 7 h) of ozone concentrations for the five seasons in the chambers, at an ambient plot and at a nearby EMEP-station, are presented in Table 6. For further details concerning air sampling and ozone monitoring, see Wallin et al. (1990).

Accumulated dose over a threshold of 40 ppb (AOT40)

Recently, a new concept of critical levels for ozone has been agreed upon (Fuhrer and Acherman 1994). For forest trees, as well as for crops, it was agreed that these levels should be based on significant reductions in growth or total biomass in plants grown under near-field conditions, i.e. the experimental facility should be either open-top chambers or open release systems. Since there are very few data published from this kind of study in Europe, the critical level suggested for forest trees is based on a very small data set. The critical level for forest trees as well as for crops is based on a dose concept, the Accumulated exposure Over a Threshold of 40 ppb (AOT40). The AOT40 for forest trees has been set to 10 ppm hours, calculated for 24 h per day over a six months growing season. However, this is to be regarded as provisional at this stage (Fuhrer and Acherman 1994).

In Table 7 the AOT40 for the seasons 1985–1989 are

Table 2. Monthly mean temperatures (°C) and precipitation (mm) at Rörvik environmental monitoring station, May 1985 – April 1990. Data from K. Sjöberg (pers. comm.).

	1985		1986		1987		1988		1989		1990	
Month	°C	mm	°C	mm	°C	mm	°C	mm	°C[1]	mm	°C	mm
J		43	−3.4	49	−7.9	7	3.3	140	5.3	7	3.4	100
F		14	−6.0	0	−1.3	42	1.6	87	5.1	54	4.3	77
M		54	1.3	79	−2.5	35	−0.1	40	5.6	72	4.4	23
A		85	4.1	28	5.5	28	–	50	6.0	34	4.9	50
M	11.8	35	11.9	82	9.6	65	–	24	11.0	28		75
J	13.6	65	15.5	10	12.3	118	17.8	13	14.9	18		92
J	16.5	54	17.0	40	15.6	108	17.5	156	17.0	95		62
A	16.2	105	15.6	98	14.7	60	16.6	72	14.6	52		70
S	11.8	153	11.2	35	12.4	78	14.2	79	12.4	45		106
O	10.4	62	10.4	85	9.7	59	8.4	92	9.0	112		74
N	1.6	43	7.5	94	4.1	83	2.8	30	5.1	19		43
D	1.2	152	1.4	60	2.6	50	2.9	50	3.0	54		79
Sum		865		660		733		833		590		851

[1] Monthly temperatures in April-October 1989 were from the nearby climate station at Säve.

presented, together with the summarized AOT40 for the five seasons for the different treatments as well as for ambient air. In these calculations we have used the vegetation seasons instead of exposure periods (Table 6), in order to have the most valid AOT values for the plants. This is the reason for the high AOT40 values for all treatments in 1985. Except for 1985, the trees in CF and NF were never exposed to > 10 ppm-h while the opposite was true for the plants in NF+. The mean concentrations of ozone in NF+ were always in the same range as the ozone concentrations measured at 5 m above ground at Rörvik EMEP station (Table 6). This clearly indicates that the ozone concentrations used in the experiment were realistic. The calculations of AOT40 in Table 7 can also be compared to the situation in Sweden 1990–1993 (Fig. 7 in Kindbom et al. 1995).

Nutrient content of needles

Needles of different age classes were sampled for nutrient analysis in December each year, except in 1985. The needle contents of P, K, Mg and Ca were analyzed using plasma emission spectrometry at SGAB-analys (Analytical Lab.), Sweden. The nitrogen content was analyzed using an elemental analyzer (CarloErba, Italy). In 1986 the N contents were low while the N-based proportions of P, K, Ca, Mg were supra optimal compared to those proposed by Ingestad (1979) (Table 8). In order to increase the nutrient status, the plants were fertilized once a week from mid-May to mid-July each year throughout the experiment, from 1987 onwards. The fertilizer was composed of macronutrients (N:P:K:Ca:Mg:S) in the following weight-based proportions 100:11.4:57:7.9:5.3: 7.5:0.6 and micro-nutrients. The amount of nutrients added each week was successively increased, with the

Table 3. Monthly means (24 h) of SO_2, NO_2 and O_3 at the EMEP (Environmental Monitoring and Evaluation Programme) station at Rörvik (5 m above ground), May 1985–April 1990[1]. All concentrations are given in $\mu g\ m^{-3}$.

	1985			1986			1987			1988			1989			1990		
Month	SO_2	NO_2	O_3	SO_2	NO_2	O_3	SO_2	NO_2	O_3	SO_2	NO_2	O_3	SO_2	NO_2	O_3	SO_2	NO_2	O_3
J				13.6	8.9	44	14.6	8.6	39	9.8	9.2	27	7.6	13.5	45	6.2	9.5	44
F				11.4	6.9	–	12.0	6.6	42	7.6	5.9	44	8.6	8.2	54	5.6	12.2	44
M				12.2	6.6	–	11.6	6.9	62	6.8	6.3	63	8.8	7.2	64	3.6	6.3	65
A				5.8	4.6	74	6.6	4.3	69	4.6	4.9	72	7.6	5.6	72	2.8	4.9	77
M	4.0	4.3	94	4.4	4.3	85	1.8	3.0	69	3.2	4.3	92	4.4	4.6	84			
J	4.2	3.0	86	4.6	3.0	85	2.2	3.0	69	2.6	3.6	81	3.6	3.9	88			
J	4.2	2.3	81	4.2	2.6	79	2.4	2.0	58	3.2	3.0	82	3.2	4.9	63			
A	3.6	2.3	81	2.2	2.6	73	2.0	3.6	53	3.6	4.6	66	3.8	3.9	61			
S	4.6	3.6	66	1.8	4.3	58	1.6	1.1	52	6.0	4.9	61	4.2	4.6	57			
O	4.4	5.9	56	7.2	6.9	48	6.0	6.6	44	6.2	7.2	47	4.4	6.9	48			
N	11.2	7.6	45	6.0	7.3	–	4.6	7.3	25	8.4	13.2	44	5.2	13.8	33			
D	5.2	8.3	43	11.4	7.6	35	4.4	8.6	35	5.8	9.2	50	3.4	10.2	38			

[1] From: Lövblad et al. 1985, 1986, 1989, 1990, Dahlberg et al. 1987, Lövblad and Sjöberg 1988.

Table 4. Gaseous and particle-bound NO_3^- concentrations, ozone concentrations, air temperatures and air humidities in open-top chambers and in ambient air during three different days in 1989. NF = non-filtered air; NF+ = non-filtered air enriched with ozone; NF++ = non-filtered air enriched with ozone (ozone > NF+); AA = ambient air. The values are given as means during the measuring periods between 11.00 and 16.00 h local time.

Day	Treatment	NO_3^- gas		NO_3^- particle		Ozone	Air temp.	RH
		μg m⁻³	% of NF	μg m⁻³	% of NF	μg m⁻³	°C	%
7 July	AA	2.38	180	1.34	172	120	27	40
	NF	1.32	100	0.78	100	124	–	–
	NF+	1.56	118	0.87	112	192	28	35
	NF++	2.00	152	0.57	73	256	–	–
11 Sept	AA	0.230	121	0.22	157	56	16	47
	NF	0.104	100	0.14	100	56	–	–
	NF+	0.146	140	0.12	89	92	17	46
	NF++	0.265	255	0.07	48	148	–	–
12 Sept	AA	0.264	194	0.37	168	66	16	58
	NF	0.136	100	0.22	100	62	–	–
	NF+	0.146	107	0.18	80	102	18	56
	NF++	0.230	169	0.19	87	160	–	–

maximum amount being added in late June each year. The total amount corresponded to a yearly supply of 20 g N m⁻². As plants were harvested each year, the total nutrient supply per chamber was not increased. Fertilization increased the amount of nitrogen in both current and 1 yr old needles to 1.8–2.0% (d.wt.) in the first year. Over the next two years the N content decreased to 1.5–1.6% (d.wt.). The range 1.5–1.8% of N in needles of Norway spruce has been considered optimal for both frost resistance and stem growth (Aronsson 1985). The high concentration in 1987 was probably caused by a lag phase in the structural response of the crown, e.g. number of shoots and needles, to the sudden increase in nutrient supply. It normally takes 6–12 months before trees with preformed initials respond to increased nutrient supply with increasing shoot and needle numbers (Linder and Rook 1984). No significant differences in the nitrogen content could be observed between treatments.

The fertilization decreased the proportion of P, K, Ca, Mg in relation to N, indicating that the N-supply was limiting before fertilization. There were no differences in nutrient ratios between treatments except for the K:N ratio, where the lowest value was always for the trees from the CF treatment. However, this was only significantly lower in the 1 yr old shoots in 1987 (p < 0.05).

Table 5. Periods of ozone exposure (NF+) and charcoal-filtration (CF) during 1985 – 1989.

Year	Ozone exposure		Filtration	
	Start	Stop	Start	Stop
1985	12 July	8 Oct	11 July	5 Nov
1986	12 June	4 Oct	25 April	17 Dec
1987	8 May	6 Nov	14 April	4 Dec
1988	22 April	2 Nov	30 March	21 Nov
1989	12 May	7 Nov	7 March	12 Dec

The Mg:N ratio was also lower in the 1 yr old shoots from the NF treatment in 1987, than in the other treatments.

It can be concluded that ozone did not alter the nutrient balance of the trees in this experiment. These results also demonstrate that the increase in the gas phase NO_3^- in NF+, relative to NF (Table 8), was of no importance for the nitrogen status of the plants.

Chlorophyll content of needles

The chlorophyll a+b content of the needles from different shoot age classes was regularly analyzed 2–5 times per year throughout the experiment (Fig. 1a–d). Chlorophyll was analyzed according to Arnon (1949).

In 1985 and 1986, the chlorophyll content of the needle flush of 1985 in trees from the CF treatment was in the range of 1.3 to 2.5 mg g⁻¹ (Fig. 1a). In July 1987, the chlorophyll content of CF- and NF-treated needles increased rapidly to c. 4.6 mg g⁻¹, probably as a consequence of the fertilization, as the chlorophyll content in needles of Norway spruce has been shown to correlate strongly with the nitrogen content (Linder 1980). The NF+-treated needles also responded to fertilization, although to a much lesser extent. During the following two years, the chlorophyll content of CF-treated needles remained around 4.5 mg g⁻¹, while the content in the NF- and NF+-treated needles started to decrease. After September 1987, the chlorophyll content of NF+-treated needles of the 1985 needle flush was significantly lower than that in NF- and CF-treated needles. After September 1988, the chlorophyll content was significantly reduced compared to CF in both NF- and NF+-treated needles, which in turn were significantly different.

The changes with time in chlorophyll content of the needle flush of 1986 (Fig. 1b) were similar to that of the flush of the previous year. After fertilization the chloro-

Table 6. 24 h and 7 h (11.00–18.00, local time) exposure period means of ozone concentrations in open- top chambers and in ambient air (AA) at Rörvik, Sweden, 1985–1989[1]. Abbreviations as in Table 1. Concentrations are given in μg m⁻³.

Treatment	1985 24 h	1985 7 h	1986 24 h	1986 7 h	1987 24 h	1987 7 h	1988 24 h	1988 7 h	1989 24 h	1989 7 h
CF	8	9	7	8	9	10	15	16	14	16
NF	47	50	52	61	48	56	59	70	51	61
NF+	59	85	63	89	63	98	70	103	64	99
AA[2]	54	62	56	66	53	62	62	74	53	63
AA[3]	70	80	68	79	56	63	72	84	69	77

[1] 1985, 15 July–5 Nov (548 h missing); 1986, 12 June–15 Nov (730 h missing); 1987, 8 May–6 Nov (527 h missing); 1988, 24 April–1 Nov (1001 h missing), 1989, 13 May–12 Nov (1246 h missing). The monthly means in April each year varied between 35 and 39 ppb and the monthly means in May 1985 and 1986 were 48 and 43 ppb.
[2] Ambient air 1 m above ground.
[3] Data from measurements in ambient air at Rörvik monitoring station, 5 m above ground (K. Sjöberg pers. comm.).

phyll content rapidly increased in all treatments. Compared to CF- and NF-treated needles, significantly lower chlorophyll contents in NF+-treated needles were first observed in September 1987. After July 1989, the chlorophyll content of NF+-treated needles was also significantly lower than that of NF-treated needles.

The changes in chlorophyll content with time in needles from the flushes of 1987 and 1988 are presented in Fig. 1c and d, respectively. The differences in chlorophyll contents between treatments were less than those in the previous flushes and only occasionally statistically significant, although the NF+-treated needles consistently had the lowest chlorophyll content.

The decreased chlorophyll contents in needles from the NF and NF+ treatments compared to that in needles from the CF treatment could not be explained by a low N or Mg supply in these treatments, as no corresponding differences were observed in the N or Mg content. We therefore suggest that the effect on the chlorophyll content is mainly attributable to toxic reactions in the needles.

Structure of needles

Visual examination of trees and needles is the most fre-

quently used method for assessing injury to trees, and estimations of needle loss are currently widely used in the field (Panzer 1988). However, apart from being a non-specific response of trees to stress, needle loss also represents the final stage of a sequence of reactions to stress. Thus, needle loss is not a useful tool for determining relationships. On the other hand, the use of light- and electron microscopy increases the possibilities of making an early diagnosis of an injury that eventually may lead to visible injury and finally to needle loss. Furthermore, information about structural changes will give indications of where to look for disturbed functions.

Changes in ultrastructure in needles from the flush of 1985 were followed from November 1985 to October 1987. In 1985, needles were sampled from the upper part of the branches from the first branch whorl from the top of the tree. Throughout the period needles were sampled from these branches, avoiding needles with visible damage. Further details on sampling dates and microscopy procedures are given by Sutinen et al. (1990).

Ozone-induced changes were first observed in the chloroplasts; the length of the chloroplasts and the size of the starch grains decreased, while the density and the granulation of the stroma increased. The second organelle to be affected was the microbody, followed by changes in the mitochondria. These changes were first observed in needles from the NF+ treatment and later in needles from

Table 7. AOT40 (Accumulated Ozone dose in ppm-hours above a Threshold of 40 ppb) in open-top chambers and in ambient air at Rörvik, Sweden, for the seasons of 1985–1989[1]. Abbreviations as in Table 1.

Treatment	1985	1986	1987	1988	1989	Σ 1985–1989
CF	15.3	8.0	1.7	1.2	3.1	29.3
NF	15.8	9.9	4.3	8.0	6.3	44.3
NF+	23.5	18.9	18.5	21.7	16.7	99.3
AA[2]	16.5	10.9	6.2	9.9	3.1	46.6

[1] 1985, 1 April–5 Nov (850 h missing); 1986, 1 April–5 Nov (1340 h missing); 1987, 1 April–6 Nov (604 h missing); 1988, 1 April–1 Nov (1024 h missing); 1989, 1 April–12 Nov (1461 h missing). The number of ppb-h between 1 April and start of exposure/filtration each season is compensated for by taking data from the nearby monitoring station at Rörvik. Data are not corrected for missing hours.
[2] Ambient air 1 m above ground.

138

Table 8. Nutrient contents in needles of different age classes from different years and ozone treatments (means ± standard deviation). Significant differences (* = p < 0.05) between treatments within each year and needle age class are indicated by different letters. Statistical tests were made with ANOVA. Abbreviations as in Table 1. C, current year needles; C+1, 1 yr old; C+2, 2 yr old shoots.

| Needle age class | Year | Treatment | N % of d.wt. | Weight proportions | | | |
				N/P	N/K	N/Ca	N/Mg
C	1986	CF	1.2±0.1	18±2.4	56±14	68±20	8.4±0.4
		NF	1.3±0.2	19±2.9	53±7	71±29	8.3±1.9
		NF+	1.2±0.0	18±0.1	58±1	66±5	8.3±1.2
			n.s.	n.s.	n.s.	n.s.	n.s.
C	1987	CF	1.9±0.0	12±1.3	39±4	30±3	3.7±0.3
		NF	2.0±0.1	11±0.0	38±0	30±3	2.9±0.4
		NF+	2.0±0.2	12±0.3	44±6	29±4	3.0±0.3
			n.s.	n.s.	n.s.	n.s.	n.s.
C	1988	CF	1.8±0.2	11±1.4	36±5	50±9	6.4±1.7
		NF	1.9±0.1	11±0.2	41±0	49±8	5.0±0.8
		NF+	1.8±0.0	12±0.4	39±3	45±8	5.6±0.3
			n.s.	n.s.	n.s.	n.s.	n.s.
C	1989	CF	1.6±0.1	12±0.8	36±3	66±10	6.8±1.1
		NF	1.6±0.1	11±1.7	38±0	60±12	6.0±1.1
		NF+	1.6±0.0	12±0.5	38±1	55±4	6.6±0.7
			n.s.	n.s.	n.s.	n.s.	n.s.
C+1	1987	CF	1.9±0.1	12±0.9	30±1a	132±4	9.3±0.3a
		NF	1.8±0.1	12±1.1	39±4b	120±16	7.0±0.5b
		NF+	1.6±0.1	14±0.1	44±3b	118±8	8.4±0.5a
			n.s.	n.s.	*	n.s.	*
C+1	1988	CF	1.5±0.0	8±0.3	36±2	81±1	5.4±0.9
		NF	1.6±0.0	7±0.6	43±3	72±28	4.2±0.6
		NF+	1.5±0.1	8±0.2	46±4	65±3	3.9±0.3
			n.s.	n.s.	n.s.	n.s.	n.s.
C+1	1989	CF	1.6±0.0	9±1.1	32±5	117±18	6.4±1.3
		NF	1.6±0.0	10±0.8	40±1	106±1	5.6±0.1
		NF+	1.6±0.2	11±1.0	41±8	100±23	5.8±0.1
			n.s.	n.s.	n.s.	n.s.	n.s.
C+2	1989	CF	1.4±0.0	8±0.7	34±6	126±27	6.1±2.0
		NF	1.3±0.1	9±0.1	40±2	112±1	5.3±0.4
		NF+	1.4±0.0	10±1.1	42±8	114±8	5.5±0.0
			n.s.	n.s.	n.s.	n.s.	n.s.

the NF-treatment. However, only the length of chloroplasts was quantified. No changes were observed in needles from the CF treatment.

The ozone-induced changes were similar to changes observed in Norway spruce exposed to ozone in climate chambers (Sutinen 1987a). However, they were different from changes occurring as a result of exposure to SO_2 and NO_2 (Costonis 1970, Percy and Riding 1981, Kärenlampi and Houpis 1986, Maier-Maercker 1986) and to frost (Soikkeli and Kärenlampi 1984). Furthermore, they also differed from changes occurring in nutrient-deficient plants (Thomson and Weier 1962, Whatley 1971, Hall et al. 1972), thus suggesting that exposure to ozone may cause specific changes in needles of Norway spruce.

The injury typically progressed within the mesophyll tissue. Structural changes occurred first in the outer cell layer facing the sky and then began to appear in the underlying mesophyll tissue (Fig. 2). Such a gradual progression of needle injury has also been observed in ozone exposed needles from young trees of Norway spruce (Sutinen 1987a) as well as in Norway spruce stands in the Taunus mountains in West Germany (Sutinen 1987b).

Shoot photosynthesis

The net photosynthesis at saturating PPFD (Photosynthetic Photon Flux Density) of shoots of different ages was measured at the end of the growing seasons (September–October) of 1987, 1988 and 1989 (Fig. 3). These studies were made by measuring the gas exchange of the shoots in a custom-built gas exchange system in 1987 and 1988 (Wallin 1990) and with a portable CO_2 porometer (Licor-6100) in 1989. The photosynthetic rates in the CF and the NF treatments were within the range that have been observed in natural stands of Norway spruce (Fuchs et al. 1977, Lange et al. 1987, Zimmerman et al. 1988). In the two- and three-year-old shoots, it was generally observed that photosynthesis declined with increasing ozone concentration and age. In 1987, a significant reduc-

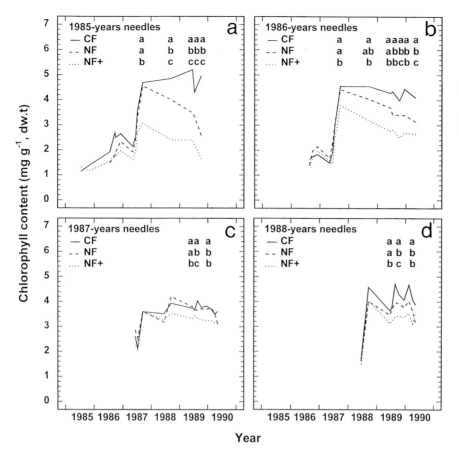

Fig. 1 a-d. The development of the chlorophyll a + b content during 1985–1990 in the needle flushes of 1985–1988 (a–d). Significant differences (p < 0.05) between treatments at each sampling occasion are indicated by different letters. Abbreviations as in Table 1.

tion in the photosynthesis of the two-year-old shoots in NF compared to CF was observed. However, at that time the sum of AOT40 for 1985–1987 was still rather similar in NF and CF (Table 7). No obvious explanation can be found, except that it could be suspected that the threshold of 40 ppb might be set too high. Furthermore, in the current-year shoots, net photosynthesis was generally lower in the CF treatment than in the NF and the NF+ treatments. Similar observations of ozone-induced stimulatory effects on net photosynthesis have been made in other studies of Norway spruce (Eamus et al. 1990) as well as in studies of other species, e.g. Ponderosa pine (Beyers et al. 1992). Inhibitory effects have been observed in older shoots in several studies and it has been suggested that ozone accelerates the process of ageing (Reich 1983, Reich and Amundson 1985).

At present there is no explanation for the change in the response of net photosynthesis in Norway spruce to ozone. However, if ozone accelerates the whole lifecycle of the needles, as illustrated in Fig. 4, the ozone-induced stimulation of net photosynthesis as well as the inhibition can be explained.

Net photosynthetic properties

Different properties of the net photosynthesis were measured in order to define the mechanisms behind the ozone effects and to determine under which conditions the effects could be of importance. The aim of these measurements was also to make modelling of the canopy photosynthesis possible. The measured properties were dark respiration, gas phase conductance and the photosynthetic light and CO_2 responses (Wallin et al. 1990, 1992a, b). The gas phase conductance, the light and CO_2 responses were used to calculate the intercellular CO_2 concentration, the apparent quantum yield and the carboxylation efficiency (Table 9). The measurements were made concurrently with those of net photosynthesis at saturating PPFD in 1987 and 1988.

The dark respiration was only affected in the current-year shoots, where it was stimulated in the NF+ treatment (Table 9). The effect on net photosynthesis could therefore not be explained by an effect on the dark respiration, either in young or in older shoots.

The intercellular CO_2 concentration (C_i) increased with increasing ozone concentration in all shoot ages, but not significantly in the 1 yr old shoots (Table 9). The stimulation of net photosynthesis in the current-year shoots could therefore at least partly be explained by the ob-

140

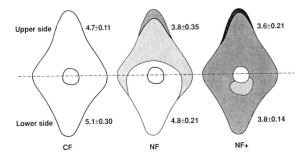

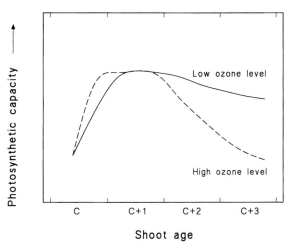

Fig. 2. Schematic illustration of the progression of tissue damage with time in the 1985 flush of needles of Norway spruce. The drawings are typical of the situation prevailing in needles on 3 April 1987. The length (mean and standard deviation) of the chloroplasts in the upper and lower half of the needle is indicated to the right of each cross section. In CF treated needles the mean length of chloroplasts in the upper and lower half of the needle was not significantly different. At this time, tissue damage had almost reached the endodermis (light grey) in NF treated needles. More tissue damage was found in the outer cell layers of the upper side (medium grey). There was a significant difference ($p < 0.05$) in chloroplast length between the upper and the lower half of the needle. In NF+-treated needles almost all parts of the needle were affected. As in the NF needles, the worst tissue damage (dark grey) was in the outer cell layers of the upper side. The length of the chloroplasts was reduced in both upper and lower sides of the needle (medium grey). The cells beneath the endodermis in the lower part of the needle were the least damaged (light grey). CF = filtered air, NF = non-filtered air, NF+ = non-filtered air + ozone. White = no damage, light grey to dark grey = increasing cell damage. From Sutinen et al. (1990).

Fig. 4. A model of the development of the photosynthetic capacity in shoots of Norway spruce with time, with and without ozone. (C = current year). Abbreviations as in Table 8.

served increase in gas phase conductance to CO_2 (Wallin et al. 1990). However, it could not explain the inhibition of net photosynthesis in the older shoots. Instead, the increase in C_i concealed some of the effects on the photosynthetic apparatus. The increase in C_i is also a result of a higher gas phase conductance relative to the photosyn-

thetic rate, suggesting that ozone will negatively influence the water economy of these needles. This was also shown in a separate study on the effects of soil moisture deficit on trees exposed to different ozone concentrations (Wallin and Skärby 1992).

Both the apparent quantum yield and the carboxylation efficiency in the older shoots were negatively affected in NF and NF+ (Table 9), with the largest decrease being in the carboxylation efficiency, which is determined mainly by the Rubisco activity. Since Wallin et al. (1992b) observed also a general over-capacity in the regeneration of RuBP, determined by photosynthesis measurements at saturating CO_2 concentrations, it was concluded that one of the major impacts of ozone was on the total activity of Rubisco. The ozone-induced reduction in the carboxylation efficiency will affect the photosynthetic rate negatively under most conditions, while the reduced apparent quantum yield will affect photosynthesis mainly under conditions of low PPFD.

Stem growth rate and biomass

The increment growth of the main stem of each tree was measured once a month during the vegetation seasons. The girth at the base and the top of each internode of the stem, and the length of the internodes, were measured. The stem volumes at the beginning and at the end of the experiment were calculated, as well as the relative growth rate for the whole growth period 1985–1989 (Table 10). In April 1990, before the vegetation period, the trees were harvested and dry weights of stems and branches, including needles, were measured (Table 10). No significant differences in growth rates or dry weight biomass were detected between the treatments, although the bio-

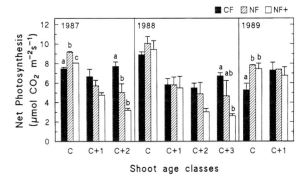

Fig. 3. The photosynthetic capacity at saturating PPFD and at approximately 330 µmol mol^{-1} CO_2 of different age classes of shoots from different treatments at the end of the growing seasons of 1987, 1988 and 1989. Bars marked with different letters (a–c) are significantly different at $p < 0.05$. Error bars shows standard deviation (n-1). Data from 1987 and 1988 are published in Wallin et al. 1990, 1992b. Abbreviations as in Table 8.

Table 9. Apparent quantum yield (Φ_i), carboxylation efficiency (CE), gas phase conductance to CO_2 (g_{gc}), intercellular CO_2 concentration (C_i) and dark respiration (R_d) of shoots of Norway spruce exposed to different ozone treatments. The values, except for gas phase conductance, are used in the modelling of the canopy photosynthesis. Symbols and abbreviations as in Table 8.

Shoot age	Treatment	Φ_i[2] (mol CO_2 (mol^{-1})	CE[3] (mmol m^{-2} s^{-1}	g_{gc}[3] (mmol m^{-2} s^{-1}	C_i[3] (μmol mol^1	R_d[1] (μmol m^{-2} s^{-1}
C	CF	0.046	0.055	104	227ab	0.82a
	NF	0.049	0.059	108	222a	0.80a
	NF+	0.047	0.054	115	236b	0.88b
		n.s.	n.s.	n.s.	*	*
C+1	CF	0.037	0.047	44	192	0.50
	NF	0.037	0.043	53	209	0.58
	NF+	0.030	0.041	46	208	0.60
		n.s.	n.s.	n.s.	n.s.	n.s.
C+2	CF	0.039a	0.042a	48	212a	0.61
	NF	0.030ab	0.030ab	50	234b	0.44
	NF+	0.022b	0.017b	42	261c	0.51
		*	*	n.s.	**	n.s.

[1] from Wallin et al. 1990
[2] from Wallin et al. 1992a
[3] from Wallin et al. 1992b

mass of trees from the CF treatment was higher than that of trees from the NF treatment, which in turn was higher than that of trees from the NF+ treatment. Because of the variation and small number of replicates, there was little possibility of detecting statistically significant growth reductions.

The effects of ozone on growth has also been studied in three American long-term experiments. There were no effects of ozone on any growth variable in red spruce exposed for two seasons (Kohut et al. 1990). Ozone exposure of Fraser fir seedlings for two and a half years did not produce any effect on growth (Seiler et al. 1994). Similarly, in loblolly pine, no ozone effects on total biomass or on diameter were evident at the end of three growing seasons (Kelly et al. 1993). However, ozone exposure resulted in a net retention of carbon in above-ground biomass of the loblolly pine and a subsequent reduction in carbon allocation to the root system.

Canopy photosynthesis – a modelling approach

The canopy photosynthesis is defined by a number of properties of the photosynthetic apparatus as well as by the structure of the canopy, which will influence light interception. In order to estimate the effects of ozone on canopy photosynthesis, a simple model was used. The model was composed of three sub-models, calculating the diurnal variation of light intensities, the light interception of the shoots in the canopy, and the shoot photosynthesis (Table 11). The modelling of the PPFD was based on the radiation geometry equations in Sesták et al. (1971). The modelling of the light interception and shoot photosynthesis was based on the equations presented by Landsberg (1986), modified to involve different shoot age classes.

The simulation of the canopy photosynthesis was made for one sunny day with 50% diffuse light and a maximum PPFD of 1600 μmol m^{-2} s^{-1}, and one cloudy day, with 100% diffuse light and a maximum PPFD of 600 μmol m^{-2} s^{-1}, in July at the same latitude and longitude as the experiment (58°N, 12°E). Three LAI values (2, 6 and 10) for a canopy with 89 yr old trees, and one LAI (2) for a canopy with 8 yr old trees, were simulated. An old forest

Table 10. Relative growth rate (RGR) of stem 1985–1989, stem biomass, and total above-ground biomass in Norway spruce exposed to different ozone concentrations (means ± standard deviation). Abbreviations as in Table 1. Seasons: see Table 4.

Treatment	RGR per 5 seasons	Stem biomass (d.wt.) (kg)	Total above-ground biomass (d.wt.) (kg)
CF	4.91±0.39	1.6±0.04	4.1±0.39
NF	4.70±0.49 (4%)	1.5±0.11 (6%)	3.6±0.46 (12%)
NF+	4.23±0.76 (14%)	1.4±0.14 (12.5%)	3.3±0.53 (20%)

Table 11. Input parameters for modelling the canopy photosynthesis in Norway spruce of different ages. The model is based on three sub-models calculating the solar geometry and light intensity, the light interception of the canopy and the shoot photosynthesis.

Solar geometry and light intensity	Canopy light interception	Shoot photosynthesis
Date	LAI of different needle ages	Quantum yield
Zone mean time (ZMT)	Diffuse light coefficient (0–1)	Carboxylation efficiency
Greenwich mean time (GMT)	Mean leaf-sun angle (60°)	Intercellular CO_2 conc.
Longitude		Dark respiration
Latitude		
Daily maximum PPFD		

dominated by Norway spruce trees in south-west Sweden typically had a LAI in the range of 6–10 (Wallin unpubl.). The LAI values of different shoot age classes were calculated from the data in Table 12. The modelling of the shoot photosynthesis was based on the net photosynthetic properties in Table 9. The data for the 2 yr old shoots were used for all age classes of that age and older. No attempt has been made to compensate for the diurnal variation in temperature or gas phase conductance. It is assumed that the intercellular CO_2 concentration is constant during the day.

The results of the simulation show that the canopy photosynthesis in an 8 yr old stand with LAI = 2 will hardly be affected by the current ambient concentrations of ozone (NF) and that an ozone concentration raised to the NF+ concentration level will only have a slight negative effect (Table 13). However, in an 89 yr old stand with the same LAI, the NF concentrations will result in a reduction of c. 10% in the canopy photosynthesis, with a further reduction to 36% at NF+ concentrations. The simulation also indicates that the negative ozone effects will increase at lower light intensities, caused either by a mutual shading of large proportions of the canopy because of high LAI or by reduced radiation as a result of cloudy weather. The increasing ozone effect on canopy photosynthesis with decreasing light intensities in the canopy can probably be attributed to the ozone-induced reduction in the apparent quantum yield. A significant reduction in the apparent quantum yield was observed only in the NF+ treatment, which suggests that a LAI-dependent effect will operate only at high ozone concentrations.

The data presented in Table 13 suggest that the effect of ozone on canopy photosynthesis of young trees will be small, while the effect of ozone on canopy photosynthesis of stands of older trees will be larger, particularly in stands with dense canopies. It is also important to consider that the ozone concentrations at the height of the canopy in a mature forest approach those in the NF+ treatment. The ozone concentration above and within a mature Norway spruce forest is very similar during daylight hours, because of vertical mixing (Pleijel, pers. comm), while the ozone concentration close to the ground at the experimental site was reduced as a result of local deposition of ozone (Table 6). Ozone concentrations measured at higher levels (5–20 m) should therefore be fed into concentration-response relationships rather than measurements of the concentrations in air entering an open top chamber, in order to predict ozone effects on older trees.

However, the model may also overestimate the negative effects of ozone, as it does not mimic the heterogeneity of the canopy, e.g. that the position of older shoots will result in more mutual shading, as well as more shading by stem and branches of older shoots, than for young shoots. In this case, older shoots would contribute less to the overall canopy photosynthesis, thus decreasing the overall effect of ozone.

Another uncertainty, especially for conifers, is whether or not ozone affects shoots on older trees to the same extent as shoots on young trees, such as those used in this study. Very limited information exists on effects of ozone on adult trees (Pye 1988). In leaves from mature trees and seedlings of red oak, exposed to ozone for two years, the photosynthesis of leaves of seedlings was less affected than that of leaves of mature trees (Hanson et al. 1994). These authors also emphasize the errors that may be introduced if responses in young seedlings are used to predict responses of mature trees, as well as by extrapolating data from one species to another. Contradictory to these results, Grulke and Miller (1994) found in a study of giant sequoia in California that shoots of older trees were less affected by ozone, because of lower stomatal conductances and thus lower ozone uptake, and also

Table 12. Relative distribution (%) of needle leaf area between different age-classes of shoots of Norway spruce trees of different ages.

	Tree age	
Shoot age-class	8 yr old[1]	89 yr old[2]
current year	72	16
1 yr old	19	15
2 yr old	7	16
3 yr old	2[3]	15
≥ 4 year		38

[1] from Wallin and Ottosson (unpubl.)
[2] from Schulze et al. (1977)
[3] ≥ 3-yr

Table 13. Results from modelling the canopy photosynthesis of Norway spruce during a single day presented as g C m^{-2} d^{-1} and in % of CF. The simulation were made for different tree ages, LAI, light conditions and ozone exposures. Abbreviations as in Table 1.

Treatment	Max. PPFD	Diff. Coef.	Canopy photosynthesis g C m^{-2} d^{-1}							
			LAI = 2				LAI = 6		LAI = 10	
			Tree age							
			8 yr		89 yr		89 yr		89 yr	
CF	1600	0.5	9.8		8.6		12.5		9.3	
NF			10.0	+2%	7.6	−12%	11.1	−11%	8.3	−11%
NF+			9.2	−6%	5.5	−36%	7.5	−40%	4.5	−52%
CF	600	1.0	7.7		6.9		8.7		5.5	
NF			8.0	+4%	6.1	−12%	7.7	−12%	5.0	−9%
NF+			7.2	−7%	4.4	−36%	4.9	−44%	2.1	−62%

because the needles of adult shoots have a different morphology, e.g. thicker mesophyll and thus lower mesophyll conductances. These shifts in the shoot properties also apply to Norway spruce trees. The importance of mesophyll conductance for the ozone effect is also supported by our ultrastructural studies, showing that the changes begin in the outer mesophyll layer of the needles (Sutinen et al. 1990). Furthermore, this study also shows that the changes begin on the side of the needles facing the sky, suggesting that the effect of ozone is influenced by the light intensity. The ozone effect may therefore be less in a closed canopy of older trees where a large proportion of the shoots are exposed to high light intensities only rarely. Since this study was carried out using eight-year-old trees, most of the shoots were frequently exposed to relatively high light intensities. However, the trees were propagated from cuttings from 30 yr old trees, with some remaining characteristics of an adult tree, such as many cell layers in the mesophyll. Furthermore, needle loss in trees in the field, which has been observed all over Sweden, will significantly increase the light intensity reaching the lower parts of the canopy, and this, in turn, will increase any light-enhanced ozone injury. However, the interaction beween light intensity and ozone is unclear, as was recently illustrated by Tjoelker et al. (1993), who showed that sugar maple was less sensitive to ozone at high than at low light intensities while the opposite was true of a hybrid poplar.

The results on ultrastructure, chlorophyll content, net photosynthesis, growth and biomass, presented in this paper, as well as the modelling study, indicate that current concentrations of ozone can cause negative effects on Swedish forests. Net photosynthesis is directly coupled to production and any effect of ozone on these processes is likely to affect growth. However, in order to assess the effect of ozone on forest stands, it is necessary to quantify the effects of ozone on whole tree carbon balance since several studies have shown that ozone affects carbon allocation to the roots (Kelly et al. 1993, Gorissen et al. 1994). If ozone disturbs the carbon alloca-

tion to roots, this will have long-term implications for tolerance to drought, for nutrient uptake, and for general vitality. Finally, to assess the effects of ozone in forest stands it is necessary also to quantify the effects on shoots of older trees of Norway spruce.

Conclusions

1) This study has shown that ozone can change the ultrastructure of needles, decrease chlorophyll content and reduce photosynthesis in shoots of Norway spruce. The extent of these effects depends on length of exposure and shoot age. 2) The results indicate that the AOT dose concept using 40 ppb as a threshold, might be too high a threshold when photosynthesis is used as the effect parameter. 3) A non-significant 6–20% decrease in aboveground biomass was observed. 4) In a model approach, using the data produced on photosynthesis in different shoot ages, it was shown that ozone has the potential to decrease canopy photosynthesis and that the decrease becomes larger with increasing tree age. 5) There is no indication that the formation of tropospheric ozone will decrease in the near future. Thus, the implications from this study are that tropospheric ozone will be a stress factor for Norway spruce in Sweden in the future. 6) In order to assess the ecological and economic impacts of this air pollution problem, research is needed on the effects of ozone on whole tree carbon allocation as well as on the effects in shoots of adult trees.

Acknowledgements – Thanks are due to B. Berglind, C. Westberg, P. Almbring and L. Kihlström for technical assistance, to L. Nestor for assistance with building the gas exchange system, to E. Knudsen, the secretary for the project, to J. Bergholm for his expertise in soil chemistry, to P. Barklund for her expertise on pathogens, to B. Ehnström for his expertise on insects. This work was supported by the Swedish Environmental Protection Agency, the Swedish Environmental Research Institute (IVL), the Swedish Council for Forestry and Agricultural Research, the Joint Committee of Power and Heating Producers on Environmental Issues (KVM), the Swedish Pulp and Paper Association, Volvo, Saab Automobile AB and Nils and Dorthi Troëdssons Research Foundation.

References

Anfossi, D., Sandroni, S. and Viarengo, S. 1991. Tropospheric ozone in the nineteenth century: the Montcalieri series. – J. Geophys. Res. 96: 349–352.

Arnon, D. I. 1949. Copper enzymes in isolated chloroplasts. Polyphenol-oxidase in Beta vulgaris. – Plant Physiol. 24: 1–15.

Aronsson, A. 1985. Indication of stress at unbalanced nutrient contents of spruce and pine. – Kungl. Skogs Lantbr. akad. Tidskr. Suppl. 17: 40–50 (in Swedish).

Beck, J. and Grennfelt, P. 1994. Estimate of ozone production and destruction over northwestern Europe. – Atmos. Environ. 28: 129–140.

Beyers, J. L., Riechers, G. H. and Temple, P. J. 1992. Effects of long-term ozone exposure and drought on the photosynthetic capacity of ponderosa pine (Pinus ponderosa Laws.). – New Phytol. 122: 81–90.

Botkin, D. B., Smith, W. H. and Carlson, R. W. 1971. Ozone suppression of white pine net photosynthesis. – J. Air Pollut. Control Ass. 21: 778–780.

Brown, K. A. and Roberts, T. M. 1988. Effects of ozone on foliar leaching in Norway spruce (Picea abies L. Karst): confounding factors due to NO_x production during ozone generation. – Environ. Pollut. 55: 55–73.

Costonis, A. C. 1970. Acute foliar injury of eastern white pine induced by sulphur dioxide and ozone. – Phytopathology 60: 994–999.

Dahlberg, K., Lövblad, G. and Steen, B. 1987. Atmosfärskemisk övervakning vid IVL:s PMK-stationer. – Swedish Environ. Protect. Agency, Report 3473 (in Swedish).

Eamus, D., Barnes, J. D., Mortensen, L., Ro-Poulsen, H. and Davison, A.W. 1990. Persistent stimulation of CO_2 assimilation and stomatal conductance by summer ozone fumigation in Norway spruce. – Environ. Pollut. 63: 365–379.

Febo, A., De Santis, F. and Perrino, C. 1987. Measurement of atmospheric nitrous and nitric acid by means of annular denuders. – In: Angeletti, G. and Restelli, G. (eds), Physico-chemical behaviour of atmospheric pollutants. D. Reidel Publ., Dordrecht, pp. 121–125.

Feister, U. and Warmbt, W. 1987. Long-term measurements of surface ozone in the German Democratic Republic. – J. Atmos. Chem. 5: 1–21.

Ferm, M. 1986. A Na_2CO_3-coated denuder and filter determination of gaseous HNO_3 and particulate NO_3^- in the atmosphere. – Atmos. Environ. 20: 1193–1201.

Fuchs, M., Schulze, E.-D. and Fuchs, M. I. 1977. Spacial distribution of photosynthetic capacity and performance in a mountain spruce forest of northern Germany. II. Climatic control of carbon dioxide uptake. – Oecologia 29: 329–340.

Fuhrer, J. and Achermann, B. (eds) 1994. Critical levels for ozone – a UNECE workshop report. Schriftenreihe der Liebefeld, No. 16. – Swiss Fed. Res. Stat. Agricult. Chem. Environ. Hygiene.

Gorissen, A., Joosten, N.N., Smeulders, S.M. and Van Veen, J.A. 1994. Effects of short-term ozone exposure and soil water availability on the carbon economy of juvenile Douglas-fir. – Tree Physiol. 14: 647–657.

Grulke, N.E. and Miller, P.R. 1994. Changes in gas exchange characteristics during the life span of giant sequoia: implications for response to current and future concentrations of atmospheric ozone. – Tree Physiol. 14: 659–668.

Hall, J.D., Barr, R., Al-Abbas, A.H. and Crane, F.L. 1972. The ultrastructure of chloroplasts in mineral-deficient maize leaves. – Plant Physiol. 50: 404–409.

Hanson, P.J., Samuelson, L.J., Wullschleger, S.D., Tabberer, T.A. and Edwards, G.S. 1994. Seasonal patterns of light-saturated photosynthesis and leaf conductance for mature and seedling Quercus rubra L. foliage: differential sensitivity to ozone exposure. – Tree Physiol. 14: 1351–1366.

Harris, G. W., Carter, W. P. L., Winer, A. M., Graham, R. A. and Pitts, Jr, J.N. 1982. Studies of trace non-ozone species produced in a corona discharge ozonizer. – J. Air Pollut. Control Ass. 32: 274–276.

Heagle, A., Body, D. and Heck, W. 1973. An open-top field chamber to assess the impact of air pollution on plants. – J. Environ. Qual. 2: 365–368.

Ingestad, T. 1979. Mineral nutrition requirements of Pinus sylvestris and Picea abies seedlings. – Physiol. Plant. 45: 373–380.

Kärenlampi, L. and Houpis, J.L.J. 1986. Structural conditions of mesophyll cells of Pinus ponderosa var. scropuorum after SO_2 fumigation. – Can. J. For. Res. 16: 1381–1385.

Kats, G., Olszyk, D. M. and Thompson, C. R. 1985. Open top experimental chambers for trees. – J. Air Pollut. Control Ass. 35: 1298–1301.

Kelly, J. M., Taylor, Jr, G. E., Edwards, N. T., Adams, M. B., Edwards, G. S. and Friend, L. 1993. Growth, physiology and nutrition of Loblolly pine seedlings stressed by ozone and acidic precipitation: A summary of the Ropis-South project. – Water Air Soil Pollut. 69: 363–391.

Kindbom, K., Lövblad, G., Peterson, K. and Grennfelt, P. 1995. Concentrations of tropospheric ozone in Sweden. – Ecol. Bull. (Copenhagen) 44: 35–42.

Kohut, R. J., Laurence, J. A., Amundson, R. G., Raba, R. M. and Melkonian, J. J. 1990. Effects of ozone and acidic precipitation on the growth and photosynthesis of red spruce after two years of exposure. – Water Air Soil Pollut. 51: 277–286.

Krause, G. H. M. 1988. Ozone-induced nitrate formation in needles and leaves of Picea abies, Fagus sylvatica and Quercus robur. – Environ. Pollut. 52: 117–130.

– and Höckel, F. E. 1988. Open-Top Kammer-Anlage Eggegebirge. Ein Beitrag zur Kausalanalyse der neuartigen Waldschäden. – Staub-Reinhalt. der Luft 48: 427–432.

Landsberg, J.J. 1986. Physiological ecology of forest production. – Academic Press, London.

Lange, O. L., Zellner, H., Gebel, J., Schramel, P., Köstner, B. and Czygan, F.-C. 1987. Photosynthetic capacity, chloroplast pigments, and mineral content of the previous year's spruce needles with and without the new flush: Analysis of the forest decline phenomenon of needle bleaching. – Oecologia 73: 351–357.

Lefohn, A. S., Shadwick, D. S., Feister U. and Mohnen, V. A. 1992. Surface-level ozone: climate change and evidence for trends. – J. Air Waste Manage. Ass. 42: 136–144.

Linder, S. 1980. Chlorophyll as an indicator of nitrogen status of coniferous seedlings. – New Zeal. J. For. Sci. 10: 166–75.

– and Rook, D. A., 1984. Effects of mineral nutrition on carbon dioxide exchange and partitioning of carbon in trees. – In: Bowen, G. D. and Nambiar, E. K. S. (eds), Nutrition of plantation forests. Academic Press, London, pp. 211–236.

Lövblad, G. and Sjöberg, K. 1988. Atmosfärskemisk övervakning vid IVL:s PMK-stationer. Rapport från verksamheten 1988 – Swedish Environ. Protect. Agency, Solna, Report 3645 (in Swedish).

– , Andreasson, K. and Dahlberg, K. 1985. Sammanställning av resultat från mätningar av luftdata vid svenska EMEP-stationer 1979–1985. – Swedish Environ. Protect. Agency, Solna. Report 3234 (in Swedish).

– , Andreasson, K. and Dahlberg, K. 1986. Atmosfärskemisk övervakning vid IVL:s PMK-stationer. – Swedish Environ. Protect. Agency, Solna. Report 3329 (in Swedish).

– , Sjöberg, K., Hållinder, C. and Peterson, K. 1989. Atmosfärskemisk övervakning vid IVL:s PMK-stationer. Rapport från verksamheten 1989. – Swedish Environ. Protect. Agency, Solna, Report 3787 (in Swedish).

– , Sjöberg, K., Hållinder-Ehrencrona, C. and Peterson, K. 1990. Atmosfärskemisk övervakning vid IVL:s PMK-stationer. Rapport från verksamheten 1990. – Swedish Environ. Protect. Agency, Solna, Report 3940 (in Swedish).

Maier-Maercker, U. 1986. Histological studies on the bundle sheath in needles of *Picea abies* (L.) Karst., diseased or fumigated with SO_2. – Eur. J. For. Pathol. 16: 352–359.

Olszyk, D. M., Dawson, P. J., Morrison, C. L. and Takemoto, B. K. 1990. Plant response to nonfiltered air vs. added ozone generated from dry air or oxygen. – J. Air Waste Manage. Assoc. 40: 77–81.

Panzer, K. F. 1988. Harmonizing surveys of visible symptoms – possibilities and constraints of international cooperation. – In: Cape, J. N. and Mathy, P. (eds), Scientific basis of forest decline symptomatology. Comm. European Commun., Brussels, Report No. 15: 217–225.

Percy, K. E. and Riding, R. T. 1981. Histology and histochemistry of elongation needles of *Pinus strobus* subjected to a long-duration, low-concentration exposure of sulfur dioxide. – Can. J. Bot. 59: 2558–2567.

Pye, J. M. 1988. Impact of ozone on the growth and yield of trees: a review. – J. Environ. Qual. 17: 347–360.

Reich, P. B. 1983. Effects of low concentrations of ozone on net photosynthesis, dark respiration, and chlorophyll contents in ageing hybrid poplar leaves. – Plant Physiol. 73: 291–296.

– , and Amundson R. G. 1985. Ambient levels of ozone reduce net photosynthesis in trees and crop species. – Science 230: 566–570.

Schulze, E.-D., Fuchs, M. I. and Fuchs, M. 1977. Spatial distribution of photosynthetic capacity and performance in a mountain spruce forest of northern Germany. I. Biomass distribution and daily CO_2 uptake in different crown layers. – Oecologia 29: 43–61.

Seiler, J. R., Tyszko, P. B. and Chevone, B. I. 1994. Effects of long-term ozone fumigations on growth and gas exchange of Fraser fir seedlings. – Environ. Pollut. 85: 265–269.

Sesták, Z., Catsky, J. and Jarvis, P. G. 1971. Plant photosynthetic production Manual of methods. – Dr W. Junk N. V. Publ., The Hague.

Seufert, G. and Arndt, U. 1985. Open-top-Kammern als Teil eines Konzepts zur ökosystemaren Untersuchung der neuartigen Waldschäden. – Allg. Forst Zeit. 40: 13–18.

Soikkeli, S. and Kärenlampi, L. 1984. The effects of nitrogen fertilization on the ultrastructure of mesophyll cells of conifer needles in northern Finland. – Eur. J. For. Pathol. 14: 129–136.

Sutinen, S. 1987a. Ultrastructure of mesophyll cells of spruce needles exposed to O_3 alone and together with SO_2. – Eur. J. For. Pathol. 17: 362–368.

– 1987b. Cytology of Norway spruce needles. II. Changes in yellowing spruces from the Taunus mountains, West Germany. – Eur. J. For. Pathol. 17: 74–85.

– , Skärby, L., Wallin, G. and Selldén, G. 1990. Long-term exposure of Norway spruce, *Picea abies* (L.) Karst., to ozone in open-top chambers. II. Effects of the ultrastructure of needles. – New Phytol. 115: 345–355.

Taylor, Jr, G. E., Owens, J. G., Grizzard, T. and Selvidge, W. J. 1993. Atmosphere × canopy interactions of nitric acid vapor in loblolly pine grown in open-top chambers. – J. Environ. Qual. 22: 70–80.

Thomson, W. W. and Weier, T. E. 1962. The fine structure of chloroplasts from mineral-deficient leaves of *Phaseolus vulgaris*. – Amer. J. Bot. 49: 1047–1055.

Tjoelker, G., Volin, J. C., Oleksyn, J. and Reich, P. B. 1993. Light environment alters response to ozone stress in seedlings of *Acer saccharum* Marsh. and hybrid *Populus* L. – New Phytol. 124: 627–636.

Volz, A. and Kley, D. 1988. Evaluation of the Montsouris series of ozone measurements made in the nineteenth century. – Nature 332: 240–242.

Wallin, G. 1990. On the impact of tropospheric ozone on photosynthesis and stomatal conductance of Norway spruce, *Picea abies* (L.) Karst. – Ph.D. thesis, Univ. of Göteborg, Sweden.

– and Skärby, L. 1992. The influence of ozone on the stomatal and non-stomatal limitation of photosynthesis in Norway spruce, *Picea abies* (L.) Karst., exposed to soil moisture deficit. – Trees 6: 128–136.

– , Skärby, L. and Selldén, G. 1990. Long-term exposure of Norway spruce, *Picea abies* (L.) Karst., to ozone in open-top chambers. I. Effects on the capacity of net photosynthesis, dark respiration and leaf conductance of shoots of different ages. – New Phytol. 115: 335–344.

– , Skärby, L. and Selldén, G. 1992a. Long-term exposure of Norway spruce, *Picea abies* (L.) Karst., to ozone in open-top chambers. III. Effects on the light response of net photosynthesis in shoots of different ages. – New Phytol. 121: 387–394.

– , Ottosson, S. and Selldén, G. 1992b. Long-term exposure of Norway spruce, *Picea abies* (L.) Karst., to ozone in open-top chambers. IV. Effects on the stomatal and non-stomatal limitation of photosynthesis and on the carboxylation efficiency. – New Phytol. 121: 395–401.

Whatley, J. M. 1971. Ultrastructural changes in chloroplasts of *Phaseolus vulgaris* during development under conditions of nutrient deficiency. – New Phytol. 70: 725–742.

Wilkinson, T. G. and Barnes, R. L. 1973. Effects of ozone on $^{14}CO_2$ fixation patterns in pine. – Can. J. Bot. 51: 1573–1578.

Zimmermann, R., Oren, R., Schulze, E.-D. and Werk, K. S. 1988. Performance of two *Picea abies* (L.) Karst. stands at different stages of decline. II. Photosynthesis and leaf conductance. – Oecologia 76: 513–518.

Ecological Bulletins 44: 147–157. Copenhagen 1995

Concentrations of mineral nutrients and arginine in needles of Picea abies trees from different areas in southern Sweden in relation to nitrogen deposition and humus form

Anders Ericsson, Mats Walheim, Lars-Gösta Nordén and Torgny Näsholm

Ericsson, A., Walheim, M., Nordén, L.-G. and Näsholm, T. 1995. Concentrations of mineral nutrients and arginine in needles of *Picea abies* trees from different areas in southern Sweden in relation to nitrogen deposition and humus form. – Ecol. Bull. (Copenhagen) 44: 147–157.

The study describes the occurrence of mineral nutrient imbalances in *Picea abies* (L.) Karst. trees in southern Sweden in relation to both deposition of nitrogen and humus form. Concentrations of mineral nutrients and free arginine were used to monitor imbalanced nutrition of trees caused by surplus nitrogen availability in the soil and/or low availability of other mineral nutrients. The major part of the investigation was performed on permanent sample plots within the Swedish National Forest Inventory (NFI) representing forests typical of the region and both mor and peat plots were sampled. A number of mor plots, laid out by the National Board of Forestry (NBF) were also sampled. The NBF plots were considered to suffer from the "new type" of forest decline in Sweden when they were selected and, therefore, represented typically damaged forests of the region. Although total nitrogen and arginine concentrations of the trees in the NFI-study were highest in the area with the highest nitrogen deposition, the concentrations were low. High arginine concentrations and low P/N ratios were found in NFI trees growing on peat plots. The NBF plots showed, in comparison with mor plots of the NFI-study, a considerably greater frequency with high needle arginine concentrations. It is concluded that, although trees on peat soils in southern Sweden tended to have greater mineral nutrient imbalances in areas with the highest airborne deposition, the humus form was more important for the development of nutrient imbalances than the deposition level. The importance of studying representative material of the region, when forest health is investigated, is discussed.

A. Ericsson and T. Näsholm, Dept of Forest Genetics and Plant Physiology, The Faculty of Forestry, The Swedish Univ. of Agricultural Sciences, S-901 83 Umeå, Sweden. – M. Walheim and L.-G. Nordén, Dept of Forest Survey, The Faculty of Forestry. The Swedish Univ. of Agricultural Sciences, S-901 83 Umeå, Sweden.

Introduction

During the last decade, the role of mineral nutrient deficiencies in forest decline has received increased attention (e.g. Schulze 1989, Huettl and Fink 1991) and it is generally believed that mineral nutrient imbalances are, at least partially, caused by the deposition of airborne substances. The components of deposition, N and S compounds in particular, cause two distinct major effects on forest soils; increased N availability and decreased base-cation supply. Deposition will therefore, in the long run, cause imbalanced nutrition with surplus of N for forest trees.

The most frequently used measure of forest health is visual assessment of needle losses (cf. Huettl and Fink 1991). However, this method has a number of weaknesses (Innes 1988). Firstly, it is prone to errors related to the subjective determination of needle loss. Secondly, it does not account for natural variation in needle cover due to differences in soil water and mineral nutrient content. Thirdly, the method is insufficiently diagnostic since loss of needles can be caused by a number of stresses, both man made and natural. There is, therefore, a great need for more informative methods for monitoring forest health especially in relation to anthropogenic impacts on forests.

If surplus N or lowered base-cation access are a major threat to forests, measurements of mineral nutrient concentrations of needles may appear to be adequate for monitoring of effects of deposition. However, interpretation of such measurements is difficult since good data on concentrations critical for growth and/or vitality are not available. Furthermore it is questionable whether such absolute values can be defined. Critical values of mineral nutrient concentrations may vary depending on genetic orgin, climatic and edaphic conditions and other factors. To overcome these problems, another method to measure nutrient imbalances based on ratios between individual elements has been proposed (e.g. Ingestad 1979, Cape et al. 1990, Linder 1995, Rosengren-Brinck and Nihlgård 1995). This approach is advantageous because problems with variations in concentrations of the different elements between and within years may be reduced and therefore the interpretation of the results might be easier. However, as for mineral nutrient concentrations, values for critical ratios are uncertain.

As an alternative to measurements of mineral nutrient concentrations, specific physiological responses to mineral nutrient imbalances might be used. One such response to imbalanced nutrition is accumulation of amino acids and it has been demonstrated that the concentration of free arginine in needles of conifers reflects the availability of nitrogen in relation to other mineral nutrient elements (e.g. van Dijk and Roelofs 1988, Näsholm and Ericsson 1990, Edfast et al. 1990, Näsholm 1991, Ericsson et al. 1993). Analysis of arginine concentrations might therefore be a first step in a chain of analyses whereby the actual mineral nutrient imbalance is identified and followed by improvments of the nutrient status of the stand.

In a previous study, Ericsson et al. (1993) evaluated the possibility of using free arginine concentrations to indicate mineral nutrient imbalances in Norway spruce trees. The study was carried out in two areas in Sweden with low and high deposition, respectively. Sites were selected from permanent plots within the Swedish National Forest Inventory (NFI; Ranneby et al. 1987). These plots have been laid out in a systematically designed sampling scheme all over Sweden. The study was restricted to plots with low fertility because it was believed that such sites would be most prone to suffer from mineral nutrient imbalances given a certain amount of deposition (cf. Edfast et al. 1990).

The present investigation describes the mineral nutrient status of Norway spruce in southern Sweden. Free arginine concentrations of needles are used as an alternative method to monitor mineral nutrient imbalances. The major part of the study is based on samples taken from mor and peat plots within the network of permanent NFI-plots and comprise three regions with different deposition levels. A number of plots belonging to the Swedish National Board of Forestry (NFB) were studied. The NFI-plots represented the range of typical forests in southern Sweden while the NFB-plots were chosen to represent forests with damage symptoms typical of the region.

Material and methods

The study was carried out in southern Sweden (Figs 1–3). The studied sites were selected from two different types of permanent monitoring plots. In one, belonging to the Swedish National Forest Inventory (NFI-plots), the plots had been laid out in a systematically designed sampling system. In the other the plots had been selected by the regional county administrations on behalf of the Swedish National Board of Forestry (NBF-plots) with the aim of covering forests with typical damage symptoms. Thus, the first type of plot can be seen as representative of all forests in this region while the second type is considered to represent damaged forests.

NFI-plots

The permanent NFI sample plots, c. 22 500 in total, were laid out in Sweden between 1983 and 1987 in a systematically designed sampling scheme (Ranneby et al. 1987). They are assessed every fifth year; stand, site and tree characteristics being recorded and described (Lundmark and Odell 1984, Anon. 1992). In the present study, plots with an age of 30 yr or older and with at least six Norway spruces (Picea abies (L.) Karst.) belonging to the dominating or co-dominating tree class were selected. Half of the plots had the humus form mor and the other half peat (including the transition form peat/mor). Average stand characteristics are given in Table 1. Needle-loss inventories carried out by NFI-teams during 1984–1987 on a limited number of trees per plot, showed that c. 12% of the trees had needle losses > 20%. Seven peat plots (NFI sample plots), known to have trees with high needle arginine concentrations from a previous study (Ericsson et al. 1993), and seven new mor plots selected close to them were also included in the investigation (see below). However, to avoid bias only the new mor plots were used for the final statistical evaluation of the results.

Close mor and peat plots were paired. For 80% of the plots there was a distance of 0.3–1.1 km between the two paired areas. For the rest, the distance varied between 1.1 and 10 km. Some of the chosen plots had to be excluded because of clearcutting. The final number of pairs evaluated was 52. The aim of the pairwise design was to determine whether differences between plots were caused by local deposition conditions or by humus forms. However, since there was no pairwise correlation between mor and peat plots with reference to arginine or mineral nutrient concentrations (data not shown) no further results of this part of the study are reported below.

The western, central and eastern areas have been estimated to have an annual total deposition of 15–30, 3–6 and 7–15 kg N ha^{-1}, respectively (Westling et al. 1992). So, before sampling, the studied area was divided into three subareas representing different levels of deposition (e.g. Fig. 1).

From each NFI-plot (20 m diam.) the six spruces with

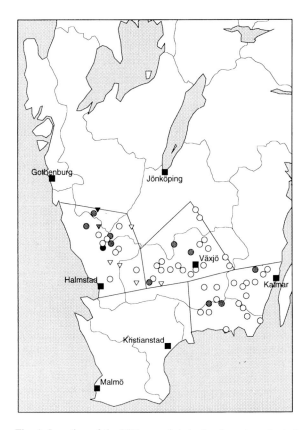

Fig. 1. Location of the NFI mor plots in the three investigated areas in southern Sweden classified in relation to needle arginine concentrations of the sampled Norway spruces. Mor plots sampled close to peat plots known to have high arginine concentrations (Ericsson et al. 1993) indicated with triangles. ○, △ ≤ 1 μmol (g dw)⁻¹; ▧, ▨ >1 <5 μmol (g dw)⁻¹; ●, ▲ ≥5 μmol (g dw)⁻¹. Results from pooled needles grown out in 1988 and 1989 and sampled in late September to mid-October 1991.

the largest diameter at breast height were selected for needle sampling. These trees, with few exceptions, belonged to the dominating or co-dominaiting height class. The trees were climbed and two exposed twigs from each of two different branches from opposite sides of the tree in the uppermost 25% of the crown were sampled using a cutter. Twigs with needles grown out in 1988 and 1989 were cut into pieces and a composite sample for each tree was immediately put into a plastic bag for instant storage in a well insulated box containing dry ice. Thereafter the needles were stored at −20°C until further analysis. The sampling was performed during late September to mid-October 1991, i.e. after the growing season had ended.

NBF-plots

The NBF plots were subjectively laid out during the period from 1984 to 1987. The plots were established to monitor the observed new forest decline in Sweden (Anon. 1984). The plots (30 × 30 m) were placed in old

and middle-aged stands with needle losses characteristic of the forest decline in the region. The instructions, given by the National Board of Forestry, directed that selected stands should preferably be located downwind of local discharge sources. Estimation of needle losses is carried out every year for all dominating and co-dominating trees on the plots, and every fifth year needles have been sampled from trees close to the border of the plots for mineral nutrient analysis. The needle-loss inventories performed by NBF-teams showed that c. 22% of the trees had needle loss >20% when the plots were laid out.

In the present study, a portion of the NBF plots (located in the counties of Älvsborg and Göteborg-Bohus) were investigated (cf. Fig. 3, Table 1). Only mor plots were chosen. Needles were sampled from four or five dominating trees just outside the plots by staff from the regional county administrations. Two twigs from the upper half of the tree were shot down using a rifle. The twigs were handled, principally, in the same way as described above for the NFI-plots, but for technical reasons, the samples sometimes were not placed in a freezer until two days after collection. Furthermore, the sampling

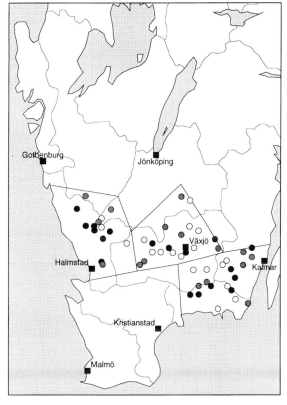

Fig. 2. Location of the NFI peat plots in the three investigated areas in southern Sweden classified in relation to needle arginine concentrations of the sampled Norway spruces. ○ ≤ 1 μmol (g dw)⁻¹; ▧ >1 <5 μmol (g dw)⁻¹; ● ≥5 μmol (g dw)⁻¹. Results from pooled needles grown out in 1988 and 1989 and sampled in late September to mid-October 1991.

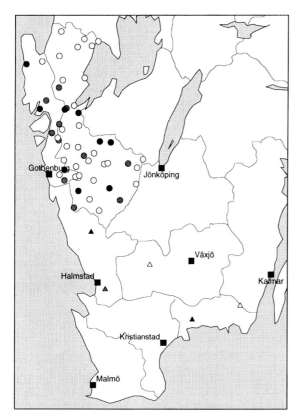

Fig. 3. Location of the NBF (circles) and IVL (triangles) mor plots in southwestern Sweden classified in relation to needle arginine concentrations of the sampled Norway spruces. ○, △ ≤ 1 μmol (g dw)⁻¹; ◍, ◮, > 1 < 5 μmol (g dw)⁻¹; ●, ▲ ≥ 5 μmol (g dw)⁻¹. Results from needles grown out in 1989 and sampled in January and February 1991 (NBF). Pooled needles grown out in 1988 and 1989 and sampled in late September to mid-October 1991 (IVL).

was performed during January and February 1991 and only needles grown out in 1989 were collected.

Table 1. Average stand characteristics (± SD) for the the different plot types and areas.

Plot type, stand area	Humus form	No. of plots	Site index, m*	Mean stand age, year	Mean height, m
NFI, all	mor	60	30±3	67±15	21±3
	peat	53	28±4	71±20	20±3
NFI, west	mor	19	31±2	68±15	22±3
	peat	14	26±5	75±27	18±3
NFI, mid	mor	21	29±3	72±17	21±3
	peat	19	27±3	73±18	21±3
NFI, east	mor	20	31±3	62±10	20±4
	peat	20	30±4	67±15	21±2
NBF	mor	47	28±3	78±17	23±3

* Height of dominating trees when 100 yr at breast height. For site classification, see Hägglund and Lundmark (1977).

IVL-plots

In addition to the NBF plots in Älvsborg and Göteborg-Bohus counties, five similar mor plots (IVL-plots) south of these counties were sampled (cf. Fig. 3). They were chosen since they, except for being a part of the NBF-series of plots, also are included in a monitoring program carried out by the Swedish Environmental Research Institute (IVL, Aneboda, Sweden). The sampling was performed by us in the same way as for the NFI-plots.

Chemical analyses

The chemical analyses, including amino acids (only data for arginine is presented) and mineral nutrients (N, P, K, Ca, Mg, Mn, Al and S), were determined by HPLC, CHN-elemental and ICP analysis, respectively, according to Ericsson et al. (1993). Samples taken from each plot were pooled before analysis.

Results
NFI plots

When comparing needles from mor and peat plots, nearly all mineral nutrient concentrations differed significantly ($p < 0.05$). Nitrogen, Mg, and S were lower in trees on mor plots, while lower concentrations of P, Mn and Al were obtained in samples from peat plots (Table 2). No significant difference was found for K or Ca between the two types of plots.

The mineral nutrient/N ratio in needles differed significantly between the mor and peat plots for four of the elements; P/N, K/N, Mn/N and Al/N all being higher in trees from mor zones (Table 2).

Nitrogen was the only element reflecting the different levels of N deposition of the three subareas (Table 3). Thus, the highest concentration was found in needles from plots located in southwestern Sweden, the area where the highest N-deposition has been recorded. There was, however, no significant difference between the western and eastern areas. Calcium, Mn and Al concentrations increased from west to east on mor plots and likewise Ca, Mg and Mn concentrations on peat plots. However, with regard to either humus form or mineral element, only two of the three areas showed statistically significant differences.

Although just partially statistically significant, the P/N, S/N and Al/N ratios of the samples from the mor plots reflected N-deposition levels. In samples from peat plots, the P/N, S/N and Al/N ratios and also the Mg/N and the Ca/N ratios showed the same pattern. That is, mean values of all these ratios were lowest for the western, highest for the central, and intermediate for the eastern areas (Table 3). The K/N ratio did not differ significantly between mor or peat plots of the three areas. Mn/N values

Table 2. Mineral nutrient concentrations and ratios in Norway spruce needles from mor and peat plots in the NFI, NBF and IVL studies. Sampling of needles grown out in 1988–89 was performed in the autumn of 1991 for the NFI and IVL material, and needles grown out in 1989 were sampled in January and February 1991 for the NBF plots. Values ($\pm$ SD) with different letters are statistically different ($p < 0.05$; Student t-test). Mineral nutrient concentrations in mg (g dw)$^{-1}$; ratios in per cent of N concentration. ND = not determined.

	NFI-plots Mor (n = 60)	Peat (n = 53)	NBF-plots Mor (n = 47)	IVL plots Mor (n = 5)
N	10.9±1.3[a]	12.0±1.9[b]	12.9±1.8	13.0±2.6
P	1.08±0.25[a]	0.90±0.22[b]	1.08±0.22	1.14±0.26
K	3.76±0.65	3.52±0.78	4.31±0.77	3.43±0.65
Ca	6.96±2.18	7.44±2.06	6.09±1.96	7.42±1.93
Mg	0.91±0.23[a]	1.04±0.28[b]	0.99±0.20	0.90±0.28
S	0.92±0.10[a]	1.02±0.19[b]	ND	0.90±0.07
Mn	1.39±0.77[a]	0.73±0.44[b]	ND	1.65±0.95
Al	0.16±0.04[a]	0.09±0.04[b]	ND	0.15±0.08
P/N	10.0±2.5[a]	7.6±1.9[b]	8.4±1.9	9.1±2.8
K/N	34.8±6.6[a]	29.9±7.9[b]	33.7±6.7	26.5±1.6
Ca/N	64.9±22.2	63.7±21.8	48.2±18.0	59.3±20.1
Mg/N	8.5±2.7	8.9±2.9	7.8±2.0	7.0±2.0
S/N	8.6±1.5	8.7±2.0	ND	7.1±1.4
Mn/N	12.9±7.4[a]	6.1±3.4[b]	ND	14.4±10.9
Al/N	1.5±0.5[a]	0.8±0.4[b]	ND	1.3±0.7

increased from west to east. Again, in all cases statistically significant differences were only found for the mean values most differing from each other.

Arginine concentrations of the needles were used to classify the NFI-plots into three groups (Figs 1, 2): plots with low (≤ 1.0 µmol (g dw)$^{-1}$), slightly increased (1–5 µmol (g dw)$^{-1}$) and high (≥ 5 µmol (g dw)$^{-1}$) levels of arginine (cf. Ericsson et al. 1993). Samples from only two mor plots showed high levels of arginine. In contrast c. 35% of the peat plots gave high arginine concentrations.

Plots giving slightly elevated arginine concentrations were more frequent when the humus form was peat. Low levels of arginine were recorded from c. 35 and 78% of the peat and mor plots, respectively.

Seven mor plots (37%; n = 19) of the western area had needle arginine concentration >1.0 µmol (g dw)$^{-1}$ and three (15%; n = 20–21) of each of the other two areas (Fig. 1). Corresponding figures for peat plots (Fig. 2) were 78, 53 and 65% of the western (n = 14), central (n = 19) and eastern (n = 20) areas, respectively. Thus, both

Table 3. Mineral nutrient concentrations in Norway spruce needles and the ratios to N in relation to regions and humus form (NFI study). Values ($\pm$ SD) with different letters and from plots with the same humus form are statistically different ($p < 0.05$; Student t-test). Mineral nutrient concentrations in mg (g dw)$^{-1}$; ratios in per cent of N concentration.

	Mor West (n = 19)	Mid (n = 21)	East (n = 20)	Peat West (n = 14)	Mid (n = 19)	East (n = 20)
N	11.5±1.3[a]	10.2±1.3[b]	11.0±1.1[a]	12.8±2.0[a]	11.0±1.3[b]	12.4±1.9[a]
P	1.09±0.21	1.11±0.26	1.04±0.28	0.88±0.14	0.87±0.18	0.95±0.28
K	3.82±0.43	3.55±0.80	3.91±0.62	3.71±0.52	3.56±0.92	3.33±0.77
Ca	5.26±1.37[a]	7.18±2.04[b]	8.33±1.92[b]	6.33±1.87[a]	7.70±2.34[ab]	7.96±1.65[b]
Mg	0.91±0.28	0.96±0.24	0.86±0.17	0.94±0.20[a]	1.01±0.30[ab]	1.13±0.29[b]
S	0.91±0.08	0.93±0.09	0.96±0.12	1.01±0.09	0.99±0.20	1.06±0.22
Mn	1.03±0.55[a]	1.21±0.82[a]	1.94±0.61[b]	0.59±0.22[a]	0.70±0.43[ab]	0.86±0.53[b]
Al	0.15±0.03[a]	0.16±0.05[ab]	0.18±0.04[b]	0.09±0.03	0.07±0.05	0.10±0.04
P/N	9.5±2.0	10.9±2.6	9.5±2.5	7.0±1.6	7.9±1.5	7.7±2.4
K/N	33.4±4.2	35.1±8.5	35.7±6.3	29.1±3.9	32.9±9.4	27.5±7.8
Ca/N	46.1±12.6[a]	70.8±20.7[b]	76.4±20.0[b]	50.7±18.9[a]	71.0±23.7[b]	65.8±18.6[b]
Mg/N	8.1±3.1[ab]	9.5±2.8[a]	7.9±1.7[b]	7.5±2.1[a]	9.3±2.8[b]	9.4±3.2[ab]
S/N	8.0±1.2[a]	9.2±1.5[b]	8.6±1.5[ab]	8.0±1.4	9.1±2.1	8.7±2.2
Mn/N	8.7±4.0[a]	12.1±8.7[a]	17.7±5.6[b]	4.7±1.8[a]	6.3±3.3[ab]	6.9±4.1[b]
Al/N	1.3±0.3[a]	1.6±0.6[ab]	1.6±0.4[b]	0.7±0.3	0.7±0.5	0.8±0.4
Arg	1.9±0.7	0.7±0.1	0.8±0.2	9.1±2.1	5.7±2.5	10.9±4.5

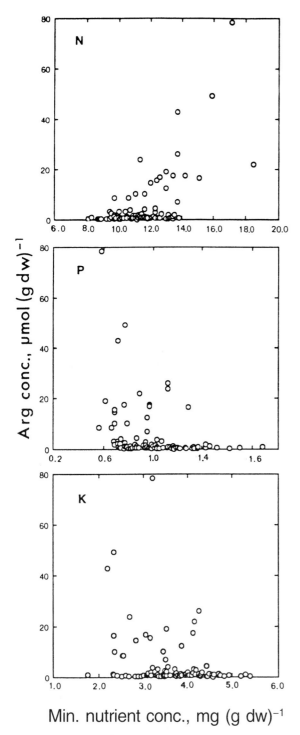

Fig. 4. The relationship between needle concentrations of arginine and corresponding concentrations of nitrogen, phosphorous and potassium. The values are for samples from both mor and peat plots of the NFI study. Results from pooled needles grown out in 1988 and 1989 and sampled in late September to mid-October 1991.

mor and peat plots having trees with elevated arginine concentrations were found to be more frequent in areas with higher N deposition.

Comparisons of arginine concentrations with mineral nutrient concentrations showed that there was a correlation between needle arginine concentration and N concentration (Fig. 4). When arginine N was subtracted from the total N pool, no such tendency remained. Regressions of concentrations of P, K or any other of the analysed elements against arginine concentration did not indicate any relationship (Fig. 4, data not shown).

Correlations of arginine concentrations with mineral nutrient/N ratios showed that arginine concentrations were strongly inversely correlated with P/N at P/N values <0.075; arginine concentrations being much higher at P/N values below this value than they are at values >0.075. This relationship remained also when arginine was plotted against ratios based on total N minus arginine N (data not shown). The arginine concentration was also markedly higher in some of the samples with a K/N ratio below c. 0.30 (Fig. 5). No correlations between any of the other mineral nutrient/N ratios and arginine concentrations were found (data not shown).

Peat plots of the NFI study with arginine concentrations ≥5 μmol (g dw)$^{-1}$ showed significantly higher N concentrations than corresponding plots with lower arginine levels (Table 4). Furthermore, the lowest P, K, and Mg concentrations were found from the peat plots with the highest arginine levels. Five of the mineral nutrient/N ratios were significantly lower for the high arginine peat plots compared to one or both of the other two groups, i.e. P/N, K/N, Ca/N, Mg/N and S/N (Table 4).

Except for the K/N ratio, no statistically significant differences between mor plots with low or intermediate arginine levels could be detected (Table 4). The high arginine plots were excluded from the Student t-test because of the low number of plots.

NBF-plots

The needles from trees on the NBF-plots, showed, on average, higher N and K concentrations than the needles from the mor plots of the NFI study (Table 2), while Ca was found in lower concentrations. Phosphorous and Mg showed approximately the same levels for the two types of plots. There were lower P/N and Ca/N ratios in needles from the NBF plots compared to those from the NFI mor plots (Table 2).

A comparison between the NBF plots and the NFI mor plots of the western subarea showed that all determined mineral elements, except for P, were highest in concentration in needles from the NBF-plots (cf. Tables 2, 3). The most pronounced difference in ratios between the above mentioned plots was the lower P/N ratio of the NBF-plots.

NBF plots with ≥5 μmol (g dw)$^{-1}$ of arginine accounted for c. 23% of the total number (Fig. 3). The

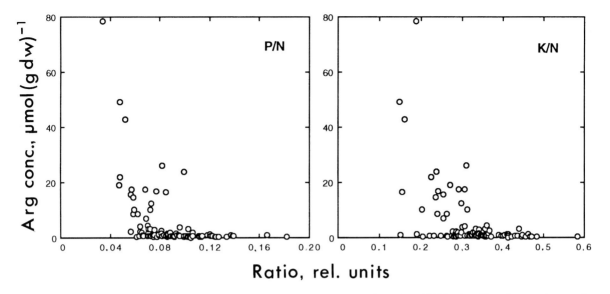

Fig. 5. The relationship between needle concentrations of arginine and corresponding P/N and K/N ratios. The values are for samples from both mor and peat plots of the NFI study. Results from pooled needles grown out in 1988 and 1989 and sampled in late September to mid-October 1991.

corresponding figure for plots with ≤ 1.0 μmol (g dw)⁻¹ was 36%. Thus, considerably more of the NBF plots had trees rich in arginine than the mor plots of the NFI-study (cf. Fig. 1).

No significant correlations between N, P or K, as well as for the P/N or K/N ratios, and arginine concentrations were found (Figs 6,7).

NBF plots with low, intermediate, and high concentrations of arginine (≤ 1, > 1 < 5, and ≥ 5 μmol (g dw)⁻¹, respectively), showed statistically significant differences in N concentration and in K/N ratio (Table 4). Plots with

high concentrations of arginine were found to have high concentration of N as well as low K/N ratio.

IVL-plots

The IVL plots were characterized by high N and low K concentrations (Table 2) compared to the other types of plots. Consequently, the lowest K/N ratios was also found for these plots. Three of the five IVL-plots had high concentrations of arginine in the needles; two with ≥ 5 μmol (g dw)⁻¹ and one with > 1 < 5 μmol (g dw)⁻¹ (Fig. 3).

Table 4. Mineral nutrient concentrations (mg (g dw)⁻¹) and ratios to N (%) in needles of Norway spruce after classification of plots in relation to arginin concentration. Values (± SD) with different letters are statistically different (p < 0.05; Student t-test). Each plot type separately tested. Mor plots of the NFI study with ≥ 5 μmol (g dw)⁻¹ not included in the statistical test. ND = not determined.

Plots Arg conc.	NFI, peat ≤ 1 (n = 17)	> 1 < 5 (n = 18)	≥ 5 (n = 18)	NFI, mor ≤ 1 (n = 46)	> 1 < 5 (n = 15)	≥ 5 (n = 2)	NBF, mor ≤ 1 (n = 15)	> 1 < 5 (n = 20)	≥ 5 (n = 12)
N	11.4±1.2ᵃ	11.3±1.1ᵃ	13.4±2.2ᵇ	10.8±1.3	11.3±1.5	11.3±0.4	12.4±1.6ᵃ	12.7±1.2ᵃᵇ	14.1±2.3ᵇ
P	0.97±0.18ᵃ	0.91±0.24ᵃᵇ	0.83±0.21ᵇ	1.11±0.26	1.00±0.19	0.83±0.06	1.01±0.19	1.13±0.19	1.08±0.29K
K	3.49±0.83ᵃᵇ	3.90±0.61ᵃ	3.16±0.72ᵇ	3.79±0.68	3.62±0.56	3.91±0.66	4.39±0.84	4.36±0.78	4.13±0.68
Ca	7.58±2.29	7.67±1.88	7.07±2.07	7.10±2.31	6.81±1.58	4.60±0.47	6.31±2.43	5.77±1.55	6.36±2.03
Mg	1.19±0.28ᵃ	0.96±0.28ᵇ	0.97±0.23ᵇ	0.94±0.23	0.82±0.24	0.69±0.06	0.96±0.19	1.01±0.20	1.00±0.22
S	1.06±0.21	1.03±0.20	0.97±0.14	0.93±0.10	0.89±0.09	0.85±0.16	ND	ND	ND
Mn	0.63±0.41	0.79±0.33	0.78±0.55	1.34±0.72	1.66±0.96	1.08±0.34	ND	ND	ND
Al	0.08±0.03	0.11±0.05	0.08±0.05	0.16±0.04	0.16±0.05	0.14±0.01	ND	ND	ND
P/N	8.5±1.4ᵃ	8.0±2.0ᵃ	6.3± .6ᵇ	10.4±2.5	8.9±2.0	7.3±0.2	8.1±1.3	9.0±2.0	7.8±2.2
K/N	31.0±7.7ᵃ	34.9±6.3ᵃ	23.8±5.1ᵇ	35.5±7.2ᵃ	32.0±3.1ᵇ	34.5±4.5	35.7±7.1ᵃ	34.5±6.0ᵃ	30.0±6.1ᵇ
Ca/N	67.6±22.2ᵃᵇ	69.1±19.6ᵃ	54.6±21.8ᵇ	66.7±23.2	61.7±17.9	40.8±5.7	52.8±24.5	45.6±11.8	46.9±17.6
Mg/N	10.6±2.7ᵃ	8.6±2.5ᵇ	7.6±2.8ᵇ	8.9±2.6	7.5±2.8	6.1±0.8	7.9±2.1	8.0±1.8	7.3±2.1
S/N	9.3±1.8ᵃ	9.2±1.9ᵃ	7.5±1.7ᵇ	8.8±1.5	8.0±1.5	7.5±1.1	ND	ND	ND
Mn/N	5.5±3.3	7.0±2.7	5.8±4.1	12.4±6.8	15.2±9.9	9.5±2.6	ND	ND	ND
Al/N	0.7±0.3ᵃ	1.0±0.4ᵇ	0.6±0.3ᵃ	1.5±0.4	1.5±0.6	1.2±0.1	ND	ND	ND

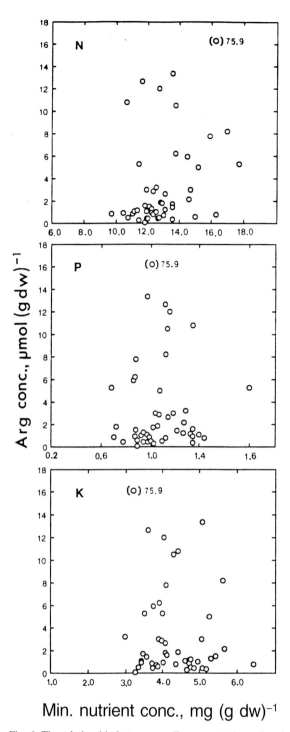

Fig. 6. The relationship between needle concentrations of arginine and corresponding concentrations of nitrogen, phosphorous and potassium. The values are for samples from the mor plots of the NBF study. Results from needles grown out in 1989 and sampled in January and February 1991.

Discussion

One aim of the present study was to determine whether or not different deposition levels were reflected in needle concentrations of mineral nutrients and arginine. The data show the range of concentrations found in forests typical of the southermost part of Sweden. A general problem with this kind of study is the lack of good data on concentrations critical for growth and/or vigour of trees. We chose to study older needles because these needles may reflect both excess nutrients and nutrient deficiencies better than younger needles (Edfast et al. unpubl.; cf. Cape et al. 1990, Rosengren-Brinck and Nihlgård 1995). Furthermore, the arginine concentration seems to increase with increasing needle age under nitrogen surplus conditions (Edfast et al. unpubl.). This choice does, however, make it more difficult to compare the results of the present study with those from previous investigations.

Concentrations of most mineral nutrients were low in the needles sampled from the NFI plots compared to levels recommended by other authors (e.g. Cape et al. 1990, Nihlgård 1990, Landmann et al. 1995). The mean N concentrations of the spruce needles were 11 and 12 mg (g dw)$^{-1}$ for mor and peat plots, respectively (Table 2), which is, according to Cape et al. (1990), lower than the limit below which visual deficiency symptoms should occur. The range of N concentrations in samples from the individual plots was 7.5–19 mg (g dw)$^{-1}$ (cf. Fig. 4). The mean P concentrations were 1.1 and 0.9 mg (g dw)$^{-1}$ for mor and peat plots, respectively, and the range for all plots was 0.6 – 1.9 mg (g dw)$^{-1}$ (cf. Fig. 4). These figures imply that most of the plots are P-limited according to the limitation value of 1.3 mg (g dw)$^{-1}$ suggested by Nihlgård (1990). Potassium ranged between 1.8 and 5.4 mg (g dw)$^{-1}$ and the mean values were 3.8 and 3.5 mg (g dw)$^{-1}$ for mor and peat plots, respectively. These mean values are lower than the 'intermediate or normal concentrations' (4.0–12.1 mg (g dw)$^{-1}$) proposed by Cape et al. (1990) and at least half of the plots have K deficient trees according to Nihlgård's limit of 3.5 mg (g dw)$^{-1}$. Mean Mg concentrations were just slightly deficient or on the limit of deficiency according to values suggested by Nihlgård (1990) but higher than 'deficiency range for growth' suggested by Cape et al. (1990). Mean Ca and S concentrations were in satisfactory amounts for both mor and peat plots as defined by the cited studies. Thus, many of the mineral elements were found in low concentrations compared to ranges given by e.g. Cape et al. (1990) in their survey of nutrient levels in stands from SW Germany to N Scotland. This difference might be a consequence of different climatic and edaphic conditions and of different levels of airborne deposition. However, one part of the difference between the results of the current study and those from other sites might be attributable to the use of older needles in our study. On the other hand, with an increase of the concentrations of the retranslocable mineral nutrients by 10%, which seems to be a general difference between the two needle-age classes

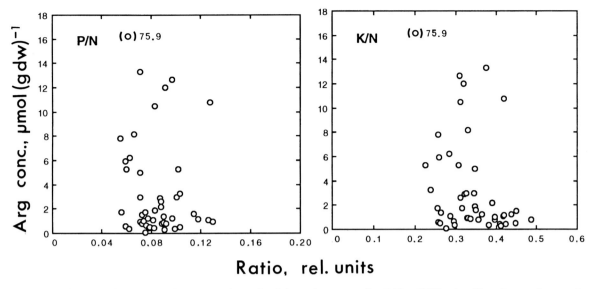

Fig. 7. The relationship between needle concentrations of arginine and corresponding P/N and K/N ratios. The values are for samples from the mor plots of the NBF study. Results from needles grown out in 1989 and sampled in January and February 1991.

(e.g. Tamm 1955, Rosengren-Brinck and Nihlgård 1995), our results would still be within reported 'normal' ranges.

An alternative way to analyse mineral nutrient data is to use ratios between N and other mineral elements (Ingestad 1979, Cape et al. 1990). The mean ratios of the NFI-study (Table 2) were in the same ranges as found in a previous study in southwestern Sweden (Ericsson et al. 1993). The similarities are most evident when comparing the ratios from the western subarea (Table 3), i.e., from the same region as used in the earlier study. Thus, the P/N ratios again indicated P deficiency, especially for peat plots. Potassium/N ratios were lower than those found in the first investigation, but this discrepancy may be explained by between-year changes in concentration of K (Aronsson 1985).

Mineral nutrient concentrations of the needles collected from the NBF plots showed, in general, higher values than was found for the needles of the mor plots in the NFI study (Tables 2, 3). This difference might be attributed to the usually higher mineral nutrient levels found in younger needles (e.g. Aronsson and Elowson 1980, Gebauer and Schulze 1991). The mean P/N ratio from the NBF plots was lower than the mean value from the NFI mor plots but was also lower than the mean value of samples from the mor plots of the western subarea (Tables 2, 3). However, for trees from all mor plots, the mean values of the P/N ratios were above the limit value (7.0–8.0) suggested by Ericsson et al. (1993). Corresponding Mg/N and K/N ratios showed similar or somewhat lower values than those obtained for the NFI study.

There is a need for methods to identify pre-visible symptoms of mineral nutrient imbalances in apparently healthy trees. Cape et al. (1990) presented a method of risk assessment based on threshold values of concentra-

tions and mineral nutrient ratios. An application of this method on our plot mean data did not indicate that any of the NFI or NBF plots were in the risk zone for damage (data not shown).

The results of the present study illustrate the difficulties of using certain limits for mineral nutrient concentrations and/or mineral nutrient ratios of needles to detect effects of factors such as air pollutants on the mineral nutrient status of conifers. Although analysis of ratios between mineral nutrients might be a useful tool in the assessment of nutrient balance, it has a number of weaknesses. As mentioned earlier, values of ratios critical for growth and/or vigour are not known. Moreover, when ratios of mineral nutrients/N are used, accumulation of N resulting from deficiency of any one nutrient or by surplus N availability, will lower all mineral nutrient/N ratios. These ratios cannot, therefore distinguish between imbalanced nutrition caused by excess N or by low availability of other nutrients. In the present study, a correlation between total N concentration and free arginine was found (Fig. 4). However, this correlation was clearly weakened when arginine was plotted against total N minus arginine N, i.e., the N dependence of arginine concentration is partly an autocorrelation. Likewise, accumulation of arginine causes increased total N concentration and by that lowered ratios of mineral nutrients/N.

Determination of free arginine concentration in conifer needles has been suggested as an alternative method for detection of N surplus in conifers (van Dijk and Roelofs 1988, Näsholm and Ericsson 1990, Edfast et al. 1990, Ericsson et al. 1993). The present study shows that trees with high arginine levels were most abundant on peat plots (Fig. 2) and they often had low P/N- and K/N ratios (Fig. 5, Table 4). These findings are in accordance with previous results by Ericsson et al. (1993), and are probably a result of the well known limitation in supply of P and K in peat soils (Clymo 1983).

A higher frequency of NBF plots than mor plots from the NFI study, had trees with high arginine concentrations (cf. Figs 1, 3). Thus, for both arginine and mineral nutrient concentrations a difference between mor plots of the NFI study and those of the NBF study was found. The difference in arginine concentrations between the two groups is probably not a result of differences in needle age because preliminary studies have demonstrated that arginine concentration increases with needle age (Edfast et al. unpubl.). The area covered by the NBF study does not over-lap the area covered by the NFI study and this might also affect the results. However, in our view these explanations do not account for the differences in needles from the two sets of plots. Instead, we believe that the differences reflect the different criteria used in selecting the plots. Since the NFI study plots represent all forests in the studied region but the NBF plots represent damaged forests, major differences are inevitable. It is clearly, therefore, extremely important to study truly representative material, with appropriate controls, when investigating forest health.

Some forms of the decline of European forests have been related to the mineral nutrient imbalances of trees (e.g. Schulze 1989). In the present study, three areas with different deposition levels were studied. Total N and arginine concentrations of needles were highest in the area with highest deposition. Moreover, the Ca and Mn concentrations and the ratios Ca/N and Mg/N were lowest in the same area. These tendencies could be signs of deposition-mediated effects. However, compared to results found in other studies, total N concentrations (Cape et al. 1990) and arginine concentrations (van Dijk and Roelofs 1988, Näsholm and Ericsson 1990) were low also in the western area. Peat soils, which are known to supply low amounts of P and K for tree growth, had high total N and arginine concentrations and low P/N and K/N ratios compared to mor soils. The more or less inherent tendencies for imbalanced nutrition on these soils might, however, have been aggravated by deposition. This might be reflected in the greater frequency of plots producing trees with high arginine concentrations in regions with high deposition. It can, however, be concluded that the humus form was more important for the development of nutrient imbalances than the deposition level. Moreover, the discrepancy between the NFI and the NBF mor plots emphasizes the importance of studying truly representative material when questions of effects of deposition on forest health are addressed.

Acknowledgements – We are sincerely grateful to all persons involved in the field sampling and which were employed by the Swedish National Board of Forestry, Älvsborg and Göteborg-Bohus county administrations, Tönnersjöheden Field Station, Dept of Forest Genetics and Plant Physiology, and Dept of Forest Survey. M. Zetterström is thanked for assistance in the laboratory. We are grateful to J. Blackwell for help with the English language. This work has been supported financially by the Swedish Environmental Protection Agency, the Swedish National Board of Forestry and the Swedish Council of Forestry and Agricultural Research.

References

Anon. 1984. Instruktion för bevakning av skogsskadors utveckling. – Utläggning och uppföljning av permanenta observationsytor. – Swedish Nat. Board of For., Jönköping, (in Swedish).

Anon. 1992. Field manual, Swedish National Forest Inventory. Dept. of Forest Survey, Swedish Univ. Agricult. Sci., Umeå, (in Swedish).

Aronsson, A. 1985. Indications of stress at unbalanced nutrient contents of spruce and pine. – Roy. Swed. Acad. Agricult. For. Suppl. 17: 40–51, (in Swedish with English summary).

– and Elowson, S. 1980. Effects of irrigation and fertilization on mineral nutrients in Scots pine needles. – Ecol. Bull. (Stockholm) 32: 219–228.

Brodin, Y.-W. and Kuylenstierna, J. C. I. 1992. Acidification and critical loads in Nordic countries: A background. – Ambio 21: 332–338.

Cape, N., Freer-Smith, P. H., Paterson, I. S., Parkinson, J. A. and Wolfenden, J. 1990. The nutritional status of *Picea abies* (L.) Karst. across Europe, and implications for 'forest decline'. – Trees 4: 211–224.

Clymo, R. S. 1983. Peat. – In: Gore, A. J. P. (ed.), Ecosystems of the world. 4A. Mires: Swamp, bog, fen and moor. Elsevier, Amsterdam, pp. 159–224.

Dijk, H. F. G. van and Roelofs, J. G. M. 1988. Effects of excessive ammonium deposition on the nutritional status and condition of pine needles. – Physiol. Plant. 73: 494–501.

Edfast, A.-B., Näsholm, T. and Ericsson, A. 1990. Free amino acid concentrations in needles of Norway spruce and Scots pine trees on different sites in areas with two levels of nitrogen deposition. – Can. J. For. Res. 20: 1132–136.

Ericsson, A., Nordén, L.-G., Näsholm, T. and Walheim, M. 1993. Mineral nutrient imbalances and arginine concentrations in needles of *Picea abies* (L.) Karst. from two areas with different levels of airborne deposition. – Trees 8: 67–74.

Gebauer, G. and Schulze, E.-D. 1991. Carbon and nitrogen isotope ratios in different compartments of a healthy and a declining *Picea abies* forest in the Fichtelgebirge, NE Bavaria. – Oecologia 87: 198–207.

Hägglund, B. and Lundmark, J.-E. 1977. Site index estimation by means of site properties of Scots pine and Norway spruce in Sweden. – Stud. For. Suec. 138: 1–38.

Huettl, R. F. and Fink, S. 1991. Pollution, nutrition and plant function. – In: Porter J. R. and D. W. Lawlor (eds), Plant growth: interactions with nutrition and environment. Cambridge Univ. Press, Cambridge, pp. 207–226.

Ingestad, T. 1979. Mineral nutrient requirements of *Pinus silvestris* and *Picea abies* seedlings. – Physiol. Plant. 45: 373–380

Innes, J. L. 1988. Forest health surveys: problems in assessing observer objectivity. – Can. J. For. Res 18: 560–565.

Landmann, G., Bonneau, M., Bouhot-Delduc, L., Fromard, F., Chéret, V., Dagnac, J. and Souchier, B. 1995. Crown damage in Norway spruce and silver fir: Relation to nutritional status and soil chemical characteristics in the French mountains. – In: Landmann, G. and Bonneau, M. (eds), Forest decline and atmospheric deposition effects in the French mountains. Springer, Berlin, (in press).

Linder, S. 1995. Foliar analysis for detecting and correcting nutrient imbalances in Norway spruce. – Ecol. Bull. (Copenhagen) 44: 178–190.

Lundmark, J.-E. and Odell, G. 1984. Field manual for the Swedish forest soil survey. – Dept Forest Survey, Swedish Univ. Agricult. Sci., Umeå, (in Swedish).

Näsholm, T. 1991. Aspects of nitrogen metabolism in Scots pine, Norway spruce and birch as influenced by the availability of nitrogen in pedosphere and atmosphere. – Thesis, Dept Forest Gen. Plant Physiol., Swedish Univ. Agricult. Sci., Umeå.

– and Ericsson, A. 1990. Seasonal changes in amino acids, protein and total nitrogen in needles of Scots pine trees. – Tree Physiol. 6: 267–281.

Nihlgård, B. 1990. The vitality of forest trees in Sweden – stress symptoms and causal relations. – In: Liljelund, L.-E., Lundmark, J.-E., Nihlgård, B., Nohrstedt, H.-Ö. and Rosén, K. (eds), Vitality fertilization – present knowledge and research needs. Swedish Environ. Protect. Board, Solna. Report 3813, pp. 45–70, (in Swedish with English summary).

Ranneby, B., Cruse, T., Hägglund, B., Jonasson, H. and Swärd, J. 1987. Designing a new national forest survey for Sweden. – Stud. For. Suec. 177: 1–29.

Rosengren-Brinck, U. and Nihlgård, B. 1995. Nutritional status in needles of Norway spruce in relation to water and nutrient supply. – Ecol. Bull. (Copenhagen) 44: 168–177.

Schulze, E.-D. 1989. Air pollution and forest decline in a spruce (*Picea abies*) forest. – Science 244: 776–783.

Tamm, C. O. 1955. Studies on forest nutrition: I. Seasonal variation in the nutrient content of conifer needles. – Medd. Skogsforskningsinst., Stockholm, 45: 1–34.

Westling, O., Hallgren Larsson, E., Sjöblad, K. and Lövblad, G. 1992. Deposition och effekter av luftföroreningar i södra och mellersta Sverige. – IVL Report B 1079, Stockholm, (in Swedish).

Ecological Bulletins 44: 158–167. Copenhagen 1995

Effects of acid deposition on tree roots

Hans Persson, Hooshang Majdi and Anna Clemensson-Lindell

Persson, H., Majdi, H. and Clemensson-Lindell, A. 1995. Effects of acid deposition on tree roots. – Ecol. Bull. (Copenhagen) 44: 158–167.

The influence of air pollution on forest trees includes direct effects on foliage as well as indirect soil-mediated effects. In nitrogen-loaded areas, many forest stands show a positive growth response to the increased nitrogen input. However, with extensive soil acidification and cation leakage, damage in forest stands is frequently observed, in particular in mature forest stands. The most important soil-mediated factors which cause a reduction in fine-root growth and mycorrhizal development are: i) high nitrogen/cation ratios and ii) aluminium (Al) toxicity, viz. elevated Al/cation ratios, leading to an increased sensitivity of the root systems to environmental stress (drought, wind-break, nutrient shortage, etc.). Extensive data on fine-root growth in response to experimental manipulation of plant nutrients in the forest soil are available from many large-scale field experiments in Sweden and elsewhere. Available data have been evaluated in order to chart the effects of nitrogen and sulphur deposition on the mineral nutrient balances in tree fine roots and to assess the risk of Al interference on cation uptake. It is concluded that Al toxicity should be considered as a predisposing factor for forest decline on SW Swedish sites, reducing root function and inhibiting nutrient uptake. A chronically high nitrogen deposition is furthermore likely to produce longer-lasting damage symptoms on fine roots and their function. Aluminium-induced deficiencies of important cations in the forest trees may contribute to forest decline. In SW Swedish forest stands, potassium deficiency is likely to be another important predisposing factor.

H. Persson, H. Majdi and A. Clemensson-Lindell, Dept of Ecology and Environmental Res., Swedish Univ. of Agricultural Sciences, Box 7072, S-750 07 Uppsala, Sweden.

Introduction

The influence of air pollution on forest trees includes direct effects on foliage as well as indirect soil-mediated effects (Aber et al. 1989, Cowling 1989, Schulze 1989, Ulrich 1990, Marschner 1991a). Large forest regions in SW Sweden have been exposed to high levels of acid deposition for many decades, causing soil acidification in forest soils (Hallbäcken and Tamm 1986). Historically, SO_2 has been the major acidification agent, but lately nitrogen compounds have become increasingly important. Nowadays nitrogen deposition has reached 15–25 kg ha^{-1} yr^{-1} in SW Swedish forests, about one half is deposited as NH_4-N and the other half as NO_3-N (Lövblad et al. 1995). The amount and chemical form of nitrogen strongly affects the pH in the rhizosphere and rhizoplane (Gijsman 1990, Marschner and Dell 1994). In forest soils, indirect pH effects are the most important for tree roots,

as regards the solubilization of nutrients and toxic elements or changes in microbial activity.

Many forest stands show a positive growth response to increased nitrogen input, even in heavily N-loaded areas (Nilsson and Wiklund 1992). Nitrogen fertilization experiments suggest that part of the increased forest production is caused by a translocation of biomass production from below-ground to above-ground parts (Ågren et al. 1980, Persson 1980, Persson 1993). At the same time fine-root growth dynamics are strongly affected by N supply (Persson 1990, Persson and Ahlström 1991, Hendricks et al. 1993, Majdi 1994). In a field experiment ammonium sulphate application tend to increase fine-root production and decrease fine-root survival, while the nitrogen-free fertilizers decreased fine-root production but improved survival (Majdi 1994).

In areas with more extensive acidification and nutrient leaching, a decline in tree vitality has been observed

158

Table 1. Chemical composition (mg g⁻¹), Ca/Al or (Ca+Mg)/Al ratios (mol/mol) in living fine roots in soil cores of various tree stands in the upper 0–30 cm of the mineral soil horizon or in 30 cm ingrowth cores (Öringe). Estimates were obtained on untreated sites (0), control plots (C), plots subjected to irrigation and liquid fertilization (IL), dolomite lime application (D), carbonate lime application in two doses (Ca1–2), application of peat/ wood ashes (P-W), application of complete solid fertilizers (FA or FB) and application of ammonium sulphate (NS), different plots in the same stand (F1, Fd, F2).

Table 1. Chemical composition (mg g^{-1}), Ca/Al or (Ca+Mg)/Al ratios (mol/mol) in living fine roots in soil cores of various tree stands in the upper 0–30 cm of the mineral soil horizon or in 30 cm ingrowth cores (Öringe). Estimates were obtained on untreated sites (0), control plots (C), plots subjected to irrigation and liquid fertilization (IL), dolomite lime application (D), carbonate lime application in two doses (Ca1–2), application of peat/ wood ashes (P-W), application of complete solid fertilizers (FA or FB) and application of ammonium sulphate (NS), different plots in the same stand (F1, Fd, F2).

Tree species/site diameter/treatment	N	K	P	Ca	Mg	Fe	Al	Ca/Al	(Ca+Mg)/ Al	Reference
Picea abies – Skogaby; Sweden										
<2 mm, 0	7.4–9.3	0.4–1.3	0.7–0.8	1.5–2.0	0.6–0.7	3.7–6.3	6.4–11.0	0.10–0.20	0.16–0.31	Majdi and Persson (1993)
<2 mm, C	5.8–9.1	0.2–0.4	0.6–0.8	1.5–1.6	0.5	2.7–5.5	5.5–10.1	0.05–0.22	0.10–0.37	Majdi and Persson
<2 mm, NS	7.8–9.3	0.2–0.4	0.7–0.9	1.3–1.7	0.5	2.8–4.4	4.7–11.7	0.06–0.30	0.09–0.43	(in press and
<2 mm, IL	7.1–9.2	0.5–0.6	0.7–0.8	2.0–2.3	0.6–0.7	3.0–5.4	4.0–13.3	0.11–0.34	0.17–0.50	unpubl.)
Picea abies – Öringe, Sweden										
<1 mm, C	13.1–13.5	0.2–0.6	1.5–1.9	1.5–5.0	0.4–0.9	1.3–1.8	6.6–9.3	0.15–0.38	0.22–0.47	Persson and
<1 mm, D	11.4–14.3	0.3–0.8	1.2–1.9	1.4–5.7	0.6–1.2	1.1–2.3	6.2–8.8	0.15–0.45	0.25–0.58	Ahlström (1994)
<1 mm, Ca1–2	11.7–14.6	0.2–0.8	1.3–1.7	2.1–6.2	0.4–1.0	1.0–2.1	5.6–8.4	0.22–0.56	0.32–0.67	
<1 mm, P-W	11.8–15.2	0.3–1.0	1.4–2.4	2.5–6.1	0.6–1.1	1.0–1.7	4.8–8.7	0.22–0.57	0.32–0.55	
Picea abies – Fäxboda, Sweden										
<1 mm, C	–	0.4–1.2	–	2.7–2.8	0.6–0.9	4.0–5.3	7.2–7.4	0.24–0.26	0.35–0.38	Clemensson-Lindell and Persson (1992)
Picea abies – Hautes Fagnes, Belgium										
<2 mm, 0	11.3	0.2	1.0	11.5	1.1	–	1.4	5.61	6.50	van Praag et al. (1988)
Picea abies – Solling, Germany										
<2 mm, F1	8.2–12.4	1.8–2.6	0.9–1.2	1.5–3.44	0.7–0.8	2.8–9.1	4.1–13.1	0.11–0.55	0.17–0.72	Murach and
<2 mm, Fd	10.4–11.4	2.6–2.9	1.0–1.1	2.3–4.1	1.2–1.3	2.6–7.2	4.4–9.8	0.17–0.63	0.31–0.96	Schünemann (1985)
Picea abies – Wank, Germany										
<2 mm, F1	7.0–10.8	1.6–1.7	0.5–0.6	7.1–12.6	1.5–3.1	0.7–1.5	1.0–2.6	3.20–4.58	4.47–6.20	Sandhage-Hofman
<2 mm, F2	7.4–9.1	1.5–2.1	0.5–0.7	5.9–10.4	0.9–1.3	1.9–2.6	1.2–4.1	1.70–3.18	2.00–3.95	and Zech (1992)
Picea engelmanni – *Abies lasicarpa* – Rocky Mountains, USA										
<1 mm, C	10.8	2.8	0.4	3.6	0.5	–	–	–	–	Arthur and Fahley (1992)
Picea glauca – *Abies lasiocarpa* – Prince George, Canada										
<6.4 mm, 0	13.3–22.2	2.9–5.4	1.9–3.2	2.4–5.8	0.9–1.4	5.3–10.0	4.8–18.6	0.09–0.81	0.15–1.14	Kimmins and Hawkes (1978)
Pseudotsuga menziesii – Oregon, USA										
<5 mm, 0	5.4	2.0	1.0	4.9	0.7	–	–	–	–	Fogel and Hunt (1983)
Pinus sylvestris – Jädraås, Sweden										
mature stand										
<2 mm, C	9.8	1.4	1.5	1.5	1.2	–	7.9	0.13	0.30	Ahlström et al.
<2 mm, IL	15.0	1.3	2.2	1.7	1.2	–	7.2	0.16	0.34	(1988)
<2 mm, FA	16.0	1.6	1.6	1.6	0.7	–	4.8	0.22	0.39	
<2 mm, FB	11.2	1.8	1.3	0.9	0.7	–	5.3	0.11	0.26	
young stand										
<2 mm, C	6.1	1.9	0.9	2.0	0.7	–	–	–	–	Persson and Staaf (unpubl.)
Pinus sylvestris – Ilomatsi, Finland										
sapling stand										
<2 mm, 0	4.4	2.3	0.9	1.3	0.9	2.4	5.6	0.15	0.33	Helmisari (1991)
pole stage stand										
<2 mm, 0	4.1	3.0	1.0	1.0	0.9	3.6	9.5	0.07	0.17	Helmisaari (1991)
mature stand										
<2 mm, 0	3.3	1.8	0.6	1.5	0.9	6.5	13.3	0.07	0.15	Helmisaari (1991)
Fagus sylvatica – Hautes Fagnes, Belgium										
<2 mm, 0	10.0	0.4	0.6	9.4	1.1	–	3.4	1.84	2.21	van Praag et al. (1988)

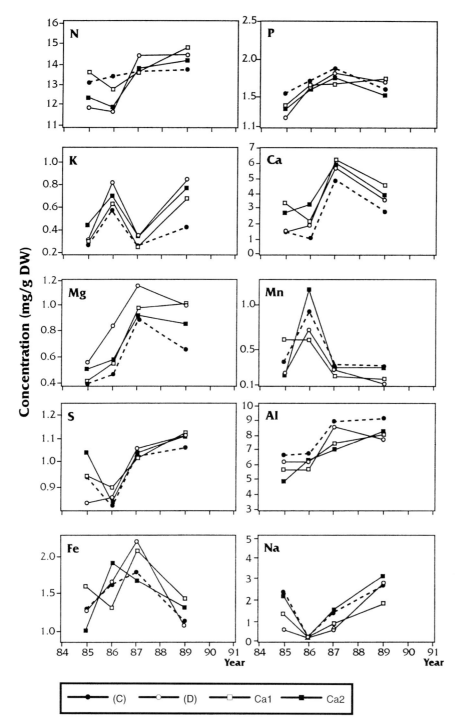

Fig. 1a. The chemical composition of living fine roots (< 1 mm in diameter) in the Öringe experiment for N, P, Na, Ca, Mg, Mn, Fe, Al, and S (mg g⁻¹) in ingrowth cores. Chemical analyses were carried out for the 1985, 1986, 1987, and 1989 sampling occasion (Persson and Ahlström 1994). The experimental treatment included besides the control (C), dolomite lime 3550 kg ha⁻¹ (D), carbonate lime in two doses of 2000 (Ca1) and 3839 kg ha⁻¹ (Ca2), respectively.

(Hüttermann and Ulrich 1984, Schulze 1989). Deficiencies of various nutrients (Mg, Ca, K, Mn and Zn) obtained from needle analyses have been reported from different *Picea abies* stands (Zech and Popp 1983, Zöttl and Mies 1983, Bosch et al. 1986, Hüttl 1989). These observations are mainly based upon studies of needle

chemistry, although attention has also been drawn to fine roots, which are directly influenced by the availability of mineral nutrients in the soil.

Root damage is often described as a decline in the amount of living fine roots, an increase in the amount of dead versus live fine roots (a lower live/dead ratio) and an

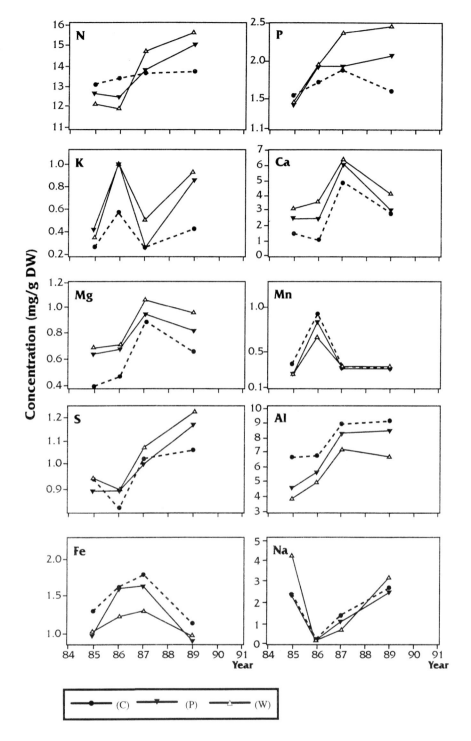

Fig. 1b. See Fig. 1a. The experimental treatment included besides the control (C), wood ash 2800 kg ha^{-1} (W) and peat ash 2800 kg ha^{-1} (P).

increasing amount of dead medium and coarse roots (Puhe et al. 1986, Majdi and Persson 1993, Persson 1993, Majdi 1994). Damage to the fine-root system may thus be expressed as a condition where the mortality rate of fine roots exceeds the regeneration rate, with the result that a net decrease in fine-root biomass may take place gra-

dually (Majdi 1994). We hypothesize that the major factors causing a reduction in fine-root growth and mycorrhizal development in acidified soils are high nitrogen/cation ratios and elevated Al/cation ratios, causing an increased sensitivity of the root systems to environmental stress (drought, wind-break, nutrient shortage, etc.). Atmospheric inputs of excess nitrogen have a significant impact on fine roots by changing plant carbon allocation patterns, production of secondary defence chemicals and storage of carbohydrates (Persson 1990, Vogt et al. 1993). Al/Ca ratios may best describe plant response to available soil Al^{3+} (Godbold et al. 1988, Schaedle et al. 1989, Gobran et al. 1993). It is generally accepted that Al affects the membrane surface by replacing cations from negatively charged head groups on membranes, thereby diminishing the cation exchange capacity of the roots (Cronan et al. 1986, Godbold et al. 1988, Clarkson 1994).

Fine-root mortality represents a substantial carbon and nutrient input into the soil (Persson 1978, Fogel and Hunt 1983, Joslin et al. 1988). The quantity of N added to the soil by root mortality is often greater than the amounts added by above-ground litterfall in certain forest ecosystems (Persson 1978, Vogt et al. 1993, Majdi and Rosengren-Brinck 1994). The importance of understanding and quantifying the processes influencing fine-root dynamics (production, longevity, and mortality) is obvious. Extensive data on fine-root growth in response to experimental manipulation of plant nutrients in forest soils are available from many large field experiments in Sweden (cf. Persson 1978, 1979, 1980, Ahlström et al. 1988, Wallander et al. 1990, Persson and Ahlström 1991, 1994, Clemensson-Lindell and Persson 1993, 1995a,b, Majdi and Persson 1993, 1994). In this article root data from a large body of literature have been compiled in order to evaluate the general pattern. Only references with original data in a tabular form have been compiled; however, a large amount of supporting evidence is available also in diagrams etc. in literature. The primary objectives of the present paper were i) to analyse available data on the effects of high nitrogen and sulphur deposition on mineral nutrient balance in tree fine roots and ii) to evaluate the risk of Al interference with cation uptake by roots.

Chemical composition of fine roots

The concentrations of N, P, K, Ca, Mg, Fe, Al, and Ca/Al, (Ca+Mg)/Al ratios of fine roots from mineral soil layers in different experimental forest sites are given in Table 1. The concentration of most nutrients in roots decrease with increasing root diameter. Although deficiency symptoms in forest trees may be reflected in nitrogen/cation ratios in fine roots, few attempts have been made to explain forest damage symptoms from fine-root chemistry. The concentrations of most macro-nutrients are generally higher in roots than in leaves of seedlings (Ericsson and Kähr 1993). However, in mature trees the opposite may be found (cf. Table 1; data from Ilomatsi; Helmisaari 1991).

The investigations at Öringe clearly demonstrate that element concentrations (Ca, P, Mg and K) in fine roots are affected by changes in soil nutrient status after liming, ash-application, etc. (see Figs 1a,b and Table 1). Further-

more, it was demonstrated from the same data that alkalizing substances increased the Ca/Al ratio of the fine roots and the pH in the soil solution. Thus, the nutritional status was generally improved, although increased pH and salt concentrations caused a negative growth response the first four years after application in the peat ash and high-dose calcium carbonate treatments.

The Ca/Al and (Ca+Mg)/Al ratios are very low in most Scandinavian sites (Table 1); however, low ratios were also found in the German Solling site and at the Canadian *Picea glauca – Abies lasiocarpa* site (Table 1). The Ca/Al and (Ca+Mg)/Al ratios in fine roots differed with soil depth (Table 1). Available data indicate that these ratios are generally lower in dead than in living fine roots (cf. references in Table 1). The concentration of Al in itself should not be expected to be critical for damage symptoms, but rather the Ca/Al molar ratio (Meiwes et al. 1986). The concentrations of Fe and Al were high in fine roots compared with large diameter roots and decreased with stand age (see Table 1; data on *Pinus sylvestris*). A low Ca/Al ratio (<0.1 mol/mol) in the fine roots should be regarded as a possible indication of root damage (cf. Meiwes et al. 1986). It has been hypothesized that nutrient deficiencies may result from an antagonistic role of Al on the uptake of different cations (cf. Godbold et al. 1988). Available data from many of the investigated forest stands indicate a low Ca/Al ratio in fine roots (often <0.1 mol/mol) and it may be concluded that Al toxicity may be a predisposing factor to forest decline (Table 1).

The data in Table 1 show that in most sites concentrations of Ca varied considerably in comparison with other nutrients, such as P, K and Mg. High Ca concentrations were found in lime rich soils viz. on several *Picea abies* and *Fagus sylvatica* sites in Germany and Belgium (Table 1). Low K concentrations were found in the fine roots of most SW Swedish forest stands. Low K concentrations in fine-roots to some extent could be explained by the fact that roots had been frozen and washed before being analyzed (cf. Clemensson-Lindell and Persson 1992; data for Fäxboda in Table 1). Nutrient losses (in particular K) usually occur after prolonged washing of roots (Evdokimova and Grishina 1968, Clemensson-Lindell and Persson 1992), but this was not necessary in most of the investigated stands. Even assuming a loss of more than half of the original potassium in the fine roots due to washing, the K concentration in fine roots would still be low in the SW Swedish forest stands (Skogaby and Öringe) and in the two forest stands in Belgium (Hautes Fagnes). In spite of the fact that all data in Table 1 are based on washed samples, variation in K between the different sites are still evident. Fairly high K concentrations were obtained in *Pinus sylvestris* stands at Jädraås and at Ilomatsi, for *Picea abies* in the German sites (Solling and Wank) and for different tree species in the sites in the USA (Rocky Mountains and Oregon) and in Canada (Prince George).

Nitrogen is the most important macro-nutrient affect-

Table 2. Nitrogen: mineral nutrient ratios in living fine roots in the mineral soil of various forest stands. For experimental treatments and references see Table 1.

Tree species/ site diameter/ treatment	N	K	P	Ca	Mg
Picea abies – Skogaby, Sweden					
<2 mm, C	100	3–4	8–11	20–21	7
<2 mm, NS	100	2–5	8–11	15–20	6
<2 mm, LI	100	6–7	9–10	25–28	7–8
Picea abies – Öringe, Sweden					
<2 mm, C	100	2–5	11–14	11–38	3–7
<2 mm, D	100	2–6	9–15	11–44	5–9
<2 mm, L1-L2	100	2–6	10–13	16–47	3–8
<2 mm, P-W	100	2–7	10–18	19–45	4–8
Picea abies – Hautes Fagnes, Belgium					
<2 mm, C	100	2	9	102	10
Picea abies – Solling, Germany					
<2 mm, F1	100	20–29	10–13	17–38	8–9
<2 mm, Fd	100	24–27	9–10	21–38	11–12
Picea abies – Wank, Germany					
<2 mm, F1	100	18–19	6–7	80–142	17–35
<2 mm, F2	100	18–25	6–8	72–126	11–16
Picea engelmanni – *Abies lasicarpa* – Rocky Mountains, USA					
<1 mm, 0	100	26	4	33	5
Picea glauca – *Abies lasiocarpa* – Prince George, Canada					
<6.4 mm, 0	100	16–30	11–18	14–33	5–8
Pseudotsuga menziesii – Oregon, USA					
<5 mm, 0	100	37	19	91	13
Pinus sylvestris – Jädraås, Sweden					
mature stand					
<2 mm, C	100	14	15	15	12
<2 mm, IL	100	9	15	11	8
<2 mm, FA	100	10	10	10	4
<2 mm, FB	100	16	12	8	6
young stand					
<2 mm, C	100	31	15	33	11
Pinus sylvestris – Ilomatsi, Finland					
sapling stand					
<2 mm, 0	100	52	20	30	20
pole stage stand					
<2 mm, 0	100	73	24	24	22
mature stand					
<2 mm, 0	100	55	18	45	27
Fagus sylvatica – Hautes Fagnes, Belgium					
<2 mm, 0	100	4	6	94	11

may be regarded as better phytoindicators of the nutritional conditions in the forest soil than foliar concentrations and give a more direct indication of experimental impacts from fertilization, liming etc. (cf. Joslin et al. 1988, Persson and Ahlström 1994). The proportions of nitrogen in relation to other macro-nutrients from mineral soil layers in different experimental forest sites have been compiled in Table 2. Some variation around the optimal proportion values should be expected without plant growth affecting. According to Ingestad (1979), the optimum weight proportions are 100 N: 45 K: 15 P: 6 Ca: 6 Mg for *Picea abies*. In field experiments where a close relationship between tree growth rate and net uptake rate of nutrients is desired, complete sets of nutrients should be applied at these levels (see Majdi and Persson 1993 and Bergholm et al. 1994 for Skogaby; Persson 1980 and Ahlström et al. 1988 for Jädraås).

It is evident from the data in Table 2 that the potassium to nitrogen ratio is low in forest stands in SW Sweden (Skogaby and Öringe), when compared with other Swedish sites (Jädraås). Low K/N ratios were also found in the two forest stands in Belgium (Hautes Fagnes), compared with other European forest stands (Solling, Wank and Ilomatsi) and forest stands in the USA (Oregon) and Canada (Prince George). Significantly lower potassium concentrations were also found in the Skogaby NS treatment in 1992 compared with the IL-treatment (Nilsson and Wiklund 1995). The data suggest that K shortage may be a reality in forest stands of southern Sweden and that accelerated acidification in combination with increased nitrogen deposition (the NS-treatment at Skogaby) increase the risk of K deficiency. K deficiency is known to depress the allocation pattern to roots in *Betula pendula* seedlings (Ericsson and Kähr 1994).

It has also been reported that K^+ can alleviate ammonium toxicity because of antagonistic effects towards NH_4^+ (Nelson and Hsieh 1971, Ingestad 1976). The uptake rates of NH_4^+ versus NO_3^- of coniferous tree roots differ considerably depending on both concentrations and proportions of the nitrogen forms supplied (Marschner 1991b).

The form in which nitrogen is absorbed largely determines the acidifying or alkalizing effect of plant nutrient uptake. Gijsman (1990) concludes from results of an experiment on rhizosphere pH along different root zones of *Pseudotsuga menziesii* that nitrate nutrition enables the plant to protect its most essential root zone from the adverse effects of strong acidity by locally raising the rhizosphere pH. The rhizospere pH in the Skogaby site in SW Sweden was significantly decreased in the ammonium sulphate treatment areas, thus indicating a high rate of ammonium uptake (Majdi and Persson 1995). As a rule, pH in the rhizosphere of SW Swedish forests did not differ from, or was lower than, in the bulk soil (Clemensson-Lindell and Persson 1993, Majdi and Persson 1994).

Shortages of macro-nutrients other than K in relation to N were generally not indicated in the data from the different forest stands in Table 2. However, we know

ing plant growth. The proportions of nitrogen to other macro-nutrients have been used to describe the plant requirements of different macro-nutrients (cf. Ingestad 1979). In trees, elemental concentrations in fine roots

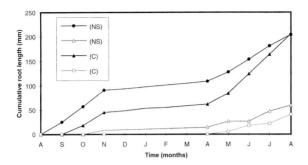

Fig. 2. Cumulative fine-root production and mortality in the control and ammonium sulphate treatment areas from August 1991 to August 1992 at a depth < 30 cm. The data were obtained in 10 minirhizotron tubes in each of three plots in the control and ammonium sulphate treatment areas. Area of observation = 1.80 × 1.35 cm at each 1.35 cm depth. Cumulative production and cumulative mortality in the control areas (C), cumulative production and cumulative mortality in the NS-treated areas (NS). Filled symbols = production, unfilled symbols mortality.

from different ecosystem studies that low levels of most cations, in combination with an increased mineral requirement due to the growth stimulating effects of nitrogen, may cause such nutrient imbalances (Zech and Popp 1983, Zöttl and Hüttl 1985, Rehfuss 1985, MacDonald et al. 1992). Concentrations of Ca and Mg are extremely variable in different forest stands, and concentrations over the optimum requirements are indicated at one of the German sites (Wank) and at the Belgium sites (Hautes Fagnes). Low P concentration was found in the Rocky Mountains site in the USA.

Effects on fine-root development

Part of the increased above-ground tree production subsequent to nitrogen deposition seems to be related to a shift in the carbon allocation pattern from below- to above-ground, caused by changes in fine-root turnover (Ågren et al. 1980, Santantonio and Hermann 1985, Persson 1993). At the same time, the growth dynamics and vitality of the fine roots are affected. Swedish root investigations in the field indicate so far that: 1) liquid fertilization (with a complete fertilizer) caused an increased fine-root production (cf. e.g. Persson 1980, Ahlström et al. 1988), 2) solid fertilizers in combination with liming generally caused a decrease in fine-root production, whilst solid fertilizers alone (ammonium nitrate and urea) in some field experiments caused a decrease in fine-root production. However, in some experiments an increase in fine-root growth was observed (Persson 1980, Persson and Ahlström 1991, Persson et al. 1995). 3) Ammonium sulphate application, viz. increased nitrogen and sulphur deposition (the NS-treatment at Skogaby), caused a decreased fine-root vitality (based on morphological characteristics), an increased concentration of fine roots in

the LFH-layer and an increased amount of dead fine roots in the soil profile (Clemensson-Lindell and Persson 1995a, Majdi and Persson 1995, Persson et al. 1995). Experimentally decreased nitrogen and sulphur deposition in a catchment area in SW Sweden (roof experiment) tend to increase fine-root vitality (Clemensson-Lindell and Persson 1995b).

In most root investigations, information about the growth dynamics of fine roots is generally obtained by destructive soil sampling, including a tremendous amount of time for processing the samples (cf. Vogt and Persson 1991, Persson 1993). Estimates of change in the amount of fine roots and nutrient concentrations were until the last decades very rare in literature (Persson 1990). Information about the role of tree roots in the carbon and mineral nutrient cycling is crucial for understanding the effect of soil acidification on tree growth in forest ecosystems. The development of a new technique, e.g. the minirhizotron technique (Majdi et al. 1992, Hendrick and Pregitzer 1993), which has been tested and further developed within the Skogaby project, makes it possible to follow the growth dynamics of the fine roots in the field with high resolution. Since the same roots are followed at the same part of the tube, the observed rate of changes in production or mortality of fine roots are directly related to fluctuations in soil chemistry and nutrient availability or environmental changes (soil temperature or moisture).

Results from the minirhizotron investigations on fine-root production and mortality in the control (C) and ammonium sulphate treatment areas (NS) in Skogaby are shown in Fig. 2. The NS-treatment areas were subjected to increased nitrogen and sulphur supply from 1988 onwards. The data reveal that fine-root production tended to increase in the NS-treated areas compared with the control areas while the fine-root survival decreased (cf. also Majdi 1994). Fine-root mortality in the NS treatment areas were higher than in the control during the whole study period. Since the amount of fine roots at any given time is governed by the balance between the production and mortality of fine roots, the NS treatment may in the longer run result in fewer fine roots and a less vital fine-root system.

Reduced vitality of fine roots in the NS treatment at

Table 3. The percentage share of fine roots (<1 mm in diameter) in the control and ammonium sulphate treatment areas at Skogaby. Number of replicates is 4, each based on 4 samples. Significant difference (Student's t-test, p > 0.05) compared with controls (*) are indicated. C = control; NS = ammonium sulphate. Data are given for the mineral soil horizon 0–5 cm. Original data are to be found in Clemensson-Lindell and Persson (1994a).

Treatment	Percentage of class		
	1	2	3
C	2±3	74± 7	23±5
NS	2±4	54±11*	44±7*

Skogaby was indicated in 1992 from data obtained from direct soil coring, and sorting the fine roots into 3 vitality classes based on morphological characteristics (Table 3, Clemensson-Lindell and Persson 1995a). Few fine roots belonged to vitality class 1 (roots less suberized, well branched, with white turgid root tips). Most fine roots belonged to class 2 (darkened roots with few white root tips) and class 3 (dead roots). The amount of living fine roots of class 2 decreased significantly in treatment areas subjected to increased nitrogen and sulphur deposition (NS), while the amount of dead roots increased. Although a 30% increase in above-ground production was shown on NS-treated areas in response to the NS-application, the vitality of the fine-root system appeared to be in a state of deterioration, as indicated by a decrease in fine-root biomass, an increased amount of dead fine roots, a decreased specific root length (fine-root length/fine-root dry weight) (Clemensson-Lindell and Persson 1995a).

Recent findings on fine-root carbon and nutrient cycling stress the importance of a more holistic approach to the ecosystem structure and function, for the assessment and prediction of disturbances to terrestrial systems (Hendricks et al. 1993). The main question to answer is to what extent fine-root and foliar carbon allocation and utilization patterns are similar across nitrogen availability gradients.

Concluding remarks

In order to study the relationships that might exist between atmospheric deposition and damage to tree roots, we compiled data on fine-root chemistry and growth response from forests experimentally supplied with sulphur and nitrogen. Available data from SW Swedish forest trials indicate Ca/Al and Ca/Al molar ratios in roots below levels indicating a high risk of root damage (<0.1; see Meiwes et al. 1986 and Puhe et al. 1986). The Al concentration in fine roots was much higher in the investigated forest stands than in the roots of seedlings grown at high Al concentrations (Göransson and Eldhuset 1987), indicating either a longer time of exposure of the fine roots to Al in the forest soil and/or frequent drought periods causing a low pH and a high Al concentration in the rhizosphere soil. The high Al concentrations in the fine roots indicate that Al should not be disregarded as a predisposing factor to root damage in the investigated Swedish forest stands.

Root damage resulting from elevated Al^{3+} concentrations in fine roots, which prevent the uptake of important mineral nutrients and result in unbalanced growth conditions, has been proposed as an explanation of forest decline (Cronan et al. 1986, Gobold et al. 1988, Marschner 1991a). Leaching of base cations, in combination with high nutrient requirements due to increased plant growth rate, may enhance nutrient imbalances in forest trees (Aber et al. 1989). The ability of the forest ecosystems to accumulate nitrogen from atmospheric depositions is limited before an imbalance between demand for and supply of cations is obtained. Chronically high nitrogen deposition is likely to produce longer-lasting effects on fine-root systems and their function. In SW Swedish sites, K deficiency is likely to be another important predisposing factor and accelerating acidification in combination with increased nitrogen deposition increases this risk.

References

Aber, J., Nadelhoffer, K. J., Steudler, P. and Melillo, J. M. 1989. Nitrogen saturation in northern forest ecosystems. – Bioscience 39: 378–386.

Ågren, G., Axelsson, B., Flower-Ellis, J. G. K., Linder, S., Persson, H., Staaf, H. and Troeng, E. 1980. Annual carbon budget for a young Scots pine. – Ecol. Bull. (Stockholm) 32: 307–313.

Ahlström, K., Persson, H. and Börjesson, I. 1988. Fertilization in a mature Scots pine (*Pinus sylvestris* L.) stand – effects on fine roots. – Plant Soil 106: 179–190.

Arthur, M. A. and Fahley, T. J. 1992. Biomass and nutrients in an Engelmann spruce – subalpine fir forest in north central Colorado: pools, annual production, and internal cycling. – Can. J. For. Res. 22: 315–325.

Bergholm, J., Jansson, P.-E., Johansson, U., Majdi, H., Nilsson, L.-O., Persson, H., Rosengren-Brinck, U. and Wiklund, K. 1994. – Air pollution, tree vitality and forest production – The Skogaby project. General description of a field experiment with Norway spruce in south Sweden. – Ecosystem Research Report 13, CEC, Brussels.

Bosch, C., Pfannkuch, E. E., Rehfuess, K. E., Runkel, K. H., Schramel, P. and Senser, M. 1986. Einfluss einer Düngung mit Magnesium und Calcium, von Ozon und sauren Nebel auf Frosthärte, Ernährungzustand und Biomasse-Produktion junger Fichten (*Picea abies* (L.) Karst.). – Forstwiss. Centralbl. 105: 218–229.

Clarkson, D. T. 1994. Ionic relations. – In: Wilkins, M.B. (ed.) Advanced plant physiology. Pitman Publ., London, pp. 218–229.

Clemensson-Lindell, A. and Persson, H. 1992. Effects of freezing on rhizosphere and root nutrient content using two soil sampling methods. – Plant Soil 139: 39–45.

– and Persson, H. 1993. Long-term effects of liming on the fine-root standing crop of *Picea abies* and *Pinus sylvestris* in relation to chemical changes in the soil. – Scand. J. For. Res. 8: 384–394.

– and Persson, H. 1995a. Fine-root vitality in a Norway spruce stand subjected to varying nutrient supplies. – Plant Soil 168–169.

– and Persson, H. 1995b. The effects of nitrogen addition and removal on fine-root vitality and distribution in three catchment areas at Gårdsjön. – Forest Ecol. Manage. 71: 123–131.

Cowling, E. B. 1989. Recent changes in chemical climate and related effects on forests in North America and Europe. – Ambio 18: 167–171.

Cronan, C. S., Walker, W. J. and Bloom, P. R. 1986. Predicting aqueous aluminium concentrations in natural waters. – Nature 324: 140–143.

Ericsson, T. and Kähr, M. 1994. Growth and nutrition of birch seedlings in relation to potassium supply rate. – Trees 7: 78–85.

Evdokimova, T. I. and Grishina, L. A. 1968. Productivity of root systems of herbaceous vegetation on flood plain meadows and methods for its study. – In: Methods of productivity studies in root systems and rhizosphere organisms. Int. Symp. USSR, USRR Acad. of Sciences, pp. 24–27.

Fogel, R. and Hunt, G. 1983. Contribution of mycorrhizae and soil fungi to nutrient cycling in a Douglas-fir ecosystem. – Can. J. For. Res. 13: 219–232.

Gijsman, A. 1990. Nitrogen nutrition and rhizosphere pH of Douglas-fir. – Ph.D. thesis, Rijksuniversitet, Gronigen.

Godbold, D. L., Fritz, E. and Hüttermann, A. 1988. Aluminium toxicity and forest decline. – Proc. Natl. Acad. Sci. (USA) 85: 3888–3892.

Gobran, G. R., Fenn, L. B., Persson, H. and Al-Windi, I. 1993. Nutrient response of Norway spruce and willow to varying levels of calcium and aluminium. – Fertilizer Res. 34: 181–189.

Göransson, A. and Eldhuset, R. D. 1987. Effects of aluminium on growth and nutrient uptake of *Betula pendula* seedlings. – Physiol. Plant. 69: 193–199.

Hallbäcken, L. and Tamm, C. O. 1986. Changes in soil acidity from 1927 to 1982–1984 in a forest area of south-west Sweden. – Scand. J. For. Res. 1: 219–232.

Helmisaari, H. S. 1991. Variation in nutrient concentrations of *Pinus sylvestris* roots. – In: McMichael, B. L. and Persson, H. (eds), Plant Roots and their Environment. Development in Agricultural and Managed-Forest Ecology 24: 204–212.

Hendrick, R. L. and Pregitzer, K. S. 1993. The dynamics of fine root length, biomass, and nitrogen content in two northern hardwood ecosystems. – Can. J. For. Res. 23: 2507–2520.

Hendricks, J. J., Nadelhoffer, K. J. and Aber, D. A. 1993. Assessing the role of fine roots in carbon and nutrient cycling. – Trends Ecol. Evol. 8: 174–178.

Hüttl, R. F. 1989. 'New types' of forest damages in central Europe. – Yale Univ. Press, New Haven, CT.

Hüttermann, A. and Ulrich B. 1984. Solid phase – solution – root interactions in soils subjected to acid deposition. – Phil. Trans. R. Soc. Lond. B 305: 353–368.

Ingestad, T. 1976. Nitrogen and cation nutrition of three ecologically different plant species. – Physiol. Plant. 38: 29–34.

– 1979. Mineral nutrient requirements of *Pinus sylvestris* and *Picea abies* seedlings. – Physiol. Plant. 45: 373–380.

Joslin, J. D., Kelly, J. M. and Wolfe, M. H. 1988. Elemental patterns in roots and foliage of mature spruce across a gradient of soil aluminium. – Water Air Soil Pollut. 40: 375–390.

Kimmins, J. P. and Hawkes, B. C. 1978. Distribution and chemistry of fine roots in a white spruce – subalpine fir stand in British Columbia: implications for management. – Can. J. For. Res. 8: 266–279.

Lövblad, G., Kindbom, K., Grennfelt, P., Hultberg, H. and Westling, O. 1995. Deposition of acidifying substances in Sweden. – Ecol. Bull. (Copenhagen) 44: 17–34.

MacDonald, N. W., Burton, A. J., Liechty, H. O., Witter, J. A., Pregitzer, K. S., Mroz, G. D. and Richter, D. D. 1992. Ion leaching in forest ecosystems along a Great Lakes air pollution gradient. – J. Env. Qual. 21: 614–623.

Majdi, H. 1994. Effects of nutrient applications on fine-root dynamics and root/rhizosphere chemistry in a Norway spruce stand. – Ph.D. thesis, Dept Ecol. Env. Res., Swed. Univ. Agr. Sci. Rep. 71: 1–83.

– and Persson, H. 1993. Spatial distribution of fine roots, rhizosphere and bulk-soil chemistry in an acidified *Picea abies* stand. – Scand. J. For. Res. 8: 147–155.

– and Persson, H. 1995. Effect of ammonium sulphate application on the chemistry of bulk soil, rhizosphere, fine roots and fine-root distribution in a *Picea abies* (L.) Karst. stand. – Plant Soil 168–169: 151–160.

– and Rosengren-Brinck, U. 1994. Effects of ammonium sulphate application on the rhizosphere, fine-root and needle chemistry in a *Picea abies* (L.) Karst. stand. – Plant Soil 162: 71–81.

–, Smucker A. J. M. and Persson, H. 1992. A comparison between minirhizotron and monolith sampling methods for measuring root growth of maize (*Zea mays* L.). – Plant Soil 147: 135–142.

Marschner, H. 1991a. Mechanisms of adaptation of plants to acid soils. – Plant Soil 134: 1–20.

– 1991b. Ammonium and nitrate uptake rates and rhizosphere pH in non-mycorrhizal roots of Norway spruce (*Picea abies* (L.) Karst.). – Trees 5: 14–21.

– and Dell, B. 1994. Nutrient uptake in mycorrhizal symbiosis. – Plant Soil 159: 89–102.

Meiwes, K. J., Khanna, P. K. and Ulrich, B. 1986. Parameters for describing soil acidification and their relevance to the stability of forest ecosystems. – For. Ecol. Manage. 15: 161–179.

Murach, D. and Schünemann, E. 1985. Reaktion der Feinwurzeln von Fichten auf Kalkungsmassnahmen. – Allg. Forst Zeitschr. 43: 1151–1154.

Nelson, P. V. and Hsieh, K.-H. 1971. Ammonium toxicity in *Chrysanthemum*: critical level and symptoms. – Commun. Soil Sci. Plant Anal. 2: 739–448.

Nilsson, L. O. and Wiklund, K. 1992. Influence of nutrient and water stress on Norway spruce production in south Sweden – the role of air pollutants. – Plant Soil 147: 251–265.

– and Wiklund, K. 1995. Nutrient balance and P, K, Ca, Mg, S and B accumulation in a Norway spruce stand following ammonium sulphate application, fertigation, irrigation, drought and N-free-fertilization. – Plant Soil 168–169: 437–446.

Persson, H. 1978. Root dynamics in a young Scots pine stand in central Sweden. – Oikos 30: 508–519.

– 1979. Fine root production, mortality and decomposition in forest ecosystems. – Vegetatio 41: 101–109.

– 1980. Fine-root dynamics in a Scots pine stand with and without near optimum nutrient and water regimes. – Acta Phytogeogr. Suec. 68: 101–110.

– (ed.) 1990. Above-and below-ground interactions in forest trees in acidified soils. – Air Pollut. Res. Rep. 32. Comm. European Comm., Brussels, Belgium.

– 1993. Factors affecting fine root dynamics of trees. – Suo 43: 163–172.

– and Ahlström, K. 1991. The effects of liming and fertilization on fine root growth. – Water Air Soil Pollut. 54: 365–379.

– and Ahlström, K. 1994. The effects of alkalizing compounds on fine-root growth in a Norway spruce stand in southwest Sweden. – J. Env. Sci. Health A29(4): 803–820.

–, von Fircks, Y., Majdi, H. and Nilsson, L. O. 1995. Root distribution in a Norway spruce (*Picea abies* (L.) Karst.) stand subjected to drought and ammonium-sulphate application. – Plant Soil 168–169: 161–165.

Puhe, J., Persson, H. and Börjesson, I. 1986. Wurzelwachtum und Wurzelshäden in Skandinavischen Nadelwäldern. – Allg. Forst Zeitschr. 20: 488–492.

Rehfuss, K. E. 1985. Vielfältige Formen der Fichtenerkrankung in Süddeutschland. – In: Niesslein, E. and Voss, G. (eds), Was wir über das Waldsterben wissen, pp. 124–130.

Sandhage-Hofmann, A. and Zech, W. 1992. Dynamik und Elementgehalte von Fichtenfeinwurzeln in Kalkdesteinsböden am Wank (Bayerische Kalkalpen). – Z. Pflanzenernähr. Bodenk. 156: 181–190.

Santantonio, D. and Hermann, R. K. 1985. Standing crop, production and turnover of fine roots on dry, moderate and wet sites of mature Douglas-fir in western Oregon. – Ann. Sci. For. 42: 113–142.

Schaedle, M., Thornton, F. C., Raynal, D. J. and Tepper, H. B. 1989. Response of tree seedlings to aluminium. – Tree Physiol. 5: 337–356.

Schulze, E. D. 1989. Air pollution and forest decline in a spruce (*Picea abies*) forest. – Science 244: 776–783.

Ulrich, B. 1990. An ecosystem approach to soil acidification. – In: Ulrich, B. and Sumner, M. E. (eds), Soil acidity. Springer, Berlin, pp. 28–79.

van Praag, H. J., Sougnez-Remy, S., Weissen, F. and Carletti, G.

1988. Root turnover in a beech and a spruce stand in the Belgian Ardennes. – Plant Soil 105: 87–103.

Wallander, H., Persson, H. and Ahlström, K. 1990. Effects of nitrogen fertilization on fungal biomass in ectomycorrhizal roots and surrounding soil – In: Persson, H. (ed.), Above and below-ground interactions in forest trees in acidified soils. Air Poll. Res. Rep. 32: 99–102.

Vogt, K. A. and Persson, H. 1991. Measuring growth and development of roots. – In: Lassoie J. P. and Hincley T. M. (eds), Techniques and approaches in forest tree ecophysiology. CRC Press, Boca Raton, pp. 477–501.

– , Publicover, D. A., Bloomfield, J., Perez, J. M., Vogt, D. J. and Silver, W. L. 1993. Belowground responses as indicators of environmental change. – Environ. Experim. Bot. 133: 189–205.

Zech, W. and Popp, E. 1983. Magnesiummangel, einer Grunde für das Fichten- und Tannensterben in NO-Bayern. – Forstwiss. Centralbl. 102: 50–55.

Zöttl, H. W. and Mies, E. 1983. Nährelementversorgung und Schadstoffbelastung von Fichtenökosystemen im Südschwartzwald unter Immisionseinfluss. – Mitt. Dtsh Bodenksl. Ges. 38: 429–434.

– and Hüttl, R. F. 1985. Schadsymptome und Ernährungszustand von Fichtenbeständen im südwestdeutschen Alpenvorland. – Allg. Forstz. 40: 197–199.

Ecological Bulletins 44: 168–177. Copenhagen 1995

Nutritional status in needles of Norway spruce in relation to water and nutrient supply

Ulrika Rosengren-Brinck and Bengt Nihlgård

Rosengren-Brinck, U. and Nihlgård, B. 1995. Nutritionl status in needles of Norway spruce in relation to water and nutrient supply. – Ecol. Bull. (Copenhagen) 44: 168–177.

The nutrient status of one- to five-year-old needles of Norway spruce *Picea abies* was analysed in the Skogaby field experiment (SW Sweden) five years after the various treatments had been started. The treatments included application of ammonium sulphate (NS), N-free fertilizer (V), liquid fertilizer with a complete set of macro- and micronutrients (IF), irrigation (I), and artificial drought (D). For diagnostic purposes, four different techniques were used: i) comparisons of concentrations of N, K, P, Ca, Mg, and S with corresponding critical concentrations; ii) comparisons of K/N, P/N, Ca/N, and Mg/N ratios with corresponding critical ratios; iii) changes in nutrient concentration and nutrient content per needle with needle age; and iv) graphical vector analysis according to Timmer and Stone. On control plots, N and P seemed to alternate as limiting nutrients. Despite an annual precipitation of c. 1100 mm, water also seemed to limit tree growth. The I treatment caused a K deficiency, partly explained by increased tree growth. Application of a total of 500 kg N ha^{-1} during five years led to relative deficiencies in P and K. An accumulation of N in older needles indicated that the NS plots were approaching N saturation. Both nutrient ratios and vector analysis diminished the dilution effects due to growth variation, thus making the diagnosis more accurate compared to using the critical concentrations concept. Retranslocation of nutrients between needles of different age was best indicated when based on the nutrient content per needle.

U. Rosengren-Brinck and B. Nihlgård, Dept of Plant Ecology, Univ. of Lund, Ecology Building, S-223 62 Lund, Sweden.

Introduction

The availabilities of nitrogen (N) and phosphorus (P) in soil are usually considered to be decisive factors regulating the growth of most tree species in the boreal zone. Scarcities of other important nutrients, like potassium (K), magnesium (Mg), and calcium (Ca) occur less commonly (Tamm 1991). Under normal conditions, the photosynthesising cells in the foliage can regulate nutrient concentrations, and maintain almost constant relations between nitrogen and other nutrients (Ingestad 1979, Linder 1995). Nutrients are translocated from accumulated reserves to ensure that concentrations remain high enough to avoid deleterious long-term effect on the plant (Chapin 1988). Deficiencies can develop in response to stress induced by factors such as drought. Under drought conditions, nutrient uptake is reduced which leads in turn to lowered concentrations of many nutrients (Evers 1986, Chapin 1991). Concentrations of some of the nutrients

are lowered to a critical level, below which plant growth is negatively affected. Although deficiency symptoms sometimes appear on such occasions, perennial plants have a remarkable ability to adapt their growth according to the availability of different nutrients, often causing the deficiency symptoms to disappear again. Irrigation and fertilization, on the other hand, can lead to a temporary increase in nutrient concentrations and nutrients may even accumulate in the plant (Chapin 1991). Thereafter nutrients are probably taken up faster than they can be used for growth, which leads to their further accumulation in the plant. In such cases, a change in supply should not affect growth. By contrast, when the nutrient concentrations in photosynthesising organs are low enough to limit growth, an extra supply of nutrients can have positive effects (Marschner 1986).

In most parts of Europe today the atmospheric deposition of N is elevated and this increased input of N has enhanced forest productivity. However, since it may be

Table 1. Descriptions of treatments used in the Skogaby field experiment and basal area growth (m^2 ha^{-1}) during 1993. Adapted from Bergholm et al. (1995).

	Treatment	no. of replicates	Description	Basal area growth*
C	Control	4	No treatment.	1.07
D1	Drought	4	Roofs placed 1–1.5 m above ground pre-	1.19
D2			vented 2/3 of the throughfall from reaching the ground in half the plot (D1) during April to September 1988–89. After two years of drought treatment (1988–89), the roofs were moved to the other half (D2) of the plot which was subjected to drought 1990–92.	0.89
I	Irrigation	4	Irrigation was performed using sprinklers. The amount of water was adjusted to avoid any water storage deficits exceeding 20 mm of water during May to September. Water was taken from Lake Råsjön.	1.73
IF	"Optimum" fertilization with irrigation	4	100 kg N ha^{-1} yr^{-1} together with a complete set of other nutrients according to the principles established by Ingestad (1988), were added from May to August 1988–92 in water solution and spread with the irrigation water as for the irrigation treatment.	2.58
NS	N-S fertilization	4	100 kg ha^{-1} of N and 114 kg ha^{-1} of S yr^{-1} were spread as solid ammonium sulphate in equal proportions on three occasions each year.	1.42
V	N-free fertilization	4	1000 kg ha^{-1} of the commercial N-free full fertilizer "Skog-vital" (manufactured by Supra, Sweden) added in solid phase during 1988–1989.	1.12

*Nilsson and Wiklund pers. comm.

difficult for trees to take up corresponding amounts of mineral elements like P, K and Mg the N-induced growth enhancement can disturb the nutritional balance in the tree (Nihlgård 1985). In addition, the "acid rain" leads to the accumulation of high levels of potentially toxic Al ions in the soil solution. Such a nutrient imbalance can predispose trees to forest damage (Schulze 1989).

In this field study we investigated the long-term (5 yr) effects of various experimental treatments, including nutrient additions, drought and irrigation on the nutritional status of Norway spruce. Our aim was to learn more about how the long-term stress induced by the ongoing

soil acidification and N deposition will affect the forest ecosystem.

Material and methods

Study site

The experimental area is situated in south-western Sweden (56°33'N, 13°13'E) c. 16 km from the coast and 95–115 m a.s.l. The site was reforested with Norway spruce (*Picea abies* L. Karst.) in 1966 after a clear-cutting. The

Table 2. Added amounts of macro nutrients (kg ha^{-1}) in different treatments in the Skogaby experiment during 1988–1992. The abbreviations are explained in Table 1. Adapted from Bergholm et al. (1995).

Treatment	N	P	K	Ca	Mg	S	Application
NS	100	0	0	0	0	114	Yearly during 1988–1992
V	0	30	50	210	48	80	1988–1989
I	1.7	0	1.1	4.7	2.9	4.6	Yearly*
IF	102	17	49	10.7	8.9	13.6	Yearly during 1988–1992**

* Based on an irrigation rate of 200 mm yr^{-1}.
** Ca was given as a single dose as ground limestone prior to the start of the irrigation. Nutrients from the irrigation are included (based on an irrigation of 200 mm yr^{-1}).

Table 3. Critical concentrations for macro nutrients (mg g⁻¹) and ratios (in per cent) of K, P, Ca and Mg, respectively, to N, set for current-year needles after the first growing season.

Reference	N	K	P	Ca	Mg	S	K/N	P/N	Ca/N	Mg/N
Hüttl and Fink (1988)	13	4.0	1.1	2.0	0.7					
Foerst et al. (1987)	13	3.5	1.3	1–2	0.6–0.7					
Nihlgård (1990)	13	3.5	1.3	1	0.7	0.6	30–40	10–12	5	5
Linder (1995)*	16						35	10	2.5	4
Ingestad (1979)*							50	16	5	5

* Optimum values.

climate is maritime, with a mean annual precipitation of c. 1100 mm. May and June are often very dry. The annual mean air temperature is c. 7.5°C.

The mean pH of the precipitation is 4.5 (1988–1991) and deposition in this area adds 20–25 kg ha⁻¹ yr⁻¹ of N and 10–15 kg ha⁻¹ yr⁻¹ of S (Bergkvist, pers. comm.). The soil is a haplic podzol according to FAO (1988), on a loamy sand till, with a $pH_{(H_2O)}$ of 3.9 in the humus layer and between 3.9 and 4.5 in the mineral soil (Bergholm et al. 1995). Physical and chemical characteristics of the soil are described by Majdi and Persson (1993). The stand had an average basal area of c. 25 m² ha⁻¹ (Nilsson and Wiklund 1992) when field plots (45 × 45 m) were established in 1987. A randomised experimental design with 10 treatments and 4 blocks was set up, and the treatments were started in spring 1988 (Tables 1 and 2). Basal area growth (m² ha⁻¹) during the experimental period (1987–1993) is shown in Table 1. For more details on the experimental design, treatments and site characteristics, see Bergholm et al. (1995).

Needle sampling

Needles were sampled in the middle of February 1993

from dominant or co-dominant trees selected on the basis of stem diameters. One tree of mean diameter and two trees with ± one standard deviation from the mean diameter were selected on each plot, i.e. 12 trees from each of the C, I, IF, NS, and V treatments. From the drought treatments only two trees with ± one standard deviation were selected, i.e. 8 trees each from the D1 and D2 treatments. On each tree, one branch pointing close to SW was manually sampled from the 7th branch whorl from the top. In the laboratory, needles were separated into age classes, i.e. current year (grown 1992) (C), one-year-older needles (C+1), two-year-older needles (C+2), three-year-older-needles (C+3) and four-year-older needles (C+4). Only green, visibly undamaged needles from nodal shoots of the first and second order of side branches were used. The needles were dried at 40–55°C until they fell off (c. 48 h). To determine concentrations of Ca, K, Mg, P and S with Inductively Coupled Plasma Emission Spectroscopy (Perkin Elmer), 1000 mg of unground needles was wet digested in 20 ml concentrated HNO₃ and heated to 125°C. Nitrogen was analysed with the Kjeldahl-method (Nihlgård 1971) using 250 mg needles, with K₂SO₄ and CuSO₄ (1:10 by weight) as catalysts. Each sample was analysed separately. Concentrations were corrected to 85°C dry weight using a 40°C/

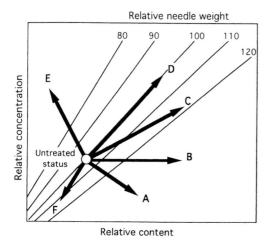

Fig. 1. Interpretation of relationships between relative nutrient concentration, nutrient content per needle and dry weight of needles following treatment (adapted from Timmer and Stone 1978, Timmer and Morrow 1984).

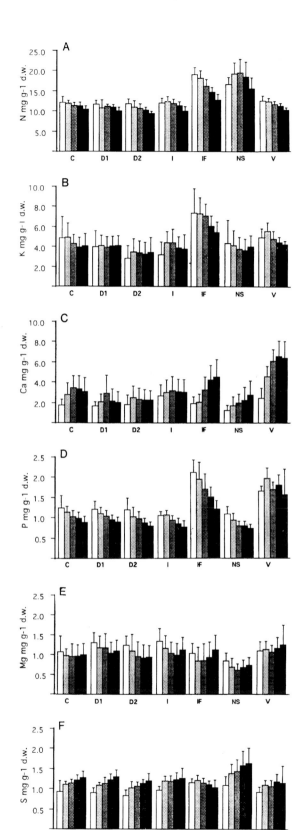

Fig. 2. Nutrient concentration (mg g^{-1} dw) in needle year-classes C (current) throughout C+4 (5 yr old) sampled in February 1993. Bars represent standard deviation, and treatment abbreviations are explained in Table 1.

85°C conversion factor. The coefficient of variation of the analysis was <5%.

Nutritional diagnosis and statistics

Foliar diagnosis was performed using different techniques. i) Needle nutrient concentrations were compared with empirical critical concentrations, i.e. concentrations in needles the growth of which had been reduced by ≥10% of the maximum level (Table 3). ii) Changes in nutrient concentrations and in nutrient contents per needle in different needle year classes were compared with corresponding changes in needles on trees in the irrigated and fertilized treatment (IF), which was assumed to have optimum water and nutrient supplies. That conditions were more optimal in this treatment was confirmed by the higher growth rate compared with the control (Table 1). iii) Proportions of K, P, Ca and Mg in relation to N (K/N, P/N, Ca/N, and Mg/N) were compared with critical values and optimum proportions (Table 3). iv) The foliar response was also examined graphically by using "graphical vector technique" as described by Timmer and Stone (1978), Timmer and Morrow (1984) and Weetman (1989). Nutrient contents per needle are plotted against nutrient concentrations and needle weights. Only data from current-year needles are used. With this technique, dilution effects, sufficiency, luxury consumption, toxicity and possible antagonism are possible to identify (Fig. 1). All nutrient concentrations and contents in the different treatments were normalised to 100 and then compared with the corresponding values in the optimum (IF) treatment. The interpretation was based on the direction and magnitude of each response vector ("vector analysis").

All statistics (analyses of variance with split plot design and Spjotvol-Stoline test) were carried out according to Sokal and Rohlf (1981) and Abacus (1989).

Results

Comparisons with critical concentration

In all treatments except those including N applications (i.e. IF and NS) needle N concentrations were lower than

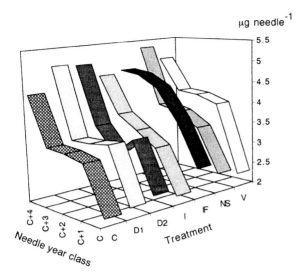

μg needle⁻¹

Fig. 3. Needle dry weight (μg needle⁻¹) for needle year-classes C (current) throughout C+4 (5 yr old) sampled in February 1993. Treatment abbreviations are explained in Table 1.

the critical level of 13 mg g⁻¹ d.w. (Fig. 2 and Table 3). For P, needles in all treatments except IF and V, which received soluble P, had concentrations below the critical P level according to Foerst et al. (1987) and Nihlgård (1990). However, concentrations lower than the critical P concentration according to Hüttl and Fink (1988) only occurred in I and NS. In the D2 and I treatments, K concentrations were below critical levels, and NS showed low Ca concentrations. Both Mg and S concentrations were well above critical concentrations in needles from all treatments (Table 3).

Changes in nutrient composition with needle age

The concentration of N decreased with needle age in all treatments except NS. In the latter treatment concentrations increased until year class C+2 whereafter they decreased (Fig. 2). Except for the D and I plots, the K concentration decreased with needle age. The P concentration decreased with age in all treatments, although the decrease was less pronounced in the treatment where the N-free fertilizer had been applied (V). In most cases, Ca concentration increased with needle age, but in the drought (D1 and D2) and irrigation (I) treatments it remained more or less constant. The variation in Mg concentration was generally small, and after 5 growing seasons the concentration was similar to that in current-year needles. S concentrations increased with needle age from C to C+4 in all treatments except IF where a small decrease with needle age was obtained (Fig. 2).

Needle dry weight increased with age in all treatments, and was highest in C+4 needles (Fig. 3). In most treatments, the needle weight largely increased from C to C+1, remained stable in C+2 and C+3, and then increased strongly between C+3 and C+4. However, when both water and nutrients were added (IF), the increase in needle weight was smooth and steady. Neither changes in water nor changes in the nutrient supply caused significant changes in needle weight, apart from C+2 and C+3 needles from the IF treatment, which were significantly heavier than needles from the control plots (p < 0.05).

The N content per needle increased with age in all treatments, but in the IF treatment the N content decreased between C+3 and C+4 (Table 4). The application of N caused significant increases in the N content per needle in the IF and NS treatments. The corresponding increases for K and P with needle age were, in most cases, proportional to the increase in needle weight. The IF treatment caused an increase in K content compared with the control which was significant (p < 0.05) for three out of five needle-year classes. The P content was significantly increased in four needle classes in the IF and V treatments compared with the control, but no other treatments differed from the control in terms of P. The Ca content showed a 3–6 fold increase with age, and the application of Ca in the V treatment caused a significant increase in all needle-year classes except C needles. Both the Mg and S contents increased with needle age in all treatments and were highest in C+4 needles, except in IF where the S content was highest in C+3 needles (Table 4).

Nutrient ratios

The ratios of K, P, Ca and Mg, respectively, to N were expressed as percentages of N (Table 5). The K/N ratios in current-year needles indicated that there were relative deficiencies of K in D2, I and NS according to critical ratios given by Nihlgård (1990) and optimum ratios given by Linder (1995), and in all treatments ratios were below the optimum values set by Ingestad (1979) (Tables 3 and 5). The P/N ratios in current needles were below critical levels according to Nihlgård (1990) and Linder (1995) in I- and NS-treated plots, whereas the P/N ratios for all treatments were below optimum according to Ingestad (1979). All treatments resulted in Ca/N and Mg/N ratios higher than the optima in current-year needles proposed by Ingestad (1979) and Linder (1995). In most cases, ratios of K, P, Ca and Mg, respectively, to N did not differ between the control and IF treatment, whereas the P/N ratios in the NS plots were significantly lower compared with both the control and the IF treatment in all needle-year classes. On the other hand, plots treated with N-free fertilizer (i.e. V) had significantly higher P/N ratios compared with the control in all needle-year classes. Neither K/N nor Mg/N showed any consistent change with needle age, whereas P/N decreased and Ca/N increased.

Table 4. Nutrient content (µg needle^{-1}) for needle year-class C (current) to C+4 (5 yr old) sampled in February 1993. Different letters denote significant differences between treatments (p < 0.05) according to ANOVA with split plot design followed by Spjotvol-Stoline range test. Treatment abbreviations are explained in Table 1.

	Treatment							
Year class	C	D1	D2	I	IF	NS	V	p-value
Nitrogen								
C	26.0[a]	26.0[a]	27.5[a]	26.7[a]	51.8[b]	40.6[ab]	29.4[a]	<0.001
C+1	38.2[a]	36.7[a]	37.2[a]	40.1[a]	67.3[a]	66.3[a]	49.3[a]	0.002
C+2	36.2[a]	38.2[a]	32.4[a]	41.8[a]	72.3[b]	72.9[b]	48.1[a]	<0.001
C+3	37.4[a]	37.3[a]	37.4[a]	43.9[a]	69.3[b]	71.6[b]	47.4[a]	<0.001
C+4	46.3[a]	51.2[a]	46.7[a]	47.3[a]	60.2[a]	79.5[b]	50.5[a]	<0.001
Potassium								
C	10.2[a]	8.65[a]	6.22[a]	7.10[a]	19.0[b]	9.22[a]	11.3[a]	<0.001
C+1	16.7[a]	14.2[a]	12.2[a]	14.4[a]	26.4[a]	13.7[a]	22.1[a]	0.012
C+2	14.5[a]	13.2[a]	10.8[a]	15.6[a]	31.1[b]	13.8[a]	20.2[a]	<0.001
C+3	13.1[a]	13.8[a]	12.2[a]	15.1[a]	28.1[b]	14.0[a]	18.8[a]	<0.001
C+4	18.0	21.6	18.2	18.2	24.5	20.7	21.1	0.41
Phosphorous								
C	2.70[a]	2.67[a]	2.42[a]	2.34[a]	5.71[b]	2.60[a]	3.82[a]	<0.001
C+1	3.69[a]	3.83[a]	3.46[a]	3.48[a]	7.04[b]	3.17[a]	7.78[b]	<0.001
C+2	3.32[a]	3.57[a]	2.91[a]	3.33[a]	7.44[b]	3.01[a]	6.96[b]	<0.001
C+3	3.24[a]	3.23[a]	3.15[a]	3.31[a]	7.10[b]	3.13[a]	7.49[b]	<0.001
C+4	3.90[a]	4.61[ab]	4.00[a]	3.80[a]	5.75[ab]	3.88[a]	7.30[b]	<0.001
Calcium								
C	3.85[ab]	3.86[ab]	3.46[ab]	6.03[b]	5.24[ab]	3.08[a]	6.16[b]	0.008
C+1	8.38[a]	7.77[a]	7.86[a]	9.89[a]	7.83[a]	5.97[a]	18.4[b]	<0.001
C+2	10.9[a]	10.6[a]	6.68[a]	11.4[a]	15.1[a]	7.44[a]	25.4[b]	<0.001
C+3	10.5[a]	7.59[a]	8.11[a]	12.0[ab]	20.3[bc]	8.61[a]	28.1[c]	<0.001
C+4	13.5[a]	11.3[a]	11.2[a]	14.6[a]	21.0[ab]	14.0[a]	32.9[b]	<0.001
Magnesium								
C	2.43	2.96	2.65	3.00	2.77	2.11	2.66	0.45
C+1	3.01[ab]	3.95[ab]	3.56[ab]	3.78[ab]	3.06[ab]	2.31[a]	4.39[b]	0.039
C+2	2.87[ab]	4.03[ab]	2.76[ab]	3.67[ab]	3.85[ab]	2.27[a]	4.33[b]	0.037
C+3	3.02[ab]	3.40[abc]	3.13[abc]	3.90[abc]	4.38[bc]	2.58[a]	4.90[c]	0.001
C+4	4.25	5.37	4.50	5.60	5.22	3.80	5.80	0.062
Sulphur								
C	2.05[ab]	2.04[ab]	1.72[a]	2.14[ab]	3.14[b]	2.61[ab]	2.14[ab]	0.005
C+1	3.58	3.84	3.51	3.89	4.43	4.68	4.38	0.55
C+2	3.68[ab]	3.88[ab]	3.24[a]	4.20[ab]	5.06[ab]	5.23[b]	4.40[ab]	0.015
C+3	4.04	4.25	4.17	4.75	5.19	5.96	4.88	0.077
C+4	5.74[ab]	6.85[ab]	6.16[ab]	6.12[ab]	4.83[a]	8.39[b]	5.34[a]	0.034

Vector analysis

Simultaneous comparisons between element concentration, element content per needle and needle weight for current needles of all treatments are presented (Fig. 4). Adding both water and nutrients (i.e. the IF-treatment) caused the needle weight to increase by 24% and the relative P concentration to increase by 110%, whereas the relative P content increased by 70%. Evaluating the direction of the vector of each nutrient indicated that the availabilities of P, N, K, S, and Ca were not adequate in the control, whereas Mg showed a slight decrease in both concentration and content in IF owing to dilution effects according to the interpretation based on Fig. 1. The magnitude of the vectors indicated that P was the most probable limiting nutrient, followed by N and K. In the irrigation treatment, K was the most likely limiting nutrient. Comparisons of the application of ammonium sulphate

(NS) or N-free fertilizer (V) with IF suggested that P was the most limiting nutrient in NS and that N was most limiting in V. For drought treatments (D1 and D2) K appeared to be the most probable limiting nutrient.

Discussion

Treatment effects

Interpretations based on critical concentrations suggested that N and possibly P (depending on the interpretation tool used) limited tree growth in the control plots at the Skogaby site, whereas interpretations based on nutrient ratios indicated that P (and possibly K) was the limiting element (Fig. 2, Tables 3 and 5). The vector analysis, where simultaneous interpretations of concentration, content and needle weight were made, suggested P to be the

Table 5. Ratios (in per cent) of K, P, Ca and Mg, respectively, to N, in needles sampled in February 1993. Different letters denote significant differences between treatments ($p < 0.05$) according to ANOVA with split plot design followed by Spjotvol-Stoline range test. Treatment abbreviations are explained in Table 1.

Year class	C	D1	D2	I	IF	NS	V	p-value
K/N								
C	41[a]	34[a]	25[a]	26[a]	38[a]	26[a]	40[a]	0.007
C+1	41[b]	40[ab]	32[ab]	36[ab]	40[ab]	22[a]	45[b]	0.012
C+2	37[b]	35[ab]	32[ab]	37[b]	44[b]	20[a]	41[b]	0.001
C+3	35[ab]	37[ab]	32[ab]	34[ab]	41[b]	21[a]	39[b]	0.012
C+4	39	41	36	37	42	26	42	0.25
P/N								
C	10[bc]	10[bc]	10[bc]	8.8[b]	11[c]	6.6[a]	13[d]	<0.001
C+1	9.6[b]	11[b]	9.4[b]	8.7[b]	11[b]	5.0[a]	16[c]	<0.001
C+2	8.9[b]	9.4[bc]	9.1[bc]	8.0[b]	10[c]	4.3[a]	14[d]	<0.001
C+3	8.8[bc]	8.7[bc]	8.6[bc]	7.5[b]	10[c]	4.5[a]	16[d]	<0.001
C+4	8.6[b]	9.0[b]	8.5[ab]	7.8[ab]	9.6[b]	4.8[a]	15[c]	<0.001
Ca/N								
C	14[ab]	15[ab]	16[ab]	22[b]	10[a]	7.7[a]	19[b]	<0.001
C+1	24[b]	21[ab]	22[ab]	24[b]	11[a]	9.4[a]	37[c]	<0.001
C+2	28[a]	26[a]	21[a]	27[a]	20[a]	11[a]	52[b]	<0.001
C+3	29[b]	19[ab]	22[ab]	27[ab]	29[b]	13[a]	58[c]	<0.001
C+4	30[a]	21[a]	24[a]	31[a]	36[a]	18[a]	63[b]	<0.001
Mg/N								
C	9.0[b]	11[b]	11[b]	11[b]	5.4[a]	5.2[a]	8.8[b]	<0.001
C+1	8.3[b]	11[b]	10[b]	9.5[b]	4.6[a]	3.7[a]	9.3[b]	<0.001
C+2	8.3[bc]	11[c]	9.0[c]	8.7[c]	5.2[ab]	3.2[a]	9.2[c]	<0.001
C+3	8.6[bc]	9.4[bc]	9.0[bc]	8.7[bc]	6.3[ab]	3.8[a]	10[c]	<0.001
C+4	9.6[b]	11[b]	9.9[b]	11[b]	8.9[b]	4.8[a]	12[b]	<0.001

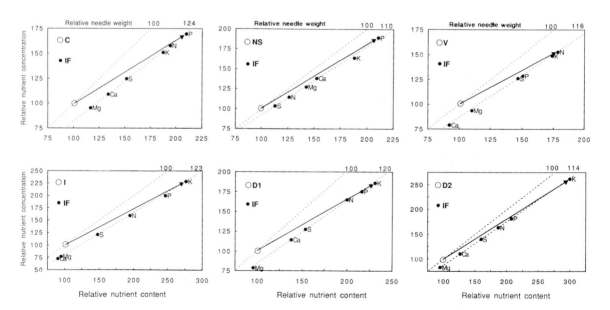

Fig. 4. Graphical vector analyses of relative nutrient content, relative nutrient concentration and relative needle weight. All three variables for each treatment (including the control) are set to 100 and compared with those in the IF treatment, here regarded as the optimum situation, with adequate water and nutrient supplies. The upper dotted line shows the relative needle weight line for respective treatment (relation 1:1 concentration:content) and the lower dotted line shows the relative needle weight line for IF, fitted through the data points. For interpretation, see Fig. 1 and for abbreviations, see Table 1.

most likely limiting element (Fig. 4). However, adding soluble P but no N (V treatment), only increased basal area growth by c. 12% (Table 1), indicating that P was not the only limiting element. Needles from the control plots in Skogaby seemed to be well supplied with Mg, which probably reflects the maritime influence from the nearby sea. By contrast, studies from Central Europe have shown that Mg deficiency is often associated with tree decline (e.g. Oren et al. 1988, Cape et al. 1990, Le Goaster et al. 1990/91, Katzensteiner et al. 1992). Our conclusion, based on these interpretations, is therefore that at Skogaby, P and N alternate as limiting nutrients. Despite the high annual precipitation, the water supply also seemed to limit tree growth, as indicated by the fact that irrigation increased tree growth by c. 60% during 1993 (Table 1). The water deficiency was most likely due to early summer droughts normally occurring in this area, but irrigation may also have stimulated N-mineralization. The increased growth in the I treatment caused a decrease in both concentrations and contents of K and P, which only partly can be explained by dilution effects since K/N and P/N ratios also decreased. This was also confirmed in the vector analysis (Fig. 4).

Providing an "optimum water and nutrient supply" (IF) increased basal area growth by almost 150% (Table 1) and there were no apparent nutrient deficiencies or unbalanced nutrient ratios. Actually, in most cases nutrient ratios in the control did not differ significantly ($p < 0.05$) from those in IF, despite large differences in growth rate. It seemed as though the dilution affected nutrient ratios less than nutrient concentrations.

In total, 500 kg N ha^{-1} was applied on the NS plots during 1988–92, which resulted in 30% more basal area growth compared with the control (Table 1). Normally, when N is scarce in the ecosystem and limiting growth, the N concentration decreases with increasing needle age (e.g. Tamm 1955, Carlyle 1986, Cape et al. 1990). Following the NS-application, the N concentration showed a maximum in C+2 needles, whereas all other treatments (as well as the control) showed decreasing N concentrations from C+1 onwards. The N content (μg needle^{-1}), on the other hand, increased with age and weight in all treatments (except in IF where the content showed a maximum in C+2 needles). Calculations in which the differences in needle dry weight were accounted for showed that the N content in C+4 needles of the NS treatment was c. 45% larger compared to the control and 20% larger compared with IF. This excess N in needles from the NS plots is most likely stored in the vacuoles as amino acids, mainly arginine (Rabe and Lovatt 1986, Van Dijk and Roelofs 1988, Gezelius and Näsholm 1993, Näsholm 1994). According to Aber et al. (1989), this may be regarded as an early indication of N saturation.

The reasons why assimilated nitrogen is stored as arginine instead of being used for growth are likely complex. However, P might be a limiting element (Fig. 4), making the P/N ratio in the needles sub-optimal (Nihlgård 1990, Linder 1995). This hypothesis is supported by Gezelius and Näsholm (1993), who showed that in Scots pine seedlings with a nutritional imbalance assimilated N was diverted to arginine when net protein synthesis was limited by factors other than N. A relative P deficiency was also found for Douglas fir in areas with high N deposition (Mohren et al. 1986, Houdijk and Roelofs 1993). This diversion of N can be considered as an increased metabolic expenditure (Bloom et al. 1985, Waring 1985). The tree must divert resources for transports to and from the site of the storage, convert them into specific storage compounds (i.e. arginine) and protect them from herbivores and pathogens by synthesising secondary metabolites. This resource consumption may reduce the competitive ability of the tree (Waring 1985).

Application of the N-free fertilizer caused a small, but non-significant, increase in basal area growth (Table 1). The needle analysis pointed out N as the most likely limiting nutrient, which again supports the interpretation of the data from the control plots, suggesting that N and P alternate as most limiting. The stand is still young and can be expected to have a surplus of cations released during the clear-cut phase.

An extended drought such as that achieved in the D1 and D2 plots affected not only the water status of the tree, but also the nutrient availability in the soil since air spaces replace water in the pores between soil particles reducing the rate of ion diffusion to the root surface (Chapin 1991). This was reflected in generally low needle concentrations (Fig. 2). The vector analysis indicated K to be limiting in D1 (drought during 1988–90) as well as in the D2 treatment (drought during 1990–92). However, the use of the vector analysis for identifying limiting nutrients in treatments exposed to drought may give a false picture, since water is likely to be the most limiting factor.

Retranslocation

Retranslocation processes can be divided into two phases i) translocation from existing foliage to the new shoots and needles in spring and early summer, and ii) translocation from senescing needles to remaining tissue during autumn (Helmisaari 1992). In most treatments at Skogaby the nutrient content per needle increased with needle age (Table 4), in contrast to findings by Nambiar and Fife (1991), Helmisaari (1992), and Millard and Proe (1992), among others. However, this does not necessarily mean that no retranslocation occurred, since translocation may have been balanced by replenishment prior to the sampling in February (Everett and Thran 1992). The IF treatment is the only exception, owing to the extremely fast growth rate, with obvious retranslocations of N, K and P from C+4 needles to younger foliage. This is consistent with the hypothesis that the internal circulation of nutrients is driven by shoot growth, making the amounts of nutrients retranslocated greater in trees with a higher growth rate (Nambiar and Fife 1991, Helmisaari

1992, Millard and Proe 1992, 1993). Removal of buds from Scots pine just before shoot and needle elongation does not cause any decrease in N concentration in C+1 needles, in contrast to trees with intact buds (Gezelius et al. 1981, Helmisaari 1992).

Estimations of retranslocation based on nutrient concentrations cannot be reliably interpreted (Nambiar and Fife 1991). In our study, estimates based on concentration indicate that N and P were retranslocated in all treatments (except N in NS). However, the needle weight increased in all needle-year classes analysed in this study, balancing the slight decreases in N, K, P, and Mg concentrations. Due to annual variations, the initial needle size for each needle year class may differ, which makes interpretations of retranslocation more difficult when analysing several year classes at one sampling (Nambiar and Fife 1991, Everett and Thran 1992). A more informative analysis of the retranslocation would be provided by following the cohort of one needle-year class over several years, starting at the current-year needle stage and ending when they are litter needles.

Evaluation of the diagnostic tools

Depending on the technique used, different conclusions can be drawn as to which factor(s) is limiting tree growth in Skogaby. Using nutrient ratios instead of nutrient concentrations to make a diagnosis is considered to decrease the problem of annual variation (Van den Driessche 1979, Linder 1995). It also seems to reduce the dilution effects (Table 5), making interpretation of the needle analysis more precise. However, when analysing mineral nutrient/N ratios we cannot distinguish between an imbalance due to excess N and an imbalance due to low availability of other nutrients (Ericsson et al. 1995). The vector analysis is based on the fact that there are links between tree growth and needle weight from the first growing season after fertilization (Timmer and Morrow 1984). In this study the plots were treated for five growing seasons, and the correlation between needle weight and tree growth may have been lower owing to a successive increase in the number of needles. Nevertheless, adjusting the relative needle weights according to the tree growth in 1993 would, in most cases, change the magnitude of the response vector rather than the direction of the vector. It has also been shown that vector analysis is applicable when interpreting responses to applied nutrients (Valentine and Allen 1990). This probably made the method difficult to use for interpretations of the drought treatments, since in these the relation between needle weight and tree growth might have weakened after five years of treatment. However, the vector analysis reduces the problem with dilution effects, and a simultaneous interpretation of several nutrients can be made, promoting a more correct interpretation.

Conclusions

Interpretations of the needle analyses from the experimentally manipulated forest plots suggested that both P and N were limiting nutrients in this stand. Also, despite high annual precipitation, the water supply also seemed to limit tree growth due to frequent summer drought periods in this area. The addition of ammonium sulphate over a five-year period led to relative deficiencies in P and K, and N saturation was indicated by the accumulation of N in older needles. Estimates of retranslocation between needles of different age, varied depending on whether nutrient concentration or nutrient content per needle was studied. Nutrient content seemed to give a better basis for retranslocation estimates. The vector analysis technique seems to provide reliable interpretations of the needle analysis for those nutrients applied in the experiment, except when drought is severe.

Acknowledgement – We thank L.-O. Nilsson and K. Wiklund for providing us with data on tree growth. We also want to thank M. Gyllin and A. Jonshagen for conducting good field work, and M.-L. Gernersson and I. Persson for their skilful laboratory work. A. Aronsson, H. Staaf, G. Tyler and S. Widell are gratefully acknowledged for critically reading the manuscript and D. Tilles for correcting the language. This work was supported by the Swedish Environmental Protection Agency (SNV) and the Foundation of Forestry Research (Stiftelsen Skogsbrukets Forskningsfond).

References

Abacus 1989. SuperANOVA. – Abacus Inc., Berkeley, CA.

Aber, J. D., Nadelhoffer, K. J., Steudler, P. and Melillo, J. M. 1989. Nitrogen saturation in northern forest ecosystems. – BioScience 39: 378–386.

Bergholm, J., Jansson P.-E., Johansson, U., Majdi, H., Nilsson, L.-O., Persson, H., Rosengren-Brinck, U. and Wiklund, K. 1995. Air pollution, tree vitality and forest production – the Skogaby project. General description of a field experiment with Norway spruce in south Sweden. – In: Nilsson, L.-O. and Hüttl, R. F. (eds), Nutrient uptake and cycling in forest ecosystems. CEC/IUFRO-symp. Ecosystem Report 13, EUR 15465, (in press).

Bloom, A. J., Chapin, F. S. and Mooney, H. M. 1985. Resource limitation in plants – an economic analogy. – Ann. Rev. Ecol. Syst. 16: 363–392.

Cape, J. L., Freer-Smith, P. H., Paterson, I. S., Parkinson, J. A. and Wolfenden, J. 1990. The nutritional status of *Picea abies* (L.) Karst. across Europe, and implications for 'forest decline'. – Trees 4: 211–234.

Carlyle, J. C. 1986. Nitrogen cycling in forested ecosystems. – For. Abstr. 47: 307–336.

Chapin, F. S. III. 1988. Ecological aspects of plant mineral nutrition. – Adv. Min. Nutr. 3: 161–191.

– 1991. Effects of multiple environmental stresses on nutrient availability and use. – In: Mooney, H. A., Winner, W. E. and Pell, E. J. (eds), Response of plants to multiple stresses. Academic Press, San Diego, pp. 67–88.

Ericsson, A., Walheim, M., Nordén, L.-G. and Näsholm, T. 1995. Concentrations of mineral nutrients and arginine in needles of *Picea abies* trees from different areas in southern Sweden in relation to nitrogen deposition and humus form. – Ecol. Bull. (Copenhagen) 44: 147–157.

Everett, R. L. and Thran, D. F. 1992. Nutrient dynamics in singleleaf pinyon (*Pinus monophylla* Torr and Frem.) needles. – Tree Physiol. 10: 59–68.

Evers, F. H. 1986. Die Blatt- und Nadelanalyse als Instrument der Bioindikation. – Allg. Forst. Z. 1986: 6–9.

FAO-Unesco 1988. Soil map of the world. Revised legend. – FAO, Rome.

Foerst, K., Sauter, U. and Neuerburg, W. 1987. Bericht zur Ernährungssituation der Wälder in Bayern und über die Anlage von Walddüngeversuchen. – Bayerische Forstliche Versuchs- und Forschungsanstalt München.

Gezelius, K. and Näsholm, T. 1993. Free amino acids and protein in Scots pine seedlings cultivated at different nutrient availabilities. – Tree Physiol. 13: 71–86.

– , Ericsson, A., Hällgren, J.-E. and Brunes, L. 1981. Effects of bud removal in Scots pine (*Pinus sylvestris*) seedlings. – Physiol. Plant. 51: 181–188.

Helmisaari, H.-S. 1992. Nutrient retranslocation within the foliage of *Pinus sylvestris*. – Tree Physiol. 10: 45–58.

Houdijk, A. L. F. M. and Roelofs, J. G. M. 1993. The effects of atmospheric nitrogen deposition and soil chemistry on the nutritional status of *Pseudotsuga menziesii, Pinus nigra* and *Pinus sylvestris*. – Environ. Pollut. 80: 79–84.

Hüttl, R. F. and Fink, S. 1988. Diagnostische Düngungsversuche zur Revitalizerung geschädigter Fichtenbestände (*Picea abies* Karst.) in Südwestdeutschland. – Forstw. Cbl. 107: 173–183.

Ingestad, T. 1979. Mineral nutrient requirements of *Pinus silvestris* and *Picea abies* seedlings. – Physiol. Plant. 45: 373–380.

– 1988. A fertilization model based on the concepts of nutrient flux density and nutrient productivity. – Scand. J. For. Res. 3: 157–173.

Katzensteiner, K., Glatzel, G., Kazda, M. and Sterba, H. 1992. Effects of air pollutants on mineral nutrition of Norway spruce and revitalisation of declining stands in Austria. – Water Air Soil Pollut. 61: 309–322.

Le Goaster, S., Dambrine, E. and Ranger, J. 1990/91. Mineral supply of healthy and declining trees of a young spruce stand. – Water Air Soil Pollut. 54: 269–280.

Linder, S. 1995. Foliar analysis for detecting and correcting nutrient imbalances in forest stands. – Ecol. Bull. (Copenhagen) 44: 178–190.

Majdi, H. and Persson, H. 1993. Spatial distribution of fine roots, rhizosphere and bulk-soil chemistry in an acidified *Picea abies* stand. – Scand. J. For. Res. 8: 147–155.

Marschner, H. 1986. Mineral nutrition of higher plants. – Academic Press, London.

Millard, P. and Proe, M. F. 1992. Storage and internal cycling of nitrogen in relation to seasonal growth of Sitka spruce. – Tree Physiol. 10: 33–43.

– and Proe, M. F. 1993. Nitrogen uptake, partitioning and internal cycling in *Picea sitchensis* (Bong.) Carr. as influenced by nitrogen supply. – New Phytol. 125: 113–119.

Mohren, G. M. J., Van den Burg, J. and Burger, F. W. 1986. Phosphorus deficiency induced by nitrogen input in Douglas fir in the Netherlands. – Plant Soil 95: 191–200.

Nambiar, E. K. S. and Fife, D. N. 1991. Nutrient retranslocation in temperate conifers. – Tree Physiol. 9: 185–207.

Näsholm, T. 1994. Removal of nitrogen during needle senescence in Scots pine (*Pinus sylvestris* L.). – Oecologia 99: 290–296.

Nihlgård, B. 1971. Pedological influence of spruce planted on former beech forest soil in Scania, south Sweden. – Oikos 22: 302–314.

– 1985. The ammonium hypothesis, an additional explanation to the forest dieback in Europe. – Ambio 14: 2–8.

– 1990. The vitality of forest trees in Sweden – stress symptoms and causal relations (in Swedish with English summary). – In: Liljelund, L.-E. (ed.), Vitality fertilization – present knowledge and research needs. Swedish Environ. Protect. Agency, Solna, Report 3813, pp. 45–70.

Nilsson, L.-O. and Wiklund, K. 1992. Influence of nutrient and water stress on Norway spruce production in south Sweden – the role of air pollutants. – Plant Soil 147: 251–265.

Oren, R., Werk, K. S., Schulze, E.-D., Meyer, J., Schneider, B. U. and Schramel, P. 1988. Performance of two *Picea abies* (L.) Karst stands at different stages of decline. VI. Nutrient concentration. – Oecologia 77: 151–162.

Rabe, E. and Lovatt, C. J. 1986. Increased arginine biosynthesis during phosphorus deficiency. – Plant Physiol. 81: 774–779.

Schulze, E.-D. 1989. Air pollution and forest decline in a spruce (*Picea abies*) forest. – Science 244: 776–783.

Sokal, R. R. and Rohlf, F. J. 1981. Biometry, 2nd ed. – W. H. Freeman, NY.

Tamm, C. O. 1955. Studies on forest nutrition: I. Seasonal variation in the nutrient content of conifer needles. – Medd. Stat. Skogsforskningsinst. Stockholm 45: 1–34.

– 1991. Nitrogen in terrestrial ecosystems – Questions of productivity, vegetational changes, and ecosystem stability. – Ecol. Stud. 81.

Timmer, V. R. and Stone, E. L. 1978. Comparative foliar analysis of young balsam fir fertilized with nitrogen, phosphorous, potassium and lime. – Soil. Sci. Soc. Am. J. 42: 125–130.

– and Morrow, L. D. 1984. Predicting fertilizer growth response and nutrient status of Jack pine by foliar diagnosis. – In: Stone, E. L. (ed.), Forest soils and treatment impacts. Proc. 6th North Amer. Forest Soils Conf. Univ. of Tennessee, Knoxville, pp. 335–351.

Valentine, D. W. and Allen, H. L. 1990. Foliar responses to fertilization identify nutrient limitation in loblolly pine. – Can. J. For. Res. 20: 144–151.

Van den Driessche, R. 1979. Estimating potential response to fertilizer based on tree tissue and litter analysis. – In: Proc. For. Fertilization Conference, Coll. For. Resour., Univ. Washington, Seattle pp. 214–220.

Van Dijk, H. F. G. and Roelofs, J. G. M. 1988. Effects of excessive ammonium deposition on the nutritional status and condition of pine needles. – Physiol. Plant. 73: 494–501.

Waring, R. H. 1985. Imbalanced forest ecosystems: Assessments and consequences. – For. Ecol. Manag. 12: 93–112.

Weetman, G. F. 1989. Graphical vector analysis technique for testing stand nutritional status. – In: Dyck, W. J. and Mees, C. A. (eds), Research strategies for long-term site productivity. Proceedings, IEA/BE A3 Workshop, Seattle, WA. IEA/BE A3 Report No. 8. For. Res. Inst., New Zealand, Bulletin 152.

Ecological Bulletins 44: 178–190. Copenhagen 1995

Foliar analysis for detecting and correcting nutrient imbalances in Norway spruce

Sune Linder

Linder, S. 1995. Foliar analysis for detecting and correcting nutrient imbalances in Norway spruce. – Ecol. Bull. (Copenhagen) 44: 178–190.

Results are presented from the first seven years of a nutrient optimisation experiment in young stands of Norway spruce in northern Sweden. The principal aim of the experiment was to eliminate water and mineral nutrients as growth-limiting factors, at the same time as leaching to the groundwater was avoided. The approach applied was the defininition of target values for the foliage concentration of each nutrient element. On the basis of repeated foliar analysis and predicted growth response the proportions and amounts of nutrients applied were adjusted annually. Imbalances in the nutrient status of the trees, induced by fertilisation as determined by foliage analysis, were successfully corrected by adjustment of the amount and composition of the fertiliser mix. Accumulation, followed by depletion, of starch in needles during summer had a pronounced effect on nutrient concentrations, thus making evaluation of nutritional status difficult. A variation in needle dry weight of up to 30% occurred during the growing season. The depletion of starch coincided with the onset of growth and was both earlier and faster in fertilised trees than in control trees. This indicates that the growth rate in non-treated stands was not limited by carbon, but rather by nutrient availability. It is recommended that for diagnostic purposes, several age-classes of foliage are sampled on a number of occasions during the season(s). If sampling is restricted to one age-class of foliage, it is recommended that one-year-old foliage is used, to reduce between-year variation and to enable sampling throughout the season. If nutrient concentrations are assessed on samples taken during the period late spring to early autumn, the carbohydrate content must be determined to allow values to be normalised. Nutrient imbalances can, however, be detected without correcting for carbohydrate reserves, by calculating the ratio between elements. Experience obtained during the first seven seasons has indicated that the nutrient quotients relative to nitrogen, based on detailed studies of plant nutrition, have been more generally valid than the concentrations regarded as optimal for Norway spruce.

S. Linder. Dept of Ecology and Environmental Res., Swedish Univ. of Agricultural Sciences. P. O. Box 7072, S-750 07 Uppsala, Sweden.

Introduction

Our basic knowledge of the effects of nutrient stress on the yield, vitality and survival of forest trees is still inadequate. Until recently, the study of forest nutrition has mainly aimed at improving yield by detecting nutrient deficiencies or at improving the fertility of poor sites to allow 'maximum' production, by supplying the limiting nutrient(s). Over the latest two decades, however, it has become increasingly evident that the 'vitality' and yield of forest ecosystems can be severely affected by anthropogenic pollution; effects described as 'new type forest decline' (Blank 1985). Even if the causal relation-

ships were unclear, it was commonly agreed at an early stage that nutritional disturbances were involved (e.g. Hüttl 1988, Oren and Schulze 1989, Schulze et al. 1989). In declining Norway spruce stands in Denmark, however, nutritional deficiencies or imbalances could not explain the observed decline (Saxe 1993), stressing the fact that a multitude of factors can result in similar symptoms. Not only anthropogenic pollution, but also new management systems, such as whole tree-harvesting and the use of slash for energy purposes, may deplete forest sites of valuable nutrient elements, which must be replaced to maintain a high 'vitality' and sustained yield (e.g. Staaf and Olsson 1994). To be able to identify and correctly treat forest stands, which are declining or which have reduced 'vitality' as an effect of nutrient disorders, we require an improvement in our diagnostic tools.

Attempts have been made to evaluate the causes and magnitude of current forest damage and decline by regional surveys of nutritional status in 'healthy' and 'declining' trees (e.g. van Praag and Weissen 1986, Kaupenjohann et al. 1989, Cape et al. 1990). Such studies give, however, mainly circumstantial evidence; the lack of reliable historical records is obvious and the need for long-term studies and experiments is emphasised (e.g. Woodman and Cowling 1987, Tamm 1995).

In forest science there is, however, a number of long-term experiments, designed for purposes other than the study of forest decline, which are of great value in the study of long-term changes occurring in forest ecosystems. Such a series of experiments, including repeated additions of nitrogen and other nutrient elements, was started in Sweden during the late 1960's to determine the primary production of forest ecosystems at optimum nutrient levels, and to study disturbances in ecosystem functions under supra-optimal nutrient regimes (Tamm 1968, 1991). This type of long-term experiment has proved important not only for establishing the relationship between tree nutrition and forest yield, but also for developing concepts and methods for assessing the impact of air pollution on forest ecosystems, as well as for establishing 'critical loads' for nitrogen and sulphur (Tamm 1989, 1995).

A further development of the experiments mentioned above is the establishment of a new type of forest nutrient optimisation trial (Linder 1990). The experiments were established in young stands of Norway spruce in the north (Flakaliden) and southeast (Asa) of Sweden, where treatments commenced in 1987 and 1988, respectively. A third experiment (Skogaby) using similar concepts, but focusing on the impacts of anthropogenic pollution, was started at the same time in southeastern Sweden (cf. Bergholm et al. 1994). The principal aim of these experiments was to eliminate, by means of combined irrigation and fertilisation, water, mineral nutrients or both, as growth-limiting factors, at the same time as leaching of nutrients to the groundwater was avoided. The supply of mineral nutrients was adjusted annually to the nutrient status of the trees and soil. The concepts and methods used were based on the following assumptions: 1) Optimal 'vitality' and conditions for growth can exist only if all essential plant nutrients are present in the correct proportions. 2) Within a wide range the concentration per se of a nutrient element is not essential to the 'vitality' of a plant; the proportions of elements relative to nitrogen are at least as important. 3) To optimise biomass production in a given climate, all essential mineral nutrients should be supplied at a rate which is adjusted to the current mineralisation and fixation rates and nutrient demand of the crop. 4) The optimal proportion between nutrient elements is similar for all higher plants and can be defined in relation to nitrogen.

These concepts were not completely new (cf. Shear et al. 1948), but had not previously been rigorously tested in long-term field experiments. Earlier combined irrigation and fertilisation experiments in forest stands had tried to create non-limiting conditions, in terms of water and nutrient availability, but without attempting to define and maintain an optimal nutrient balance in the stand (e.g. Aronsson and Elowson 1980, Linder et al. 1987, Pereira et al. 1989).

The method used was to set target values for the concentration of each individual nutrient element in the foliage. The target values were based on results from detailed studies of plant nutrient requirements established in laboratory studies, and experience from long-term forest nutrient experiments. On the basis of foliar analysis and predicted growth response, the amount of element to be applied was estimated relative to the target figure. Subsequent analysis of nutrients in foliage and soil water showed the extent to which the target had been met and the information was fed back to the proportions and amounts of nutrient to be added on the next occasion (cf. Ericsson et al. 1992, Linder and Flower-Ellis 1992). For this approach to be meaningful, it was important to standardise the procedures for foliar sampling and analysis.

In the present paper, results are presented from the first seven years of nutrient optimisation at the northern site (Flakaliden), using diagnostic foliar analysis to determine composition and amount of nutrient supply, to create and maintain optimal internal nutrient status in the trees.

Materials and methods
Site

The results presented were obtained in a current nutrient optimisation experiment at Flakaliden (64°07'N; 19°27'E; alt. 310 m a.s.l.) in northern Sweden (Linder 1990). The principal aim of the experiment was to demonstrate the potential yield of Norway spruce, under given climatic conditions and non-limiting soil water, by optimising the nutritional status of the stands, at the same time as leaching of nutrients to the groundwater was avoided. The experiment was laid out in a young Norway spruce stand, planted in 1963 after clear-felling. Two nutrient optimisation treatments were included. The first treatment (IL) was a complete nutrient solution, which was injected into the irrigation water and supplied every day during the growing season (June to mid-August). The second treatment (F) was a solid fertiliser mix, which was applied in early June each year. Controls (C) and plots with irrigation (I) were also included. The treatments, which began in 1987, were replicated four times, and each replicate consisted initially of a double plot, made up of two 50 × 50 m plots. Some of the subplots were later used for new treatments, but without reducing the number of true replicates. The total area of the experiment was 8.25 ha.

Table 1. Target values used as 'optimal' in one-year-old foliage of Norway spruce. Nitrogen [N] is given in mg g^{-1} (structural weight) and other values in per cent of nitrogen (by weight) during the first seven years of treatment. To convert [N] from structural dry weight to total dry weight, for periods when starch is not stored in the foliage, the value should be decreased by c. 10%. For further details see text.

		Macronutrients						Micronutrients					
	N	P	K	Ca	S	Mg	Mn	Fe	Zn	Cu	B	Mo
1987	[20]	15	45	7	6	5	0.4	0.7	0.03	0.03	0.05	0.007
1990	[18]	10	35	2.5	5	4	0.4	0.7	0.03	0.03	0.05	0.007
1993	[18]	10	35	2.5	5	4	0.05	0.2	0.05	0.03	0.05	0.007

Nutrient treatment

An 'optimal' nutritional status to be obtained in the foliage of the trees was defined in terms of target needle concentrations for each individual nutrient element. On the basis of foliar analysis and predicted growth response it was possible to estimate applications of mineral fertiliser appropriate to the target values (cf. Ericsson et al. 1992, Linder and Flower-Ellis 1992).

The initial target values (Table 1) for macronutrients were derived from studies of 'optimal' nutrition of Norway spruce in laboratory (Ingestad 1959, 1979, 1982, Ingestad and Kähr 1985) and field experiments (Tamm 1968, Ingestad et al. 1981, Aronsson 1983, 1985, Linder 1987, Tamm 1991). For micronutrients, values recommended by Ingestad (1967) for optimal fertilisation of tree seedlings were used. Based on experience from the first three years of treatment and new results from detailed laboratory experiments (Ericsson and Ingestad 1988, Ericsson and Kähr 1993) target values for macronutrients were revised in 1990. A revision of the target values for manganese, iron and zinc occurred in 1993 and was based on results by Göransson (1993, 1994) and Göransson and McDonald (1993).

The initial composition of the nutrient solution and fertiliser mix was as recommended by Ingestad (1967, 1979). The amount and composition of the nutrient solution and fertiliser mix was later changed each year in relation to the results obtained (Table 2).

Sampling and analyses

Following a complete inventory of all plots (> 18 000 trees), and detailed structural analysis of 32 trees in October 1986, four height classes of tree were selected for serial branch sampling in IL, F and C plots. The sample trees were distributed over all replicate plots of the C, F and IL treatments so that every plot had approximately the same number. In total 70 trees per treatment were utilized. Ten of those, selected across replicates, were sampled on each occasion, and the same group of trees was used only once a year. A detailed description of the selection of sample trees, branch sampling and subsequent growth analysis has been given by Flower-Ellis (1993). To cover the snow-free season (c. May-October) adequately, it was estimated that seven sampling occasions would be needed. These were concentrated to the period of most active growth, with longer intervals at the beginning and end of the season. From the end of April 1987, serial branch samples were collected from whorls 4, 7 and 10 (counted from the top) on every occasion. During the winter of 1991/92 an extra five sampling occasions were included.

The sample branches were dissected by age-classes and orders before being frozen (−20°C) to await further processing. Subsamples to be used for nutrient and carbohydrate analysis were immediately immersed in liquid nitrogen before being stored at −20°C until analysed. The subsamples were dried in a ventilated oven (85°C, 48 h), then separated into the fractions needles and shoot axes, which were counted and measured. The fractions were pooled by age-classes within treatments. For the diagnostic nutrient and carbohydrate analysis, needles from

Table 2. Summary of macronutrients (N, P, K, S, Ca, Mg) supplied, kg ha^{-1}, during the first seven years, to irrigated-fertilised (IL) and fertilised (F) plots, respectively. For further details see text.

	Nitrogen		Phosphorus		Potassium		Calcium		Sulphur		Magnesium	
	IL	F	IL	F	IL	F	IL	F	IL	F	IL	F
1987	100	100	17	17	48	48	53	9	8	12	18	6
1988	100	100	17	17	48	48	–	9	8	12	–	6
1989	100	100	21	29	55	58	–	10	6	8	–	5
1990	75	75	16	22	41	48	15	15	6	10	14	10
1991	75	75	13	22	41	48	–	11	–	4	5	10
1992	75	75	11	15	34	35	–	8	–	4	5	10
1993	75	75	10	10	30	30	–	5	–	8	8	8

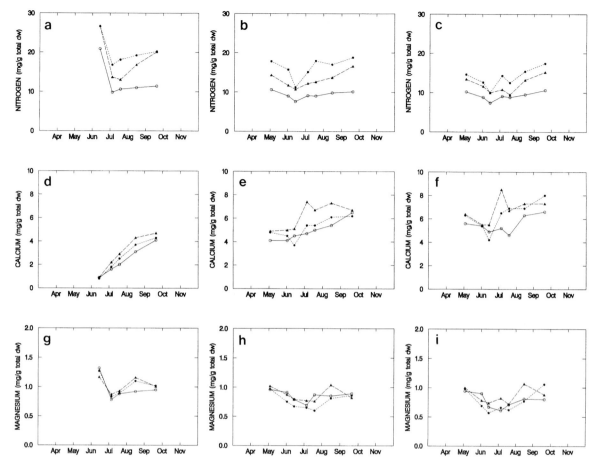

Fig. 1. The seasonal variation in nitrogen, (a–c), magnesium (d–f), and calcium (g–i) during 1988 in current (C: a, d, g), one-year-old (C+1: b, e, h) and two-year-old (C+2: c, f, i) needles on whorl 4 of young Norway spruce trees subjected to different treatments. The concentrations are expressed as mg g^{-1} (total dry weight). Symbols: Control: open circle; fertilised: filled circle; irrigated-fertilised: filled triangle.

second order shoots from the 4th whorl, were used. After milling to pass a 0.12 mm sieve, the material was dried under vacuum before being analysed. Three age-classes of needle (C: current; C+1: one-year-old; C+2: two-year-old) were analysed on each occasion.

Total nitrogen (N) was analysed with an elemental combustion analyzer (Carlo Erba, NA 1500, Carlo Erba Strumentatzione, Milan, Italy) and other elements (K, P, Ca, S, Mg, Mn, Fe, Zn, B, Cu, Al) by an inductively coupled plasma atomic emission spectrometer (Jobin Yvon 70+, Longjumeau, France). Starch and soluble sugars were analysed enzymatically, according to Steen and Larsson (1986), slightly modified (50 mg sample extracted 40 min at 60°C, Termamyl 120L). All samples from one season were analysed at the same time and on each occasion a 'standard' sample was included. The 'standard' was prepared by sampling all one-year-old shoots on a control tree during summer when the carbohydrate content was high (29% of dry weight). To determine the accuracy of the analysis, ten individual samples

from this batch were analysed for nutrients, starch and soluble sugars. The standard deviation for the individual nutrient elements, starch and total carbohydrate, was < 5 per cent.

Results and discussion
Sampling for diagnostic foliage analysis

For discussion of the usefulness of diagnostic needle analysis and the possibility of detecting and correcting nutrient imbalances in forest stands, the presentation of results has been restricted mainly to results from autumn samplings during the period 1987–1993. To illustrate and discuss seasonal variation in foliage nutrition, results from two out of the seven years studied are presented (1988 and 1993). A further restriction is to concentrate mainly on four (N, P, K, Mg) of the 11 nutrient elements analysed.

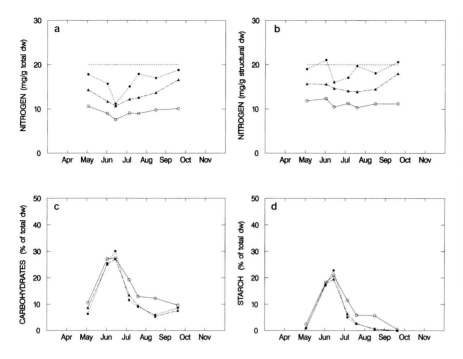

Fig. 2. The seasonal variation in nitrogen concentration (mg g^{-1}) in one-year-old (C+1) needles on whorl 4 of Norway spruce during 1988, before (a) and after (b) correction for variations in carbohydrate content. The seasonal variation in the amount (% of total dry weight) of total carbohydrates and starch during the same period is shown in diagrams c and d, respectively. The horizontal lines in the diagrams show the target value of the nitrogen concentration used up to 1990. Symbols: Control: open circle; fertilised: filled circle; irrigated-fertilised: filled triangle.

For foliar analysis to be meaningful, particularly in comparative studies, it is important to standardise procedures for foliar sampling and analysis, taking into account temporal and spatial variation (e.g. Wehrmann 1959, Tamm 1964, van den Driessche 1974). In the present study, comparability of successive samples was ensured by always sampling shoots from the fourth whorl, and the effect of between-tree variation was reduced by pooling the samples from ten trees.

Seasonal variation of nutrient concentration

In agreement with earlier studies in conifers (e.g. Wehrmann 1959, Höhne 1964, Evers 1972, Aronsson and Elowson 1980, Oren et al. 1988b) a pronounced seasonal variation was found for all nutrient elements analysed, independent of age-class of foliage or treatment. However, the magnitude and pattern of seasonal variation varied between elements and age-class of foliage (Fig. 1). Nitrogen (Fig. 1a–c), calcium (Fig. 1d–f), and magnesium (Fig. 1g–i) were chosen to represent the main patterns. The example was taken from the second year of treatment (1988), but the general pattern was similar during the seven seasons studied (1987–1993). Except for Ca (Fig. 1d) the concentration (mg g^{-1} total dry weight) of other macronutrients (N, P, K, S, Mg) decreased rapidly in current (C) foliage, reaching a minimum in early July, whereafter the concentrations increased once more. The magnitude of the variation in one-year-old (C+1) and two-year-old (C+2) foliage was smaller, but still pronounced. With the exception of Ca and Mn, element concentrations decreased with increasing needle age.

These results agree in general with a number of recent reports on Norway spruce nutrition (e.g. Oren et al. 1988b, Cape et al. 1990, Rosengren-Brinck and Nihlgård 1994, 1995).

Even on fertilised plots, there was a clear decrease in the mineral nutrient concentration, during the growing season, in needles older than the current season. Part of this decrease can be explained by the resorption of nutrients and their transport to the developing needles and shoots (e.g. Fife and Nambiar 1982, 1984, Helmisaari 1992), but most of the variation was caused by the seasonal variation in the carbohydrate reserves, principally starch (Fig. 2c–d). If the effect of variations in the carbohydrate content was eliminated, needle residual dry weight (structural weight) at constant length increased more or less linearly with age (Flower-Ellis 1993). In Norway spruce, this amounts to 0.18 mg yr^{-1} per single needle, at a standardised length of 10 mm, i.e. c. 5%. The development of new phloem cells in the vascular tissue may lie behind the increase in needle weight (cf. Ewers 1982), but other substances unaccounted for in the carbohydrate analyses may also be involved. The 'dilution effect' on needle nutrient concentrations, associated with the increase in structural weight, is thus c. 5% in a year. Neglect of this leads to overestimation of the extent of nutrient resorption from foliage. As an effect of between-year variation in needle size, the method of expressing nutrient contents 'per needle' or per '1000 needles' is not to be recommended, unless estimated at a standardised length (cf. Flower-Ellis 1993).

The general recommendation that foliage should be sampled for diagnostic purposes in autumn or winter (e.g. Wehrmann 1959, Tamm 1964, Höhne 1964), reduces the influence of stored carbohydrates, but includes an improvement in nutrient status, which can occur after growth has ceased in autumn. The autumn uptake of nutrients can depend on the weather conditions and hence add to the between-year variation. The results from seven years of autumn sampling showed a considerable between-year variation in nutrient concentrations (s.d. = 10 to 20% of the means), which is in agreement with earlier reports (cf. van den Driessche 1974). The magnitude of the variation depended on age-class of foliage, treatment and nutrient element. Across treatments, the extent of variation for all elements, except Mg, was smaller in one-year-old than in current foliage, and higher in treated than in non-treated stands (F > IL > C). Of all macro-elements, Mg had the lowest between-year variation in current foliage, but the highest in older foliage. Ranking of the variation between other macroelements was similar across treatments and age-classes of foliage (K > Ca > N > S > P). The between-year variation in carbohydrate concentration of the autumn samples was higher (s.d. > 20%) than for the macronutrients, but was independent of age-class of foliage or treatment. Normalising the element concentrations with regard to carbohydrate reserves had a minor effect on the variation and no effect on the ranking between elements or age-classes of foliage. The highest between-year variation of all elements was found for Fe (c. 40% s.d. of the mean), while other micronutrients had a variation similar to that found for macroelements. The ranking between micronutrients was Fe > B > Zn > Mn > Cu.

Seasonal variation of starch reserves

Most of the seasonal variation in nutrient concentrations could be explained by variations in the amount of starch stored in the needles (Fig. 2d). The maximum starch content attained by needles appeared to be independent of treatment, whereas the rate of breakdown of the starch reserves was far greater in fertilised than in unfertilised trees (Figs 2d and 3). The rapid breakdown of the starch reserve in fertilised (F, IL) trees was also in good agreement with earlier bud-break and more rapid shoot extension of such trees (Linder and Flower-Ellis 1992). This agrees with earlier findings in young stands of Scots pine (Ericsson 1979), and indicates that the availability of nutrients, rather than of carbon, limits yield in these northern conifer stands.

The starch content of previous year's and older needles varied from 0 to >30% of their dry weight, while the concentration of soluble sugars varied relatively little (c. 10% of dry weight), irrespective of treatment. Up to c. 40% of the needle's dry weight may thus consist of non-structural carbohydrates during early summer.

The size and dynamics of the starch reserve in all age-classes of foliage exhibited a wide range of variation during the seven seasons for which it was measured (cf. Figs 2d and 3). The course of accumulation of the starch

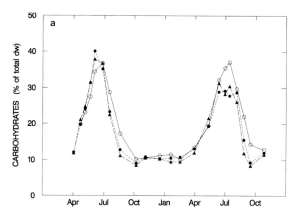

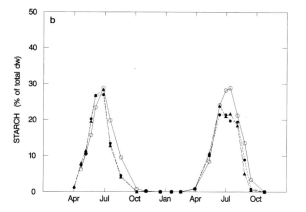

Fig. 3. The seasonal variation in the concentration (% of total dry weight) of carbohydrates (a) and starch (b), during 1991 and 1992, in one-year-old (C+1) needles on whorl 4 of Norway spruce trees subjected to different treatments. Symbols: Control: open circle; fertilised: filled circle; irrigated-fertilised: filled triangle.

reserve in the early part of the growing season, and its breakdown in the latter part of the summer, differed markedly. In an "average" year, starch has begun to accumulate by mid-April at the latest, peak values are reached in mid-June, and by mid-October the starch reserves have been completely emptied. The beginning of the decline in the starch reserve coincides with the start of extension growth, when the daily demand for carbohydrates normally exceeds net assimilation.

After the nutrient concentrations (mg g^{-1} total dry weight) were normalised with regard to carbohydrate reserves and expressed as mg g^{-1} structural dry weight, the seasonal variation decreased, but some variation remained (cf. Fig. 2a–b).

When defining the level of nutrient stress in trees, there is reason to believe that the minimum concentrations found during the period of active growth are more pertinent than peak values found during the non-active part of the year. Furthermore, the period when sampling is possible without disturbance from starch reserves in the

Table 3. Nutrient concentrations in one-year-old foliage of young Norway spruce trees subjected to different treatments. Values shown are from needles on whorl 4, sampled in the autumn, during the first seven years of treatment. The nutrient concentrations in May 1987, one month before treatments were commenced, are given for comparison. These spring values are normalised in relation to their content of starch, but not soluble sugars. The concentrations are given as mg g^{-1} (N, P, K, Ca, S, Mg, Mn) or µg g^{-1} (Fe, Zn, Cu, B) of total dry weight. The treatments were irrigation-fertilisation (IL), fertilisation (F), and control (C). For further details see text.

Date	Plot	N	P	K	Ca	S	Mg	Mn	Fe	Zn	Cu	B
19 May 1987	C	10.4	1.42	5.9	3.9	0.69	1.03	0.69	31.5	38.7	1.3	8.9
19 May 1987	F	10.9	1.52	5.7	4.0	0.68	0.96	0.82	33.0	38.2	2.3	7.5
19 May 1987	IL	10.8	1.58	6.1	3.9	0.71	1.02	0.65	26.9	40.7	2.3	6.6
15 Sept 1987	C	10.0	1.47	5.0	5.2	0.70	0.81	1.13	31.7	45.8	2.4	8.5
15 Sept 1987	F	12.7	1.64	5.5	5.1	0.80	0.93	0.90	34.6	40.1	2.8	10.5
15 Sept 1987	IL	14.4	1.73	6.5	5.5	0.93	0.85	0.96	34.5	44.2	3.2	7.9
20 Sept 1988	C	10.1	1.94	6.0	6.5	0.91	0.89	1.23	70.4	72.8	3.0	9.6
20 Sept 1988	F	18.8	2.10	6.7	6.2	1.10	0.86	1.31	77.9	54.9	3.8	12.2
20 Sept 1988	IL	16.6	2.27	7.1	6.7	1.14	0.82	1.18	82.6	58.6	4.1	11.6
4 Oct 1989	C	10.5	1.95	6.0	5.9	0.85	0.75	1.28	61.6	65.7	2.5	8.3
4 Oct 1989	F	21.8	2.42	7.3	6.5	1.04	0.77	1.50	55.4	64.8	3.1	11.5
4 Oct 1989	IL	18.4	2.65	8.0	6.2	1.07	0.65	1.29	52.8	64.8	3.4	10.1
4 Oct 1990	C	10.3	1.60	3.7	5.0	0.82	0.73	1.02	45.0	56.1	2.6	6.3
4 Oct 1990	F	16.2	1.86	5.0	5.9	0.96	0.50	1.05	38.5	49.2	2.9	9.1
4 Oct 1990	IL	15.6	2.02	5.4	5.8	1.02	0.54	1.10	37.8	63.4	3.0	9.6
7 Oct 1991	C	10.8	1.44	5.3	6.5	0.78	0.88	1.33	40.4	67.2	3.5	5.7
7 Oct 1991	F	14.4	1.92	6.7	7.1	0.91	0.70	1.37	37.9	54.4	3.9	7.9
7 Oct 1991	IL	14.4	2.25	6.8	6.8	0.93	0.72	1.28	37.0	62.3	3.8	9.0
27 Oct 1992	C	9.9	1.45	4.7	4.7	0.65	0.70	0.83	32.1	50.3	2.3	5.3
27 Oct 1992	F	15.2	2.08	7.1	4.8	0.83	0.58	0.99	30.3	43.0	2.9	7.7
27 Oct 1992	IL	13.3	2.12	7.5	5.1	0.86	0.66	1.00	27.0	70.4	3.1	6.9
28 Sept 1993	C	10.4	1.59	5.1	4.4	0.75	0.62	0.97	38.2	47.2	2.1	6.4
28 Oct 1993	F	16.2	2.14	7.7	4.3	0.95	0.54	0.97	40.5	39.4	3.0	8.7
28 Oct 1993	IL	15.4	2.25	8.1	4.9	0.98	0.57	1.01	42.9	46.6	2.9	7.8

foliage is rather short (Fig. 3). Results by Oren et al. (1988a) indicate that the starch accumulation in spring starts earlier at lower latitudes, making the suitable sampling period even shorter.

Recommendations

Based on the present results, and earlier studies in Scots pine (cf. Aronsson and Elowson 1980, Linder 1990), it is recommended that several age-classes of foliage are sampled on a number of occasions during the season(s). The reason for repeated sampling is that without knowledge of within and between-year variation in element concentrations the value of a single sample for evaluation of nutrient status is limited (Evers 1972). If the sampling is restricted to one age-class of foliage, it is recommended that one-year-old (C+1) foliage is used, to reduce between-year variation and to enable sampling throughout the season. If nutrient concentrations are assessed on samples taken during the period late spring to early autumn, the carbohydrate content should be determined to allow values to be normalised. Nutrient imbalances can, however, be detected without correcting for carbohydrate reserves, by calculating the ratio between elements (cf. Hüttl 1990, Linder 1990, Ericsson et al. 1992). An al-

ternative method is to express nutrient contents on a leaf area basis (e.g. van den Driessche 1974), but for small conifer needles this method may introduce new errors and uncertainties into the estimates.

Use of diagnostic foliage analysis to estimate nutrient supply

In May 1987, the initial nutrient status of the trees at the Flakaliden site was poor (Table 3). More than half of the element concentrations in the foliage were below or close to values considered to be deficient or in the range of critical levels in Norway spruce (e.g. Cape et al. 1990, Hüttl 1990, Nihlgård 1990). Only three elements (Ca, Mn, Zn) were consistently above the target values considered as optimal (Table 1) and remained so throughout the study (Tables 3 and 4). They are therefore given less attention in the further presentation and discussion of the results.

Already two weeks after the treatments were commenced, the effect of nutrient supply could be seen in the needle concentrations of all nutrients (data not shown). In all age-classes of foliage the highest element concentrations were found in fertilised plots (F and IL). In current foliage (C) the concentration of all elements, except Mn,

Table 4. The ratio between nitrogen and other nutrient elements in one-year-old foliage of young Norway spruce trees subjected to different treatments. The values are expressed in per cent of nitrogen (by weight). Values shown are from needles on whorl 4, sampled in the autumn, during the first seven years of treatment. The ratios in May 1987, one month before treatments were commenced, are given for comparison. The concentrations of nitrogen for each occasion are given in Table 3 and the target values for each individual element are found in Table 1. For further details see text.

Date	Plot	P:N	K:N	Ca:N	S:N	Mg:N	Mn:N	Fe:N	Zn:N	Cu:N	B:N
19 May 1987	C	13.7	56.7	37.8	6.6	9.9	6.6	0.30	0.37	0.013	0.084
15 Sept 1987	C	14.7	50.0	52.0	7.0	8.1	11.3	0.32	0.46	0.024	0.085
20 Sept 1988	C	19.2	59.4	64.4	9.0	8.8	12.2	0.70	0.72	0.030	0.095
4 Oct 1989	C	18.6	57.1	56.2	8.1	7.1	12.2	0.59	0.63	0.024	0.079
4 Oct 1990	C	15.5	35.9	48.5	8.0	7.1	9.9	0.44	0.55	0.025	0.061
7 Oct 1991	C	13.3	49.1	60.2	7.2	8.1	12.3	0.37	0.62	0.032	0.053
27 Oct 1992	C	14.6	47.5	47.5	6.6	7.1	8.4	0.32	0.51	0.024	0.054
28 Sept 1993	C	15.3	49.0	42.3	7.2	6.0	9.3	0.37	0.45	0.021	0.062
19 May 1987	F	13.9	52.1	37.2	6.2	8.8	7.6	0.30	0.35	0.021	0.069
15 Sept 1987	F	12.9	43.3	40.2	6.3	7.3	7.1	0.27	0.32	0.022	0.083
20 Sept 1988	F	11.2	35.6	33.0	5.9	4.6	7.0	0.41	0.29	0.020	0.065
4 Oct 1989	F	11.1	33.5	29.8	4.8	3.5	6.9	0.25	0.30	0.014	0.053
4 Oct 1990	F	11.5	30.9	36.4	5.9	3.1	6.5	0.24	0.30	0.018	0.056
7 Oct 1991	F	13.3	46.5	49.3	6.3	4.9	9.5	0.26	0.38	0.027	0.055
27 Oct 1992	F	13.7	46.7	31.6	5.5	3.8	6.5	0.20	0.28	0.019	0.051
28 Sept 1993	F	13.2	47.5	26.5	5.9	3.3	6.0	0.25	0.24	0.018	0.054
19 May 1987	IL	14.6	57.0	36.6	6.6	9.5	6.0	0.25	0.38	0.022	0.061
15 Sept 1987	IL	12.0	45.1	38.2	6.5	5.9	6.7	0.24	0.31	0.022	0.055
20 Sept 1988	IL	13.7	42.8	40.4	6.9	4.9	7.1	0.50	0.35	0.025	0.070
4 Oct 1989	IL	14.4	43.5	33.7	5.8	3.5	7.0	0.29	0.35	0.019	0.055
4 Oct 1990	IL	12.9	34.6	37.2	6.5	3.5	7.1	0.24	0.41	0.019	0.062
7 Oct 1991	IL	15.6	47.2	47.2	6.5	5.0	8.9	0.26	0.43	0.026	0.063
27 Oct 1992	IL	15.9	56.4	38.3	6.5	5.0	7.5	0.20	0.53	0.023	0.052
28 Sept 1993	IL	14.6	52.6	31.8	6.4	3.7	6.6	0.28	0.30	0.019	0.051

was highest in the IL plots, and in older foliage (C+1, C+2) the highest concentrations of all elements, except K, were found in F-plots. At the end of the season, however, the situation had changed and in current foliage only N, Fe, and B concentrations were higher in F- than IL-plots. In C+1 and C+2 foliage Mg and B were also higher in foliage from F- than IL-plots. In spite of a fast response to treatment, in terms of concentration of foliar nutrients, no increase was found in biomass production (Linder and Flower-Ellis 1992, Flower-Ellis 1993). The lack of growth response during the first year of fertilisation is, however, the normal pattern found for boreal and temperate conifers (cf. Linder and Rook 1984).

Although the nutrients supplied, during the first two seasons, were the same in amount and composition in both the IL and the F treatment, the method of application resulted in differences in needle nutrient concentrations (Figs 1, 2 and 4; Table 3). In general, the uptake was higher for all elements, except nitrogen, when the supply was distributed throughout the season. This difference in uptake of nitrogen probably reflects improved conditions for competition from the field and bottom-layer vegetation, when nutrients were supplied "daily" instead of once a year (Kellner 1993). A similar effect was initially found in an irrigation and fertilisation experiment in young Scots pine (cf. Aronsson and Elowson 1980, Linder 1987).

At the end of 1988, the second season of treatment, the

absolute variation between treatments in foliage concentration of nutrient elements was not large, except for N (Table 3). The N concentration in C+1 needles had increased markedly, foliage from trees on F-plots having the highest N concentration (Fig. 2). This resulted in decreasing and non-optimal P:N and K:N ratios, especially in trees on F-plots (Fig. 4, Table 4). To improve the P:N and K:N ratios, the amounts of P and K in relation to N in the fertiliser mix were increased in the following year (Table 2). The foliage concentrations of both P and K increased during the following season (Table 3), but since N increased even more the P:N and K:N ratios continued to decrease, as did the ratio Mg:N (Table 4). The fast response to altered supply rate of P and K is in agreement with earlier studies (e.g. Hüttl 1989, 1990). The foliage analysis indicated that the N concentration in autumn and spring was sufficiently close to the target value for nitrogen supply to be reduced in 1990 (Table 2). With the exception of Ca, the concentration of all macronutrient elements still lay below the levels initially considered optimal (Table 3).

In the light of new results from laboratory experiments (cf. Ericsson et al. 1992), the target values for macronutrients were revised before the season of 1990 (Table 1). The new ratios for P:N (10%) and K:N (35%) are similar to those recommended by Ericsson et al. (1993, 1995), based on studies of nutrient imbalances and arginine accumulation in Norway spruce. The new results

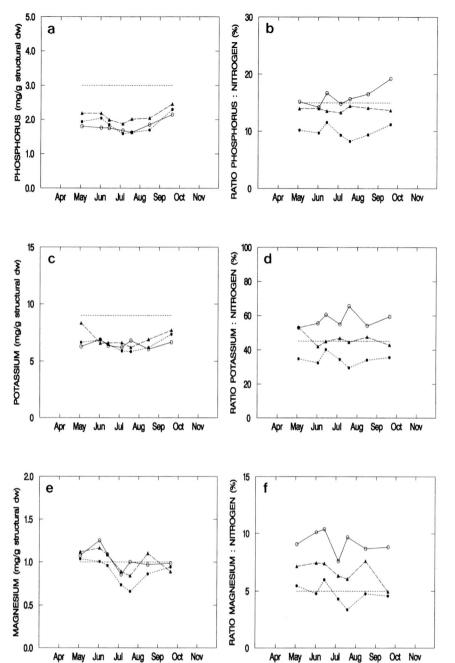

Fig. 4. The seasonal variation in needle concentration of phosphorus (a), potassium (c), and magnesium (e) in relation to structural carbon, and the ratio (%) between phosphorus and nitrogen (b), potassium and nitrogen (d), and magnesium and nitrogen (f) in one-year-old (C+1) needles of young Norway spruce trees subjected to different treatments. The horizontal lines in the diagrams show the target values used up to 1990. Symbols: Control: open circle; fertilised: filled circle; irrigated-fertilised: filled triangle. Values shown are from 1988, which was the second season of treatment.

revealed that in general the concentration of macronutrients had been sufficient, both in terms of concentration and when expressed as a ratio to nitrogen (Tables 3 and 4). One exception was that, in both F and IL treatments, the Mg:N quotient approached levels which have been reported from damaged Norway spruce stands in Germany (e.g. Kaupenjohann et al. 1989, Hüttl 1990).

Based on the foliar analysis and new target values, the supply of most nutrients was reduced, during 1990, by 25%. Extra Mg was supplied in the form of dolomite, but

the effect on foliage concentrations was not seen until the following season (Table 3). Over the next three years the supply of nitrogen was maintained at a level of 75 kg N ha^{-1} yr^{-1}, while the rate of other elements was decreased (Table 2).

During the summer of 1993 all macronutrients, except Ca and Mg, stayed above the old and close to the new target values, when expressed as quotients to N (Fig. 5). Calcium, however, was one order of magnitude above the target value at all autumn samplings (Table 3), and since

Fig. 5. The seasonal variation in needle concentration of phosphorus (a), potassium (c), and magnesium (e) in relation to structural carbon, and the ratio (%) between phosphorus and nitrogen (b), potassium and nitrogen (d), and magnesium and nitrogen (f) in one-year-old (C+1) needles of young Norway spruce trees subjected to different treatments. The horizontal lines in the diagrams show the target values used up to 1990 (dashed line) and thereafter (solid line). Symbols: Control: open circle; fertilised: filled circle; irrigated-fertilised: filled triangle. Values shown are from 1993, which was the 7th season of treatment.

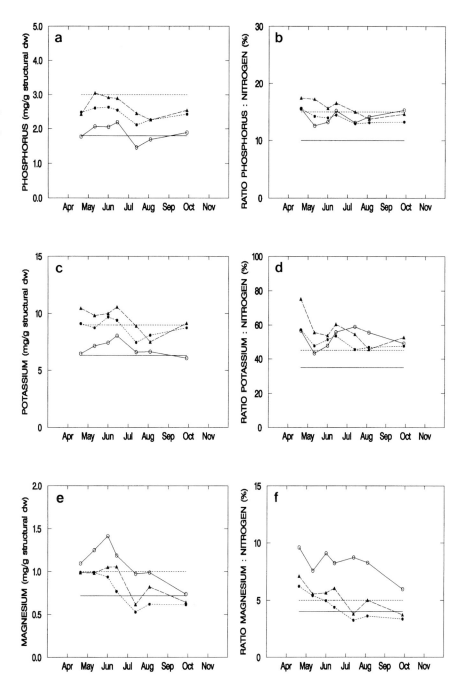

Ca like Mn and Al accumulates in the foliage over time, constant concentrations or ratios cannot be expected (cf. Fig. 1).

Among the analysed micronutrients (Mn, Fe, Zn, Cu, B) two elements (Mn, Zn) were consistently above the target values and two below (Fe, Cu). Boron concentrations varied around the target value when expressed in terms of foliage concentration (Table 3), but the B:N ratios were always above the set value and independent of treatment (Table 4).

In 1993, new target values were set for Mn, Fe, and Zn. The values were based on new findings regarding nutrient requirements in birch seedlings. In the field experiment foliage concentrations of Mn were always found to be at least one order of magnitude above the old target value (Tables 3 and 4). The initial target value was, however, found to be another order of magnitude above the optimal requirement for growth (Göransson 1994). There was also a major difference between the old and new target values for Fe:N, which had to be adjusted from 0.7 to

0.2% (Göransson 1993). An effect of this reduction in target value was that foliar concentrations of Fe and Fe:N ratios, which had been considered too low, had in fact been sufficient throughout the experimental period (Tables 3 and 4). The minor increase in recommended levels of Zn:N (Göransson and McDonald 1993), from 0.03 to 0.05, had no effect on the interpretation of the results.

The foliage concentrations of Cu and the Cu:N ratios were always low in relation to the target value (Table 3 and 4), but surprisingly constant over time and between treatments, in spite of large differences in biomass production (Linder and Flower-Ellis 1992). Since no 'dilution effect' was detected, this could be an indication that Cu, and some of the other micronutrients were being taken up in relation to demand rather than to supply. Similar results were reported from a study where micronutrients in tree leaves were analysed across a range of site fertilities (Ahrens 1964). The assumption that the target value for Cu could have been set too high is supported by results from *Pinus radiata* where 4 µg g^{-1} of total dry weight was recommended as satisfactory (Will 1985). This lower value agrees with the concentrations found by Ahrens (1964), and would give a Cu:N ratio of 0.02, which is in good agreement with the values normally found in the present study (Table 4).

Based on experience from long-term field experiments (Aronsson 1983, 1985), the target value for B could probably be adjusted downwards. This should, however, be based on the same criteria as for other nutrient elements, i.e. definition of the critical level at which foliage element concentration begins to limit growth. For micronutrients such as B and Cu, such experiments are technically very difficult (Göransson pers. comm.), and until better information is available we have to rely upon current experience from long-term field experiments on forest nutrition.

Conclusions

By combining diagnostic foliar analysis and annual adjustment of proportion and amount of nutrient supply in young stands of Norway spruce, it was possible to attain and maintain optimal nutrient status, defined in terms of target values for each nutrient element in the foliage. Nutrient imbalances, which initially were induced by fertilisation, could be detected by means of foliar analyses and corrected by adjusted nutrient applications.

The results indicate that it is possible, by means of diagnostic foliar analyses, to detect and if neccessary correct nutrient imbalances, which occur as a result of anthropogenic pollution or unsuitable management systems. For this approach to be meaningful, particularly in comparative studies, it is important to standardise the procedures for foliar sampling and analysis and to define target values for individual nutrient elements. The method should, however, be tested and evaluated on a

'practical' level before being recommended for large-scale 'vitality-fertilisation' of forests.

Acknowledgements – The nutrient optimisation experiment at the Flakaliden research site was established with support from The Swedish Forestry Research Foundation (SSFf). Major support for operational costs and research has later been obtained from The Swedish Environmental Protection Agency (SNV), The Swedish Council of Forestry and Agricultural Research (SJFR), The Swedish National Board for Industrial and Technical Development (NUTEK), and the former Swedish State Power Board (Vattenfall AB). I am most grateful to A. Flower-Ellis for her devoted work in developing and carrying out the carbohydrate analysis. Many thanks are due also to B.-O. Wigren, U. Nylander, L. Olsson and G. Moen for their skilful technical help in the field. I finally would like to thank C. O. Tamm, A. Aronsson and J. Flower-Ellis for continuous help and valuable discussions during the study and preparation of the manuscript.

References

Ahrens, E. 1964. Untersuchung über den Gehalt von Blättern und Nadeln verschiedener Baumarten an Kupfer, Zink, Bor, Molybdän and Mangan. – Allg. Forst- u. Jagdztg. 135: 8–16.

Aronsson, A. 1983. Growth disturbances caused by boron deficiency in some fertilized pine and spruce stands on mineral soil. – Comm. Inst. For. Fenn. 116: 116–122.

– 1985. Indications of stress at unbalanced nutrient contents of spruce and pine. – J. Royal Swed. Acad. Agric. For. Suppl. 17: 40–51 (in Swedish, English summary).

– and Elowson, S. 1980. Effects of irrigation and fertilization on mineral nutrients in Scots pine needles. – Ecol. Bull. (Stockholm) 32: 219–228.

Bergholm, J., Jansson, P.-E., Johansson, U., Majdi, H., Nilsson, L.-O., Persson, H., Rosengren-Brinck, U. and Wiklund, K. 1994. Air pollution, tree vitality and forest production – the Skogaby project. General description of a field experiment. – In: Hüttl, R. F. and Nilsson, L.-O. (eds), Nutrient uptake and cycling and uptake in forest ecosystems. Ecosystem Report No. 13, EUR 15465.

Blank, I. W. 1985. A new type of forest decline in Germany. – Nature 314: 311–314.

Cape, J. N., Freer-Smith, P. H., Paterson, I. S., Parkinson, J. A. and Wolfenden, J. 1990. The nutritional status of *Picea abies* (L.) Karst. across Europe, and implications for 'forest decline'. – Trees 4: 211–224.

Ericsson, A. 1979. Effects of fertilization and irrigation on the seasonal changes of carbohydrate reserves in different age-classes of needle on Scots pine trees (*Pinus sylvestris*). – Physiol. Plant. 45: 270–280.

– , Norden, L.-G., Näsholm, T. and Walheim, M. 1993. Mineral nutrient imbalances and arginine concentrations in needles of *Picea abies* (L.) Karst. from areas with different levels of airborne deposition. – Trees 8: 67–74.

– , Walheim, M., Norden, L.-G. and Näsholm, T. 1995. Concentrations of mineral nutrients and arginine in needles of *Picea abies* trees from different areas in southern Sweden in relation to nitrogen deposition and humus form. – Ecol. Bull. (Copenhagen) 44: 147–157.

Ericsson, T. and Ingestad, T. 1988. Nutrition and growth of birch seedlings at varied relative phosphorus addition rates. – Physiol. Plant. 72: 227–235.

– and Kähr, M. 1993. Growth and nutrition of birch seedlings in relation to potassium supply rate. – Trees 7: 78–85.

– , Rytter, L. and Linder, S. 1992. Nutritional dynamics and requirements of short rotation forests. – In: Mitchell, C.P., Ford-Robertson, J. B., Hinckley, T. and Sennerby-Forsse, L.

(eds), Ecophysiology of short rotation forest crops. Elsevier Applied Science, London, pp. 35–65.

Evers, F.H. 1972. Die jahrweisen Fluktuationen der Nährelementkonzentrationen in Fichtennadeln und ihre Bedeutung für die Interpretation nadelanalytischer Befunde. – Allg. Forst- u. J.-Ztg. 143: 68–74.

Ewers, F.W. 1982. Secondary growth in needle leaves of *Pinus longaeva* (bristlecone pine) and other conifers: quantitative data. – Amer. J. Bot. 69: 1552–1559.

Fife, D.N. and Nambiar, E.K.S. 1982. Accumulation and retranslocation of nutrients in developing needles in relation to seasonal growth of young radiata pine trees. – Ann. Bot. 50: 817–829.

– and Nambiar, E. K. S. 1984. Movement of nutrients in radiata pine needles in relation to the growth of shoots. – Ann. Bot. 54: 303–314.

Flower-Ellis, J. G. K. 1993. Dry-matter allocation in Norway spruce branches: a demographic approach. – Stud. For. Suec. 191: 51–73.

Göransson, A. 1993. Growth and nutrition of small *Betula pendula* plants growing at different relative addition rates of iron. – Trees 8: 33–38.

– 1994. Growth and nutrition of small *Betula pendula* plants growing at different relative addition rates of manganese. – Tree Physiol. 14: 375–388.

– and McDonald, A.J.S. 1993. Growth and nutrition of small *Betula pendula* plants at different relative addition rates of iron, manganese and zinc. – Plant Soil 155/156: 469–472.

Helmisaari, H.-S. 1992. Nutrient retranslocation within the foliage of *Pinus sylvestris*. – Tree Physiol. 10: 45–58.

Höhne, H. 1964. Untersuchungen über die jahreszeitlichen Veränderungen des Gewichtes und Elementgehaltes von Fichtennadeln in jüngeren Beständen des Osterzgebirges. – Arch. f. Forstwes. 13: 747–774.

Hüttl, R. F. 1988. 'New Type' forest declines and restabilisation/revitalization strategies. A programme focus. – Water Air Soil Pollut. 41: 95–111.

– 1989. Liming and fertilization as mitigation tools in declining forest ecosystems. – Water Air Soil Pollut. 44: 93–118.

– 1990. Nutrient supply and fertilizer experiments in view of N saturation. – Plant Soil 128: 45–58.

Ingestad, T. 1959. Studies on the nutrition of forest tree seedlings. II. Mineral nutrition of spruce. – Physiol. Plant. 12: 568–593.

– 1967. Methods of uniform optimum fertilization of forest tree plants. – In: Proc. XIVth IUFRO Congress, Münich, Section 22: 265–269.

– 1979. Mineral nutrient requirements in *Pinus sylvestris* and *Picea abies* seedlings. – Physiol. Plant. 45: 373–380.

– 1982. Relative addition rate and external concentration: driving variables used in plant nutrition research. – Plant Cell Environ. 5: 443–453.

– and Kähr, M. 1985. Nutrition and growth of coniferous seedlings at varied relative nitrogen addition rate. – Physiol. Plant. 65: 109–116.

– , Aronsson, A. and Ågren, G.I. 1981. Nutrient flux density model of mineral nutrition in conifer ecosystems. – Stud. For. Suec. 160: 61–71.

Kaupenjohann, M., Zech, W., Hantschel, R., Horn, R. and Schneider, B. U. 1989. Mineral nutrition of forest trees: A regional survey. – In: Schulze, E.-D., Lange, O. L. and Oren, R. (eds), Forest decline and air pollution. A study of spruce (*Picea abies*) on acid soils. Ecol. Stud. 77: 280–296.

Kellner, O. 1993. Effects of fertilization on forest flora and vegetation. – Comprehensive summaries of Uppsala dissertations faculty science 464, Acta universitatis Upsaliensis, Uppsala 1993.

Linder, S. 1987. Responses to water and nutrition in coniferous ecosystems. – In: Schulze, E.-D. and Zwölfer, H. (eds), Potentials and limitations of ecosystem analysis. Ecol. Stud. 61: 180–202.

– 1990. Nutritional control of forest yield. – In: Nutrition in

trees. The Marcus Wallenberg Foundation Symposia Proc. 6: 62–87.

– and Rook, D. A. 1984. Effects of mineral nutrition on carbon dioxide exchange and partitioning of carbon in trees. – In: Bowen, G. D. and Nambiar, E. K. S. (eds), Nutrition of plantation forests. Academic Press, London, pp. 211–236.

– and Flower-Ellis, J.G.K. 1992. Environmental and physiological constraints to forest yield. – In: Teller, A., Mathy, P. and Jeffers, J.N.R. (eds), Responses of forest ecosystems to environmental changes. Elsevier Applied Science, London, pp. 149–164.

– , Benson, M.L., Myers, B.J. and Raison, R.J. 1987. Canopy dynamics and growth of *Pinus radiata*. I. Effects of irrigation and fertilisation during a drought. – Can. J. For. Res. 10: 1157–1165.

Nihlgård, B. 1990. The vitality of forest trees in Sweden – stress symptoms and causal relations. – In: Liljelund, L.-E., Lundmark, J.-E., Nihlgård, B., Nohrstedt, H.-Ö. and Rosen, K. (eds), Vitality fertilization – present knowledge and research needs. Swedish Environ. Protect. Agency, Solna, Report 3813, pp. 45–70 (in Swedish, English summary).

Oren, R. and Schulze, E.-D. 1989. Nutritional disharmony and forest decline: A conceptual model. – In: Schulze, E.-D., Lange, O. L. and Oren, R. (eds), Forest decline and air pollution. A study of spruce (*Picea abies*) on acid soils. Ecol. Stud. 77: 425–443.

– , Schulze, E.-D., Meyer, J., Schneider, B. U. and Heilmeier, H. 1988a. Performance of two *Picea abies* (L.) Karst. stands at different stage of decline I. Carbon relations and stand growth. – Oecologia 75: 25–37.

– , Werk, K. S., Schulze, E.-D., Meyer, J., Schneider, B. U. and Schramel, P. 1988b. Performance of two *Picea abies* (L.) Karst. stands at different stage of decline VI. Nutrient concentration. – Oecologia 77: 151–162.

Pereira, J.S., Linder, S., Araujo, M.C., Pereira, H., Ericsson, T., Borallho, N. and Leal, L. 1989. Optimization of biomass production in *Eucalyptus globulus* – a case study. – In: Pereira, J.S. and Landsberg, J.J. (eds), Biomass production by fast-growing trees. Kluwer Academic Publ., pp. 101–121.

Rosengren-Brinck, U. and Nihlgård, B. 1994. Nutritional status in needles of Norway spruce (*Picea abies*) in relation to water and nutrient supply. – In: Rosengren-Brinck, U. (ed.), The influence of nitrogen on the nutrient status in Norway spruce (*Picea abies* L. Karst). Ph.D. thesis, Lund Univ. 1994, pp. 55–75.

– and Nihlgård, B. 1995. Nutritional status in needles of Norway spruce (*Picea abies*) in relation to water and nutrient supply. – Ecol. Bull. (Copenhagen) 44: 168–177.

Saxe, H. 1993. Triggering and predisposing factors in the "red" decline syndrome of Norway spruce (*Picea abies*). – Trees 8: 39–48.

Schulze, E.-D., Oren, R. and Lange, O.L. 1989. Nutrient relations of trees in healthy and declining Norway spruce stands. – In: Schulze, E.-D., Lange, O. L. and Oren, R. (eds), Forest decline and air pollution. A study of spruce (*Picea abies*) on acid soils. Ecol. Stud. 77: 392–417.

Shear, C. B., Crane, H. L. and Myers, A. T. 1948. Nutrient element balance: Application of the concept to the interpretation of foliar analysis. – Proc. Amer. Soc. Horticult. Sci. 51: 319–326.

Staaf, H. and Olsson, B.A. 1994. Effects of slash removal and stump harvesting on soil water chemistry in a clearcutting in SW Sweden. – Scand. J. For. Res. 9: 305–310.

Steen, E. and Larsson, K. 1986. Carbohydrates in roots and rhizomes of perennial grasses. – New Phytol. 104: 339–346.

Tamm, C.O. 1955. Studies on forest nutrition. I. Seasonal variation in the nutrient content of conifer needles. – Medd. Stat. Skogsforskn. Inst. 45: 1–34.

– 1964. Determination of nutrient requirements of forest stands. – Int. Rev. For. Res. 1: 115–170.

– 1968. An attempt to assess the optimum nitrogen level in Norway spruce under field conditions. – Stud. For. Suec. 61.

- 1989. The role of field experiments in the study of soil mediated effects of air pollution on forest trees. – Comm. Norw. For. Res. Inst. 42: 179–196.
- 1991. Nitrogen in terrestrial ecosystems – Questions of productivity, vegetational changes, and ecosystem stability. Ecol. Stud. 81, Springer-Verlag.
- and Popovic, B. 1995. Long-term field experiments simulating increased deposition of sulphur and nitrogen to forest plots. – Ecol. Bull. (Copenhagen) 44: 301–321.
Van den Driessche, R. 1974. Prediction of mineral nutrient status of trees by foliar analysis. – Bot. Rev. 40: 347–394.

Van Praag, H.J. and Weissen, F. 1986. Foliar mineral composition, fertilisation and dieback of Norway spruce in the Belgian Ardennes. – Tree Physiol. 1: 169–176.
Wehrmann, J. 1959. Methodische Untersuchungen zur Durchführung von Nadelanalysen in Kiefernbeständen. – Forstwiss. Cbl. 78: 77–97.
Will, G. 1985. Nutrient deficiencies and fertiliser use in New Zealand exotic forests. – Forest Res. Inst., Rotorua, New Zealand, Bull. 97.
Woodman, J.N. and Cowling, E.B. 1987. Airborne chemicals and forest health. – Environ. Sci. Technol. 21: 120–126.

Ecological Bulletins 44: 191–196. Copenhagen 1995

Interactions between aluminium, calcium and magnesium – Impacts on nutrition and growth of forest trees

Tom Ericsson, Anders Göransson, Harmke Van Oene and George Gobran

Ericsson, T., Göransson, A., Van Oene, H. and Gobran, G. 1995. Interactions between aluminium, calcium and magnesium – Impacts on nutrition and growth of forest trees. – Ecol. Bull. (Copenhagen) 44: 191–196.

Aluminium in the form of Al^{3+} has a negative impact on uptake of base cations and hence on growth of plants. In this chapter we will examine how this ion interferes with the root surface chemistry and thereby affects the uptake of Ca^{2+} and Mg^{2+}. We show that Al^{3+} competes effectively with cations on the root surface, hence limiting their uptake. Uptake of cations is related to the cation exchange capacity of the roots and that this root property is strongly influenced by Al^{3+} as well as the amounts and degree of adsorption of Ca^{2+} and Mg^{2+}. Ca and Mg act as antagonists by negatively affecting each other's uptake. The consequence of these interactions on nutrition and growth of forest trees is discussed.

T. Ericsson, A. Göransson, H. Van Oene and G. Gobran, Dept of Ecology and Environmental Research, Swedish Univ. of Agricultural Sciences, P.O. Box 7072, S-750 07 Uppsala, Sweden.

Introduction

The forest decline in central Europe and eastern North America has been a matter of intensive debate since the late 1970's. The number of tree species associated with the forest die-back has gradually increased and includes today both conifers and broad-leaved trees (e.g. Hüttl 1989). Needle yellowing and premature losses of older needles in conifers, particularly Norway spruce *Picea abies* (L.) Karst., are commonly observed deficiency symptoms in declining stands together with decreased contents of mineral nutrients in the tissues (e.g. Hüttl 1993). A number of possible causes for the observed growth disturbances have been suggested. Exposure to elevated concentrations of gaseous air pollutants such as SO_2, NO_2, O_3 (Prinz et al. 1982, Amundson et al. 1986, Kozlowski and Constantinidou 1986) and to acid mist (Prinz et al. 1982, Wisniewski 1982) can damage the foliar tissue, particularly the membranes. This may result in pigment destruction and accelerated crown leaching of physiologically mobile and exchangeable nutrients (K, Mg, Ca, Zn, Mn). However, artificial exposure of trees to these pollutants has failed to reproduce the typical needle yellowing symptoms and associated changes in needle structure (Fink 1988) and mineral composition which normally occur in damaged forest stands.

Leaching of base cations, as an indirect effect of acid rain, in combination with an increased mineral requirement resulting from stimulated plant growth from nitrogen deposition, has also been proposed as a probable explanation of the observed nutrient imbalances in forest trees. Analyses of Norway spruce needles collected from different sites in Germany have revealed deficiencies of Mg (Zech and Popp 1983), Ca (Zech and Popp 1983, Zöttl and Mies 1983, Bosch et al. 1986), K (Zech et al. 1985, Zöttl and Hüttl 1985), Mn (Hüttl 1985, Rehfuess 1985) and Zn (Zech et al. 1985, Zöttl and Hüttl 1985). Root diseases or antagonists of the rhizosphere flora or mycorrhiza may also contribute to inhibit mineral uptake from the soil (Kandler 1990). Shortages of Mg, in particular, (Ericsson and Kähr 1995) and Mn (Göransson 1994), but also of K (Ericsson and Kähr 1993) are likely to worsen an already imbalanced nutrient situation, since such shortages depress the allocation of dry matter to roots.

Root damage, and consequent reduced nutrient acquisition, resulting from elevated Al^{3+} concentrations in the soil solution, has been proposed as an explanation of decreased forest growth (Ulrich 1981, Rost-Siebert 1983, Cronan et al. 1989). However, direct toxicity of aluminium, as the mechanism behind forest die back, has been questioned by Göransson and Eldhuset (1987, 1991,

Table 1. Influence of aluminium (Al^{3+}) on the relative growth rate, as a percent of the control, of tree seedlings grown in culture solution with free access to all nutrients (FA) or under nutrient-limiting conditions (LA). Data from Göransson and Eldhuset (1987, 1991). The calculated Ca/Al ratio was < 0.2 in all experiments.

Al^{3+} mM	Betula pendula		Picea abies	Pinus sylvestris	Ectomycorrhizal Pinus sylvestris
	FA	LA	FA	FA	FA
0.2	–	–	93	–	–
0.3	–	–	89	–	–
1	95	93	55	87	–
3	88	46	39	97	91
6	61	38	36	88	92
10	70	30	27	60	78
15	–	26	37	50	51
30	–	–	–	39	48

1993). They demonstrated that Norway spruce, Scots pine and birch were damaged at external Al concentrations of 0.2–1.0 mM, which are high compared to those found in the soil solution of acid boreal forest soils (< 0.1 mM, Nilsson and Bergkvist 1983, Berdén et al. 1987, Table 1). However, in their experiments the uptake of Ca and Mg was already reduced at low aluminium concentrations in the culture solution, at which there were no growth disturbances (Göransson and Eldhuset 1993). A negative impact of Al on uptake of Ca and Mg is commonly observed in other tree species, as well as in herbaceous plants (Rost-Siebert 1983, Jorns and Hecht-Buchholtz 1985, Thornton et al. 1986, Andersson 1988, Asp et al. 1988, Bengtsson et al. 1988, Godbold et al. 1988a,b).

The aim of this paper is to examine how Al^{3+} interferes with the root surface chemistry, thereby affecting the uptake of Ca^{2+} and Mg^{2+}, and tree growth.

Influence of aluminium on root surface chemistry

Evidence of a positive relationship between plant growth and the root cation exchange capacity (CEC_r) has accumulated. The review by Haynes (1980) indicates, for example, that the relative absorption of Ca^{2+} and Mg^{2+} increased with an increase in the CEC_r. Although the CEC_r has obvious physiological and ecological effects, the fundamental mechanisms in which it is involved are not yet understood (Frejat et al. 1967, Haynes 1980, Koedam and Buscher 1983, Dufey and Braun 1986).

It is generally accepted that Al interferes with the membrane surface by displacing Ca^{2+} from negatively charged head groups on membranes (Clarkson 1984), and by affecting the cation exchange capacity of the roots (Cronan 1991). The level of Al saturation on the nutrient-absorbing surfaces increases when Ca is in short supply, and decreases at high Ca availabilities. This explains why Ulrich (1983) suggested a relationship between the Ca^{2+}/Al^{3+} in the soil solution and the growth of tree roots. According to Rost-Siebert (1983, 1985), growth and nu-

trient uptake of European beech *Fagus sylvatica* and Norway spruce are negatively affected at Ca/Al molar ratios below 1. Gobran et al. (1993) examined data from several independent greenhouse and field experiments on tree species and found that CAB (CAB = log (Ca/Al), where Ca^{2+} and Al^{3+} are expressed on an equivalent basis) can potentially describe the degree to which plant growth is supported or hindered in an acid soil. These authors also concluded that application of calcium in amounts exceeding the requirements for growth can be recommended as a means of ameliorating the negative impact of Al on nutrient acquisition.

A similar approach to that advocated by Ulrich (1983) for determining the effects of soil acidification on growth of trees has been suggested by Sverdrup and Warfvinge (1993). In examining published data pertaining to growth responses of trees, grasses and herbs to Al, they used the log (Ca + Mg + K)/Al molar ratio as the diagnostic tool and obtained a positive correlation between this ratio and plant growth. However, although both CAB and the Ca + Mg + K/Al terms correlate well with plant growth, these ratios do not offer any deeper physiological understanding of the observed nutrient imbalances in forest trees.

Interactions between aluminium, calcium and magnesium

The impact of Al^{3+} on uptake of Ca^{2+} and Mg^{2+} has been extensively studied in small birch *Betula pendula* plants under laboratory conditions by Ericsson et al. (unpubl.). This study was complemented by examining a possible relationship between nutrient uptake and exchange processes occurring on the root surface, and modelling and simulating the results. The experiments were carried out under low Mg or Ca conditions which allowed the plants to express only c. 70% of their growth potential. This was achieved by either supplying Mg or Ca once per hour in exponentially increasing amounts at a sub-optimum relative addition rate of 0.15 g g^{-1} d^{-1} (see Ingestad 1982, Ingestad and Lund 1986, Ingestad and Ågren 1992). The

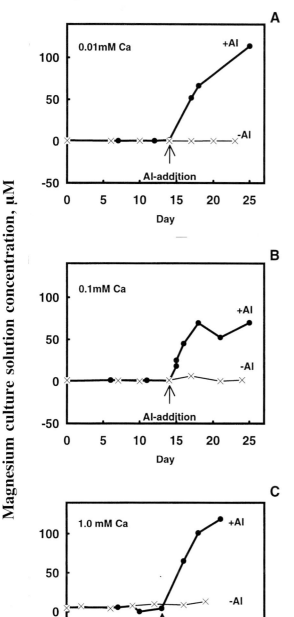

Fig. 1. Influence of 0.5 mM Al³⁺ on uptake of Mg, here represented by the residual Mg concentration in the culture solution immediately before the next Mg addition, in Mg-limited birch seedlings grown at 0.01 (A), 0.1 (B) or 1.0 (C) mM Ca in the culture solution. Each data point represents a mean of two replicates.

interactions between Mg and Ca, and the reported beneficial influence of Ca on nutrient acquisition in presence of Al, were studied by keeping the culture solution concentration of Ca in the low Mg treatments, or of Mg in the low Ca treatments, at either 0.01, 0.1 or 1.0 mM. Aluminium was added to the culture solution once the growth of the birch seedlings had acclimated to the sub-optimum supply rate of either Ca or Mg. Disturbances in nutrient uptake resulting from root damage were avoided by selecting an Al concentration shown to be harmless to root growth (0.5 mM, see Göransson and Eldhuset 1987).

A residual Mg concentration of c. 1, 3 or 10 μM was measured in the culture solution immediately before the next Mg addition, in the treatments with a constant Ca^{2+} concentration of 0.01, 0.1 or 1.0 mM, respectively (Fig. 1). Addition of Al immediately decreased the efficiency by which the plants were able to take up Mg (Fig. 1a–c). A Mg concentration in the culture solution of 100–180 μM, depending on the Ca concentration, was required to prevent a decrease in the uptake rate of Mg in the presence of 0.5 mM Al^{3+}. The influence of Al on the uptake efficiency of Ca and Mg was associated with a large

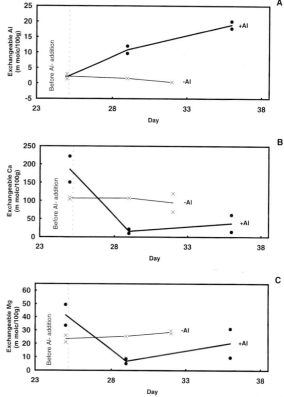

Fig. 2. Exchangeable Al (A), Ca (B) and Mg (C) from roots of Mg-limited birch seedlings, before and after addition of 0.5 mM Al^{3+} to the culture solution, grown at a constant Ca^{2+} concentration of 0.01 mM in the culture solution. The adsorbed Al, Ca and Mg upon addition of Al are presented in meq. 100 g⁻¹ dry matter. The data points are from two independent experiments.

Table 2. Measured and simulated nutrient (Ca and Mg) uptake rate per unit root growth rate (dn/dW$_r$) after Al addition, expressed as a percentage of dn/dW$_r$ before Al addition, for *Betula* plants grown with a relative growth rate (R$_G$) of 21.5% or 10% d^{-1} (from Van Oene (1994)).

R$_G$ %	dCa/dW$_r$		dMg/dW$_r$	
	measured	simulated	measured	simulated
21.5	50	55	60	55
10.0	150	98	130	98

change in the adsorbed cations on the root surface. While adsorbed Al^{3+} increased sharply (Fig. 2a), the adsorbed Ca^{2+} and Mg^{2+} decreased (Fig. 2b,c). This cannot be explained by physiological processes, i.e. processes requiring metabolic energy, since it occurred almost instantaneously. Gobran et al. (unpubl.) attributed the immediate effect of Al^{3+} in increasing Ca^{2+} and Mg^{2+} concentrations in the culture solution to the cation exchange processes of the root surface, which are non-metabolic processes.

Ericsson et al. (unpubl.) did not observe any beneficial influence of Ca on cation uptake in presence of Al. On the contrary, Ca given in excess of growth requirements reduced the uptake of Mg still further. A similar trend was observed in the experiments in which Ca was the growth-limiting nutrient. When Mg was given in excess of growth requirements the negative impact on Ca uptake was less pronounced, whether Al was present or absent in the culture solution.

Because of the high affinity of negatively charged surfaces for Al^{3+}, CEC$_r$ has a possible mediating role in Al toxicity to tree species. The competitive power of Al^{3+} over Ca^{2+} and Mg^{2+} adsorption by roots of several plant species has been reported for annual crops and for tree species (Dufey et al. 1991); in Norway spruce (Schröder et al. 1988), red spruce (*Picea rubens*, Cronan 1991) and ryegrass (*Lolium multiflorum*, Rengel and Robinson 1989).

The study by Ericsson et al. (unpubl.) clearly demonstrated that Al^{3+} effectively competes with cations on the root surface and hence limits their uptake simultaneously. Gobran et al. (unpubl.) hypothesized that uptake of cations is related to CEC$_r$, and to amount and degree of adsorbed cations on the root surface. This hypothesis has been tested by a model developed by Van Oene (1994), which describes the effects of Al^{3+} on the uptake of base cations. The model is based on the membrane and cell wall properties of the roots. The uptake of cations is described as a function of CEC$_r$, exchangeable fractions of these cations of the CEC$_r$, and root growth. The uptake capacity of the roots is represented by the CEC$_r$, the magnitude of which is affected by chemical reactions of H, Al and Ca with the negative sites on the root surface; these reactions are not exchange reactions. This model was calibrated using the exchange parameters measured

by Gobran et al. (unpubl.) and applied to the experiments of Ericsson et al. (unpubl.). There was a close agreement between simulated and observed plant nutrient concentrations and solution concentrations. The application of the model to an independent data set from a similar type of experiment (Göransson and Eldhuset 1993) showed a good agreement between the measured and simulated nutrient uptake rates per unit root growth rate (Table 2). Thus the model showed that it is possible to relate uptake of nutrients to the cation exchange properties of the roots. Also revealed in the model are the ameliorating influence of higher cation concentrations (H, Ca, and Mg) on Al effects. However, Ca and Mg acted as antagonists by negatively affecting each other's uptake. The model suggested that uptake of Ca and Mg is dependent on the ratio of their concentrations in the solution. Antagonism between Ca and Mg has also been reported by Tan et al. (1992) and Mengel and Kirkby (1987). Clearly, this antagonism between Ca and Mg stresses the importance of not over-dosing Ca in the composition of so-called "vitality fertilizers" designed for use in areas with a high rate of N- and S-deposition and in soils low in pH and Mg-delivering capacity. It is also clear from the model, as well as from the laboratory study of Ericsson et al. (unpubl.), that concentrations of Mg in the soil solution which are adequate for vigorous plant growth may become strongly growth-limiting in the presence of Al or when there is large excess of Ca in relation to Mg.

According to the principles of exchange processes, divalent cations are more effective in ameliorating Al^{3+} toxicity than monovalent cations. Therefore, Ca^{2+} and Mg^{2+} are well-known to be much more effective than K$^+$ and Na$^+$ (Kinraide et al. 1985). At comparable ionic strengths, in principle, Ca^{2+} should be more effective than Mg^{2+} because of greater binding selectivity on root and soil surfaces. However, increasing ionic strengths of solution resulting from large applications of Ca^{2+} or Mg^{2+} on the exchange sites (root and soil), results in imbalanced conditions. It is therefore, necessary to consider the quantitative relationship between important nutrient cations such as Ca^{2+} and Mg^{2+}, and Al^{3+}, in solution and adsorbed on the exchange complex (soil and root surface), which is important for the response of plants to soil- and water acidification. Understanding the significance of the processes involved has great practical applications. Further research is necessary in order to increase understanding of the uptake mechanisms of nutrients, the interactions between nutrients and their relation to root characteristics.

References

Amundson, R. G., Walker, R. B. and Legge, A. H. 1986. Sulphur gas emission in boreal forest. VII. Pine tree physiology. – Water Air Soil Pollut. 29: 129–147.

Andersson, M. 1988. Toxicity and tolerance of aluminium in vascular plants. A literature review. – Water Air Soil Pollut. 39: 439–462.

Asp, H., Bengtsson, B. and Jensén, P. 1988. Growth and cation uptake in spruce (*Picea abies* Karst.) grown in sand culture with various aluminium contents. – Plant Soil 111: 127–133.

Bengtsson, B., Asp, H., Jensén, P. and Berggren, D. 1988. Influences of aluminium on phosphate and calcium uptake in beech (*Fagus sylvatica*) grown in nutrient solution and soil solution. – Physiol. Plant. 74: 299–305.

Berdén, M., Nilsson, S.I., Rosén, K. and Tyler, G. 1987. Soil acidification-extent, causes and consequences. An evaluation of literature information and current research. – Swedish Environmental Protection Board, Solna, Sweden, Report 3292.

Bosch, C., Pfannkuch, E., Rehfuess, K.E., Runkel, K.H., Schramel, P. and Senser, M. 1986. Einfluss einer Düngung mit Magnesium und Calcium, von Ozon und saurem Nebel auf Frosthärte, Ernährungszustand und Biomasse-Produktion junger Fichten (*Picea abies* (L.) Karst.). – Forstwiss. Centralbl. 105: 218–229.

Clarkson, D.T. 1984. Ionic relations. – In: Wilkins, M.B. (ed.), Advanced plant physiology. Pitman Publishing, London, pp. 319–354.

Cronan, C.S. 1991. Differential adsorption of Al, Ca, and Mg by roots of red spruce (*Picea rubens* Sarg.). – Tree Physiol. 8: 227–237.

– , April, R., Bartlett, R.J., Bloom, P.R., Driscoll, C.T., Gherini, S.A., Henderson, G.S., Joslin, J.D., Kelly, J.M., Newton, R.M., Parnell, R.A., Patterson, H.H., Raynal, D.J., Schaedle, M., Schofield, C.L., Sucoff, E.I., Tepper, H.B. and Thornton, F.C. 1989. Aluminum toxicity in forests exposed to acidic deposition: The Albios result. – Water Air Soil Pollut. 48: 181–192.

Dufey, J.E. and Braun, R. 1986. Cation exchange capacity of roots: titration, sum of exchangeable cations, copper adsorption. – J. Plant Nutrit. 9: 1147–1155.

– , Drimmer, D., Lambert, I. and Dupont, P.H. 1991. Composition of root exchange sites in acidic solutions. – In: McMichael, B.L. and Persson, H. (eds), Plant roots and their environment. Elsevier, Amsterdam, pp. 31–38.

Ericsson, T. and Kähr, M. 1993. Growth and nutrition of birch seedlings in relation to potassium supply rate. – Trees 7: 78–85.

– and Kähr, M. In press. Growth and nutrition of birch seedlings at varied relative addition rates of magnesium. – Tree Physiol.

Fink, S. 1988. Histological and cytological changes caused by air pollutants and other abiotic factors. – In: Schulte-Hostede, S., Darrall, N.M., Blank, L.W. and Wellburn, A.R. (eds), Air pollution and plant metabolism. Elsevier Applied Sciences, Oxford. pp. 36–54.

Frejat, A., Ansett, A. and Lemaire, F. 1967. Capacité d'échange de cations des systèmes radiculaires et des sols, et leurs relations avec l'alimentation minérale. – Ann. Agron. 18: 31–64.

Gobran, G.R., Fenn, L.B., Persson, H. and Al Windi, I. 1993. Nutrition response of Norway spruce and willow to varying levels of calcium and aluminium. – Fert. Res. 34: 181–189.

Godbold, D.L., Fritz, E. and Hüttermann, A. 1988a. Aluminium toxicity and forest decline. – Proc. Natl. Acad. Sci. 85: 3888–3892.

– , Dictus, K. and Hüttermann, A. 1988b. Influence of aluminium and nitrate on root growth and mineral nutrition of Norway spruce (*Picea abies*) seedlings. – Can. J. For. Res. 18: 1167–1171.

Göransson, A. 1994. Growth and nutrition of small *Betula pendula* plants at different relative rates of manganese. – Tree Physiol. 14: 375–388.

– and Eldhuset, T. 1987. Effects of aluminium on growth and nutrient uptake of *Betula pendula* seedlings. – Physiol. Plant. 69: 193–199.

– and Eldhuset, T. 1991. Effects of aluminium on growth and nutrient uptake of small *Picea abies* and *Pinus sylvestris* plants. – Trees 5: 136–142.

– and Eldhuset, T. 1993. Effects of aluminium on calcium, magnesium and nitrogen nutrition in small *Betula pendula* plants growing at high and low nutrient supply rates. – Ph.D. thesis, Swed. Univ. Agric. Sci., Uppsala, Rep. No. 60.

Haynes, R.J. 1980. Ion exchange properties of roots and ionic interactions within the root apoplasm: their role in ion accumulation by plants. – Bot. Rev. 46: 75–99.

Hüttl, R.F. 1985. "Neuartige" Waldschäden und Nährelementversorgung von Fichtenbeständen (*Picea abies* Karst.) in Südwestdeutschland. – Freib. Bodenkdl. Abh. 16.

– 1989. Liming and fertilization as migitation tools in declining forest ecosystems. – Water Air Soil Pollut. 44: 93–118.

– 1993. Mg deficiency – A "new" phenomenon in declining forests – Symptoms and effects, causes, recuperation. – In: Hüttl, R.F. and Mueller-Dombois (eds), Forest decline in the Atlantic and Pacific region. Springer, Berlin, pp. 97–114.

Ingestad, T. 1982. Relative addition rate and external concentration; Driving variables used in plant nutrition research. – Plant Cell Environ. 5: 443- 453.

– and Lund, A.-B. 1986. Theory and techniques for steady state mineral nutrition and growth of plants. – Scand. J. For. Res. 1: 439–453.

– and Ågren, G.I. 1992. Theories and methods on plant nutrition and growth. – Physiol. Plant. 84: 177–184.

Jorns, C.A. and Hecht-Buchholz, C. 1985. Aluminiuminduzierter Magnesium- und Calciummangel im Laborversuch bei Fichtensämlingen. – Allg. Forstz. 46: 1248–1252.

Kandler, O. 1990. Epidemiological evaluation of the development of Waldsterben in Germany. – Plant Disease 74: 4–12.

Kinraide, T.B., Arnold, R.C. and Baligar, V.C. 1985. A rapid assay for aluminium phytotoxicity at submicromolar concentrations. – Physiol. Plant. 65: 245–250.

Koedam, N. and Buscher, P. 1983. Studies on the possible role of cation exchange capacity in the soil preference of mosses. – Plant Soil 70: 77–93.

Kozlowski, T.T. and Constantinidou, H.A. 1986. Responses of woody plants to environmental pollution. Part 1. Sources and types of pollutants and plant responses. – For. Abstr. 47: 5–51.

Mengel, K. and Kirkby, E.A. 1987. Principles of plant nutrition. 4th ed. – Int. Potash Inst., Bern, Switerland.

Nilsson, S.I. and Bergkvist, B. 1983. Aluminium chemistry and acidification processes in a shallow podzol on the Swedish westcoast. – Water Air Soil Pollut. 20: 311–329.

Prinz, B., Krause G.H.M. and Stratmann, H. 1982. Waldschäden in der Bundesrepublik Deutschland. – LIS-Berichte 28, LIS-Essen.

Rehfuss, K.E. 1985. Vielfältige Formen der Fichtenerkrankung in Süddeutschland. – In: Niesslein, E. and Voss, G. (eds), Was wir über das Waldsterben wissen. Deutscher Instituts Verlag, Köln, pp. 124–130.

Rengel, Z. and Robinson, D.L. 1989. Competitive Al^{3+} inhibition of net Mg^{2+} uptake by intact *Lolium multiflorum* roots. – Plant Physiol. 91: 1407–1413.

Rost-Siebert, K. 1983. Aluminium – Toxizität und – Toleranz an Keimplanzen von Fichte (*Picea abies* Karst.) und Buche (*Fagus sylvatica* L.). – Allg. Forstz. 38: 686–689.

– 1985. Untersuchungen zur H- und Al-ionen-toxizität an Keimplanzen von Fichte (*Picea abies* Karst.) und Buche (*Fagus sylvatica* L.) in Lösungskultur. – Ber. Forsch. Waldökosyst./Waldsterben, Göttingen, Bd. 12.

Schröder, W.H., Bauch, J. and Endeward, R. 1988. Microbeam analysis of Ca exchange and uptake in the fine roots of spruce: influence of pH and aluminium. – Trees 2: 96–103.

Sverdrup, H. and Warfvinge, P. 1993. The effect of soil acidification on growth of trees, grass and herbs as expressed by the (Ca + Mg + K)/Al ratio. – Lund Univ., Dept Chemical Engin. II, Reports in Ecology and Environmental Engineering 1993: 2.

Tan, K., Keltjens, W.G. and Findenegg, G.R. 1992. Calcium-induced modification of aluminium toxicity in *Sorghum* genotypes. – J. Plant Nutr. 15: 1395–1405.

Thornton, F.C., Schaedle, M., Raynal, D.J. and Zipperer, C. 1986. Effects of aluminium on Honeylocust (*Gleditsia triacanthos* L.) seedlings in solution culture. – J. Exp. Bot. 37: 775–785.

Ulrich, B. 1981. Eine ökosystemare Hypothese über die Ursachen des Tannensterbens (*Abies alba* Mill.). – Forstwiss. Centalbl. 100: 228–236.

– 1983. Soil acidity and its relations to acid deposition. – In: Ulrich, B and Panchrath, J. (eds), Effects of accumulation of air pollutants in forest ecosystems. Reidel Publishing, Boston, pp. 127–146.

Van Oene, H. 1994. Modelling the impacts of sulphur and nitrogen deposition on forests. – Ph.D. thesis, Swed. Univ. Agric. Sci., Uppsala, Rep. No. 68.

Wisniewski, J. 1982. The potential acidity associated with dews, frosts, and fogs. – Water Air Soil Pollut. 17: 361–377.

Zech, W. and Popp, E. 1983. Magnesiummangel, einer Grunde für das Fichten- und Tannensterben in NO-Bayern. – Forstwiss. Centralbl. 102: 50–55.

–, Suttner, T.H. and Popp, E. 1985. Elemental analysis and physiological response of forest trees in SO_2 polluted areas of N.E. Bavaria. – Water Air Soil Pollut. 25: 175–183.

Zöttl, H.W. and Mies, E. 1983. Nährelementversorgung und Schadstoffbelastung von Fichtenökosystemen im Südschwarzwald unter Immissionseinfluss. – Mitt. Deutsch. Bodenkdl. Ges. 38: 429–434.

– and Hüttl, R.F. 1985. Schadsymptome und Ernährungszustand von Fichtenbeständen im südwestdeutschen Alpenvorland. – Allg. Forstz 40: 197–199.

Ecological Bulletins 44: 197–214. Copenhagen 1995

Interactions between soil acidification, plant growth and nutrient uptake in ectomycorrhizal associations of forest trees

R. D. Finlay

Finlay, R.D. 1995. Interactions between soil acidification, plant growth and nutrient uptake in ectomycorrhizal associations of forest trees. – Ecol. Bull. (Copenhagen) 44: 197–214.

Interactions between soil acidification and ectomycorrhizal associations of forest trees are reviewed, placing special emphasis, where appropriate, on recent Swedish research. Mycorrhizal fungi represent an important component of the biodiversity of forest ecosystems and are themselves subject to the influence of acidification. In addition, these symbiotic fungi also possess the capacity to mediate the extent to which their host plants are influenced by soil acidification. The ways in which different mycorrhizal fungi respond to acidification are still very poorly understood. Ectomycorrhizal fungi growing in symbiotic association with different hosts may respond in very different ways from that of a pure fungal culture and the overriding problems of identifying and quantifying them in soil have so far restricted progress in understanding changes occurring in the field. Controlled laboratory experiments have improved our knowledge of interactions involving identified species. Acidification can influence ectomycorrhizas either directly through altered soil pH, or indirectly through changes in soil nutrient availability, metal solubility, or carbon flow. Each of these factors can operate directly on the individual symbionts or exert its effect indirectly on one symbiont by directly influencing the other symbiont. The ability of ectomycorrhizal fungi to mediate the effects of acidification on forest trees is of special significance when discussing changes in mycorrhizal community structure and the possible effects of a given level of soil acidification. This mediation is achieved in a number of ways including modification of the soil chemical environment, altering patterns of plant nutrient uptake and increasing tolerance to, or detoxifying, the increased levels of heavy metals or aluminium often associated with soil acidification. The possible mechanisms behind these interactions are discussed with special reference to increased aluminium concentrations. Methodological problems associated with demonstration of different mechanisms are also discussed. Improved knowledge of the functional basis of observed mycorrhizal responses and effects is a pre-requisite for better understanding and prediction of the dynamics of forest ecosystems and their responses to anthropogenically generated stresses.

R.D. Finlay, Microbial Ecology, Dept of Ecology, Univ. of Lund, Ecology Building, S-223 62 Lund, Sweden.

Introduction

The predominant concern of this book is the impact of soil acidification on the various biotic components of terrestrial ecosystems, or on the chemical processes occurring within them. This chapter is somewhat different from the others in that the biotic components under discussion, ectomycorrhizal fungi, may also have a significant influence themselves. While mycorrhizal fungi are themselves subject to the influence of acidification, they also appear to have an important mediating influence on the ability of forest trees to withstand some of the

effects of acidification. Each of the important factors involved is therefore dealt with in two ways, considering both the effects of acidification on ectomycorrhiza and effects of ectomycorrhizal fungi on trees exposed to acidification.

The widespread problem of acid deposition has provided the stimulus for a number of national research programmes throughout Europe, the results of which have been recently summarised in a number of review articles (e.g. Schulze et al. 1989, Kauppi et al. 1990, Landmann and Bonneau 1994). Foremost amongst the effects of acidic atmospheric deposition has been the role of

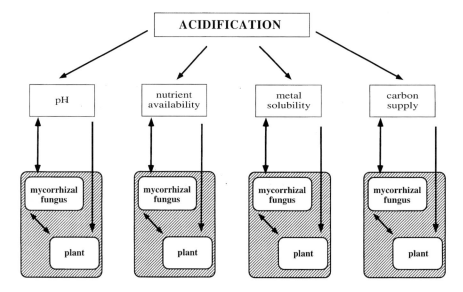

Fig. 1. Principal types of interaction between soil acidification and ectomycorrhizal associations. Effects on pH, nutrient availability, metal solubility or carbon supply may be exerted directly on either of the symbiotic partners, or indirectly through interaction between the symbionts.

soil acidification in forest decline. This paper examines the interactions between soil acidification and tree growth and nutrient uptake with particular reference to the role played by ectomycorrhizal associations. Work on the interactions involving ectomycorrhizal fungi has lagged behind other studies of acidification because of severe methodological problems which are outlined below. One of the most recent reviews of the subject (Dighton and Jansen 1991) reflects the scale of the problem in its title 'Atmospheric pollutants and ectomycorrhizae – more questions than answers?'

Ectomycorrhizal fungi are of interest in two respects. Firstly they represent an important component of the biodiversity which is threatened by acidification of forest ecosystems. Almost 1000 species of ectomycorrhizal fungi have been recorded in Swedish forests (Hallingbäck 1994) but we still know little about their function or the impact of anthropogenic influences on their distribution and abundance. Secondly, and possibly more importantly, there is increasing evidence to suggest that these symbiotic associations play an important role in mediating the effects of soil acidification on forest trees, influencing the ability of host plant species to tolerate a range of different stresses normally associated with increased soil acidity. In this chapter this evidence is reviewed, with special emphasis on recent Swedish experiments.

Mycorrhizal interactions

Ectomycorrhizal associations occur more or less ubiquitously in temperate and boreal forest ecosystems where they form the dominant nutrient absorbing organs of most trees. In undisturbed forests colonisation of tree roots by these fungi is the rule rather than the exception. The

intimate association of mycorrhizal fungi with plant roots, the primary producers of the soil ecosystem, places them in the unique position of having direct access to energy rich plant assimilates and they play an important role in the absorption and translocation of nutrients from soil to their hosts. Mycorrhiza thus play a central role in the interactions between soil microbial communities and plants, influencing soil biological activity, patterns of nutrient cycling and vegetation dynamics (Read 1991, Finlay and Söderström 1992).

Ectomycorrhizal fungi can be directly or indirectly influenced themselves by soil acidification but, through their symbiotic associations with forest trees, are additionally able to mediate the effects of soil acidification on their host plants. The possible types of interaction are outlined in Fig. 1. Soil acidification may result in effects on pH, availability of soil nutrients and solubility of metals which may influence plants or ectomycorrhizal fungi directly. These factors may also exert an indirect influence on the fungi by virtue of their effects on the host plants and one factor which may be of importance in this respect is the effect on translocation of carbon to tree roots.

The influence of mycorrhizal associations on the ability of the plants to tolerate soil acidification includes their possible effects on rhizosphere pH buffering, nutrient uptake and storage, weathering of minerals and heavy metal and aluminium tolerance. In addition to the above interactions it should be stressed that mycorrhizal fungi form a dominant component of the microbial biomass in most forest soils and, through their direct access to plant assimilates and rapid turnover they are probably responsible for a major input of carbon into the below ground ecosystem and play a major role in nutrient cycling.

In the following sections experimental evidence for these possible interactions will be considered in more detail.

Soil pH

Effects on ectomycorrhiza

A large body of research exists on effects of acid precipitation, in particular on forest decline. Controlled experiments on the effects on ectomycorrhiza are difficult to perform and interpret however and the direct effects of pH changes are often difficult or impossible to distinguish from indirect effects of pH on metal solubility or nutrient availability. Several studies (Reich et al. 1985, 1986, Stroo and Alexander 1985, Dighton and Skeffington 1987, Stroo et al. 1988, Deans et al. 1990) indicate an overall decrease in levels of ectomycorrhizal infection in response to simulated acid rain. Other studies have shown both stimulation and inhibition of mycorrhiza formation (Shafer et al. 1985, Stroo and Alexander, 1985) and effects on the relative abundance of different mycorrhiza types (Meier et al. 1989a,b, Deans et al. 1990).

Studies of fruitbody occurrence have shown an overall decrease in the numbers of fruitbodies and shifts in the species structure (Jansen and de Nie 1988, Jansen and de Vries 1988, Arnolds 1990). These data must be treated with caution though since the correlation between changes in mycorrhizal morphotypes and fruitbodies is often weak. In an experiment by Shaw et al. (1992, 1993) to examine the effects of SO_2 and ozone on young Pinus sylvestris and Picea abies trees no effects were found on the mycorrhizal community but there was a succession of fungal species which was accelerated by low levels of available P. No correlation between fruitbodies and mycorrhizal morphotypes was found in this experiment. Danielson and Visser (1989), in a study of Pinus contorta and Phleum pratense growing in acidified soil, concluded that the better survival of the tree species compared to the understorey species was due to ectomycorrhizal infection which was more tolerant of low pH than the VA endophytes. A large number of ectomycorrhizal fungi have pH optima of c. 3.0 when tested in pure culture (Hung and Trappe 1983) but very little is known about their activity or efficiency at different pH values when growing in symbiotic association.

Effects of liming

Attempts to counteract natural or anthropogenic acidification of forest soils by liming will almost certainly influence mycorrhizal fungi but our knowledge of these effects is still far from complete. This topic has been discussed elsewhere and will be discussed only briefly here in as much as it is relevant to pH responses. Since the possible effects on microbial communities were reviewed in 1984 (Söderström 1984) a number of studies have been conducted on the effects of liming on populations of ectomycorrhizal fungi (see Erland 1992). Many of these have suffered from the fact that identification of mycorrhizal species from roots is at best difficult and at

worst impossible. Quantitative studies of mycorrhizal abundance are therefore partly or wholly restricted to mycorrhizal morphotypes rather than species. Changes in the composition of ectomycorrhizal communities have often been reported in response to liming (Lehto 1984, Erland and Söderström 1990, 1991a,b, Antibus and Linkins 1992). Different results have been obtained in relation to the effect of liming on overall mycorrhizal abundance. Erland and Söderström (1990) found an increase in the +1 percentage mycorrhizal colonization from 70% at pH 4 to almost 100% at pH 5 and thereafter a decrease to < 40% at pH 7.5. In later field experiments Erland and Söderström (1991b) demonstrated an increase in total numbers of mycorrhizas per unit root length in response to a lime application of 5 tons ha^{-1}. Semjonova (1990) found no correlation between application of lime and the overall level of mycorrhization but possible changes in the species composition of fungi involved were not examined.

Several studies suggest that liming may have particular effects on specific fungi. Rapp (1992) showed that liming can reduce the abundance of Lactarius subdulcis but increase the abundance of Laccaria amethystina, while Taylor and Brand (1992) found three mycorrhiza types, 'Piceirhiza nigra', Amphinema byssoides and a Tuber species which all increased in abundance in limed plots. Effects of liming on N mineralization (Lyngstad 1992) may indirectly influence mycorrhizal species composition but this has so far not been examined. Although a number of studies have examined pH optima of mycorrhizal fungi in pure culture (e.g. Hung and Trappe 1983) growth rates of the fungus may be different when grown in association with the host plants (Erland et al. 1990) and little is known of the differences in pH sensitivity of different intact associations. Recent experiments by Andersson and Söderström (1995) suggest that effects of liming on mycorrhizal associations can be highly variable. Although liming appears to have an overall negative effect on mycorrhizal fruitbody production, the colonization of roots by mycorrhizal fungi may be strongly influenced by other factors and further studies of liming interactions in different soil types are still needed. There have been almost no physiological studies of how changes in pH affect the functioning of established mycorrhizal associations but experiments by Erland et al. (1991) indicated that translocation of ^{14}C-labelled carbon to the fungal mycelium was reduced by a pH increase from 3.8 to 5.2. The possible significance of these results is discussed later in the section dealing with carbon allocation.

pH regulation by ectomycorrhiza

The possible role of mycorrhizal fungi as rhizosphere buffers has been discussed by Rygiewicz et al. (1984a,b) who showed that mycorrhizal roots released fewer hydrogen ions per ammonium ion taken up than non-mycorrhizal roots. The general significance of this possible effect has not been widely tested but it is likely that mycorrhizal fungi influence soil pH in a number of other ways, including the production of organic acids, interactions with bacteria and weathering of minerals (e.g. Leyval and Berthelin 1989, 1991, 1993, Garbaye et al. 1992, Griffiths et al. 1994). Further research into these types of interaction is thus necessary and the role of organic acids in particular is discussed in the following section.

Nutrient availability

Soil acidification changes the availability of a number of soil nutrients in different ways. Increased solubility of metals such as Al and Fe, and Mn may result in complexation of P, reducing the concentrations that are readily available for uptake. Increased solubility of metals such as Al can constitute a problem in itself and will be dealt with separately in the section dealing with elevated metal concentrations. Acidic depositions provide an input of nitrogen (as ammonium or nitrate) and sulphur (mainly as sulphate) (Lövblad et al. 1992) and the biological effects of these depend on the extent to which these elements are limiting at a given place or time. Proton loading of a particular soil is logarithmically related to the pH of the incident rainfall and the capacity of a particular soil to buffer a given input of protons will depend on a number of factors. Exchange between protons, largely in the form of hydronium ions (H_3O^+), and base cations can lead to subsequent leaching of base cations from the soil, reduced base saturation and deficiencies of elements such as K, Ca and Mg. Interactions involving changes in nutrient availability are thus complex and difficult to separate from other factors. Ectomycorrhizal fungi themselves can be directly influenced by increased concentrations of nutrients such as nitrogen but also play an important role in sequestering nutrients which may be in short supply as a consequence of acidification.

Effects on ectomycorrhiza

Fertilisation of plants occurs both as a deliberate human activity and an unintended consequence of atmospheric nitrogen deposition arising from pollution. Large inputs of nitrogen as frequently practised in forest fertilisation often result in reductions in ectomycorrhizal abundance (Alexander and Fairley 1983) but the effects of long term nitrogen deposition at lower levels still require further research. Reductions in overall mycorrhizal abundance (Jentschke et al. 1991, Arnebrant and Söderström 1992) and changes in population structure (Arnebrant and Söderström 1992) have both been found. Other studies (Gorissen et al. 1991) report increases in ectomycorrhizal frequency at intermediate ammonium levels. Using a semi-hydroponic system Wallander and Nylund (1991, 1992) showed that high nitrogen availability did not sig-

nificantly affect internal sugar concentrations in *Pinus sylvestris* but that development of the extramatrical mycelium was restricted by excess N. Mycelial biomass on the roots, within the mantle and Hartig net was much less affected. These observations are consistent with those showing a reduction in production of mycorrhizal sporocarps following N fertilisation (Wästerlund 1982).

Methods which have been applied to investigate the above processes are broadly similar to those used to investigate other types of anthropogenic disturbance. These include both semi-descriptive studies of changes occurring in the field (Alexander and Fairley 1983, Arnebrant and Söderström 1992), and laboratory studies using known species (Gorissen et al. 1991, Wallander and Nylund 1992). Better knowledge of the functional basis underlying changes observed in the field is still required before predictive tools can be properly developed. In particular more research is required to distinguish the effects of long term nitrogen fertilisation at low levels from those of large single doses and to examine the differential susceptibility of different mycorrhizal populations to these two types of nitrogen supply (Högberg 1989). Methods based on determination of ^{15}N-enrichment (Högberg 1991) may provide convenient tools with which to study nitrogen saturation, critical loads of N and effects of revitalisation fertilisation (Högberg et al. 1992). Mycorrhiza-related differences in discrimination between ^{14}N and ^{15}N may also occur during nutrient uptake or because of differences in utilisation of organic nitrogen (Högberg 1990).

Little is known about the influence of different levels of nitrogen deposition on populations of mycorrhizal fungi and methods with which to study identified fungal species on root tips (Henrion et al. 1992, Erland et al. 1994) have only recently become available. Previous studies have relied on fruitbody data or studies of morphotypes (Arnebrant and Söderström 1992, Tyler et al. 1992) and although shifts in species composition have been shown, little attention has been paid to possible functional differences between the observed fungi and fruitbody distribution may not accurately reflect patterns of root infection. Differences in the apparent ability of ectomycorrhizal fungi to tolerate nitrogen additions may be related to differences in the extent to which they rely upon organic nitrogen sources or to differences in their physiology such as the ability to accumulate arginine (Finlay 1992, Finlay et al. 1992). Immobilisation of N is likely to require large amounts of labile, microbially available carbon and root exudation, organic acid production and rapid turnover of mycorrhizal hyphae are all likely to be important in influencing pools of dissolved organic carbon (Finlay and Söderström 1992, Griffiths et al. 1994). Further, more detailed studies of ectomycorrhizal populations and their nitrogen metabolism and interactions with organic matter and other soil microbial populations are now necessary to improve our understanding of the role they play in immobilisation and turnover of nitrogen in soil.

Effects of ectomycorrhiza

The role ectomycorrhizal associations in uptake and translocation of nutrients has received much attention and there have been numerous demonstrations of the improved nutritional status of mycorrhizal plants (see Harley and Smith 1983). There have been fewer direct demonstrations of uptake and translocation of nutrients by ectomycorrhizal mycelia but the original experiments of Melin et al. (1958 and references therein) have now been refined and extended to include a range of different mycorrhizal associations growing under more natural conditions (see Finlay 1992). Most of these experiments have studied ^{32}P-phosphorus (Finlay and Read 1986a, Finlay 1989), ^{14}C-carbon (Finlay and Read 1986b) or mineral nitrogen labelled with ^{15}N (Finlay et al. 1988, 1989).

Ectomycorrhiza have both a quantitative and a qualitative influence on nutrient availability. It is now widely accepted that the improved nutrient uptake of mycorrhizal plants arises, in part, from an altered spatial geometry of the root system in which surface area available for nutrient absorption is massively increased, at minimal synthetic cost to the plant, by the presence of a more or less extensive mycorrhizal mycelium. In addition to this physical extension of the root system, the mycorrhizal mycelium is better able to penetrate nutrient rich soil micro-sites which may be unavailable to plant roots. This quantitative increase in nutrient absorbing capacity may be important in situations where supply of nutrients is limited by low soil mobility or where acid depositions cause leaching of base cations from the soil and result in reduced availability of Mg^{2+}, Ca^{2+} and K^+. Uptake of ^{45}Ca and ^{86}Rb (a potassium analogue) by the mycelium of *Paxillus involutus,* and translocation of these elements to connected mycorrhizal *Fagus sylvatica* plants has been shown by Finlay et al. (1995c). Mycelial uptake and translocation of ^{45}Ca and ^{22}Na have also been shown in other mycorrhizal associations, grown under sterile conditions (Melin et al. 1958). These experiments are supported by observations that mycorrhizal plants exposed to simulated acid rain or elevated aluminium levels have relatively higher Mg and K concentrations than non-mycorrhizal plants (Finlay et al. 1995a,b). This suggests that ectomycorrhizal plants may be better able to withstand the reductions in availability of these nutrients resulting from loss by leaching.

Renewed base saturation may occur as a result of mineral weathering and the buffering capacity of a soil and the rate at which base cations are released will depend upon will depend on the parent material and the stage of evolution of the soil. The extent to which ectomycorrhizal fungi contribute to weathering by accelerating the release of base cations is still uncertain. Much of the recent evidence comes from *Pseudotsuga menziesii* forest soils in the Pacific Northwest. Oxalic acid is produced in large amounts by some ectomycorrhizal fungi (Graustein et al. 1977, Cromack et al. 1979, Lapeyrie et

al. 1987) and appears to have a significant weathering effect, influencing the availability of a number of nutrients in the soil solution associated with mycelial mats (Griffiths et al. 1994). Fungi such as *Hysterangium setchellii* and *Gauteria monticola* form extensive mycelial mats and are able to accumulate most nutrients important for growth and productivity (Entry et al. 1992). In addition to sequestration of mineral elements such as phosphate anions organic acids such as oxalate and citrate are able to form soluble organic complexes with metals, reducing their toxicity to plants (Hue et al. 1986). Recent experiments in my laboratory have shown production of significant amounts of organic acids such as oxalate and citrate by a number of ectomycorrhizal fungi, both in pure culture and in symbiotic association with different tree species. Further experiments are now needed to determine extent of variation in production the factors influencing the level of production.

Another qualitative effect of ectomycorrhizal fungi is their key role in utilisation of less recalcitrant forms of soil organic nitrogen such as proteins and amino acids. An alternative view of nitrogen cycling in coniferous forest ecosystems based on the presence of proteolytic and peptidolytic activity in some ectomycorrhizal fungi was proposed by Abuzinadah et al. in 1986. The early stages of litter decomposition have been shown to be associated with increases in absolute amounts of N (Berg and Söderström 1979, Berg and Staaf 1981) caused by accumulation in the microbial biomass. This immobilisation in microbial tissues leads to higher concentrations of N than in plant tissue. Figures as high as 6.7% (Abuzinadah and Read 1989) have been reported. Proteolytic ability in mycorrhizal fungi will enable utilisation of these organic resources before the nitrogen is re-immobilised in tissues of decomposer organisms or complexed with carboxylic acids. Abuzinadah et al. (1986) suggested that this would lead to tighter cycling of nutrients and rapid transfer of nitrogen to the mycorrhizal plants. Mycorrhizal fungi are known to differ in their proteolytic activity (Abuzinadah and Read 1986, Finlay et al. 1992) and also in their ability to transfer assimilated nitrogen to their host plants (Abuzinadah and Read 1989). These processes are likely to be of significance in undisturbed forest ecosystems where nitrogen is limiting. In forests polluted by nitrogen deposition there may be a progressive loss of this proteolytic activity and selection for different mycorrhizal fungi with other properties. However little is known about shifts in mycorrhizal community structure at the functional level and more research is needed.

Elevated metal concentrations

Acidification of forest soils and the accompanying increased solubility of metals have led to an increased interest in interactions between ectomycorrhizas and metals. The effects of many heavy metals may often be confined to local gradients associated with point sources of pollution such as smelters (Rühling et al. 1984, Ohtonen et al. 1990) whereas increased solubility of Al represents a more general problem. Reductions in the pH of the forest soil solution lead to more than exponential increase in the concentration of Al, accompanied by loss of nutrients such as Ca, Mg and K from the exchangeable pool. Decreases of between 0.5 and 1 pH unit have been recorded in the past 50 yr in forests in Sweden and Germany and concentrations of Al as high as 0.7 mM have been measured in forest soil solutions with a pH of 4.0–4.2 (Hallbäcken and Tamm 1986, Falkengren-Grerup 1987, Falkengren-Grerup et al. 1987, Tyler et al. 1987). The possible toxic influence of this aluminium on growth and nutrient acquisition by vascular plants (Andersson 1988, Marschner 1991) has now become an area of widespread research interest and much research has been focused on the physiological basis of Al tolerance (see Taylor 1991). In particular, there has been much discussion of the effects of Al on base cation uptake and the use of models based on soil solution ratios of base cations (such as Ca^{2+} and Mg^{2+}) to Al^{3+} (e.g. Sverdrup et al. 1992). Although divalent cations are able to modify Al toxicity in solution culture experiments, the use of Ca:Al ratios as a general measure of Al toxicity seems problematic and the direct applicability of data from simple solution culture experiments to the field is questionable. In a study of natural plant communities growing under a range of field conditions found Ca:Al ratios to have a lower predictive value than a range of other parameters. The use of simple models based on ratios of base cations to aluminium has also been criticised by Högberg and Jensén (1994) since these usually ignore the role of ion channels or specific carriers and pay scant regard to a range of rhizosphere effects such as pH gradients, exudate production and symbiotic associations with mycorrhizal fungi. Further study of the biological complexity of these rhizosphere interactions will play a key role in helping to understand and predict responses to toxic metals. In particular the extent to which uptake of base cations is constrained by low ratios of base cations to Al is highly likely to be greatly modified by the presence of ectomycorrhizal fungi.

Effects on ectomycorrhiza

Ectomycorrhizal fungi display inter- and intra-specific differences in resistance to different metals (Rühling et al. 1984, Brown and Wilkins 1985, Jones and Hutchinson 1986, Colpaert and Van Assche 1987, 1992, 1993, Rühling and Söderström 1990). Different experimental approaches have been adopted to study the effects of metals on mycorrhiza. Field studies have included observations of changes in the numbers of fruitbodies in relation to localised heavy metal contamination (Rühling et al. 1984, Rühling and Söderstöm 1990) and measurements of the

numbers of mycorrhizal species forming sporophores, as well as numbers of different types of mycorrhizal root tips (Ohtonen et al. 1990). Other studies have attempted to measure metal contents of fruitbodies (Tyler 1980) to determine whether bioexclusion or bioconcentration of metals occurs. As suggested by this author, this method may indicate the capacity of the mycelium to take up and translocate metals from the substrate but it will not necessarily accurately reflect metal concentrations in the mycelium or mycorrhizal short roots.

Studies of the fungi in pure culture suggest that many ectomycorrhizal species tolerate very high concentrations of Al (Hintikka 1988, Väre 1990) but that there is interspecific (Jongbloed and Borst-Pauwels 1992, Jongbloed et al. 1992) and intra-specific variation in their sensitivity. Field studies of mycorrhizal abundance along gradients of metal contamination have shown a generally depressive effect of raised metal concentrations (Ohtonen et al. 1990) although numbers of *Laccaria laccata* fruitbodies have actually been shown to increase along such gradients (Rühling et al. 1984). Although some of the evidence is conflicting (Denny and Wilkins 1987b) there is often a relationship between the degree of metal tolerance of different fungal isolates and the metal concentrations at their sites of origin (Colpaert and Van Assche 1987, 1992).

Methods for studying effects of Al on plants typically involve hydroponic growth systems in which plants are exposed to culture solutions containing different concentrations of Al (Bengtsson et al. 1988, Arovaara and Ilvesniemi 1990a,b, Bengtsson 1992). Although this type of approach permits better control of Al concentrations it is not appropriate for intact mycorrhizal systems where the mycelium requires a solid medium providing physical support and some aeration. Most experiments have used sand (Cumming and Weinstein 1990a,b, Browning and Hutchinson 1991, Jentschke et al. 1991, McQuattie and Schier 1992, Hentschel et al. 1993) or perlite (Wilkins and Hodson 1989, Hodson and Wilkins 1991). Alternative systems include steady state nutrition growth units (Göransson and Eldhuset 1991) or simpler semi-hydroponic systems (Finlay et al. 1995a) of the type described by Nylund and Wallander (1989). Levels of aluminium used in these experiments vary widely from 35 μM (Finlay et al. 1995c) to 30 mM (Göransson and Eldhuset 1991) but the concentrations used in solution culture experiments usually exceed the concentrations of phytotoxic Al^{3+} which can be measured in the humus layer of forest soils. In many forests levels of exchangeable Al in the organic A horizon may not exceed 2% of total cation exchange capacity due to complexation of Al with organic matter (Tyler 1987) while concentrations in the soil solution of the mineral B horizon may be between 0.3 mM and 1 mM (Bergkvist 1987). The usual justification for using concentrations in excess of those usually found in the field is that localised concentrations of aluminium in soil microsites may often exceed those of the bulk soil solution, especially under conditions of drought.

Table 1. Influence of different soil solution concentrations of Al^{3+} on plant dry weight and root concentrations of K, Mg, Ca and P in mycorrhizal (Myc) and non mycorrhizal (Non-myc) *Pinus contorta* and *Fagus sylvatica* plants grown in peat microcosms.

Experiment	Al level	Species	Results	Myc	Non-Myc
				Myc	Non-Myc
		Pinus contorta			
1	control ≈ 15 μM	*Paxillus involutus*	Δ_{Al} – dw	−33%	−86%
	Al ≈ 250 μM	*Suillus variegatus*	Δ_{Al} – dw	−50%	−81%
				Myc	
2		*Paxillus involutus*			
		Fagus sylvatica	Δ_{Al} – dw	−45%	
		Fagus sylvatica	Δ_{Al} – Al	+264%	
	control ≈ 11 μM	*Fagus sylvatica*	Δ_{Al} – K	−31%	
	Al ≈ 650 μM	*Fagus sylvatica*	Δ_{Al} – Mg	0% NS	
		Fagus sylvatica	Δ_{Al} – Ca	−9% NS	
		Pinus contorta	Δ_{Al} – dw	+20% NS	
		Pinus contorta	Δ_{Al} – Al	−7% NS	
		Pinus contorta	Δ_{Al} – K	−4% NS	
		Pinus contorta	Δ_{Al} – Mg	−30% NS	
		Pinus contorta	Δ_{Al} – Ca	−23% NS	
				Myc	Non-Myc
	control ≈ 35 μM	*Paxillus involuts*			
3	≈ 80 μM	*Fagus sylvatica*	Δ_{Al} – dw	−18%	−20%
	≈ 400 μM	*Fagus sylvatica*	Δ_{Al} – dw	−51%	−46%
3	≈ 80 μM	*Pinus contorta*	Δ_{Al} – dw	+15%	−43%
	≈ 400 μM	*Pinus contorta*	Δ_{Al} – dw	−2% NS	−65%
3	≈ 80 μM	*Pinus contorta*	Δ_{Al} – P	+27%	+46%
	≈ 400 μM	*Pinus contorta*	Δ_{Al} – P	+94%	+40%
3	≈ 80 μM	*Pinus contorta*	Δ_{Al} – K	0% NS	0% NS
	≈ 400 μM	*Pinus contorta*	Δ_{Al} – K	+12% NS	−41%
3	≈ 80 μM	*Pinus contorta*	Δ_{Al} – Mg	−43%	+39%
	≈ 400 μM	*Pinus contorta*	Δ_{Al} – Mg	−21%	+65%

Δ_{Al} percentage figures indicate the % increase or decrease in aluminium treatments relative to controls. All differences are statistically significant ($p < 0.05$ or less) unless otherwise indicated (NS). Levels of monomeric Al^{3+} were determined by complexation with 8-hydroxyquinoline and extraction of the resulting complex into methyl isobutyl ketone (MIBK) (Barnes 1975).

Table 2. Influence of 0.7 mM Al Cl_3 on total plant dry weight and root concentrations of K, Mg, Ca and P in mycorrhizal (Myc) and non mycorrhizal (Non-myc) *Pinus contorta*, *Betula pendula* and *Fagus sylvatica* plants grown in semi-hydroponic culture. Mycorrhizal plants were colonised by the ectomycorrhizal fungus *Laccaria bicolor*.

Experiment	Al level	Species	Results	Myc	Non-Myc
1	≈ 0.7 mM	*Pinus contorta*	Δ_{Al} – dw	−25%	−44%
		Laccaria bicolor	Δ_{Al} – K	−9%	−38%
			Δ_{Al} – Mg	+13% NS	−80%
			Δ_{Al} – Ca	−59%	−75%
			Δ_{Al} – P	−30%	−54%
3	≈ 0.7 mM	*Betula pendula*	Δ_{Al} – dw	−45%	−41%
		Laccaria bicolor			
5	≈ 0.7 mM	*Fagus sylvatica*	Δ_{Al} – dw	−32%	−22%
		Laccaria bicolor			

Δ_{Al} percentage figures indicate the % increase or decrease in aluminium treatments relative to controls. All differences are statistically significant ($p < 0.05$ or less) unless otherwise indicated (NS).

Table 3. Influence of ectomycorrhizal colonisation on plant dry weight and root concentrations of K, Mg, Ca and P in *Pinus contorta* and *Fagus sylvatica* plants grown in peat microcosms with different soil solution concentrations of Al^{3+}. Data are from the experiment referred to as Experiment 3 in Table 1.

Species	Results			
		35 µM	80 µM	400 µM
	Paxillus involutus			
Fagus sylvatica	Δ_{myc} – dw	–3% NS	–0.5% NS	–13%
Pinus contorta	Δ_{myc} – dw	+95%	+295%	+441%
Pinus contorta	Δ_{myc} – P	+130%	+100%	+220%
Pinus contorta	Δ_{myc} – K	+14%	+14%	+116%
Pinus contorta	Δ_{myc} – Mg	+424%	+113%	+153%

Δ_{myc} percentage figures indicate the % increase or decrease in mycorrhizal treatments relative to non-mycorrhizal controls. All differences are statistically significant (p < 0.05 or less) unless otherwise indicated (NS). Levels of monomeric Al^{3+} were determined by complexation with 8-hydroxyquinoline and extraction of the resulting complex into methyl isobutyl ketone (MIBK) (Barnes 1975).

Reductions in growth (Arovaara and Ilvesniemi 1990a,b), Ca and Mg uptake (Bengtsson et al. 1988, Göransson and Eldhuset 1991, Bengtsson 1992, Hentschel et al. 1993) and P uptake (Bengtsson et al 1988, Finlay et al. 1995a) and both increases (Finlay et al. 1995c) and decreases (Asp et al. 1988) in K uptake have been shown in a range of plant species in response to exposure to elevated Al concentrations (Tables 1–4). Effects of Al on the growth of ectomycorrhizal fungi growing in symbiosis may depend on the particular tree species involved and its own sensitivity to the metal. Finlay et al. (1995c) found no effects of short term exposure to

Al on translocation of K (^{86}Rb), ^{32}P and ^{45}Ca through ectomycorrhizal mycelium of *Paxillus involutus*. Longer term exposure to high (0.08–0.7 mM) concentrations of Al resulted in negative effects on ectomycorrhizal colonisation of *Fagus sylvatica* and *Pinus contorta* plants by *P. involutus* measured in terms of ergosterol concentrations (Finlay et al. 1995b). This effect seemed to be related to direct effects on the plant roots and a reduction in root tips available for fungal colonisation than to any direct effect of the Al on the fungus. Mycorrhizal pine plants appeared better able to withstand raised Al concentrations but there was no great effect of *P. involutus* on

Table 4. Influence of ectomycorrhizal colonisation by the fungus *Laccaria bicolor* on total plant dry weight and root concentrations of K, Mg, Ca and P in *Pinus contorta*, *Betula pendula* and *Fagus sylvatica* plants grown in semi-hydroponic culture with and without 0.7 mM AlCl$_3$.

Experiment	Al level	Species	Results		
				control	Al
1	≈ 0.7 mM	*Pinus contorta*	Δ_{myc} – dw	+23%	+63%
		Laccaria bicolor	Δ_{myc} – K	–18%	+22%
			Δ_{myc} – Mg	–38% NS	+260%
			Δ_{myc} – Ca	–4% NS	+61%
			Δ_{myc} – P	+67%	+156%
				5 µM P	20 µM P
2*	≈ 0.7 mM	*Pinus contorta*	Δ_{myc} – dw	+45%	+58%
		Laccaria bicolor			
				control	Al
3	≈ 0.7 mM	*Betula pendula*	Δ_{myc} – dw	–24%	–30%
		Laccaria bicolor			
					Al
4	≈ 0.7 mM	*Pinus contorta*	Δ_{myc} – dw		+60%
	≈ 0.7 mM	*Betula pendula*	Δ_{myc} – dw		+18% NS
				control	Al
5	≈ 0.7 mM	*Fagus sylvatica*	Δ_{myc} – dw	+8% NS	–55% NS

Δ_{myc} percentage figures indicate the % increase or decrease in mycorrhizal treatments relative to non-mycorrhizal controls. All differences are statistically significant (p < 0.05 or less) unless otherwise indicated (NS). * In Experiment 2 all treatments contained AlCl$_3$ but the P concentration of the nutrient solution was changed from the normal value of 20 µM to 5 µM.

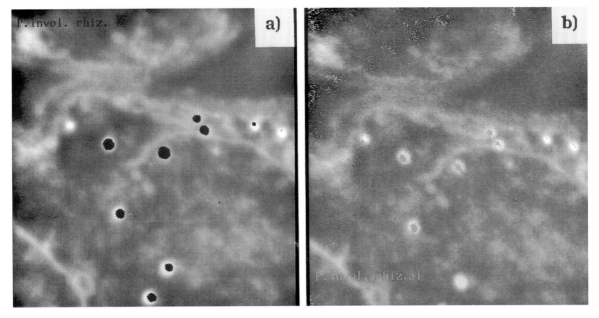

Fig. 2. Electron spectroscopic imaging micrograph of *Paxillus involutus* mycorrhizal mycelium associated with *Fagus sylvatica* plants exposed to 0.3–0.7 mM Al. a) elemental map showing distribution of P and accumulation in polyphosphate droplets in the mycelium. b) elemental map of Al showing association with polyP droplets.

beech plants (Tables 1 and 3). In studies using semi-hydroponic systems (Finlay et al. 1995a) colonisation of root systems of *Pinus contorta*, *Betula pendula* and *Fagus sylvatica* by the ectomycorrhizal fungus *Laccaria bicolor*, as measured by ergosterol concentrations, was unaffected by raised Al concentrations of 0.7–1.0 mM Al. Dry weight yield and root concentrations of mycorrhizal plants were less well affected than those of non-mycorrhizal pine plants (Tables 2 and 4).

Effects of ectomycorrhiza

The role of ectomycorrhizal infection in ameliorating the adverse effects of elevated metal concentrations has been widely discussed (see Wilkins 1991) but so far there have been few detailed experiments with fully intact, identified mycorrhizal associations. Different experimental approaches have been adopted and the results have generally shown that ectomycorrhizal plants tolerate higher metal concentrations than non-mycorrhizal plants and have lower shoot concentrations of the metal.

Experiments with plants have mainly involved culture in sand, perlite or vermiculite and exposure to metals such as zinc (Denny and Wilkins 1987a, b), copper (Colpaert and Van Assche 1992), nickel (Jones and Hutchinson 1986), cadmium (Colpaert and Van Assche 1993) or aluminium (Jentschke et al. 1991, Hentschel et al. 1993). The latter two studies were unable to demonstrate any convincing effect of mycorrhizal infection on the ability of *Picea abies* seedlings to withstand the negative effects

of Al exposure. However the other studies mentioned above showed largely positive effects of the mycorrhizal plants on the ability of the plants to tolerate metal pollution. Experiments by Cumming and Weinstein (1990a, b) using *Pinus rigida* and the ectomycorrhizal fungus *Pisolithus tinctorius* showed that the fungal symbiont was able to modulate ionic relations in the rhizosphere, reducing Al-P precipitation, Al uptake and subsequent exposure of tissue to Al whilst maintaining foliar P levels. Association of pitch pine seedlings with *P. tinctorius* also enabled the trees to utilise P from insoluble $AlPO_4$ which was otherwise unavailable to non-mycorrhizal plants (Cumming and Weinstein 1990c) and to maintain foliar K levels in the presence of Al.

Other experiments performed in my laboratory, or with colleagues at the Swedish Univ. of Agricultural Sci. (Finlay et al. 1995a, b) have shown strong ameliorative effects of ectomycorrhizal fungi on the ability of plants to withstand elevated Al levels (see Tables 1–4 for details). Growth reductions of *Pinus contorta* occurring in response to elevated Al levels were minimised by the ectomycorrhizal fungi *Paxillus involutus* and *Laccaria bicolor* and the fungi helped to maintain foliar concentrations of K, P and Mg. The interactions of the same fungi with *Fagus sylvatica* were much less strong and had no significant ameliorative effect. The effects of mycorrhizal infection were strongly dependent on the stage of development at which seedlings were exposed to high Al concentrations and seedlings which were already fully mycorrhizal before exposure were much less severely affected than younger seedlings exposed to Al before mycorrhizal infection had been established (Table 1, Experiment 1).

Current theories attribute these ameliorative effects to immobilisation of the metal through complex formation at the mycelial surface (Väre 1990) or in the root cortex (Hodson and Wilkins 1991) or sheath (Eeckhaoudt et al. 1992). The idea that ectomycorrhizal fungi reduce the toxic effects of aluminium and heavy metals by binding them within their own tissue and reducing exposure to the plant has led to a number of studies attempting to localise the metals within mycorrhizal roots. Techniques for localisation include EDAX [Energy Dispersive Analysis with X-rays] (Denny and Wilkins 1987b, Donner and Heyser 1989), PIXE [Particle Induced X-ray Emission] (Hult et al. 1992), LAMMA [Laser Microprobe Mass Analysis] (Eeckhaoudt et al. 1992), and EELS/ESI [Element Energy Loss Spectroscopy/Electron Spectroscopic Imaging] (Kottke 1991). Nuclear magnetic resonance [NMR] spectroscopy (Martin et al. 1994) has also been used in conjunction with EELS (Holopainen et al. 1992, Kottke and Martin 1994) to examine patterns of polyphosphate accumulation with fungal mycelium exposed to aluminium and it has been suggested that this polyphosphate can act as a detoxifying agent by binding metals and reducing exposure to the plant root. An advantage of NMR studies is that living tissue can be used, avoiding artefacts of sample preparation (see Methodological problems and research priorities), however relatively large amounts of tissue are required.

The idea that polyphosphate is involved in detoxification of Al is indirectly supported by a number of other studies which show either accumulation or maintenance of phosphorus concentrations within mycorrhizal roots exposed to metals (Finlay et al. 1995a,b). EELS studies of material from these experiments showed some co-localisation of polyphosphate and aluminium (Fig. 2) in hyphae of *Paxillus involutus* associated with *Pinus contorta* plants exposed to 0.7 mM Al suggesting that polyphosphate may have played some role in the detoxification of the aluminium. The quantitative significance of the effect was difficult to determine however and further studies are still required. Further indirect support is provided by a cryo-microanalytical study of element distribution in *Eucalyptus rudis* roots (Egerton-Warburton et al. 1993) showing accumulation of Al and P in the mycorrhizal sheath of plants exposed to Al. Although the experiments of Martin et al. (1994) were carried out with pure cultures of *Laccaria bicolor* grown at artificially high P concentrations NMR studies by McFall et al. (1992) have shown that the ectomycorrhizal fungus *Hebeloma arenosa* is able to accumulate significant amounts of phosphorus as poly P even at low, growth-limiting soil P concentrations. Turnau et al. (1993) have used EELS and ESI methods to examine element distribution in mycorrhizal roots collected from heavily polluted soils and found some evidence of metal accumulation in vacuoles in association with phosphate. Other substances which may sequester metals include organic acids such as oxalate (Cromack et al. 1979, Hue et al. 1986) or inducible metallothionein-like proteins which bind heavy metals (Morselt et al. 1986). Further studies of field material collected from sites with defined pollution characteristics or of known mycorrhizal associations grown under controlled laboratory conditions will assist in determining the general significance of these possible detoxification mechanisms and establishing the extent of inter- and intra-specific variation in detoxification capacity.

Studies with intact mycorrhizal systems, including the host plant are necessary in order to obtain an accurate picture of the effects of metals and are essential partly because effects of the interaction on the plants can be measured and partly because pure fungal cultures may behave differently from fungal mycelium growing in symbiotic association with a plant host. An example of this phenomenon is provided by experiments on zinc exposure by Denny and Wilkins (1987b) who found that extramatrical hyphae of the ectomycorrhizal fungus *Paxillus involutus* growing in association with *Betula* spp. contained higher concentrations (192 µmol g^{-1}) of zinc than the mantle (44.4 µmol g^{-1}) whereas there was no such accumulation in mycelium growing in the absence of *Betula* seedlings.

It is also evident that detoxification is differently effective in different species and strains and that there are also different mechanisms involved for different metals; for example the mechanisms for zinc and copper tolerance appear to be completely different (Colpaert and Van Assche 1987). It is thus important to avoid extrapolation from one metal to another. Further physiological examination of identified strains and species, in combination with the range of sophisticated localisation techniques discussed above, will no doubt greatly improve our existing knowledge and application of ectomycorrhizal fungi in relation to metal toxicity. Ectomycorrhizal fungi are important in recolonisation of derelict or degraded soils resulting from anthropogenic sources such as mine spoils, which often contain high concentrations of heavy metals. Reductions in plant uptake of elements such as manganese have been attributed to the presence of mycorrhizal fungi such as *Pisolithus tinctorius* (Marx and Artman 1979, Berry 1982). Future use of ectomycorrhizal fungi as a management tool in environmental amelioration now requires further, more detailed studies of the processes involved and of the fungi occurring naturally in such environments.

Carbon allocation

Effects of changes in carbon allocation to ectomycorrhiza

The direct effects of pH on the fungi are compounded by

the indirect effects of reduced pH which act through changes in plant metabolism. One effect of acidification may be a reduction in carbohydrate supply to the roots and a consequent reduction in carbon available to the mycorrhizal fungi (Stroo et al. 1988). Carbohydrate physiology can be an indicator of physiological dysfunction and photosynthesis and allocation processes are sensitive indicators of physiological stress, reflecting the relationship between pollutant sensitivity of plants and availability of carbon (energy) for compensation and repair processes. The physiology of carbon allocation has thus been suggested by Andersen and Rygiewicz (1991) as a suitable framework within which to examine the effect of different anthropogenic factors. The study of carbon physiology and allocation has the advantage that it allows examination of the way in which plants integrate multiple stresses at the organism level. Since mycorrhizal fungi are an integral part of most naturally occurring root systems any anthropogenic factor which alters carbon allocation will affect the vigour of established mycorrhizas and colonization and development of new associations.

Detailed information about the above processes is still lacking although a number of studies are now in progress in relation to the effects of CO_2 enrichment (e.g. Norby et al. 1987, O'Neill et al. 1987). Studies of the effects of altered pH on carbon translocation by Erland et al. (1991) showed no differences in carbon assimilation by *Pinus sylvestris* plants but loss of labelled carbon from mycorrhizal roots systems was more rapid at reduced pH values. Conversely recent experiments on the uptake and respiration of ^{14}C-labelled glutamate, glutamine and alanine in our laboratory (Chalot et al. 1994a,b) have shown increases in respiration of these substrates at higher pH values (Chalot et al. 1995). Further more detailed measurements of these processes are still required.

Effects of ectomycorrhiza on below ground carbon allocation and cycling

Ectomycorrhizal associations may play an important role in influencing below ground allocation of carbon and some of the relevant features of this carbon cycling have been reviewed by Finlay and Söderström (1992). Cycling of carbon compounds is ultimately of significance to studies of soil acidification and immobilisation of pollutants such as nitrogen. Ectomycorrhizal mycelial respiration was estimated as 30% of below ground respiration by Söderström and Read (1987) but they were not able to calculate a specific rate because it was not possible to estimate the mycelial biomass. A complete carbon budget has now been produced by Rygiewicz and Andersen (1994) for *Pinus ponderosa* seedlings colonised by the ectomycorrhizal fungus *Hebeloma crustuliniforme*. In this study hyphal respiration was estimated to be 19.4% of total belowground respiration and hyphae had the

highest respiration rates of any seedling fraction. Total below ground respiration of mycorrhizal seedlings was 2.1 times higher than that of uninoculated seedlings. This was due to three factors, 1) the high specific respiration rate of hyphae, 2) the inherently high respiration rates of colonised roots, and 3) the greater percentage of active, fine roots in mycorrhizal plants. Mycorrhizal plants thus lost photosynthetically fixed carbon at a higher rate than control plants. The above observations are consistent with those of Reid et al. (1983) who showed that *Pinus taeda* and *Pinus contorta* colonised by *Pisolithus tinctorius* and *Suillus granulatus* lost a greater proportion of assimilated carbon through root respiration than did non-mycorrhizal plants. Abuzinadah and Read (1989) suggested that ectomycorrhizal plants could utilise carbon from heterotrophically assimilated organic nitrogen sources and estimated that up to 9% of the photosynthetically fixed carbon could be assimilated in this way. In our present studies (Chalot et al. 1994b and unpubl.) we have observed that amino acids such as alanine can be used as a respiratory substrate by mycorrhizal fungi and that mycorrhizal plants have a higher respiratory loss of ^{14}C than non-mycorrhizal plants. Taken together these studies indicate that ectomycorrhizal associations may have a strong influence on patterns of carbon partitioning, altering both the quality and quantity of carbon allocated below ground. Anthropogenic stresses which alter ectomycorrhizal colonisation in forests may thus indirectly alter the role of forests in carbon sequestration.

Methodological problems and research priorities

The foregoing discussion has dealt with some of the major ways in which processes of acidification influence, and are influenced by, ectomycorrhizal associations. The list of topics dealt with above is not exhaustive and many other types of interaction have not been mentioned. For example the role of other microorganisms, such as rhizosphere microfungi, in forest decline has been discussed by Devêvre et al. (1994) and it is highly likely that ectomycorrhizal fungi interact strongly with other biotic components of the rhizosphere. There is a growing awareness amongst plant ecologists and foresters of the potential importance of mycorrhizal fungi in the tolerance of anthrogenically generated stress. However there is still a widespread ignorance of the methods involved in their study and a reluctance to view the mycorrhizal habit as a complex of evolutionary adaptations to different edaphic conditions. For this reason a number of the methodological problems associated with the study of ectomycorrhizal associations are discussed below and root symbionts are discussed in relation to their potential importance as indicators of anthropogenic stress.

Methodological problems

Studies involving ectomycorrhizal fungi are complicated by most of the general problems that affect other studies of acidification. Lack of suitable controls and difficulty in distinguishing primary effects of pH changes from secondarily induced changes in nutrient availability metal solubility are typical problems. The study of local pollution gradients, on the one hand, and regional or national effects of diffuse, transboundary pollutants on the other, require different approaches.

Additional problems stem from the fact that mycorrhizal fungi are difficult to quantify and identify when growing on plant roots. Traditional studies involving the counting of fruitbodies may be adequate when performed repeatedly over a long time period and in the same place. The fact that many ectomycorrhizal species do not form macrocarps and that the relationship between fruitbody abundance and the frequency of root colonisation may not be stable over time make total reliance on fruitbody data unwise, however. Although identification of ectomycorrhiza from infected roots is possible (Agerer 1991) the methods are time consuming, difficult to learn and so far only applied to a small fraction of known mycorrhizal species. Fortunately the rapid development of molecular methods (e.g. Gardes et al. 1991, Gardes and Bruns 1993) has meant that identification of individual mycorrhizal roots is becoming an increasingly realistic possibility. There is still no reliable method with which to quantify the biomass of mycorrhizal fungi against the background of non-mycorrhizal fungi but this is less of a problem since it is likely that many anthropogenic stresses will lead to changes in species composition but not necessarily changes in overall biomass. Since ectomycorrhizal fungi form discrete infection units, identification and counting of these, using molecular methods, should allow changes in community structure to be monitored. The difficulties of working with mixed communities of unidentified mycorrhizal fungi in the field necessitate the parallel development of more detailed laboratory studies under controlled conditions and using known fungal species.

With regard to some of the more specific problems involved with acidification and metal toxicity, continued development of reliable methods to measure biologically active forms of pollutants such as Al (e.g. Clarke et al. 1992) is required to enable meaningful measurements of pollutant stresses. Better methods of specimen preparation and improved element localisation are also desirable in attempts to elucidate tolerance mechanisms. Structural integrity of membranes is not always preserved using traditional ethanol dehydration and chemical fixation methods and secondary binding to substances such as polyphosphate may occur as a result of leakage. In such cases the cation composition will simply reflect the most efficient ion exchanger and not an *in vivo* detoxification mechanism. These problems have been discussed by Orlovich and Ashford (1993) who point out that polyphosphate granules in the vacuoles of *Pisolithus tinctorius* are an artefact of conventional specimen preparation, formed when membranes become leaky, allowing influx of precipitating ions such as calcium, and that vacuoles of freeze-substituted hyphae contain polyphosphate in a liquid form. Complementary studies with NMR using living tissue support the role of polyphosphate but further studies employing freeze substitution are clearly desirable.

Root symbionts as indicators of anthropogenic influence

Mycorrhizal fungi play a key role in a number of situations where anthropogenic effects are of interest. As well as being influenced themselves by different anthropogenic factors, ectomycorrhizas also influence the ability of their host plants to tolerate different anthropogenically generated stresses.

Vogt et al. (1993) have identified root symbionts as a key component of the belowground responses which may be used as indicators of environmental change. Their suitability as indicators of change arises from several properties. 1) Since mycorrhizal roots are in direct contact with the soil they will be sensitive to and reflect the response of the system to any anthropogenic stress that directly impacts on the physical or chemical characteristics of the soil. Changes may thus be initially measurable in roots before aboveground effects are noticed. 2) Ectomycorrhizal roots have evolved to buffer or ameliorate changes so that above ground effects are minimised: – sequestration of toxic metals prevents transport to shoots, storage of polyphosphate buffers aboveground tissues from seasonal fluctuations in soil P supply. Direct measurements of mycorrhizal roots, rather than shoot material may thus reveal changes at an earlier stage. 3) Root symbionts (ectomycorrhizas) are the first filter that plants have to counteract negative effects of the soil physical or chemical environment on their growth. Mycorrhizal fungi differ in their tolerance of toxic trace elements and their composition on root systems may be altered due to stress. Loss of dominant mycorrhizal species in areas where diversity is already low may thus lead to increased susceptibility to stress. 4) Fine roots and mycorrhizas have a relatively short life span and their turnover appears to be strongly controlled by environmental factors; they are thus an extremely dynamic component of the belowground ecosystem and can respond rapidly to stress. Whilst they may also be more sensitive to random fluctuations they may reflect the current state of the environment more clearly than other ecosystem components. 5) Stress affects the total amount of carbon fixed by plants and modifies carbon allocation to biomass, symbionts and secondary metabolites. Mycorrhizal fungi are dependent for their growth on the supply of host assimilates and stresses which shift the allocation of carbon reserves to the production of new leaves at the expense of supporting root tissues (Winner and Atkinson 1986) will be rapidly reflected in decreased fine root and mycorrhiza biomass.

That changes in ectomycorrhizal abundance, species composition and activity do occur is not in question; a growing body of experimental evidence exists to suggest that this is so. Unfortunately, in many studies, the functional basis underlying observed changes is often unclear or not even examined at all. Descriptive field studies of changes in mycorrhizal abundance or species composition are still necessary as methods of measuring the integrated effects of complex stress interactions. However complementary physiological studies, under more controlled conditions, are also needed to better describe the range of fungal variation, the physiological limits to tolerance of different stresses and the mechanisms by which mycorrhizal fungi are affected, or themselves affect the ability of plants to withstand different anthropogenic stresses. The challenge which remains is to integrate laboratory and field studies in such a way that current theories regarding the role of ectomycorrhiza in various anthropogenic interactions can be critically tested and extended to predict changes occurring in the field.

Key problems and research priorities

A number of key problems which are relevant to mycorrhizal interactions and soil acidification are outlined below and discussed in relation to suitable techniques which are currently available and research needs which should be given priority. Many of the ideas outlined below have also been proposed by Bledsoe (1992).

Distinguishing mycorrhizal biomass and activity from that of the other soil microorganisms

There are still no really adequate methods to distinguish ectomycorrhizal fungal biomass in the soil from that of other soil fungi. Biomass determinations are thus based on quantification of infected roots. Ectomycorrhiza are easy to enumerate since they form functionally discrete infection units. Determination of the total abundance of these roots, although time consuming using visual methods, is not difficult. Chemical methods, such as the quantification of ergosterol, are not specific for ectomycorrhizal fungi but may be useful in situations where the non-mycorrhizal fungal background can be determined separately. More data is still needed, however, on how ergosterol content differs with mycorrhizal species and age.

Activity of mycorrhizal fungi is easier to distinguish from that of other microorganisms by virtue of the rather well defined interactions and processes of chemical exchange which take place between the fungi and their host plants. A key distinction that needs to be made is that between direct effects on mycorrhizal activity and those which operate indirectly through influence on the host plant.

Distinguishing different mycorrhizal fungi from each other (identification)

Studies of anthropogenic influence on ectomycorrhizal fungi have often reported changes in species composition without overall changes in abundance. Even in situations where overall changes of abundance occur it is essential to know the extent to which different components (species) of the mycorrhizal community contribute to the overall change. It is thus evident that future monitoring methods should not rely totally on overall changes in biomass and activity but should discriminate between different fungi. Molecular methods to do this are now being developed and they will greatly increase the resolution with which changes in mycorrhizal communities can be monitored. The taxonomic resolution appropriate to study of different anthropogenic influences can be varied by using a hierarchical approach involving different methods including sequence analysis, taxon-specific oligonucleotide probes and RFLP pattern analysis and it is likely that as these methods are further developed they will replace traditional immunological methods such as ELISA. Processing of the large numbers of samples involved in field experiments needs to be made more rapid and simple: dot-blot probes and automated DNA profiling of PCR-amplified rRNA gene fragments using capillary electrophoresis (Martin et al. 1993) appear to be promising methods in this respect. The above methods have already been used with some success to monitor the spread and persistence of introduced ectomycorrhizal isolates in forest nurseries and natural forest ecosystems.

Modern molecular methods have a great potential for identification of specific fungi and will undoubtedly be useful in monitoring the distribution and competitive ability of introduced fungi, as well as responses of natural populations to various anthropogenic disturbances. Our ecological and physiological knowledge of these fungi is however still too restricted to interpret or predict changes in community structure occurring as a result of soil acidification. Production of descriptive inventories of fungi with these (expensive) methods is of little interest without more basic information about their physiology. Combined use of new molecular methods with well defined fungal strains or species with known physiological properties is thus an important priority in future work.

Physiological studies of stress tolerance

Detailed physiological experiments under controlled conditions are needed as a complement to field studies in order to be able to identify the functional basis of observed responses to anthropogenic stresses. Lack of standardisation has been a central problem and standardised test systems are needed to evaluate the effects of specific anthropogenic pollutant stresses on different types of mycorrhizal association.

Identification of key soil biological and chemical variables and cataloguing the extent of possible mycorrhizal changes to these under defined conditions will provide a theoretical framework within which to predict future responses to possible anthropogenic impacts. Such a framework will also provide a means of structuring and standardising future field observations. Collection of data in a central database would aid comparison of results. Intraspecific variation in certain parameters is likely to be high and it is therefore essential to relate edaphic characteristics of different sites to the particular strains of fungi occupying them. Responses of pure cultures grown under laboratory conditions will not necessarily reflect those of fungi growing in symbiosis with host plants and in situ tests with different hosts should be developed where possible. Examples of specific variables which can be usefully measured include tolerance of different metal ions, ability to utilise different inorganic and organic substrates and responses to different pesticides and levels of N fertilisation.

Large-scale manipulative experiments

Characterisation of physiological features of different fungi under standardised soil conditions and with defined host plants should permit better prediction of likely responses to different anthropogenic stresses. Competition between fungal strains for root infection sites will be governed by intrinsic factors but also by interactions with other soil microorganisms. Detailed study of these interactions is therefore an important priority in future studies. The biological background of soil microorganisms is likely to be very variable and large-scale manipulative experiments are therefore likely to be useful in evaluating the range of responses to particular stresses shown in different sites.

Selective management of indigenous and introduced fungi

One of the ultimate goals of mycorrhizal research should be the selective management of indigenous and introduced ectomycorrhizal fungi to improve tree growth and minimise the detrimental effects of anthropogenic stresses. At present little is known about why particular mycorrhizal strains are more efficient at stimulating growth than others. Genetic selection of fungal strains which is now in progress in Canada and France should make an important contribution to future application of efficient fungal strains in the field. Further, more detailed knowledge about the functional mechanisms underpinning responses of different fungal strains to different anthropogenic stresses is a pre-requisite for development of breeding programmes and application of genetically engineered ectomycorrhizal strains in the field, as well as better management of indigenous fungal populations. Ultimately more information is required about the factors promoting and maintaining ecological diversity of mycorrhizal fungi in natural stands and once the functional relationships of these fungi are better understood the consequences of disappearance of particular species for tree growth and nutrition will be easier to predict.

Concluding remarks

Recently developed techniques are now beginning to open up new possibilities for identification and manipulation of hitherto unidentified mycorrhizal species. Ectomycorrhizal associations play an essential role in the nutrition of economically important tree species and are an integral component of all natural forest ecosystems. As well as being influenced themselves by anthropogenic effects such as acidification, ectomycorrhizal fungi appear to have an important influence on the ability of their host trees to withstand the stresses associated with pollution. The functional basis of the nutritional interactions between individual plants and ectomycorrhizal fungi is relatively well understood in comparison to most other rhizosphere organisms but our understanding of physiological differences which exist between different mycorrhizal fungi is still poor. Although new molecular tools will help us to observe changes in fungal community structure, parallel physiological studies of the fungi are required to help interpret the functional basis of the changes and to enable us to predict responses to particular pollutant loads. The biological interactions between the symbiotic partners are well understood at the individual plant level. More knowledge is required concerning the significance of these symbiotic interactions at the level of the plant community and the way in which they influence, or are influenced by, other components of the ecosystem or anthropogenic stresses. Progress in the above areas is an essential step before we are better able to describe and predict the dynamics of terrestrial ecosystems and their responses to anthropogenically generated stresses.

Lack of standardisation has been a serious problem in previous studies and standardised tests, in which defined processes can be studied under controlled conditions, with model plants, soils and mycobionts, must be developed without delay.

Acknowledgements – The Swedish Environmental Protection Agency provided financial support for many of the studies of mine referred to in this review. I. Kottke (Eberhard-Karls-Univ. Tübingen, Germany) kindly assisted me with the EELS analysis and her help is also gratefully acknowledged.

References

Abuzinadah, R. A. and Read, D. J. 1986. The role of proteins in the nitrogen nutrition of ectomycorrhizal plants. II. Utilization of protein by mycorrhizal plants of *Pinus contorta*. – New Phytol. 103: 495–506.
– and Read, D. J. 1989. Carbon transfer associated with assimilation of organic nitrogen sources by silver birch (*Betula pendula* Roth.) – Trees 3: 17–23.
– , Finlay, R. D. and Read, D. J. 1986. The role of proteins in the nitrogen nutrition of ectomycorrhizal plants. I. Utilization of peptides and proteins by ectomycorrhizal fungi. – New Phytol. 103: 481–493.
Agerer, R. 1991. Characterization of ectomycorrhizae. – In: Norris, J. R., Read, D. J. and Varma, A. K. (eds), Methods in

microbiology Vol. 23. Techniques for the study of mycorrhizae. Academic Press, London, pp. 25–73.

Alexander, I. J. and Fairley, R. I. 1983. Effects of N fertilisation on populations of fine roots and mycorrhizas in spruce humus. – Plant Soil 71: 49–53.

Andersen, C. P. and Rygiewicz, P. T. 1991. Stress interactions and mycorrhizal plant response, understanding carbon allocation priorities. – Environ. Pollut. 73: 217–244.

Andersson, M. 1988. Toxicity and tolerance of aluminium in vascular plants. – Water Air Soil Pollut. 39: 439–462.

Andersson, S. and Söderström, B. 1995. Effects of lime on ectomycorrhizal colonization of *Picea abies* (L.) Karst. seedlings planted in a spruce forest. – Scand. J. For. Res. 00: 000–000.

Antibus, R. K. and Linkins, A. E. 1992. Effects of liming a red pine forest floor on mycorrhizal numbers and mycorrhizal and soil acid phosphatase activities. – Soil Biol. Biochem. 24: 479–487.

Arnebrant, K. and Söderström, B. 1992. Effects of different fertilizer treatments on ectomycorrhizal colonization potential in two Scots pine forests in Sweden. – For. Ecol. Manage. 53: 77–89.

Arnolds, E. 1990. Decline of ectomycorrhizal fungi in Europe. – Agricult. Ecosyst. Environ. 29: 209–244.

Arovaara, H. and Ilvesniemi, H. 1990a. Effects of soluble inorganic aluminium on the growth and nutrient concentrations of *Pinus sylvestris* and *Picea abies* seedlings. – Scand. J. For. Res. 5: 49–57.

– and Ilvesniemi, H. 1990b. The effects of soluble inorganic aluminium and nutrient imbalances on *Pinus sylvestris* and *Picea abies* seedlings. – In: Kauppi, P., Anttila, P. and Kenttamies, K. (eds), Acidification in Finland. Springer, Berlin, pp. 715–733.

Asp, H., Bengtsson, B. and Jensén, P. 1988. Growth and cation uptake in spruce (*Picea abies* Karst.) grown in sand culture with various aluminium contents. – Plant Soil 111: 127–133.

Barnes, R. B. 1975. The determination of specific forms of aluminium in natural water. – Chem. Geol. 15: 177–191.

Bengtsson, B. 1992. Influence of aluminium and nitrogen on uptake and distribution of minerals in beech roots (*Fagus sylvatica*). – Vegetatio 101: 35–41.

– , Asp, H., Jensén, P. and Berggren, D. 1988. Influence of aluminium on phosphate and calcium uptake in beech (*Fagus sylvatica*) grown in nutrient solution and soil solution. – Physiol. Plant. 74: 299–305.

Berg, B. and Söderström, B. 1979. Fungal biomass and nitrogen in decomposing Scots pine needle litter. – Soil Biol. Biochem. 11: 339–341.

– and Staaf, H. 1981. Leaching accumulation and release of nitrogen in decomposing forest litter. – In: Clarke, F. E. and Rosswall, T. (eds), Terrestrial nitrogen cycles. Ecol. Bull. (Copenhagen) 33: 163–178.

Bergkvist, B. 1987. Soil solution chemistry and metal budgets of spruce forest ecosystems in south Sweden. – Water Air Soil Pollut. 33: 131–154.

Berry, C. R. 1982. Survival and growth of pine hybrid seedlings with *Pisolithus* ectomycorrhizae on coal spoils in Alabama and Tennessee. – J. Environ. Qual. 11: 709–715.

Bledsoe, C. S. 1992. Physiological ecology of ectomycorrhizae: implications for field application. – In: Allen, M. F. (ed.), Mycorrhizal functioning. Chapman and Hall, London, pp. 424–437.

Brown, M. T. and Wilkins, D. A. 1985. Zinc tolerance of *Amanita* and *Paxillus*. – Trans. Brit. Mycol. Soc. 84: 367–369.

Browning, M. H. R. and Hutchinson, T. C. 1991. The effects of aluminium and calcium on the growth and nutrition of selected ectomycorrhizal fungi of jack pine. – Can. J. Bot. 69: 1691–1699.

Chalot, M., Brun, A., Finlay, R. D. and Söderström, B. 1994a. Metabolism of [¹⁴C]glutamate and [¹⁴C]glutamine by the ectomycorrhizal fungus *Paxillus involutus*. – Microbiology 140: 1641–1649.

– , Brun, A., Finlay, R. D. and Söderström, B. 1994b. Respiration of [¹⁴C]alanine by the ectomycorrhizal fungus *Paxillus involutus*. – FEMS Microbiol. Lett. 121: 87–92.

– , Kytöviita, M. M., Finlay, R. D. and Söderström, B. 1995. Factors affecting amino acid uptake in the ectomycorrhizal fungus *Paxillus involutus*. – Mycol. Res. 00: 000–000.

Clarke, N., Danielsson, L.-G. and Sparén, A. 1992. The determination of quickly reacting aluminium in natural waters by kinetic discrimination in a flow system. – Int. J. Anal. Chem. 48: 77–100.

Colpaert, J. V. and Van Assche, J. A. 1987. Heavy metal tolerance in some ectomycorrhizal fungi. – Funct. Ecol. 1: 415–421.

– and Van Assche, J. A. 1992. The effects of cadmium and the cadmium-zinc interaction on the axenic growth of ectomycorrhizal fungi. – Plant Soil 145: 237–243.

– and Van Assche, J. A. 1993. The effects of cadmium on ectomycorrhizal *Pinus sylvestris* L. – New Phytol. 123: 325–333.

Cromack, K., Sollins, P., Graustein, W. C., Speidel, K., Todd, A. W., Spycher, G., Li, C. Y. and Todd, R. L. 1979. Calcium oxalate accumulation and soil weathering in mats of the hypogeous fungus *Hysterangium crassum*. – Soil Biol. Biochem. 11: 463–468.

Cumming, J. R. and Weinstein, L. H. 1990a. Aluminium-mycorrhiza interactions in the physiology of pitch pine seedlings. – Plant Soil 125: 7–18.

– and Weinstein, L. H. 1990b. Nitrogen source effects on Al toxicity in nonmycorrhizal and mycorrhizal pitch pine (*Pinus rigida*) seedlings. I. Growth and nutrition. – Can. J. Bot. 68: 2644–2652.

– and Weinstein, L. H. 1990c. Utilization of AlPO₄ as a phosphorus source by ectomycorrhizal (*Pinus rigida*) seedlings. – New Phytol. 116: 99–106.

Danielson, R. M. and Visser, S. 1989. Effects of forest soil acidification on ectomycorrhizal and vesicular-arbuscular mycorrhizal development. – New Phytol. 112: 41–47.

Deans, J. D., Leith, I. D., Sheppard, J. N., Cape, J. N., Fowler, D., Murray, M. B. and Mason, P. A. 1990. The influence of acid mists on growth, dry matter partitioning, nutrient concentrations and mycorrhizal fruiting bodies in red spruce seedlings. – New Phytol. 115: 459–464.

Denny, H. J. and Wilkins, D. A. 1987a. Zinc tolerance in *Betula* spp. III. Variation in response to zinc among ectomycorrhizal associates. – New Phytol. 106: 535–544.

– and Wilkins, D. A. 1987b. Zinc tolerance in *Betula* spp. IV. The mechanism of ectomycorrhizal amelioration of zinc toxicity. – New Phytol. 106: 545–553.

Devêvre, O., Garbaye, J., Le Tacon, F., Perrin, R. and Estivalet, D. 1994. Role of rhizosphere microfungi in the decline of Norway spruce in acidic soils. – In: Landmann, G. and Bonneau, M. (eds), Forest decline and atmospheric deposition effects in the french mountains. Springer, Berlin, pp. 000–000.

Dighton, J. and Skeffington, R. A. 1987. Effects of artificial acid precipitation on the mycorrhizas of Scots pine seedlings. – New Phytol. 107: 191–202.

– and Jansen, A.-E. 1991. Atmospheric pollutants and ectomycorrhizae – more questions than answers. – Environ. Pollut. 73: 179–204.

Donner, V. B. and Heyser, W. 1989. Buchenmycorrhizen, Möglichkeiten der Elementselektion unter besonderer Berücksichtigung einiger Schwermetalle. – Forstw. Cbl. 108: 150–163.

Eeckhaoudt, S., Vandeputte, D., Van Praag, H., Van Grieken, R. and Jacob, W. 1992. Laser microprobe mass analysis (LAMMA) of aluminium and lead in fine roots and their ectomycorrhizal mantles of Norway spruce (*Picea abies* (L.) Karst.). – Tree Physiol. 10: 209–215.

Entry, J. A., Rose, C. L. and Cromack, K. 1992. Microbial biomass and nutrient concentrations in hyphal mats of the ectomycorrhizal fungus *Hysterangium setchellii* in a coniferous forest soil. – Soil Biol. Biochem. 24: 447–453.

Egerton-Warburton, L.M., Kuo, J., Griffin, B.J. and Lamont, B.B. 1993. The effect of aluminium on the distribution of calcium, magnesium and phosphorus in mycorrhizal and non-mycorrhizal seedlings of *Eucalyptus rudis*: a cryomicroanalytical study. – Plant Soil 155/156: 481–484.

Erland, S. 1992. Effects of liming on pine ectomycorrhiza. – In: Read, D.J., Lewis, D.H., Fitter, A.H. and Alexander, I.J. (eds), Mycorrhizas in ecosystems. C.A.B. Intern., Wallingford, pp. 113–118.

– and Söderström, B. 1990. Effects of liming on ectomycorrhizal fungi infecting *Pinus sylvestris* L. I. Mycorrhizal infection in limed humus in the laboratory and isolation of fungi from mycorrhizal roots. – New Phytol. 115: 675–682.

– and Söderström, B. 1991a. Effects of lime and ash treatments on ectomycorrhizal infection of *Pinus sylvestris* L. seedlings planted in a pine forest. – Scand. J. For. Res. 6: 519–525.

– and Söderström, B. 1991b. Effects of liming on ectomycorrhizal fungi infecting *Pinus sylvestris* L. III. Saprophytic growth and host plant infection at different pH values in unsterile humus. – New Phytol. 117: 405–411.

–, Söderström, B. and Andersson, S. 1990. Effects of liming on ectomycorrhizal fungi infecting *Pinus sylvestris* II. Growth rates in pure culture at different pH values compared to growth rates in symbiosis with the host plant. – New Phytol. 115: 683–688.

–, Finlay, R.D. and Söderström, B. 1991. The influence of substrate pH on carbon translocation in ectomycorrhizal and non-mycorrhizal pine seedlings. – New Phytol. 119: 235–242.

–, Henrion, B., Martin, F., Glover, L.A. and Alexander I.J. 1994. Identification of the mycorrhizal basidiomycete *Tylospora fibrillosa* Donk. by RFLP analysis of parts of the PCR-amplified ITS and IGS regions of ribosomal DNA. – New Phytol. 126: 525–532.

Falkengren-Grerup, U. 1987. Long-term changes in pH of forest soils in southern Sweden. – Environ. Pollut. (Ser. A) 43: 79–90.

–, Linnermark, N. and Tyler, G. 1987. Changes in acidity and cation pools of south Swedish soils between 1949 and 1985. – Chemosphere 16: 2239–2248.

Finlay, R.D. 1989. Functional aspects of phosphorus uptake and carbon translocation in incompatible ectomycorrhizal associations between *Pinus sylvestris* and *Suillus grevillei* and *Boletinus cavipes*. – New Phytol. 112: 185–192.

– 1992. Uptake and mycelial translocation of nutrients by ectomycorrhizal fungi. – In: Read, D.J., Lewis, D.H., Fitter, A.H. and Alexander, I.J. (eds), Mycorrhizas in ecosystems. C.A.B. International, Wallingford, pp. 91–97.

– and Read, D.J. 1986a. The structure and function of the vegetative mycelium of ectomycorrhizal plants. II. The uptake and distribution of phosphorus by mycelial strands interconnecting host plants. – New Phytol. 103: 157–165.

– and Read, D.J. 1986b. The structure and function of the vegetative mycelium of ectomycorrhizal plants. I. Translocation of ^{14}C-labelled carbon between plants interconnected by a common mycelium. – New Phytol. 103: 143–156.

– and Söderström, B. 1992. Mycorrhiza and carbon flow to soil. – In: Allen, M.F. (ed.), Mycorrhizal functioning. Chapman and Hall, London, pp. 134–160.

–, Ek, H., Odham, G. and Söderström, B. 1988. Mycelial uptake, translocation and assimilation of nitrogen from ^{15}N-labelled ammonium by *Pinus sylvestris* plants infected with four different ectomycorrhizal fungi. – New Phytol. 110: 59–66.

–, Ek, H., Odham, G. and Söderström, B. 1989. Uptake, translocation and assimilation of nitrogen from ^{15}N-labelled ammonium and nitrate sources by intact ectomycorrhizal

systems of *Fagus sylvatica* infected with *Paxillus involutus*. – New Phytol. 113: 47–55.

–, Frostegård, Å. and Sonnerfelt, A.-M. 1992. Utilization of organic and inorganic nitrogen sources by ectomycorrhizal fungi of different successional stages grown in pure culture and in symbiosis with *Pinus contorta* (Dougl. ex Loud). – New Phytol. 120: 105–115.

–, Bengtsson, B., Jensén, P. and Söderström, B. 1995a. Interactions between exposure to raised aluminium concentrations and ectomycorrhizal infection in *Pinus contorta*, and *Fagus sylvatica* plants grown in a semi-hydroponic culture system. – Tree Physiol. 00: 000–000.

–, Jensén, P. and Söderström, B. 1995b. Effects of elevated soil aluminium on growth and nutrient uptake by mycorrhizal and non-mycorrhizal *Fagus sylvatica* and *Pinus contorta* plants. – Plant Soil 00: 000–000.

–, Söderström, B. and Jensén, P. 1995c. Mycelial uptake and translocation of [45]Ca, ^{86}Rb and ^{32}P in ectomycorrhizal associations between *Fagus sylvatica* and *Paxillus involutus* exposed to different levels of soil aluminium. – Plant Soil 00: 000–000.

Garbaye, J., Churin, J.-L. and Duponnois, R. 1992. Effects of substrate sterilization, fungicide treatment and mycorrhization helper bacteria on ectomycorrhizal formation of pedunculate oak (*Quercus robur*) innoculated with *Laccaria laccata* in two peat bare-root nurseries. – Biol. Fert. Soils 13: 55–57.

Gardes, M. and Bruns, T.D. 1993. ITS primers with enhanced specificity for basidiomycetes – application to the identification of mycorrhizae and rusts. – Mol. Ecol. 2: 113–118.

–, White, T.J., Fortin, J.A., Bruns, T.D. and Taylor, J.W. 1991. Identification of indigenous and introduced symbiotic fungi in ectomycorrhizae by amplification of nuclear and mitochondrial ribosomal DNA. – Can. J. Bot. 69: 180–190.

Göransson, A. and Eldhuset, T.D. 1991. Effects of aluminium on growth and nutrient uptake of small *Picea abies* and *Pinus sylvestris* plants. – Trees 5: 136–142.

Gorissen, A., Joosten, N.N. and Jansen, A.E. 1991. Effects of ozone and ammonium sulphate on carbon partitioning to mycorrhizal roots of juvenile Douglas fir. – New Phytol. 119: 243–250.

Graustein, W.C., Cromack, Jr. K. and Sollins, P. 1977. Calcium oxalate: Occurrence in soils and effect on nutrient and geochemical cycles. – Science 198: 1252–1254.

Griffiths, R.P., Baham, J.E. and Caldwell, B.A. 1994. Soil solution chemistry of ectomycorrhizal mats in forest soil. – Soil Biol. Biochem. 26: 331–337.

Hallbäcken, L. and Tamm, C.O. 1986. Changes in soil acidity from 1927 to 1982–1984 in a forest area of south-west Sweden. – Scand. J. For. Res. 1: 219–232.

Hallingbäck, T. 1994. The macrofungi of Sweden. – Swed. Environ. Protect. Agency, Solna, Report No. 4313, (in Swedish).

Harley, J.H. and Smith, S.E. 1983. Mycorrhizal symbiosis. – Academic Press, London.

Henrion, B., Le Tacon, F. and Martin, F. 1992. Rapid identification of genetic variation of ectomycorrhizal fungi by amplification of ribosomal RNA genes. – New Phytol. 122: 289–298.

Hentschel, E., Godbold, D.L., Marschner, P., Schlegel, H. and Jentschke, G. 1993. The effects of *Paxillus involutus* Fr. on aluminium sensitivity of Norway spruce seedlings. – Tree Physiol. 12: 379–390.

Hintikka, V. 1988. High aluminium tolerance among ectomycorrhizal fungi. – Karstenia 28: 41–44.

Hodson, M.J. and Wilkins, D.A. 1991. Localization of aluminium in the roots of Norway spruce [*Picea abies* (L.) Karst.] inoculated with *Paxillus involutus* Fr. – New Phytol. 118: 273–278.

Högberg, P. 1989. Growth and nitrogen inflow rates in mycorrhizal and non-mycorrhizal seedlings of *Pinus sylvestris*. – For. Ecol. Manage. 28: 7–17.

– 1990. ^{15}N natural abundance as a possible marker of the ectomycorrhizal habit of trees in mixed African woodlands. – New Phytol. 115: 483–486.

– 1991. Development of ^{15}N enrichment in a nitrogen-fertilized forest soil-plant system. – Soil Biol. Biochem. 24: 335–338.

– and Jensén, P. 1994. Aluminium and uptake of base cations by tree roots: A critique of the model proposed by Sverdrup et al. – Water Air Soil Pollut. 75: 121–125.

–, Tamm, C.O. and Högberg, M. 1992. Variations in ^{15}N abundance in a forest fertilization trial – Critical loads of N, N-saturation, contamination and effects of revitalization fertilization. – Plant Soil 142: 211–219.

Holopainen, T., Kottke, I., Martin, F. and Turnau, K. 1992. Polyphosphates as detoxifying agents in mycorrhizas. – Proc. int. symp. manage. mycorrhizas Agricult., Horticult. Forestry, Univ. Western Australia, Nedlands.

Hue, N.V., Craddock, G.R. and Adams, F. 1986. Effect of organic acids on aluminium toxicity in subsoils. – Soil Sci. Soc. Am. J. 50: 28–34.

Hult, M., Bengtsson, B., Larsson, N.P. and Yang, C. 1992. Particle-induced X-ray emission microanalysis of root samples from beech (*Fagus sylvatica*). – Scanning Microsc. 6: 581–592.

Hung, L. and Trappe, J.M. 1983. Growth variation between and within species of ectomycorrhizal fungi in response to pH in vitro. – Mycologia 75: 234–241.

Jansen, A.E. and de Nie, H.W. 1988. Relations between mycorrhizas and fruitbodies of mycorrhizal fungi in Douglas fir plantations in The Netherlands. – Acta Bot. Neerl. 37: 243–250.

– and de Vries, F.W. 1988. Qualitative and quantitative research on the relation between ectomycorrhizae of *Pseudotsuga menziesii*, vitality of the host and acid rain. – Dutch priority program. acid., Report 25–02, Wageningen Agricult. Univ., The Netherlands.

Jentschke, G., Godbold, D.L. and Hüttermann, A. 1991. Culture of mycorrhizal tree seedlings under controlled conditions: Effects of nitrogen and aluminium. – Physiol. Plant. 81: 408–416.

Jones, M.D. and Hutchinson, T.C. 1986. The effect of mycorrhizal infection on the response of *Betula papyrifera* to nickel and copper. – New Phytol. 102: 429–442.

Jongbloed, R.H. and Borst-Pauwels, G.W.F. 1992. Effects of aluminium and pH on growth and potassium uptake by three ectomycorrhizal fungi in liquid culture. – Plant Soil 140: 157–165.

–, Tosserams, M.W.A. and Borst-Pauwels, G.W.F. 1992. The effect of aluminium on phosphate uptake by three isolated ectomycorrhizal fungi. – Plant Soil 140: 167–174.

Kauppi, P., Anttila, P. and Kenttamies, K. (eds) 1990. Acidification in Finland. – Springer, New York.

Kottke, I. 1991. Electron energy loss spectroscopy and imaging techniques for subcellular localization of elements in mycorrhiza. – In: Norris, J.R., Read, D.J. and Varma, A.K. (eds), Methods in microbiology Vol. 23, Techniques for the study of mycorrhizae. Academic Press, London, pp. 369–382.

– and Martin, F. 1994. Demonstration of aluminium in polyphosphate of *Laccaria amethystea* (Bolt. ex Hooker) Murr. by means of electron energy-loss spectroscopy. – J. Microsc. 174: 225–232.

Landmann, G. and Bonneau, M. (eds) 1994. Forest decline and atmospheric deposition effects in the French mountains. – Springer, Berlin.

Lapeyrie, F., Chilvers, G.A. and Behm, C.A. 1987. Oxalic acid synthesis by the mycorrhizal fungus *Paxillus involutus* (Batsch. ex. Fr.) Fr. – New Phytol. 106: 139–146.

Lehto, T. 1984. The effect of liming on the mycorrhizae of Scots Pine. – Fol. Forest. 609: 1–20.

Leyval, C. and Berthelin, J. 1989. Interactions between *Laccaria laccata*, *Agrobacterium radiobacter* and beech roots: Influence on P, K, Mg and Fe mobilization from minerals and plant growth. – Plant Soil 117: 103–110.

– and Berthelin, J. 1991. Weathering of a mica by roots and rhizospheric microorganisms of pine. – Soil Sci. Soc. Am. J. 55: 1009–1016.

– and Berthelin, J. 1993. Rhizodeposition and net release of soluble organic compounds by pine and beech seedlings inoculated with rhizobacteria and ectomycorrhizal fungi. – Biol. Fertil. Soils 15: 259–267.

Lövblad, G., Amann, M., Andersen, B., Hovmand, S.J. and Pedersen, U. 1992. Deposition of sulfur and nitrogen in the Nordic countries: present and future. – Ambio 21: 339–347.

Lyngstad, I. 1992. Effect of liming on mineralization of soil nitrogen as measured by plant uptake and nitrogen released during incubation. – Plant Soil 144: 247–253.

Martin, F., Vairelles, D. and Henrion, B. 1993. Automated ribosomal DNA fingerprinting by capillary electrophoresis of PCR products. – Anal. Biochem. 214: 182–189.

–, Rubini, P., Coté, R. and Kottke, I. 1994. Aluminium polyphosphate complexes in the mycorrhizal basidiomycete *Laccaria bicolor*: a ^{27}Al-nuclear magnetic resonance study. – Planta 194: 241–246.

Marschner, H. 1991. Mechanisms of adaptation of plants to acid soils. – Plant Soil 134: 1–20.

Marx, D.H. and Artman, J.D. 1979. *Pisolithus tinctorius* ectomycorrhiza improve survival and growth of pine seedlings on acid coal spoils in Kentucky and Virginia. – Reclamation Review 2: 23–31.

McFall, J.S., Slack, S.A. and Wehrli, S. 1992. Phosphorus distribution in red pine roots and the ectomycorrhizal fungus *Hebeloma arenosa*. – Plant Physiol. 100: 713–717.

McQuattie, C.J. and Schier, G.A. 1992. Effect of ozone and aluminium on pitch pine seedlings: anatomy of mycorrhizae. – Can. J. For. Res. 22: 1901–1916.

Meier, S., Robarge, W.P., Bruck, R.I. and Grand, L.F. 1989a. Effects of simulated rain acidity on ectomycorrhizae of red spruce seedlings potted in natural soil. – Environ. Pollut. 59: 205–216.

–, Robarge, W.P., Bruck, R.I. and Grand, L.F. 1989b. Effects of simulated rain acidity on ectomycorrhizae of red spruce seedlings potted in natural soil. – Environ. Pollut. 59: 315–24.

Melin, E., Nilsson, H. and Hacskaylo, E. 1958. Translocation of cations to seedlings of *Pinus virginiana* through mycorrhizal mycelium. – Bot. Gaz. 119: 243–246.

Morselt, A.F.W., Smits, W.T.M. and Limonard, T. 1986. Histochemical demonstration of heavy metal tolerance in ectomycorrhizal fungi. – Plant Soil 96: 417–420.

Norby, R.J., O'Neill, E.G., Gregory Hood, W. and Luxmoore, R.J. 1987. Carbon allocation, root exudation and mycorrhizal colonization of *Pinus echinata* seedlings grown under CO_2 enrichment. – Tree Physiol. 3: 203–210.

Nylund, J. and Wallander, H. 1989. Effects of ectomycorrhiza on host growth and carbon balance in a semi-hydroponic cultivation system. – New Phytol. 112: 389–398.

Ohtonen, R., Markkola, A.M., Heinonen-Tanski, H. and Fritze, H. 1990. Soil biological parameters as indicators of change in Scots pine forests (*Pinus sylvestris* L.) caused by air pollution. – In: Kauppi, P., Anttila, P. and Kenttämies, K. (eds), Acidification in Finland. Springer, Berlin, pp. 373–393.

O'Neill, E.G., Luxmoore, R.J. and Norby, R.J. 1987. Increases in mycorrhizal colonization and seedling growth in *Pinus echinata* and *Quercus alba* in an enriched carbon dioxide atmosphere. – Can. J. For. Res. 17: 878–883.

Orlovich, D.A. and Ashford, A.E. 1993. Polyphosphate granules are an artefact of specimen preparation in the ectomycorrhizal fungus *Pisolithus tinctorius*. – Protoplasma 173: 91–102.

Rapp, C. 1992. Effects of liming and N-fertilization on ectomycorrhizas in a mature beech stand in the Solling area Germany. – In: Read, D.J., Lewis, D.H., Fitter, A.H. and

Alexander, I. J. (eds), Mycorrhizas in ecosystems. C.A.B. Intern., Wallingford, pp. 397–???.

Read, D. J. 1991. Mycorrhizas in ecosystems. – Experientia 47: 376–391.

Reich, P. B., Schoettle, A. W., Stroo, H. F., Troiano, J. and Amundson, R. G. 1985. Effects of O_3, SO_2 and acidic rain on mycorrhizal infection in northern red oak seedlings. – Can. J. Bot. 63: 2049–2055.

– , Schoettle, A. W., Stroo, H. F. and Amundson, R. G. 1986. Acid rain and ozone influence mycorrhizal infection in tree seedlings. – J. Air Poll. Cont. Ass. 36: 724–726.

Reid, C. P. P., Kidd, F. A. and Ekwebelam, S. A. 1983. Nitrogen nutrition, photosynthesis and carbon allocation in ectomycorrhizal pine. – Plant Soil 71: 415–431.

Rygiewicz, P. T. and Andersen, C. P. 1994. Mycorrhizae alter quality and quantity of carbon allocated below ground. – Nature 369: 58–60.

– , Bledsoe, C. S. and Zasoski, R. J. 1984a. Effects of ectomycorrhizae and solution pH on (^{15}N) ammonium uptake by coniferous seedlings. – Can. J. For. Res. 14: 885–892.

– , Bledsoe, C. S. and Zasoski, R. J. 1984b. Effects of ectomycorrhizae and solution pH on (^{15}N) nitrate uptake by coniferous seedlings. – Can. J. For. Res. 14: 893–899.

Rühling, Å. and Söderström, B. 1990. Changes in fruitbody production of mycorrhizal and litter decomposing macromycetes in heavy metal polluted coniferous forests in north Sweden. – Water Air Soil Pollut. 49: 375–387.

– , Bååth, E., Nordgren, A. and Söderström, B. 1984. Fungi in metal-contaminated soil near the Gusum brass mill, Sweden. – Ambio 13: 34–36.

Schulze, E.-D., Lange, O. L. and Oren, R. (eds) 1989. Ecological studies Vol. 77. – Sringer, Berlin.

Semjonova, L. 1990. Effect of liming on mycorrhiza formation in *Pinus sylvestris* and *Picea abies*. – Agricult. Ecosyst. Environ. 28: 459–462.

Shafer, S. R., Grand, L. F., Bruck, R. I. and Heagle, A. S. 1985. Formation of ectomycorrhizae on *Pinus taeda* seedlings exposed to simulated acidic rain. – Can. J. For. Res. 15: 66–71.

Shaw, P. J. A., Dighton, J., Poskitt, J. and McLeod, A. R. 1992. The effects of sulphur dioxide and ozone on mycorrhizas of Scots pine and Norway spruce in a field fumigation system. – Mycol. Res. 96: 785–791.

– , Dighton, J. and Poskitt, J. 1993. Studies on the mycorrhizal community infecting trees in the Liphook Forest fumigation experiment. – Agricult. Econ. 47: 185–191.

Stroo, H. F. and Alexander, M. 1985. Effect of simulated acid rain on mycorrhizal infection of *Pinus strobus* L. – Water Air Soil Pollut. 25: 107–114.

– , Reich, P. B. and Schoettle, A. W. 1988. Effects of ozone and acid rain on white pine (*Pinus strobus*) seedlings grown in five soils. II. Mycorrhizal infection. – Can. J. Bot. 66: 1510–1516.

Sverdrup, H., Warfvinge, P. and Rosén, K. 1992. A model for the impact of soil solution Ca:Al ratio, soil moisture and temperature on tree base cation uptake. – Water Air Soil Pollut. 61: 365–383.

Söderström, B. 1984. Some microbial data on acidification and liming of forest soils. A literature review. – Swed. Nat. Environ. Protect. Board Report. No. PM 1860, Solna, Sweden.

– and Read, D. J. 1987. Respiratory activity of intact and excised ectomycorrhizal mycelial systems growing in unsterilized soil. – Soil Biol. Biochem. 19: 231–237.

Taylor, A. F. S. and Brand, F. 1992. Reaction of the natural Norway spruce mycorrhizal flora to liming and acid irrigation. – In: Read, D. J., Lewis, D. H., Fitter, A. H. and Alexander, I. J. (eds), Mycorrhizas in ecosystems. C.A.B. Intern., Wallingford, pp. 404.

Taylor, G. J. 1991. Current views of the aluminum stress response; the physiological basis of tolerance. – Curr. Top. Plant Biochem. Physiol. 10: 57–93.

Turnau, K., Kottke, I. and Oberwinkler, F. 1993. *Paxillus involutus-Pinus sylvestris* mycorrhizae from heavily polluted forest. I. Element localization using electron energy loss spectroscopy and imaging. – Bot. Acta 106: 213–219.

Tyler, G. 1980. Metals in sporophores of basidiomycetes. – Trans. Brit. Mycol. Soc. 74: 41–49.

– 1987. Acidification and chemical properties of *Fagus sylvatica* L. forest soils. – Scand. J. For. Res. 2: 263–271.

– , Berggren, D., Bergkvist, B., Falkengren-Grerup, U., Folkeson, L. and Ruhling, Å. 1987. Soil acidification and metal solubility in forests of southern Sweden. – In: Hutchinson, T. C. and Meema, K. M. (eds), Effects of atmospheric pollutants on forests, wetlands and agricultural ecosystems, NATO ASI Ser., Vol. G16: Springer, Berlin, pp. 347–359.

– , Balsberg-Påhlsson, A.-M., Bergkvist, B., Falkengren-Grerup, U., Folkeson, L., Nihlgård, B., Rühling, Å. and Stjernquist, I. 1992. Chemical and biological effects of artificially increased nitrogen deposition to the ground in a Swedish beech forest. – Scand. J. For. Res. 7: 515–532.

Vogt, K. A., Publicover, D. A., Bloomfield, J., Perez, J. M., Vogt, D. J. and Silver, W. L. 1993. Belowground responses as indicators of environmental change. – Environ. Exp. Bot. 33: 189–205.

Väre, H. 1990. Aluminium polyphosphate in the ectomycorrhizal fungus *Suillus variegatus* (Fr.) O. Kunze as revealed by energy dispersive spectrometry. – New Phytol. 116: 663–668.

Wallander, H. and Nylund, J.-E. 1991. Effects of excess nitrogen on carbohydrate concentration and mycorrhizal development of *Pinus sylvestris* L. seedlings. – New Phytol. 119: 405–411.

– and Nylund, J.-E. 1992. Effects of excess nitrogen and phosphorus starvation on the extramatrical mycelium of ectomycorrhizas of *Pinus sylvestris* L. – New Phytol. 120: 495–503.

Wästerlund, I. 1982. Försvinner tallens mykorrhizasvampar vid gödsling? – Svensk Bot. Tidskr. 76: 411–417, (in Swedish).

Wilkins, D. A. 1991. The influence of sheathing (ecto-)mycorrhizas of trees on the uptake and toxicity of metals. – Agricult. Ecosyst. Environ. 35: 245–260.

– and Hodson, M. J. 1989. The effects of aluminium and *Paxillus involutus* Fr. on the growth of Norway spruce [*Picea abies* (L.) Karst.]. – New Phytol. 113: 225–232.

Winner, W. E. and Atkinson, C. J. 1986. Absorption of air pollution by plants and consequences for growth. – Trends Ecol. Evol. 1: 15–18.

Ecological Bulletins 44: 215–226. Copenhagen 1995

Long-term changes in flora and vegetation in deciduous forests of southern Sweden

Ursula Falkengren-Grerup

Falkengren-Grerup, U. 1995. Long-term changes in flora and vegetation in deciduous forests of southern Sweden. – Ecol. Bull. (Copenhagen) 44: 215–226.

This article focuses on south Swedish research on soil acidification and its impact on field-layer vegetation in beech *Fagus sylvatica*, oak *Quercus robur* and hornbeam *Carpinus betulus* forests. Soil acidification has occurred since at least the 1950's and is of such magnitude that it has been detected over a period of only 10 yr. Acidification can be quantified as a lowering of soil pH, most pronounced in soils with originally high pH. The cations are leached out of the soil and the pools available to plants decrease gradually. The average annual decrease in recent decades has been 1%.

Changes in the occurrence of vascular plants have been investigated mainly by repeating earlier vegetation analyses after 10, 30, 40 and 60 yr. Many typical field-layer species of deciduous forests have been negatively affected, both quantitatively and qualitatively, during recent decades, especially in soils which have become too acid for them. The more intense management of beech forests today has, among other things, allowed certain weed species to establish themselves. If management trends continue in this direction, these species may become part of the "natural" species composition of these forests.

The doubled N-deposition since the 1950's is another factor that, until now, seems to have led to a higher diversity and abundance of several species in deciduous forests. Nowadays N is available in excess in many south Swedish forest soils and NO_3 may leach out of the soil, leading to further acidification and accelerated leaching of other nutrients. The form of inorganic N available to plants may limit growth. Acid soils have a lower nitrification potential and NH_4 as the sole N-form may be inadequate for plants usually growing in less acid soils. For some species the hydrogen ion concentration per se seems to be crucial, but decreased growth and survival may also result from a high Al concentration or Al/Ca ratio.

U. Falkengren-Grerup, Soil-Plant Res., Lund Univ., Ecology Building, S-223 62 Lund, Sweden

Introduction

At the beginning of the 1980's there was little evidence in forest ecosystems of changes, in soil chemistry and in vegetation, related to anthropogenic deposition of acidifying substances. Soil acidification was subsequently observed in central and north Europe (see Nilsson and Tyler 1995) and during the second half of the 1980's the impact of soil acidification and nitrogen deposition on vegetation became apparent in many European countries.

By repeating older studies it was shown that the field-layer vegetation had changed in a way that suggested the cause to be a more acid root environment. Examples of such studies, except for Sweden which are given below, are for Germany (Ballach et al. 1985, Wittig et al. 1985,

Rost-Siebert and Jahn 1988), Switzerland (Kuhn et al. 1987), Great Britain (Tregenza 1986), Norway (Bjørnstad 1991), France (Becker et al. 1992) and the Netherlands (van Breemen and van Dijk 1988). However, many of these studies also indicate some influence, or even a major influence, of an increased nitrogen deposition in recent decades.

Because the management of forests has changed considerably in central Europe, the effects of a changed chemical environment may be difficult to separate from those of new management practices. Raking of litter has been abandoned, as has the coppicing of trees: the combined effects of such changes have been discussed by Wilmans and Bogenrieder 1986, Kuhn et al. 1987 and Wilmans 1989, for example.

Table 1. Mean pH-changes in the humus layer of deciduous forests determined by making repeated measurements over three different time periods or by comparing forests of similar type at two different times (calculated from Falkengren-Grerup 1986, 1990a, Falkengren-Grerup et al. 1987, Falkengren-Grerup and Eriksson 1990).

First study	Second study	No. of years between studies	No. of sites	pH-change	Extraction
1980	1988	8	80	+0.03	0.2 M KCl
1965/70	1985	15–20	23	−0.20	H_2O
1949/54	1984	30–35	61	−0.75	H_2O
1929/54	1979–88	30–60	105,132	−0.65	H_2O

Another way to indicate environmental changes was used by Ellenberg (1985) who compared indices of soil chemistry, temperature and light for threatened species with those for non-threatened species. Increased nitrogen and soil acidity were found to be among the threats.

Laboratory experiments have long been used to explain physiological and ecological amplitudes of species, e.g. with respect to pH and to various metals. Recent studies have focused on pH, Al and Mn, usually at field relevant soil solution concentrations. The importance of other elements (N-form, base cations) in modifying toxic effects, of Al for example, have also been considered (Rorison 1986, Kroetze et al. 1989, Runge and Rode 1991). Field experiments with varying amounts and different forms of nitrogen have also been performed (see Kellner and Redbo-Torstensson 1995).

The aims of this article are to review the research on effects of soil acidification and nitrogen deposition on the field-layer vegetation in south Swedish forest ecosystems, mainly deciduous forests (beech *Fagus sylvatica*, oak *Quercus robur* and hornbeam *Carpinus betulus*). Repeated field studies over several time periods are described, as well as experiments aiming at explaining the importance of single chemical compounds in the acidified and N-enriched soils. The reported experiments are designed to be relevant for south Swedish conditions. Forest management during recent centuries is described, in order to elucidate its possible effects on the vegetation. Some results on soil acidification are given as a background for the development of vegetation.

History of deciduous forests

In the southernmost province, Skåne, the areas covered by forests and open fields are about the same today as they were at the end of the 17th century, seen in early maps of forest distribution. The most fertile soils, mainly the moraine clays on the coastal plains, were already completely cultivated. There were vast beech forests on the nutrient-poor moraines (inland), usually common land belonging to the farmers who extracted some timber and allowed their cattle to graze in the forests. The tall-stemmed beech forests owned by the noble families were often fenced for game, horses or cattle, and the rearing of

pigs which grazed the beech mast was of great economic importance (Campbell 1928, Bergendorff and Emanuelsson 1982). Intensive felling, top cutting and grazing reduced the forested area during the 18th century.

The fields close to the villages provided hay and leaves for the cattle together with firewood and fencing material, from oak, hornbeam, lime *Tilia cordata*, ash *Fraxinus excelsior*, hazel *Corylus avellana* and birch *Betula* spp., for example. These wooded meadows covered large areas in the 18th century (Campbell 1928, Bergendorff and Emanuelsson 1982). Beech was, however, a rare species in these meadows as it does not tolerate repeated cutting of twigs and branches (Pott 1985). In the 19th century most of the wooded meadows were abandoned, as a result of new laws of land ownership and of new cultivation methods. Agriculture took over and only a minor part of the land became forested (Anon. 1982). The forested area in historical times was smallest between 1860 and 1880 (Bergman 1960, Persson 1971).

The grazing in forests ceased gradually during the second half of the 19th century and timber production became the main interest (Petersens 1932, Bergman 1960). The area covered by deciduous forests increased, with the dominant species being beech, which covered 80 000 ha in southern Sweden between 1920 and 1950. The deciduous forests were not profitable in the 1960's and the beech forest area was reduced by 25% (Anon. 1971, 1982) and mainly replanted with spruce. The area of spruce increased by 1000 ha yr^{-1} between 1920 and 1985 (Lindgren and Malmer 1970, Anon. 1990). Spruce was also planted on former pastures, fields and arable land. A law was introduced in 1974 to maintain the beech forests, and the area today is almost stable. Another law was introduced in 1984 to save the remaining areas of hardwoods.

Changes in soil chemistry

The review by Nilsson and Tyler (1995) is concerned with studies of soil acidification. Data for south Sweden, which were collected mainly to describe soil-vegetation dynamics during recent decades, will be summarized below, to facilitate understanding of causal relationships.

The pH in the humus layer has decreased since the

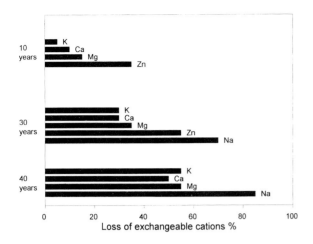

Fig. 1. Loss of exchangeable cations (%) in forest soils analyzed at 10-, 30- and 40-yr intervals. Base cations were extracted using 1 M NH₄Ac, pH 7.0 (topsoil) or pH 4.8 (other depths). The 10-yr study was made on the humus layer in beech forests (n = 264); the 30-year study included soil profiles of 0–100 cm depth from forests with different tree species (n = 10); and the 40-yr study was made on the C-horizons (70–80 cm depth) in beech and oak forests (n = 19; Falkengren-Grerup et al. 1987, Falkengren-Grerup and Eriksson 1990, Falkengren-Grerup and Tyler 1991a).

1950's, especially in soils with a relatively high initial pH (Falkengren-Grerup 1986, 1987). The decrease was largest over an interval of 30–60 yr between the studies (Table 1), while no change was found over a 8 yr period in spite of decreasing base cation pools (Fig. 1). The changes were in, as well as below, the rooting zone (Falkengren-Grerup et al. 1987).

The greatest losses of exchangeable base cations were found for the longest period studied and in soils with inititally large pools of exchangeable ions. There was a large amount of leaching of most elements and only a small proportion was accumulated in the vegetation (Warfvinge et al. 1993). There were also measurable losses over 10 yr (Fig. 1). The exchangeable base cation pools decreased by an average of 1% per year over the whole period and were usually replaced by H- or Al-ions (Falkengren-Grerup et al. 1987, Berggren et al. 1990, Falkengren-Grerup and Tyler 1992a).

Another important effect of a pH decrease is the change in solubility of many trace elements. The solubility of most cationic elements increases substantially between pH 5 and pH 4 in the soil solution. The soil pool of these elements may be strongly reduced when pH decreases below 4, while the soil solution concentrations increase temporarily. This is true especially for pH-labile oxides, hydroxides, exchangeable ions and some organic complexes with Zn, Cd, Ni, Co, Al and Mn (Tyler 1981).

The solubility of Al is especially important for the vegetation. When the concentration of organic compounds in the soil solution is low, the solubility of Al is nearly completely determined by pH. Below pH 4.5, any further decrease in pH results in a higher concentration of Al in the soil solution (Bergkvist 1987). The dominant Al-form in these circumstances is free Al^{3+}, which is potentially toxic to plants in concentrations which are common today (Andersson 1988, Bengtsson et al. 1988).

Bacteria and other microbes in the soil are often affected by soil acidity. Nitrifiers, which transfer ammonium to nitrate, are rarely active in mor and are active only to a small extent in mineral soils with a low pH. Nitrogen mineralization is often assessed under laboratory conditions with favourable temperature and water regimes. In such an experiment, the nitrate production of beech and oak forest soils increased similarly with pH at 0–5 and 5–10 cm soil depth, while the total amount of mineralized nitrogen did not change with pH (Fig. 2). The

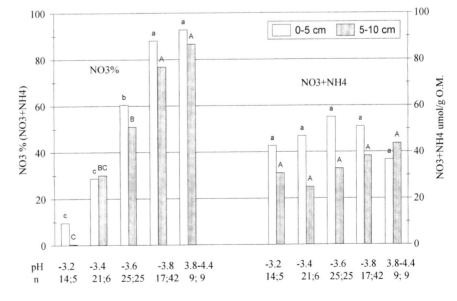

Fig. 2. Nitrate production capacity (NO₃ % (NH₄+NO₃)) and net mineralization (µmol g⁻¹ organic matter, O.M., closely correlated with µmol cm⁻³; r = 0.8–0.9, n = 100) of two layers of beech forest topsoils (0–5 and 5–10 cm, litter excluded) as related to pH(KCl) of the soil. The data were obtained in an 8 wk laboratory incubation experiment at 20°C and 60% water holding capacity. Means within one soil layer which lack common letters differ between pH classes (ONEWAY ANOVA followed by Tukey's test).

Table 2. Distribution of 65 forest species in relation to the median site pH of the humus layer. The number in each species group (herbs, grasses/sedges, ferns, shrubs) is given for three pH(KCl) classes, based on 526 sample sites in deciduous forests studied from 1978. Species present in ≥ 20 of the sites are included (see Table 3).

	No. of species in pH-classes			Total no. of species
	3.3–3.5	3.6–4.0	4.1–4.5	
Herbs	4	17	20	41
Grasses/sedges	5	8	4	17
Ferns	3	2	0	5
Shrubs	2	0	0	2

form of plant-available nitrogen in the soil may be one factor that causes vegetation changes in acidified soils, as discussed below.

Present flora

The composition of the vegetation of a large number (c. 500) of oak, beech and hornbeam forest sites has been recorded since 1978 in south Sweden. There were twice as many herbs as grasses and sedges and only a few fern and shrub species (Table 2). Most of the herbs grew on less acid soils than the grasses and ferns, half of the species being found on soils in the highest pH class. The sampling of sites is of course of great importance to the

Table 3. Field-layer species (herbs, grasses/sedges, ferns) and saplings of woody plants in 526 oak, beech and hornbeam sites in south Sweden. The 10th and 50th percentiles are given for the relative frequencies in 11 classes of 0.2 pH unit measured in 0.2 M KCl (< 2.9, 2.9–3.1, ..., 4.5–4.7, > 4.7. n = number of sites in which the species is present; (n > 20).

Species	n	10 perc	50 perc	Species	n	10 perc	50 perc
Vaccinium myrtillus	86	2.8	3.1	*Lamium galeobdolon*	237	3.4	4.1
Carex pilulifera	210	2.8	3.1	*Melica nutans*	35	3.6	4.1
Trientalis europaea	146	2.8	3.2	*Galium aparine*	44	3.6	4.1
Deschampsia flexuosa	321	2.8	3.3	*Circaea lutetiana*	38	3.7	4.1
Pteridium aquilinum	36	2.8	3.3	*Stachys sylvatica*	27	3.7	4.1
Vaccinium vitis-idaea	29	3.0	3.3	*Stellaria media*	99	3.6	4.2
Molinia caerulea	21	3.1	3.3	*Anthriscus sylvestris*	48	3.6	4.2
Dryopteris dilatata	37	2.9	3.4	*Urtica dioica*	99	3.6	4.2
Galium saxatile	90	3.1	3.4	*Polygonatum verticillatum*	28	3.4	4.3
Dryopteris carthusiana	122	3.1	3.4	*Polygonatum multiflorum*	85	3.5	4.3
Luzula pilosa	212	2.9	3.5	*Lactuca muralis*	70	3.6	4.3
Melampyrum pratense	126	3.0	3.5	*Mercurialis perennis*	125	3.6	4.3
Holcus mollis	80	3.0	3.5	*Festuca gigantea*	46	3.7	4.3
Potentilla erecta	44	3.1	3.5	*Carex sylvatica*	37	3.7	4.3
Maianthemum bifolium	271	3.0	3.6	*Geranium robertianum*	44	3.7	4.3
Epilobium angustifolium	75	3.0	3.6	*Veronica chamaedrys*	81	3.7	4.3
Athyrium filix-femina	88	3.0	3.7	*Aegopodium podagraria*	91	3.8	4.3
Juncus effusus	41	3.1	3.7	*Taraxacum vulgare*	47	3.8	4.3
Agrostis capillaris	196	3.1	3.8	*Fragaria vesca*	66	3.7	4.4
Galeopsis spp.	220	3.2	3.8	*Geum urbanum*	105	3.8	4.4
Melica uniflora	109	3.3	3.8	*Pulmonaria obscura*	41	3.9	4.4
Convallaria majalis	203	3.1	3.9	*Hepatica nobilis*	70	3.8	4.5
Oxalis acetosella	375	3.1	3.9	**Saplings of woody plants**			
Milium effusum	145	3.2	3.9	*Frangula alnus*	102	2.9	3.2
Rubus idaeus	281	3.2	3.9	*Betula pubescens*	26	2.9	3.2
Moehringia trinervia	106	3.3	3.9	*Betula pendula*	53	2.9	3.5
Stellaria holostea	156	3.3	3.9	*Sorbus aucuparia*	277	2.9	3.6
Stellaria nemorum	135	3.3	3.9	*Populus tremula*	53	3.1	3.6
Deschampsia caespitosa	157	3.3	3.9	*Acer pseudoplatanus*	38	3.2	3.6
Anthoxantum odoratum	74	3.3	3.9	*Rubus fructicosus*	80	3.0	3.7
Poa pratensis	48	3.5	3.9	*Sambucus* spp.	50	3.2	3.7
Dryopteris filix-mas	91	3.3	4.0	*Malus sylvestris*	57	3.2	3.7
Dactylis glomerata	144	3.4	4.0	*Tilia cordata*	37	3.2	3.9
Rubus saxatilis	64	3.4	4.0	*Corylus avellana*	173	3.2	4.0
Veronica officinalis	61	3.4	4.0	*Prunus avium*	117	3.3	4.0
Galium odoratum	66	3.4	4.0	*Lonicera periclymenum*	35	3.2	4.1
Viola riviniana/				*Fraxinus excelsior*	134	3.3	4.2
reichenbachiana	290	3.4	4.0	*Acer platanoides*	138	3.4	4.2
Hypericum spp.	59	3.5	4.0	*Viburnum opulus*	63	3.5	4.2
Campanula rotundifolia	36	3.5	4.0	*Prunus padus*	47	3.5	4.3
Rumex acetosella	53	3.6	4.0	*Crataegus* spp.	72	3.6	4.3
Campanula persicifolia	27	3.7	4.0	*Ulmus glabra*	69	3.7	4.3
Poa nemoralis	227	3.3	4.1	*Hedera helix*	34	3.7	4.3
Lathyrus linifolius	113	3.4	4.1	*Euonymus europaeus*	20	3.7	4.4

Table 4. Summary of frequency changes of field-layer species in deciduous forests during recent decades. The studies comprise three periods: 1979/81–1989/90 (95 sites); 1947/70–1984/88 (74 sites); 1929/54–1979/88 (105 and 132 sites, respectively). The signs represent: increase (+), decrease (−), no change (0; meaning <10% change). One sign represents 10–50% change, two signs >50% change. Species with no signs were not investigated (from Falkengren-Grerup 1986, 1990a, Falkengren-Grerup and Tyler 1991b).

| | Time period | | |
	10 yr	15–40 yr	30–60 yr
Decreasing			
Pulmonaria obscura		− −	
Dentaria bulbifera		− −	
Polygonatum multiflorum	−	− −	
Mercurialis perennis	−	− − c	0
Viola riviniana/			
reichenbachiana	−		+
Galium odoratum	−	0 a	0
Unchanged			
Lamium galeobdolon	−	0 c	+
Luzula pilosa	+	−	
Convallaria majalis	0	+	
Maianthemum bifolium	0	0	+ +
Oxalis acetosella	0	0 a	+
Deschampsia caespitosa	+	0	
Deschampsia flexuosa	0 b	0 b	+ +
Increasing			
Aegopodium podagraria		+ +	
Poa nemoralis	0	+ +	+ +
Melica uniflora	0 a	+ +	+ +
Stellaria holostea	0	+	+ +
Stellaria nemorum	0 b	+ +	+ +
Milium effusum	+	+ +	+ +
Dryopteris filix-mas,			
Athyrium filix-femina	+	+	
Carex sylvatica	+	+ +	
Lactuca muralis	+	+ +	
Epilobium angustifolium	+ +	+ +	
Galeopsis tetrahit	+ +		
Rubus idaeus	+	+ +	+ +
Urtica dioica	+ +	+ +	

a=decreased cover; b= increased cover; c= decreased cover in the most acid soils.

results, as the most acid soils are usually underrepresented in studies considering a large number of species. The validity of the comparison between species groups within a pH class is independent of choice of sites.

Deciduous forest soils nowadays have a relatively narrow pH range. The 50% percentile of occurrence of the above species varied between pH (KCl) 3.1 and 4.5 (Table 3). *Deschampsia flexuosa*, *Maianthemum bifolium* and *Oxalis acetosella* are common species in very acid soils. *Poa nemoralis*, *Lamium galeobdolon* and *Viola riviniana/reichenbachiana* are also common but found in somewhat less acid soils. *Geum urbanum*, *Mercurialis perennis* and *Aegopodium podagraria* are quite frequent in the least acid soils, the two latter being able to dominate the field-layer vegetation totally (Lindgren 1970, Brunet 1991).

The woody species are often saplings of tree other than

those dominant in the canopy. The shrub layer is more developed in the oak forests than in the beech and hornbeam forest, because of the less dense canopy and different management practice. *Corylus avellana* is frequent in oak forests. The most common species in all studied sites was *Sorbus aucuparia*, which can grow on quite acid soil, as can *Frangula alnus* (Table 3). *Euonymus europaeus*, *Hedera helix* and *Viburnum opulus* are non-tree species which grow in less acid soils.

Changes in vegetation

Several studies of the forest flora in Skåne demonstrate changes during recent decades. Vegetation and soil analyses in semi-permanent plots were repeated 10–40 yr later, in stands usually having the same tree generation (Falkengren-Grerup 1986, 1990a, Falkengren-Grerup and Eriksson 1990, Falkengren-Grerup and Tyler 1991b). Deciduous forests, especially of beech, were studied over 10, 15–40 and 30–60 yr and the different data sets were compared. The changes for the two shorter periods were quite similar (Table 4). One group of species decreased, a somewhat larger group was unchanged and the largest group increased in frequency. In spite of the simultaneous decreases in soil pH and base cation pools, the number of field-layer species per sample area increased (Falkengren-Grerup 1986). Grasses and sedges, in particular, became more frequent, e.g. *Milium effusum, Poa nemora-*

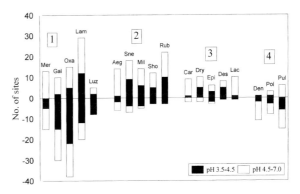

Fig. 3. Number of sites where the cover of a species decreased (−) or increased (+) over 15–40 yr, as related to two classes of pH(H$_2$O) in the humus layer, measured in the second study. Only sites where a species changed considerably are included (from Falkengren-Grerup 1986). The underlined letters of the names are used as labels.
1 = species decreasing in the most acid soils (*Mercurialis perennis*, *Galium odoratum*, *Oxalis acetosella*, *Lamium galeobdolon*, *Luzula pilosa*); 2 = increasing species with a tendency to decrease in the most acid soils (*Aegopodium podagraria*, *Stellaria nemorum*, *Milium effusum*, *Stellaria holostea*, *Rubus idaeus*); 3 = species mainly increasing over the whole pH-range (*Carex sylvatica*, *Dryopteris filix-mas*, *Epilobium angustifolium*, *Deschampsia flexuosa*, *Lactuca muralis*); 4 = species mainly decreasing over the whole pH-range (*Dentaria bulbifera*, *Polygonatum multiflorum*, *Pulmonaria obscura*).

Table 5. Examples of changes in frequency of field-layer species between 1980 and 1990. The species are divided into groups and into typical and occasional species (from Falkengren-Grerup and Tyler 1991b).

Typical species		Occasional species	
increase	decrease	increase	decrease
Woody species		**Woody species**	**Woody species**
Fagus sylvatica		*Quercus robur*	*Acer platanoides*
		Carpinus betulus	
		Sambucus spp.	
		Fraxinus excelsior	
		Acer pseudoplatanus	
		Sorbus aucuparia	
Grasses/sedges	**Grasses**	**Grasses/sedges**	
Deschampsia flexuosa	*Melica uniflora*	*Agrostis capillaris*	
Milium effusum		*Deschampsia caespitosa*	
Carex sylvatica		*Dactylis* spp.	
Carex pilulifera		*Juncus effusus*	
Herbs	**Herbs**	**Herbs**	
Rubus idaeus	*Galium odoratum*	*Epilobium angustifolium*	
Stellaria nemorum	*Viola riviniana/reichenb.*	*Urtica dioica*	
Moehringia trinervia	*Polygonatum multiflorum*	*Galeopsis tetrahit*	
Lactuca muralis	*Mercurialis perennis*	*Veronica officinalis*	
	Lamium galeobdolon		
	Oxalis acetosella		

lis, Melica uniflora and *Carex sylvatica,* and none of the studied species decreased. Other species typical of the beech forest flora also increased, e.g. *Rubus idaeus, Lactuca muralis* and *Stellaria nemorum,* as well as more occasional species like *Epilobium angustifolium* and *Urtica dioica.* In common for the declining species is that they are restricted to less acid soils while the opposite is true for many of the unchanged species.

Changes in cover, rather than changes in frequency, may reveal more gradual and differentiated changes within a pH-range. For sites at which a species was present on one or both occasions, a definite pH dependence was found for some species (Fig. 3: species group 1 and partly 2). An increase was mainly found at a high pH and a decrease at a low pH. *Galium odoratum* was a typical example of this pattern. It was present on the same number of sites as before but the cover had decreased markedly in the most acid soils and increased somewhat in the less acid soils. Other examples were *Mercurialis perennis, Lamium galeobdolon, Oxalis acetosella* and *Luzula pilosa.* The rest of the species mainly increased (group 3) or decreased (group 4) over the whole pH-range.

The number of species in the 10 yr study had increased from 14 to 17 500 m^{-2} plot in spite of continuing soil acidification (Falkengren-Grerup and Tyler 1991b). Among the species increasing were grasses/sedges, tree seedlings and herbs of a more occasional, "weedy" character (Table 5). The decreasing species were mainly herbs characteristic of less acid soils. The increase was usually largest in sites with the highest soil pH (Fig. 4A), or at high management intensity (thinning, soil disturbance, slash; Fig. 4B) where it was quite pronounced. Woody species were an exception in both cases. The "weedy"

species (usually not encountered in undisturbed sites) had become more frequent, usually by colonizing sites with some soil disturbance in combination with substantial thinning. Some of these species are common on clear-felled areas and occur sparsely in the forests. The decrease of *Lamium galeobdolon* and *Oxalis acetosella,* characteristic species of undisturbed forests, was also highest at the most intensive management.

In the study covering the longest time period a comparison was made between deciduous forest sites analyzed in 1929–54 and sites analyzed in 1979–88. Out of thirteen species reported in the older study all but two, which remained unchanged, were more frequent in the 1980's (Table 4). This study was thus not quite consistent with those involving shorter time periods. It is possible that the lower occurrence in the 1930's is due to lower light intensity in unmanaged, dense forests (Lindquist 1931). The deposition of acidifying substances and soil acidification was small before the 1950's (Warfvinge et al. 1993) and deposition probably had little immediate effect on the less acid and buffered soils of the time.

Possible causes of the vegetation changes

Variables commonly used to describe soil acidity in relation to vegetation are pH and exchangeable concentrations of Ca, Mg, Al and Mn. It is difficult to find causal relationships between single elements and plant distributions in field studies, as many soil variables are closely related, such as pH and base cation concentration, for example.

In acidified soils, with decreasing pH and nutrient

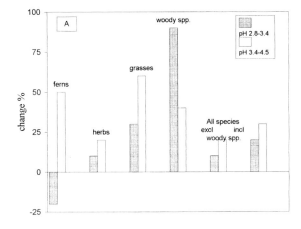

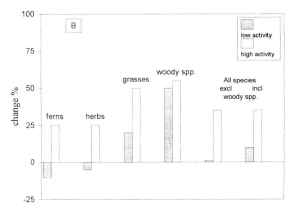

Fig. 4. Change (%) in number of species between 1980 and 1990, studied in beech sites of 500 m². The change is given for five species groups and is related to, A: two classes of pH(KCl) of the humus layer and, B: low or high intensity of forest management. Low intensity means that no or few trees were cut during the 10 yr. High intensity means more cutting, some soil cultivation or machine damage and some slash remaining (Falkengren-Grerup and Tyler 1991b).

pools and increasing soluble and exchangeable Al concentrations, root environment would be expected to be less favourable to most plants, which have to avoid or adapt to the new conditions. Each species has its specific physiological amplitude. It is probable that many species react slowly to environmental changes, especially long-lived species, which are not all dependent on sexual reproduction. Individual plants of *Hepatica nobilis* may, for example, be 100 yr old (Inghe and Tamm 1985). If there is a delay in reaction resulting from the lesser susceptibility of adult individuals than of seedlings, changes in vegetation may become more pronounced in the future and may continue even if there is no further soil acidification.

Soil acidification may also affect the vegetation indirectly by a negative effect on other organisms in the soil, such as earthworms, which improve soil porosity

and decomposition of organic material. The abundance of earthworms decreases drastically with increased soil acidity and the species composition changes towards smaller species which penetrate a more superficial part of the soil profile (Nordström and Rundgren 1974, Staaf 1987). A laboratory experiment showed that growth and reproduction decreased in more acidic soils, probably because of inadequate forage (Bengtsson et al. 1986), as did the digging activity (Laverack 1961). Liming increased both species diversity and abundance (Edwards and Lofty 1977, Lofs-Holmin 1983).

H-ion concentration

The occurrence of many species is clearly related to the soil chemistry. Growth and flowering of *Mercurialis perennis*, *Galium odoratum*, *Stellaria nemorum* and *Maianthemum bifolium* were studied within ten forest sites representing their total pH-range (Falkengren-Grerup 1990b). The first three species were restricted to less acid soils while *Maianthemum bifolium* did not seem to be able to compete with other species at higher pH. The biomass was positively related to pH and exchangeable base cations.

In a pot experiment, using soil from a beech forest on quite acid dystric cambisol with little vegetation, it was shown that only when pH was raised to values corresponding to field conditions for the 13 tested species was survival positively affected and growth possible (Falkengren-Grerup and Tyler 1993c). The pH rise per se seemed to be more important than base cation supply, as soil treated with $SrCO_3$ gave similar results as to that treated with $CaCO_3+MgCO_3$, and Sr is not considered an essential nutrient. The Ca and Mg concentrations of the soil solution even decreased in the $SrCO_3$ treatment but growth was still similar in the two treatments. The concomitant increase in mineralization rate and the decrease in Al concentration was probably also favourable for growth. Comparable effects of $SrCO_3$ were found for four rare forest grasses (Brunet and Neymark 1992). In a combined soil and solution experiment, using soil or synthetic solution derived from mor (raw humus) of a beech forest, the eight studied species responded to the increased pH with a higher growth rate (Falkengren-Grerup and Tyler 1993b). It was concluded that inability of most forest plants to grow in mor (characterized by insignificant concentrations of reactive Al in the soil solution) was mainly attributable to the high H-ion concentration.

The species' reaction to pH has been measured in several other laboratory experiments using soil solution composed as in the field, keeping all other factors under control. Acid mull and mor may have a pH in the soil solution as low as 3.3, which is too acid for most field-layer species to survive. *Galium odoratum* has a low growth potential already at a solution pH of 4.0, root growth being more sensitive than shoot growth (Anders-

Table 6. Frequency (%) of vascular plants in 116 sampling sites in acid mor and mull in beech and oak forests (base saturation <15%). Soils with an organic matter content >25% are designated as mor (n = 34), otherwise as mull (n = 82). Sample depth is 0–5 cm.

Species	Mor	Mull
Preference for mull		
Melica uniflora	0	17
Stellaria nemorum	0	16
Urtica dioica	0	13
Lamium galeobdolon	3	32
Viola riviniana/reichenbachiana	9	46
Poa nemoralis	9	40
Polygonatum multiflorum	3	13
Dryopteris filix-mas	12	50
Stellaria holostea	6	23
Deschampsia caespitosa	9	30
Rubus idaeus	18	60
Galeopsis tetrahit	15	38
Athyrium filix-femina	9	21
Milium effusum	9	21
Convallaria majalis	29	48
No decided preference		
Oxalis acetosella	53	71
Maianthemum bifolium	59	61
Preference for mor		
Luzula pilosa	82	45
Carex pilulifera	71	44
Trientalis europaea	65	45
Deschampsia flexuosa	97	72

son 1992). Similar results were found for *Bromus benekenii* (Andersson and Brunet 1993).

Base cations

Exchangeable pools of macronutrients have decreased by c. 1% yr^{-1} during recent decades (Falkengren-Grerup and Tyler 1991a). It is possible that this decrease in itself, regardless of other changes caused by soil acidification, could affect the distribution of plants. However, additon of Ca and Mg to a fairly acid dystric cambisol did not counteract other negative properties of the soil, nor did a complete fertilizer, if it did not increase soil pH (Brunet and Neymark 1992, Falkengren-Grerup and Tyler 1992b).

Similar results were found in a field experiment with addition of six times the exchangeable pools of Ca, Mg and K in the topsoil (Falkengren-Grerup 1995). Out of seven transplanted species, survival of *Mercurialis perennis* and *Galium odoratum* was very poor in the acid soil (pH (H_2O) 4.1) in spite of the base cation application. *Melica uniflora* and *Viola reichenbachiana,* species naturally occurring in slightly more acid soils, were favoured by Mg application.

Soil acidification initially raises the soil solution concentrations of several nutrients. However, it is questionable whether this has a positive effect on plants, because concentrations of H- and Al-ions are also high. Field studies of soil acidity gradients, caused by large amounts of acid water flowing down beech stems, showed that the soil acidity had increased and a flora of more acid-tolerant species had developed. This occurs in spite of the 2–3 times higher concentration of Ca, Mg and K in the stemflow water than in the throughfall water; the H concentration had increased at the same rate (Falkengren-Grerup 1989).

The relation between base cation and H-ion concentration in the soil solution is, however, of importance for the toxicity at critical H-levels. When Ca+Mg+K was varied, within normal field concentrations, an increase of the base cations favoured growth of *Bromus benekenii* and *Hordelymus europaeus* (Brunet 1994).

Aluminium and manganese

One of the elements potentially most toxic to plants is Al. The monomeric, inorganic Al^{3+} is the form most often taken up by plants in acid soils, Al chelated to organic compounds being mainly unavailable (Asp and Berggren 1990). Most analyses of Al in soil solutions and extracts include the organic complexes and chelates, as Al speciation has only recently become susceptible to routine analysis. The proportion of Al^{3+} is difficult to estimate from analyses of total Al. It may vary between 10 and 50% in the rhizosphere solution of acid beech forests in south Sweden (Andersson and Brunet 1993) and between 3 and 75% in soils of open land (Tyler 1993). The concentration of Al^{3+} may amount to 80 µM in the most acid soils of these forests and to 250 µM in the open land soils.

Many agricultural and acidifuge wild plants are sensitive to Al but the concentration in south Swedish forest soils is now high enough also to damage many forest species of acidic soils. The growth of *Allium ursinum* and *Bromus benekenii* was retarded at 20 µM Al (Andersson 1993, Andersson and Brunet 1993), a concentration commonly found in acid forest soil solutions but rarely in sites where these species grow. (The Ca concentration in this and the following experiments was 0.2 mM) *Galium odoratum* seemed to tolerate 70 µM Al (Falkengren-Grerup and Tyler 1992b). An Al concentration of 40 µM decreased growth of *Bromus benekenii* and *Hordelymus europaeus* as compared to 20 µM, but when Ca was increased threefold the negative effects of Al were reduced (Brunet 1994). A better characterization of Al toxicity may sometimes be given by the Ca/Al ratio (Rengel 1992).

Manganese is another metal with a high solubility in acidified soils (Falkengren-Grerup and Tyler 1991a, 1993a). Concentrations in the soil solution are occasionally as high as 0.2 mM in south Swedish, well-drained forests. The toxicity of Mn has been tested on agricultural plants and some forest plants (Hackett 1964, Kroeze et al. 1989). Many species can tolerate concentrations as high as 2 mM, and there may even be a small positive effect on growth at 0.1–0.2 mM. *Galium odoratum* was not negatively affected by 0.5 mM Mn (Falkengren-Grerup and Tyler 1992b). Thus the Mn concentrations in these forest soils, and possibly in most acidified soils, seem to be of less importance to plant performance than the H and Al concentration.

In acid soils, lacking deep-digging worms, a superficial layer of organic material may develop into a mor layer. Most species growing in acid soils prefer mull, being absent or rare in mor, and only a few species are more abundant in mor (Table 6). The organic layer has a very high H-ion concentration but only low concentrations of Al, particularly reactive Al, in the soil solution (Falkengren-Grerup and Tyler 1993b). Seeds of four forest species seemed to germinate equally well in mor and mull, while seedlings and transplanted mature plants did not survive in mor unless the soil was limed (Staaf 1992). Glasshouse experiments with mor soil or soil solution indicated that the high H-concentration was the main chemical growth-limiting factor for eight forest species studied with the exception of the most acid-tolerant species in the study, *Deschampsia flexuosa* (Falkengren-Grerup and Tyler 1993b).

Phosphorous

The role of P for species and vegetation changes in acidified soils has been little studied. It has low mobility in the soil. Most forms of P in soils, such as Al-, Fe- and Ca-phosphates and organic P, are not readily available to plants, but P availability is extremely dependent on the root physiology of the species and on the type of mycorrhizal infection. It is possible that shortage of P may limit growth in acidic south Swedish soils with high concentrations of Al. In a field study, addition of phosphates to an acid beech forest soil increased biomass production of three of the six field-layer species studied (*Galium odoratum*, *Lamium galeobdolon* and *Stellaria holostea*) (Falkengren-Grerup et al. 1994).

Nitrogen

Deposition of N has increased during recent decades. The amounts have doubled in south Sweden since the 1950's and may amount to 20 kg ha^{-1} yr^{-1} in deciduous forests, with equal amounts of NH_4 and NO_3 (Anon. 1986). The increased deposition probably raises the total available N in the soil which may affect the decomposition rate and the occurrence of various decomposer species (bacteria and fungi). The N deposition may also induce soil acidification in nitrifying soils if NO_3 formed is not utilized by organisms but is leached out of the soil (N saturation), a situation which exists in south Swedish forest soils today (Anon. 1991, Tyler et al. 1992, Bergkvist and Folkeson 1995).

An increased N availability in forest soils is often considered to favour diversity and growth of plants. Nitrogen has until now been considered growth-limiting in most forest soils. It should therefore have positive effects as long as no other element, such as P or other nutrients, becomes limiting. Diversity and abundance of field-layer species may change, as different species have quite different N requirements. New species may establish from the seed bank or from the surrounding vegetation. Altered competition may reduce the occurrence of other species, as is often found in N fertilized fields, where some grasses attain dominance (Thurston et al. 1976, Heil and Diemont 1983, Bobbink 1991).

The repeated studies over recent decades showed that half of the studied species had increased in cover, while less than 5% of the species had decreased (Falkengren-Grerup 1986, Falkengren-Grerup and Eriksson 1990). Among the increasing species were *Aegopodium podagraria*, *Epilobium angustifolium*, *Rubus idaeus*, *Stellaria nemorum* and *Urtica dioica*, all considered to be nitrophilous species (Ellenberg et al. 1991). This indicates that the higher N availability in forest soils has favoured establishment of some species and increased the abundance of other nitrophilous species. This is also indicated by a comparison of the field-layer vegetation in oak forests in two Swedish provinces (Skåne, Småland) with different N deposition rates. The province with the highest deposition (Skåne) had a larger number and a greater share of nitrophilous species (Rühling and Tyler 1986, Tyler 1987).

These effects of increased N are valid as long as N is a limiting nutrient, but south Swedish forest soils may now be N-saturated. This was apparent from a 5 yr field experiment in two beech forests, where the current N deposition was increased artificially by 3 or 9 times (Tyler et al. 1992). *Rubus idaeus* and *Impatiens parviflora*, present in one of the forests, increased somewhat in cover. These species are common in intensively managed forests. Frequency of all other field-layer species either remained unchanged or decreased. The species which decreased were *Anemone nemorosa*, *Maianthemum bifolium*, *Oxalis acetosella*, *Viola riviniana/reichenbachiana* (Falkengren-Grerup 1993). When competition from tree roots was excluded by trenching, biomass of the studied field-layer species increased, indicating that water or other nutrients limited growth in the intact system.

The importance of different N forms to plant performance can be studied in laboratory experiments. Ammonium is the main form available in the most acid soils while the proportion of NO_3 increases in less acid soils (Fig. 2). In experiments with NH_4 and/or NO_3 in synthetic acid forest soil solutions none of the seven studied species grew better when N was present as NH_4 only, whereas most species grew best when both forms were present (Falkengren-Grerup and Lakkenborg-Kristensen 1994). Lack of NO_3 may therefore limit plant distribution in none-nitrifying soils. *Deschampsia flexuosa* was one of the species studied, and the same results were also found in some older studies (Hackett 1965, Gigon and Rorison 1972). Older studies have often used high nutrient concentrations and high pH (e.g. Gigon and Rorison 1972). The observed preference for NO_3 of calcicole species and for NH_4 of calcifuge species may therefore not be applicable to acid forest soils of today.

Forest management

In forests established after clear-felling there is often a succession of field-layer species in the different development phases of the stand. Some species are present in all phases, others only initially or in the mature phase. Light affects the distribution of most species. In oak and hornbeam forests, in which canopy closure varied between 60 and 140%, light increased the occurrence of *Rubus idaeus, Vaccinium myrtillus* and grasses, for example, while *Convallaria majalis* and *Mercurialis perennis* were more dominant in less light forests (Tyler 1989).

The continuous changes in management practice of the mature forests may also be of great importance to the vegetation and to the soil properties. Such changes have taken place in south Sweden during the 20th century. In the 1930's half the beech stands in Skåne had > 5% of daylight penetrating to the ground, while ca 10% had only 2%, because of lack of thinning (Lindquist 1931). Management has been quite intense during the last decade with a high degree of thinning and soil scarification. This change has enabled "weedy" species to establish. Species typical of less disturbed forest stands have decreased, because of competition and other factors attributable to a more open canopy. As a result of mechanical disturbance of the humus layer, seeds of species represented in the soil bank are exposed and can germinate. These are often species other than those established in the vegetation, e.g. *Rubus idaeus, Juncus effusus* and *Agrostis capillaris* (Staaf et al. 1987, Falkengren-Grerup and Tyler 1991a).

Some plants require forests with a long continuity, e.g. *Actaea spicata, Bromus benekenii, Bromus ramosus, Festuca altissima, Hedera helix, Hordelymus europaeus* and *Veronica montana* (Brunet 1993, Brunet et al. 1993). One explanation for this may be a requirement for high and even humidity and a long growth period. Another explanation may be the restricted dispersal ability in the fragmented landscape.

Soil chemistry may be distinctly modified by the tree species. Beech and oak tend to decrease soil pH more than hornbeam and much more than maple and lime (Nordén 1992). A successional change from oak to elm probably counteracted soil acidification in a deciduous forest in Skåne (Persson et al. 1987). The more acidic soil under spruce as compared to beech and birch is an effect of the greater H-ion deposition onto the ground (Bergkvist and Folkeson 1995).

Summary and conclusions

It is most probable that soil acidification will continue in the near future in a similar way to that in recent decades. The exchangeable pools of Ca, Mg and K will decrease further in large parts of the soil profile and reach values which may be too low even for the highly acidophilous forest flora. The concentrations of potentially toxic elements like H, Al and Mn will increase even in sites which still retain a fairly high base saturation and may, often in combination with the lowered cation concentrations, negatively affect the growth potential of many species. The high or increasing N deposition rates will promote soil acidification, as many forest soils will be N saturated.

Forest management may contribute to changes in the flora in the short term as well as in the long term as it continues to change in response to the economy and to consideration of soil chemistry and environment. Thinning and harvesting practices may change over a few years, while choice of tree species is made over a longer term. The threat from inappropriate management is therefore less than from soil acidification caused by acid deposition, which is a slower process, not easily reversed and only marginally possible to control, regionally or nationally.

Several of the characteristic forest species are threatened by the current changes towards a more acidic root environment with toxic effects of H- and Al-ions. The decreasing amounts of most nutrients except N may enhance the toxic effects by reducing antagonistic interactions between elements. Nutrient deficiency per se may also become of importance.

References

Andersson, M. 1988. Toxicity and tolerance of aluminium in vascular plants. – Water Air Soil Pollut. 39: 439–462.
– 1992. Effects of pH and aluminium on growth of *Galium odoratum* (L.) Scop. in flowing solution culture – Environ. Exp. Bot. 32: 497–504.
– 1993. Aluminium toxicity as a factor limiting the distribution of *Allium ursinum* (L.). – Ann. Bot. 72: 607–611.
– and Brunet J. 1993. Sensitivity to H and Al ions limiting growth and distribution of the woodland grass *Bromus benekenii*. – Plant Soil 153: 243–254.
Anon. 1971. Bokskogens bevarande. – Jordbruksdepartementet, Stockholm. SOU 1971:71.
– 1982. Ädellövskog – Förslag till skydd och vård. – Swedish Environ. Protect. Agency, Solna, PM 1587.
– 1986. Sura och försurade vatten. – Monitor. Swedish Environ. Protection Agency, Solna.
– 1990. Skog 2000. Skogligt program för Kristianstad och Malmöhus län. – Länsstyrelsen L/M län.
– 1991. Miljöatlas, Resultat från IVLs undersökningar i miljön. – Swedish Environ. Res. Inst., Stockholm.
Asp, H. and Berggren, D. 1990. Phosphate and calcium uptake in beech *Fagus sylvatica* in the presence of aluminium and natural fulvic acids. – Physiol. Plant. 78: 79–84.
Ballach, H.-J., Greven, H. and Wittig, R. 1985. Biomonitoring in Waldgebieten Nordrhein-Westfalens. – Staub Reinhaltung der Luft 45: 567–573.
Becker, M., Bonneau, M. and Le Tacon, F. 1992. Long-term

vegetation changes in an *Abies alba* forest: natural development compared with response to fertilization. – J. Veg. Sci. 3: 467–474.

Bengtsson, B., Asp, H., Jensén, P. and Berggren, D. 1988. Influence of aluminium on phosphate and calcium uptake in beech *Fagus sylvatica* grown in nutrient solution and soil solution. – Physiol. Plant. 74: 299–305.

Bengtsson, G., Gunnarsson, T. and Rundgren, S. 1986. Effects of metal pollution on the earthworm *Dendrobaena rubida* (Sav.) in acidified soils. – Water Air Soil Pollut. 28: 361–383.

Bergendorff, C. and Emanuelsson, U. 1982. Skottskogen – en försummad del av vårt kulturlandskap. – Svensk Bot. Tidskr. 76: 91–100.

Berggren, D., Bergkvist, B., Falkengren-Grerup, U., Folkeson, L. and Tyler, G. 1990. Metal solubility and pathways in acidified forest ecosystems of south Sweden. – Sci. Tot. Environ. 96: 103–114.

Bergkvist, B. 1987. Leaching of metals from forest soils as influenced by tree species and management. – For. Ecol. Manage. 22: 29–56.

– and Folkeson, L. 1995. The influence on acid deposition, proton budgets and element fluxes in south Swedish forest ecosystems. – Ecol. Bull. (Copenhagen) 44: 90–99.

Bergman, F. A. 1960. Skånes skogar. – Skånes Natur 47: 199–222.

Bjørnstad, O. N. 1991. Changes in forest soils and vegetation in Søgne, southern Norway, during a 20 year period. – Holarct. Ecol. 14: 234–244.

Bobbink, R. 1991. Effects of nutrient enrichment in Dutch chalk grassland. – J. Appl. Ecol. 28: 28–41.

Brunet, J. 1991. Vegetationen i Skånes alm- och askskogar. – Svensk Bot. Tidskr. 85: 377–384.

– 1993. Environmental and historical factors limiting the distribution of rare forest grasses in south Sweden. – For. Ecol. Manage. 61: 263–275.

– 1994. Interacting effects of pH, aluminium and base cations on growth and mineral composition of the woodland grasses *Bromus benekenii* and *Hordelymus europaeus*. – Plant Soil 161: 157–166.

–, and Neymark, M. 1992. Importance of soil acidity to the distribution of rare forest grasses in south Sweden. – Flora 187: 317–326.

–, Andersson, P.-A. and Weimarck, G. 1993. Faktorer som påverkar utbredningen av Skånes ädellövskogsflora. – Svensk Bot. Tidskr. 87: 177–186.

Campbell, Å. 1928. Skånska bygder under förra hälften av 1700-talet. Etnografisk studie över den skånska allmogens äldre odlingar, hägnader och byggnader. – Thesis, Uppsala Univ.

Edwards, C. A. and Lofty, J. R. 1977. Biology of earthworms. – Chapman and Hall, London.

Ellenberg, H. 1985. Veränderungen der Flora Mitteleuropas unter dem Einfluss von Düngung und Immisionen. – Schweiz. Z. Forstwes. 136: 19–39.

– 1991. Zeigerwerte von Pflanzen in Mitteleuropa – Scripta Geobot. 18: 1–248.

Falkengren-Grerup, U. 1986. Soil acidification and vegetation changes in deciduous forest in southern Sweden. – Oecologia 70: 339–347.

– 1987. Long-term changes in pH of forest soils in southern Sweden. – Environ. Pollut. 43: 79–90.

– 1989. Effect of stemflow on beech forest soils and vegetation in southern Sweden. – J. Appl. Ecol. 26: 341–352.

– 1990a. Distribution of field layer species in Swedish deciduous forests in 1929–54 and 1979–88 as related to soil pH. – Vegetatio 86: 143–150.

– 1990b. Biometric and chemical analysis of five herbs in a regional acid-base gradient in Swedish beech forest soils. – Acta Oecol. 11: 755–766.

– 1993. Effects on beech forest species of experimentally enhanced nitrogen deposition. – Flora 188: 85–91.

– 1995. Replacement of nutrient losses caused by acidification of a beech forest soil and its effects on transplanted field-layer species. – Plant Soil 168–169, in press.

– and Eriksson, H. 1990. Changes in soil, vegetation and forest yield between 1947 and 1988 in beech and oak sites of southern Sweden. – For. Ecol. Manage. 38: 37–53.

–, Linnermark, N. and Tyler, G. 1987. Changes in acidity and cation pools of south Swedish soils between 1949 and 1985. – Chemosphere 16: 2239–2248.

– and Lakkenborg-Kristensen, H. 1994. Importance of ammonium and nitrate to the performance of forest herbs and grasses. – Environ. Exp. Bot. 34: 31–38.

– and Tyler, G. 1991a. Changes of cation pools of the topsoil in south Swedish beech forests between 1979 and 1989. – Scand. J. For. Res. 6: 145–152.

– and Tyler, G. 1991b. Dynamic floristic changes of Swedish beech forest in relation to soil acidity and stand management. – Vegetatio 95: 149–158.

– and Tyler, G. 1992a. Changes since 1950 of mineral pools in the upper C-horizon of Swedish deciduous forest soils. – Water Air Soil Pollut. 64: 495–501.

– and Tyler, G. 1992b. Chemical conditions limiting survival and growth of *Galium odoratum* (L.) Scop. in acid forest soil. – Acta Oecol. 13: 169–180.

– and Tyler, G. 1993a. The importance of soil acidity, moisture, exchangeable cation pools and organic matter solubility to the cationic composition of beech forest (*Fagus sylvatica* L.) soil solution. – Z. Pflanzenernähr. Bodenkd. 156: 365–370.

– and Tyler, G. 1993b. Soil chemical properties excluding field-layer species from beech forest mor. – Plant Soil 148: 185–191 (errata in 150, 323).

– and Tyler, G. 1993c. Experimental evidence for the relative sensitivity of deciduous forest plants to high soil acidity. – For. Ecol. Manage. 60: 311–326.

–, Rühling, Å. and Tyler, G. 1994. Effects of phosphorus application on vascular plants and macrofungi in an acid beech forest soil. – Sci. Tot. Environ. 151: 125–130.

Gigon, A. and Rorison, I. H. 1972. The response of some ecologically distinct plant species to nitrate- and to ammonium-nitrogen. – J. Ecol. 60: 93–102.

Hackett, C. 1964. Ecological aspects of the nutrition of *Deschampsia flexuosa* (L.) Trin. – J. Ecol. 52: 159–167.

– 1965. Ecological aspects of the nutrition of *Deschampsia flexuosa* (L.) Trin. II. The effects of Al, Ca, Fe, K, Mn, N, P and pH on the growth of seedlings and established plants. – J. Ecol. 53: 315–333.

Heil, G. W. and Diemont W. H. 1983. Raised nutrient levels change heathland into grassland. – Vegetatio 53: 113–120.

Inghe, O. and Tamm, C. O. 1985. Survival and flowering of perennial herbs. IV. The behaviour of *Hepatica nobilis* and *Sanicula europaea* on permanent plots during 1943–1981. – Oikos 45: 400–420.

Kellner, O. and Redbo-Torstensson, P. 1995. Effects of elevated nitrogen deposition on the field layer vegetation in coniferous forests. – Ecol. Bull. (Copenhagen) 44: 227–237.

Kroeze, C., Pegtel, D. M. and Blom, C. J. 1989. An experimental comparison of aluminium and manganese susceptibility in *Antennaria dioica*, *Arnica montana*, *Viola canina*, *Filago minima* and *Deschampsia flexuosa*. – Acta Bot. Neerl. 38: 165–172.

Kuhn, N., Amiet, R. and Hufschmid, N. 1987. Veränderungen in der Waldvegetation der Schweiz infolge Nährstoffanreicherungen aus der Atmosphäre. – Allg. Forst. Jagdz. 158: 77–84.

Laverack, M. S. 1961. Tactile and chemical perception in earthworms. II. Responses to acid pH solutions. – Comp. Biochem. Physiol. 2: 22–34.

Lindgren, L. 1970. Beech forest vegetation in Sweden – a survey. – Bot. Notiser 123: 401–424.

– and Malmer, N. 1970. Skogsutvecklingen i Skåne under fyra decennier. – Skånes Natur 57: 2–8.

Lindquist, B. 1931. Den skandinaviska bokskogens biologi. – Sv. skogsv. för tidskr. 3: 179–532.

Lofs-Holmin, A. 1983. Influence of agricultural practices on earthworms (Lumbricidae). – Acta Agr. Scand. 33: 225–234.

Nilsson, S.I. and Tyler, G. 1995. Acidification-induced chemical changes of forests soils during recent decades – a review. – Ecol. Bull. (Copenhagen) 44: 54–64.

Nordén, U. 1992. Soil acidification and element fluxes as related to tree species in deciduous forests of south Sweden. – Ph. D. thesis, Lund Univ.

Nordström, S. and Rundgren, S. 1974. Environmental factors and lumbricid associations in southern Sweden. – Pedobiologia 14: 1–27.

Persson, S., Malmer, N. and Wallén, B. 1987. Leaf litter fall and soil acidity during half a century of secondary succession in a temperate deciduous forest. – Vegetatio 73: 31–45.

Persson, Å. 1971. Vegetations- och landskapshistoria på Söderåsen. – Skånes Natur 58: 29–44.

Petersens, F. af. 1932. Skogshushållning. – Skrifter utgivna av de skånska hushållningssällskapen med anledning av deras hundraårsjubileum år 1914, X.

Pott, R. 1985. Vegetationsgeschichtliche und pflanzensoziologische Untersuchungen zur Niederwaldwirtschaft in Westfalen. – Abh. Westfälisches Mus. f. Naturk. 47: 1–75.

Rengel, Z. 1992. Role of calcium in aluminium toxicity. – New Phytol. 121: 499–513.

Rorison, I.H. 1986. The response of plants to acid soils. – Experientia 42: 357–362.

Rost-Siebert, K. and Jahn, G. 1988. Veränderungen der Waldbodenvegetation während der letzten Jahrzehnte – Eignung zur Bioindikation von Immisionswirkungen? – Forst und Holz 43: 75–81.

Runge, M. and Rode, M.W. 1991. Effects of soil acidity on plant associations. – In: Ulrich, B. and Sumner, M.E. (eds), Soil acidity. Springer, Berlin, pp. 183–203.

Rühling, Å. and Tyler, G. 1986. Vetetationen i sydsvenska ekskogar – en regional jämförelse. – Svensk Bot. Tidskr. 80: 133–143.

Staaf, H. 1987. Foliage litter turnover and earthworm populations in three beech forests of contrasting soil and vegetation types. – Oecologia 72: 58–64.

– 1992. Performance of some field-layer vegetation species introduced into an acid beech forest with mor soil. – Acta Oecol. 13: 753–765.

–, Jonsson, M. and Olsén, L.-G. 1987. Buried germinative seeds in mature beech forests with different herbaceous vegetation and soil types. – Holarct. Ecol. 10: 268–277.

Thurston, J.M., William, E.D. and Johnston, A.E. 1976. Modern developments in an experiment on permanent grassland started in 1856: Effects of fertilizers and lime on botanical composition and crop and soil analysis. – Ann. Agron. 27: 1043–1082.

Tregenza, N.J.C. 1986. Acidification effects on the flora of Cornwall. – Cornwall Trust Nature Cons., Cornwall Cons. Review, Report 1: 1–21.

Tyler, G. 1981. Leaching of metals from the A-horizon of a spruce forest soil. – Water Air Soil Pollut. 15: 353–369.

– 1987. Probable effects of soil acidification and nitrogen deposition on the floristic composition of oak (*Quercus robur* L.) forest. – Flora (Jena) 179: 165–170.

– 1989. The interacting effects of soil acidity and canopy cover on the species composition of field layer vegetation in oak-hornbeam forest. – For. Ecol. Manage. 28: 101–114.

– 1993. Soil solution chemistry controlling the field distribution of *Melica ciliata* L. – Ann. Bot. 71: 295–301.

–, Balsberg Påhlsson, A-M., Bergkvist, B., Falkengren-Grerup, U., Folkeson, L., Nihlgård, B., Rühling, Å. and Stjernkvist, I. 1992. Chemical and biological effects of simulated nitrogen deposition to the ground in a Swedish beech forest. – Scand. J. For. Res. 7: 515–532.

van Breemen, N. and van Dijk, H.F.G. 1988. Ecosystem effects of atmospheric deposition of nitrogen in the Netherlands. – Environ. Pollut. 54: 249–274.

Warfvinge, P., Falkengren-Grerup, U., Andersen B. and Sverdrup, H. 1993. Modelling long-term cation supply in acidified forest stands. – Environ. Pollut. 80: 209–221.

Wilmans, O. 1989. Zur Frage der Reaktion der Waldboden-Vegetation auf Stoffeintrag durch Regen, eine Studie auf der schwäbischen Alp. – Allg. Forst. Jagdz. 160: 165–175.

Wilmans, O. and Bogenrieder, A. 1986. Veränderungen der Buchenwälder des Kaiserstuhls im Laufe von vier Jahrzehnten und ihre Interpretation – pflanzensoziologische Tabellen als Dokumente. – Abhandlungen 48: 55–79.

Wittig, R., Ballach, H.-J. and Brandt, C.J. 1985. Increase of number of acid indicators in the herb layer of the millet grass-beech forest of the Westphalian Bight. – Angew. Bot. 59: 219–232.

Ecological Bulletins 44: 227–237. Copenhagen 1995

Effects of elevated nitrogen deposition on the field-layer vegetation in coniferous forests

Olle Kellner[1] and Peter Redbo-Torstensson

Kellner, O. and Redbo-Torstensson, P. 1995. Effects of elevated nitrogen deposition on the field-layer vegetation in coniferous forests. – Ecol. Bull. (Copenhagen) 44: 227–237.

Time series studies on vegetation changes in European coniferous forests are reviewed in relation to nitrogen deposition. In areas with high N deposition, grasses had increased together with mean indicator values for nitrogen. The same type of changes have been recorded in nitrogen fertilization experiments. In order to investigate whether such vegetation changes only take place at high N deposition levels, we performed a five-year experiment in which nitrogen was applied to the field layer in a coniferous forest in central Sweden. There were five different fertilization treatments: i) control with no extra nitrogen, ii) 0.5, iii) 1.0, iv) 2.0, and v) 4.0 g N m^{-2} yr^{-1}.
Six species were present in sufficient amounts to make analyses of changes possible: *Deschampsia flexuosa*, *Linnaea borealis*, *Luzula pilosa*, *Trientalis europaea*, *Vaccinium myrtillus* and *V. vitis-idaea*. For only two of these species (*D. flexuosa* and *T. europaea*) was there a significant effect of the nitrogen treatment. A third species (*V. myrtillus*) changed significantly through the years, but showed no effect of nitrogen. For *D. flexuosa*, the magnitude of the increase in shoot density after five years was closely related to the amount of nitrogen added. In control plots its shoot density remained fairly constant, whereas in the range from the lowest to the highest nitrogen treatment, the shoot density increased 1.7, 3.5, 5.3 and 8.7 times.
Ramet density of *T. europaea* more than doubled in the two treatments with the lowest amounts of added nitrogen, and increased more than five times in the two highest treatments. In the control treatment ramet density was fairly constant.
Vaccinium myrtillus increased in density during the five years, from 36 ramets m^{-2} in 1988 to 46 m^{-2} 1992. The increase in ramet density was observed in all treatments and we detected no significant effect of nitrogen treatments.
Our results indicate that changes in the field layer are induced already at low deposition rates (0.5 g N m^{-2} yr^{-1} in through-fall). The types of changes observed are similar to those observed in areas with elevated nitrogen deposition. From the reviewed time-series studies and fertilization experiments it is evident that nitrogen generally has the most pronounced effects at moderately poor sites, whereas the poorest sites change their vegetation only slowly, and the richest site are more or less unaffected.

O. Kellner, Dept of Ecology and Environmental Research, Swedish Univ. of Agricultural Science, P.O. Box 7072, S-750 07 Uppsala, Sweden. – P. Redbo-Torstensson, Dept of Ecological Botany, Uppsala Univ., Villavägen 14, S-752 36 Uppsala, Sweden.

Introduction

During the last 30 to 40 years, terrestrial ecosystems in Europe have been subjected to an increasing atmospheric deposition of nitrogen compounds (Galloway and Likens 1981, Grennfelt and Hultberg 1986). Availability of nutrients is an important factor in determining the species composition of plant communities. Nitrogen is the element that limits plant growth in most temperate and boreal forests (Tamm 1991), and many plant species have their distribution restricted to habitats within limited ranges of nitrogen availability. Eutrophication of terrestrial ecosystems due to the deposition of nitrogen is considered a major threat to the flora of many European

---

[1]Authors in alphabetical order

Table 1. Retrospective studies of vegetation change in coniferous forest.

Geographical area	Current nitrogen deposition rate in forest stand ($g\ N\ m^{-2}\ yr^{-1}$)	Dominating tree species	Original vegetation and number of studied sites	Year/years of first observation	Year/years of last observation	Vegetation changes	References and comments
Continental Europe Lowland in eastern Germany	2–3 (?)	*Pinus sylvestris*	lichen (*Cladina*) type (1 site), lichen-rich type (1 site), *Vaccinium myrtillus* types (2 sites), *Deschampsia flexuosa – Rubus idaea* type (1 site)	1964	1985	Vegetation of the poor to intermediate sites changed to a more fertile vegetation type: lichen type → lichen-rich type, lichen-rich type → *Vaccinium myrtillus* type. *V. myrtillus* type → *Deschampsia flexuosa* type. Vegetation of the richest site (*D. flexuosa – Rubus idaeus* type) did not change. No good data on species number.	1
Schwarzwald, south-west Germany	1–5	*Abies alba/Picea abies*	various types (c. 30 stands)	1940 or 1960 or 1970	1985–86	Decrease of typical forest species, increase of clearing species, increase of mean indicator values (Ellenberg et al. 1991) for nitrogen and light, no decrease in value for soil reaction. Increase in species number.	2
North-east Bavaria, Germany	1.5–2.5	*Pinus sylvestris*	*Vaccinium myrtillus* type (32 stands)	1946 or 1967 or 1971	1987–88	Decrease in *Calluna vulgaris*, *Vaccinium vitis-idaea*, *Cladina* spp. and *Cetraria islandica*. Increase in *D. flexuosa* and of *Quercus*-seedlings. Increase in species number.	3
Vosges mountains, east France	1.7	*Abies alba*	*Vaccinium myrtillus* type (1 stand)	1969	1989	Increase in mean indicator values for nitrogen and soil reaction. Increase in graminoids (e.g. *Deschampsia* and *Luzula*), herbs (e.g. *Oxalis*), *Rubus*, ferns and some deciduous trees. Increase in species number.	4
Fennoscandia South Finland	c. 1	*Pinus sylvestris*	*Calluna – Vaccinium* types (60 stands)	1950–56	1983–86	Increase in *Epilobium angustifolium*, *Calamagrostis arundinacea*, *Maianthemum bifolium* and *Trientalis europaea*. Decrease in *V. vitis-idaea*. No significant change in *V. myrtillus* or *Deschampsia flexuosa*. No change in species number.	5
South Sweden	1–2	mainly *Pinus sylvestris* and *Picea abies*	*Vaccinium* types (many stands)	1973–77	1983–87	Many sites changed to grass type (mainly dominated by *D. flexuosa*). No data on species number.	6

cont.

References and comments:
1. Hofmann (1972, 1987) and Hofmann et al. (1990). Through-fall deposition is said to vary between 1.6 and 15 g N m^{-2} yr^{-1} in the lowlands of former East Germany (data from Simon and Westendorff, referred to in Hofmann et al. 1990). Deposition rates were not measured in the study sites, but Hofmann et al. (1990) appear to assume that they were in the lower part of the regional range. The study comprised repeated observations during the period, with assessments of the cover of each species in experimental plots in fertilization trials (including control plots). Unfortunately, vegetation data were not fully presented in the papers, which makes comparisons for single species difficult. Hofmann et al. (1990) stated that where nitrogen deposition is c. 5 g N m^{-2} yr^{-1}, the diversity of vegetation types decreased, as poor to intermediate sites (lichen types, *Vaccinium* types and *Deschampsia* type) changed to *Calamagrostis epigeios* dominance, while the richest (herb types) changed to *Rubus idaeus* – *Urtica dioica* dominance. No data were presented to support this statement, but there were data on such conversions in long term fertilization trials (Hofmann 1987, Hofmann et al. 1990).

2. Bürger (1991). Through-fall deposition data from Adam et al., referred to by Bürger (1991). Deposition was highest on the west slopes of the Schwarzwald and decreased gradually eastwards of the mountain crest. A large number of sites in various types of forests were investigated by describing relevés according to phytosociological methods. These descriptions were compared with older site descriptions. The relevés were not located at exactly the same places as the older ones, so direct comparisons of each relevé were not possible. In stead, Bürger (1991) made comparisons of whole forest types, as described by many relevés. Abundance changes of single species were not analysed, only the change in number of species from different ecological groups. The study included broadleaved forests as well as fir and spruce forests, but only the results for coniferous forests are included in the table. The forests in the area have suffered from thinning of the tree crowns, and Bürger (1991) concluded that this thinning, together with nitrogen deposition, had caused the changes.

3. Rodenkirchen (1991, 1992). Through-fall deposition data from Hüser, Rehfuess and Sauter, referred to by Rodenkirchen (1992). Among a large number of sites with old vegetation descriptions, those sites with no changes in light regime were chosen for new descriptions, using the Braun-Blanquet (relevé) method. In nitrogen fertilization trials within the area, there were vegetation changes similar to those recorded in the time series study (Rodenkirchen 1991). Experiments with full (NPK) fertilizer and/or lime within the same area were also reported (Rodenkirchen 1991, 1992).

4. Becker et al. (1992). Through-fall deposition data from Probst et al., referred to by Becker et al. (1992). The vegetation in 65 experimental plots was described according to the Braun-Blanquet (relevé) method at the start of a liming/fertilization experiment, and the description was repeated twenty years later. In the first description, only the twelve most common species at the site were investigated, but in the last description all species were included. Liming in 1969 had caused even more pronounced changes towards a nitrophilic flora, with colonization of more species, further increase of some nitrophilic species and decrease of some low-demanding (e.g. *V. myrtillus*) and intermediate (e.g. *D. flexuosa*) species. The liming caused an improvement of humus type from mor/moder to acid mull/mull-moder, which probably improved the nitrogen availability. The stand suffered from thinning of the tree crowns in the 1970's and early 1980's.

5. Nieppola (1992). Data on total nitrogen deposition to forest from large-scale estimates by Lövblad et al. (1992). Vegetation descriptions made in permanent plots in mature stands. The stands were thinned during the period. Similar results were obtained in a study in 16 unthinned stands (Lähde and Nieppola 1987).

6. Rosén et al. (1992). Data on total nitrogen deposition to forest from large-scale estimates by Lövblad et al. (1992). Vegetation data were from the Swedish National Forest Survey, in which 20 000 plots all over Sweden were classified into vegetation types according to the system by Hägglund and Lundmark (1977). There were no data on single species abundances. The change in vegetation type took place only in the south-western part of the country, and the deposition given in the table refers to that area. In the central and northern parts of the country, where nitrogen deposition is well below 1 g N m^{-2} yr^{-1}, there was no significant change in vegetation type.

countries (Ellenberg 1985). There are, already, several examples indicating that deposition of nitrogen is causing vegetational changes. For example, many heathlands of north-west Europe have changed into grasslands (Heil and Diemont 1983, van Breemen and van Dijk 1988); chalk grasslands have become dominated by *Brachypodium pinnatum* (L.) Beauv. (Bobbink 1991); broadleaved forest vegetation has changed into more nitrogen-rich types in both central Europe (Wittig 1988) and southern Sweden (Falkengren-Grerup 1986, 1992, Tyler 1987), and coniferous forest (reviewed below). Plants and vegetation adapted to nitrogen-poor soils are considered to be the most susceptible to increased nitrogen deposition (Ellenberg 1985, 1988, Nilsson and Grennfelt 1988). Ombrotrophic bogs, for instance, are assumed to be particularly sensitive (Nilsson and Grennfelt 1988, Morris 1991, Redbo-Torstensson 1994).

Coniferous forests are the most common forest types in northern Europe. In Sweden, they cover c. 60% of the land-area below the alpine zone. Many of these forests grow in nitrogen-poor habitats and are inhabited by several species adapted to nitrogen-poor soils. We may, therefore, expect their flora to be sensitive to increased atmospheric nitrogen deposition. In this paper we review the evidence for effects on field-layer vegetation of fairly high nitrogen deposition levels in coniferous forests, and then we present a study on the effects of only slightly elevated nitrogen deposition.

A source of information on nitrogen effects are the studies of nitrogen fertilization experiments in coniferous forests. Such studies have consistently reported an increase of grasses, if present (e.g. Mälkönen et al. 1980, Rodenkirchen 1991, Kellner 1993a, van Dobben et al. 1993). For dwarf shrubs, both positive and negative responses have been reported. However, when the initial site conditions at the experimental sites are taken into consideration, it is apparent that dwarf shrubs responded positively only at the poorer sites, where there was no strong competition from grasses or herbs (Rodenkirchen 1991, Kellner 1993a). Thus, it seems as if the response of vascular plants to nitrogen fertilization was generally determined by changes in the competitive relationships. The changes in field-layer vegetation after fertilization could often be measured as increased mean scores of the nitrogen indicator value according to Ellenberg et al. 1991 (see Rodenkirchen 1991 and van Dobben et al. 1993).

A drawback of most fertilization experiments is that nitrogen has not been added in a way that is equivalent to atmospheric deposition. Usually nitrogen has been added in high doses (10 g N m^{-2} or more) at intervals of several years. At best, nitrogen has been added once a year in doses comparable to the nitrogen deposition levels in central Europe and south-west Scandinavia (van Dobben et al. 1993). The study by van Dobben et al. (1993) is interesting also because treatments with artificial acidification (sulphuric acid) and liming are linked with a nitrogen dosage experiment.

Studies of actual changes of field-layer vegetation at current levels of air-borne nitrogen deposition may be of two types; time-series observations and regional comparisons. Both these approaches have drawbacks: in time-series it may be hard to establish that nitrogen deposition has in fact caused the changes; natural succession and changes in management practices are alternative possibilities. Similarly, regional comparisons suffer from the difficulty of separating deposition gradients from other geographical gradients.

We have searched the literature for time-series studies of field-layer vegetation in European coniferous forests, spanning at least ten years and in which the last observation was made after 1980. Studies made in broad-leaved forests were not included, since these topics are covered by Falkengren-Grerup (1995). We found six relevant studies or groups of studies (Table 1). The studies were made with different methods and the studied areas differ in nitrogen loads as well as in site properties. In two of the studies the stands had suffered from crown thinning. Nevertheless, there are similarities in the vegetation changes observed at sites with a deposition higher than 1 g N m^{-2}. A consistent observation is an increase in grasses and nitrophilic herbs together with an increase in the mean score of the nitrogen indicator value (according to Ellenberg et al. 1991). Species number increased where any change was observed. Thus, the changes are similar to those observed in nitrogen fertilization experiments, and it seems that nitrogen enrichment is a major factor behind the changes. The studies by Hofmann (1987, Hofmann et al. 1990) and Rodenkirchen (1991, 1992) include fertilization experiments linked to the retrospective studies, which strengthen this conclusion. In south Finland, where N deposition was the lowest of the studied areas, *Deschampsia flexuosa* did not increase, whereas it did so in all the other *Vaccinium*-type sites.

Nitrogen enrichment of these systems is caused by a combination of increased nitrogen deposition and changed land use. Before the beginning of this century, more nitrogen was lost from the systems, through litter removal (in central Europe), live-stock grazing and more frequent forest fires (in Fennoscandia). The study by Rosén et al. (1992) was made along a deposition gradient, which strengthens the inference that deposition was the main cause of the changes.

In the following, we report the results of a five-year fertilization experiment in which nitrogen was applied to permanent plots at rates relevant to the current deposition rates in Fennoscandia. In the Discussion we will first consider the implications of this experiment and then discuss more general aspects of the effects of nitrogen deposition on the field layer in coniferous forests.

Material and methods

The study site is located on a small island (2 km^2), named Grimskär, in the archipelago off Söderhamn (61°13'50"N, 17°14'30"E). Annual mean temperature at the meteorological station at Söderhamn is 4.3°C, and mean annual precipitation is 630 mm (means for 1931–1960). The study was performed in a mixed mature coniferous forest dominated by *Picea abies* (L.) Karst. and *Pinus sylvestris* L. Dominant trees are 80–100 yr old at breast height (1.3 m above ground), and basal area at breast height is 34 m^2 ha^{-1}; 16 m^2 ha^{-1} for *P. abies* and 18 m^2 ha^{-1} for *P. sylvestris*. The soil is a washed till, stony at the surface, with a well developed mor layer. The ground is covered by a moss carpet dominated by *Pleurozium schreberi* (Brid.) Mitt., *Hylocomium splendens* (Hedw.) B.S.G. and *Ptilium crista-castrensis* (Hedw.) De Not. The field layer is dominated by *V. myrtillus*. According to the species composition, site productivity can be classified as moderately poor (*Vaccinium myrtillus* type, Hägglund and Lundmark 1977).

Within this forest we established 25 plots (2.0 × 2.0 m) in May 1988. We chose the plots subjectively in an attempt to minimize the differences in vegetation. Eight field-layer species were present in the plots: *V. myrtillus*, *V. vitis-idaea*, *Trientalis europaea* L., *Linnaea borealis* L., *Goodyera repens* (L.) R. Br., *D. flexuosa*, *Luzula pilosa* (L.) Willd. and *Maianthemum bifolium* (L.) F. W. Schm. Each year in September, from 1988 until 1992, we counted the number of ramets of each species in the various plots, disregarding *D. flexuosa*. Each plant part with no connection above ground to any other plant part was defined as a ramet. Because of the large number of shoots, we estimated the density of *D. flexuosa* by counting the number of shoots in ten 0.316 × 0.316 m (0.1 m^2) frames laid out at random in each plot.

To simulate increased nitrogen deposition, five different fertilization treatments were employed: i) control with no extra nitrogen, ii) 0.5, iii) 1.0, iv) 2.0, and v) 4.0 g N m^{-2} yr^{-1}. Ammonium nitrate was used as the nitrogen source. The rates chosen correspond to the range of nitrogen deposition in different parts of Sweden (Lövblad et al. 1992). We randomized the five treatments in five replicates over the 25 2.0 × 2.0 m plots. Application of nitrogen started in May 1988 and continued until September 1991. The nitrogen fertilizer was added as powder in five applications per year, once a month between May and September, to avoid any problem associated with snow cover.

We analysed the effect of treatment on each species by an Anova with a repeated measured design with the nitrogen treatment as a grouping factor (Wilkinson 1990). A repeated measured design has more than one measurement on each subject. In this analysis, the subject was the ramet density in the respective plots,which were repeatedly measured over five years. The test makes it possible to detect changes in density over time together with any interaction of these changes with the nitrogen treatment. For those species where a significant interaction effect of nitrogen was detected, we performed a one-way repeated Anova for each nitrogen treatment and

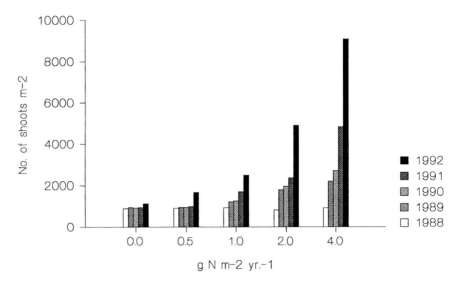

Fig. 1. The yearly density of *Deschampsia flexuosa* shoots m^{-2} from 1988 to 1992, in five different nitrogen treatments.

a single degree-of-freedom polynomial test. This procedure makes it possible to detect the nitrogen application rate at which there was a significant effect on ramet density, and also to detect any linear trend over time (Wilkinson 1990). For *D. flexuosa* and *T. europaea*, the data were log transformed to achieve equal variance within cells. The univariate F test and the Huynh-Feldt and multivariate F statistics lead to the same conclusions. Thus, we assume that it is safe to trust the results given by the traditional univariate F test.

Results

Of the eight field-layer species present in the 25 plots, only six were present in all the plots in such a number that it was possible to analyse the variation in abundance statistically. We detected a significant variation over years for three of the six species studied. For only two of these species was there a significant interaction between the changes over years and the nitrogen treatment.

Table 2. Analyses of variance in density of *Deschampsia flexuosa* over a period of five years, resulting from nitrogen treatment. The analyses were performed by an Anova with a repeated measured design with the nitrogen treatment as a grouping factor. Five different fertilization treatments were employed: i) control with no extra nitrogen, ii) 0.5, iii) 1.0, iv) 2.0, and v) 4.0 g N m^{-2} yr^{-1}.

Source	SS	df	F	p
Year	18.9	4	56.5	<0.001
Year * Nitrogen	8.1	16	6.06	<0.001
Error	3.3	80		

Deschampsia flexuosa

The shoots of *D. flexuosa* were evenly distributed in the area. At the beginning of the study the density of shoots was similar in all treatments, ranging between 820 to 900 shoots m^{-2}. In control plots the density of shoots remained fairly constant over the five years, unlike all the fertilized plots, in which the density increased (Fig. 1). In the highest nitrogen treatment, the shoot density had increased > 8-fold after five years and the grass was beginning to become tufted. In the fifth year, there were even a small number of inflorescences in the highest nitrogen treatment.

We found a significant interaction between nitrogen and time in the increase in shoot density (Table 2). The one-way repeated measure analysis indicated that the increase in shoot density was significant for all nitrogen treatments (Table 3). For all nitrogen treatments, except the lowest, the polynomial test indicated a significant linear trend in ramet density with time. A comparison of the sum of squares of the one-way repeated measure design and the single degree-of-freedom polynomial design contrast showed that the linear trend accounted for 90 to 98% of the variability over the five years in the three highest nitrogen treatments.

The magnitude of the increase in ramet density after five years was closely related to the amount of added nitrogen. In the range from the lowest up to the highest nitrogen treatment the shoot density increased 1.7, 3.5, 5.3 and 8.7 times.

Trientalis europaea

At the beginning of the study, in 1988, the ramet density of *T. europaea* varied from 12 to 27 m^{-2} between plots. In 1992, ramet density varied between 26 and 123 m^{-2}. The

Table 3. Analyses of variance in density of *Deschampsia flexuosa* over a period of five years within different nitrogen treatments. The analyses were performed with an one-way repeated Anova and a single degree-of-freedom polynomial contrast for each nitrogen treatment.

g N m^{-2} yr^{-1}	One way repeated ANOVA					Single degree of freedom polynomial contrast			
	Source	SS	df	F	p	SS	df	F	p
0	Year	0.186	4	1.20	NS				
	Error	0.617	16						
0.5	Year	1.046	4	6.43	0.003	0.644	1	5.63	0.077
	Error	0.651	16			0.457	4		
1.0	Year	5.277	4	12.9	<0.001	5.111	1	23.9	0.008
	Error	1.639	16			0.856	4		
2.0	Year	7.373	4	26.0	<0.001	6.671	1	74.8	0.001
	Error	1.132	16			0.357	4		
4.0	Year	13.137	4	19.8	<0.001	12.814	1	32.5	0.005
	Error	2.656	16			1.576	4		

increase in ramet density was most pronounced in the fertilized plots (Fig. 2).

The statistical analysis showed that the increase in ramet density over time was significant and that the change in density interacted with the nitrogen treatment (Table 4). The one-way repeated measure design indicated that the increase in shoot density was significant in all treatments (Table 5). In all nitrogen treatments the polynomial test indicated a significant linear trend in ramet density over time. A comparison of the sum of squares of the one-way repeated measure design and the single degree-of-freedom polynomial design contrast showed that the linear trend accounted for 86 to 97% of the variability over the five years in the nitrogen treatments. In the control treatment, however, there was no linear trend with time.

The magnitude of the increase in ramet density after five years was related to the amount of added nitrogen. The density of ramets more than doubled in the two treatments with the lowest amounts of added nitrogen and increased more than five times in the two highest nitrogen treatments.

Vaccinium myrtillus

Vaccinium myrtillus increased slightly in density over the five years. The density increased from 36 ramets m^{-2} in 1988 to 46 m^{-2} in 1992. The variation with time was significant ($F_{4,80} = 5.21$; $p = 0.001$). The increase in ramet density was observed in all treatments and we detected no significant interaction between the increase in ramet density over the period and nitrogen treatments ($F_{16,80} = 0.877$; NS). The single degree-of-freedom polynomial design contrast of the difference between years was significant ($F_{1,4} = 9.41$; $p = 0.006$) and linear trend accounted for 65% of the variability over the five years.

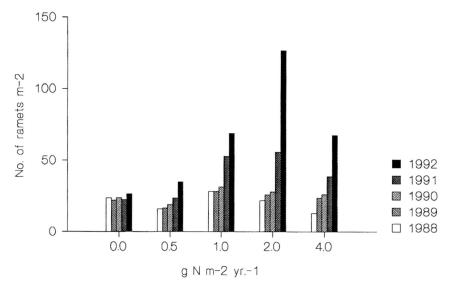

Fig. 2. The yearly density of *Trientalis europaea* ramets m^{-2} from 1988 to 1992, in five different nitrogen treatments.

Table 4. Analyses of variance in density of *Trientalis europaea* over a period of five years, resulting from nitrogen treatment. The analyses were performed by an Anova with a repeated measured design with the nitrogen treatment as a grouping factor. Five different fertilization treatments were employed: i) control with no extra nitrogen, ii) 0.5, iii) 1.0, iv) 2.0, and v) 4.0 g N m^{-2} yr^{-1}.

Source	SS	df	F	p
Year	16.5	4	79.6	<0.001
Year * Nitrogen	6.4	16	7.68	<0.001
Error	4.2	80		

Luzula pilosa, Linnaea borealis and *Vaccinium vitis-idea*

In the three species *L. pilosa*, *L. borealis* and *V. vitis-idaea*, there was no significant variation in the abundance of ramets between years in any plot and no effect of nitrogen treatment.

Discussion

The experiment

In the context of the critical load for nitrogen and the extent to which nitrogen deposition may cause changes in vegetation composition, two significant results were obtained in this fertilization experiment. First, the capacity to utilize an increased supply of nitrogen in growth differs between species. Second, this difference between species is already apparent at very low levels of nitrogen fertilization.

That species differ in ability to utilize an increased supply of nitrogen is a well-known fact from several studies (Chapin 1980). This variation is of great importance since it usually results in changes in the competitive relationship between species when nitrogen supply increases (e.g. Berendse and Elberse 1990). In our study,

this variation is observed at comparatively low rates of nitrogen supply. The result indicates that a long-term change in coniferous forest vegetation composition as a result of nitrogen deposition is possible within the current range of deposition over most of Sweden.

Only two of six species in the field layer were able to utilize the increased supply of nitrogen to give an increase in abundance. Many forest fertilization experiments indicate that the response of a single species usually depends on the current type of vegetation. In an open stand on a nutrient-poor soil with dwarf-shrub vegetation and with no graminoid species present, the dominant dwarf shrub will increase if nitrogen is added (Tälli and Veermets 1979, Faltynowicz 1986, Kellner 1993a). However, if there are faster-growing species present, these species will increase and the dwarf shrubs will decrease after nitrogen fertilization (Mälkönen et al. 1980, van Dobben et al. 1993, Kellner 1993a,b). In our experiment, *Deschampsia flexuosa* and *Trientalis europaea* increased and *Vaccinium myrtillus* remained unchanged. Apparently the competition from *Deschampsia* and *Trientalis* was not intense enough to cause *V. myrtillus* to decrease. Like the forest at our experimental site, most of the coniferous forests in relatively unpolluted areas of northern Europe are dwarf-shrub dominated forest types, usually containing scattered plants of grasses (usually *D. flexuosa*, but also *Festuca ovina* L., *Agrostis capillaris* L., *Calamagrostis arundinacea* (L.) Roth and others) and forbs (e.g. *T. europaea*). In such forests, addition of nitrogen will typically lead to expansion of *D. flexuosa* or some other grass species, whereas the dwarf shrubs will decline (Hofmann et al. 1990, Rodenkirchen 1992, Rosén et al. 1992, van Dobben et al. 1993). Our study indicates that such vegetational changes are already initiated at low deposition rates. However, the results should be interpreted with some caution since the interaction with the tree layer is not the same when nitrogen is added only to small plots as when it is added to whole sites. It has been suggested that nitrogen deposition promotes a denser tree canopy, which might lead to light limitation of field-layer species (Högbom and Högberg 1991).

Table 5. Analyses of variance in density of *Trientalis europaea* over a period of five years within different nitrogen treatments. The analyses were performed with an one-way repeated Anova and a single degree-of-freedom polynomial contrast for each nitrogen treatment.

g N m^{-2} yr^{-1}	One way repeated ANOVA					Single degree of freedom polynomial contrast			
	Source	SS	df	F	p	SS	df	F	p
0	Year	0.116	4	4.94	0.009	0.034	1	4.58	NS
	Error	0.094	16			0.029	4		
0.5	Year	1.541	4	4.86	0.009	1.316	1	10.9	0.030
	Error	1.270	16			0.484	4		
1.0	Year	3.372	4	21.6	<0.001	2.885	1	40.7	0.003
	Error	0.623	16			0.284	4		
2.0	Year	10.363	4	59.9	<0.001	8.886	1	175	<0.001
	Error	0.692	16			0.204	4		
4.0	Year	7.530	4	20.4	<0.001	7.296	1	45.7	0.002
	Error	1.477	16			0.638	4		

Of the species in our experiment, *Deschampsia flexuosa* increased the most, and it is also the species that has increased most after nitrogen addition in the majority of studies. However, taller and more broad-leaved grasses, if present, may outcompete *D. flexuosa* (Hofmann 1987, Hester et al. 1991b, Kellner 1993b). Other species typical of forest clearings also increase in abundance after nitrogen addition, e.g. *Epilobium angustifolium,* and *Rubus idaeus* L. (Hofmann 1987, Bürger 1991, Rodenkirchen 1992, van Dobben et al. 1993). Under an extremely high nitrogen load, tall herbs (*E. angustifolium, Urtica dioica* L.) or semi-shrubs (*Rubus* spp.) may come to dominate the vegetation (Hofmann 1987, Hofmann et al. 1990).

General

Mechanisms of vegetation changes resulting from nitrogen deposition

The main mechanism in N-limited forests for vegetation changes after increased nitrogen deposition seems to be the change in competitive relationships that is caused by increased nitrogen availability. Evidence from heathlands (Berendse and Aerts 1984, Heil and Bruggink 1987, Aerts et al. 1990) and successional birch woodland (Hester et al. 1991b) confirm this conclusion. However, other mechanisms are also possible and evidence relating to their importance in coniferous forests is briefly reviewed below.

Coniferous forests in north Europe usually grow on podsolic soils with acid mor-layers, in which NH_4^+ is the prevailing form of inorganic nitrogen. In areas with high N-deposition, the concentration of NO_3^- increases because of increased nitrification (Högbom and Högberg 1991). Many of the species typical of coniferous forests prefer ammonium as their nitrogen source (Haynes 1986), and may even lack the enzyme nitrate reductase, which is needed in the assimilation of nitrate (Högberg et al. 1990). In contrast, many weedy species and species occurring in richer habitats prefer nitrate as their nitrogen source (Haynes 1986), which may contribute to the changed competitive relationships. High nitrate concentrations may also contribute to break seed dormancy of some species.

Nitrogen addition can lead to a denser tree canopy, which may have a decisive influence on the field-layer vegetation. It is often difficult to separate the effects of competition within the field layer from the effects of shading and nutrient uptake by the tree layer. However, a study in successional birch woodland (Hester et al. 1991a,b) indicates that *Calluna vulgaris* (L.) Hull was limited mainly by tree canopy shading, in contrast to *Vaccinium myrtillus*, which was limited mainly by competing grass species. Similarly, an early decrease of *C. vulgaris* in a fertilized and irrigated old pine forest was interpreted as an effect of shading from the tree canopy (Kellner and Mårshagen 1991).

The patterns of herbivory may often be changed by nitrogen addition. For example, in a forest fertilization experiment, the vegetation changed from *Vaccinium myrtillus* dominance to *Epilobium angustifolium* dominance when herbivores were fenced out, but to *Calamagrostis epigeios* (L.) Roth dominance when herbivores were present (Hofmann 1987). Ellenberg (1986, 1988) argued that the population of roe-deer *Capreolus capreolus* has increased in Germany because of the increased field-layer production resulting from nitrogen deposition and thinning of tree canopies, and that the increased herbivore population has caused a shift in vegetation towards less palatable species. The effect of nitrogen addition on mycorrhizal fungi is probably another important aspect of indirect effects on vascular plants, though the effect on field-layer species is little known (Turnau et al. 1992, Tyler et al. 1992, Wallander and Nylund 1992).

Nutrient imbalance can occur on sites which approach nitrogen saturation. For example, in a central European spruce stand with high nitrogen deposition, *Oxalis acetosella* L., which is considered a nitrophilic plant (Ellenberg et al. 1991), did not increase in control plots but increased in plots receiving $CaCO_3$ or Ca SO_4 (Rodenkirchen 1993). Similarly, many species in broad-leaved forests in south Sweden are constrained by the low pH-levels and high Al-concentrations prevailing, and do not respond positively to nitrogen addition (Falkengren-Grerup 1993, Falkengren-Grerup and Tyler 1993). However, most coniferous forests in Fennoscandia are still nitrogen-limited. The effect of soil acidification on the field-layer vegetation in coniferous forests would be expected to be less than in broad-leaved forests, since the soils of coniferous forests are fairly acid also at more pristine sites, and the species growing there are adapted to those more acid conditions. Wittig (1988) proposed this as the explanation why mean indicator values for soil reaction (Ellenberg et al. 1991) have not decreased in some central European forest sites (Table 1).

An important aspect of nitrogen effects on forest vegetation is what happens during the regenerative phase of the forest. Hypothetically, we would expect the effects on the field-layer vegetation in a clearing to be more pronounced than in a closed stand, both because there is no tree stand that takes up part of the nitrogen and because the field layer is in a dynamic phase, with more disturbed soil available for colonization, etc. However, there are no published studies of this aspect.

In conclusion, changed competitive relationships within the field layer seem to be the most important mechanism behind vegetation changes caused by nitrogen deposition. Other mechanisms might be important at certain sites and/or for certain species. Of special interest are the influences of grazing and of tree canopy shading, since these factors can be manipulated through hunting and forest management, respectively. Unfortunately, current knowledge of the interplay between nitrogen deposition, tree canopy density, herbivores and field-layer vegetation is extremely scarce (but see Ellenberg 1988 for some conclusions based on circumstantial evidence).

Effects on different vegetation types

The effects of nitrogen deposition on the coniferous forest field-layer vegetation seem to be dependent on soil fertility. Intuitively, one would expect the naturally most fertile sites to be least influenced by nitrogen deposition. Indeed, this seems to be correct, since grass and herb forest types (forest types according to Hägglund and Lundmark 1977) were less changed than dwarf shrub types in a 20-yr study made in several forest types with pine in north-east Germany (Hofmann 1987, Hofmann et al. 1990). Similarly, in a fertilization experiment on different sites in central Sweden, a herb forest type site was hardly affected at all by nitrogen fertilization (15 g N m^{-2}), whereas a *Vaccinium vitis-idaea* type site changed considerably (Kellner 1993b).

However, the poorest sites usually have fewer changes in their field-layer vegetation than intermediate sites, because they lack species (usually graminoids) that are able to respond to the nitrogen addition with substantially increased growth. This was the case in fertilization experiments in central and northern Sweden, where a lichen type site and a lichen-rich type site were less changed than *Vaccinium myrtillus* sites by repeated nitrogen fertilization (12–60 g N m^{-2} repeated at 5–7 yr intervals for 20 yr; Kellner 1993a). Similarly, in the above-mentioned study in north-east Germany, a lichen type site which lacked a field-layer in 1963 had not attained any field-layer cover in 1985, whereas *V. myrtillus* sites changed to *Deschampsia flexuosa* dominance over the same period (Hofmann 1972, 1987, Hofmann et al. 1990). At very dry sites, water limitation may contribute towards weaker response to nitrogen deposition.

The position of the forest stand in the landscape is of great importance. Not only does it affect deposition rates, but also the availability of seed sources for nitrophilic plants that might colonize in the stand. Forest edges are more liable to vegetation changes both because of higher deposition rates and because of the presence of nearby sources of ruderal plant seeds (Sougnez and Weissen 1977, Hasselrot and Grennfelt 1987).

Effects on species richness

Species richness in Nordic coniferous forest is generally considered to be positively correlated with soil pH (Helliwell 1978) or "soil fertility" (Tonteri et al. 1990). "Soil fertility" is associated both with relative nitrogen content, pH and availability of base cations in the soil; variables that are usually strongly intercorrelated (Dahl et al. 1967). Similar positive relationships between species richness and soil fertility or pH have been found in temperate deciduous forests (Grubb 1987, Peet and Christensen 1988, Palmer 1990), exceptions being spruce plantations on former arable land and the most fertile broad-leaved forests, in which light availability governs species richness of the field layer (Helliwell 1978, Tyler 1989). As far as we know, there are no studies in which the influence of pH and nitrogen content on species richness have been separated.

Three retrospective studies in central European coniferous forests have shown that species numbers had increased during the past 20 yr or more (Bürger 1991, Rodenkirchen 1991, 1992, Becker et al. 1992). Fertilization with only nitrogen had very little impact on the number of vascular plant species in coniferous forests, although species composition changed (Rodenkirchen 1991, Kellner 1993a,b). However, fertilization with large amounts of NPK or complete nutrient mixtures resulted in substantial increases in number of species, especially when combined with liming (Guzikowa et al. 1976, Persson 1981, Hofmann 1987, Kellner and Mårshagen 1991, Rodenkirchen 1991, Becker et al. 1992). However, if the canopy closes beyond the point where light becomes limiting, species number will decrease again (Tamm 1991).

Conclusion

Our experiment shows that even low levels of nitrogen deposition result in changes in field-layer vegetation. The experience from this study and others is that nitrogen deposition usually results in an increase of the most competitive species at the site, whereas less competitive species may decrease. For the common *Vaccinium myrtillus* forest type, this usually means that *Deschampsia flexuosa* increases while *V. myrtillus* and *V. vitis-idaea* decrease.

We recommend that future research should be directed at: 1) interaction effects of nitrogen deposition and tree canopy density on the field-layer; 2) interactions between nitrogen deposition, field-layer vegetation and herbivores; and 3) the influence of nitrogen deposition on field-layer succession during forest regeneration. Information on these three issues should have important implications for forest management in a situation with chronic nitrogen deposition.

Acknowledgements – The work was financed by the Swedish Environmental Protection Board.

References

Aerts, R., Berendse, F., de Caluwe, H. and Schmitz, M. 1990. Competition in heathland along an experimental gradient of nutrient availability. – Oikos 57: 310–318.

Becker, M., Bonneau, M. and Le Tacon, F. 1992. Long-term vegetation changes in an *Abies alba* forest: natural development compared with response to fertilization. – J. Veg. Sci. 3: 467–474.

Berendse, F. and Aerts, R. 1984. Competition between *Erica tetralix* L. and *Molinia caerulea* (L.) Moench as affected by the availability of nutrients. – Acta Oecol. 5: 3–14.

– and Elberse, W. T. 1990. Competition and nutrient availability in heathland and grassland ecosystems. – In: Grace, J. B.

and Tilman, D. (ed.), Perspectives on plant competition. Academic Press Inc., San Diego. pp. 93–116.

Bobbink, R. 1991. Effects of nutrient enrichment in Dutch chalk grassland. – J. Appl. Ecol. 28: 28–41.

Bürger, R. 1991. Immissionen und Kronenverlichtung als Ursachen für Veränderungen der Waldbodenvegetation im Schwarzwald. – Tuexenia 11: 407–424.

Chapin, S. F. III. 1980. The mineral nutrition of wild plants. – Ann. Rev. Ecol. Syst. 11: 233–260.

Dahl, E., Gjems, O. and Kielland-Lund, J. 1967. On the vegetation types of Norwegian conifer forests in relation to the chemical properties of the humus layer. – Meddelser fra Det norske skogforsøksvesen 23: 503–531.

Ellenberg, H. jun. 1985. Veränderungen der Flora Mitteleuropas unter dem Einfluss von Düngung und Imissionen. – Schweiz. Z. Forstwes. 136: 19–39.

– 1986. Immissionen – Produktivität der Krautschicht – Populationsdynamik des Rehwilds: ein Versuch zum Verständnis ökologischer Zusammenhänge. – Natur Landsch. 61: 335–340.

– 1988. Eutrophierung – Veränderungen der Waldvegetation – Folgen für den Reh-Wildverbiss und dessen Rückwirkungen auf die Vegetation. – Schweiz. Z. Forstwes. 139: 261–282.

Ellenberg, H. sen., Weber, H. E., Düll, R., Wirth, V., Werner, W. and Paulissen, D. 1991. Zeigerwerte von Pflanzen in Mitteleuropa. Scripta Geobotanica 18. – Verlag Erich Goltze KG, Göttingen.

Falkengren-Grerup, U. 1986. Soil acidification and vegetation changes in deciduous forest in southern Sweden. – Oecologia 70: 339–347.

– 1992. Mark- och floraförändringar i sydsvensk ädellövskog (Summary: Soil and floral changes in hardwood forests of southern Sweden). – Swedish Environmental Protection Board, Report 4061.

– 1993. Effects on beech forest species of experimentally enhanced nitrogen deposition. – Flora 188: 85–91.

– and Tyler, G. 1993. Soil chemical properties excluding field-layer species from beech forest mor. – Plant Soil 148: 185–191.

Faltynowicz, W. 1986. The dynamics and role of lichens in a managed *Cladonia*-pine forest (*Cladonia-Pinetum*). – Monographiae botanicae 69: 1–97.

Galloway, J. N. and Likens, G. E. 1981. Acid precipitation; the importance of nitric acid. – Atm. Environ. 15: 1081–1085.

Grennfelt, P. and Hultberg, H. 1986. Effects of nitrogen deposition on the acidification of terrestrial and aquatic ecosystems. – Water Air Soil Pollut. 31: 945–964.

Grubb, P. J. 1987. Global trends in species-richness in terrestrial vegetation: a view from the northern hemisphere. – In: Gee, J. H. R. and Giller, P. S. (eds), Organization of communities, past and present. Symp. Brit. Ecol. Soc. 27. Blackwell Sci. Publ., Oxford. pp. 99–118.

Guzikowa, M., Latocha, E., Pancer-Kotewoja, E. and Zarzycki, K. 1976. The effect of fertilization on a pine forest ecosystem in an industrial region. – Ekol. Pol. 24: 307–318.

Hägglund, B. and Lundmark, J.-E. 1977. Site index estimation by means of site properties. Scots pine and Norway spruce in Sweden. – Stud. For. Suec. 138: 1–38.

Hasselrot, B. and Grennfelt, P. 1987. Deposition of air pollutants in a wind-exposed forest edge. – Water Air Soil Pollut. 34: 135–143.

Haynes, R. J. 1986. Uptake and assimilation of mineral nitrogen by plants. – In: Haynes, R. J. (ed.), Mineral nitrogen in the plant-soil system. Academic Press, Orlando. pp. 303–378.

Heil, G. W. and Diemont, W. H. 1983. Raised nutrient levels change heathland into grassland. – Vegetatio 53: 113–120.

– and Bruggink, M. 1987. Competition for nutrients between *Calluna vulgaris* (L.) Hull and *Molinia caerulea* (L.) Moench. – Oecologia 73: 105–107.

Helliwell, D. R. 1978. Floristic diversity in some central Swedish forests. – Forestry 51: 151–161.

Hester, A. J., Miles, J. and Gimingham, C. H. 1991a. Succession from heather moorland to birch woodland. I. Experimental alteration of specific environmental conditions in the field. – J. Ecol. 79: 303–316.

– , Miles, J. and Gimingham, C. H. 1991b. Succession from heather moorland to birch woodland. II. Growth and competition between *Vaccinium myrtillus*, *Deschampsia flexuosa* and *Agrostis capillaris*. – J. Ecol. 79: 317–327.

Högberg, P., Johannisson, C., Nicklasson, H. and Högbom, L. 1990. Shoot nitrate reductase activities of field-layer species in different forest types. – Scand. J. For. Res. 5: 449–456.

Högbom, L. and Högberg, P. 1991. Nitrate nutrition of *Deschampsia flexuosa* (L.) Trin. in relation to nitrogen deposition in Sweden. – Oecologia 87: 488–494.

Hofmann, G. 1972. Vegetationsveränderungen in Kiefernbeständen durch Mineraldüngungen und Möglichkeiten zur Nutzanwendung der Ergebnisse für biologische Leistungsprüfungen. – Beitr. Forstwirtsch. 2: 29–36.

– 1987. Vegetationsänderungen in Kiefernbeständen durch Mineraldüngung. – Hercynia 24: 271–278.

– , Heinsdorf, D. and Krauss, H.-H. 1990. Wirkung atmogener Stickstoffeinträge auf Produktivität und Stabilität von Kiefern-Forstökosystemen. – Beitr. Forstwirtsch. 24: 59–73.

Kellner, O. 1993a. Effects on associated flora of sylvicultural nitrogen fertilization repeated at long intervals. – J. Appl. Ecol. 30: 563–574.

– 1993b. Effects of fertilization on forest flora and vegetation. – Acta Univ. Ups. 464: 1–32.

– and Mårshagen, M. 1991. Effects of irrigation and fertilization on the ground vegetation in a 130-year-old stand of Scots pine. – Can. J. For. Res. 21: 733–738.

Lähde, E. and Nieppola, J. 1987. Vegetation changes in old stands of *Pinus sylvestris* L. in southern Finland. – Scand. J. For. Res. 2: 369–377.

Lövblad, G., Amann, M., Andersen, B., Hovmand, M., Joffre, S. and Pedersen, U. 1992. Deposition of sulfur and nitrogen in the Nordic countries: present and future. – Ambio 21: 339–347.

Mälkönen, E., Kellomäki, S. and Holm, J. 1980. Effect of nitrogen, phosphorus and potassium fertilization on ground vegetation in Norway spruce stands (in Finnish with summary in English). – Metsäntutkimuslaitoksen Julkaisuja 98: 3: 1–35.

Morris, J. T. 1991. Effects of nitrogen loading on wetland ecosystems with particular reference to atmospheric deposition. – Ann. Rev. Ecol. Syst. 22: 257–279.

Nieppola, J. 1992. Long-term vegetation changes in stands of *Pinus sylvestris* in southern Finland. – J. Veg. Sci. 3: 475–484.

Nilsson, J. and Grennfelt, P. (eds) 1988. Critical loads for sulphur and nitrogen. Miljørapport 1988: 15. – Nordic council of ministers, Copenhagen.

Palmer, M. W. 1990. Spatial scale and patterns of vegetation, flora and species richness on hardwood forests of the North Carolina Piedmont. – Coenoses 5: 89–96.

Peet, R. K. and Christensen, N. L. 1988. Changes in species diversity during secondary succession on the North Carolina piedmont. – In: During, H. J., Werger, M. J. A. and Willems, J. H. (eds), Diversity and pattern in plant communities. SPB Academic Publishing, The Hague. pp. 233–245.

Persson, H. 1981. The effect of fertilization and irrigation on the vegetation dynamics of a pine-heath ecosystem. – Vegetatio 46: 181–192.

Redbo-Torstensson, P. 1994. The demographic consequences of nitrogen fertilization of a population of sundew, *Drosera rotundifolia*. – Acta. Bot. Neerl. 43: 175–188.

Rodenkirchen, H. 1991. Der Wandel der Bodenvegetation in den Kiefernforsten des Staatlichen Forstamtes Waldsassen infolge anthropogen bedingter Bodenveränderung. – Bayreuther Bodenkundliche Berichte 17: 261–273.

– 1992. Effects of acidic precipitation, fertilization and liming on the ground vegetation in coniferous forests of southern Germany. – Water Air Soil Pollut. 61: 279–294.

– 1993. Wirkungen von Luftverunreinigung und künstlichem sauren Regen auf die Bodenvegetation in Koniferenwäldern. – Forstw. Cbl. 112: 70–75.

Rosén, K., Gundersen, P., Tegnhammar, L., Johansson, M. and Frogner, T. 1992. Nitrogen enrichment of Nordic forest ecosystems, the concept of critical loads. – Ambio 21: 364–368.

Sougnez, N. and Weissen, F. 1977. Evolution de la couverture morte et de la couverture vivante après fumure en vieille forêt d'épicéa commun. – Bull. Rech. Agron. Gembloux 12: 233–248.

Tälli, P. and Veermets, M. 1979. Über die Reagierbarkeit der Bodenvegetation der *Vaccinium*-Kiefernbestände auf die Mineraldünger. – Metsanduslikud Uurimused, Estonian SSR 14: 119–133.

Tamm, C.O. 1991. Nitrogen in terrestrial ecosystems. Ecological Studies 81. – Springer, Berlin.

Tonteri, T., Mikkola, K. and Lahti, T. 1990. Compositional gradients in the forest vegetation of Finland. – J. Veg. Sci. 1: 691–698.

Turnau, K., Mitka, J. and Kędzierska, A. 1992. Mycorrhizal status of herb-layer plants in a fertilized oak-pine forest. – Plant Soil 143: 148–152.

Tyler, G. 1987. Probable effects of soil acidification and nitrogen deposition on the floristic composition of oak (*Quercus robur* L.) forest. – Flora 179: 165–170.

– 1989. Interacting effects of soil acidity and canopy cover on the species composition of field-layer vegetation in oak/hornbeam forests. – For. Ecol. Manage. 28: 101–114.

– , Balsberg Pålsson, A.-M., Bergkvist, B., Falkengren-Grerup, U., Folkesson, L., Nihlgård, B., Rühling, Å. and Stjernquist, I. 1992. Chemical and biological effects of artificially increased nitrogen deposition to the ground in a Swedish beech forest. – Scand. J. For. Res. 7: 515–532.

van Breemen, N. and van Dijk, H.F.G. 1988. Ecosystem effects of atmospheric deposition of nitrogen in the Netherlands. – Environ. Pollut. 54: 249–274.

van Dobben, H.F., ter Braak, C.J.F. and Tamm, C.O. 1993. Forest undergrowth as a biomonitor for deposition of nutrients and acidity. – Thesis, Univ. of Utrecht. pp. 113–138.

Wallander, H. and Nylund, J.-E. 1992. Effects of excess nitrogen and phosphorus starvation on the extramatrical mycelium of ectomycorhizas of *Pinus sylvestris* L. – New Phytol. 120: 495–503.

Wilkinson, L. 1990. SYSTAT: The system for statistics. – SYSTAT Inc., Evanston, IL.

Wittig, R. 1988. Retrospective studies of changes in central European forests by means of repeating phytosociological surveys. – In: Salbitano, F. (ed.), Human influence on forest ecosystems development in Europe. ESF FERN-CNR, Pigatora Editrice, Bologna. pp. 139–147.

Ecological Bulletins 44: 238–247. Copenhagen 1995

Changes in epiphytic lichen and moss flora in some beech forests in southern Sweden during 15 years

Kurt Olsson

Olsson, K. 1995. Changes in epiphytic lichen and moss flora in some beech forests in southern Sweden during 15 years. – Ecol. Bull. (Copenhagen) 44: 238–247.

Epiphytic vegetation (lichens, bryophytes and algae) in 12 beech forests in the provinces of Skåne, Halland and Småland was analysed at the end of the 1970's and again at the beginning of the 1990's. The number of species per plot did not change during the investigation period, although the species turnover was between 15% and 35%, mostly involving species with a sparse abundance. In the Skåne sites, *Lepraria incana* and *Chrysotrix candelaris* increased while *Pertusaria amara, P. pertusa* and *Phlyctis argena* decreased. In the Halland sites, the changes were small. In the inland area (Småland), with a richer epiphytic flora, *L. incana, Scoliciosporum chlorococcum, Lecanora conizaeiodes, Isothecium myurum, Pertusaria flavida* and *Ochrolechia androgyna* increased. Although there was a great variation in the epiphytic vegetation among different trees in the sites, an ordination of the vegetation shows that on three sites a significant change in the epiphytic vegetation had occurred. At two of these sites the forest had become less illuminated as a result of a more closed canopy, while at the third one only the differences in the epiphytic vegetation among trees had increased. The vegetational changes found in the investigation were either non directional inter-annual variation or could be related to a change in the canopy cover in the forest causing a decrease in light and an increase in moisture.

K. Olsson, Dept of Ecology, Univ. of Lund, Ecology Building, S-223 62 Lund, Sweden.

Introduction

In recent decades, there have been numerous reports of a decline in species number of epiphytic lichen and moss floras both in urban and in rural areas (Ferry et al. 1973, Hawksworth and Rose 1976). The main causes of the decline are assumed to be related to air pollution, especially sulphur dioxide (Ferry et al. 1973). However, several later reports have shown that the epiphytic flora is recovering in areas that were formerly heavily polluted (Gilbert 1992). In Western Europe the fallout of SO_2 has decreased substantially during the last 15 years (Hawksworth and McManus 1989, Lövblad 1990), while in eastern Europe there has been no such decrease (Lövblad 1990).

For a long period of time there have been reports from southern Sweden about a decline in species number of the lichens in urban areas (Almborn 1943). In recent decades, several authors (Skye 1979, Ekman 1990, Arup and Ekman 1991) have claimed that the epiphytic flora in south-ern Sweden has changed, even in rural areas, as a result of the effects of airborne pollutants.

The present study compares relevées of the crypto-gamic vegetation on beech stems in 12 beech forests in southern Sweden carried out with an interval of 15 years, 1977–79 and 1991–92. There are two specific goals: firstly, to record quantitatively the changes and species turnover of the epiphytic cryptogamic vegetation (lichens, mosses, algae) on the beech stems on these sites; secondly, to study the extent to which the documented changes are directed, and thus considered different from inter-annual changes caused by inter-annual climatic variations. The trends are then correlated with changes in the environment not related to airborne pollutants. This analysis is assisted by ordination using correspondence analysis (CA, Ter Braak 1988, Ter Braak and Prentice 1988). It is suggested that any unexplained trends are related either to an increase or to a decrease in the load of airborne pollutants.

Table 1. Study sites.

	Position (lat long)	Altitude (m)	Area of site (ha)	Dist to sea (km)	Annual mean temp (C°)	Annual mean precip (mm)	Humidity of veg. period 1 (mm)	Humidity of summer 3 (Martonne)	Potential aq balance 4 (mm)	Temp. mean amplitude 5 (C°)	Cont. index 6	Mean temp coldest mon 8 (C°)	Conc. SO_4 in precip. 9 (μekv l^{-1})
Skåne													
Öved	55.70N,13.64E	60	20	32	7.2	660	85	36	-50	18	16.6	-1.6	85
Eriksdal	55.58N,13.08E	65	25	18	7.3	660	75	34	-50	18	16.7	-1.2	85
Högestad	55.57N,13.91E	60	60	15	7.3	660	75	34	-50	18	16.7	-1.2	85
Anklam	55.61N,13.08E	100	9	20	7.2	660	75	35	-50	18	16.7	-1.6	85
Kagarp	55.78N,13.93E	175	7	16	6.8	820	100	35	-40	15.5	15.6	-2.1	80
Maglehem	55.75N,14.12E	85	150	4	6.9	590	50	29	-50	17	14.6	-1.7	80
Halland													
Grimeton	57.09N,12.48E	125	25	14	7	790	175	40	50	18.3	16.7	-2	65
Sibbarp	57.08N,12.55E	100	42	14	7	790	200	41	50	18.3	16.7	-2	65
Öströö	57.07N,12.53E	75	25	12	7	790	200	40	50	18.3	16.7	-2	65
Småland													
Hinnsjöhult	56.89N,14.93E	200	12	85	5.5	650	25	36	-40	10	18.2	-3.9	63
Helgö	56.95N,14.76E	170	6	88	5.5	650	25	36	-40	19	18.1	-3.9	63
Nydala	57.30N,14.30E	200	20	125	5.4	690	75	40	-20	19	18.0	-3.9	57

1. Humidity of vegetation period (precipitation – evaporation) (Eriksson 1986). 3. Humidity June to August (according to Martonne) (Eriksson 1986). 4. Potential water balance of vegetation period (Eriksson 1986). 5. Amplitude of monthly mean temperature (Eriksson 1982). 6. Continentality index according to Gorczynski (Mattsson 1971). Data from Eriksson (1982). 8. Mean temperature of coldest month, mean 1931–1960 (Eriksson 1982). 9. Concentration of SO_4 in precipitation, mean 1983–1987 (Granat 1988).

Study sites and methods

Investigation area

The climate in the coastal regions of southern Sweden is more oceanic than that of the inner part of Småland (Table 1, Ångström 1968). The precipitation and humidity decreases from west to east. The mean temperature of the warmest month is about the same, while the mean temperature of the coldest month is lower and the vegetation period shorter in the more continental areas.

Most of southern Sweden belongs to the boreo-nemoral vegetation zone, except for the main part of the province of Skåne and the coastal parts of the province of Halland, which belong to the nemoral zone (Sjörs 1971). Spontaneous beech *Fagus sylvatica* L. forests occur throughout the nemoral zone and in the south-western part of the boreo-nemoral zone up to a line from Kungsbacka on the west coast to Kalmar on the Baltic coast (Lindgren 1970). The southern part of Skåne is covered by calcareous till, with cambisols soils predominating in the deciduous forests. The rest of the area is characterized by soils of podzolic type on non-calcareous till (Lundegårdh et al. 1970, Troedsson and Nykvist 1973).

Sixteen beech forest sites in southern Sweden were selected in 1977 for an investigation of the epiphytic cryptogamic vegetation on beech stems. The sites were chosen to represent three different gradients, viz. one with decreasing oceanicity, one from nemoral to boreonemoral conditions and one from calcareous to acid soils. Each site was a closed beech forest (canopy cover ≥ 75%) of at least 5 ha. The trees in the forests should be well grown, with an estimated age of c. 100 yr. The sites should not be exposed to local pollution sources like factories, pig farms and cities.

The sites were first investigated in the years 1977–1979 and then 1991–1992. Four of the sites were excluded in the second investigation because they had been heavily thinned in the meantime and now had a dense scrub layer of young beech trees.

Field work

In the first investigation, a sample plot of 20 × 20 m was selected in each site. The plot was situated on even ground ≥100 m from the nearest forest edge. Neither the plots nor individual trees were permanently marked, but the plots were referenced to coordinates and to distinct landmarks. In the second investigation, each new sample plot was established < 50 m from the earlier one. On both occasions the canopy cover in the sample plot was estimated visually.

Seven to ten trees were selected in each plot. The selected trees were well grown with no big branches below 5 m and with a maximum inclination of 5°. A measuring tape was stretched around the stem at five heights above ground (60, 90, 120, 150 and 180 cm) with zero placed to the north. The presence of all epiphytic species was recorded along every one centimetre interval of the transect formed by the edge of the tape. Species occurring on the stem, but not hit by the transect, were also recorded. These species are not included in the data for numerical treatment, because they cannot be given an appropriate cover value. However, they are taken into account when looking at the species turnover. Thus, a species is not considered lost or gained if it appeared outside the transect line in one sampling year, but was hit by the line in the other sampling year. Species appearing only outside transect lines have been treated as neither lost nor gained.

Nomenclature

The names of the lichens follow Santesson (1993). *Lepraria incana* is treated in a collective sense. *Scoliciosporum chlorococcum* was not distinguished from coccoid green algae, they are together treated as *S. chlorococcum* (Ahti and Vitikainen 1974). The mosses were named according to Corley et al. (1981). The fungus *Ascodichaena rugosa* was named according to Hawksworth (1983).

Data treatments

The presence/absence data from each one centimetre along the four upper transects were computed to percentage transect cover, both for each tree and for the sample plot as a whole. In this investigation the lowest transect line is excluded, to avoid influence from the ground.

Indicator values (Ellenberg et al. 1991, Hultengren et al. 1991) were used in the interpretation of species changes. The indicator values in Ellenberg et al. (1991) were also used to compute characteristic indicator values (CIV) for each tree and each site (Persson 1981) according to the following formula:

$$CIV_{jk} = \frac{\Sigma_i(C_{ij}Z_{ik})}{\Sigma_i(C_{ij}); \ Z_{ik}k \neq 0}$$

C_{ij} = relative cover of species i in plot j
Z_{ik} = indicator value k for species i

The program TABORD (van der Maarel et al. 1978) was used to classify the vegetation.

The program CANOCO (Ter Braak 1988) was used for the CA analyses. The input data were the square-root-transformed transect cover values. The transformation reduces the strong weight that would otherwise accrue to a few common and dominating species.

Climatic variables obtained from Eriksson (1982 and 1986) (Table 1) were used as environmental variables in the ordinations. The mean temperature of the coldest month (usually February) was used to represent continentality. The humidity of the summer represents the humidity of the site. The concentration of SO_4^{2-} in precipitation

(Granat 1988) is used as a proxy for SO_2 in the air. The characteristic indicator values were also used as environmental variables.

A method based on Bloom (1980) was used to test the significance of changes between the years. The ordination scores on the first three axes of a correspondence analysis ordination (Ter Braak 1988) of transect cover values for every tree were used to compute the Euclidean distance (d) between all trees within the sites the first year and the Euclidean distances between all trees in the sites between the two years.

$$d_{ik} = (\Sigma_j(x_{ij} - x_{kj})^2)^{1/2}$$

x_{ij} = ordination score of tree i on axis j (j = 1,3)
x_{kj} = ordination score of tree k on axis j (j = 1,3)

A t-test (SYSTAT 1990) was used to test whether the inter-year distances were longer than the intra-year distances.

Results
Species composition

There are clear differences between the three regions with respect both to species composition and abundance (Table 2). The number of species per site in the regions increases in the order Skåne < Halland < Småland.

The epiphytic vegetation at the sites in Skåne is dominated by *Scoliciosporum chlorococcum* and *Lepraria incana* (Table 2). *Lepraria incana* increased during the investigation period. It now has between 25% and 50% transect cover at all sites. Another increasing species is *Chrysotrix candelaris,* which appeared at only one site in the end of the 1970's and is now present at all but one site. The other changes in Skåne are decreasing species. *Pertusaria amara* has decreased at several sites but is still present, whereas *Pertusaria pertusa* and *Phlyctis argena* have disappeared from most of the sites. The mean number of species per site has decreased from 17±4 in the first investigation to 16±2 in the second. This change is not significant.

In the Halland group of sites, *Scoliciosporum chlorococcum* and *Hypnum cupressiforme* are the dominant species. There are small changes in the species composition at these sites. *Parmelia saxatilis* was found at two of the sites in 1978 but has now disappeared and *Pertusaria hemisphaerica* has disappeared from two of the three sites. The mean number of species has decreased from 24±6 to 22±3 but this change is not significant.

The sites in Småland have no common dominating species, in contrast to Skåne and Halland. The most abundant species are *Hypogymnia physodes, Hypnum cupressiforme, Lepraria incana, Pertusaria amara* and *P. pertusa.* *L. incana* has increased at two of the sites. *Scoliciosporum chlorococcum* and *Lecanora conizae-*

Table 2. Vegetation table

Site	Skåne												Halland						Småland					
Investigation	a 1	n 2	k 1	a 2	e 1	r 2	m 1	a 2	o 1	v 2	h 1	o 2	o 1	s 2	g 1	r 2	s 1	i 2	h 1	e 2	h 1	i 2	n 1	y 2
Lecanora conizaeoides	4	3	1	2	2	4	2	2	5	5	2	2	–	–	–	–	–	–	–	3	–	–	–	5
Hypogymnia physodes	5	5	5	3	2	4	7	6	4	4	4	3	4	4	3	2	2	2	7	8	8	7	7	2
Lepraria incana	8	8	6	8	8	8	7	8	7	8	8	8	6	6	6	7	7	7	6	8	7	7	4	7
Parmelia sulcata	2	2	4	2	1	2	4	2	–	2	1	–	5	5	2	4	2	2	4	5	2	3	5	4
Parmeliopsis hyperopta	5	5	4	5	–	4	4	3	2	4	2	3	3	4	2	2	1	3	–	2	1	–	2	–
Pertusaria amara	2	2	4	2	2	2	5	1	–	1	1	–	6	5	3	2	2	2	5	4	6	6	6	5
Ascodichaena rugosa	4	5	6	3	7	6	5	7	5	6	5	2	3	4	4	4	7	5	2	2	2	4	3	2
Scoliciosporum chlorococcum	9	9	9	9	8	9	9	9	9	9	9	8	9	8	8	8	9	9	4	5	4	5	7	7
Melanelia subaurifera	2	1	1	1	2	2	2	2	–	2	2	2	2	2	2	2	4	2	2	2	2	2	4	2
Dicranum scoparium	1	2	1	2	1	–	1	–	–	–	1	1	2	2	3	2	2	2	2	2	2	2	1	2
Loxospora elatina	2	2	3	2	2	3	2	2	2	2	2	3	2	3	2	2	2	5	1	–	2	2	2	–
Hypnum cupressiforme	3	2	5	5	4	5	2	–	–	2	6	7	7	8	8	9	8	7	8	7	4	5	6	8
Evernia prunastri	1	2	1	1	–	2	4	2	–	1	1	–	2	4	2	2	1	1	1	2	2	2	2	–
Pertusaria spp.	–	–	3	2	2	1	2	2	–	–	2	2	2	–	–	2	2	2	2	2	–	–	4	–
Phlyctis argena	2	–	–	–	2	–	2	2	–	–	2	–	–	2	2	2	–	1	2	2	2	3	5	5
Ulota crispa	2	4	1	2	2	–	–	2	–	–	–	1	2	2	2	1	3	2	2	2	2	2	2	2
Pertusaria coccodes	2	–	–	2	1	–	2	–	–	–	–	–	3	3	2	–	2	2	4	4	4	4	4	2
Pertusaria pertusa	1	–	2	–	2	–	2	1	–	–	2	2	2	2	2	2	1	2	4	5	3	5	5	4
Cladonia coniocraea	–	2	2	2	2	1	–	–	–	–	1	2	2	2	2	2	2	2	3	2	2	2	2	2
Ochrolechia subviridis	–	–	1	–	–	–	2	–	–	–	–	–	1	1	–	–	1	2	2	–	2	2	–	1
Pertusaria hemisphaerica	–	–	2	–	2	–	1	2	–	–	–	–	2	2	2	–	2	–	2	2	2	2	3	2
Lecanora argentata	–	–	1	–	1	–	2	–	2	2	1	1	1	2	2	2	–	1	2	2	4	2	5	2
Graphis scripta	–	–	–	–	–	–	–	–	–	–	1	–	2	1	2	2	–	2	2	–	3	2	1	–
Radula complanata	–	–	–	–	–	–	–	–	–	–	–	–	1	2	2	–	2	–	1	–	2	–	2	2
Platismatia glauca	1	–	1	1	–	2	–	–	–	–	–	–	1	–	1	–	–	–	2	2	4	4	2	–
Ptilidium pulcherrimum	2	2	2	1	1	1	–	–	–	–	–	–	–	1	2	–	–	1	–	2	2	2	2	1
Thelotrema lepadinum	–	–	–	–	–	–	–	–	–	–	–	–	–	–	–	–	–	–	3	1	4	4	–	–
Usnea subfloridana	–	–	–	–	–	–	–	–	–	–	–	–	–	–	–	–	–	–	2	2	2	1	1	–
Metzgeria furcata	–	–	–	–	–	–	–	–	2	1	–	–	–	2	3	2	–	–	2	2	1	1	2	2
Opegrapha spp.	–	–	–	–	–	–	–	–	–	–	–	–	–	–	2	1	–	–	2	2	–	2	–	1
Pyrenula nitida	–	–	–	–	–	–	–	–	–	–	–	–	–	2	–	–	–	–	1	4	4	2	–	–
Frullania dilatata	–	–	–	–	–	–	–	–	–	–	–	–	2	2	1	–	–	–	2	–	4	3	2	2
Pertusaria hymenia	–	–	–	–	1	–	–	–	–	–	1	–	2	–	1	–	–	–	2	2	1	–	2	–
Lecanora intumescens	–	–	–	2	–	–	–	–	–	–	–	–	2	–	–	–	–	–	1	–	1	2	2	–
Pseudevernia furfuracea	–	–	–	1	–	–	–	–	–	–	–	–	1	2	–	–	–	–	–	2	2	2	1	–
Cetraria spp.	–	–	–	–	–	–	–	–	–	–	–	–	–	–	–	–	–	1	–	–	–	–	–	–
Geocalyx graveolens	–	–	–	–	–	–	–	–	–	–	–	–	–	–	–	–	–	–	–	–	–	–	–	1
Lecanora circumborealis	–	–	–	–	–	–	–	–	–	–	–	–	–	–	–	–	–	–	–	–	–	–	–	1
Lejeunea cavifolia	–	–	–	–	–	–	–	–	–	–	–	–	–	–	–	–	–	–	–	–	–	–	–	2
Sanionia uncinata	–	–	–	–	–	–	–	–	–	–	–	–	–	–	–	–	–	–	–	–	–	–	–	4
Ochrolechia microstictoides	–	–	–	–	–	–	–	–	–	–	–	–	–	–	–	–	–	–	–	–	–	–	4	–
Buellia punctata	–	–	–	–	–	–	–	–	–	–	–	–	–	–	–	–	–	–	–	–	–	–	2	2
Lecidella euphorea	–	–	–	–	–	–	–	–	–	–	–	–	–	–	–	–	–	–	–	–	–	–	1	–
Lophocolea heterophylla	–	–	–	–	–	–	–	–	–	–	–	2	–	–	–	–	–	–	–	–	–	–	–	–
Orthotrichum lyellii	–	–	–	–	–	–	–	–	1	–	–	–	–	–	–	–	–	–	–	–	–	–	2	2
Neckera complanata	–	–	–	–	–	–	–	–	–	–	–	–	–	–	2	–	–	–	–	–	–	–	–	2
Lecanora spp.	–	–	–	–	–	–	–	–	–	–	–	–	–	–	2	–	–	–	–	–	–	2	–	–
Plagiothecium denticulatum	–	–	–	–	–	–	–	–	–	–	–	–	–	–	2	–	–	–	–	–	–	–	–	–
Athelia spp.	–	–	–	–	–	–	–	–	2	–	3	–	–	–	–	–	–	–	–	–	–	–	–	–
Xanthoria candelaria	–	–	–	–	–	–	–	–	1	–	–	–	–	–	–	–	–	–	–	–	–	–	–	–
Bryoria spp.	–	–	–	–	–	–	–	–	1	–	–	–	–	–	–	–	–	–	–	–	–	–	–	–
Usnea spp.	–	–	–	–	–	1	–	–	–	–	–	–	–	–	–	–	–	–	–	–	–	–	–	–
Lecanora carpinea	–	–	–	–	–	–	–	–	–	–	–	–	–	–	–	–	–	–	–	–	–	1	–	–

cont.

oides have also increased. Other species which have increased are *Isothecium myurum*, *Pertusaria flavida* and *Ochrolechia androgyna*. The mean number of species per site has decreased from 34±2 to 33±5. Again, there is no significant change.

At most of the sites, the net changes in number of

Table 2. Continued

	Skåne												Halland						Småland					
Site	a	n	k	a	e	r	m	a	o	v	h	o	o	s	g	r	s	i	h	e	h	i	n	y
Investigation	1	2	1	2	1	2	1	2	1	2	1	2	1	2	1	2	1	2	1	2	1	2	1	2
Sphaerophorus globosus	–	–	–	–	–	–	–	–	–	–	–	–	–	–	–	–	–	–	–	–	1	–	–	–
Buellia griseovirens	–	–	–	–	–	–	–	–	–	–	–	–	–	–	2	–	–	–	–	–	2	–	–	–
Lecidea nylanderi	–	–	–	–	–	–	–	–	–	–	–	–	–	–	–	–	–	–	–	–	3	–	–	–
Antitrichia curtipendula	–	–	–	–	–	–	–	–	–	–	–	–	–	–	–	–	–	–	–	–	3	4	–	–
Hypogymnia farinacea	–	–	–	–	–	–	–	–	–	–	–	–	–	–	–	–	–	–	–	–	2	3	–	–
Cetraria sepincola	–	–	–	–	–	–	–	–	–	–	–	–	–	–	–	–	1	–	–	–	2	–	–	–
Porina carpinea	–	–	–	2	–	–	–	–	2	–	–	–	–	–	–	–	–	–	–	–	–	–	1	–
Fuscidea cyathoides	–	–	–	–	–	–	–	–	–	–	–	–	–	1	–	–	–	–	–	–	–	–	–	–
Pertusaria leucostoma	–	–	–	–	–	–	–	–	–	–	–	–	–	2	–	2	–	–	–	–	–	–	–	–
Parmelia spp.	–	–	–	–	–	–	–	–	–	–	–	–	2	–	–	–	–	–	–	–	–	–	–	–
Lecanora pulicaris	–	–	1	–	2	–	1	–	2	–	–	–	–	–	–	–	–	1	–	–	–	–	–	–
Cavernularia hultenii	–	–	2	–	–	–	–	–	–	–	–	–	–	–	–	–	–	–	–	–	–	–	–	–
Pertusaria glomerata	–	–	2	–	–	–	–	–	–	–	–	–	–	–	–	–	–	–	–	–	–	–	–	–
Hypogymnia spp.	–	–	1	–	–	–	–	–	–	–	–	–	–	–	–	–	–	–	–	–	–	–	–	–
Micarea prasina	–	–	2	–	–	–	–	–	–	–	–	–	–	–	–	–	–	–	–	–	–	–	–	–
Ochrolechia turneri	–	–	1	–	–	–	–	–	–	–	–	–	–	–	–	–	–	–	–	–	–	–	–	–
Physcia spp.	–	–	2	–	–	–	–	–	–	–	–	–	–	–	–	–	–	–	–	–	–	–	–	–
Parmeliella triptophylla	–	–	2	2	–	–	–	–	–	–	–	–	–	–	–	–	–	–	–	–	–	–	–	–
Parmelia saxatilis	–	–	4	4	–	–	1	–	–	1	–	–	4	1	2	–	–	–	–	–	–	1	2	–
Ramalina spp.	–	–	–	–	–	–	–	–	–	–	–	–	–	–	–	–	–	–	1	–	–	–	–	–
Isothecium myurum	–	–	–	–	–	–	–	–	–	–	–	–	–	–	–	–	–	–	2	–	2	–	–	5
Biatora efflorescens	–	–	–	–	–	–	–	–	–	–	–	–	–	–	–	–	–	–	2	–	–	–	–	–
Pertusaria coronata	–	–	–	–	–	–	–	–	–	–	–	–	–	–	–	–	–	–	4	–	–	2	3	–
Pertusaria flavida	–	–	–	–	–	–	–	–	–	–	–	–	–	–	1	–	–	–	2	–	2	–	–	–
Pylaisia polyantha	–	–	–	–	–	–	–	–	–	–	–	–	–	–	–	–	–	3	2	–	–	–	–	2
Sphinctrina turbinata	–	–	–	–	–	–	–	–	–	–	–	–	–	–	–	–	–	–	2	1	2	–	–	–
Ochrolechia androgyna	–	–	–	–	–	–	–	–	–	–	–	–	–	–	–	–	–	–	2	–	–	4	–	2
Pertusaria leioplaca	–	–	–	–	–	–	–	–	–	–	–	–	–	–	–	–	–	–	1	–	–	2	–	–
Lecanora albescens	–	–	–	–	–	2	–	2	–	1	–	–	2	–	2	–	–	–	2	–	–	–	–	–
Lecanora expallens	–	–	–	–	–	–	–	–	–	–	–	–	–	–	–	–	–	–	3	–	2	–	–	–
Orthotrichum spp.	–	–	–	–	–	–	–	–	–	–	–	–	2	–	–	–	–	2	2	–	–	–	–	–
Homalothecium sericeum	–	–	–	–	–	–	–	–	–	–	–	–	–	–	–	–	–	–	2	–	–	–	–	–
Hypogymnia tubulosa	–	–	1	–	–	–	2	–	–	–	–	–	–	–	–	–	–	–	1	–	1	–	2	–
Lecidea spp.	–	–	–	–	–	–	–	–	–	–	–	–	–	–	–	–	–	–	2	–	–	–	–	–
Alectoria spp.	–	–	–	–	–	–	–	–	–	–	–	–	–	–	–	–	–	–	2	–	1	–	1	–
Chrysothrix candelaris	–	4	–	2	–	4	–	–	4	2	–	5	–	–	–	–	4	1	–	–	–	–	–	–
Hypocenomyce scalaris	1	–	–	–	–	–	–	–	3	2	1	–	–	–	–	–	–	–	–	–	–	–	–	–
Pertusaria albescens	2	–	2	–	–	–	–	–	–	–	–	1	2	–	–	–	–	2	4	2	–	–	2	–

Abbreviated site names: Kagarp (k a), Högestad (h o), Anklam (a n), Eriksdal (e r), Maglehem (m a), Öved (o v), Öströö (o s), Grimeton (g r), Sibbarp (s i), Helgö (h e), Hinnsjöhult (h i), Nydala (n y). Transect cover scale: >50% (9), 25%–50% (8), 12.5%–25% (7), 6.2%–12.5% (6), 3.1%–6.2% (5), 1%–3.1% (4), 0.5%–1% (3), <0.5% (2), present but not hit by line (1). Investigation no 1 was in 1977–79 and no 2 in 1991–92.

species are small, than less 4 species (Table 3). The number of sites with an increase in the number of species is the same as the number with a decrease. At three of the sites (Nydala, Grimeton and Maglehem) the number has decreased by 6 to 7 species (13% to 20%). The highest turnover among the species occurred at Nydala and Helgö, with 22 and 20 species, respectively (c. 30%). Other sites with high turnover in the number of species are Grimeton, Öströö, Hinnsjöhult and Kagarp. However, Hinnsjöhult had the most persistent flora, with only 14% (10 species) turnover. The species lost or gained are usually the sparse ones in the vegetation (Table 2). In 86% of the sites, the species' average transect cover is <0.5%. In Skåne it is only *Chrysothrix candelaris* that shows major changes. In 1977, it was found at Öved with 2% average transect cover. In 1992, it had decreased at Öved to 0.28% but appeared as a new species on all the sites in Skåne other than Maglehem (Högestad 4.7%, Anklam 2%, Eriksdal 2% and Kagarp 0.28%).

In Småland there are many more species with high turnover rates. *Lecanora conizaeoides* was not present at the first investigation while in 1992 it was found both at Nydala and Helgö with 4.6% and 0.75% average transect cover, respectively. Other new species appearing at Nydala are *Isothecium myurum* (4.7% average transect cover) and *Sanionia uncinata* (2%). Disappeared species from Nydala are *Pertusaria* spp. (2%) and *Ochrolechia microstictoides* (2%). At Hinnsjöhult, *Ochrolechia androgyna* has appeared and now has 2% average transect cover and the same is true of *Pertusaria coronata* at

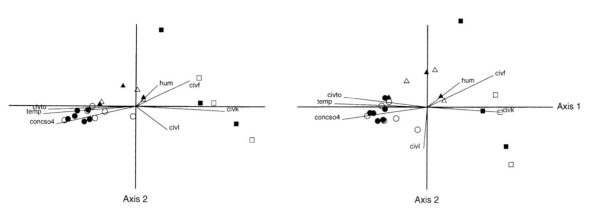

Entire sites   **Single trees, centroids**

Fig. 1. A sites environment bi-plot resulting from correspondence analyses (Ter Braak 1988). In the left plot, mean transect covers for entire sites are used. In the right plot, the transect covers for single trees are used, the centroids representing the sites are plotted. Squares represent the sites in Småland, triangles those in Halland and circles those in Skåne. Open symbols represent investigation 1 (1977–79), filled investigation 2 (1991–92). The arrows show the direction of maximum variation of the environmental variables (Table 1). temp = mean temperature of the coldest month, concso4 = concentration of SO_4^{2-} in precipitation, hum = humidity of the summer, civk = characteristic indicator value for light, civto = characteristic indicator value for toxitolerance, civl = characteristic indicator value for light and civf = characteristic indicator value for moisture.

Helgö. Other turnover rates are small. In Halland, there was no turnover of species with an average transect cover > 1%.

Ordination

The first ordination, CA (Ter Braak 1988), of mean values of all the sites, divides the sites into the three groups seen in Table 2, viz. the group of sites in Skåne, the group in Halland and the group in Småland (Fig. 1). However, Sibbarp in Halland comes rather close to the sites in Skåne. The main between-site variation is along axis 1 (eigenvalue 0.325). Axes 2, 3 and 4 have eigenvalues of 0.177, 0.132 and 0.104, respectively.

The transect cover for the individual trees was used in another CA ordination (Fig. 2). The eigenvalues of the four first axes are 0.408, 0.266, 0.170 and 0.166, respectively. In this ordination, the between-site variation is also along axis 1, while axis 2 mostly represents the within-site variation. The centroids of the sites are located according to the same pattern as obtained when using the mean values for the sites, as in the first ordination (Fig. 1). Plotting the individual trees (Fig. 2) shows that there are great differences in the epiphytic vegetation among trees within a site. The sites in Småland and Halland are more heterogeneous than those in Skåne. This is mainly because a larger number of species creates a greater variation within a site.

Using the Euclidean distances between trees in the first three ordination axes of the second ordination to test the significance of the changes in the sites, I found that only three of the sites have changed significantly during the investigation period (Table 4). The greatest change was at

Nydala, with somewhat smaller ones at Maglehem and Grimeton. The other changes are insignificant. At Nydala and Maglehem, there has been a large increase in the canopy cover at the same time (Table 3). No such change was observed at Grimeton.

In the first ordination (Fig. 1), the environmental variables (Table 1) with strongest correlation to axis 1 are mean temperature of the coldest month (r = −0.91) and concentration of SO_4^{2-} in precipitation (r = −0.78). A multiple regression shows that temperature and SO_4^{2-} have a significant contribution to the explanation of axis 1 (regr. coeff. = −0.59*** and −0.51***, respectively). The characteristic indicator values for continentality and toxitolerance (Table 5a) also both show high correlation to axis 1 (r = 0.86 and r = −0.86, respectively). The variable with the strongest correlation to axis 2 is Martonne's humidity value of the summer (r = 0.54), but the regression gives no significance.

In the ordination of single trees (Fig. 2), the same environmental variables (Table 1) have strong correlations to axis 1 as in the first ordination, viz. mean temperature of the coldest month and concentration of SO_4^{2-} in precipitation (r = −0.84 and r = −0.77, respectively). A multiple regression shows a significant contribution for these two variables (regr. coeff. −0.49*** and −0.52***, respectively). The characteristic indicator value for continentality and for toxitolerance also shows correlation with axis 1 (r = 0.69, −0.76 respectively). In this ordination, the characteristic indicator value for moisture is also correlated to axis 1 (r = 0.62). Two variables are correlated to axis 2, characteristic indicator value for light and for moisture (r = −0.73, r = 0.57, respectively). None of the environmental variables is correlated to axis 3 and 4.

In both ordinations, the sites in Skåne form a close

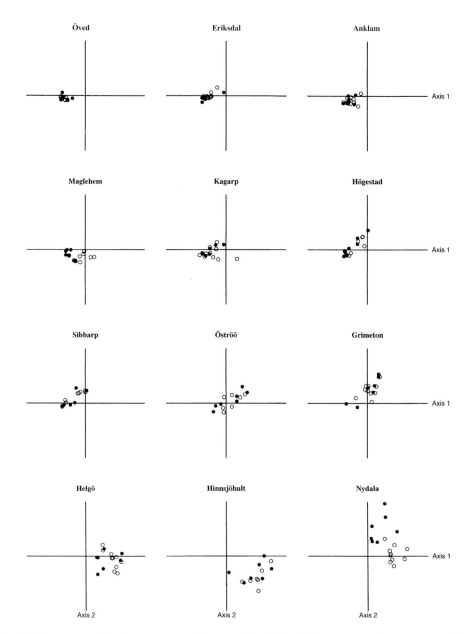

Fig. 2. A plot of single trees resulting from a correspondence analysis. Each subplot comprises the trees at one site. Open circles represent investigation 1 (1977–79), filled investigation 2 (1991–92).

group in that part of the diagram where the heads of the vectors for oceanic conditions (high temperature of the coldest month) and pollution (high SO_4-concentration) are also found. The Småland sites are positioned in the opposite part of axis 1, characterized by more continental conditions (opposite to mild coldest month) and by low pollution. The Småland sites also shows a great variation along axis 2, interpreted as a variation in light and moisture conditions.

Discussion

The main variation in the data separates the sites geographically into three groups, with the Skåne sites at one end, the Småland sites at the other, and the Halland sites in between (Fig. 1). This represents a plant geographic gradient, described e.g. by Degelius (1935) and Almborn (1948). Nowadays the pollution parallels this gradient, from the most polluted sites in Skåne, because of influence from local sources and from the Baltic countries, Poland, Germany and Denmark (Lövblad 1990), to the

Table 3. Number of species and canopy cover.

Site	Number of species 77–79	91–92	Diff	Diff %	Species turnover Lost	New	Turn-over	Turn-over %	Canopy cover 77–79	91–92	Diff
Skåne											
Öved	11	15	4	15	1	5	6	23	75	70	–5
Eriksdal	16	14	–2	–7	5	4	9	30	95	85	–10
Högestad	14	15	1	3	2	3	5	17	90	85	–5
Anklam	16	16	0	0	3	2	5	16	75	95	20
Kagarp	21	19	–2	–5	7	3	10	25	90	50	–40
Maglehem	21	14	–7	–20	7	1	8	23	80	95	15
Halland											
Grimeton	28	22	–6	–12	9	4	13	26	90	90	0
Sibbarp	18	19	1	3	4	2	6	16	95	85	–10
Öströö	27	25	–2	–4	7	4	11	21	90	60	–30
Småland											
Hinnsjöhult	33	36	3	4	4	6	10	14	80	80	0
Helgö	33	36	3	4	9	11	20	29	85	85	0
Nydala	36	28	–8	–13	15	7	22	34	75	95	20

rural, forested areas in Småland. The distant influences, in particular, may be greater in Halland than in Småland.

In the ordination diagrams based on mean values over the entire sites (Fig. 1), there are changes, even if small, in almost all sites during the 1980's. However, at all sites there are also differences in the epiphytic vegetation among the trees (Fig. 2). If this within-site variation among trees is taken into account, there are only three sites with significant changes. At two of these sites, viz. Nydala and Maglehem, the canopy cover has increased (Table 3), which makes the forest less illuminated and the microclimate moister for the epiphytic vegetation. Nydala has changed (Table 5b) from a site with rather light-demanding species (characteristic indicator value for light 5.69) into a site with more shade-tolerant and moisture-demanding species (characteristic indicator

value for light 4.79 and for moisture 4.15). In 1992, Nydala had the lowest light value of all the sites (4.79) and the highest moisture value (4.15). The characteristic indicator value for toxitolerance has increased from the lowest value of all sites to the same value as the other two sites in Småland. This is mainly because the abundant shade-tolerant species also have high toxitolerance. Still, the characteristic indicator value for toxitolerance is low, compared to the sites in Skåne. Light-demanding species have also decreased at Maglehem, but the difference is smaller than in Nydala. The loss of species is considerable, most of them are sparse and have high light indicator value. Grimeton shows significant changes, too, but these result from greater variation in the epiphytic vegetation in 1992 than in 1978/79 (Fig. 2). The canopy cover has not changed.

Table 4. Mean intra site distances between trees 1977–79 and mean inter sites distances between trees 1977/79–1991/92. Scores from the first three axes of the correspondence analyses in Fig. 2 are used.

	1977/1979 Mean	SD	1977/79–1991/92 Mean	SD	Diff	Significance
Skåne						
Öved	0.259	0.108	0.292	0.15	0.03	
Eriksdal	0.302	0.186	0.378	0.216	0.08	
Högestad	0.63	0.371	0.686	0.5	0.06	
Anklam	0.375	0.164	0.369	0.181	–0.01	
Kagarp	0.671	0.304	0.561	0.303	–0.11	
Maglehem	0.412	0.173	0.511	0.235	0.10	*
Halland						
Öströö	0.624	0.255	0.691	0.329	0.07	
Grimeton	0.556	0.251	0.78	0.406	0.22	***
Sibbarp	0.491	0.344	0.678	0.368	0.19	
Småland						
Hinnsjöhult	0.91	0.597	0.948	0.516	0.04	
Helgö	0.657	0.324	0.72	0.25	0.06	
Nydala	0.735	0.282	1.615	0.492	0.88	***

Table 5a. Characteristic indicator values for the sites.

	Light	Continentality	Moisture	Toxi-tolerance
Skåne				
Mean 77–79	5.47	4.08	3.16	8.19
SD	0.32	0.31	0.10	0.22
Mean 91–92	5.31	4.16	3.22	8.25
SD	0.27	0.28	0.15	0.03
Halland				
Mean 77–79	5.14	4.20	3.74	7.94
SD	0.30	0.20	0.22	0.25
Mean 91–92	5.14	4.20	3.74	7.96
SD	0.40	0.36	0.39	0.27
Småland				
Mean 77–79	5.66	5.13	3.65	7.10
SD	0.39	0.55	0.25	0.30
Mean 91–92	5.22	5.16	3.80	7.60
SD	0.42	0.42	0.30	0.37

The indicator values range from 1 to 9. Light: 1 – species growing in deep shade, 9 – species growing in full light. Continentality: 1 – euoceanic species, 9 – continental species. Moisture: 1 – species favoured by driest conditions, 9 – species common in very humid sites. Toxitolerance: 1 – species extremely sensitive to pollution. 9 – species very toxitolerant.

Table 5b. Changes in characteristic indicator values, in SD units. For explanations, see Table 5a.

	Light	Continentality	Moisture	Toxi-tolerance
Skåne				
Öved	−0.03	0.11	0.12	0.11
Anklam	−0.29	0.00	0.15	−0.23
Eriksdal	0.38	−0.20	0.29	−0.32
Högestad	−0.93	0.69	0.38	−0.11
Kagarp	−0.93	0.71	0.38	0.68
Maglehem	−1.02	−0.41	−0.26	0.73
Halland				
Öströö	−0.29	0.07	0.44	0.05
Grimeton	−0.46	0.24	0.38	0.27
Sibbarp	0.73	−0.35	−0.76	−0.14
Småland				
Hinnsjöhult	−1.22	−0.38	0.73	−0.34
Helgö	−0.03	0.15	−0.90	1.42
Nydala	−2.61	0.30	1.48	2.37

The changes in the mean number of species are small, one or two species, but the species turnover is high at all sites (Table 3). Most of the species involved in the turnover are sparse and there are hardly any significant trends in the turnover changes (Table 2). There seems to be a great component of chance in whether an existing rare species will survive or a new species will establish (Arup and Ekman 1991).

There are small changes in the species composition (Table 2) at the sites in Halland. At the sites in Skåne, the changes are greater but still rather small. Above all, the changes indicate that the sites have become darker. Shade-tolerant species which have increased are *Lepraria*

incana and *Chrysotrix candelaris*, while *Pertusaria amara, P. pertusa* and *Phlyctis argena* are decreasing light-demanding species. At Kagarp, the canopy cover has decreased by 40%. This has resulted from a thinning in 1990 but the flora showed no evidence of the increase in light at the investigation in 1992. The shade-tolerant species have increased, e.g. *Lepraria incana* and *Chrysotrix candelaris,* and some light-demanding species have decreased, e.g. *Hypogymnia physodes* and *Pertusaria amara,* reflecting the darker conditions before the thinning.

At the sites in Småland, there are also rather small changes, except at Nydala. *Scoliciosporum chlorococcum* has increased at Hinnsjöhult and Helgö. This species is nitrophilous (Ahti and Vitikainen 1974) but also light-demanding. The increase, in spite of the absence of increase in light may, indicate that the availability of nitrogen has increased.

At the sites in Skåne and Halland, there are no changes, or very small ones in the characteristic indicator values. At the three sites in Småland, the light value has decreased and the moisture and toxitolerance values have increased (Table 5a). These changes mainly result from the large changes at Nydala.

In conclusion, there is no general trend in the changes found during the investigation period. At two of the three sites, where there have been significant changes in the vegetation, the changes may be an effect of a marked increase in the canopy cover resulting in a darker and moister microclimate.

Although the other sites do not show significant changes, the turnover of species has been considerable. At most of the sites, 20–35% of the species are either lost or new during the period (Table 3). The turnover may be even higher because all rare species at a site are not included when the analysis is done in a small sample plot. There is a potential risk that a weak trend of rare recolonizers will be missed. It seems unlikely, though, that such a trend would be missed if it appears in a whole area covered by several sample plots. Most of the species involved are sparse, and the changes then seem to be related to chance. However, in Skåne the turnover of species indicates that the studied forests in general may have grown darker. In Småland, the toxitolerant species seem to have increased but this is an effect of the colonization and expansion of some shade-tolerant species which also have high values for toxitolerance. Ekman (1990) found that the light and moisture climate of the forest plays a major role in determining the epiphytic flora. In the present investigation, also, changes in canopy cover, causing changes in air moisture and light, are more important factors than pollution. In this investigation there are no clear indications of any effects of an increasing or decreasing load of pollutants during the last 15 years.

Acknowledgments – The study was partly financed by grants from The Swedish Environmental Protection Board. N. Malmer has given great support and important criticism during the work. I.C. Prentice has also given valuable comments.

References

Ångström, A. 1968. Sveriges klimat. – Generalstabens litografiska anstalts förlag, Stockholm.

Ahti, T. and Vitikainen, O. 1974. *Bacidia chlorococca*, a common toxitolerant lichen in Finland. – Mem. Soc. Fauna Flora Fennica 49: 95–100.

Almborn, O. 1943. Lavfloran i botaniska trädgården i Lund. – Bot. Notiser 96: 167–177.

– 1948. Distribution and ecology of some south Scandinavian lichens. – Bot. Notiser, Suppl. 1 (2): 1–252.

Arup, U. and Ekman, S. 1991. Lavfloran på Hallands Väderö. (The lichen flora of Hallands Väderö, S. Sweden – changes over a 40-year period) – Svensk Bot. Tidskr. 85: 225–320.

Bloom, S.A. 1980. Multivariate quantification of community recovery. – In: Cairns, J. Jr. (ed.), The recovery process in damaged ecosystems. Ann Arbor Science Publishers Ann Arbor pp. 141–151.

Corley, M.F.V., Crundwell, A.C., Düll, R., Hill, M.O. and Smith, A.J.E. 1981. Mosses of Europe and the Azores; an annotated list of species, with synonyms from the recent literature. – J. Bryol. 11: 609–689.

Degelius, G. 1935. Das ozeanische element der Strauch- und Laubflechtenflora von Skandinavien. – Acta Phytogeogr. Suec. 7: 1–411.

Ekman, U. 1990. Lavfloran i Dalby Söderskog (The lichen flora of Dalby Söderskog National Park). – Svensk Bot. Tidskr. 84: 61–240.

Ellenberg, H., Weber, H.E., Düll, R., Wirth, V., Werner, W. and Paulissen, D. 1991. Zeigerwerte von Pflanzen in Mitteleuropa. – Script Geobot. 18: 175–237.

Eriksson, B. 1982. Data rörande Sveriges temperaturklimat (Data concerning the air temperature climate of Sweden). – SMHI Rep. Meteoro. Climato. 39: 1–34.

– 1986. Nederbörds- och humiditetsklimatet i Sverige under vegetationsperioden (The precipitation and humidity climate in Sweden during the vegetation period). – SMHI Rep. Meteoro. Climatolo. 46: 1–79.

Ferry, B.W., Baddeley, M.S. and Hawksworth, D.L. 1973. Air pollution and lichens. – The Athlone Press, London.

Gilbert, O.L. 1992. Lichen reinvasion with declining air pollution. – In: Bates, J.W. and Farmer, A.M. (eds), Bryophytes and lichens in a changing environment. Clarendon Press, Oxford, pp. 159–177.

Granat, L. 1988. Luft- och nederbördskemiska stationsnätet inom PMK. Rapport från verksamheten 1988. (The net of stations for air and precipitation chemistry in Swedish program for environmental monitoring (PMK). Report of activities 1988). – Swedish Environ. Protection Agency, Solna, Report 3649.

Hawksworth, D.L. 1983. The nomenclature of the beech bark fungus: A solution of the complex case of *Ascodichaena, Dichaena* and *Polymorphum*. – Taxon 32: 212–217.

– and Rose, F. 1976. Lichens as air pollution monitors. – Edward Arnold, London.

– and McManus, P.M. 1989. Lichen recolonization in London under conditions of rapidly falling sulphur dioxide levels, and the concept of zone skipping. – Bot. J. Linn. Soc. 100: 99–109.

Hultengren, S., Martinsson, P.-O. and Stenström, J. 1991. Lavar och luftföroreningar. Känslighetsklassning och indexberäkning av epifytiska lavar (Lichens and air pollution. Classification of sensitivity and calculation of indices in epiphytic lichens). – Swedish Environ. Protection Agency, Solna, Report 3967.

Lindgren, L. 1970. Beech forest vegetation in Sweden – a survey. – Bot. Notiser 123: 401–424.

Lövblad, G. 1990. Luftföroreningshalter och deposition i bakgrundsluft. – Swedish Environ. Protection Agency, Solna, Report 3812: 1–65.

Lundegårdh, P.H., Lundqvist, J. and Lindström, M. 1970. Berg och jord i Sverige. – Almqvist & Wiksell, Stockholm.

Maarel, E. van der, Janssen, J.G.M. and Louppen, J.M.W. 1978. TABORD, a program for structuring phytosociological tables. – Vegetatio 38: 143–156.

Mattsson, J. O. 1971. Väderlekslära och klimatologi. – CWK Gleerups, Lund.

Persson, S. 1981. Ecological indicator values as an aid in the interpretation of ordination diagrams. – J. Ecol. 69: 71–84.

Santesson, R. 1993. The lichens and lichenicolous fungi of Sweden and Norway. – SBT-förlaget, Lund.

Skye, E. 1979. Lichens as biological indicators of air pollution. – Ann. Rev. Phytopathol. 17: 325–341.

Sjörs, H. 1971. Ekologisk botanik. – Almqvist & Wiksell, Stockholm.

SYSTAT 1990. Software for statistics and graphics. – SYSTAT Inc., Evanston.

Ter Braak, C.J.F. 1988. CANOCO – a FORTRAN program for canocical community ordination by [partial] [detrended] [canonical] correspondence analysis, principal components analysis and redundancy analysis (version 2.1). – TNO Inst. Appl. Computer Sci., Wageningen.

– and Prentice, I.C. 1988. A theory of gradient analysis – Adv. Ecol. Res. 18: 271–314.

Troedsson, T. and Nykvist, N. 1973. Marklära och markvård. – Almqvist & Wiksell, Stockholm.

Ecological Bulletins 44: 248–258. Copenhagen 1995

Arthropods and passerine birds in coniferous forest: the impact of acidification and needle-loss

Bengt Gunnarsson

Gunnarsson, B. 1995. Arthropods and passerines birds in coniferous forest: the impact of acidification and needle-loss. – Ecol. Bull. (Copenhagen) 44: 248–258.

The micro-habitat structure on coniferous trees changes as a result of needle-loss. This structural change in the vegetation may affect arthropods living in spruce *Picea abies* by indirect mechanisms, e.g. altered relations between prey and predators. The impact of acidification and needle-loss on some tree-living arthropods and passerine birds is reviewed. New information about the taxonomic composition of spiders in relation to needle density in a field experiment is reported. The main combined findings from the review and field experiments are: 1) Acid precipitation may be toxic because of high H^+ concentrations. However, simulated acid rain (pH 4.0) did not reduce the growth rate of a spruce-living spider. There is at present no evidence of toxic effects on arthropods at this level of pH. 2) Experiments in the field and laboratory and data from natural populations suggested that spruce-living arthropods are affected by the needle density on branches. These data showed a positive correlation between needle density and spider abundance. However, a large-scale field experiment could not confirm this relationship. 3) The interaction between bird predation and needle density was examined in a large-scale field experiment. There were strong negative effects of bird predation on arthropod abundance. Moreover, the taxonomic composition among spiders changed as a result of bird predation: raptorial spiders increased their relative abundance whereas sheetweb spiders decreased their relative abundance when bird predation was excluded. There were also some cases of bird predation/needle density interactions: arthropod abundance was affected by bird predation on needle-dense branches in some cases and on needle-sparse branches in others. In the absence of bird predation, the needle density affected the spider size distribution: large spiders were more common on needle-sparse branches than on needle-dense ones. The species composition was affected by similar interactions, e.g. bird predation effects on crab spiders (Thomisidae) were found on needle-sparse branches only. 4) The foraging behaviour of willow tits *Parus montanus* was studied at two sites differing in the needle density of the coniferous trees. The birds at the site with relatively needle-sparse branches spent proportionally more time scanning for predators and less time for prey handling than at the site with needle-dense trees. It is concluded that needle density in coniferous trees affects the behaviour of both spruce-living spiders and passerine birds. The effects on the arthropod population density may, however, be weaker than regulating factors, e.g. predation. It seems probable that the relationship between vegetation structure and the arthropod community is complex. Effects of needle density may interact with bird predation or it may be masked by other biotic factors.

B. Gunnarsson, Sect. of Animal Ecology, Dept of Zoology, Göteborg Univ., Medicinareg. 18, S-413 90 Göteborg, Sweden.

Introduction

In Scandinavian coniferous forest, there are several studies showing remarkable species richness among arthropods (e.g. Ehnström and Waldén 1986, Biström and Väisänen 1988). However, during the 20th century rapid changes have occurred in European coniferous forests.

Some of these changes are the effect of air pollution. Direct or indirect effects of air pollution on the population dynamics of forest arthropods are to be expected, but reliable data have been lacking (Führer 1985).

Coniferous forests in several parts of central and northern Europe (Blank 1985, Andersson 1986, Schulze 1989) are severely affected by air pollution. This may be at-

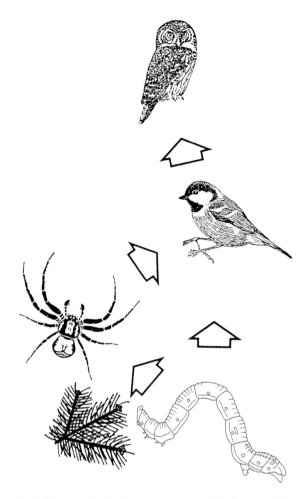

Fig. 1. The trophic relations between some animal groups living in coniferous trees. The insects include herbivores, predators and parasites. Spiders are intermediate level predators. Passerine birds are partly dependent on arthropods as prey. The top predators are owls and hawks.

and diversity of plant-living arthropods (May 1978, Southwood 1978, Lawton 1983, 1986, Morse et al. 1985).

However, if there are changes in the arthropod community resulting from altered vegetation structure, this may affect other trophic levels as well. Passerine birds foraging in coniferous forest use tree-living arthropods as an important food resource (Palmgren 1932, Askenmo et al. 1977, Jansson and von Brömssen 1981, Jansson 1982, Gunnarsson 1983, Hogstad 1984, Suhonen et al. 1992). A micro-habitat change caused by needle-loss may affect the interaction between the birds and their arthropod prey. For instance, the foraging efficiency of birds should be related to the vegetation structure, since the number of prey refuges may change. Moreover, the birds are subjected to predation pressure from e.g. owls (Ekman 1986, Suhonen 1993), and their vulnerability to predators may be related to needle density in coniferous trees causing variation in visibility.

Here, I review studies on the impact of acid precipitation and needle density on tree-living arthropods and passerine birds. I also present new data on the influence of bird predation and needle density on the spider community in spruce.

The effects of acidification are in two major categories: direct and indirect. Arthropods living in the vegetation in areas with air pollution may be affected directly by chemical substances. The toxic effect of acid precipitation has been studied on spiders living in coniferous forest. Earlier investigations show that many species could be found on spruce branches in all seasons (Askenmo et al. 1977, Jansson and von Brömssen 1981, Gunnarsson 1983, 1985). This means that spiders living in coniferous trees may be exposed to "acid rain" during their entire life-cycle.

In a polluted area, organisms may not only be affected by the toxicity of different pollutants but also through changed interactions between the biotic and abiotic components of the environment. Focussing on the needle density on branches of coniferous trees, I examined the importance of changes in vegetation structure in several trophic levels. Three such levels will be considered (Fig. 1): 1) arthropods other than spiders, including herbivores, predators, and some parasites, all being potential prey of spiders and birds; 2) spiders, i.e. intermediate level predators, which are potential prey of passerine birds; 3) birds, i.e. vertebrate insectivores, partly dependent on insects and spiders as prey. The top predators in this systems are not included here, but they are still important and their role will be discussed below.

The prediction in these studies is that the needle density affects the interactions between organisms within or between the trophic levels. The specific hypotheses are presented in the context of each study. Results for arthropods and birds are separated in the presentation.

tributable to direct deposition of pollutants on the trees (Lindberg et al. 1986) or to indirect damage, e.g. soil acidification, which causes disturbance in the uptake of nutrients (Berdén et al. 1987). In coniferous forests, the intensity of needle-loss has often been used as a measure of the condition of the trees (Blank 1985). The accelerated needle-loss in polluted areas has been interpreted as a symptom of increasing stress caused by air pollution, drought and other weather conditions and this may also affect the future growth negatively. For instance, in *Picea abies*, a declining stand produced on average approximately 65% of the wood per area unit in comparison with an undamaged stand (Oren et al. 1988). Apart from physiological changes, the architecture of the trees is altered by a decreasing needle density on the branches (Westman and Lesinski 1986). These structural changes may affect the tree-living fauna. Vegetation structure has been suggested as a major factor in determining the abundance

Study sites

The experiments and observations were made at three coniferous forest sites in south-west Sweden. One of the sites was situated c. 40 km east of Göteborg (site Ulasjö; 'U'), and another c. 20 km south-east of Kungsbacka (site Hultåsvägen; 'H'). A third site was situated c. 25 km east of Göteborg. The environmental conditions and tree composition were similar to the other two sites. Although sites 'U' and 'H' were separated by c. 25 km, the forests had similar composition of trees and were at about the same elevation above sea level. In both sites, spruce *Picea abies* was predominating but there were also scattered stands of pine *Pinus sylvestris* and birch *Betula verrucosa* trees. However, there was a significant difference between the two sites in the proportion of coniferous trees with noticeable needle-loss (Hake 1991). In site H, the proportions of spruces and pines with needle-loss >20% were 0.53 and 0.67, respectively: in site U, the corresponding proportions were 0.42 and 0.37 (spruce: $p < 0.001$; pine: $p < 0.001$). There are at least two possible explanations for the difference. First, since site H was closest to the sea, it should be exposed to higher concentrations of chloride, since the prevailing winds from south-west will bring in salt from the sea. Secondly, since the coniferous forest at site H is the first one at a higher altitude than the coastal areas, there is considerable deposition of long distance air pollutants on the branches, especially along forest edges (Hasselrot and Grennfelt 1987).

Arthropods

Direct effects

The toxic effects of acid water on the common sheetweb spider *Pityohyphantes phrygianus* were studied. Spiders gain free water by adsorption and/or drinking (Pulz 1987). Consequently, rain was simulated in the laboratory by gently spraying water of different acidity on spiders and their webs in experimental vials (Gunnarsson and Johnsson 1989). Three categories were used: i) control, water of pH ≈ 7; ii) simulated acid rain, water of pH 4.0; iii) water of pH 2.2. Juvenile spiders were allowed to grow under identical conditions, except for the acidity of the provided water. The growth rates in the three treatments were compared after one month. There was no difference in growth rate between the spiders in control and those given simulated acid rain, but the spiders provided with pH 2.2 water showed a significantly lower growth rate. A second comparison was made after a further 1.5 months. The result was similar, but the specimens in the pH 2.2 treatment now showed a negative growth rate. The conclusions were that *P. phrygianus* was remarkably resistant to acid water and the rain in polluted areas could not influence the growth rate, but the possibility of long-term effects could not be excluded. The experiment was conducted with juvenile specimens, so effects on adult spiders are unknown. However, studies on the egg hatching under the same conditions as described above suggested that the conclusions hold, at least for this sheetweb spider (Gunnarsson unpubl.).

There are several other studies on direct effects of toxic pollutants on arthropods (Alstad et al. 1982, Heliövaara and Väisänen 1993). However, they often focus on heavy metals (e.g. Clausen 1984a), or gaseous compounds such as sulphur dioxide (e.g. Ginevan and Lane 1978, Feir and Hale 1983). In general, there are few conclusions from these studies that can be applied to systems other than the investigated one. It thus seems that studies on the direct action of toxic pollutants on terrestrial arthropods are not good indicators of general effects in the polluted ecosystem.

Indirect effects

Changes in the habitat structure may have large consequences for animals. In arthropods, for instance, the vegetation structure affects both numbers and body size distribution (Morse et al. 1985, Lawton 1986). In short, the argument is as follows. Several studies indicate that the edges of plants are fractal structures. This means that the edge length increases as the measurement unit becomes smaller. Assuming that the arthropods are the 'rulers', different animal size distributions are expected on plants, or habitats, with different fractal dimensions. Plants with a fractal dimension of c. 2 (i.e. highly irregular shapes), should house relatively more small specimens than plants with a fractal dimension of c. 1 (smooth leaves, etc.). To put it in other words, fine lobes and complicated leaf shapes should provide relatively more space for small animals than for larger ones (Lawton 1986). Support for these ideas has been provided by studies on different systems with plants and arthropods (Morse et al. 1985, Shorrocks et al. 1991, Gunnarsson 1992).

Consequently, variation in the vegetation structure may in itself cause changes in the arthropod community. In the present case, reducing needle density may lead to altered micro-habitat structure that affects interspecific relations among arthropods. For instance, the number of niches available on the branches may decline. If there are vertebrate predators affecting the arthropod populations, an altered needle density can be accompanied by a change in the predator-prey relationship. This can include variation in the enemy-free space on branches since a reduction in the number of refuges is conceivable if the needle density is reduced.

A review of results obtained in studies of arthropod population dynamics in spruces with different needle density is given below. There are also new data on the taxonomic composition of spruce-living spiders affected by bird predation and variation in needle density.

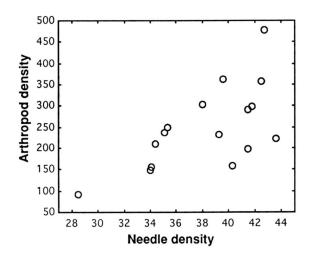

Fig. 2. Arthropod abundance (numbers kg^{-1} dry twig-mass) in relation to the needle density (% dry weight) in a natural population in south-west Sweden (p = 0.019).

Population dynamics

The major prediction is that variation in needle density on spruce branches affects arthropod abundance. If there is covariation between needle density and arthropod abundance the relationship may be caused by at least two processes: a) the needle density in itself affecting the arthropods, e.g., by impact on the number of available niches; b) the needle density is affecting the interactions between or within trophic levels. For instance, the abundance of certain arthropods may change if interspecific competition is affected by needle density, or predators may increase their foraging efficiency in needle-sparse branches. This was tested by field and laboratory experiments, and by studies of natural populations under contrasting conditions.

(1) In a descriptive study, two nearby spruce stands at site H were compared, one stand having a relatively high and one a relatively low percentage of needle-loss (Gunnarsson 1988). On average, there was 33% needles (based on dry weights of needles and twigs) on branches in the stand in an exposed position, and 40% in the stand in a sheltered position (p < 0.01). Since there is higher deposition of air pollutants at forest edges (Hasselrot and Grennfelt 1987), it seems likely that part of the needle density difference between the stands can be attributed to air pollution. The abundance of autumn spider populations were assessed in the two stands. There was no difference in observed total spider density. However, the specimens were categorised into two sizes (Askenmo et al. 1977, Norberg 1978, Gunnarsson 1983), and the abundance of large spiders (body length ≥ 2.5 mm) was 5.2 (numbers kg^{-1} dry branch-mass) in the stand with low needle density and 10.5 in the stand with a higher needle density (p < 0.02). There was no difference in abundance of small spiders (< 2.5 mm). There was a significant difference (p < 0.001) between the total size distribution

in the two stands; the mean length was 1.55 mm in the stand with low needle density and 1.84 mm in the stand with a higher needle density (Gunnarsson 1990).

This suggested that large spiders were affected negatively by low needle density. Alternatively, spatial differences may have caused the observed pattern, i.e. other conditions other than needle density, such as microclimate or colonization rate, may explain why large spiders were so few on branches in the low-needle density stand.

(2) The spider abundance in relation to the proportion of needles in each branch was examined at site U (Gunnarsson 1990). In an autumn sample, the spider density was positively correlated with needle density (p < 0.02). The spider density was given as numbers kg^{-1} dry twig-mass (twig-mass = total branch-mass – needle-mass), so the amount of needles in the branches did not influence the density estimates. Unpublished results show a similar positive correlation for arthropods other than spiders (Spearman rank correlation test, r_s = 0.539, N = 16, p = 0.037), and for the total arthropod community on the branches (Fig. 2, r_s = 0.603, N = 16, p = 0.019).

The hypothesis that needle density has an impact on spider abundance was further tested in laboratory and field experiments (Gunnarsson 1990). First, in the laboratory, spiders were allowed to choose between staying on the branches in terraria or not. After being used with natural needle densities, branches were used a second time with an average of 37% of the needles removed. The mean number of spiders found on unaltered branches was 23.3 versus 19.4 specimens on needle-sparse ones (p < 0.01). Some of the alternative explanations could be ruled out since interspecific predation and effects of the experimental sequence causing the difference in spider numbers were controlled. Secondly, in the field experiment, two branches were randomly selected in each of 13 experimental trees. One of the branches in each tree was used with its natural needle density and in the second branch an average of 30% of the needles were removed. After two months the branches were cut down and examined in the laboratory. The mean spider abundance in needle-sparse branches was 65.5 (spiders kg^{-1} dry twig-mass), compared with 102.3 on unaltered branches (p < 0.05).

There were no differences in mean size of spiders on needle-sparse and unaltered branches in either the laboratory or the field experiment.

To summarize, in a natural population, arthropod abundance increased with needle density, and in laboratory and field experiments, reduced needle density resulted in decreased spider abundance. Thus, these results strongly suggest a difference in the carrying capacity of arthropods in branches with high and low needle density.

(3) In a field experiment, the abundance of spruce-living spiders was examined in relation to needle density (Sundberg and Gunnarsson 1995). The study site was part of a vast coniferous forest situated c. 25 km east of Göteborg in south-west Sweden. The abundance and size distribution of spiders were recorded on spruce branches at

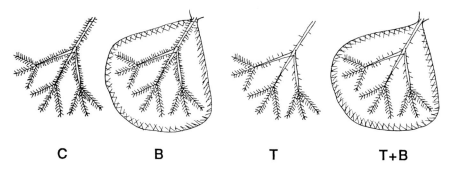

C   B   T   T+B

Fig. 3. Experimental design in a large-scale field experiment. Control (C) and three manipulations were used: branches without bird predation, i.e. enclosed with nets (B); needle-sparse branches (T); branches without bird predation and needle-sparse (T+B). Effects of the experimental treatment were evaluated by multiple comparisons within each sample.

C ≠ B and T ≠ T+B — Bird predation effects irrespective of needle density

C ≠ T and B ≠ T+B — Needle density effects irrespective of bird predation

C ≠ B or T ≠ T+B — Interaction: bird predation effects dependent on needle density

C ≠ T or B ≠ T+B — Interaction: needle density effects dependent on bird predation

the beginning and the end of the experiment. All spiders and c. 24% of the needles were removed from the experimental branches. The branches were enclosed in a plastic sack and shaken vigorously, then it was carefully checked that no spiders remained on the branches. The needles were removed by hand picking. In the control branches, the needle density was left unaltered but the spiders were removed.

There were no initial differences in the spider community on the selected branches. Spiders were collected after seven weeks of colonization on control and needle-sparse branches. The mean density of spiders was significantly lower on needle-sparse than on control branches (p < 0.047). This was shown to be an effect of reduced density of large (length ≥2.5 mm) spiders, but not of small (<2.5 mm) ones. However, the total size distributions did not differ between the branch categories. Thus, the field experiment confirmed earlier results in the laboratory studies and natural populations.

(4) A long-term field experiment at site U was conducted to elucidate the effects of bird predation and needle density on population dynamics of spruce-living arthropods (Gunnarsson unpubl.). One control and three manipulated branch categories were used in each of 70 experimental spruces. The three experimental categories were (Fig. 3): needle-sparse branches (needles were removed by hand picking and scraping); bird predation-free branches (coarse-meshed nets, mesh size 10 mm, prevented birds from foraging on the branches); branches that were needle-sparse and bird predation-free. Experimental samples were taken four times, each autumn and spring for two years. The abundance of spiders and of arthropods was tested separately. Any effects of interactions between bird predation and needle density were evaluated by multiple comparisons between the control and experimental categories (Fig. 3).

The analyses showed that strong effects of bird predation could be found on spiders as well as other arthropods. For instance, the median densities of spiders and

other arthropods on bird predation-free branches were between 2.1 and 4.7 times the control densities. There were also some interactions: in the autumn samples, spiders were affected by bird predation on needle-sparse branches in 1989 and on needle-dense branches in 1990. In other arthropods, similar interactions were found in spring 1990 and spring 1991.

The impact of bird predation on the spider size distribution was investigated by comparisons of the pooled spider samples within each of the four experimental samples. The most relevant test is to compare controls with bird predation-free branches. The median body lengths in the controls were shorter (68%–91%) than on predation-free branches in all experimental samples.

The effects of needle density on spruce-living arthropods were subtle compared to bird predation effects. There was not a single significant difference in arthropod abundance between branches with unaltered and reduced needle density. This was irrespective of the bird predation influence. However, there were spider size differences that could be attributed to the needle density. In the first autumn sample, the median spider length on bird predation-free branches with unaltered needle density was larger than on bird predation-free and needle-sparse branches. However, in the following three samples there was a shift in the direction of the size difference. Spiders on needle-sparse branches were larger than specimens on branches with unaltered needle density.

In this field experiment over two years, the major effects were caused by bird predation. Obviously passerine birds have a strong impact on the arthropod community in all seasons in coniferous forest in south-west Sweden. The absence of needle density effects on arthropod abundance is remarkable and hard to explain. Nevertheless the results suggested that bird predation effects sometimes depended on needle density. However, the relationship appeared complex and varied between different years. Clear effects of needle density on spider size distribution was found on bird predation-free branches.

Large spiders were more common on needle-sparse branches than on branches with unaltered needle density, except in the first experimental sample. These results support the idea that small arthropods prefer a fine-grained habitat, as suggested by other experiments (e.g. Gunnarsson 1992).

Spider community

(1) The descriptive study at site H, reported above, also revealed that there may be taxonomic differences related to the needle density (Gunnarsson 1988). In the stand with needle-dense branches, there was a positive correlation between needle density and the relative abundance of raptorial spiders (p < 0.05). The major family among raptorial spiders was Thomisidae (16% by number in both stands), but this category also included Clubionidae, Anyphaenidae, and Salticidae. Moreover, there were positive correlations between relative twig-mass (i.e. total branch-mass – needle-mass) and the relative abundance of sheetweb spiders (Linyphiidae; c. 60% of the spiders). This means that there were negative correlations between needle density and relative linyphiid abundance in the stand with needle-sparse branches (p < 0.01), and in the stand with higher needle density (p ≈ 0.05).

The observed pattern in taxonomic composition in relation to needle density might be explained by interaction between the micro-habitat structure and bird predation. First, the high relative abundance of raptorial spiders may be associated with needle-dense branches, since such branches should offer good hiding-places for the spiders between the needles. Secondly, sheetweb spiders may not be so severely affected by needle-loss since there will still be possibilities for attaching the webs. Hence, the relative increase of sheetweb spiders may be caused by a frequency reduction of other families. An alternative explanation could be that needle-sparse branches may have a higher abundance of flying insects, favouring web-building spiders. However, this explanation was not supported by available data (see e.g. Fig. 2).

The relation between the taxonomic composition and needle density was tested in a field experiment.

(2) In the field experiment at site U, reviewed above, the spiders were determined to family level. Consequently, it was possible to examine the influence of bird predation and needle density separately by comparisons between the control and the three experimental categories.

Experimental procedure and sampling

In autumn 1988, 70 spruces were selected at random within the experimental area. In each experimental tree, four branches were selected between 1 and 4 m above the ground, using random numbers to determine the direction of the branch. The treatment of each branch was randomized.

One control and three manipulated branch categories were used in each tree (Fig. 3): needle-sparse branches (needles artifically removed); bird predation-free branches (enclosed with coarse-meshed nets, mesh size 10 mm, the arthropods were free to move within and between branches); branches that were needle-sparse and bird predation-free. The needle density was reduced in September to December 1988. In spring 1989, the needle-thinned branches were checked. Seven trees with one or more branches with dry parts and yellow needles were replaced by new ones. In April 1989, the experiment was started and those branches selected to be protected from bird predation were enclosed in net-sacks. There was additional reduction of needle density in summer 1990, to compensate for the growth of the branches.

The first experimental sample, consisting of 15 trees, was taken in late September 1989. The trees included in each experimental sample were randomly selected. It was carefully checked that the net-sacks were intact and the branches were undamaged, otherwise the tree was omitted from the sample. In mid-March 1990, 18 trees were sampled. The autumn 1990 and spring 1991 samples, both of 15 trees, were taken on similar dates to the previous year. The branches were cut into plastic sacks which were sealed and brought to the laboratory where they were stored at +4°C until examination (Gunnarsson 1983).

The branches were cut into small pieces and each piece was shaken and carefully examined over a white bowl (Askenmo et al. 1977, Gunnarsson 1983), and arthropods were collected from the branch pieces. Spiders were stored in 70% ethanol, and identified to family level. For each branch, the relative abundances of the families were recorded. The relative abundance was calculated as the proportion of the total number of individuals per branch. This does not give any information about spider density, but makes it possible to compare the family composition in different treatments. Further details on the experimental procedure and results on arthropod abundance will be presented elsewhere (Gunnarsson unpubl.).

Results

The family composition of spiders on the control and the experimental branches are shown in the Appendix. The predominating family was Linyphiidae, sheetweb spiders, which produce hunting-webs consisting of a dense sheet in the horizontal plane, supported by threads above and below the sheet. The vertical threads above the sheet work as filter. The flying insects hitting the threads will be knocked down to the sheet, where the spider attacks its prey. The linyphiids were founds at mean relative abundances between 0.219 and 0.748 (median of means: 0.524). Another family present at high relative abundances was Thomisidae (including Philodrominae). They are raptorial spiders, which do not produce any hunting-webs. The mean relative abundance of Thomisidae varied between 0.052 and 0.229 (median of means: 0.128).

The impact of bird predation and of needle density was

Table 1. Relative abundances of Clubionidae, Thomisidae, and Linyphiidae in control and experimental branches. Statistical differences between control (C), needle-sparse (T), bird predation-free (B), and needle-sparse plus bird predation-free (T+B) branches were evaluated by the Kruskal-Wallis one-way ANOVA in the September 1989 sample, and the Friedman two-way ANOVA in the other samples (see the text). The multiple comparisons were made at the 5% level.

Date	Spider family	p	Statistical significance Multiple comparisons
Sep 1989	Clubionidae	0.0001	C<B; C<T+B; T<B; T<T+B
	Thomisidae	0.22	
	Linyphiidae	0.0017	B<C; B<T; T+B<C
Mar 1990	Clubionidae	0.0001	C<B; C<T+B; T<B; T<T+B
	Thomisidae	0.0005	T<B; T<T+B
	Linyphiidae	0.29	
Sep 1990	Clubionidae	0.0001	C<B; C<T+B; T<B; T<T+B
	Thomisidae	0.42	
	Linyphiidae	0.0002	B<C; B<T
Mar 1991	Clubionidae	0.0001	C<T+B; T<T+B
	Thomisidae	0.0003	T<B; T<T+B
	Linyphiidae	0.0062	B<T; T+B<T

examined by comparing the relative abundances of the controls and the experimental treatments. Three families were used in the analyses: Clubionidae, Thomisidae, and Linyphiidae. Thomisidae and Linyphiidae represent raptorial and web-building spiders, respectively. Clubionidae also are raptorial spiders, in the present system represented by a few, very large species only (especially *Clubiona subsultans*). The mean relative abundance of *Clubiona* was very variable, ranging from 0 to 0.229, in the present data set (Appendix). These large variations suggested a strong response to experimental treatment, or temporal factors, and justified the inclusion in the analysis.

In the analysis of the autumn 1989 sample, the bird predation-free category consisted of 12 branches only since unfortunately the spiders from three branches were destroyed before determination was made. The other experimental categories and control consisted of 15 branches. There were strong bird predation effects on Clubionidae and Linyphiidae (Table 1). The comparison between the control and the net-enclosed branches reflected the bird predation effect. There was a ten-fold increase of Clubionidae in net-enclosed branches. In Linyphiidae, the trend was opposite to that in Clubionidae, i.e. the relative abundance was lower in bird predation-free branches than in the control. The relative abundance of Linyphiidae in the control was 1.8 times the abundance in net-enclosed branches with unaltered needle density. However, no effect of bird predation was found on needle-sparse branches, suggesting an interaction: bird predation favoured Linyphiidae on needle dense branches only (Fig. 3). There was no evidence of any needle density effects on the relative abundance of the three families (this comparison refers to control vs needle-sparse and bird predation-free vs bird predation-free plus needle-sparse branches, Fig. 3).

In spring 1990, the relative abundance of Clubionidae was positively affected by the absence of bird predation (Table 1). Thomisidae were favoured on needle-sparse branches protected from bird predation only (Fig. 3). There were no effects of needle density between the categories.

The autumn 1990 samples showed differences similar to those of the autumn 1989 samples (Table 1). The relative abundance of Clubionidae in bird predation-free branches was 8.4 times the control. In the Linyphiidae, the relative abundance in the control was 2.2 times the bird predation-free category but no effects were found on needle-sparse branches (Fig. 3). There were no needle density effects.

In spring 1991, there were significant bird predation effects on needle-sparse branches (Fig. 3) in all three families. Clubionidae and Thomisidae increased and Linyphiidae decreased in the absence of bird predation. No needle density effects were observed (Table 1).

The effect of needle density on the relative abundance of sheetweb spiders, the most common web-builders, was investigated by correlation analyses among each branch category, including the control, in the experimental samples. There were two significant correlations among the 16 data sets: a positive correlation between needle density and the relative abundance of Linyphiidae in the control group in autumn 1989 (Spearman rank correlation test, $r_s = 0.629$, $p = 0.019$), and a negative correlation in bird predation-free branches in spring 1990 ($r_s = -0.515$, $p = 0.034$).

Discussion

There is clear evidence for bird predation effects on the taxonomic composition of the spider community in spruce. In many cases, however, interactions with needle density were involved. The free-hunting spider families Clubionidae and Thomisidae increased their relative abundance on branches protected from bird predation. The sheetweb spiders Linyphiidae, however, decreased their relative abundance on these branches. This suggests that competition and/or interspecific predation between spiders have strong effects that are reversed or absent in the presence of bird predation. The foraging of passerines in spruce branches is favouring the linyphiids. Competition effects are rare among spiders and this has been attributed to strong vertebrate predation pressure on spiders, for example (Wise 1993).

The demonstrated effect of needle density variation on spider relative abundance is rather weak in the present experiment. However, in combination with bird predation there were effects, especially on Thomisidae on needle-sparse branches (see above). In the absence of bird predation there was a negative relationship between needle density and linyphiid relative abundance. This was similar to the observed relation in the descriptive study at site H (Gunnarsson 1988). The positive correlation in the control showed that the pattern between needle density and the relative abundance of Linyphiidae is variable. The relationship may be altered in presence of strong bird predation pressure.

Norberg (1978) reported on the taxonomic composition of spruce-living spiders from the same area as the present study. In his study, Thomisidae (including Philodrominae) made up 37–45% by number, and Linyphiidae 27%. This suggests that the spider community has changed during the last 15 years. However, the present study was restricted to the lower part of the spruces, but Norberg's study was not.

Why did the experiment not reveal effects of changing micro-habitat structure on the spider taxonomic composition? There are several studies showing that changing habitat structure will lead to an altered spider community (Uetz 1991, Wise 1993). It is possible that the procedure with a mixture of needle-sparse and needle dense branches in the experiment may have contributed to the result. For instance, migration between branches will result in dilution of any needle density effect. It is known that spiders actively search for a suitable micro-habitat (Riechert and Gillespie 1986). However, the bird predation effect was clearly apparent, meaning that any needle density effect is much weaker.

Birds

In winter, mixed-species flocks of passerine birds forage in coniferous forests in the south part of Scandinavia (Alerstam et al. 1974, Ekman 1979). These flocks may include coal tits *Parus ater*, willow tits *P. montanus*, crested tits *P. cristatus*, goldcrests *Regulus regulus*, treecreepers *Certhia familiaris*, and occasionally also great tits *P. major*, blue tits *P. caeruleus* and marsh tits *P. palustris*. Two factors crucial to winter survival of wintering birds are food availability and the risk of being killed by predators such as sparrowhawks *Accipiter nisus* and pygmy owls *Glaucidium passerinum* (Jansson et al. 1981, Ekman 1984, 1986, Suhonen 1993).

Earlier experimental and descriptive studies have shown that birds in coniferous forest significantly reduce their non-renewable food resources, mainly tree-living spiders, during winter (Askenmo et al. 1977, Norberg 1978, Jansson and von Brömssen 1981, Gunnarsson 1983, Suhonen et al. 1992). Later experiments showed that these results are applicable to arthropods in general, both during summer and winter (Gunnarsson unpubl.). Field experiments conducted at the U site showed that the absence of bird predation during winter increased spider survival by c. 20% (Askenmo et al. 1977, Gunnarsson 1983). In particular, large spiders (body length ≥ 2.5 mm) were affected negatively by bird predation. The needle density on spruce branches may affect the foraging efficiency of birds when searching for prey (Hake and Eriksson 1990). For instance, the visibility of spiders may be higher on needle-sparse branches, making the foraging

success partly dependent on the branch structure. There was some support for this hypothesis in a field experiment (see above; Gunnarsson, unpubl.).

There are at least two factors related to needle density which could influence the behaviour of wintering birds. First, needle-loss may affect the prey populations by reducing densities (Gunnarsson 1990). Secondly, the risks of predation from sparrowhawks and pygmy owls may increase as a result of thinning of the tree canopies. If there is a balance between starvation and predation risks, the birds may be expected to change their behaviour as a result of altered structure of the trees.

In a comparative study, the foraging behaviour of willow tits was investigated at the H and U sites (Hake 1991). In winter, the willow tits spend all daytime foraging, and the time available is almost exclusively devoted to searching for food, handling prey, and scanning for predators. The proportions of these activities make up the daily time budget of the birds (Ekman 1987). Time budgets of willow tits foraging in pine were studied during December and January at the H and U sites. A significant difference (p = 0.0043) in the proportion of time spent 'scanning' was observed. At the H site, where needle density was lower, willow tits spent 40% of the time 'scanning', compared with 31% at the U site. As a compensation for increased 'scanning' time, birds spent only 17% of the time 'handling' at the H site, compared to 24% at the U site (p < 0.026).

To reduce predation risk, willow tits join mixed-species flocks during winter. These flocks were, on average, significantly larger at the H site, containing c. 29 individuals compared with 20 at the U site (p < 0.05). Moreover, the mean number of species in the flocks differed between the areas: 6 species at the H site, and 4 at the U site (p < 0.02).

Thus, the behaviour of birds at the H site indicated that they were affected by one or more habitat characteristics that differed between sites (Hake 1991). A likely explanation is the difference in needle density on branches of coniferous trees. In December and January, the birds almost exclusively forage in the upper half of mature pines (Jansson 1982, Ekman 1987, Hake pers. comm.), and the reduced protection caused by needle-loss should make the birds at the H site especially vulnerable to predation. The needle-loss should facilitate the detection of owls and hawks by the passerines. However, the increased "scanning" time may not be large enough to balance the risks. It is not possible to reduce foraging, i.e. searching and handling, below a critical level of starvation.

The increased flock size in the area with low needle density, H, supports the idea that the risk of predation was higher. By joining a mixed-species flock, birds may maintain protection from predators without having to reduce the time spent foraging, as individual scanning time decreases when flock size becomes larger (Ekman 1987, Hogstad 1988). However, foraging success is probably decreased when flock size increases, as tits live on limited and dispersed food during winter (Askenmo et al. 1977, Jansson and von Brömssen 1981, Gunnarsson 1983). This may reduce the probability of winter survival and consequently population size of tits in coniferous forests (Hake 1991).

General discussion and conclusion

In general, indirect effects of acidification seem to be of importance in the present system. There are clear effects on spider and bird behaviour related to needle density variations. Spiders changed their colonizing rate in relation to the vegetation structure both in the laboratory and in the field. Birds in coniferous forest with high or low needle density, may be exposed to differences in the risk of predation, and they adjusted their foraging behaviour accordingly. Thus, the animals perceive the change in habitat structure and they respond by changing their behaviour. The results are providing further support to the idea that the habitat structure is important to animal populations since the biotic relations can be affected (May 1978, Southwood 1978, Bell et al. 1991). Epiphytes, such as lichens, may be important to the arthropod community in spruce (Pettersson et al. 1995). However, epiphytic lichens were not considered in the present study since the abundance was very low in the study sites. Negative correlations between the number of spider species on trees and SO_2-concentrations have been attributed to declining epiphytic flora (Clausen 1984b, 1986).

The interaction between bird predation and needle density on the spruce-living arthropod population appears to be complex. There is no doubt that there is a strong impact of bird predation on the arthropod numbers in spruce branches. This has been established in field experiments in the present forest system (Askenmo et al. 1977, Gunnarsson 1983), and similar effects are known in several other contrasting systems (Holmes et al. 1979, Pacala and Roughgarden 1984, Spiller and Schoener 1988, Riechert and Hedrick 1990). There is also good evidence for shifts in the taxonomic composition of the spider community resulting from bird predation. Considering the earlier experimental results on spider numbers and needle density, it was surprising that no effects of vegetation structure on the abundance or taxonomy were observed in the large-scale field experiment. However, it is apparent that bird predation effects can interact with needle density, but there were no consistent trends. Sometimes bird predation effects were found on needle dense branches, and in other samples on needle-sparse ones. There are at least two possible explanations for the experimental results. Firstly, any effect of needle density is dependent on factors not controlled in the experiment. For instance, the experimental winters were much milder than average. Spider survival on branches may have been much higher than in a cold winter. Increasing population densities in spruce is probably interfering with the vegetation structure since there may be no "low-spider-density" patches to move to. Secondly, the manipulations may not have produced a "critical experiment". There was no way to control exactly the amount of needles removed from the branches and in some branches the needle removal did not produce the planned difference (Gunnarsson unpubl.). However, needle density in the field experiment clearly affected the size distribution of spiders, suggesting that interspecific relations may be changed as a result of vegetation structure.

In conclusion, effects of variation in needle density can affect both abundance and taxonomic composition of spruce-living arthropods but other community-controlling factors, i.e. vertebrate predation, may be stronger. In areas with moderate needle-loss, between 20% and 30%, the effects of changed micro-habitat structure in itself on arthropod populations are probably small.

Acknowledgements – This study would have been impossible without field and laboratory assistance by the following persons: S. Blid, K. Carlsson, U. Carlsson, A. Christensen, P. Fält, J. Grudemo, M. Hellsten, J. Johnsson, R. Neergaard, P. Stålhandske, I. Sundberg and P. Wingård. I am most grateful to M. Hake and J. Johnsson for comments on this paper. Continuous financial support from the National Swedish Environment Protection Board made this study possible.

References

Alerstam, T., Nilsson, S.G. and Ulfstrand, S. 1974. Niche differentiation during winter in woodland birds in southern Sweden and the island of Gotland. – Oikos 25: 321–330.

Alstad, D.N., Edmunds, G.F. Jr. and Weinstein, L.H. 1982. Effects of air pollutants on insect populations. – Annu. Rev. Entomol. 27: 369–384.

Andersson, F. 1986. Acidic deposition and its effects on the forests of Nordic Europe. – Water Air Soil Pollut. 30: 17–29.

Askenmo, C., von Brömssen, A., Ekman, J. and Jansson, C. 1977. Impact of some wintering birds on spider abundance in spruce. – Oikos 28: 90–94.

Bell, S. S., McCoy, E. D. and Mushinsky, H. R. (eds) 1991. Habitat structure. The physical arrangement of objects in space. – Chapman and Hall, London.

Berdén, M., Nilsson, S., Rosén, K. and Tyler, G. 1987. Soil acidification – Extent, causes and consequences. – Nat. Swedish Environm. Protect. Board, Solna, Report No. 3292.

Blank, L.W. 1985. A new type of forest decline in Germany. – Nature 314: 311–314.

Biström, O. and Väisänen, R. 1988. Ancient-forest invertebrates of the Pyhän-Häkki national park in central Finland. – Acta Zool. Fennica 185: 1–69.

Clausen, I.H.S. 1984a. Lead (Pb) in spiders: A possible measure of atmospheric Pb pollution. – Environ. Pollut. (Ser. B) 8: 217–230.

– 1984b. Notes on the impact of air pollution (SO_2 & Pb) on spider (Araneae) populations in North Zealand, Denmark. – Ent. Meddr. 52: 32–39.

– 1986. The use of spiders (Araneae) as ecological indicators. – Bull. Br. Arachnol. Soc. 7: 83 – 86.

Ekman, J. 1979. Coherence, composition and territories of winter social groups of the willow tit *Parus montanus* and the crested tit *P. cristatus*. – Ornis Scand. 10: 56–58.

– 1984. Density-dependent seasonal mortality and population fluctuations of the temperate-zone willow tit (*Parus montanus*). – J. Anim. Ecol. 53: 119–134.

– 1986. Tree use and predator vulnerability of wintering passerines. – Ornis Scand. 17: 261–267.

– 1987. Exposure and time use in willow tit flocks: the cost of subordination. – Anim. Behav. 35: 445–452.

Ehnström, B. and Waldén, H. W. 1986. Faunavård i skogsbruket. Del 2 – den lägre faunan. – Skogsstyrelsen, Jönköping, (in Swedish).

Feir, D. and Hale, R. 1983. Responses of the large milkweed bug, *Oncopeltus fasciatus* (Hemiptera: Lygaeidae), to high levels of air pollutants. – Int. J. Environ. Studies 20: 269–273.

Führer, E. 1985. Air pollution and the incidence of forest insect problems. – Z. ang. Ent. 99: 371–377.

Ginevan, M. E. and Lane, D. D. 1978. Effects of sulphur dioxide in air on the fruit fly, *Drosophila melanogaster*. – Environ. Sci. Technol. 12: 828–831.

Gunnarsson, B. 1983. Winter mortality of spruce-living spiders: effect of spider interactions and bird predation. – Oikos 40: 226–233.

– 1985. Interspecific predation as a mortality factor among overwintering spiders. – Oecologia 65: 498–502.

– 1988. Spruce-living spiders and forest decline; the importance of needle-loss. – Biol. Conserv. 43: 309–319.

– 1990. Vegetation structure and the abundance and size distribution of spruce-living spiders. – J. Anim. Ecol. 59: 743–752.

– 1992. Fractal dimension of plants and body size distribution in spiders. – Funct. Ecol. 6: 636–641.

– and Johnsson, J. 1989. Effects of simulated acid rain on growth rate in a spruce-living spider. – Environ. Pollut. 56: 311–317.

Hake, M. 1991. The effects of needle loss in coniferous forests in south-west Sweden on the winter foraging behaviour of willow tits *Parus montanus*. – Biol. Conserv. 58: 357–366.

– and Eriksson, M. O. G. 1990. Passerine birds and their prey affected by forest die-back: a tentative work model. – Fauna norv. Ser. C, Cinclus, Suppl. 1: 13–16.

Hasselrot, B. and Grennfelt, P. 1987. Deposition of air pollutants in a windexposed forest edge. – Water Air Soil Pollut. 34: 135–143.

Heliövaara, K. and Väisänen, R. 1993. Insects and pollution. – CRC Press, Boca Raton.

Hogstad, O. 1984. Variation in numbers, territoriality and flock size of a goldcrest *Regulus regulus* population in winter. – Ibis 126: 296–306.

– 1988. Advantages of social foraging in willow tits *Parus montanus*. – Ibis 126: 296–306.

Holmes, R. T., Schultz, J. C. and Nothnagle, P. 1979. Bird predation on forest insects: an exclosure experiment. – Science 206: 462–463.

Jansson, C. 1982. Food supply, foraging, diet and winter mortality in two coniferous forest tit species. – Ph.D. thesis, Univ. of Göteborg.

– and von Brömssen, A. 1981. Winter decline of spiders and insects in spruce *Picea abies* and its relation to predation by birds. – Holarct. Ecol. 4: 82–93.

– , Ekman, J. and von Brömssen, A. 1981. Winter mortality and food supply in tits *Parus* spp. – Oikos 37: 313–322.

Lawton, J. H. 1983. Plant architecture and the diversity of phytophagous insects. – Annu. Rev. Entomol. 28: 23–39.

– 1986. Surface availability and insect community structure: the effects of architecture and fractal dimension of plants. – In: Juniper, B. E. and Southwood, T. R. E. (eds), Insects and the plant surface. Edward Arnold, London, pp. 317–331.

Lindberg, S. E., Lovett, G. M., Richter, D. D. and Johnson, D. W. 1986. Atmospheric deposition and canopy interactions of major ions in a forest. – Science 231: 141–145.

May, R. M. 1978. The dynamics and diversity of insect faunas. – In: Mound, L. A. and Waloff, N. (eds), Diversity of insect faunas. Blackwell, Oxford, pp. 188–204.

Morse, D. R., Lawton, J. H., Dodson, M. M. and Williamson, M. H. 1985. Fractal dimension of vegetation and the distribution of arthropod body lengths. – Nature 314: 731–733.

Norberg, R. Å. 1978. Energy content of some spiders and insects on branches of spruce (*Picea abies*) in winter; prey of certain passerine birds. – Oikos 31: 222–229.

Oren, R., Schultze, E.-D., Werk, K. S., Meyer, J., Schneider, B. U. and Heilmeier, H. 1988. Performance of two *Picea abies* (L.) Karst. stands at different stages of forest decline. I. Carbon relations and stand growth. – Oecologia 75: 25–37.

Pacala, S. and Roughgarden, J. 1984. Control of arthropod abundance by *Anolis* lizards on St. Eustatius (Neth. Antilles). – Oecologia 64: 160–162.

Palmgren, P. 1932. Zur Biologie von *Regulus r. regulus* (L.) und *Parus atricapillus borealis* Selys. Eine vergleichend-ökologische Untersuchung. – Acta Zool. Fennica 14: 1–113.

Pettersson, R. B., Ball, J. P., Renhorn, K.-E., Esseen, P.-A. and Sjöberg, K. 1995. Invertebrate communities in boreal forest canopies as influenced by forestry and lichens with implications for passerine birds. – Biol. Conserv. (in press).

Pulz, R. 1987. Thermal and water relations. – In: Nentwig, W. (ed.), Ecophysiology of spiders. Springer, Berlin, pp. 26–55.

Riechert, S. E. and Gillespie, R. G. 1986. Habitat choice and utilization in web-building spiders. – In: Shear, W. A. (ed.), Spiders. Web, behavior, and evolution. Stanford Univ. Press, Stanford, pp. 23–48.

– and Hedrick, A. V. 1990. Levels of predation and genetically based anti-predator behaviour in the spider, *Agelenopsis aperta*. – Anim. Behav. 40: 679–687.

Schulze, E.-D. 1989. Air pollution and forest decline in a spruce (*Picea abies*) forest. – Science 244: 776–83.

Shorrocks, B., Marsters, J., Ward, I. and Evennett, P. J. 1991. The fractal dimension of lichens and the distribution of arthropod body lengths. – Funct. Ecol. 5: 457–460.

Southwood, T. R. E. 1978. The components of diversity. – In: Mound, L. A. and Waloff, N. (eds), Diversity of insect faunas. Blackwell, Oxford, pp. 19–40.

Spiller, D. A. and Schoener, T. W. 1988. An experimental study of the effects of lizards on web-spider communities. – Ecol. Monogr. 58: 57–77.

Suhonen, J. 1993. Predation risk influences the use of foraging sites by tits. – Ecology 74: 1197–1203.

– , Alatalo, R. V., Carlson, A. and Höglund, J. 1992. Food resource distribution and the organization of the *Parus* guild in a spruce forest. – Ornis Scand. 23: 467–474.

Sundberg, I. and Gunnarsson, B. 1995. Spider abundance in relation to needle density in spruce. – J. Arachnol. 23: (in press).

Uetz, G. W. 1991. Habitat structure and spider foraging. – In: Bell, S. S., McCoy, E. D. and Mushinsky, H. R. (eds), Habitat structure. The physical arrangement of objects in space. Chapman and Hall, London, pp. 325–348.

Westman, L. and Lesinski, J. 1986. Kronutglesning och andra förändringar i grankronan. Morfologisk beskrivning. – Nat. Swedish Environ. Protect. Board, Solna, Report No. 3262, (in Swedish).

Wise, D. H. 1993. Spiders in ecological webs. – Cambridge Univ. Press, Cambridge.

Appendix. The relative abundance of spider families in control (C), needle-sparse branches (T), branches without bird predation (B), and needle-sparse plus bird predation-free branches (T+B), in four experimental samples. Relative abundance, by number, given as mean ± S.D. proportion in each category. The spider families are: Dictynidae (DIC), Uloboridae (ULO), Clubionidae (CLU), Anyphaenidae (ANY), Thomisidae (THO), Agelenidae (AGE), Theridiidae (THE), Tetragnathidae (TET), Metidae (MET), Araneidae (ARA), and Linyphiidae (LIN). N denotes number of branches.

Date and category	DIC	ULO	CLU	ANY	THO	AGE	THE	TET	MET	ARA	LIN
September 1989											
C (N = 15)	0.003 (±0.009)	0	0.016 (±0.020)	0	0.088 (±0.088)	0.032 (±0.057)	0.060 (±0.055)	0.018 (±0.030)	0.014 (±0.023)	0.100 (±0.065)	0.668 (±0.170)
T (N = 15)	0	0.005 (±0.020)	0.020 (±0.040)	0	0.074 (±0.079)	0.046 (±0.119)	0.061 (±0.068)	0.017 (±0.023)	0.096 (±0.255)	0.048 (±0.050)	0.611 (±0.229)
B (N = 12)	0	0	0.160 (±0.097)	0	0.126 (±0.072)	0.137 (±0.121)	0.100 (±0.061)	0.018 (±0.035)	0	0.078 (±0.051)	0.370 (±0.186)
T+B (N = 15)	0.023 (±0.077)	0	0.117 (±0.073)	0	0.086 (±0.056)	0.230 (±0.164)	0.065 (±0.080)	0.004 (±0.014)	0	0.043 (±0.050)	0.431 (±0.232)
March 1990											
C (N = 18)	0.005 (±0.020)	0.002 (±0.008)	0.006 (±0.017)	0	0.078 (±0.081)	0	0.052 (±0.117)	0.045 (±0.073)	0	0.111 (±0.128)	0.702 (±0.196)
T (N = 18)	0	0.008 (±0.034)	0.007 (±0.029)	0	0.052 (±0.129)	0	0.068 (±0.123)	0.019 (±0.060)	0.012 (±0.034)	0.154 (±0.141)	0.679 (±0.290)
B (N = 18)	0.001 (±0.002)	0.001 (±0.004)	0.036 (±0.023)	0.002 (±0.008)	0.183 (±0.162)	0.004 (±0.009)	0.133 (±0.100)	0	0.005 (±0.017)	0.049 (±0.050)	0.586 (±0.202)
T+B (N = 18)	0.005 (±0.013)	0.001 (±0.006)	0.061 (±0.046)	0	0.130 (±0.072)	0.006 (±0.011)	0.137 (±0.110)	0.004 (±0.012)	0	0.038 (±0.042)	0.616 (±0.131)
September 1990											
C (N = 15)	0.015 (±0.038)	0	0.019 (±0.035)	0	0.195 (±0.120)	0.028 (±0.070)	0.050 (±0.071)	0.043 (±0.070)	0.056 (±0.086)	0.069 (±0.056)	0.490 (±0.100)
T (N = 15)	0	0	0.034 (±0.103)	0	0.121 (±0.104)	0.032 (±0.053)	0.046 (±0.059)	0.044 (±0.063)	0.035 (±0.071)	0.097 (±0.095)	0.483 (±0.198)
B (N = 15)	0.006 (±0.016)	0	0.159 (±0.122)	0	0.166 (±0.147)	0.152 (±0.137)	0.127 (±0.104)	0.007 (±0.020)	0	0.018 (±0.029)	0.219 (±0.143)
T+B (N = 15)	0.016 (±0.052)	0	0.229 (±0.145)	0	0.134 (±0.137)	0.121 (±0.140)	0.110 (±0.112)	0.008 (±0.024)	0.003 (±0.008)	0.039 (±0.056)	0.334 (±0.189)
March 1991											
C (N = 15)	0.004 (±0.015)	0	0.004 (±0.016)	0	0.226 (±0.281)	0	0.033 (±0.044)	0.061 (±0.097)	0.003 (±0.012)	0.098 (±0.105)	0.558 (±0.295)
T (N = 15)	0	0	0	0	0.014 (±0.037)	0	0.048 (±0.096)	0.051 (±0.133)	0.008 (±0.032)	0.123 (±0.181)	0.748 (±0.234)
B (N = 15)	0	0	0.030 (±0.035)	0	0.229 (±0.137)	0.006 (±0.023)	0.272 (±0.148)	0.004 (±0.010)	0.002 (±0.007)	0.039 (±0.051)	0.415 (±0.222)
T+B (N = 15)	0	0	0.054 (±0.056)	0.003 (±0.011)	0.228 (±0.105)	0	0.217 (±0.166)	0.011 (±0.037)	0.025 (±0.076)	0.070 (±0.080)	0.389 (±0.197)

Ecological Bulletins 44: 259–270. Copenhagen 1995

Effects of soil acidification on forest land snails

Ulf Gärdenfors, Henrik W. Waldén and Ingvar Wäreborn

Gärdenfors, U., Waldén, H. W. and Wäreborn, I. 1995. Effects of soil acidification on forest land snails. – Ecol. Bull. (Copenhagen) 44: 259–270.

The objectives of our investigation were to determine whether on-going acidification in Sweden affects forest land snail populations and to study the relationships between snail density and the complex of chemical parameters related to calcium. We approached the problems through: i) re-inventories of snail populations in localities originally investigated some decades earlier, ii) experimental liming in forests, and iii) examination of elemental composition in snail shells from different environmental situations. The re-inventories were made at 57 forest sites in southern and central Sweden, originally sampled between 14 and 46 yr earlier. In general, snail density clearly decreased over the 14–46 yr periods, particularly at localities with low base saturation and low pH. In some oligotrophic coniferous forest sites in southern Sweden the snail fauna even seemed to have died out. Parallel to this a decrease in calcium concentration in litter has occurred. According to rarefraction analyses the number of species had decreased significantly only at six sites in SW Sweden. Experimental liming in beech forests resulted in an increase of snail density by a factor of 10–90, indicating that liming can counteract depauperization of snail populations caused by soil acidification. Analysis of the results from the retrospective studies and liming experiments showed highly significant correlations between snail density and calcium concentration, pH, base saturation and base cation concentration of the litter. These abiotic factors were strongly inter-correlated. Analyses by instrumental neutron activation and other methods of shell elemental composition indicated that concentrations of certain elements were dependent on acidity in the environment in which the snails lived. This suggests that shells from different localities and time periods can be used as bio-indicators and environmental archives.

U. Gärdenfors, Dept Wildlife Ecology, Swedish Univ. of Agricultural Sci., P.O. Box 7002, S-750 07 Uppsala, Sweden. – H. W. Waldén, Natural History Museum, S-402 35 Göteborg, Sweden. – I. Wäreborn, Univ. Linköping, Dept of Teacher Education, S-581 83 Linköping, Sweden.

Introduction

Land snails are part of the decomposition biota in most terrestrial ecosystems (Lindquist 1941, Frömming 1958, Mason 1974). They are preyed upon by many animals, and at least the larger species, function as an important calcium source for birds (Schifferli 1977, Graveland et al. 1994).

Malacologists know by experience that, in general, land snails are positively influenced (in abundance as well as in species richness) by the presence of lime, in litter and soil. This has been confirmed in laboratory breedings (Schmidt 1955, Voelker 1959, Crowell 1973) and field studies (Wäreborn 1969, 1982, Graveland et al. 1994). The snails need calcium for shell construction, reproduction and various physiological processes (Kün-

kel 1929, Robertson 1941, Tompa 1976, Chétail and Krampitz 1982, Wäreborn 1982). Several other studies have also demonstrated a positive correlation between snail density and pH in litter and soil (Burch 1955, Valovirta 1968, Wäreborn 1970, 1982, Waldén 1981). This correlation is not surprising since pH is strongly correlated with calcium concentration.

In the light of the knowledge mentioned above, Gärdenfors (1987) listed several land snail species suspected of being susceptible to acidification in terrestrial habitats. Snails living in oligo- and mesotrophic forests or marshes with low or intermediate levels of base saturation, as well as in habitats disturbed by pollution or other human activities, were considered to be more affected than others.

In the 1970's a re-inventory of 148 localities in the

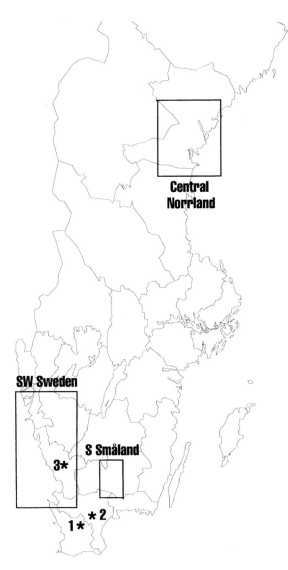

Fig. 1. Map showing the investigated areas and the localities where liming experiments were conducted in Sweden; 1) Ekeröd, 2) Nösdala, and 3) Hult.

Göteborg region, originally investigated in 1921–22, gave some indications that the mollusc fauna had been negatively affected by acidification (Waldén 1992). When Graveland et al. (1994) re-sampled old study sites in the Netherlands, they found that snail abundance had seriously declined in forests on poor soils but had remained unaffected in forests on calcareous sandy soils.

The detailed effects on snail populations of ongoing acidification in terrestrial habitats, including effects of changes in soil chemistry, shell chemical composition, and regional trends, have not been investigated. We chose to approach the topic in three ways: 1) Retrospective studies of snail populations in localities originally investigated some decades earlier. We wanted to relate any observed population changes to changes in soil chemistry

during the elapsed period and to current chemical conditions. 2) Experimental liming in forests to quantify reactions of snail populations to calcium and related parameters. 3) Analyses of the elemental composition of snail shells from sites representing different environmental conditions.

Some of the results have already been published by Gärdenfors et al. (1988, 1995), Gärdenfors (1992), Waldén et al. (1992), and Wäreborn (1992). In this paper additional data are presented and the results from the separate parts of the investigation are analyzed and evaluated together.

Material and methods

Retrospective studies

The sites where the first inventories of snails had been made (between 1941 and 1974) were located using markings on maps, original map sketches and field notes (including position, topography and vegetation). The re-inventories were performed by the same persons who had made the original inventory (except for 10 sites in SW Sweden investigated during the 1940's and 1950's by H. Lohmander), which facilitated the location of the exact positions. Sampled litter, including the uppermost part of the F-layer (cf. Bridges 1978) and mosses, was sifted in the field using a sieve with a 10×10 mm mesh. After air-drying, the samples were passed through sieves with finer meshes, whereupon the snails were extracted by hand, from the finer fractions under a magnifying glass. Three regions in Sweden (Fig. 1) were studied: 1) Southern Småland (in the southern part of the province of Småland, vicinity of Lake Möckeln; boreo-nemoral zone) originally investigated by Wäreborn in 1965–66 (Wäreborn 1969, 1982). Twenty localities were sampled in 1987–88; 16 localities in 1987 and 4 in 1988, representing 8 from oligotrophic mixed forests (mainly Norway spruce and birch), 2 from oligotrophic oak forests and 10 from mesotrophic broadleaved forests. Two of the latter were re-sampled a second time in 1988. Approximately 3 l of sifted litter (corresponding to 20–25 l of unsifted litter) were subjectively collected, mainly along three longitudinal transects, within the original 10×20 m sample areas. 2) South-western Sweden (from NW Skåne to the provinces of Bohuslän and Västergötland; nemoral and boreo-nemoral zones) originally investigated by Lohmander and Waldén in 1941–1966 (Waldén 1965, 1986). Twenty-five localities (mainly with broadleaved mixed forest; but a few with coniferous forests with deciduous trees) were sampled in 1987–90 (13 localities in 1987, 11 in 1988 and 1 in 1990); 6 of those investigated in 1987 were re-sampled a second time in 1989. Twenty litres (before sifting) of litter were subjectively collected and sifted within each sample area (representing < 1 ha). Besides these 25 localities, two reference localities (investigated yearly in 1972–81 by Waldén unpubl.) were re-sampled in 1988; Hördalen (province of Halland) and St. Mollungen (province of Västergötland). 3) Central Norrland (provinces of Medelpad and Ångermanland; boreal zone) originally investigated by Waldén in 1954–1974. Twelve localities (conifer forests mixed with deciduous trees) were sampled in 1988–1991 (3 localities in 1988, 6 in 1990 and 3 in 1991) in the same way as 2).

Of all the originally investigated sites, 55, 46 and 45 localities in the three regions respectively were considered for re-inventories. Of these 20, 25 and 12 localities were selected using the criteria of minimum structural and/or anthropogenic changes since the original inventory. Certain successional changes was of course inevitable.

In southern Småland the re-inventories in 1987–88 were done within ±7 days of the original sampling dates in 1964–66. However, the second re-inventory (1988) was performed 15 and 24 days later than the original inventory (1965). In SW Sweden the deviation from the original sampling dates was 0–3 days for 9 localities, in the other cases wider, in the most extreme case 74 days. In central Norrland the deviation varied between 5 and 62 days.

The effects of weather conditions on snail density were studied: i) at two reference localities in SW Sweden where the snail fauna was sampled yearly from 1972–81, and ii) by comparing rainfall and temperature in the years when the first inventory was made with the conditions during the re-inventories, looking for extreme deviations which could be suspected to affect snail populations.

Liming experiments

Three mature beech forests already limed for other purposes were utilized for the experimental liming investigation. The liming experiments were established using two or three replicates 1–10 yr before sampling. Within 10×10 m plots a number of squares were randomly distributed and the litter within them was collected and sifted. The forests were (Fig. 1): 1) Ekeröd (province of Skåne), limed in 1983, sampled in 1987 using 36 0.1 m^2 squares in 3×4 experimental plots treated with 0, 2, 5 or 10 tons of dolomitic limestone ha^{-1}. 2) Nösdala (province of Skåne), limed in 1987, sampled in 1988 using 20, 0.1 m^2 squares in 2×2 plots treated with 0 or 2 tons of dolomitic limestone ha^{-1}. 3) Hult (province of Halland), limed in 1979, sampled in 1987 and 1988 using 26, 0.25 m^2 squares in 3×2 plots treated with 0 and 2 tons of finely ground limestome ("supra") ha^{-1}.

Chemical analyses

For most of the re-investigated localities colorimetrical pH measurements on litter were available from the first inventory. Samples from Småland had also been analyzed for total calcium concentration and ash content of the litter.

In connection with the re-inventories and liming experiments, chemical analyses were performed on the sampled litter. In fresh litter the pH was measured both colorimetrically and electrometrically (pH H$_2$O). In air-dried, ground and homogenized litter the following variables were measured: pH (H$_2$O), determined electrometrically; total calcium (Ca-tot); base cations (S) (Nömmik 1974); base saturation (S/T) at pH 7; total (Kjeldahl) nitrogen (N-tot); ash-content and easily soluble calcium (Ca-AL); potassium (K-AL) and phosphorus (P-AL) extracted in ammonium-acetate-lactate solution (Egnér et al. 1960). All calculations (except for pH) were made on variables calculated on an ash-free basis (Sjörs 1961).

Shell analyses

Shells of five snail species, viz. *Cepaea hortensis* (Müller), *C. nemoralis* (L.), *Arianta arbustorum* (L.), *Cochlodina laminata* (Montagu) and *Clausilia bidentata* (Ström), were collected in various localities and habitats. Fifty samples (one or two shells each) – 21, 4, 5, 13 and 7 samples, respectively, of these species – were analyzed to determine elemental concentrations. Instrumental neutron activation analysis (INAA) was applied to most of the samples; for several samples atomic absorption, ICP and other special methods were also used (Gärdenfors et al. 1988, 1995). In many cases, litter samples were collected together with the snails, and less often soil and vegetation samples but these samples have not yet been analysed.

Statistical methods

Most statistical analyses (tests specified in the text) were conducted with the program StatView. One exception was the rarefaction analyses which were conducted with the program RAREFACT, using data on live specimens (Krebs 1991, see Krebs 1989).

Results
Total density

The total number of live snails found had decreased in all three re-investigated regions. In southern Småland (21–24 yr after the first inventory) we found that there had been an average total decrease of 64% (p < 0.01, n = 10; paired t-test) at mesotrophic localities, and of 80% (p < 0.01, n = 10) at oligotrophic localities. In SW Sweden (22–46 yr between inventories) there was a decrease of 27% (p = 0.02, n = 25). In central Norrland (14–36 yr between inventories) there had been an average decrease of 38% of litter-dwelling individuals, but the change was not significant (p = 0.13, n = 12).

At two of the most oligotrophic localities in southern Småland not a single snail specimen was found in 1987,

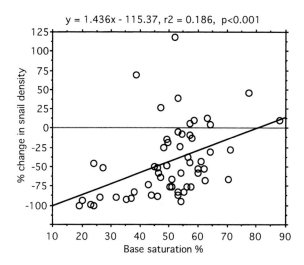

y = 1.436x - 115.37, r2 = 0.186, p<0.001

Fig. 2. Percentage change (over 14–46 yr) in snail density in all three re-investigated regions in relation to base saturation (S/T) at the time of re-inventory (1987–1990). N = 57, Pearson linear regression.

whereas 66 and 74 specimens were found in the original inventory. At the two mesotrophic localities in this area, which were re-investigated twice with some seasonal delay in the third inventory, observed decreases in density were 82% and 88% from 1965 to 1987 and 65% and 66% from 1965 to 1988.

In SW Sweden, six localities (two from 1943, four from 1961–63) were re-investigated twice, both in 1987 and 1989. Replacing the data from 1987 from these six localities with the corresponding data from 1989, gave an average decrease in all 25 localities of 24% (p = 0.01, paired t-test) instead of 27% up to 1987 (cf. above). For these six sites the observed mean decrease in density was 26% from the first to the second (1987) inventory, and 15% from the first to the third (1989) inventory (neither decrease was statistically significant; paired t-test).

For sites in SW Sweden originally sampled during 1941–53 (n = 10) and at those sampled in 1961–66 (n = 15), snail abundance at the end of the 1980's tended to have decreased by 40% (p = 0.11; Wilcoxon signed-rank test) and 20% (p = 0.07), respectively. This comparison suggests that the decrease in snail density in SW Sweden might have begun early, well before 1960.

In general, the percentage decrease in snail density in southern Sweden (southern Småland and SW Sweden) was negatively correlated with base saturation (southern Småland r = -0.576, n = 20, p < 0.01; SW Sweden r = -0.414, n = 25, p < 0.05), total calcium concentration (southern Småland, r = -0.461, n = 20, p < 0.05; SW Sweden r = -0.359, n = 25, p = 0.08) and pH measured electrometrically (southern Småland r = -0.512, n = 20, p = 0.02; SW Sweden r = -0.399, n = 24, p = 0.05). The chemical parameters were measured during the re-inventory. In central Norrland, the same pattern was seen, although the change was not significant (for base saturation, S/T:

r = -0.566, n = 11, p = 0.07, one locality with very few specimens was excluded from the analysis). Considering all three regions together, the percentual decrease was significantly correlated to base saturation (r = -0.432, n = 57, p < 0.001; Fig. 2) and pH measured electrometrically (r = -0.321, n = 53, p < 0.05), but not with calcium concentration (r = -0.211, n = 57, p = 0.12).

In southern Småland, the percentage of empty shells (dead specimens) was on average 9.7% higher in 1987 than in 1964–66 (z = -2.84, n = 15, p < 0.01, Wilcoxon signed-rank test), suggesting a possible effect of the cold winter of 1986–87 (see below). The percentage of empty shells was not, however, correlated with a decrease in the density in corresponding samples (r = -0.144, n = 15, p = 0.41).

In the two localities sampled annually during 1972–1981, total snail densities had coefficients of variation of 34% (St. Mollungen) and 45% (Hördalen) (Waldén unpubl.). In 1988, when these localities were re-sampled, the densities were 92% and 61%, respectively, as high as those found on average in 1972–81.

Species richness and density

With the exception of mesotrophic forests in southern Småland, there was a good correlation between the current number of species and the complex of inter-correlated chemical parameters related to calcium. This is illustrated in Fig. 3 for base saturation in SW Sweden.

In southern Småland, the average number of species in the samples had decreased from 6.0 to 3.5 at oligotrophic localities and from 17.1 to 15.5 at mesotrophic localities. In SW Sweden, the average number of species in the samples had decreased from 13.2 to 11.3. In central Norrland, where only litter-dwelling species could be used for comparisons, the average number of encountered species had decreased from 8.3 to 7.5. However, rarefaction analysis (considering only live specimens)

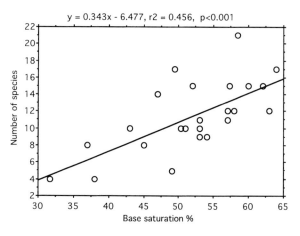

y = 0.343x - 6.477, r2 = 0.456, p<0.001

Fig. 3. Relationship between base saturation (S/T) and number of species in SW Sweden, 1987–1990. N = 25.

Table 1. Number of records (localities) and changes in density for snail species in three re-investigated regions in Sweden. Probability (p) of reduced density was tested by Wilcoxon signed-ranked test; N tot = no. of localities where the species (live specimens) was present in either first, second or both samples. No. of records includes observations of empty shells (indicating presence of the species at the locality) as well as live specimens.

Region	Southern Småland 1964–1966 and 1987–1988. 20 localities.				Southwest Sweden 1941–1966 and 1987–1990. 25 localities.				Central Norrland 1954–1974 and 1988–1991. 12 localities.			
Species	Change in density (%)	No. of records at 1st and 2nd invest.	N tot	p (density)	Change in density (%)	No. of records at 1st and 2nd invest.	N tot	p (density)	Change in density (%)	No. of records at 1st and 2nd invest.	N tot	p (density)
Group I: Species with statistically significant decrease in density in at least two of the studied regions.												
Carychium tridentatum (Risso)	–48	8, 6	8	<0.05	–88	8, 3	8	0.01	–[1]	–	–	–
Clausilia bidentata (Ström)	–79	11, 11	11	<0.01	–79	17, 15	20	<0.05	–[3]	7, 7	10	–
Cochlicopa lubrica (Müller)	–80	12, 11	12	<0.01	–91	13, 10	18	<0.05	–55	9, 5	9	<0.05
Cochlicopa lubricella (Porro)	–45	5, 5	5	<0.05	–93	13, 3	12	<0.05	–90	2, 1	2	0.18
Discus ruderatus (Férussac)	–86	13, 7	11	<0.01	–84	9, 7	10	<0.02	–[3]	12, 12	12	<0.02
Euconulus fulvus (Müller)	–71	20, 16	20	0.001	–43	25, 25	25	<0.05	–65	12, 12	12	<0.02
Nesovitrea hammonis (Ström)	–60	20, 17	20	<0.001	–31	25, 24	25	<0.05	–32	12, 12	12	0.07
Vitrea contracta (Westerlund)	–63	5, 5	5	<0.05	–100	5, 0	4	<0.05	–	–	–	–
Group II: Species with statistically significant decrease in density in one of the studied regions												
Columella aspera Waldén[2]	–79	20, 15	20	<0.01	+33	24, 23	25	0.98	–[3]	3, 4	5	–
Discus rotundatus (Müller)	–78	10, 11	10	0.18	–62	22, 18	22	0.01	–	–	–	–
Nesovitrea petronella (Pfeiffer)	–88	9, 5	9	<0.01	–81	4, 1	4	0.07	–39	12, 12	12	0.12
Vertigo pusilla Müller	–71	6, 4	6	<0.05	+25	15, 12	16	0.44	–54	5, 6	6	0.49
Vertigo substriata (Jeffreys)	–84	19, 10	19	<0.001	–0.4	22, 20	23	0.48	–52	9, 8	9	0.12
Vitrea crystallina (Müller)	–74	5, 4	5	<0.05	–72	8, 6	8	0.13	–[1]	–	–	–

Besides listed species, the following species were found during the investigations but had either not changed significantly in any of the regions or the observations were to few to allow any statistical tests: *Acanthinula aculeata* (Müller), *Aegopinella nitidula* (Drap.), *Aegopinella pura* (Alder), *Arianta arbustorum* (L.), *Balea perversa* (L.), *Bradybaena fruticum* (Müller), *Cepaea hortensis* (Müller), *Clausilia pumila* Pfeiffer, *Clausilia cruciata* Studer, *Cochlodina laminata* (Montagu), *Columella edentula* (Drap.), *Euomphalia strigella* (Drap.), *Helicigona lapicida* (L.), *Lauria cylindracea* (Da Costa), *Macrogastra plicatula* (Drap.), *Oxychilus alliarius* (Müller), *Oxychilus cellarius* (Müller), *Punctum pygmaeum* (Drap.), *Spermodea lamellata* (Jeffreys), *Succinea putris* (L.), *Trichia hispida* (L.), *Vallonia costata* (Müller), *Vallonia excentrica* Sterki, *Vertigo alpestris* Alder, *Vertigo angustior* Jeffreys, *Vertigo ronnebyensis* (Westerlund), *Vitrina pellucida* (Müller), *Zoogenetes harpa* (Say).

1) Occurs in the region but was not included in this survey.
2) A considerable number of the snails of this species climb the lower vegetation, especially *Vaccinium* and particularly during the summer. Consequently it is difficult to estimate the populations with the methods used in our investigation.
3) Not adequately quantified during the first investigation.

Table 2. Lowest values for total calcium content (ash free weight), pH (measured electrometrically using air-dried litter in a H_2O-suspension) and base saturation (at pH 7.0) in litter where the listed snail species (in order of base saturation) have been recorded in our investigations in southern Sweden.

	Base satur. %	Calcium ‰	pH
Nesovitrea hammonis	23	3.7	3.8
Euconulus fulvus	23	4.2[1]	3.8
Columella aspera	23	4.4	3.8
Vertigo substriata	27	4.3	3.8
Punctum pygmaeum	32	4.8	4.4
Vitrina pellucida	32	4.8	4.4
Clausilia bidentata	37	4.8	4.7
Cochlicopa lubrica	37	4.8	4.7
Discus rotundatus	37	4.8	4.7
Oxychilus alliarius	38	5.6	4.5
Discus ruderatus	38	5.9[1]	4.5
Carychium tridentatum	40	8.9	4.7
Aegopinella pura	44	9.7	4.8
Cochlodina laminata	44	9.9	4.9
Vitrea crystallina	44	9.9	4.9
Bradybaena fruticum	46	8.9	5.2
Vertigo pusilla	46	7.0	4.9
Cochlicopa lubricella	48	11.2	4.9
Nesovitrea petronella	48	12.5	4.9
Vitrea contracta	48	12.5	4.9

1) The species was also observed on a stump in one locality with a Ca-concentration of 3.7‰ in the sampled litter.

showed that a real loss of species had only occurred in 6 localities in SW Sweden, while the species number in the remaining 19 localities was as expected from the overall density. In the 20 localities in southern Småland we found more species than expected in one and as expected in 19 localities. In central Norrland the number of species found was as expected in all 12 localities.

When comparing changes in density of individual species, we found that several species showed a similar reaction in all studied regions, while some showed a more complex or contradictory pattern (Table 1). *Carychium tridentatum* (Risso), *Clausilia bidentata*, *Cochlicopa lubrica* (Müller), *C. lubricella* (Porro), *Discus ruderatus* (Férussac), *Euconulus fulvus* (Müller), *Nesovitrea hammonis* (Ström) and *Vitrea contracta* (Westerlund) all appeared to have decreased significantly in the samples from at least two of the three regions compared with the original inventory. Certain other species had apparently decreased substantially as well, although in most cases the decrease was not statistically significant.

Abiotic variables

Table 2 shows the lowest values found for some calcium-related variables in the litter of sites where individual snail species were recorded during our investigations.

The only abiotic variables available for retrospective comparisons were pH (measured colorimetrically) and – from southern Småland only – total calcium. There were

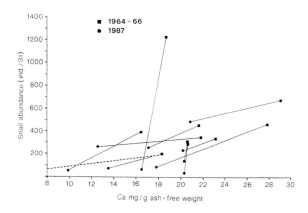

Fig. 4. Snail density plotted against calcium content at 10 mesotrophic forest localities in southern Småland in 1964–66 and in 1987; no significant regressions on calcium were found here at any of these sampling periods. The broken line indicates a Pearson regression from a larger series of oligotrophic forest localities in the same area, sampled 1964–66 (y = 12.37x–37.42; n = 31, r = 0.741, p < 0.001). It is presented here for comparison and is valid for a Ca range of 3–18‰ (Wäreborn 1982). It is obvious that the density level of mesotrophic localities in 1987 was similar to the level that was recorded for the oligothrophic ones in 1964–66. Modified after Wäreborn 1992.

only small changes in pH: –0.2 pH units (n.s.) in southern Småland, almost no change in SW Sweden, and –0.7 pH units (n.s.) in central Norrland. The total calcium concentration (calculated on the organic fraction – ash free weight, AFW) in southern Småland had decreased during the 21–24 yr in mesotrophic forests (p < 0.01, n = 10; paired t-test) from, on average, 21.8 to 16.9‰, i.e. by 22%. The corresponding decrease in oligotrophic forests (p < 0.001, n = 10) was from 9.6 to 5.8‰, i.e. 40%. The overall average decrease was 31% (see Figs 4 and 5).

Certain chemical properties of the litter, measured in connection with the re-inventories and liming experi-

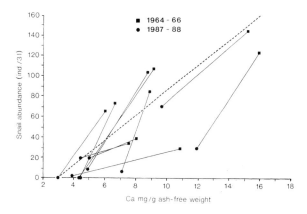

Fig. 5. Snail density related to calcium content in 10 oligotrophic forests (two oak forests to the right in the figure and 8 Norway spruce forests) in southern Småland in 1964–66 and in 1987–88. The broken line is the same linear regression line from 1964–66 as found in Fig. 4. Modified after Wäreborn (1992).

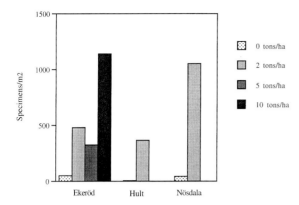

Fig. 6. Effects shown as average total snail densities in plots of liming with 0, 2, 5 and 10 tons of limestone ha^{-1} in three beech woods in southern Sweden. In Ekeröd and Hult each bar represents an average of 3 replicates, in Nösdala 2.

ments, were strongly correlated. Thus, for example, in all regions correlations were found for the calcium-related variables ($p < 0.05$–0.001), viz. calcium and base cation concentrations, base saturation and pH.

A thorough investigation of soil moisture at the mesotrophic forest localities in southern Småland was made in 1966 (Wäreborn 1969). After some days of dry weather, three samples per site were taken simultaneously (19 August) for soil moisture analyses. We assumed that those values were representative for the soil moisture conditions during the re-inventories of these forests. It was found that soil moisture determined in this way was better correlated with snail density ($r = 0.801$, $n = 10$, $p < 0.01$) than were any of the calcium-related parameters.

When comparing the weather data (precipitation and temperature) during the first inventory years (and the year before) with the corresponding later inventory years we found only a few cases where the weather differed substantially between the two inventories. During 1986 and 1987 (preceding 1987 and 1988 when the majority of the re-sampling was conducted) precipitation was near or above average in all three studied regions. The average winter temperature in 1986–87 (preceding 16 of the 20 re-inventory samplings in southern Småland, 13 of 25 in SW Sweden, but none of the 12 in central Norrland) was well below average (although snow came early and a good snow cover was present throughout the winter). In contrast, the winter of 1987–88 was mild (temperature above average), with sparse snow cover in southern Sweden. The summer temperature was below average in 1987 and above average in 1988.

At one of the two studied reference localities (Hördalen, which is situated on the western margin of the area of re-investigated localities in SW Sweden) there was a positive correlation ($r = 0.647$, $n = 10$, $p < 0.05$) between the number of snail specimens in the samples (sampled during spring) and the amount of precipitation in the preceding year at the nearest weather station (Göteborg, c. 20 km N of Hördalen; average precipitation from 1961–1990 was 791 mm). At the second reference locality (St. Mollungen, just outside the region in SW Sweden where the re-inventory was done) there was no such correlation (nearest weather station Borås, c. 30 km SW of St. Mollungen; average precipitation 1961–1990 was 976 mm).

Effects of liming

The snail populations in the investigated beech forests as a whole reacted very strongly to artificial liming (Fig. 6). Two tons of limestone ha^{-1} increased populations by 10–90 times compared with untreated plots, with an overall average of > 20 times ($p < 0.01$, paired t-test). Several plots had a small number of specimens in spite of being limed, which suggests that the fauna had been very impoverished before the treatment.

At one locality (Ekeröd) the experiment also included treatments with 5 and 10 tons ha^{-1}. Here the recorded numbers of snails were 50, 480, 325 and 1140 m^{-2} in the treatments 0, 2, 5 and 10 tons ha^{-1}, respectively ($p = 0.47$, $n = 3 \times 4$; one factor ANOVA). *Punctum pygmaeum* (Drap.) accounted for a large proportion of the total number; excluding that species the numbers were 28, 67, 120 and 360 m^{-2} ($p < 0.05$, $n = 3 \times 4$, one factor ANOVA; untreated plots and plots treated with two tons of limestone ha^{-1} both differed from plots treated with 10 tons ha^{-1} at the 95% significance level).

In general, there was a strong correlation between snail density and calcium-related variables. Figure 7 illustrates the density of the species *Euconulus fulvus* in relation to pH in litter in the sampled squares.

In all three beech forests a total of 14 snail species was found in the investigated plots. Nine species, viz. *Acan-*

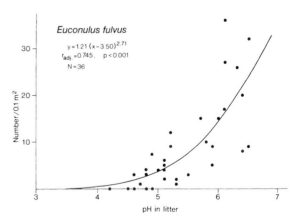

Fig. 7. Numbers of *Euconulus fulvus* in relation to the litter pH (pH H$_2$O in fresh litter) of individual sampling squares (0.1 m^2; $n = 36$) in a beech wood at Ekeröd, Skåne, treated with different dosages (0, 2, 5 or 10 tons ha^{-1}) of lime. After Waldén et al. (1992).

Table 3. Pearson correlation coefficients between snail abundance and chemical properties in litter from SW Sweden, southern Småland, central Norrland and three liming (0, 2, 5 or 10 tons ha⁻¹, where each treatment series was considered as one plot) experimental areas in southern Sweden.

	specimen index n = 65	Ca-tot AFW n = 65	pH (H$_2$O) electrom. n = 61[1]	S/T n = 65	S AFW n = 65	N-tot AFW n = 65	P (AL) AFW n = 61[2]	K (AL) AFW n = 65
specim. index	1							
Ca-tot AFW	0.617***	1						
pH electrometrical[1]	0.751***	0.750***	1					
S/T	0.829***	0.784***	0.885***	1				
S AFW	0.758***	0.848***	0.850***	0.877***	1			
N-tot AFW	0.358**	0.262*	0.455***	0.296*	0.513***	1		
P (AL) AFW[2]	−0.015 n.s.	−0.278*	−0.106 n.s.	−0.015*	−0.193	0.238 n.s.	1	
K (AL) AFW	0.270*	0.062 n.s.	0.294*	0.338**	0.274*	0.385**	0.494***	1

1) pH-values are lacking from four localities in SW Sweden.
2) All Ekeröd plots were excluded in the correlation with phosphorus as parts of that area had been treated experimentally with phosphorus in a system independent of the liming.

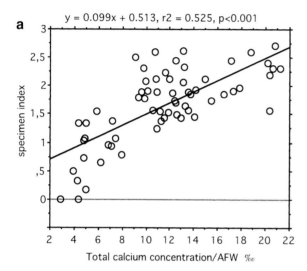

a

y = 0.099x + 0.513, r2 = 0.525, p<0.001

Total calcium concentration/AFW ‰

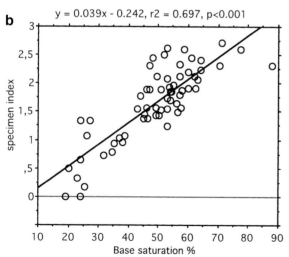

b

y = 0.039x - 0.242, r2 = 0.697, p<0.001

Base saturation %

Fig. 8. Relationship between chemical variables and snail density (specimen index) based on data from all four investigations (three re-inventories plus liming experiments), a. calcium concentration ‰ (ash-free weight), b. base saturation (%). N = 65.

thinula aculeata (Müller), *Oxychilus alliarius* (Miller), *Discus rotundatus, Clausilia bidentata, Cochlodina laminata, Cochlicopa lubrica, Vitrina pellucida* (Müller), *Aegopinella pura* (Alder) and *Spermodea lamellata* (Jeffreys), were found only in the limed plots. With the exception of *O. alliarius*, they are all suspected to be sensitive to soil acidification (Gärdenfors 1987), and most of them had decreased in southern Sweden according to our re-inventories (Table 1).

In addition to the randomly distributed squares, untreated plots in Hult and Ekeröd were subjected to selective searches, using the same sampling methods as in the re-inventories. Using this procedure we found four of the six species that had only been found in sample squares from limed plots and under bark, also the species (*Discus ruderatus*) at these localities.

Synthesis of re-inventories and liming experiments

A direct comparison of the results from the different parts of our investigation was not possible without a transformation of snail numbers. The re-inventories had to adopt the same specific (volume-based) methodologies used in the original samplings, while in the liming experiments strictly quantitative, area-based samples could be chosen. By relating the number of specimens (a) in each sample to the mean number (n) of snails encountered within each of the four sub-investigations (three retrospective investigations plus the liming experiments) a specimen index log(1+a100/n) could be calculated for each sample. In the liming experiments, each treatment series (comprising 9–13 squares) was regarded as one sample.

Following this procedure, there were highly significant ($p \leq 0.001$, n = 61–65) correlations between snail density (specimen index) and the variables calcium, pH, base saturation and base cation concentration (Table 3, Figs 8a–b). The specimen index was also correlated to a cer-

Table 4. Pearson correlation coefficients between snail abundance (specimen index) and chemical properties in litter from SW Sweden, southern Småland, central Norrland and three liming (0 and 2 tons ha^{-1}) experimental areas in southern Sweden, after grouping data according to calcium concentration in litter (Ca-tot/ash-free weight).

	Ca-tot AFW	pH (H$_2$O) electrom.[1]	S/T	S AFW	N-tot AFW	P (AL) AFW[2]	K (AL) AFW
Ca-conc. <12‰ specimen index	0.753*** n = 34	0.754*** n = 33	0.840*** n = 34	0.771*** n = 34	0.507** n = 34	0.316 n.s. n = 33	0.453** n = 34
Ca-conc. ≥12‰ specimen index	0.425* n = 29	0.508** n = 26	0.573** n = 29	0.638*** n = 29	0.344 n.s. n = 29	0.244 n.s. n = 28	−0.116 n.s. n = 29

1) pH values are lacking from 4 localities in SW Sweden.
2) The two (0 and 2 tons ha^{-1}) Ekeröd plots were excluded in the correlation with phosphorus as parts of the Ekeröd area had experimentally been treated with phosphorus in a system independent of the liming.

tain extent with N (r = 0.358, n = 65, p < 0.01) and K (r = 0.27, n = 65; p < 0.05). A multiple linear regression analysis revealed that the seven analyzed chemical properties of the litter explained 85% of the specimen index variation. Base saturation showed the highest regression coefficient.

The material was divided in two groups according to calcium concentration (< 12‰ and ≥ 12‰), a level chosen because it divided the material in roughly two equal halves. After grouping a better correlation between snail density and most abiotic parameters was generally seen at lower calcium concentrations (Table 4). Among the calcium-related parameters this was particularly apparent for Ca.

Shell analyses

In the analyses of the composition of shells, 50 different elements were detected. Together with C, O and H, which were not included in our analysis program, as well as Cl (see Ndiokwere 1983) and F (see Hammer and Brunson 1975) a total of 55 elements have been identified from land snail shells. On average, Ca made up 37.7–38.5% of the shell matter. Na, K, Sr, and in three of the five analyzed species, Mn, each had a concentration of 1–0.1‰. The remaining elements constituted lower fractions, in some cases < 0.001‰ (see Gärdenfors et al. 1995 for further details). An analyzed subfossil (a few thousand years old) shell of Cepaea hortensis had, with the exception of U, Eu and Br (indicating a former contact with sea water due to a higher sea level), a very similar chemical composition to that of shells collected alive.

In C. hortensis the concentrations of Ca (z = −1.889, n = 17, p = 0.06; Mann-Whitney U-test) and Sr (z = −1.716, n = 15, p = 0.09) seemed to be higher in adult shells than in juvenile ones, whereas the opposite was observed for K (z = −2.219, n = 16, p = 0.03), Ag (z = −2.097, n = 16, p = 0.04), Zn (z = −2.515, n = 15, p = 0.01) and Se (z = −2.329, n = 16, p = 0.02) and possibly also Rb (z = −1.837, n = 15, p = 0.07), Sn

(z = −1.846, n = 9, p = 0.06) and Br (z = −1.869, n = 16, p = 0.06).

The lower the Ca concentration in the shell the higher were the concentrations of Zn (r = −0.778, n = 15, p < 0.001; simple regression analysis), Rb (r = −0.762 n = 15, p < 0.001), K (r = −0.672, n = 16, p < 0.01), Ag (r = −0.685, n = 16, p < 0.01), Br (r = −0.648, n = 16, p < 0.01), Se (r = −0.592, n = 16, p < 0.02), Cr (r = −0.619, n = 15, p < 0.02), Cs (r = −0.588, n = 15, p = 0.02) and Cd (r = −0.625, n = 12, p < 0.05).

In Cochlodina laminata, concentrations of Sr, Mn, Ag, Au, Zn, Cd, Sb, and possibly of K and Cs, seemed (no significant correlations due to the small number of observations) to be higher in shells from regions with more acid rain. In some shells of Clausilia bidentata and Cochlodina laminata found in an area with a high levels of acid deposition the concentration of Mn (up to 4.7 and 1.5‰, respectively) was the second (after Ca) or third highest (after Ca and N), respectively, among the elements analyzed.

Discussion

Do the results reflect reality?

The land snail density found in field studies performed during 1987–91 was, on average, 27–80% lower than that recorded in the original inventories made 14–46 yr earlier. There could be several reasons for this: 1) differences in the methodologies used, 2) populations measured in the latest inventories could have been temporarily low owing to unfavourable preceding weather conditions, 3) habitat structure could have changed owing to sucession, etc, 4) long-term changes in chemical properties of the habitats.

The original inventories and re-inventories were made using identical sampling methodology and, except for 10 of the first localities investigated in SW Sweden, by the same persons.

It is known, by experience, that snail populations vary

less in size compared with e.g. most species of insects. In many insect species high temperatures and dry weather are required to reach maximum activity and development. By contrast, Boag (1985) found that maximum activity in land snails occurred at 6–15°C under highly humid conditions. At 16–20°C or higher, activity generally decreased, even at high humidity. Naturally, weather effects must be accounted for in a study of the present kind. In the long-term (10 yr) studies at the two reference localities, coefficients of variation of 34% and 45% were found for snail density, and annual precipitation played a major role in at least one of the localities. This locality is on the coast and has a generally drier weather regime, but with more variation in amount of rain, compared with the other, inland site. The latter generally receives more precipitation; thus snail activity is probably not as often limited by lack of moisture.

A comparison of weather conditions during the original investigation with those during the re-investigation at the different localities generally did not reveal any notable differences between the first and second inventories (with a few exceptions, influencing few of the samples). During 1986 and 1987 (preceding 1987 and 1988 when most of the re-sampling was conducted) all three studied regions received average or above-average precipitation.

The low winter temperature in 1986–87 may have affected snail populations, which could partly explain why average densities were higher in 1988–89 than in 1987 at the eight localities re-sampled twice. However, in a comparison of densities at 10 mesotrophic localities sampled after the cold winter of 1965–66, when weather conditions were very similar to those in 1986–87, with densities at comparable localities sampled after the mild winter of 1964–65, the average difference was only 4.6% ($z = -1.94$, n.s.; Mann-Whitney U-test) (Wäreborn 1992). The fact that the re-inventory was spread over four years (1987, 29 localities; 1988, 18; 1990, 7; 1991, 3) should also have moderated the weather effects.

In choosing the 57 localities used for re-inventories from among the 146 localities considered, emphasis was placed on finding sites where a minimum degree of structural and/or anthropogenic change had taken place. A certain degree of successional change was inevitable, however.

Re-analysis of the litter and surface soil at the re-investigated localities in southern Småland revealed that there had been an average decrease in total calcium of 31% over 21–24 yr. Falkengren-Grerup et al. (1987) found an average calcium decrease in surface soil in Skåne (southernmost Sweden) of 30% over 35 yr. The re-inventories as well as the liming experiments clearly showed a strong correlation between calcium concentration (as well as base saturation and pH) and snail density. For example, according to the equation obtained from the regression of specimen index on calcium concentration ($y = 0.10x + 0.51$), we should expect a 37% decrease in snail density when calcium concentration decreases from, e.g. 12‰ to 10‰ (−20%), and a 39% decrease when calcium decreases from 7‰ to 5‰ (−29%). Such a response of snail density to a decrease in calcium concentration was observed de facto in the southern Småland material (the only investigated region where calcium was measured both in the original inventory and in the re-inventory; Figs 4 and 5).

In general, there was a negative correlation between the degree of decrease in snail density and the base saturation (Fig. 2), pH and (in southern Sweden) Ca concentration of the localities. In other words, the density decreased more in localities with poor soils, which are less buffered and therefore more affected by cation leaching resulting from H^+-deposition. It also suggests that changes in the chemical environment have played a larger role for the snail density compared with weather effects and sampling errors. It should also be noted that several species showed a similar pattern of response in different regions (Table 1).

Together, these facts support the conclusion that the observed decreases in snail density are not attributable mainly to temporary environmental variations or to sampling errors but that the really reflect a long-term change in the chemical properties of the environment, owing to acidification and cation leaching from soil and litter.

Species density and richness

A decrease in snail density resulted in a parallel decrease in the number of species recorded in the samples, owing to the lower probability of finding individual species. Except for seven localities in SW Sweden, the observed decrease in number of species in our samples was of the magnitude that would be expected based on the decrease in overall density (according to rarefaction analyses). Thus, except for SW Sweden, there was no statistical evidence that any species had been lost in the studied regions. Still, it should be kept in mind that when the density changes, the rarefaction analysis actually compares samples from different populations. If local densities approach zero, e.g. as in some plots in southern Småland, species will inevitably disappear, even though this cannot be revealed by a rarefaction analysis. Also, when chemical properties fall well below observed minimum values (Table 2), one would expect species to die out, although such local extinctions cannot always be confirmed by a rarefaction analysis owing to a parallel decrease in density.

In our study, densities of certain species had evidently decreased (Table 1). Some of these, particularly *Nesovitrea hammonis*, *Euconulus fulvus*, *Discus ruderatus* and *Cochlicopa lubrica*, are ubiquitous and still widely distributed in Sweden. In addition to the species in Table 1, several other more stenotopic species were found in lower numbers (or not at all) in the second inventory than at the first inventory, but they occurred in too few samples to allow any statistical analyses.

Influence of abiotic conditions

From Table 2 it is apparent that species differ in their environmental requirements. The figures should not be treated as definite limits. The relevant minimum values will vary depending on other variables in the litter. For instance, a snail stressed by drought will probably not tolerate as low a pH as a snail in a moist habitat. Moreover, the measured chemical values represent a homogenized sample composed of several subsamples. In the field, individual snails may thrive and survive in microhabitats with more favourable conditions.

The capacity to survive in favourable microhabitats was indirectly illustrated in connection with the liming experiments. In addition to the randomly distributed squares, untreated plots in Hult and Ekeröd were subjected to selective searches in different microhabitats. Six species *(Discus rotundatus, Cochlodina laminata, Clausilia bidentata, Acanthinula aculeata, Oxychilus alliarius and Vitrina pellucida)* out of 11 had been found only in squares in limed plots. Now, the first four species were also found in litter in untreated plots.

Shell formation and physiological processes require relatively high levels of calcium availability. The positive effect of a supply of extra calcium was obvious in the liming experiments, where two tons of limestone per ha increased the snail density by a factor of 10–90. However, the variation between the plots was high, probably because the density before liming may have been close to zero in some plots.

Above a certain level the Ca concentration may have less influence on snail density. In samples from the retrospective and liming studies the correlation between calcium concentration and snail density was less good at localities where the calcium concentration was $>12‰$ than at sites where it was $<12‰$ (Table 4). The same pattern was seen on the habitat scale: In southern Småland, snail density was correlated with the calcium concentration in oligotrophic forests ($p < 0.05$, $r = 0.714$, $n = 10$) but not in mesotrophic ones. Graveland et al. (1994) obtained similar results.

Moisture in soil and litter is an important factor regulating land snail density (Wäreborn 1969). In our investigations soil moisture was measured in mesotrophic forests in southern Småland, although only during the original inventory. Here, soil moisture was correlated with snail density, but not with calcium or pH. Thus, much of the unexplained variation (15%) in the previously mentioned (section Synthesis of re-inventories and liming experiments) multiple regression analysis might be attributable to soil moisture.

It may seem paradoxical that the results indicate that the average decrease in snail density in SW Sweden has been lower than in southern Småland, and on the same level as in central Norrland, even though SW Sweden receives the most acid rain. Part of the explanation may be found in the higher deposition of airborne cations by winds from the sea and from industrial emissions in the south. The most affected area in SW Sweden annually receives c. 100 meq m^{-2} of base cations. This level is higher than in any other region in Sweden, being two and eight times higher than in southern Småland and central Norrland, respectively (Granat 1989, Sjöberg et al. 1992). As a result, the level of base saturation may remain high and to some extent counteract the effects of acidification.

Another factor contributing to the extent of the decrease in snail popolations may be when acidification, and the accompanying decrease in snail density, started in the different regions. Odén (1968) demonstrated that the acidification advanced gradually during the period 1958–1965, from a centre over the Netherlands, in a north-east direction. There is also evidence of early effects in Sweden, and already during the period 1910–1948 the roach, *Leuciscus rutilus* L., gradually decreased and even disappeared from lakes in SW Sweden (Almer 1972, Fleisher et al. 1993). In southern Småland a decrease of the species was first observed around 1960 (Varenius 1990), which may indicate that also soil acidification occurred later in Småland. Thus, at many sites at the time of the original inventories the initial snail density level in SW Sweden may have already been affected. Such a pattern was also indicated in our material from SW Sweden when comparing localities inventoried early (1941–1953) with those inventoried later (after 1960) in the first investigation period.

Environmental reflection in shell chemical composition

Before a snail population reaches its lower tolerance limit at a site exposed to ongoing acidification, the chemical constitution of the shell seems to be altered. Our investigations in this field have only just begun, but the results so far suggest that concentrations of several elements in shells change as the environment becomes more acid. Thus, it may be possible to use snail shells as environmental archives by analysing old museum specimens or even sub-fossils. Our results indicate that when using the elemental composition of shells as a bioindicator, comparisons should preferably be made between conspecific shells of similar ontogenetical age. In cases of calcium deficiency, the element may be substituted by other chemically similar elements. Thus, we anticipate that the most distinct results will be obtained when studying shells, for instance in a time series, from regions poor in lime.

Acknowledgements – The investigation was mainly funded by the Swedish Environmental Protection Agency. The inventory in central Norrland also received financial support from Sundsvall Municipality and the Västernorrland County administration. Funds for the shell analyses were obtained from Tryggers Stiftelse. The shell analyses were made in co-operation with T. Westermark and co-workers. Valuable comments from H. Staaf improved the manuscript substantially.

References

Almer, B. 1972. Effects of acidification on fish stocks in lakes on the west coast of Sweden. – Inform. Inst. Freshwater Res. Drottningholm. No. 12 (in Swedish, with English summary).

Boag, D. A. 1985. Microdistribution of three genera of small terrestrial snails (Stylommatophora: Pulmonata). – Can. J. Zool. 63: 1089–1095.

Bridges, E. M. 1978. World soils. – Cambridge Univ. Press, Cambridge.

Burch, J. B. 1955. Some ecological factors of the soil affecting the distribution and abundance of land snails in eastern Virginia. – The Nautilus 69: 62–69.

Chétail, M. and Krampitz, G. 1982. Calcium and skeletal structures in molluscs: concluding remarks. – Malacologia 22: 337–339.

Crowell, H. H. 1973. Laboratory study of calcium requirements of the brown garden snail, *Helix aspersa* Müller. – Proc. Malacol. Soc. Lond. 40: 491–503.

Egnér, H., Riehm, H. and Domingo, W. R. 1960. Beurteilung des Nährstoffzustandes der Böden. II. Chemische Extraktionsmethoden zur Phosphor- und Kaliumbestimmung. – Ann. Roy. Agric. Coll. Sweden, Stockholm, 26: 199–216.

Falkengren-Grerup, U., Linnemark, N. and Tyler, G. 1987. Changes in soil acidity and cation pools of south Swedish soils between 1949 and 1985. – Chemosphere 16: 2239–2248.

Fleischer, S., Andersson, G., Brodin, Y., Dickson, W., Herrmann, J. and Muniz, I. 1993. Acid water research in Sweden – Knowledge for tomorrow? – Ambio 22: 258–263.

Frömming, E. 1958. Die Rolle unsere Landschnecken bei der Stoffumwandlung und Humusbildung. – Z. Angew. Zool. 45: 341–350.

Gärdenfors, U. 1987. Impact of airborne pollution on terrestrial invertebrates with particular reference to molluscs. – Swedish Environment Protect. Board, Solna, Report No. 3362.

– 1992. Effects of artificial liming on land snail populations. – J. Appl. Ecol. 29: 50–54.

–, Westermark, T., Carell, B., Forberg, S., Emanuelsson, U., Mutvei, H. and Waldén, H. W. 1988. Use of land-snail shells as environmental archives: Preliminary results. – Ambio 17: 347–349.

–, Bignert, A., Carell, B., Forberg, S., Mutvei, H. and Westermark, T. 1995. Elemental composition of some land snail shells (Mollusca, Gastropoda) and observations of environmental interest. – Bull. l'Inst. Océanogr., Monaco, Num. Spec. 14 (in press).

Granat, L. 1989. Luft- och nederbördskemiska stationsnätet inom PMK. Rapport från verksamheten 1988. – Swedish Environ. Protect. Agency, Solna, Report 3679 (in Swedish).

Graveland, J., van der Wal, R., van Balen, J. H. and van Noordwijk, A. J. 1994. Poor reproduction in forest passerines from decline of snail abundance on acidified soils. – Nature 368: 446–448.

Hammer, W. P. and Brunson, R. B. 1975. Fluoride accumulation in the land snail *Oreohelix subrudis* from Western Montana. – The Nautilus 89: 65–89.

Krebs, C. J. 1989. Ecological methodology. – Harper & Row, New York.

– 1991. Fortran programs for ecological methodology. – Exeter software, Exeter.

Künkel, K. 1929. Experimentelle Studie über *Vitrina brevis* Férussac. – Zool. Jahrb. Abt. Allg. Zool. Physiol. Tiere 46: 575–626.

Lindquist, B. 1941. Experimentelle Untersuchungen über die Bedeutung einiger Landmollusken für die Zersetzung der Waldstreu. – Kung. Fysiogr. Sällsk. Lund Förh. 11: 144–156.

Mason, C. F. 1974. Mollusca. – In: Dickinson, C. H. and Pugh, G. J. F. (eds), Biology of plant litter decomposition. Academic Press, London, Vol. 2: 555–591.

Ndiokwere, C. L. 1983. Arsenic, antimony, gold and mercury levels in the soft tissues of intertidal and terrestrial molluscs and trace element composition of their shells. – Radioisotopes 32: 117–120.

Nömmik, H. 1974. Ammonium chloride – Imidazole extraction procedure for determining titrable acidity, exchangeable base cations, and cation exchange capacity in soils. – Soil Sci. 113: 254–262.

Odén, S. 1968. The acidification of air and precipitation and its consequences for the natural environment. – Bull. Ecol. Comm., Stockholm, No. 1 (in Swedish, English translation published by Translation Consultants Ltd, Arlington, Va).

Robertson, J. D. 1941. The function and metabolism of calcium in the Invertebrata. – Biol. Rev. 16: 106–133.

Schifferli, L. 1977. Bruchstücke von Schneckenhäuschen als Calciumquelle für die Bildung der Eischale beim Haussperling *Passer domesticus*. – Arch. Orn. Beob. 74: 71–74.

Schmidt, H. A. 1955. Zur Abhängigkeit der Entwicklung von Gehäuseschnecken vom Kalkgehalt des Bodens. Dargestellt bei *Oxychilus draparnaldi*. – Arch. Moll. 84: 167–177.

Sjöberg, K., Persson, K., Lövblad, G. and Granat, L. 1992. Luft- och nederbördskemiska stationsnätet inom PMK. Rapport från verksamheten 1991. – Swedish Environment. Protect. Agency, Solna, Report 4090 (in Swedish).

Sjörs, H. 1961. Some chemical properties of the humus layer in Swedish natural soils. – Bull. Roy. School Forestry, Stockholm 37: 1–51.

Tompa, A. 1976. A comparative study of the ultrastructure and mineralogy of calcified land snail eggs (Pulmonata: Stylommatophora). – J. Morph. 150: 861–871.

Valovirta, I. 1968. Land molluscs in relation to acidity on hyperite hills in central Finland. – Ann. Zool. Fennica 5: 245–253.

Varenius, B. (ed.) 1990. Miljöfakta i Kronobergs län (Environmental data in Kronoberg county). – Länsstyrelsen i Kronobergs län 1990: 4 (in Swedish).

Voelker, J. 1959. Der chemische Einfluss von Kalziumkarbonat auf Wachstum, Entwicklung und Gehäusebau von *Achatina fulica* Bowd. (Pulmonata). – Mitt. Hamburg Zool. Mus. Inst. 57: 37–78.

Waldén, H. W. 1965. Terrestrial faunistic studies in Sweden. – Proc. First European Malac. Congress, pp. 95–109.

– 1981. Communities and diversity of land molluscs in Scandinavian woodlands. I. High diversity communities in taluses and boulder slopes in SW Sweden. – J. Conchol. 30: 351–372.

– 1986. The 1921–1981 survey on the distribution and ecology of land Mollusca in southern and central Sweden. – Proc. VIII Int. Malacol. Congr., Budapest, pp. 329–336.

– 1992. Changes in a terrestrial mollusc fauna (Sweden: Göteborg region) over 50 years by human impact and natural succession. – Proc. IX Int. Malacol. Congr. Edinburgh, pp. 387–402.

–, Gärdenfors, U. and Wäreborn, I. 1992. The impact of acid rain and heavy metals on the terrestrial mollusc fauna. – Proc. X Int. Malacol. Congr., Tübingen, Part 2, pp. 425–435.

Wäreborn, I. 1969. Land molluscs and their environments in an oligotrophic area in southern Sweden. – Oikos 20: 461–479.

– 1970. Environmental factors influencing the distribution of land molluscs of an oligotrophic area in southern Sweden. – Oikos 21: 285–291.

– 1982. Environments and molluscs in a non-calcareous forest area in southern Sweden. – Ph.D. thesis, Lund Univ.

– 1992. Changes in the land mollusc fauna and soil chemistry in an inland district in southern Sweden. – Ecography 15: 62–69.

Ecological Bulletins 44: 271–300. Copenhagen 1995

Acidification of groundwater in forested till areas

Gert Knutsson, Sten Bergström, Lars-Göran Danielsson, Gunnar Jacks, Lars Lundin, Lena Maxe, Per Sandén, Harald Sverdrup and Per Warfvinge

Knutsson, G., Bergström, S., Danielsson, L.-G., Jacks, G., Lundin, L., Maxe, L., Sverdrup, H. and Warfvinge, P. 1995. Acidification of groundwater in forested till areas. – Ecol. Bull. (Copenhagen) 44: 271–300.

An integrated study of the chemical and hydrological processes involved in the acidification of soil and groundwater has been conducted in order to estimate future changes in the groundwater chemistry caused by changes in acid deposition in relation to other acidifying processes. The project included field studies, development of analytical methods and hydrological as well as hydrochemical modelling. An improved method for the measurement of aluminium speciation was developed. The aluminium species of the greatest toxicological importance – "quickly reacting aluminium" – are analyzed by kinetic discrimination in a flow system (the Flow Injection Analysis-method). The field studies were carried out in five areas (three small catchments, one hill slope and three springs), representing different types of till and with various amounts of acid load. An integrated dynamic model was developed by combining a modification of the existing hydrological model PULSE with two new hydrochemical models, SAFE and PROFILE. The integrated model was used for prediction of the response in the groundwater chemistry to different future acid deposition scenarios. Critical loads for shallow groundwater were calculated using the PROFILE model. Comparing the field areas with past surveys of soil and water acidification in Scandinavia shows that they represent the main features of till areas. Cross-combinations of the studied areas and the acid deposition pattern are commonly found. It seems likely, that as the acidification of the soil profile has reached c. 3 m depth in Fårahall, southwestern Sweden, many wells in this part of the country with a shallower groundwater table have seen a drastic acidification during the last decades. A catchment with the characteristics of the Masbyn site with its surficial water pathways transplanted to southwestern Sweden would have experienced an early acidification of the surface water and is representative of the catchments of many long-time acidified lakes. A special, though common, case is that of catchments with very shallow soils on rocks found along the west coast of Sweden, as well as in Sörlandet in Norway, where the breakthrough to the surface waters resulted in the first acidified lakes as early as in the 50's and 60's. Such environments are present in the upper reaches of Stubbetorp. Even the groundwater in the underlying rock may then be acidified. The field investigations and the modelling give a picture of the status and development of the acidification of soil and shallow groundwater in till areas. The budgets compiled from the fieldwork indicate that the weathering is insufficient to maintain the buffering in the investigated areas. At current deposition rates stores of base cations will be almost depleted after a few decades. The base cation stores at Fårahall and Masbyn are sufficient for buffering for only 10–20 yr, while at Stubbetorp and Risfallet they may last for c. 50 yr. The scenarios developed through modelling are equally bleak, unless there are drastic reductions in acidic depositions. The acidification models have been validated by independent estimates of weathering rates (PROFILE) and by comparison with historical data (SAFE). The comparison with the past acidification history in Scandinavia confirms that the main features of the predictions made in this work can be relied upon.

G. Knutsson, G. Jacks and L. Maxe, Div. of Land and Water Resources, Royal Inst. of Technology, S-100 44 Stockholm, Sweden. – S. Bergström and P. Sandén, Swedish Meteorological and Hydrological Inst., S-601 76 Norrköping, Sweden. – L.-G. Danielsson, Div. of Analytical Chemistry, Royal Inst. of Technology, S-100 44 Stockholm, Sweden. – L. Lundin, Dept of Forest Soils, Swedish Univ. of Agricultural Sciences, S-750 07 Uppsala, Sweden. – H. Sverdrup and P. Warfvinge, Dept of Chemical Engineering II, Lund Inst. of Technology, S-221 00 Lund, Sweden.

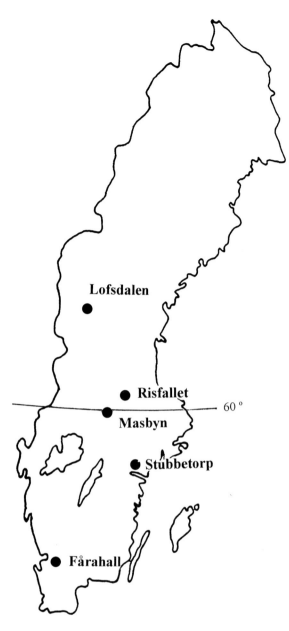

Fig. 1. Geographical location of the field research areas.

Introduction

Problems

Effects of acid deposition on surface water were observed in Sweden during the 1960's but for forest soils and groundwater not for another twenty years. Studies at the beginning of the 1980's (Jacks and Knutsson 1981, Jacks et al. 1984) showed that significant changes in the groundwater chemistry, for example a decrease in alkalinity and an increase in hardness, were caused by increased acid load and that there was a risk for acid-

ification of shallow groundwater in sensitive regions. About 300 000 dug wells are used for water supply in permanent households and c. 650 000 wells in holiday houses. Many of these are in a critical situation. Today 50% of the dug wells have a pH < 6.3 and an alkalinity of < 0.5 meq l^{-1} (Bertills et al. 1989). Shallow wells in coarse textured soils are the most affected.

In Sweden the dominating overburden on the bedrock is glacial till. Sufficient groundwater yields for wells are often found in medium or coarse textured tills with fairly short water pathways. In these tills the water in wells are sensitive to acidification. The most serious risk with such an acidification is the potential release of heavy metals from soils and water pipe systems as well as the corrosion of constructions. A deterioration of groundwater will also be transmitted to surface water such as streams and lakes. The discharge of acid groundwater containing high concentrations of metals can have a strongly negative, sometimes toxic, impact on biota in surface water.

It is of great importance to predict future effects of acidification on the groundwater quality. To do this there is a need for increased and comprehensive knowledge of the hydrology and hydrochemistry involved in the acidification of soil and groundwater together with soil physical and chemical properties in different geological environments. This information obtained must be put into hydrochemical and acidification models to elucidate future development. However, an evaluation of the models hitherto used in acidification studies showed the need for a development of new models (Bergström and Lindström 1989). The existing models were found to have an insufficient description of water residence time, which limited the possibilities for an advanced chemical subroutine. As the rate of weathering is the most important parameter to determine the resilience of both soils and water against acidification, it was decided to develop a weathering rate submodel instead of using weathering as a calibration parameter. Furthermore, there was a need for an improved method for speciation of metals, especially of aluminium, iron and manganese.

The Swedish Integrated Groundwater Acidification Project was therefore initiated by the Swedish Environmental Protection Agency to provide information about soil and groundwater acidification as a basis for policy decisions. The integrated project consists of seven subprojects concerned with theoretical studies, development of hydrological and hydrochemical models as well as a method for aluminium speciation and field investigations. Results from the project have earlier been presented (Sandén and Warfvinge 1992, Maxe 1995).

Objectives

The overall objective of the integrated project was to obtain a better understanding of the chemical and hydrological processes involved in the acidification of soil and groundwater in order to estimate future changes of the

Table 1. Total deposition at the field research areas (not Lofsdalen). Estimated from the compilation made by IVL, presented
in Lövblad et al. (1992).

Element Catchment	Na	K	Ca	Mg	NH₄	NO₃	SO₄	Cl
					meq m⁻² yr⁻¹			
Fårahall	151	13	54	45	75	91	197	193
Stubbetorp	16	2	9	5	26	24	52	20
Masbyn	23	4	11	8	31	29	63	28
Risfallet	8	1	5	3	18	19	35	9

groundwater chemistry caused by changes in acid deposition in relation to other acidifying processes. The tasks
for the modelling work were: 1) to determine the time
required to acidify groundwater of different sensitivity at
different locations within Sweden. 2) To quantify response in the chemical composition of groundwater to
different future acid deposition scenarios and to estimate
critical loads.

The goals of the analytical subproject were: 1) to
develop an improved method for the measurement of
aluminium speciation in natural waters. 2) To supervise
and control the laboratories performing analysis of field
samples.

The purposes of the field investigations were: 1) to
study the chemical and hydrological processes in small
catchments and hillslopes along catenas from upslope
recharge areas to downslope discharge areas close to the
discharging stream. 2) To determine the quantitative
characteristics of the hillslope geology and hydrology,
especially the chemistry of soil, soil water, groundwater
and stream water, as well as the soil moisture conditions,
the flow paths and the residence time of water. 3) To
produce relevant data as an input to the hydrological and
hydrochemical modelling work.

The research was based on the currently accepted concepts in hydrology, for example that groundwater is a
dominating part of the runoff even during periods of
snow melt and heavy rains.

Field investigations and analytical methods

The field investigations began in the autumn of 1986 and
finished at the end of 1990. Five field research areas were
selected (Fig. 1). These areas represent hydrogeologically
and lithologically different types of non-calcareous till.
The research areas were selected on the following main
criteria a) moderate or high acid deposition, b) till overburden, c) normal salic soils, and d) presence of drained
soils with a normal groundwater level within a few
metres depth. The aim was to use established field research areas where other investigations of the soil, water
and vegetation were already going on. The main investigations were performed on hill slopes that were

considered to be representative of the small catchment
within which they were located.

The geology was documented by mapping, sampling,
boring and geophysics. The research included studies of
amount and chemical composition of precipitation, soil
water, groundwater and runoff, in order to calculate water
balance and hydrochemical budgets. Special attention
was paid to percolation, groundwater recharge and water
turnover. Evapotranspiration was not measured but was
determined as a difference.

The hydrochemical budgets were calculated from determinations within the investigated areas and from estimations from national compilations. The variables considered are atmospheric deposition (Dep), weathering
(Weath), net vegetational uptake (VegUp) and output
from the catchment by leaching (Leach). The difference
between the variables is taken as equal to the changes in
the soil chemical exchange pool (IonEx). An expression
of this balance would be

$$Dep + Weath - (VegUp + Leach) = IonEx.$$

Deposition rates were derived from the compilation made
by Lövblad (Lövblad et al. 1992) but adjusted according
to the actual precipitation amounts measured in the catchments. The dry deposition was scaled to give a balance
between input of chloride by precipitation and dry deposition and output of chloride by stream discharge (Table
1).

Weathering was calculated by models based on soil
chemical determinations from profiles of the upper metre
of the soil at a few sites within the slopes. The pits dug
for sampling are located in upslope areas with drained
profiles, as the models consider vertical percolation.
Transforming these weathering determinations from profiles to mean values for the whole catchment introduces
an element of uncertainty as lateral water flows and
limited percolation occur in the lower parts of the slopes.
This is especially evident when considering a catchment
where >50% of the area is moist or wet with imperfectly
drained profiles.

Several methods have been employed. Historical
weathering (Olsson and Melkerud 1991) was calculated
from base cation losses during the Holocene period using
zirconium as reference (Schutzel et al. 1963). The calcium weathering rate was investigated using strontium as
an analogue and differentiating between atmospherically
deposited strontium and weathering-derived strontium by
the use of the stable isotopes ^{87}Sr and ^{86}Sr (Jacks et al.
1989). Recent weathering was calculated by modelling,
using primary mineral composition, soil specific surface,
pH, CO_2, organic content and water conditions as the
main input variables (Sverdrup and Warfvinge 1988).

Vegetational uptake was calculated from measurements at other sites of element distribution in different
parts of the trees (Rosén 1982, Olsson et al. 1993). These
estimations are transformed to the catchment stands. Both
the long-term mean nutrient uptake over a hundred years

and the present uptake, taking current annual increment into account were calculated.

Physical and chemical properties of soil profiles along the slope were also investigated. Measurements included texture, organic content, compact and bulk densities, porosity, soil specific surface, water retention characteristics and hydraulic conductivity of the upper soil horizons (usually to the depth of one metre). The chemical analysis was mainly focused on the exchangeable cations and pH. Total content of elements in the soil and mineralogy was usually only determined for a few samples from each research area.

Along the hillslopes, determinations were made of groundwater levels in piezometers to several depths and of runoff at V-notch weirs with recording gauges. Water pathways and turnover were investigated by tracer techniques, mainly with tritium as tracer. Soil water and groundwater chemistry were studied at several locations and the chemical composition was analyzed in soil water, groundwater and stream water. The analysis included acids, alkalinity, organic anions, base cations, aluminium, sulphate, nitrogen and in some areas also fluoride and organic carbon. Water analyses were usually performed according to Swedish standards (SIS 3120, 1986; Maxe 1995).

Runoff and chemical composition of stream water were combined in calculating the outflow of elements from the catchment. Variations in water chemistry over time were studied. This aimed to reveal the influence of climatic conditions, in particular the dampening effect of ionic exchange during the water flow through the soil, and to show where equilibrium exists. Not all of the measurements were made in all the catchments. Especially in Lofsdalen the measurements were less extensive.

Analytical quality control

The analytical quality control of water analysis within the project was set up to comprise both ring tests and the regular submission of control samples to participating laboratories. Parameters studied were, first and foremost; pH, alkalinity/acidity, Ca, Mg, K, Na, SO_4 and Cl. Four ring tests have been undertaken with five or six laboratories. Samples for the ring tests, two for each round, were obtained from the study areas of the project or from sources of representative natural waters. The first round revealed some serious problems, but for the next three, coefficients of variation of 1.8–14% were obtained for the different parameters, with the exception of pH and alkalinity, for which discrepancies remained. Ring tests two and three, two years apart, were run with the same two samples, which were later also used as control samples. Sample stability was excellent: 12 and 20% decrease for K was the only discernible change.

Development of a method for aluminium-speciation

The primary objectives of the method development were to avoid the uncertainty about which species are included in a certain measurement, inherent in the use of operational definitions, and to increase the speed of analysis.

The chemistry of the method is based on the use of 8-hydroxyquinoline (Oxine) as the reagent. This reagent forms a coloured complex with aluminium extractable into chloroform. A pH of 5.0 was chosen for the reaction as this would minimize the change in pH experienced by the natural water samples during analysis. Oxine is not specific to aluminium, but reacts with a range of metals. A masking step using hydroxylamine/1.10 phenanthroline was therefore included to avoid disturbances from those metals.

The reaction and subsequent extraction is performed in a flow system (FIA). Fractionation is obtained through kinetic discrimination in that the reaction between Oxine and aluminium is allowed to proceed for only 2.3 s. Thus, only kinetically labile forms will be included. The uncertainty in this operational definition of included species is minimised by the use of a flow system which facilitates close control of reaction time (Hanning 1988, Clarke et al. 1992).

We launched an ambitious characterization program trying to elucidate what is actually measured. In this work we have used pure substances with known complexation constants for aluminium as well as humic and fulvic acids extracted from river water. The results of this characterization are summarized in Table 2 together with similar results for the common Driscoll procedure (Driscoll 1984). Equilibrium dialysis experiments showed that complexes formed with humic and fulvic acid in some natural surface water samples from Masbyn were not included either (Clarke et al. 1991, 1992, 1995, Berdén et al. 1994).

Table 2. A comparison of the inclusion of various forms of aluminium in our method with reaction time of about 2.3 s (Al_{qr}) and in the more conventional Barnes-Driscoll method (Al_i).

Complex	Included in	
	Al_{qr}	Al_i
Al^{3+}, $Al(OH)^{2+}$, $Al(OH)^{2+}$, $Al(OH)_3$ (aq)	yes	yes
Sulfato complexes	yes	–
Carbonato and silicato complexes	yes	–
$Al(OH)_4$	partially	–
Phospato complexes	partially	–
Glycinato and oxalato complexes	partially	–
Acetato complexes	partially	yes
Salicylato complexes	partially	yes
Large polymeric hydroxo complexes	no	no
Fluoro complexes	no	yes
Citrato complexes	no	no
Complexes with extracted humus	no	usually not
Complexes with natural humus	no	usually not

The procedure that has been used for most of our analysis of natural waters has the following characteristics: Sample volume, 250 µl, sampling rate 1 min⁻¹, 4 injections sample⁻¹ yielding a detection limit of 5–10 µg l⁻¹. Taking the time needed for calibration into account this adds up to a sample capacity of c. 10 samples h⁻¹. For a number of natural water samples having concentrations in the range 50–500 µg l⁻¹ and with varying concentrations of organic matter the repeatability was 2–9% r.s.d. The within-lab reproducibility was determined using a standard containing 0.25 mg l⁻¹ Al and found to be ±0.0049 mg l⁻¹ (2.0%). We can therefore state that the FIA method shows adequate stability over time (Clarke et al. 1992, Berdén et al. 1994).

The most obvious advantage of the FIA method is that it provides a direct measurement of aluminium species of great toxicological importance. Driscoll's method requires two measurements of aluminium fractions combined with a determination of flouride activity, used in correcting for the inclusion of Al-fluoro complexes, to obtain a comparable figure (Driscoll 1984). The FIA method also has a much lager sample capacity than manual methods of speciation, and the sample consumption is much lower.

Field research areas

The physiographical data of the field research areas are summarized in Table 3. A comprehensive description of the field research areas, as well as of the results of the field investigations are given in a separate report (Maxe 1995).

Fårahall

The southernmost field research area, Fårahall, consisting of a single hillslope, is located in the south-western part of Sweden, close to the sea, on the northern escarpment of the horst Hallandsåsen. It is exposed to a high acid load (Fig. 1). The area is above the highest shore line. The slope has a thick till soil with depth to a fractured bedrock surface of 3–6 m (Fig. 2). There is a significant content of stones and boulders in the soil. Apart from these coarse grains the soil is well graded, though with a comparatively low content of silt. In an international comparison the soil would be considered a coarse till.

The developed soil profile is a haplic podzol. The soil chemistry reveals a pH only slowly increasing with soil

Table 3. Physiographical characteristics of the field research areas.

Field research area	Fårahall	Stubbetorp	Masbyn	Risfallet	Lofsdalen
Location	56°22'N; 13°7'E	58°44'N; 16°21'E	59°55'N; 15°15'E	60°21'N; 16°14'E	62°01'N; 13°30'E
Area (km²)	0.12	0.87	0.186	0.45	0.23
Elevation (m a.s.l.)	145–180	80–130	320–360	215–270	600–700
Bedrock	Gneisses	Granites	Granites	Gneissic granites	Porphyries
Soil cover (%)					
Till	94	46	90	53	100
Gravel/sand	0	13	0	0	
Silt/clay	0	0	0	0	
Organic soil	6	5	3	7	
<0.5 m soil*	0	36	7**	40**	
Precipitation (mm)[1&2]	1150	690	950	620	830
(% snow***)	(20)	(30)	(40)	(35)	(45)
Runoff (mm)	450	230	440	250	500
Temperature (°C)[3]					
Annual mean	+6.5	+5.7	+4.2	+4.3	+0.5
January mean	− 1	− 4	− 6	− 6	−11
July mean	+16	+16	+15	+16	+12
Temperature sum[1] (day-degrees)	1500	1300	1250	1200	750
Vegetation period[1] (days, temp >5°C)	230	205	180	180	153
Snow cover (days)[4]	60	90	140	140	180
Land use (%)					
Forest					100
Pine	0	69	49	76	
Spruce	90	11	39	9	
Deciduous	0	3	10	9	
Impediment					
Exposed bedrock	0	10	1	2	
Mire	10	7	1	4	

Climatological parameters refer to long-term mean regional values: [1]) Odin et al. (1983), [2]) Eriksson (1983), [3]) Alexandersson et al. (1991), [4]) Svenska sällskapet för antropologi och geografi (1953).
*Including exposed bedrock, ** <0.7 m depth, ***Including 50% of reported sleet.

Table 4. Chemical conditions of the soil horizons in upslope parts of the Fårahall field research area. Values from composite samples (c. 20 subsamples) from one location. Exchangeable base cations and cation exchange capacity in 1 M NH₄Cl extract, acidity, Al+H, in 1 M KCl and base saturation, BS.

Horizon	Depth cm	Na	K	Ca	Mg μeq g⁻¹	Al+H	CEC	BS %
OH	0–8	0	3.9	36	15	66	121	45
A1	8–11	0.5	0.8	2.3	1.1	42	47	10
E	11–18	0.2	0.3	0.0	0.2	29	30	2
B1	18–28	0.2	0.4	0.0	0.2	17	18	4
B2	28–48	0.1	0.1	0.0	0.1	6.6	7	3
BC	48–68	0.0	0.1	0.0	0.1	3.7	4	5
C1	68–88	0.1	0.0	0.0	0.1	3.7	4	4
C2	88–108	0	0.0	0.1	0.2	3.0	3	8

depth, with pH_{H_2O} 3.7 in the upper soil layers but only 4.3 in the upper C-horizon (Table 4). Higher pH, near 5, is found at depths of c. 3 m.

Stubbetorp

The field research area is located in the eastern part of south Sweden, within the elongated hilly area Kolmården just below the highest shore line (Fig. 3). The actions of the waves have strongly influenced the soil cover within the area. In the exposed high parts of the catchment bedrock outcrops and areas with very thin soil layers dominate. At lower altitudes wave washed sandy tills occur together with small areas covered by sorted sediments of fine sand, sand and gravel. A few small fens with peat has formed along the main stream, at the outlet of a small spring, and in a few depressions on the hills.

The soils that occur are lithosols in areas with thin soil covers, histosols in the peatlands and podzols in > 50% of the catchment. Mean values from sampling of the upslope

areas in Stubbetorp are given in Table 5. Samples of downslope areas show a higher degree of base saturation; 87% and 74% in the organic horizons, 60% in the A-horizon and 50% in the E-horizon (Maxe 1995).

Masbyn

The field research area is located in the central part of south Sweden (Bergslagen) at a regionally high altitude, on the south-western slope of the highland and therefore exposed to acid load. The catchment is situated above the highest shore line. The most frequently developed soils are podzols, with less distinct horizons in downslope locations than in the upslope ones. Upslope there are mostly drained profiles (Table 6). Porosity decreases with soil depth from ~68% in the E-horizon to only c. 30% at a depth of 1 m. The organic content in the upper B-horizon is 8–12% and the grain size distribution in the parent material is dominated by silt, fine sand and sand.

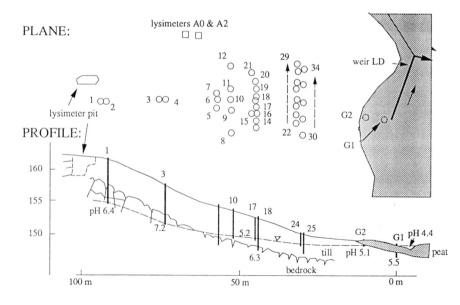

Fig. 2. The hillslope investigation area Fårahall with the installed piezometers and a profile through the slope.

Table 5. Chemical conditions of the soil horizons in upslope parts of the Stubbetorp field research area. Mean values. Loss on ignition, LOI, exchangeable base cations in NH₄Ac extracts pH 7, acidity, Al+H, in 1 M KCl, cation exchange capacity, CEC and base saturation, BS.

Horizon	Depth cm	LOI %	pH H₂O	Na	K	Ca	Mg	Al+H	CEC	BS %
						μeq g⁻¹				
F	0–4	88	4.3	1.8	26	164	27	40	259	82
H	4–7	69	4.0	1.4	9.2	110	18	85	223	59
A	7–11	8.4	4.2	0.3	1.2	10	1.8	27	40	31
E	11–16	2.4	4.5	0.3	0.5	5.1	1.1	16	23	23
B1	16–31	4.3	4.7	0.2	0.5	3.4	0.6	16	21	22
B2	31–68	1.6	5.4	0.1	0.2	1.5	0.3	3.0	5.1	40
BC	68–83	1.0	5.1	0.1	0.1	0.3	0.1	2.7	3.3	29
C	83–107	0.9	5.4	0.2	0.3	3.6	0.7	3.5	8.3	35

Risfallet

This field research area is also located in the central part of south Sweden. It is situated in the same highland region as the Masbyn area but is on the north-eastern side and is therefore partly protected from the heavy acid load. The Risfallet catchment is formed in a gentle depression just above the highest shore line of the region. The till is heterogeneous but the matrix is normally dominated by silt, fine sand and sand (60%), when soil material < 20 mm is considered. However, this apparently high content of fine material should be seen in view of a high content of stones and boulders of 35 vol-% in the top one metre of the soil, leaving usually only 20–25 vol-% of material finer than gravel as the porosity is ~40%. Bulk densities increase with soil depth from c. 1.0 g cm⁻³ in the E-horizon and upper B-horizon to almost 2.1 g cm⁻³ in the C-horizon at 1–2 m depth. Accordingly, porosity decreases with depth from c. 50% in the upper 0.2 m to c. 20% at 1.5–2 m depth. Organic content in the upper B-horizon is up to 7.5%.

In the upper soil layers pH$_{H_2O}$ is 4.3–4.9, with a lowest value of 3.9 in the humic horizon. In the C-horizon pH reaches 5.5–5.7. Base saturation decreases with depth from 60–80% in the organic layers to c. 10% in the C-horizon, but with values up to 30% in the C-horizon at 1.2 m depth (Table 7). This is probably caused by groundwater with high calcium content.

Lofsdalen

The northernmost field research area, at the south-eastern edge of the High Mountains of Sweden, is the Lofsdalen springs, with catchments consisting of small, 2–18 ha, till areas. The soil is a stony, gravelly or sandy till, composed of porphyry and quartzite. The soil profile is characterised by strongly developed E-and B-horizons at sites up- and midslope. There are often 10–20 cm leached eluvial horizons with 20–30 cm reddish-brownish B-horizons beneath. These are heavily enriched by aluminium-iron-manganese oxides combined with organic matter. Organic content is fairly high in the B-horizon, almost 12%. In the E-horizon pH$_{H_2O}$ is 4.4, increasing with depth to 5.7 in

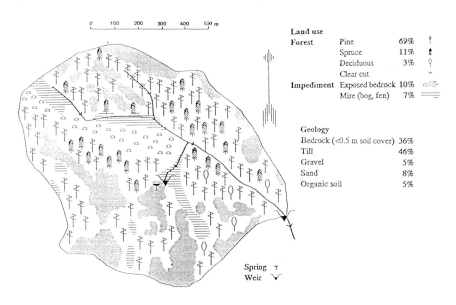

Fig. 3. Site types and forest vegetation of the Stubbetorp catchment.

Table 6. Physical and chemical conditions of the soil horizons (mean values of two subsamples) in an upslope drained site at the Masbyn catchment. Soil bulk densities, γ, porosity, P, loss on ignition, LOI, exchangeable cations in 1 M NH$_4$Ac and acidity, Al+H pH 7, in 1 M KCl, cation exchange capacity, CEC, and base saturation, BS.

Horizon	Depth cm	γ g cm^{-3}	P %	LOI %	pH H$_2$O	Na	K	Ca	Mg µeq g^{-1}	Al+H	CEC	BS %
F	0–2	–	–	–	3.9	1.5	22	80	20	59	183	68
H	2–5	–	–	–	3.7	1.0	14	67	28	96	206	53
Ah	5–7	0.72	–	29	4.1	0.3	1.8	3.7	1.7	21	28	21
E	7–14	1.15	68	3.5	4.4	0.3	0.5	0.7	0.3	12	14	13
Bs1	14–23	1.04	53	12	4.8	0.3	1.3	0.8	0.7	24	27	11
Bs2	23–29	1.12	48	8.4	4.8	0.3	0.5	0.1	0.2	6	7	14
Bs3	29–40	1.21	44	4.6	4.9	0.1	0.2	0.1	0.1	4	4	11
BC	40–53	1.35	39	2.3	–	–	–	–	–	–	–	–
C	53–95	1.60	33	0.9	5.0	0.1	0.2	0.0	0.0	3	3	10

the C-horizon (Table 8). At downslope sites the C-horizon has gleyic properties.

Climate

The climate of the areas ranges from a mild maritime type in Fårahall to a fairly cold continental type in Lofsdalen. The temperature sum in day-degrees at Fårahall is double of that at Lofsdalen while the duration of the snow cover is only a third of that at Lofsdalen. The climate of Stubbetorp is somewhat similar to Fårahall as regards temperature sum and snow cover days, but is much drier. The two other areas, Masbyn and Risfallet, are very alike in many aspects but differ in precipitation and runoff – Masbyn is the wetter of the two. Precipitation is enhanced westwards and by higher altitude. The hydrological conditions of the investigation period is shown in Fig. 4.

The acid load is of the same magnitude at Stubbetorp and Risfallet, somewhat higher at Masbyn, and highest at Fårahall, and according to new findings-quite surprising at Lofsdalen (Ahlström et al. 1995). The differences in climate are reflected in the amount and type of runoff and the pattern of the annual fluctuations of the groundwater levels, which also indicate periods of groundwater recharge. The main period of groundwater recharge in Fårahall is during the mild winter, and there is very

seldom recharge during the summer and never surface runoff. In Lofsdalen there is – in contrast – no recharge during the half-year-long winter, but an intense, short period of recharge during and after the snowmelt in May, when surface runoff can also occur. Minor periods of groundwater recharge occur during the autumn and sometimes after heavy rains during the summer. In the other three areas there are two periods of recharge, in the spring and in the autumn. Surface runoff may occur at Masbyn during periods when the groundwater level is very close to the surface of the ground.

Geology

The geology is dominated by crystalline acid rocks (granites, gneisses and porphyries) and coarse tills in all of the field research areas (Table 3). However, there are some differences in the lithological composition of the soil cover between the areas due to the transport of material by the land-ice. The till of Fårahall has, for example, a higher content of feldspars (especially plagioclase) as well as of biotite and hornblende, than the acid gneiss of the area. The composition is probably influenced by the metabasites, which are found close to Fårahall. Thus the till has a better buffer capacity than expected in contrast to the till of Lofsdalen, in which the quartz-

Table 7. Physical and chemical conditions of the soil horizons at the Risfallet catchment with soil bulk densities, γ, porosity, P, loss on ignition, LOI, exchangeable cations in 1 M NH$_4$Ac and acidity, Al+H pH 7, in 1 M KCl, cation exchange capacity, CEC, and base saturation, BS.

Horizon	Depth cm	γ g cm^{-3}	LOI %	P %	pH H$_2$O	Na	K	Ca	Mg µeq g^{-1}	Al+H	CEC	BS %
F	0–3	0.11	71	–	4.7	2.1	32	90	21	25	171	85
H	3–5	0.20	51	–	3.9	1.0	14	101	20	49	185	73
Ah	5–7	0.37	14	–	4.3	0.4	4.0	23	4.7	21	53	61
E	7–12	1.18	3	55	4.5	0.1	1.0	4.8	1.0	16	23	30
Bs1	12–17	0.97	7	52	4.9	0.2	1.0	2.9	0.7	23	28	17
Bs2	24–29	1.11	5	46	5.1	0.1	0.5	1.0	0.3	7.3	9	21
C1	65–80	1.65	2	27	5.5	0.2	0.7	0.5	0.1	15	16	9
C4	120–130	2.04	1	21	5.7	0.2	0.7	0.3	0.1	3.1	4	30

Table 8. Chemical conditions of the podzolic soil on sandy till, spring no 4, Lofsdalen. Loss on ignition, LOI, exchangeable base cations determined in 1M NH₄Ac extract (pH 7), Fe and Al in 0.2 M NH₄-oxalate extract (pH 3). Acidity (H+Al) determined in 1 M KCl (unbuffered).

Horizon	Depth cm	LOI %	pH H₂O	Na	K	Ca	Mg μeq g^{-1}	H+Al	Fe	Al	BS %
O	0–7	95.0	3.97	0.9	15	180	33	82	16	78	73
E	7–26	1.5	4.38	0.2	0.9	2.1	0.5	22	5	39	14
B	26–45	11.6	5.27	0.2	0.8	4.6	0.5	19	730	2390	24
C	45+	1.7	5.69	0.1	0.6	1.2	0.2	7	120	400	23

porphyry is mixed with >50% quartzite, transported by the land-ice. The lithological composition of the tills of the other three field research areas is more "normal" (for Sweden) and in most areas dominated by hard crystalline rocks of granitic type (Table 9).

The textural (= grain size) composition of the tills is related to the lithology, and at Lofsdalen it is mostly coarser (more gravelly) than the tills of the other field research areas. The till of Fårahall is also a stony gravelly-sandy type, whereas that of Risfallet is normally silty-sandy but is very heterogeneous, sometimes with lenses of sand and gravel and with a high frequency of stones and boulders. The most fine-grained and compact till is the silty-sandy basal till of Masbyn, but the topsoil is coarser and looser. The composition of the till of Stubbetorp has changed in grain-size by wave-washing, which has depleted the till on the slopes of finer particles and formed sandy or silty sediments in the depressions. None of the other areas has been affected by wave-washing.

Vegetation

The vegetation on the Fårahall slope is a 60–80 yr old spruce stand developed on a former *Calluna* heath. There are mesic mosses in the stand, mainly *Hylocomium splendens*. The vegetational cover of the Stubbetorp catchment is dominated by coniferous forests of different ages, mostly Scots pine. Also the Masbyn catchment is dominated by coniferous forests with ages between 15 and 70

Table 9. Mineralogical composition of soil (<2 mm) in C-horizons.

Mineral	Fårahall	Stubbetorp	Masbyn	Risfallet
Quartz %	32	52	48	39
K-Feldspar	28	19**	24**	25**
Ca-Na-Feldspar	28	23	22	26
Hornblende	4	2.2	4	5
Biotite	4*	–	–	–
Chlorite	–	1.5	0.2	2
Opaque	4	–	–	–
Amorphous	–	–	2	3

*Including chlorite, **Including K-mica.

yr. The forest sites are of the bilberry type. On the main area of the Risfallet catchment a young pine stand (22 yr old) has developed. The forest type can be classified as a narrow-leafed grass type. Lofsdalen is also a coniferous forest area. The age of most stands is 80–90 yr.

Hydrology

The hydrology of most of the research areas is characterised by naturally drained soils. The soils are fairly permeable and thick tills with relatively deep ground-water levels, at least in up- and midslope locations. Comparably low precipitation and high evapotranspiration in the eastern catchments, Stubbetorp and Risfallet, also

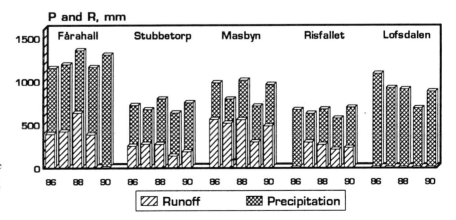

Fig. 4. Precipitation (P) and runoff (R) (mm) from the field research areas during the investigation period 1986–90. Runoff values are missing for Fårahall 1990, Risfallet 1986 and Lofsdalen 1986–90.

Table 10. Water chemistry at a depth of c. 1 m (Fårahall, Stubbetorp and Masbyn), and at a depth of c. 3 m (Risfallet); soil or ground water (SW/GW), in upslope and downslope locations. Mean values. Balance of cations and anions. Calculated concentrations of base cations not coming from atmospheric deposition.

Research area	Upslope location				Downslope location			
	Fårahall	Stubbetorp	Masbyn	Risfallet	Fårahall	Stubbetorp	Masbyn	Risfallet
SW/GW	SW	GW	GW	GW	GW	GW	GW	GW
Sampling point	SW 1.8	22, 23, 24	4	7:15	G1	32, 34	31	6:12
Sampling depth cm	180	83–135	80–120	380	60	70–119	80–120	300
pH lab		5.2	5.4	6.1		6.7	5.7	6.2
pH field	4.4	5.2	5.3	5.8	5.4	6.4	5.7	5.9
Cond mS m^{-1}		3.7	2.6	3.1	11.5	9.8	4.4	3.9
H eq l^{-1}	38	9	4	1	4	0.3	2	0.7
Na	191	119	68	68	402	170	160	92
K	46	13	15	21	14	18	28	15
Ca	205	63	86	137	256	564	178	195
Mg	214	62	32	45	245	182	114	71
NH$_4$	6	1	3	1	1	7	3	1
NO$_3$ eq l^{-1}	41	8	3	15	11	2	3	6
Cl	246	95	33	23	391	80	48	31
SO$_4$	692	184	115	151	470	33	133	151
Alk/Ac		−7	10	60	35	799	174	138
Al μM	177	19	11	5	3	5	24	4
Sum +	(694)	344	205	272	921	934	482	374
Sum −	(979)	371	151	249	907	914	358	326
Diff		27	54	23	14	20	124	48
Na ex	−1	45	41	49	97	108	121	67
K ex	29	3	11	18	−13	9	22	11
Ca ex	136	22	73	125	147	530	159	178
Mg ex	155	35	22	38	151	160	100	62
Sum ex	319	105	147	230	382	807	402	318

Sum + includes H, Na, K, Ca, Mg. Sum − includes NO$_3$, Cl, SO$_4$, Alk (if positive).
Na ex etc; calculated concentration of cations not coming from atmospheric deposition.

influence the soil water content. At Risfallet this coincides with deep till layers, strong hydraulic gradients and comparable high saturated conductivities throughout the soil profile (10^{-5}–10^{-4} m s^{-1}). Stable and deep groundwater levels have developed at Risfallet and as the discharge mainly originates from deep groundwater it is persistent and smooth. In a few small areas there are moist and wet soil types, partly constituting peatlands. In Stubbetorp, discharge is usually low during the vegetation period and in dry periods, as after the dry summers 1989 and 1990, the runoff even ceases. During dry periods the groundwater table is located in the bedrock in large areas of the catchment. In downslope locations and valley bottoms there are often higher groundwater levels, i.e. mostly in the upper 0.5 m of the soil. At the stream the piezometric head is above the ground surface, indicating an upward groundwater flow.

In Fårahall, the fairly steep slope and thick permeable till, together with high evapotranspiration, results in deep groundwater levels and deep unsaturated water percolation despite the high precipitation.

In contrast, Lofsdalen and Masbyn have comparatively low evapotranspiration and very shallow groundwater. These conditions are fairly representative of till slopes in the northern part of Sweden. In the Masbyn catchment, hydraulic conditions and groundwater levels were thoroughly investigated (Lundin 1982). Saturated hydraulic conductivity is stratified with depth, with high conductivities (10^{-4} m s^{-1}) in the upper soil layers and comparably low conductivities ($< 10^{-7}$ m s^{-1}) at depths of more than one metre. This is a common feature of till soils (Lind and Lundin 1990). With the existing high water input, high water flows occur in the top soil layers and groundwater levels are also often high. Moisture content in the unsaturated zone is often close to saturation and even small water inputs affect the groundwater level considerably. Most groundwater discharge passes the upper soil layers laterally and only small amounts of water percolate to deeper groundwater discharge. Only during dry periods there is deep percolation with outflow of deep groundwater dominating the surface water formation.

Hydrochemistry

The hydrochemical properties of the sites are reflected in Table 10 giving the groundwater composition in two typical terrain locations, upslope and downslope. Fårahall is exposed to atmospheric deposition of sea salts as well as of sulphur and nitrogen components of anthropogenic origin. The chloride content of the Fårahall groundwaters is about four times higher than at Stubbetorp and ten times higher than at Masbyn and Risfallet. The sulphur

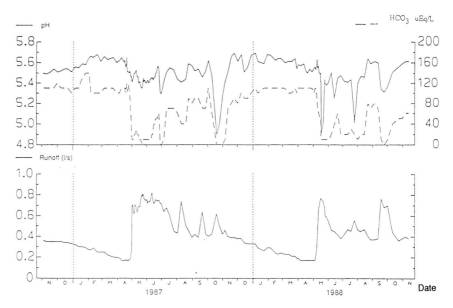

Fig. 5. The fluctuations of pH and alkalinity (HCO$_3$) (above) and of the runoff (below) in spring no. 2, Djursvallen, Lofsdalen (from Knutsson 1992).

deposition does not have an equally strong gradient: the concentration of sulphate is about three times greater at Fårahall than at the other catchments. The nitrogen deposition at the different catchments is not mirrored in the groundwater composition, as nitrogen retention in the plant-soil system is still intact in all sites. However, the slightly higher nitrate concentration in the upslope location at Fårahall is likely to be the effect of leaching of deposited nitrate in winter.

The downslope groundwater at Stubbetorp is well buffered and has a very high concentration of calcium. This might be an effect of calcite present in the fine sediments in the fracture valleys, a phenomenon which can be observed at Södertörn south of Stockholm (Jacks 1992). However, it might also be the effect of the wave washing of the till soils, leaving the coarse fractions on the hillslopes and accumulating the fine sand in the valleys. It is clear from the low sulphate concentration, indicating sulphate reduction, that the groundwater in the downslope locations has a slow turnover. Rather stagnant conditions are required for sulphate reduction, as the amount of organics in the soil is not high.

The Stubbetorp catchment is situated below the highest marine shoreline, while all the others lie above. There is little difference in the till texture along the slopes in these sites. Fårahall and Risfallet have rather coarse textures which permits a deep drainage of the water and gives a better buffered water in most positions than at Masbyn with its fine textured till. The Masbyn till is a good example of the hydrochemical effects of a loose surface till lying on more impervious lower till strata (Lundin 1982). This results in export of acidity during high flow events like snowmelt. The years of observation had generally mild winters with little snow cover and there were no pronounced acid surges in runoff water during snowmelt except in the Lofsdalen springs. The rapid snowmelt and heavy autumn rains saturate the soil in the coarse near the surface sections and cause dramatic drops in pH and alkalinity of the water in these springs (Fig. 5).

The combined effects of the soil factors, hydrology and deposition is reflected in the stores of exchangeable cations down to 1 m (Table 11). Fårahall is most depleted in base cations, even magnesium is low in spite of the proximity to the sea.

Table 11. Sum of exchangeable cations to a depth of 1 m and base saturation (%). Upslope locations (as in Tables 5–7).

Research area	Na	K	Ca	Mg	Mn	H	Al	Sum of bases	CEC	BS %
					meq m^{-2}					
Fårahall[1]	84	237	698	412	25	2621	9712	1431	13764	10
Stubbetorp*[2]	140	430	3650	660	n.d.	1830	5730	4880	12440	39
Masbyn[2]	172	636	1391	515	111	7353		2714	10067	27
Risfallet[2]	151	719	2067	449	126	8419		3386	11805	29

[1] Exchangeable base cations in the soil (<2 mm) are determined in 1M NH$_4$Cl extracts, exchangeable H and Al in 1M neutral KCl extracts. [2] Exchangeable base cations in the soil (<2 mm) are determined in 1M NH$_4$Ac extracts at pH 7.00; exchangeable H and Al in 1M neutral KCl extracts. *Upslope and midslope locations: values reduced 10% in order to account for soil material >2 mm. n.d. not determined.

Modelling structure and data base

The hydrological and the hydrochemical modelling was carried out separately by two research teams. The team at the Swedish Meteorological and Hydrological Inst. in Norrköping was responsible for the hydrological PULSE model and the team at the department of Chemical Engineering, Lund Inst. of Technology, developed the chemical models, PROFILE and SAFE. These three models were then linked according to Fig. 6.

The modelling efforts were restricted to vertical processes in the soil and thus to the recharge of the aquifer, since the processes in the upper parts of the soil were considered to be the most important for analysis of the susceptibility of shallow groundwater to acid deposition. Sensitivity analyses were carried out both in order to analyse the total effect of uncertain assumptions and to check if the complexities of the models were appropriate. This analysis is based on the uncertainty that both modelling groups identified in their model development.

The subroutines of the chemical models require time series of soil moisture storage in four horizons, water flux between these horizons, and soil temperatures. These data were delivered by the hydrological model, which was run on daily time steps, with precipitation, air temperatures and monthly estimates of the potential evapotranspiration as input. A full description of the modelling process and its results are given in a separate report (Sandén and Warfvinge 1992). The following presentation is limited to brief model descriptions, a sample of results and the most important conclusions.

The performance of the hydrological and hydrochemical models was controlled by a large data base consisting of data collected within this and other research projects. For the purpose of comparison, the groundwater acidification model MAGIC (Cosby et al. 1985) was tested at the Stubbetorp catchment (Sandén et al. 1992). MAGIC differs from SAFE in a number of important aspects. It is a process-orientated model, by which long-term trends in soil and water acidification can be reconstructed and predicted at a catchment scale.

The performance of the PULSE model was compared to that of the more physically based SOIL-model (Jansson and Halldin 1979). This work supported the use of the modified PULSE model for basinwide and regionalized simulations of the effect of acid deposition, which is the aim of the project (Espeby and Sandén 1992).

Hydrological modelling

The hydrological model structure is based on the semi-empirical concept of the PULSE model (Bergström et al. 1985) and the earlier HBV model (Bergström 1976, 1992). Semi-empirical means that the models have a main structure based on simplified physical considerations, but that some of the processes are modelled by

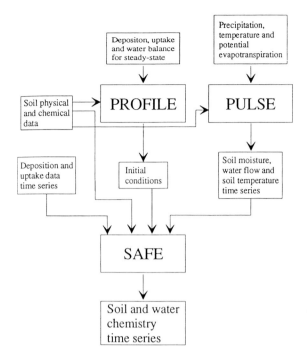

Fig. 6. The internal structure of the integrated groundwater acidification models. The flow chart shows how input data, models and output are coupled.

parameters that do not have a physical meaning and thus are subject to calibration.

Four horizons were used: the organic (O), eluvium (E), illuvium (B) and parent (C) horizons. There is, however, no theoretical limit for the vertical distribution of the model (Fig. 7). Special consideration was given to the modelling of the water pathways in each horizon. After comparison with isotope data from soil lysimeters, it was concluded that total mixing in the upper 40 cm layer and piston flow in the deeper layers is the best approach at a point (Fig. 8) (Lindström and Rodhe 1992a). Lysimeters do, however, represent a very homogeneous case. Considering the increasing model complexity of a model based on piston flow and the great heterogeneity of a recharge area, it was decided to use total mixing in all horizons when applying the model to real aquifers.

The snow routine and the groundwater drainage to the stream were calibrated against recorded runoff. However, parameters for the soil moisture routine were estimated from water retention characteristics and mean groundwater table, leaving this routine uncalibrated. Although simulations of snow or runoff were not primary tasks of this project they helped to check that the simulated water balance was within reasonable limits. An estimate of the soil temperatures at the different levels of the soil profile was obtained by the development of a separate subroutine based on filtered air temperatures.

One important simplification was that the hydrological model did not account for the dynamics of the ground-

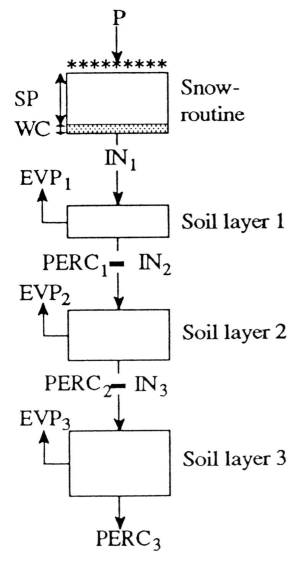

SP
WC
P
Snow-routine
EVP$_1$
IN$_1$
Soil layer 1
PERC$_1$ – IN$_2$
EVP$_2$
Soil layer 2
PERC$_2$ – IN$_3$
EVP$_3$
Soil layer 3
PERC$_3$

Fig. 7. General structure of the modified PULSE model used for simulation of the hydrological conditions in the soil profile.

water table, but that an average depth to the groundwater table had to be assumed. This simplification was subject to sensitivity analysis and the results show that the model is sensitive to this assumption.

Examples of model simulations of ^{18}O are shown in Fig. 8. This is just one example of results taken from the main report (Sandén and Warfvinge 1992), which show that the dynamics of the hydrological model are well under control with respect to a number of independent variables.

Hydrochemical modelling

Two different hydrochemical models were developed within the Swedish Integrated Groundwater Acidification Project. The model PROFILE is used to calculate the steady-state conditions of the soil profile, while SAFE is a dynamic model that simulates how the chemistry of the profile is developing over time. In this project the SAFE model is used for scenario simulations while the PROFILE model is used to provide initial conditions for this simulation and to estimate critical loads.

The data needed to run the SAFE model fall into the following three categories: 1) Boundary conditions: i) Hydrological data provided by the modified PULSE model. ii) Biomass data based on stand characteristics. iii) Deposition data for the whole simulation period. 2) Soil properties: i) Soil morphology expressed as thickness of each horizon. ii) Capacity and intensity factors, including cation exchange capacity, soil texture, mineralogy and CO_2 pressure. iii) Equilibrium coefficients, kinetic and mass transfer rate constants. 3) Initial conditions: Acid neutralizing capacity (ANC) and base cation concentration (calculated by PROFILE).

The general process descriptions are identical for the SAFE and PROFILE models. They are based on the stratification of the soil and submodels for the chemical processes and mass balance in each one of these. The chemical subroutines describe soil solution equilibrium, weathering rates, nitrogen reactions, base cation uptake and cation exchange reactions. The most important simplifying assumptions are: 1) each soil compartment is chemically isotropic and the soil solution is perfectly mixed, 2) sulphur reactions do not serve as a net sink or source of the acid neutralizing capacity, 3) only the net effect of all nitrogen reactions is included, 4) internal element cycling, such as of potassium in the upper soil layers, can be neglected, 5) dissolved organic matter can be specified as input data, and 6) organic complexing of metals, such as aluminium, can be neglected.

Control of the performance of the PROFILE and SAFE models

The performance of the PROFILE model was controlled by data from the Gårdsjön research site, on the Swedish west coast, where profiles of hydrochemical properties are available (Melkerud 1983, Giesler and Lundström 1991). One example of a simulated pH profile is shown in Fig. 9. The modelled weathering rates are further controlled by data from a large number of international sources (Sandén and Warfvinge 1992). The performance of the SAFE-model was controlled by data from sites in southern Sweden where historical data of exchangeable bases are available (Falkengren-Grerup et al. 1987). One sample of results is shown in Fig. 10.

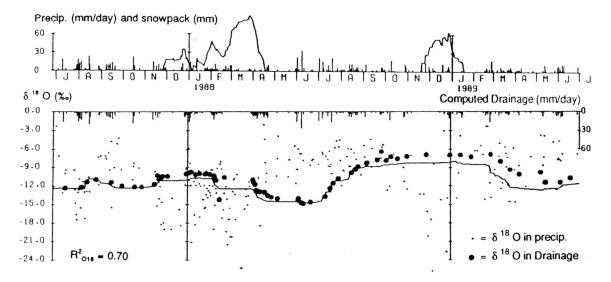

Fig. 8. ¹⁸O simulations for lysimeters 15, 40 and 80 cm deep in the Stubbetorp field research area. The figures represent from top to bottom: precipitation and modelled water equivalents of the snow, computed drainage from the 80 cm lysimeter, observed ¹⁸O concentrations in precititation (small dots), observed ¹⁸O in drainage from the 80 cm lysimeter (larger dots) and simulated concentrations of ¹⁸O from the 80 cm lysimeter (full line). The simulation is made under the assumption of complete mixing in the upper 40 cm and piston flow below this level. (From Lindström and Rodhe 1992a).

Critical loads

The PROFILE model was used in the calculation of critical load maps for the shallow groundwater in Sweden. The definition of critical loads for groundwater was based on the following limits: The acidity of the water should stay > pH 6, equivalent to an air-equilibrated pH of c. pH 6.5, and to an alkalinity > 100 μeq l⁻¹ at 2.0 m.

Water becomes more corrosive when it is acidified to < pH 6, while high alkalinity and hardness tend to decrease the rate of corrosion. If the groundwater is buffered to at least 100 μeq l⁻¹ at 2 m depth, then the deeper parts of the aquifer will be well enough buffered to prevent corrosion.

The calculation of critical loads for groundwater is done by balancing of acidity and alkalinity for the soil column down to the specified depth. Within this volume, only sinks and sources of acidity that are effective over a long time period should be included. This excludes processes where the pool is exhaustible, such as cation exchange and anion adsorption. The sources of acidity considered are, besides the acid deposition, base cation uptake by vegetation and alkalinity leaching while the sinks are the chemical weathering and uptake of nitrogen by

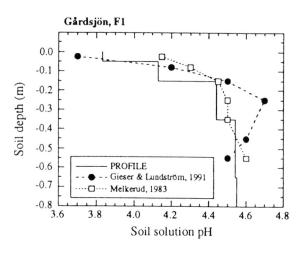

Fig. 9. Comparison between soil solution pH calculated by PROFILE and observations in the Gårdsjön basin.

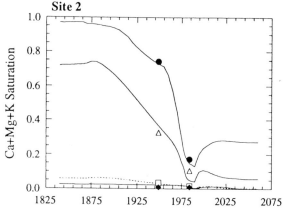

Fig. 10. Comparison between Ca+Mg+K saturation as modelled by the SAFE model for four soil layers and as observed by Falkengren-Grerup et al. (1987) in the south of Sweden. (From Warfvinge et al. 1993). Reprinted from, copyright, Environmental Pollution 80 (1993): 216 g with kind permission from Elsevier Science Ltd, UK.

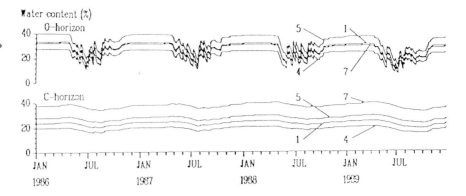

Fig. 11. Model sensitivity to change in the field capacity (FC) parameter. Base simulation (1), −20% FC-WP (4), + 20% FC-WP (5) and variable pF value (7), Stubbetorp. PULSE model.

vegetation. The critical load is by definition the acid load (D) that brings the sinks and sources to balance. ANC leaching is defined by the chemical criteria. To quantify the remaining terms, different approaches can be taken. In this study, the uptake of nitrogen (N_u) and of base cations (BC_u) by vegetation was input data, while the weathering rate (W) was calculated using the PROFILE model.

In order to calculate critical loads of groundwater with the PROFILE model, a considerable amount of data were collected. All data in this regional critical load assessment are based on information collected within the framework of the Swedish National Forest Survey. For this study, a subsample of 1395 forest sites was selected to utilise the soil samples that were available down to the C-layer at c. 60 cm soil depth. Samples at these sites were used to obtain the mineralogy data for the PROFILE model. Other input data were derived in accordance with Sverdrup et al. (1990).

Sensitivity analysis

The basis for the sensitivity analysis of the hydrological model was the determination of reasonable ranges of variability of the hydrological conditions. A major working hypothesis within the project has been that hydrological processes affect the rate of soil acidification, and hence the sensitivity of groundwater to acid deposition. Intuitively, it appears reasonable that the magnitude and temporal dynamics of the water flux should affect the fluxes and concentrations of species dissolved in water.

There is, however, only a limited number of parameters and input data in the hydrological model that can introduce uncertainty. The soil moisture routine of the modified PULSE model has only three parameters (field capacity, wilting point and the α-parameter in the exponential reduction of potential evapotranspiration). These parameters can be obtained from measured properties of the soil profile. Precipitation is the most important boundary condition to the model. Recalculation of field capacity (FC) using a variable pF with depth has a large impact on the amount of water in C and E-horizons (not

shown), while the variation pattern stays the same (Fig. 11). Changing the plant available water by 20% has a fairly large effect on the soil moisture content. Changing the precipitation volume by 30%, or using a different function for reducing evapotranspiration with depth, has a fairly small effect on the soil moisture content. The largest difference can be seen for the dry summer of 1988.

The dynamic variation in soil moisture content is completely driven by model structure and input data. Changes in the parameters can bring about only small differences in the pattern. The soil moisture content varies with the pF-value. The pF-value is dependent on the groundwater level, which varies in time and space, and the soil properties, which vary in space. The overall variation within a basin or a region will therefore be large. For the regionalization of the results this uncertainty has to be taken into account in forecasting the acidification of groundwater.

In order to study the impact of hydrology on the hydrochemical model the ranges in the parameter values used in the sensitivity analysis were summarized in a scheme of 14 hydrological simulations.

A 30 yr climatic series has been repeated several times in order to show the impact of seasonal variation and the variation of climate. In Fig. 12, the results from 14 runs within this sensitivity analysis are shown. The calculations show that an accurate hydrological submodel and good estimates of parameters are necessary for the credibility of the model and its output. From the perspective of assessing environmental damage caused by acidic water, either to groundwater or surface water, there is a significant difference if the ANC in 2020 turns out to be 0 or 120 µeq l^{-1}.

The sensitivity analysis also revealed that the time step in the hydrological input to the hydrochemical model had a profound effect on the predicted ANC. Short-term variations in concentrations of major constituents of soil water and runoff are dominated by hydrological fluxes and pathways. The decline in ANC proceeded faster the shorter the time step was set (Fig. 13).

To the uncertainty introduced by the hydrological description should then also be added the uncertainty in estimates attributable to hydrochemical model formula-

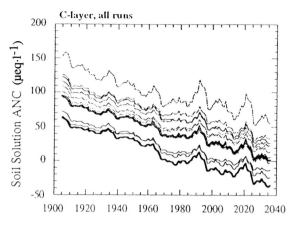

Fig. 12. Range of predictions in groundwater ANC for simulations with the entire range of hydrological parameters, Stubbetorp. SAFE model.

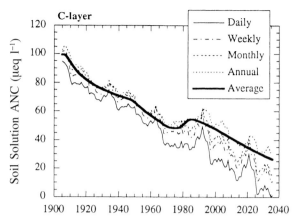

Fig. 13. Dependence of simulated trends on the time resolution in hydrological input data, Stubbetorp. SAFE model.

tion, and other data and parameters in the SAFE/PROFILE models. The parameters selected for the sensitivity analysis, along with the range of variation, were specific mineral surface area, (50%, 80%, 125% and 200% of base case values), cation exchange capacity, CEC (80% and 125% of base case values), cation exchange rate coefficient (km) (20% and 500% of base case values), CO_2 partial pressure (50% and 200% of base case values), soil density (80% and 125% of base case values).

The results from the runs within this sensitivity analysis show that the model is dynamic with respect to some input parameters. In Fig. 14, the range of predictions of change in ANC from 1903 to 1985/90 in the shallow groundwater, at a depth of 80–100 cm is shown. With the exception of one run, the predictions of change in ANC fall within ±25 µeq l^{-1} as compared with base case simulation. There are no differences in conclusions regarding if the groundwater will acidify further at this site, but to what level.

Finally, Fig. 15 illustrates the importance of chemical weathering in determining the prediction of ANC in groundwater. The predicted value is basically proportional to the weathering rate, while the decrease in ANC depends on other factors, such as cation exchange. As the base saturation decreases, however, the importance of chemical weathering increases.

Results and discussion

Flow paths and residence time

The overall interpretation of the water turnover in catchments on till includes input by precipitation, infiltration into the soil, vertical percolation through the unsaturated soil layers, groundwater recharge and a lateral groundwater flow to discharge in downslope areas with consecutive surface water formation (Gustafsson 1946). In the upslope areas, the groundwater level is often at a few

metres depth, allowing unsaturated percolation during a major part of the year. In downslope discharge areas, conditions for such percolation only occur during short periods. Most of the time, the groundwater level is high and unsaturated percolation only concerns the top half metre of the soil. At the groundwater table, the water percolation joins groundwater flow and is drawn off laterally.

Unsaturated conditions

Water added to the soil at one event constitutes only a minor percentage of the total water content (Lundin 1982,

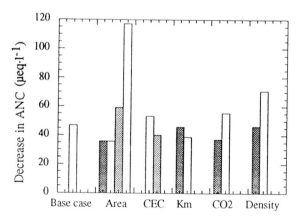

Fig. 14. The parameters selected for the sensitivity analysis, along with the range of variation, were specific mineral surface area, (50%, 80%, 125% and 200% of base case values), cation exchange capacity, CEC (80% and 125% of base case values), cation exchange rate coefficient (km) (20% and 500% of base case values), CO_2 partial pressure (50% and 200% of base case values), soil density (80% and 125% of base case values).The change in ANC differs most for runs made with large variations in exposed area, soil density and CO_2 pressure, Stubbetorp. SAFE model.

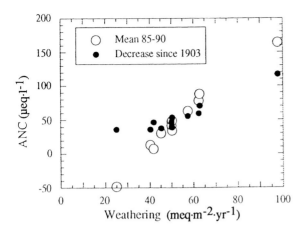

Fig. 15. Influence of weathering rate on modelled ANC and modelled decrease in ANC, Stubbetorp. SAFE model.

Bishop 1992). Under unsaturated conditions, water content in tills is still mainly high. Examples of relevant values are at 0.5 m depth a porosity of c. 40%, a water content mostly c. 30 vol-% resulting in a deficit to saturation of 10 vol-%. (Lundin 1982). In the 0.2 m layer, the deficit is 10–20 vol-% at a moisture content of 30–40 vol-% and at 1.0 m depth it is 1–2 vol-% at a moisture content of 20–25 vol-%.

In the tracer experiments, rapid flows were observed both in lysimeters and in the two percolation studies. There were early tracer detections observed in advance of the major total water movement. The particle velocity of this preferential flow was estimated at up to ten times the major piston flow, which was 1–2 m yr⁻¹. The piston flow velocity was in accordance with findings by Andersen and Sevel (1974), Saxena (1987) and Lindström and Rodhe (1992b). Percolation was also influenced by the amount of water input. This emphasized periods such as those of spring snowmelt and autumn rains. In between percolation was small.

Water turnover was studied by lysimeters in the Stubbetorp catchment, where the flow pattern in the 0–0.40 m upper soil layers coincided with an ideally mixed reservoir, while piston flow was a good approximation in the 0.4–0.8 m layers (Lindström and Rodhe 1992a). Lysimeter experiments at the Masbyn catchment showed at 0.3 m depth a first tracer leaching 2–3 months after supply which corresponded to 30–50% of the total outflow and was considered as a preferential flow. At 0.6 m depth this flow amounted to 15–20% of the total flow. The duration of the total flow was 17–29 months.

The distribution of tracer concentrations over time showed a duration of 12–24 months of tracer leaching after the time of maximum concentrations. In comparison with the preferential flow occurring after one or two months, the receding tracer elucidates the mixing of water as it passes through the soil.

Consequently, stored soil water would strongly affect the chemical composition of the water added to the soil.

The influence depends on the residence time. With large water input and especially in moisture conditions close to saturation, water passes through the soil without complete mixing. Such preferential flow occurs in the macropores and has been mentioned by several authors (e.g. Whipkey 1965, Knutsson 1971, Beven and Germann 1982, Lundin 1982, Espeby 1989). In silt and clay soils, shrinkage cracks develop in dry periods, and together with earthworm activities these allow macropore flow (Cammeraat 1992). Such preferential flows also occur in till soils. The textural composition and the existence of layers with coarser soil material provide possibilities for rapid flow. In soils with large content of stones and boulders, such as the Risfallet catchment, and in areas with frequently occurring bedrock surfaces close to the ground surface, such as the Stubbetorp catchment, water preferentially moves along the rock and stone surfaces (Troedsson 1955, Espeby 1989). Water flow could also occur in fractured bedrock, which exists in both the Stubbetorp and Risfallet catchments.

Tracer experiments showed that water added to the soil mixes with already existing soil moisture. Initially there is partly a preferential flow but during periods between water inputs, movements are small and mixing within layers takes place. Water then enters into small voids and is partly withdrawn from the large water turnover and instead incorporated into the total soil water storage. In downslope locations, deep percolation was slow and occurred during short periods, with only small amounts of added water reaching deep soil layers (>0.7 m). This resulted in comparatively low tracer concentrations, with fairly long residence times. The turnover of all the soil water to a depth of one metre took about two years and corresponded to a water throughflow of about three times the soil water content.

Saturated conditions

Full scale tracer experiments were performed in small subcatchments, taking into account the total soil column from soil surface down to the bedrock. These studies elucidated the importance of the prevailing hydrological conditions on the water turnover. The importance of moist and wet soil types close to the discharging stream was evident. During the first month after tracer supply, 26% of the output was leached by only 2% of the total water throughflow needed to leach all of the recovered tracer. High groundwater levels and 57 mm precipitation facilitated this leaching. Large outputs of tracer occurred during periods of large input of water. In a five-month period, two to six months after tracer supply, there was only 4% of tracer leaching with a runoff of 25 mm. During the following six month period, 7–13 months after tracer supply, there was a 43% of tracer leaching with a runoff of 353 mm (Fig. 16). However, only 10% and 23% of the tracer supplied in two experiments was

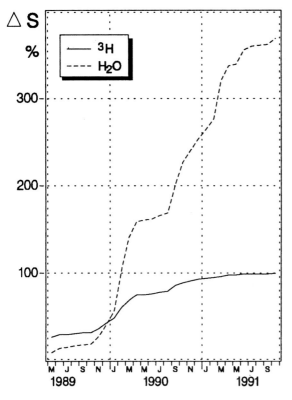

Fig. 16. Distribution of tracer (^{3}H) outflow and theoretical water turnover, runoff/storage, during a 30-month period, the Masbyn area. ΔS H$_2$O is runoff in percent of stored mean soil water content. ΔS ^{3}H is leached tracer in percent of total tracer leaching.

recovered in the stream outflow. This was in agreement with other similar studies (cf. Nyström 1985).

In discharge areas characterized by high groundwater levels, added water rapidly passes the soil and forms surface runoff. During precipitation, not only the direct input from the atmosphere enters these areas but also incoming groundwater from upslope areas. This flow partly follows pathways from somewhat deeper soil layers, so the direction of the flow is more or less upward. In the layers close to the groundwater table, this flow meets the percolating water. In these upper soil layers, high hydraulic conductivities provide a large lateral water flow. Tracer experiments showed rapid outflow from such areas, in which the groundwater of the upper soil layers contributed the major share of discharge, and exerted a considerable influence on the surface water.

The saturated lateral water movement incorporated a rapid flow concerning only a minor part of the groundwater. The velocity of this movement in the Masbyn catchment, with very shallow groundwater flow in basal till, was 0.25 m d^{-1} at 1.0 m and 0.5 m d^{-1} at 0.5 m, in comparison to average velocities of the total water movement of 0.1 m d^{-1} at 1.0 m and 0.2 m d^{-1} at 0.5 m (Lundin 1982). In this process an additional impact was exerted

by the fluctuating groundwater level, temporarily providing considerable water flow in the upper soil layers. Precipitation causes this rise in groundwater level and results in a flow pattern with a pulsating lapse. Then leaching becomes comprehensive. However, evapotranspiration was the dominating process during dry periods with low groundwater levels. Then there are only small groundwater flows, mainly concerning deep groundwater, resulting in low outflow. Rapid flow of groundwater may also occur deeper down in till types with lenses or layers of sorted sand and gravel. Flow velocities of the magnitude of 0.5–1.0 m d^{-1} in "arteries" within lodgement till have been documented by tracer tests (Knutsson 1971).

Tracer concentrations in the groundwater were determined at different times after application. After two months, tracer was mainly in the upper 0.5 m of the soil, with the largest concentrations in the 0.2 m layer. The tracer concentration in the runoff was similar to the concentrations in the 0.2–0.4 m layers. As has already been mentioned, a large part of the leaching was during the first month, originating from the upper groundwater layers. Six months after application, tracer concentrations in the uppermost groundwater layers were lower and the highest concentrations were at 0.6 m depth. At this time tracer had reached to below 0.8 m depth. Again the concentrations in the runoff were similar to those in the 0.2–0.4 m groundwater.

One year after application, tracer was found throughout the whole soil profile. Concentrations were low in the uppermost layers, increasing with depth to a largest concentration at 0.8 m. Later on, concentrations continued to decrease at all depths but tracer was found in the deeper groundwater even after two and a half years. However, tracer in runoff was then no longer discernible (Fig. 17). This revealed the obvious influence of the upper groundwater layers on the major water turnover and a very slow turnover in deeper groundwater layers.

We conclude that a minor part of the infiltrated water reached the soil layers deeper than one metre and that the residence time in those layers was several years. Furthermore, the experiments showed that a large part of the water input was drained in the upper soil layers and runoff formation was mainly generated in the moist and wet downslope discharge areas.

In the catchments studied, the dominating flow conditions vary, but hydrological conditions characteristic of the sites are distinct. In the Lofsdalen and Masbyn catchments most of the water turnover occurs during periods with high groundwater levels, for example in the upper 0.5 m soil layers. Such periods are extensive in these areas. There are low groundwater levels only for short periods, and then there is little water percolation. In the other three catchments, Fårahall, Stubbetorp and Risfallet, the groundwater level is low, with considerable unsaturated percolation but also a persistent moderate to low groundwater flow. However, there are also moist and wet soil types in these catchments, constituting discharge

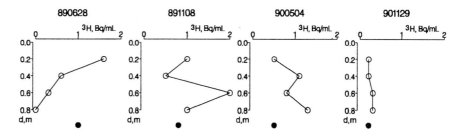

Fig. 17. Stratifications of mean tracer (^{3}H) concentrations in the groundwater with depth on four occasions and tracer concentration in streamwater (solid dots), Masbyn.

areas, which contribute both to the water yield and to influences on the runoff hydrochemistry.

Significance for acidification

The impact of hydrology on groundwater acidification varies according to the site. In areas with deep well-drained soils, acidic input is neutralized during percolation but instead there is subsequent soil acidification and in the long run the total soil column may be acidified. By then, preferential water flow has supplied the groundwater with acid compounds which already have brought about a deterioration of the groundwater. However, so far the buffering capacity of the soil and groundwater manage to provide a non-acidic groundwater.

In areas with poorly drained soils and frequent high groundwater levels, impacts of acidic input mainly affect the upper soil layers, for example to depths of 0.5–1.0 m. At these depths water is drawn off laterally and acidification is exported to the surface waters. The deep groundwater in these areas is protected against present acidification but will in the long run be affected as a consequence of the upslope acidification. At present there is an input by deep discharge of groundwater with alkalinity, which maintains a non-acidic groundwater in these layers.

Buffering processes in soils

The current acidification of forest soils is caused mainly by two factors, acidic deposition and uptake of base cations by the forest. In the southern part of Sweden about half of the acidic input to forest soils comes from forest growth and subsequent harvesting (Berdén et al. 1987). There are a number of processes that buffer against this acidification. The most important are cation exchange, anion exchange, plant uptake of nitrate, denitrification and weathering.

The cation exchange can provide buffering for periods from decades to hundreds of years, depending on the relation between the acid deposition and the stores of exchangeable cations. It is apparent that Scandinavian forest soils in general had base cation stores sufficient for only a few decades of buffering. About 20 000 acidified lakes in Sweden bear clear witness to this.

Anion exchange concerns sulphate which is adsorbed and exchanged for hydroxyl ions in the B-horizon of podzols. The more mature the soil profiles, the more anion exchange sites are available. In the Adirondacks, in the USA, there is a very efficient buffering by anion exchange in the south, where soils are older than from the last glaciation (Fuller et al. 1985). In Scandinavia, the sulphur budgets for catchments in general are found to be well balanced, meaning that what comes in by deposition also goes out with runoff (Hultberg and Grennfelt 1992). Thus generally the anion exchange provides only marginal buffering. When the soils are acidified, the number of positively charged sites increases and these can be occupied by sulphate. However, in a retrospective study of soil samples from the early 50's compared with samples from the late 80's, Gustafsson and Jacks (1993) have shown that these sites seem to be occupied by organic anions rather than sulphate. The organic anions are more competitive for the exchange sites. Possibly the growth in standing forests and the increase in the amount of organic matter in the forest floor contributed to larger amounts of organic anions being mobilized. Only in northern Sweden does there seem to be the possibility of sulphate uptake (Karltun and Gustafsson 1993). Thus, for the Fårahall field research area, and for the areas it represents, there seems to be no reason to expect anion exchange to be an important buffer. For the other sites there is not enough data to form an opinion.

Uptake of ammonium and nitrate ions by plants buffers against the acid nitrogen deposition as long as nitrogen is in short supply to the vegetation, ammonia is exchanged for hydrogen ions and nitrate for hydroxyl ions on entering the root membrane (van Breemen 1985). Currently, nitrogen deposition is in excess of the possible plant uptake in many medium quality forest areas in the southern half of Sweden (Rosén et al. 1992), though the organic matter in the forest floor can still adsorb nitrogen, lowering the C/N ratio. This is possible until a C/N ratio of c. 25 is reached; only then is there a risk of a general breakthrough of nitrate (Kriebitzsch 1978). The current C/N ratio in the forest floor in the vicinity of Fårahall is 25–30 (n = 10). From modelling the south Swedish forest ecosystem, Ågren and Bosatta (1988) have estimated the respite until nitrogen saturation occurs in southern Sweden at a few decades. It is expected to occur first in the low quality areas where the forest uptake is least and where the forest floor organic matter stores are generally also minimal.

Denitrification could provide a buffer against acid N deposition as the conversion of nitrate to nitrogen also consumes hydrogen ions. Although few data are available on the spatial distribution of denitrification, it seems unlikely that it is an important process in well drained upland soils. It is efficient in wetlands in agricultural areas (Fleischer and Stibe 1989) and also in forest wetlands where there is enough substrate in the form of nitrate for the denitrifiers (Jacks et al. 1994). However, this will not protect the groundwater from acidification once nitrogen saturation occurs, only the surface water.

Weathering is the only long-term buffer against both acid N and S deposition in groundwater recharge areas with an aerated soil zone. Scandinavian soils, being young, have enough weatherable minerals but the weathering rate is slow (Nilsson 1986). The two main determinants for the weathering rate are the mineral assemblage present and the specific surface area of the soil. Sverdrup (1990) has come to the conclusion that the amount of minerals such as epidote, pyroxenes and amphiboles is of special importance in Swedish soils, which are commonly derived from granites and gneisses. The more abundant plagioclase feldspars and potassium feldspars weather at a rate which is one or two orders of magnitude slower. The weathering rate depends on the acidity of the soil solution. In general, a more acid soil solution results in a higher weathering rate although there are exceptions such as hornblende, the weathering of which seems to be hampered by raised concentrations of dissolved aluminium (Sverdrup 1990). The overall relation between weathering rate and acidity seems to be an exponential one where the rate is proportional to the concentration of hydrogen ions raised to an exponent smaller than unity or in the order of c. 0.5 (Schnoor and Stumm 1986). The amount of organic ligands in the soil water is also important. Lundström and Öhman (1990) have found that, under laboratory conditions, the addition of a humus extract increases the weathering rate by about three times. Generally, however, it seems that increased acidity tends to decrease dissolved organic matter by precipitating it with aluminium, which serves to lower the weathering rate. Thus, on the whole, it is most unlikely that increasing acidity in soil profiles will provoke a weathering rate high enough to maintain bicarbonate buffering in the soil water and the groundwater. It is likely that weathering has decreased over time. A Scottish study of sandy soils of different age (80–13 000 yr) using the zirconium method, showed that the areal weathering rate decreases with the amount of weatherable material in the upper horizons (Bain et al. 1993). It appeared that even though the decrease in weathering is largest when the soils are newly formed, the effect seems to be significant at least up to an age of 5000 yr.

Acidification in the field research areas

The base cation uptake by the forest in the field research areas gives a substantial acidic input. However, it is in all the sites largely balanced by the combined input from deposition of base cations and weathering. Even weathering covers the need for base cations in most cases with some exceptions. The uptake of calcium exceeds the weathering rates in Fårahall, Stubbetorp and Risfallet by 140, 40 and 60%, respectively.

The cation exchange provides a considerable portion of the current buffering in all the studied catchments. In Fårahall, 115 meq m^{-2} are taken from the exchangeable store every year. The total store in the upper metre of the soil is of the order of 1400 meq m^{-2}, which is equivalent to no more than c. 10–15 yr at current rates of loss. The element in shortest supply in relation to the stores is magnesium. This is a little surprising in view of the proximity to the sea. Potassium, which would be suspected to be in short supply, is actually accumulating. The budget for sodium is very nearly balanced, which supports the reliability of the budget, as sodium is a minor component in the exchangeable pool. The Stubbetorp catchment looses c. 60 meq m^{-2} annually with stores of c. 5000 meq m^{-2}. The element in shortest supply is again magnesium. The buffer stores should last about five decades under the present conditions of deposition. The Masbyn catchment has similar losses to the Stubbetorp catchment but smaller stores. The stores of both calcium and magnesium should under current conditions last for c. 20 yr. The Risfallet catchment has exchangeable stores sufficient to sustain c. 60 yr of losses at the present rate, with magnesium in shortest supply. For the Lofsdalen springs we do not have a detailed budget. However, magnesium deficiency has been diagnosed in pines in the vicinity (Popovic pers. comm.). The fact that the stores in some of the catchments are rather small in relation to current losses does not mean that the stores will actually be depleted in 10–15 yr (Fårahall) or 20 yr (Masbyn). The increased acidity will bring about an increase in weathering rate and there will also be an export of acidity to the groundwater/surface water, causing acidification. The Fårahall and Masbyn areas are those facing water acidification in the near future. The groundwater in the Masbyn catchment may already have been considerably acidified while the Fårahall groundwater so far has been protected by the unusually thick unsaturated zone. The acidification of the soil profile has reached a depth of c. 3 m in Fårahall. Many wells in this part of the country with a relatively shallow groundwater table have become drastically acidified during the last few decades. However, the preferential flowpaths demonstrated in this and other investigations (Lundin 1982, Wels et al. 1990) may export a considerable amount of acidity directly to the surface water and this will delay the acidification of the soil and groundwater as a whole.

As mentioned at the beginning of this section, anion exchange is unlikely to provide any buffering. The sulphate budgets are more or less balanced for the Fårahall, Stubbetorp, Masbyn and Risfallet catchments, with in/out amounts of 197/200, 52/49, 63/60 and 35/37 meq m^{-2} respectively). This supports the view that the soils are in equilibrium with current sulphur deposition. In Stubbetorp sulphate reduction has been observed locally although not to the extent that it affects the budget.

The nitrogen deposition in Fårahall is 21 kg ha^{-1} yr^{-1}. The current growth rate has not been measured but the site quality would indicate an uptake in the stand of c. 6 kg ha^{-1} yr^{-1}. With the losses of 2–3 kg ha^{-1} yr^{-1} this leaves almost 10 kg ha^{-1} yr^{-1} currently bound up in the humus layer. The C/N ratio in the humus layer in the area is c. 28 and a rough extrapolation with the current excess deposition suggests that the humus layer will reach a critical C/N ratio of 25 in c. 20 yr. However, clearfelling would cause a considerable loss of base cations in a limited time and this is likely to take place before the nitrogen saturation, as the stand is approaching maturity. In the other catchments, the nitrogen deposition is 9 kg ha^{-1} yr^{-1} for the Masbyn area, 7 kg ha^{-1} yr^{-1} for Stubbetorp and 5 kg ha^{-1} yr^{-1} for Risfallet. For Masbyn the runoff losses are 2–3 kg ha^{-1} yr^{-1} and the uptake is 5 kg ha^{-1} yr^{-1} leaving only a marginal quantity for accumulation in the forest floor. The situation is similar in the other areas. Thus nitrogen saturation within the next few decades is a real threat only in the Fårahall catchment. These conclusions are in good agreement with modelling by Rosén et al. (1992), based on a large number of forest inventory sites all over Sweden. Denitrification is likely to be small in well drained upland soil, such as in the coarse Fårahall till, it will not offer any protection either, against acidification of the upland soil or of the groundwater.

A major portion of the work has been devoted to producing data for modelling of the weathering rates by the PROFILE model and independent estimates for the verification of the modelling. The estimates by PROFILE, the historical weathering rate, and the rate estimated by the use of stable strontium isotopes all fall within quite a narrow range for the Fårahall catchment. These data suggests a bleak scenario for the Fårahall area and areas with similar characteristics and deposition, which are common in southern and southwestern Sweden.

For the Stubbetorp catchment, current weathering has been estimated by the PROFILE model and by the MAGIC model. The PROFILE model gave 34 meq m^{-2} yr^{-1} for the upper soil metre or 44 meq m^{-2} yr^{-1} if adjusted to the catchment scale, and the MAGIC model gave 30–40 meq m^{-2} yr^{-1}, thus showing a good agreement. The Stubbetorp catchment represents areas below the highest marine shoreline with coarse wave-washed till soils. Wave washing of till soils is clearly shown to affect the weathering rate along hillslopes in the Stockholm area, where the weathering rate as assessed using historical lake chemistry data was found to be 20–50 meq m^{-2} yr^{-1} in sections of the terrain similar to the Stubbetorp catchment (Jacks 1992).

The weathering rate estimates differed considerably for the Masbyn catchment. Historical weathering rate estimates were 12–15 meq m^{-2} yr^{-1}, whereas the strontium method gave 20 meq m^{-2} yr^{-1} for the nearby Buskbäcken catchment (Jacks et al. 1989). The PROFILE model estimate was as much as 40 meq m^{-2} yr^{-1} for the Masbyn catchment. It is possible that the PROFILE model overestimates the weathering rate because the water pathways are very shallow in this catchment. Particularly during the wet seasons many streamlines may never reach the depth of one metre, the depth to which the PROFILE model was run. The model predicts that 21 meq m^{-2} yr^{-1} of the weathering are produced in the section 0.5–1.0 m in depth. This explanation is supported by an investigation in the neighbouring Buskbäcken catchment using the "Paces approach" (Paces 1986), in an unusually thick unsaturated zone of 2 m, which yielded 55 meq m^{-2} yr^{-1}. For the Risfallet catchment, a current weathering rate of 40 meq m^{-2} yr^{-1} was calculated by the PROFILE method. The historical weathering rate estimate was 49 meq m^{-2} yr^{-1}.

The different weathering estimates do not show any major discrepancies. The agreement is good enough to consider that the long-term buffering through weathering is insufficient in all the catchments studied. The Fårahall catchment in particular, which represents considerable areas of similar characteristics in southwestern Sweden, seems to be losing buffering strength rapidly. As major losses of base cations are related to excess sulphate coming from deposition as a counterion, a decrease in the acidic S deposition is necessary if further soil and groundwater acidification is to be avoided. The Fårahall catchment would need a reduction to almost zero.

The soil chemistry of upland sites in the investigated areas is a result of the past history with respect to the processes discussed above. The simplest but still most descriptive parameter is the soil pH (Fig. 18). The Fårahall profile is the most acidified, and the acidification is relatively recent. The Masbyn soil is rather acid, as is the Lofsdalen soil close to the surface. The Risfallet soil profile is the least acidified.

Acidic groundwater has been found only near the discharge area in Fårahall. Most of the groundwater is still protected by the deep unsaturated zone (Fig. 19). However, the lower part of the hillslope is acidified to the extent that the runoff is permanently acidic. The acidification in Stubbetorp has so far only affected the upper well-drained soils and runoff still has bicarbonate alkalinity. However, in the upper sections of the terrain below the bare rock areas there is permanently acidic groundwater, such as in the spring. The Masbyn catchment has a groundwater circulation near the surface throughout the year, which means that the acidification front has reached the groundwater in some areas. Moreover, rains and snowmelt will seasonally cause the groundwater table to rise so that a considerable export of acidity occurs. The deeper groundwater, >2 m, is to some extent protected from acidification through the shallow water pathways.

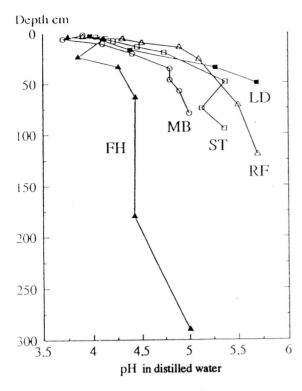

Depth cm

pH in distilled water

Fig. 18. pH in upslope soil profiles in the Fårahall (FH), Masbyn (MB), Stubbetorp (ST), Risfallet (RF) and Lofsdalen (LD) areas.

The Risfallet catchment is geologically rather similar to the Fårahall catchment, coarse till soil is found in both places. However, the lower acidic load over several decades means that the acidification is still at a shallow depth, and both groundwater and surface water have more or less permanent bicarbonate alkalinity. The Lofsdalen springs are rather similar to the Masbyn catchment, in hydrological character as the seasonal variations in water flow cause acid surges. In Masbyn the acid periods are longer, while so far they are episodic in Lofsdalen, mainly restricted to periods of snowmelt or heavy autumn rains. In addition to the shallow water pathways, the very coarse and weathering-resistant material is a main reason for the acid surges in Lofsdalen. Acid deposition seems to play an important role alongside the geological conditions in Lofsdalen (Ahlström et al. 1995), in contrast to findings further north in Sweden. The acidic episodes observed in the Svartberget catchment near Umeå could mainly be attributed to organic acids (Bishop 1992). The Lofsdalen springs are low in DOC, having c. 5 mg l^{-1}.

Cross-combinations of the hydrogeology found in the studied sites and acid deposition are common in Scandinavia. In southwestern Sweden or southern Norway, the Masbyn catchment would have suffered the early acidification of surface groundwater observed in these areas. Shallow soils such as those in the upper reaches of Stubbetorp are abundantly present in these high deposition

areas (Statens Forurensningstilsyn 1993). Even the groundwater in the rock may be acidified under such circumstances (Lång and Swedberg 1990).

Scenarios for groundwater chemistry

The model calculations show the response to four deposition scenarios: no reduction; reduction of deposition by 30% S for sulphuric acid deposition and 10% N for total nitrogen deposition; reduction of 60% S and 30% N; and reduction of 85% S and 50% N relative to the acid deposition 1988. While the predictions based on the critical loads methods are sufficient to determine the long-term consequences of deposition reductions, dynamic models are necessary to predict the rate of recovery of different receptors to changes in chemical climate.

In Stubbetorp all reduction scenarios result in significant improvement as compared to the no reduction situation. With the latter, the model predicts that the rate of acidification will increase, leading to ANC levels below 0 before 2050 (Fig. 20). The results from the steady-state calculations are shown in Table 12. It is clear that the acidification at Stubbetorp will develop to quite acid conditions unless the 60/30 scenario is put into effect. The 30/10 scenario results in a reduced rate of acidification. The N reduction has no effect on the simulations since it is assumed that all N is immobilized in the biomass under all circumstances. The 85/50 scenario is the only scenario that leads to acceptable ANC in shallow groundwater. Fig. 21 shows the relative importance of weathering and cation exchange for neutralizing acid inputs. The weathering rate modelled for Stubbetorp is high if compared with the two other catchments. This is at least partly due to differences in the method of determination of mineral surface areas. Perculation results in a weathering rate similar to that of Masbyn, c. 35 meq m^{-2} yr^{-1}.

The SAFE and MAGIC simulation both lead to the same overall conclusion that a drastic reduction in acidifying substances must be achieved if the groundwater at Stubbetorp is to be protected against acidification (Sandén et al. 1992).

In Fig. 22 the results from SAFE simulations with current deposition are shown, horizon by horizon. The soil solution pH is compared with pH measured in soil-water suspension. The base saturation is expected to continue to decrease in all the modelled catchments. The calculations for Masbyn show that only 85/50 reductions will give a significant alkalinity in the groundwater, while no reductions will lead to a damaged groundwater resource (Table 12). The sudden increase in weathering rate, from 39 meq m^{-2} yr^{-1} to 46 meq m^{-2} yr^{-1} is caused by changes in hydrology after clearcutting in the mid-1970's (Fig. 21).

Risfallet is the poorest site modelled in this study, the weathering, as determined by SAFE, amounts to only 28

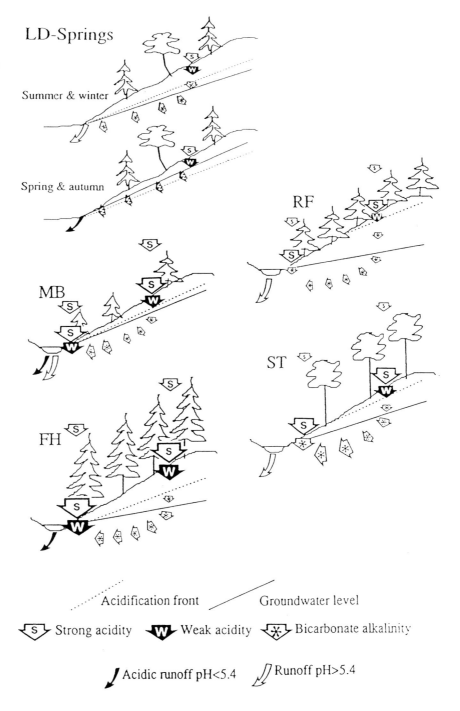

Fig. 19. Conceptual models of the acidification status in groundwater and surface water in the five field research areas, Fårahall (FH), Masbyn (MB), Stubbetorp (ST), Risfallet (RF) and Lofsdalen (LD).

LD-Springs

Summer & winter

Spring & autumn

RF

MB

ST

FH

.... Acidification front   / Groundwater level

⟨S⟩ Strong acidity   ⟨W⟩ Weak acidity   ⟨✳⟩ Bicarbonate alkalinity

⤵ Acidic runoff pH<5.4   ⤴ Runoff pH>5.4

meq m^{-2} yr^{-1}. Still, the acidification has not proceeded very far compared with that at Masbyn, because of the stand characteristics. At Risfallet, a new pine generation was planted c. 1970. Thus, during the years with the highest deposition rate, the stand only received a small amount of dry deposition. Also, a pine stand receives only 2/3 of the deposition of a mature spruce stand (Lövblad et al. 1992). The simulated pH values are close

to measured pH$_{H_2O}$ (Fig. 21). Following the clearcutting, the ANC in the groundwater increases as a result of reduced deposition. The base saturation continues to decline, indicating that the load during this phase exceeds the critical load of the system. As the stand grows, the deposition will increase, as will the rate of acidification. This leads to an accelerated rate of decline in groundwater ANC and in base saturation in the C-layer. The

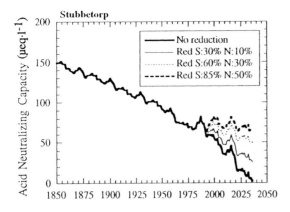

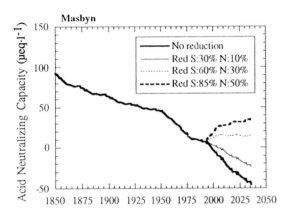

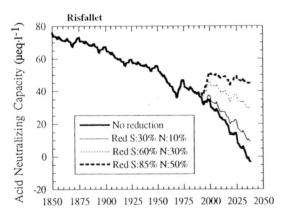

Fig. 20. Calculated ANC at the one-metre level for different future deposition scenarios, SAFE model.

The calculations show that the response of soil solution ANC is relatively rapid in all the modelled areas. This result can be explained by the relative role of different buffering mechanisms during different phases of acidification, as shown in Fig. 21. It shows that the cation exchange reaction is most important during the period of most intense increase in acid deposition. As the soil pH and the base saturation decreases, the cation exchange buffering is reduced, allowing groundwater pH to be affected. This phenomenon can only be simulated with multilayer models, where the exchangeable cations in the different soil layers are exhausted in sequence.

Critical load maps

Critical loads for groundwater were calculated with the PROFILE model, applying the critical limit of 100 µeq l⁻¹ at a depth of 2 m, and areas where these critical loads are exceeded at the current deposition rate have been calculated. The results are shown in Fig. 23. It can be seen that groundwater is very sensitive to acidification in certain areas of Sweden. The locations of these areas are closely related to the weathering rate, by far the most important source of alkalinity. Especially sensitive is a large area in middle Sweden, with the poorest mineralogy in Sweden, and the lowest weathering rates. The eastern part of Skåne also has a small area of very sensitive groundwater, corresponding to the area with quartzitic sandstone occurring locally. On average, the critical load for the groundwater is c. 0.7 keq ha⁻¹ yr⁻¹ of acidity, if all is expressed as sulphur, 11 kg ha⁻¹ yr⁻¹. Exceedance maps illustrate where the critical loads are exceeded, and by how much. On average, critical loads for groundwater have been exceeded in 2/3 of the forested land area of Sweden. This implies that 1/3 of the aquifers are at present not being negatively affected by acidification. This can be seen from the right-hand map. A major part of the protected area is located in northern Sweden. In central Sweden, in the province of Härjedalen, the present deposition largely exceeds the critical load as has been verified by the results from the field study in Lofsdalen. Also in all of southern Swedeen are the critical loads exceeded, despite the high weathering rates.

Steady-state chemistry

The PROFILE model may also be used to calculate the steady-state chemistry in response to a particular deposition input. Thereby, the long-term consequences of different deposition scenarios can be assessed. The steady-state chemistry for groundwater at 2 m depth has been calculated regionally for different scenarios. Pre-industrial deposition was assumed to be to 5%/5%/15% of current deposition for sulphur, nitrate and ammonium, respectively. The net forest uptake of base cations and N

final state will not be reached until well into the next century. The base saturation will continue to decrease even with the 85/50 scenario, but at a slower and slower rate. This is explained by comparing the modelled ANC values with the steady-state chemistry as shown in Table 12.

Table 12. Steady State Mass balance for groundwater at a depth of 2 m. D = Deposition, W = Weathering, Nu = N uptake and BCu = Base cation uptake.

		ANC Supply (meq m^{-2} yr^{-1})				[ANC]ss (μeq l^{-1})
		D	W	N$_u$	BC$_u$	
No reduction:	Stubbetorp	−118	127	68	−35	240
	Masbyn	−104	115	28	−27	24
	Risfallet	−87	70	45	−35	−25
30/10 reductions:	Stubbetorp	−93	127	61	−35	240
	Masbyn	−81	115	28	−27	70
	Risfallet	−67	70	45	−35	46
60/30 reductions:	Stubbetorp	−62	127	48	−35	339
	Masbyn	−50	115	28	−27	132
	Risfallet	−42	70	43	−35	128
85/50 reductions:	Stubbetorp	−34	127	34	−33	408
	Masbyn	−24	115	28	−27	184
	Risfallet	−20	70	23	−22	182

was set at 40% of the 1990 rate. The state of the groundwater in the past, the natural alkalinity, was calculated in order to establish a point of reference. Fig. 24 shows the cumulative distributions based on all calculation points, including the pre-industrial deposition. It shows that in 70% of Sweden the water at 2 m may have negative ANC if the deposition continues at the 1990 rate. The 85/50 scenario will yield a result similar to the pre-industrial level, with a general shift in ANC of c. 50 μeq l^{-1}. The map in Fig. 23 shows the average weathering down to a soil depth of 2.0 m, relevant to water recharging to the groundwater aquifers. There is a general decrease in the weathering rate from south to north, mostly attributable to a decline in temperature. Three minerals in the soil are of major controlling importance for the weathering rate in the field: feldspars, epidote and hornblende.

Groundwater acidification sensitivity

Calculations of groundwater quality using SAFE at Stubbetorp, Masbyn and Risfallet show that continuation of 1990 rates of acidic deposition will cause groundwater ANC to fall below 100 μeq l^{-1} at all sites. The calculations indicate that for the no reduction scenario, the ANC will reach the projected steady-state level well after 2050. For the reduction scenarios, the ANC approaches the new steady state within decades after the reductions are achieved. This strongly emphasizes the need for deposition reductions to take place as soon as possible in order to prevent further acidification. The observed differences between the sites arise from differences in soil weathering rate and forest growth history. To keep ANC >0 μeq l^{-1}, reductions of 85% S and 50% N are required at all sites. The critical loads maps are consistent with the results from the calculations with the dynamic model.

Conclusions

The project is based on field investigations in a small number of field research areas combined with modelling to generalize the results. The representativeness of the field areas is of course of crucial importance. Previous experience of soil and water acidification in Scandinavia shows that the main features of till areas are represented in the field areas. The general conclusions from the modelling work is that the developed models, covering both hydrological and hydrochemical processes, are generally applicable for the simulation of the sensitivity of groundwater in most parts of Sweden. The hydrological model (PULSE) emphasizes vertical processes. Extensive controls show that the soil temperature, water balance and vertical flux of water is described in a reasonable way. As it has been shown that the soil moisture routine can be run on field data without calibration of any empirical parameters, there seems to be little justification for a more complex model formulation where more detailed input data and more extensive parameter assessment would be necessary.

The chemical models (PROFILE, SAFE) emphasize the kinetic nature of all processes involved. Tests against field data for weathering rates and soil profile major chemistry indicate that the weathering sub-model functions with relatively high accuracy. The observed patterns in terms of soil water pH, ANC, base cation concentration and base saturation can be simultaneously reproduced for the different soil layers, using field properties. The chemical models have very dynamic and non-linear response to changes in some of the input data and model parameters. The dynamic hydrochemical model (SAFE) shows a considerable dependence on the time step used in its hydrological input and therefore a dynamic description of temperatures, recharge and moisture is needed, rather than simple annual mean values, even for the long-term

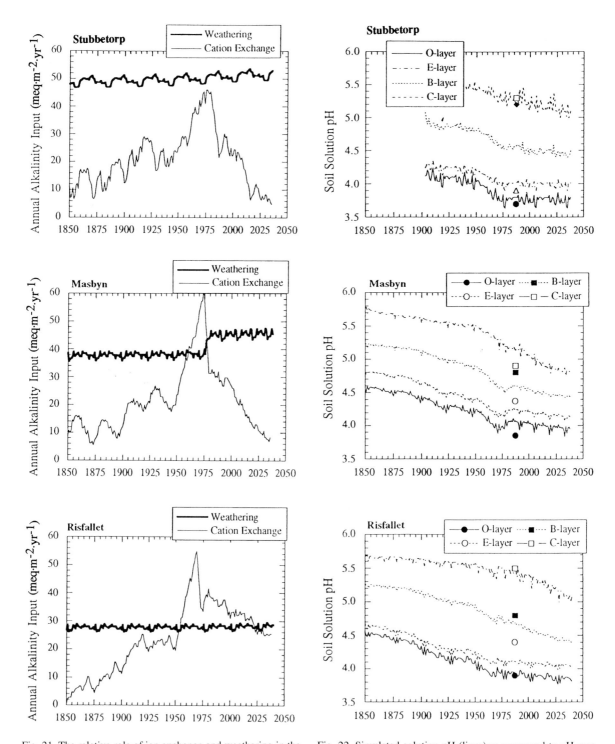

Fig. 21. The relative role of ion exchange and weathering in the three areas studied, SAFE model.

Fig. 22. Simulated solution pH (lines) as compared to pH measured in soil-water solution (dots), SAFE model.

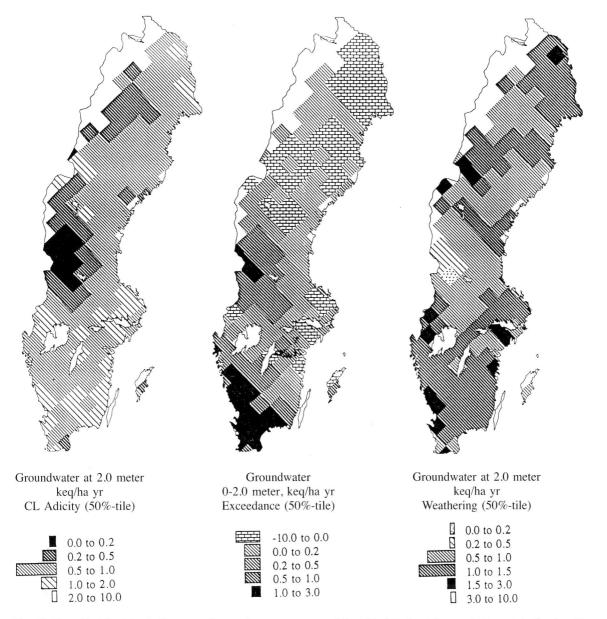

Groundwater at 2.0 meter
keq/ha yr
CL Adicity (50%-tile)

■ 0.0 to 0.2
▨ 0.2 to 0.5
▨ 0.5 to 1.0
▨ 1.0 to 2.0
▢ 2.0 to 10.0

Groundwater
0-2.0 meter, keq/ha yr
Exceedance (50%-tile)

▤ -10.0 to 0.0
▨ 0.0 to 0.2
▨ 0.2 to 0.5
▨ 0.5 to 1.0
■ 1.0 to 3.0

Groundwater at 2.0 meter
keq/ha yr
Weathering (50%-tile)

▫ 0.0 to 0.2
▨ 0.2 to 0.5
▨ 0.5 to 1.0
▨ 1.0 to 1.5
■ 1.5 to 3.0
▢ 3.0 to 10.0

Fig. 23. The critical load for shallow groundwater, the present excess of the critical load and the weathering rate in Sweden. The maps illustrate the critical limit at a depth of 2 m and the median value (50%-tile) of the sites within each grid. The critical load is defined as the largest acid deposition possible if the set quality criteria for groundwater, 100 µeq l^{-1} is to be maintained in an area (from Sandén and Warfvinge 1992).

predictions. It is also necessary to use a multi-layered model for modelling groundwater response to acidic deposition. One and two-layer models give unrealistic residence time distributions for soil water and unrealistic averages of soil chemical parameters.

The input data requirements, in the form of geochemical, hydrological and climatological data can usually be met. However, the sensitivity analysis showed that care has to be taken in the selection of values for the parameters. The exposed mineral surface area, for instance, is

largely a function of the soil texture. Even within seemingly homogenous till areas the texture may show a wide range of variation. A minor content of silt will give a substantial increase of the surface area of sandy till while a high content of stones and boulders gives lower values. The density of the soil is also of importance but is seldom determined for large volumes of soil. The hydrochemical model shows a strong response to changes in the weathering rate of minerals. Since independent tests indicate that the weathering rate can be calculated with high absolute

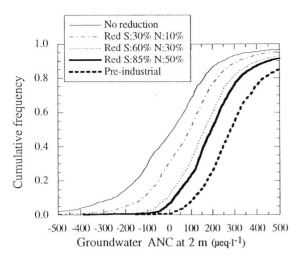

Fig. 24. The cumulative distribution of ANC in the groundwater at a depth of 2 metres under different deposition scenarios.

accuracy, errors introduced by possible deficiencies in the weathering model formulation may be less of a problem than the usual soil variation in texture within an area, as the mineralogical composition probably is more constant. Uncertainty in the main input data leads to uncertainty in the levels of ANC in the groundwater, but the rates of change and timing of the acidification processes are stable.

Acid groundwater has been found near the discharge area in Fårahall, the investigated hillslope in southwestern Sweden with the highest acid load. Most of the groundwater is still protected by the deep unsaturated zone. However, the lower part of the hillslope is evidently acidified to the extent that resulting runoff is permanently acidic. As the acidification of the soil has reached a depth of about three metres in Fårahall, it seems likely that many wells dug in this part of the country, where the groundwater table is shallow, have experienced drastic acidification during the last decade.

The acidification in the research area in southeastern Sweden (Stubbetorp) has so far only affected the upper well-drained soils and the most shallow groundwater, while both the "deeper" groundwater and the runoff still have bicarbonate alkalinity.

The Masbyn catchment in the central part of Sweden has a groundwater circulation near the surface throughout the year, which means that the acidification front has reached the groundwater in some areas. Moreover, rains and snowmelt will bring about a seasonal rise of the groundwater table so that considerable export of acidity occurs. The "deeper" groundwater (>2 m) is to some extent protected from acidification through the shallow pathways. The other catchment in the same part of Sweden (Risfallet) is geologically rather similar to the Fårahall hillslope, with coarse till soil found in both places. However, the lower acid load over several decades means that the acidification is still at a shallow depth and both

groundwater and surface water possess bicarbonate alkalinity more or less permanently.

The Lofsdalen springs in the southernmost part of the High Mountains are in hydrologic character rather similar to the Masbyn catchment, as the seasonal variations in water flow cause acid surges with fairly high concentrations of aluminium. In Masbyn the acid periods are longer, while so far they are quite episodic in Lofsdalen, mainly being restricted to periods of snowmelt or heavy autumn rains. In addition to the shallow water pathways, the very coarse and weathering-resistant material is a main reason for the acid surges in Lofsdalen. Acid deposition seems to play an important role along with the geological conditions in Lofsdalen, which contrast to the findings further northwards at Svartberget near Umeå, where the acidic episodes could mainly be attributed to organic acids.

Cross-combination of the studied areas and the acid depositions pattern are commonly found. A catchment with the characteristics of Masbyn, with its surficial water pathways, in southwestern Sweden would have experienced an early acidification of the surface water and is representative of the catchments of many long-acidified lakes. A special, though common case, is that of catchments with very shallow soils on rock, found along the coasts of Sweden, as well as in Sörlandet in Norway. Many lakes in these areas seem to have suffered acidification in the 50's and 60's (Overrein et al. 1980, Jacks 1992). Similar areas are present in the upper reaches of Stubbetorp. In such cases even the groundwater in the underlying rock may be acidic.

The field investigations and the modelling give a coherent picture of the status and development of the acidification of soil and shallow groundwater in till areas. Based on the modelled trends in soil solution chemistry, we may conclude that soil and groundwater acidification is mainly caused by acidic deposition, and the predicted future water quality depends largely on the deposition scenario. The cation uptake by forest is covered by the base cation uptake and the weathering in combination. With the exception of calcium in Fårahall, Stubbetorp and Risfallet, weathering alone covers the need of the trees. However, the leaching of base cations, mainly with sulphate as the anion, threatens to deplete the exchangeable base cation stores within a few decades. The base cation stores in Fårahall and Masbyn are sufficient for buffering for only 10–20 yr, while those for Stubbetorp and Risfallet will last for c. 50 yr. The scenarios developed through modelling are equally bleak, unless drastic reductions in acidic deposition are achieved (80% S and 50% N to halt and reverse the acidification). The weathering rates are crucial in these predictions. The verifications of the PROFILE model with independent assessments supports the confidence in the modelling. Historical data are rather few, however not contradicting the SAFE modelling. Further work in this connection is certainly needed, but there is no reason to question the general conclusions reached in the project through field work and modelling. The comparisons with past acidification history in Scandinavia confirms that the main features of the predictions made in this work can be relied upon. The results are of considerable economic importance as the acidification and the exceedance of the critical loads occur in the southern parts of the country where most of the population live and most of the dug wells are situated.

Acknowledgements – The Integrated Groundwater Acidification Project received its main funding from the Swedish Environmental Protection Agency but also valuable economic support from the departments and institutes involved. All such support is very much appreciated. The authors would also like to thank all these people, who helped us with the field work, sampling, laboratory analyses, drawing, typing and language correction. We also thank all colleagues and the agency officers who took part in field excursions, seminars and discussions and those who supplied us with data.

References

Ågren, G. I. and Bosatta, E. 1988. Nitrogen saturation of terrestrial ecosystems. – Environ. Pollut. 54: 185–197.

Ahlström, J., Degerman, E., Lindgreen, G. and Lingnell, P.-E. 1995. Försurning av små vattendrag i Norrland. – Swedish Environ. Protect. Agency, Solna, Report 4344 (in Swedish).

Alexandersson, H., Karlström, C. and Larsson-Mc Cann, S. 1991. Temperature and precipitation in Sweden 1961–90. Reference normals. – Swedish Meteorol. and Hydrol. Inst., Norrköping, No. 81 (in Swedish with English abstract).

Andersen, L.-J. and Sevel, T. 1974. Six years environmental tritium profiles in the unsaturated and saturated zones, Grönhöj, Denmark. – In: Anon. (ed.), Isotope techniques in groundwater hydrology I. IAEA, Vienna, pp. 3–20.

Bain, D. C., Meller, A., Robertson-Rintoul, M. S. E. and Buchland, S. T. 1993. Variation in weathering processes and rates with time in a chronosequence of soils from Glen Feshie Scotland. – Geoderma 57: 275–293.

Berdén, M., Nilsson, S. I., Rosén, K. and Tyler, G. 1987. Soil acidification – extent, causes and consequencies. – Swedish Environ. Protect. Agency, Solna, Report 3292.

– , Clarke, N., Danielsson, L-G and Sparén, A. 1994. Aluminium speciation: Variations caused by sample storage and the choice of analytical method. – Water Air Soil Pollut. 72: 213–233.

Bergström, S. 1976. Development and application of a conceptual runoff model for Scandinavian catchments. – Swedish Meteorol. Hydrol. Inst., Norrköping, Reports RHO, No. 7.

– 1992. The HBV model – its structure and applications. – Swedish Meteorol. Hydrol. Inst., Norrköping, Reports RH, No. 4.

– and Lindström, G. 1989. Models for analysis of groundwater and surface water acidification – A review. – Swedish Environ. Protect. Agency, Stockholm, Report 3601.

– , Carlsson, B., Sandberg, G. and Maxe, L. 1985. Integrated modelling of runoff, alkalinity and pH on a daily basis. – Nordic Hydrol. 16: 89–104.

Bertills, U., von Brömssen, U. and Saar, M. 1989. Försurningsläget i enskilda vattentäkter i Sverige. – Swedish Environ. Protect. Agency, Solna, Report 3567 (in Swedish with English summary).

Beven, K. and Germann, P. 1982. Macropores and water flow in soils. – Water Resour. Res. 18: 1311–1325.

Bishop, K. 1992. Episodic increases in stream acidity, catchment flow pathways and hydrograph separation. – Cambridge Univ. Press, Cambridge.

Breemen, N. van 1985. Transformations of nitrogen and sulphur, and acidification of soils and waters. – In: Johansson, I. (ed.), Hydrological and hydrogeochemical mechanisms and model approaches to the acidification of ecological systems. Nordic Hydrological Programme, Stockholm, Report No. 10: 131–136.

Cammeraat, L. H. 1992. Hydro-geomorphologic processes in a small forested sub catchment: Preferred flow-paths of water. – Ph.D. thesis, Dept of Physical Geography and Soil Science, Amsterdam Univ.

Clarke, N., Danielsson, L.-G. and Sparén, A. 1991. Organic bound aluminium and its interaction with a new method for aluminium speciation. – Finnish Humus News 3: 253–258.

– , Danielsson, L.-G. and Sparén, A. 1992. The determination of quickly reactive aluminium in natural waters by kinetic discrimination in a flow system. – Int. J. Environ. Anal. Chem. 48: 77–100.

– , Danielsson, L.-G. and Sparén, A. 1995. Studies of aluminium complexation to humic and fulvic acids utilizing a method for determination of quickly reacting aluminium. – Water Air Soil Poll. (in press).

Cosby, B. J., Hornberger, G. M., Galloway, J. N. and Wright, R. F. 1985. Modelling the effects of acid deposition: Assessment of a lumped-parameter model of soilwater and streamwater chemistry. – Water Resour. Res. 21: 51–63.

Driscoll, C. T. 1984. A procedure for the fractionation of aqueuos aluminium in dilute acidic waters. – Int. J. Environ. Anal. Chem. 16: 267–283.

Eriksson, B. 1983. Data rörande Sveriges nederbördsklimat. Normalvärden för perioden 1951–80. – Swedish Meteorolog. Hydrol. Inst., Norrköping, Report 1983: 28 (in Swedish).

Espeby, B. 1989. Water flow in a forested till slope. Field studies and physically based modelling. – Techn. doct. thesis, dissertation. Dept of Land and Water Resources. Royal Inst. of Technology, Stockholm. Report Trita-Kut No. 1052.

– and Sandén, P. 1992. Comparison with the SOIL-model. – In: Sandén, P. and Warfvinge, P. (eds), Modelling groundwater response to acidification. – Swedish Meteorol. and Hydrol. Inst., Norrköping, RH No. 5, pp. 66–72.

Falkengren-Grerup, U., Linnermark, N. and Tyler, G. 1987. Changes in soil acidity and cation pools of south Swedish soils between 1949 and 1985. – Chemosphere 16: 2239–2248.

Fleicher, S. and Stibe, L. 1989. Agriculture kills marine fish. – Ambio 18: 346–349.

Fuller, R. D., David, M. B. and Driscoll, C. T. 1985. Sulfate adsorption relationships in forested podsols of the northeastern USA. – Soil Sci. Soc. Am. J. 49: 1034–1040.

Geisler, R. and Lundström, U. 1991. The chemistry of several podsol profile soil solutes extracted with centrifuge drainge technique. – In: Rosén, K. (ed.), Chemical weathering under field conditions. Swedish Univ. of Agric. Sci., Dept of forest soils, Uppsala. Reports in forest ecology and forest soils No 63. pp. 157–165.

Gustafsson, J. P. and Jacks, G. 1993. Sulphur status in some Swedish podzols as influenced by acidic deposition and extractable organic carbon. – Environ. Poll. 81: 185–191.

Gustafsson, Y. 1946. Untersuchungen über die Strömungsverhältnisse in gedräntem Boden. – Acta Agric. Suec., Stockholm II: 1.

Hanning, A. 1988. Extraction in liquid-liquid segmented flow applied to kinetic speciation of aluminium in natural waters. – Licenciate thesis, The Royal Inst. of Technology, Stockholm, Trita -AWK-1208.

Hultberg, H. and Grennfelt, P. 1992. Sulphur and seasalt deposition as reflected by throughfall and runoff chemistry in forested catchments. – Environ. Poll. 75: 215–222.

Jacks, G. 1992. Acidification of soil and water below the highest holocene shoreline on Södertörn, central eastern Sweden. – Swed. Geol. Surv. Ser. Ca. 81: 145–148.

– and Knutsson, G. 1981. Känsligheten för grundvattenförsurning i olika delar av landet (förstudie), Projekt KHM. Kol-Hälsa-Miljö, Teknisk rapport. – Statens Vattenfallsverk (in Swedish with English abstract).

–, Knutsson, G., Maxe, L. and Fylkner, A. 1984. Effect of acid rain on soil and groundwater in Sweden. – In: Yaron, B., Dagan, G. and Goldshmid, J. (eds), Pollutants in porous media. Ecol. Stud. 47: 94–114.

–, Åberg, G. and Hamilton, P.J. 1989. Calcium budgets for catchments as interpretad by strontium isotopes. – Nordic Hydrol. 20: 85–96.

–, Joelsson, A. and Fleischer, S. 1994. Nitrogen retention in forest wetlands. – Ambio 23: 358–362.

Jansson, P.-E. and Halldin, S. 1979. Model for annual water and energy flow in a layered soil. – In: Halldin, S. (ed.), Comparison of forest water and energy exchange models. Int. Soc. for Ecol. Model., Copenhagen, pp. 145–163.

Karltun, E. and Gustafsson, J. P. 1993. Interference by organic complexation of Fe and Al on the SO_4^{2-}-adsorption in spodic B-horizons in Sweden. – J. Soil Sci. 44: 625–632.

Knutsson, G. 1971. Groundwater flow in till soils. – Geologiska Föreningen i Stockholm, Förhandlingar (GFF) 93: 553–573.

– 1992. Studies of some acid springs in till, Lofsdalen, Sweden. – Nordic Hydrol. Program, NHP Report nr 30, Oslo, pp. 73–84.

Kriebitzsch, W.U. 1978. Stickstoffnachlieferung in sauren Waldböden Nordwest-Deutschlands. – Scripta Geobot. 14: 1–66.

Lång, L.-O. and Swedberg, S. 1990. Occurence of acidic groundwater in Precambrian crystalline bedrock aquifers, southeastern Sweden. – Water Air Soil Pollut. 49: 315–328.

Lind, B. and Lundin, L. 1990. Saturated hydraulic conductivites of Scandinavian tills. – Nordic Hydrol. 21: 107–118.

Lindström, G. and Rodhe, A. 1992a. Transit times. – In: Sandén, P. and Warfinge, P. (eds), Modelling groundwater response to acidification. – Swedish Meteorol. and Hydrol. Inst., Norrköping, RH No. 5, pp. 48–58.

– and Rodhe, A. 1992b. Transit times of water in soil lysimeters from modelling of oxygen-18. – Water Air Soil Pollut. 65: 83–100.

Lövblad, G., Amann, M., Andersen, B., Hovmand, M., Joffre, S. and Pedersen, U. 1992. Deposition of sulphur and nitrogen in the Nordic countries: Present and future. – Ambio 21: 339–347.

Lundin, L. 1982. Mark- och grundvatten i moränmark och marktypens betydelse för avrinningen. Soil moisture and groundwater in till soil and the significance of soil type for runoff. – Uppsala Univ., Dept of Physical Geography Report No. 56 (in Swedish with English summary).

Lundström, U. and Öhman, L. O. 1990. Dissolution of feldspars in the presence of natural, organic solutes. – J. Soil Sci. 41: 359–369.

Maxe, L. (ed.) 1995. Effects of acidification on groundwater in Sweden – Hydrological and hydrochemical processes. – Swedish Environ. Protect. Agency, Solna, Report 4388.

Melkerud, P. A. 1983. Quartenary deposits and bedrock outcrops in an area around Gårdsjön, south-west Sweden, with physical, mineralogical geochemical investigation. – Swedish Univ. of Agric. Sci., Uppsala. Reports on forest ecology and forest soils No. 40.

Nilsson J. (ed.) 1986. Critical loads for nitrogen and sulphur. – Nordic Council of Ministers, Copenhagen. Miljørapport 1986: 11.

Nyström, U. 1985. Transit time distributions of water in two small forested catchments. – In: Andersson, F. and Olsson, B. (eds), Lake Gårdsjön. An acid forest lake and its catchment. – Ecol. Bull. (Copenhagen) 37: 97–100.

Odin, H., Eriksson, B. and Perttu, K. 1983. Temperature climate maps for Swedish forestry. – Swedish Univ. of Agric. Sci..

Reports in forest ecology and forest soils No. 45 (in Swedish with English abstract).

Olsson, M. and Melkerud, P-A 1991. Determination of weathering rates based on geochemical properties of the soil. – Geol. Survey of Finland, Special paper 9: 67- 78.

–, Rosén, K. and Melkerud, P.-A. 1993. Regional modelling of base cation losses from Swedish forest soils due to whole-tree harvesting. – In: Hitchon, B. and Fuge, R. (eds), Applied geochemistry. Environmental geochemistry. Selected papers 2nd Int. Symp., Uppsala, Sweden, pp. 189–194.

Overrein, L. N., Seip, H. M. and Tollan, A. 1980. Acid precipitation – effects on forest and fish. – Report of the SNSF-project 1972–1980, Oslo, Norway.

Paces, T. 1986. Weathering rates of granites and depletion of extractable cation in soils under environmental acidification. – J. Geol. Soc. Lond. 143: 673–677.

Rosén, K. 1982. Supply, loss and distribution of nutrients in three coniferous forest watersheds in central Sweden. – Swedish Univ. of Agric. Sci., Dept of Forest soils, Upppsala, Report 1.

–, Gundersen, P., Tegnhammar, L., Johansson, M. and Frogner, T. 1992. Nitrogen enrichment of Nordic forest ecosystems. – Ambio 21: 364–368.

Sandén, P. and Warfvinge, P. (eds) 1992. Modelling groundwater response to acidification. – Swedish Meteorol. and Hydrol. Inst., Norrköping, Reports RH, No. 5.

–, Maxe, L. and Wright, R.F. 1992. Scenario modelling with the MAGIC model. – In: Sandén, P. and Warfvinge, P. (eds), Modelling groundwater response to acidification. Swedish Meteorol. and Hydrol. Inst., Norrköping, RH No. 5, pp. 157–169.

Saxena, R.K. 1987. Oxygen-18 fractionation in nature and estimation of groundwater recharge. – Ph.D. thesis Uppsala Univ., Dept of Physical Geography, Upppsala, Report Series A No. 40.

Schnoor, J. L. and Stumm, W. 1986. The role of chemical weathering in the neutralization of acidic depostion. – Schweiz. Z. Hydrol. 48: 176–195.

Schutzel, H., Kutschke, D. and Wildner, G. 1963. Zur Problematik der Genese der Grauen Gneise des Sächsischen Erzgebirges (Zirkonstatistische Untersuchungen). – Freiberger Forschungsh., C 159, Mineralogie, pp. 1–65.

Statens Forurensningstilsyn 1993. Overvåking av langtransportert forurenset luft og nedbör. – Report 533/93, Oslo (in Norwegian).

Svenska sällskapet för antropologi och geografi 1952–1971. Atlas över Sverige. – Generalstabens litografiska anstalts förlag, Stockholm.

Sverdrup, H. U. 1990. The kinetics of base cation release due to chemical weathering. – Lund Univ. Press, Lund.

– and Warfvinge, P. 1988. Weathering of primary silicate minerals in the natural soil environment in relation to a chemical weathering model. – Water Air Soil Pollut. 38: 387–408.

–, de Vries, W. and Henriksen, A. 1990. Mapping critical loads. A guidence to the criteria, calculations, data collection and mapping of critical loads. – Nordic Council of Ministers, Copenhagen, Nord Report 1990: 98.

Troedsson, T. 1955. Vattnet i skogsmarken. – Kungl. Skogshögsk. Skrifter 19, Stockholm.

Warfvinge, P., Falkengren-Grerup, U., Sverdrup, H. and Andersen, B. 1993. Modelling long-term cation supply in acidified forest stands. – Environ. Poll. 80: 209–221.

Wels, C., Cornett, R.J. and LaZerte, B.D. 1990. Groundwater and wetland contribution to stream acidification: an isotopic analysis. – Water Resour. Res. 26: 2993–3003.

Whipkey, R.Z. 1965. Subsurface flow on forested slopes. – Bull. Int. Ass. Sci. Hydrol. 10: 74–85.

Ecological Bulletins 44: 301–321. Copenhagen 1995

Long-term field experiments simulating increased deposition of sulphur and nitrogen to forest plots

Carl Olof Tamm and Budimir Popovic

Tamm, C. O. and Popovic, B. 1995. Long-term field experiments simulating increased depostion of sulphur and nitrogen to forest plots. – Ecol. Bull. (Copenhagen) 44: 301–321.

In 1968, experiments with application of sulphuric acid were laid out as a supplement to the project "Optimum nutrition experiments in Swedish forest stands". For a long time the trees appeared little affected by the rather crude treatment but several soil chemical and biological properties were changed by the treatments, both in the field and when simulated in laboratory experiments. The experiments were therefore continued and extended, for example by using elementary sulphur as acidifying agent. Independent research groups have found changes in soil biology and vegetation. The lower nitrogen treatments in the optimum nutrition experiments were in the same range as the annual deposition in areas with intensive agriculture in middle Europe. The observed changes in soil chemistry and biology, as well as in tree growth and vegetation, are briefly described. It is concluded that several changes, including a case of short-term increase in tree growth, might be ascribed to shock effects of the sulphuric acid applications. On the other hand, reduced soil respiration, persisting up to 11 yr after the last acid application, is an example of a chronic effect. In many cases elementary sulphur applications had effects similar to those of sulphuric acid, but weaker, and shock effects were less likely here. In all cases nitrogen applications increased tree growth, but the larger amounts (60–90 kg N ha^{-1} yr^{-1}) became supraoptimum with time. Nitrogen compounds (ammonium nitrate and urea) increased pH in the humus layer, but acidified the mineral soil, especially at higher rates of application. As the soil and vegetation effects of nitrogen application operate in the same direction as those observed after natural disturbances of the ecosystem, it is concluded that possible "shock" effects fall within a range naturally occurring in boreal forests. Two results of particular interest: i) factorial combinations of acid and nitrogen compounds were more effective acidifiers of the mineral soil than either agent alone, ii) much of the soil acidification can be explained by the "mobile anion" concept, and that the soil is depleted of, i.a., magnesium and calcium, which are replaced to a large extent by aluminium.

C. O. Tamm and B. Popovic, Dept of Ecology and Environmental Research, Swedish Univ. of Agricultural Sciences, Box 7072 S-751 07 Uppsala, Sweden.

Introduction

In his paper of 1968, Odén showed that contents of both sulphur and nitrogen compounds in precipitation had increased markedly during the period in which the Scandinavian, later European, network for atmospheric chemistry measurements had been in operation (1950–1965). Odén concluded that these changes were likely to have ecological effects and among the hypotheses set up were: i) that forest growth was likely to increase because of increasing supply of nitrogen, known to be a growth-limiting factor in Swedish forest, and ii) that forest growth might be negatively affected by the acidifying effect of the sulphur compounds. It was commonly considered a fact that under otherwise similar conditions soils with higher pH and more exchangeable calcium ions had better tree growth. The two hypotheses contradicted each other, and Odén could not conclusively decide which of them was correct, but the available knowledge appeared to be in favour of the hypothesis of a negative effect of acidification.

Odén's hypotheses had the merit of being testable. A variety of attempts to test them have been made. The first tests were built upon comparisons between data or data trends. Experimental manipulation offers another possibility but requires long periods before conclusive results

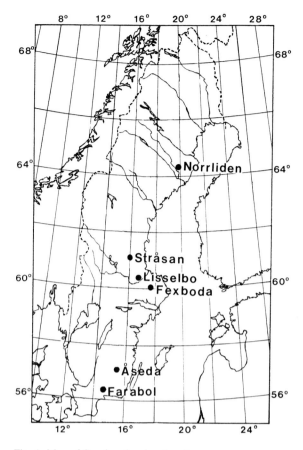

Fig. 1. Map of Sweden showing the sites.

ments were to find out whether moderate increases in acid supply (10–30 times the annual wet deposition in Sweden in the 1960's) had any impact on forest ecosystems, in particular whether they changed soil acidity and tree growth. A negative answer to that question would certainly have influenced the debate.

Plots with acid addition were included in the recently started project "Optimum nutrition experiments in forest stands" (Tamm 1985, Tamm et al. 1995), which focussed on forest nutrition (including possible negative effects on tree growth and environment of chronically high concentrations of nutrients). The first experiment with factorial additions of fertilizers had just begun within this project (E26A Stråsan 1967). Some undesirable effects of high applications (soil acidification, disturbances of tree growth, losses of nitrogen from the site) had been apparent in previous experiments (Tamm and Popovic 1974, Tamm 1980).

Both in our acidification experiments and in the optimum nutrition experiments there has been an interaction between field work and model experiments in the laboratory. Many studies have been made by independent research groups and references will be given to such work.

In this paper we shall concentrate on changes in pH_{H_2O} when describing effects in soils, although this is an incomplete measure of the acidification process. Further data can be found in the references quoted. An extensive report on the optimum nutrition experiments at Norrliden and Lisselbo can be found in Tamm et al. (1995). Readers interested in the results of lysimeter experiments running parallel with the plot experiment E42 Lisselbo 1972–1978 can find them in Farrell et al. (1984) and references given there.

can be obtained in forest ecosystems. The field experiments described here, with addition of sulphur compounds, were laid out as a response to Odén's acidification hypothesis. Long-term field experiments with forest plots receiving repeated additions of nitrogen (ammonium nitrate or urea) had already been initiated and could be used to test the effects of increased deposition of nitrogen compounds.

The immediate objectives of the acidification experi-

Experimental areas and field treatments

The first acidification experiment (E42 Lisselbo) begun in 1969. It was meant as a rough test of the feasibility of using a rather crude method for acidification (dilute but fairly strong sulphuric acid). This first experiment was then followed by others, E57 Norrliden 1971, (sulphuric

Table 1. Geographical and climatic data for the experimental sites. Mean annual temperature and precipitation from Helmisaari and Helmisaari (1992) and Andersson et al. (1994), mean total deposition of sulphur and nitrogen as estimated by Westling et al. (1992) for dense forest (Norway spruce) and open forest (Scots pine). The sites are arranged in order from north to south.

Site and tree species	Lat. N.	Long. E	Altitude, m	Mean ann. temp. °C	Mean prec. mm	Total annual deposition kg ha^{-1}	
						S	N
Norrliden, pine	64°21'	19°45'	260–275	1.2	595	5–7.5	2.5–5
Stråsan, spruce	61°10'	16°00'	360–410	3.1	745	7.5–10	6–8
Lisselbo, pine	60°28'	16°57'	80–85	4.8	593	5–10	2.5–5
Fexboda, spruce	60°08'	17°29'	45	5.4	540	10–12.5	6–8
Åseda, spruce	57°06'	15°29'	225	6.4	596	15–17.5	12–14
Farabol, spruce	56°26'	14°35'	130	6.4	669	17.5–20	14–16

Table 2. Overview of treatments in the experiments. The figures stand for the number of addition levels tested, including zero, and the column heads mark the treatments occurring in a factorial design. Figures within paranthesis mark cases where the treatments occur but not in the factorial design. In experiment E26A Stråsan, the K applications also contained Mg and a mixture of the micronutrients B, Co, Cu, Mn, Mo and Zn.

Expt. No. and year	Site	NPK/0	N levels	PK/0	P levels	K/0	Acid levels	Lime/no lime
			Treatment alternatives					
E55 1971	Norrliden		4	2				
E57 1971	Norrliden	2					4	2
E26A 1967	Stråsan		4		3	2		(2)
E40 1969	Lisselbo		4	2				2
E42 1969	Lisselbo	2					3	2
E67 1980	Fexboda						2	(2)
E63 1974	Åseda		2	2			2	(2)
1976	Farabol		2				3	(2)

acid), E63 Åseda 1974 (elementary sulphur) and E67 Fexboda 1980 (sulphuric acid and elementary sulphur).

An experiment, Farabol, was laid out in 1976 by the Forest Owners' Association in south Sweden, in close co-operation with the Sect. of Forest Ecology of the College of Forestry (now part of the Dept of Ecology and Environmental Research of the Swedish Univ. of Agricultural Sciences, SLU) and the Inst. of Forest Improvement (now the Forestry Research Inst. of Sweden), and has been described by Andersson et al. (1994). Soil acidification experiments have also been laid out on small plots by the Dept of Forest Soils, SLU (Berg 1986c). An experiment with a minicatchment acidified with elementary sulphur has been described by Hultberg et al. (1990).

Our experiments (except E67 Fexboda) were planned with both acidification and fertilization in a factorial mixture. The number of treatments had to be limited, because of costs and because of the difficulty of finding reasonably homogeneous areas of sufficient size. Therefore either nutrient combinations or amounts of acid had to be restricted. However, on most sites acidified plots could be compared with plots fertilized in different ways.

The experimental areas are situated in different parts of Sweden (Fig. 1) and consequently have different climates and deposition rates of nitrogen and sulphur (Table 1).

The experimental areas at Stråsan and Åseda were established in young stands of Norway spruce *Picea abies*, while those at Lisselbo and Norrliden were in young stands of Scots pine *Pinus sylvestris*. Fexboda and Farabol carried spruce stands in the upper middle age.

Stråsan and Norrliden are both sites typical for Swedish boreal forests on glacial till. Lisselbo and Fexboda are both situated on sandy soils, Lisselbo on the very gentle slope of a glaciofluvial esker, with many of the plots on redeposited outwash sand, Fexboda on almost flat outwash sand and fine sand from a nearby esker. At Fexboda the groundwater table was < 2 m down most of the year, while the Lisselbo area was well-drained to excessively drained. Åseda and Farabol were both established on till, but at Åseda the soil depth and the site quality was lower than normal for the region. The soil profiles have been characterized as haplic podzols in the FAO-Unesco system (FAO-Unesco 1989), on all sites where the required analyses are available. The sites not yet analysed for this purpose, E63 Åseda and E67 Fexboda, have also podzolized soils.

Table 3. Total amounts of nitrogen, sulphur and calcium added during the period specified, elements in kg ha⁻¹. For sulphur, only the amount given as sulphuric acid or elementary sulphur is noted; for calcium, only that in the form of carbonate. The sulphuric acid was diluted to about 0.8% by weight, and after application the same amount of water was applied. Nitrogen was given as ammonium nitrate in most experiments, but as urea in three of the six blocks in expt E55 and in expts E63 and Farabol.

Expt.	Nitrogen		Sulphur		Lime	
	kg N ha⁻¹	Period	kg S ha⁻¹	Period	Ca kg ha⁻¹	Year
E55 Norrliden	720–2160	1971–1990	–		–	
E57 Norrliden	1140	1971–1985	98–294 acid	1971–1976	2000	1971
E26A Stråsan	790–2340	1967–1987	–		4000	1967
E40 Lisselbo	580–1740	1969–1988	–		–	
E42 Lisselbo	1040	1969–1985	114–245 acid	1969–1976	2000	1969
E67 Fexboda	–		256 elementary or acid	1980–1983	412 +312 Mg	1980
E63 Åseda	700	1974–1989	300 elementary	1974–1976		
Farabol	600	1976–1985	600–1200 elementary	1976–1987	2160	1976–1987

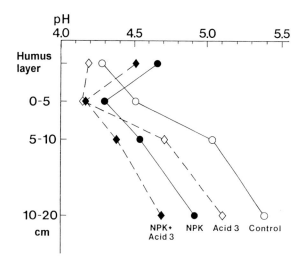

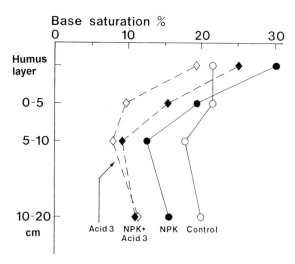

Fig. 2. Variation in pH_{H_2O} and base saturation with soil depth and treatments in expt E57 Norrliden. Sampling in June 1985, when acid 3 plots had received, per hectare, 294 kg H_2SO_4 (1971–1976), NPK plots 1080 kg nitrogen (14 applications 1971–1984), 200 kg phosphorus and 384 kg potassium (in five applications 1971–1983). (From Tamm and Popovic 1989).

The design of the acidification experiments was simple, randomized blocks with one, two or three acidification levels, given with or without fertilizers. Usually the experiments were supplemented by limed plots. The optimum nutrition experiments, in the strict sense, also consisted of randomized blocks with 2–4 replicates, with different nitrogen levels in factorial combination with other nutrients. The combinations in the experiments discussed here are shown in Table 2 and the amounts of elements added in Table 3.

A variety of methods has been used for sampling soil and vegetation (including measuring vegetation biomass and annual sampling of needles), chemical analyses of soils and plants, measuring tree growth, nitrogen mineralization, etc., not to speak of the methods used by independent research groups for studies of soil organisms and their activities. The reader can find descriptions of most of the methods for our own studies of soils and stands in Tamm and Popovic (1989). Where necessary, references will be given to other papers where method descriptions are found, and some details about the calculations of stand growth will be given in the text. Detailed descriptions of the designs and early developments of the experiments at Stråsan, Lisselbo and Norrliden have been given by Tamm et al. (1974a,b) and Holmen et al. (1976).

Results

Chemical soil changes

The first requirement, if studies of experimental soil acidification and its consequences are to be meaningful, is of course that the treatments have acidified the soil, and that this change persists for some period. Such changes are here demonstrated in expt E57 Norrliden (Fig. 2), expt E67 Fexboda (Fig. 3) and expt E63 Åseda (Table 4). In addition it is shown that fertilization also affected soil acidity (Fig. 2 and Table 4). Soil pH increased in the humus layer (O) and decreased in the mineral soil. PK fertilization lowered the mineral soil pH (statistically significant in expt E63 Åseda), but increased base saturation as a consequence of the supply of calcium and potassium with the fertilizer. In expt E57 Norrliden, N and PK were given together, but in expt E55 Norrliden, the two sources of nitrogen (urea and ammonium nitrate) were added in separate blocks, in both cases to plots with and without PK. Again there was a pH increase in the humus layer for both N and PK application and a decrease in all mineral horizons (Fig. 4). In most cases the changes were statistically significant (Tamm et al. 1995), but it is not clear whether the lowest rate of urea (N1) really affected the mineral soil. Figure 3 gives an example of the time trends in pH_{H_2O} in the humus layer and in mineral soil and also shows that sulphuric acid had a greater effect on pH than elementary sulphur. The return to pH values similar to those of the controls was more rapid after elementary sulphur application, being complete in the mineral horizons in 1989, six years after the last application. Sulphuric acid plots still had slightly lower pH values in 1989. We believe that the reason is the high acidity of the acid used but cannot exclude the possibility that the added elementary sulphur may partly follow other pathways than sulphuric acid, e.g. volatilization.

Data are also available from the other experiments, showing a stronger or weaker acidification of the mineral soil after application of acidifiers (expt E42 Lisselbo: Tamm and Popovic 1989, Farabol: Andersson et al. 1994).

304

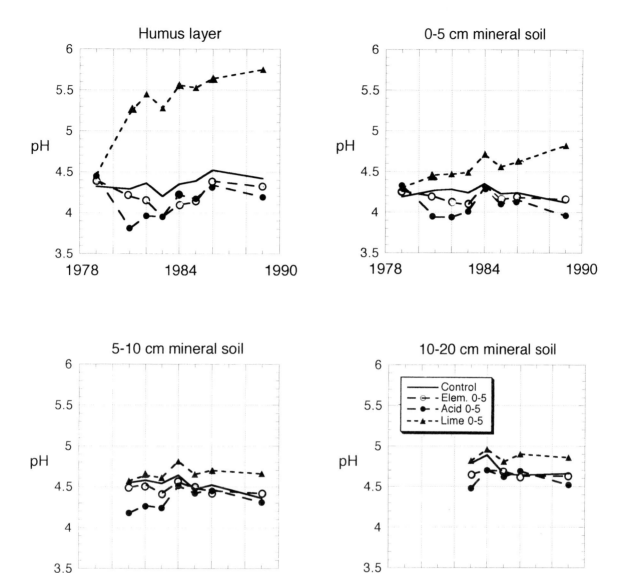

Fig. 3. Changes of pH_{H_2O} with time in different soil horizons in expt E67 Fexboda. First samples in December 1979, all other samples in October. Treatments began in May 1980. See also Table 2. Each point represents arithmetic means for four replicate plots. Legend: solid line without symbols = control; broken line with filled dots = sulphuric acid; broken line with open dots = sulphur powder; broken line with filled triangles = lime.

As the time-scale for recovery from acidification, discussed above for pH data in expt E67 Fexboda, is an important issue, it should be mentioned that some indirect information is available from expts E57 Norrliden and E42 Lisselbo. Annual needle analyses from both experiments for six plant nutrients are available (Tamm and Popovic 1989, Figs 3.12–13 and 4.3–6). The most interesting of the analysed elements is magnesium, of which half the store in the mineral soil had been lost at treatment acid 3 in 1984 at Norrliden (Tamm and Popovic 1989). In both experiments there is a tendency for needle

concentrations of magnesium to fall below control values for much of the period observed, 1976–1986 at Norrliden and 1970–1982 at Lisselbo. At the end of the observation series (1987 at Norrliden, 1984 at Lisselbo) there is very little difference between acidified plots and controls. As differences in growth-rate between acidified plots and controls were small and irregular, after the initial reaction on the treatments, we believe that the foliar concentrations reflect differences in magnesium availability in the soil. However, magnesium concentrations in needles never reached critically low values on acidified plots, in

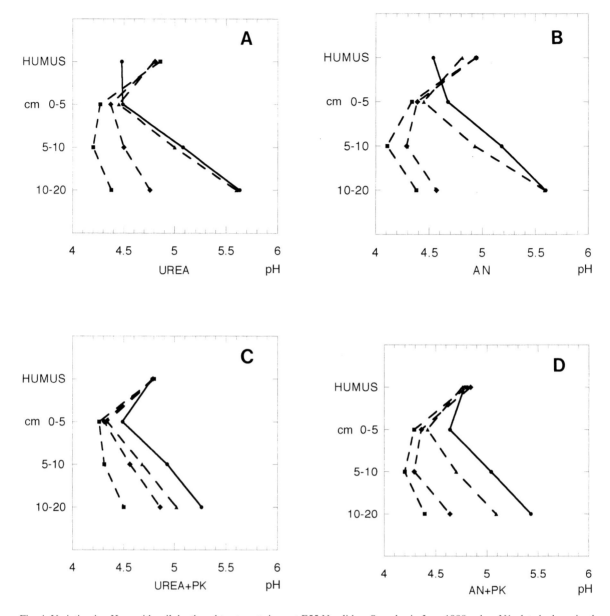

Fig. 4. Variation in pH$_{H_2O}$ with soil depth and treatments in expt E55 Norrliden. Samples in June 1988, when N1 plots had received, per hectare, 630 kg N (17 applications 1971–1987) and PK plots 240 kg P and 462 kg K (in six applications 1971–1986). Source of nitrogen: urea in subdiagrams A and C, ammonium nitrate in B and D. Arithmetic means for three plots (A and B, treatments without NPK) or two plots (C and D, treatments with NPK). Legend: solid lines = no nitrogen; broken lines = nitrogen added, triangles = N1, diamonds = N2, and quadrats = N3. From Tamm et al. (1995).

contrast to some fertilized plots, where the combined effect of magnesium leaching and biological dilution in the better growing trees resulted in very low values.

Lohm (1989) found the E horizon in expt E57 Norrliden to be deeper in acidified plots than in controls, but he did not sample all treatments and our tests in 1988 (Tamm and Popovic 1989, Appendix Table 8.1) have not confirmed his result.

From the work reviewed above, we may conclude the following:

1) The applied amounts of acid or elementary sulphur, at least in the higher dosages used, have had an acidifying effect on the soil profile. The effect has usually been comparatively small, 0.1–0.4 pH units, measured 5–10 yr after the last treatment. Where time series of soil measurements after the applications are available they

306

Table 4. Treatment effects on soil acidity (pH$_{water}$ and base saturation) in expt E63 Åseda, as measured by soil samples in 1978, 1981 and 1986. Data from 1978 and 1981 from Popovic 1984. Base saturation values from 1978 may not be fully comparable with later values because of methodical difficulties, but can be compared between treatments. One, two or three asterisks mark differences significant at the p levels 0.05, 0.01 and 0.001.

Treatment	Soil horizon and year						
	Humus layer (O)			Mineral soil			
	1978	1981	1986	0–5 cm		5–10 cm	10–20 cm
				1981	1986	1986	1986
pH							
0	4.3	4.2	4.3	4.2	4.4	4.7	4.9
S	3.9	3.8	4.2	4.1	4.2	4.6	4.8
PK	4.2	–	4.2	–	4.3	4.6	4.8
PKS	4.0	–	4.3	–	4.2	4.5	4.7
N	4.4	4.2	4.6	4.3	4.3	4.7	4.9
NS	4.0	4.1	4.5	4.1	4.2	4.6	4.8
NPK	4.3	–	4.4	–	4.1	4.3	4.5
NPKS	3.8	–	4.4	–	4.0	4.3	4.4
Difference							
+S/–S			–0.08		–0.12 **	–0.08 *	–0.08 *
+N/–N			+0.24 ***		–0.12 **	–0.17 ***	–0.17 ***
+PK/–PK			–0.06		–0.11 **	–0.21 ***	–0.26 ***
Base saturation V%							
0	33	24	20	17	9	8	7
S	20	16	14	11	6	5	(9)
PK	38	–	26	–	19	16	15
PKS	26	–	27	–	18	14	12
N	38	26	20	21	9	7	6
NS	23	18	16	14	7	8	8
NPK	36	–	29	–	18	13	10
NPKS	22	–	29	–	16	12	10
Difference							
+S/–S			–1.6	–	–2.3	–1.3	–0.9
+N/–N			+2.5	–	–0.5	–0.7	–1.2
+PK/–PK			+10.7 ***	–	+9.6 ***	+6.6 ***	+5.0 ***

show a tendency to recovery in the later measurements. Some pH and magnesium data suggest a recovery period of about a decade for the sites in question. In laboratory studies, and in the only field experiment (E67 Fexboda) in which effects of sulphuric acid and elementary sulphur can be compared, the acid appears to have acted more effectively and more rapidly than the sulphur powder.

2) Annual additions of nitrogen as urea or ammonium nitrate have acidified the mineral soil horizons considerably, except at very low rates (regime N1, corresponding to 20 or 30 kg N ha^{-1} yr^{-1} most of the time). When comparing the effects of S and N it should, however, be kept in mind that in all cases except Farabol the amounts of nitrogen have been much larger than those of sulphur,

whether expressed as weights of elements or as moles of charge.

3) Addition of PK fertilizer may also decrease the pH$_{H_2O}$ of the mineral soil, but normally increases the base saturation.

Changes in biological soil processes

Soil respiration and nitrogen turnover can be studied in model systems in the laboratory or in the field without detailed knowledge of the organisms responsible. This was the first approach to test whether the increasing soil acidity discussed in the previous section had any biolog-

307

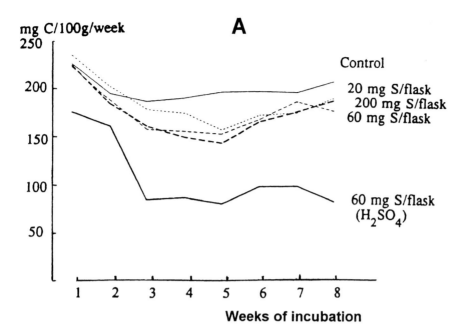

mg C/100g/week

A

Control

20 mg S/flask
200 mg S/flask
60 mg S/flask

60 mg S/flask
(H_2SO_4)

Weeks of incubation

Fig. 5. Carbon dioxide release and mineral nitrogen accumulation in incubation experiments with humus samples from untreated plots from expt E63 Åseda. Samples treated at the beginning of the incubation with the amounts of elementary sulphur or sulphuric acid noted on the curves. All values are means of three replicates. A. Carbon dioxide release in mg C w^{-1} 100 g^{-1} dry weight of sample. B. Accumulated amounts of mineral nitrogen in the same incubation experiment sampled at intervals during the incubation. Values expressed as μg N g^{-1} dry weight of sample. (From Tamm et al. 1977).

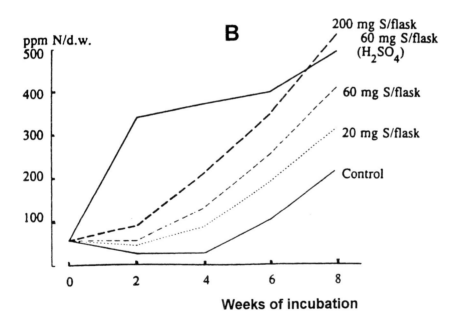

ppm N/d.w.

B

200 mg S/flask
60 mg S/flask
(H_2SO_4)

60 mg S/flask

20 mg S/flask

Control

Weeks of incubation

ical effects. Among the first experiments were incubations of humus samples from untreated plots in the experimental areas but acidified in the laboratory. Figure 5 from Tamm et al. 1977 presents results from one of the early experiments. The decrease in soil respiration upon acidification has been confirmed in a number of studies (see below), but the extent and duration of the effect depends on the method of acidification and the soil mate-

rial. The increase in accumulation of ammonium nitrogen (Fig. 5B) was at the time unexpected, and our tentative explanation was a selective harmful effect on soil microorganisms, affecting other organisms in a more negative way than those responsible for ammonium release.

The next logical step was to test whether material treated in the field behaved in the same way, and whether any effect of the field treatments remained over extended periods of time. The first tests on humus samples taken when the acid treatment was still going on, from expts E57 Norrliden (October 1974) and E42 Lisselbo (May 1976), gave no clear picture. In incubations with material from expt E63 Åseda, taken in 1978 and 1981 after the end of the sulphur applications, there were tendencies to reductions of nitrogen mineralization, but no differences were statistically significant except for a depression of the nitrification (Popovic 1984).

Further tests have been made by Lohm et al. (1984) on Norrliden samples taken in September 1978, two years after the end of the acidification treatment, which showed that soil respiration was still depressed in samples from treatment acid 3 as compared with the control. Total accumulation of mineral nitrogen did not differ between acid 3 and control, but it could be shown by an isotope dilution technique that the rate of microbial nitrogen turnover was markedly decreased in acid 3 samples. Bååth et al. (1984) also measured reduced soil respiration in 1980 in samples from acidified plots. Persson and Wirén (1993) made laboratory tests both with acidification in the incubation flasks and with material from acidified plots in expt E57 Norrliden and E67 Fexboda. They found a long-term decrease of soil respiration in all their tests. In samples acidified at the beginning of the incubation, increased accumulation of mineral nitrogen occurred but only in the initial phase of their long-term incubations.

Another important soil process is litter decomposition, which has been studied by Berg with a litter-bag technique in several of the experiments. The first tests (see Bååth et al. 1980) were made in expt E57 Norrliden October 1976 – October 1978, i.e. starting in the autumn after the last application of acid, and included bags with standard pine needle litter placed in treatments control, acid 1 and acid 3. The tests demonstrated a retarded decomposition on acidified plots, but no or small differences in chemical composition of the decaying litter between treatments.

Berg has made further litter bag tests at acidified and non-acidified plots at Fexboda (Berg 1986a), and also from small plot acidification experiments in different parts of Sweden (Berg 1986c). His general conclusion is that application of acidifying substances retards litter decomposition and that the effect can persist for a considerable period after the end of the treatment. As the needle litter quality is little affected by acid application (unlike by nitrogen fertilizer addition, Berg 1986b), the results indicate changes in microorganism activity, still persisting six months or more after the treatments.

Soil organisms (including roots and mycorrhiza)

Soil organism populations were studied on several occasions in expt E57 Norrliden, but in most cases only in selected treatments and for selected groups of organisms. Some studies have also been made on other sites.

The first study, with samples taken in 1973 and 1975, (Lundkvist 1977) concerned enchytraeid species and indicated that they decreased with an increase in acid application. The species Cognettia sphagnetorum dominated the group, so the differences found could be explained by the reaction of this single species.

A second and more comprehensive study was done in the autumn of 1976 (Bååth et al. 1980). The last acid application had been given in the early summer of that year. There was a further decrease of Cognettia sphagnetorum and of some mite species on acidified plots, but an increase in the Collembola species Tullbergia krausbaueri. As well as litter decomposition being retarded, the abundance and biomass of bacteria and of active fungal hyphae (according to FDA tests) were also negatively affected. Factor analysis revealed functional differences in the bacterial populations between controls and acid 3 plots. It was of course impossible to decide whether the effects observed were attributable to the relatively moderate decreases in pH caused by the treatments or to shock effects from the acid applied. In October 1980, four years later, a new study of the fungal populations and the soil respiration rate was made (Bååth et al. 1984). At that time soil respiration was still depressed on acidified plots, and the length of FDA active hyphae was also lower than on the control. Species composition was also affected by the treatments, the diversity being less on acidified plots, but almost all identified species occurred at all levels of acidity, though in variable frequency.

A later study (Persson et al. 1987) showed that six years after the last application, the Cognettia sphagnetorum population had completely recovered and was in fact larger on acid 3 plots than on controls at Norrliden. Nematode populations were (insignificantly) larger on acid 3 plots. Lumbricids were almost non-existant in acid 3, while the controls had a small population of Dendrobaena octaedra. The same authors also sampled E67 Fexboda (October 1982, the acid treatment still going on) and found consistently lower populations of all the three types of organisms in sulphur acid treated plots than in the controls.

Studies of root development and rhizosphere chemistry have been made in the Farabol experiment (Andersson et al. 1994). Fine root frequency was adversely affected by the sulphur treatment.

Counts of fungal fruiting bodies at Lisselbo and Norrliden have shown decreases in frequencies of most mycorrhizal species with acid addition, even many years after the end of the treatments (Gahne pers. comm.). One species, Lactarius rufus, increased in number and appeared to have a considerable tolerance to acidification as well as to moderate amounts of fertilizers at Lisselbo (see also Wästerlund 1982).

Table 5. Statistical analysis of basal area growth under bark 1972–1984 (BAG), depending on acid applications (ACID, with the levels 0, 1, 2, and 3) and pretreatment growth (BAGpre), separately for plots without NPK fertilizer (A) or with fertilizer (B). From Tamm and Popovic (1989), modified.

Source of variation	Degrees of freedom	Sum of squares	F	p
A.-NPK				
Total variation	11	23.04		
Model	3	14.89	4.66	0.036*
BAGpre	1	6.04	5.68	0.044*
ACID	1	1.38	1.30	0.287
ACID*ACID	1	7.47	7.02	0.029*
Error	8	8.51		
B.+NPK				
Total variation	11	14.70		
Model	2	9.39	7.94	0.010*
BAGpre	1	3.05	5.16	0.049*
ACID	1	6.34	10.73	0.0096**
Error	9	5.32		

Equation A: BAG=5.91+1.9(BAGpre)+2.79 (ACID) −1.03 (ACID)^2, R^2=0.64
Equation B: BAG=15.38+0.88 (BAGpre) −0.81 (ACID), R^2=0.64

Tree growth

Methodological remarks

Measurements of tree growth have been made in all acidification experiments, as well as in the other experiments on the sites indicated in Fig. 1. In all cases standing trees have been measured periodically for height and diameter over bark at breast height (dbh, 1.3 m above ground). The precision of such measurements is not very high for short periods or for plots with a limited number of trees, partly because tree growth during a given period may be more closely related to tree vitality at the beginning of that period than to treatments. Therefore we have tried to increase precision in our estimates of treatment effects by analysis of covariance with some expression for the initial conditions as adjustment variable. Initial volume or basal area growth can often be used for adjustment of periodical increment in volume or in basal area. In several of our experimental stands, however, the treatments began before full canopy closure, in some cases even before all trees had attained breast height. It has been shown by Bjørgung (1968) that, in a young

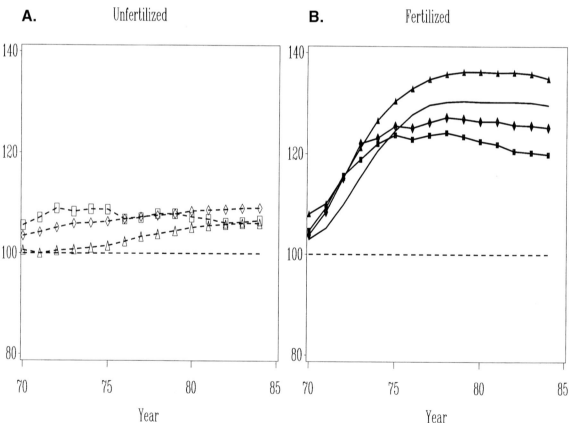

A. Unfertilized   **B.** Fertilized

Year   Year

Fig. 6. Relative basal area under bark from expt E57 Norrliden 1970 (before treatments) to 1984, measured on increment cores. A. Means of three replicates for three levels of acid, acid 1–3, without NPK. B. Means of three replicates with NPK, for each of the levels acid 0–3. (From Tamm and Popovic 1989; a version with some errors in the legend occurs in Tamm 1989). Legend: Broken line: without NPK (without symbols = Control); Open symbols: △ acid 1, ◇ acid 2, □ acid 3; Solid line and filled symbols: with NPK.

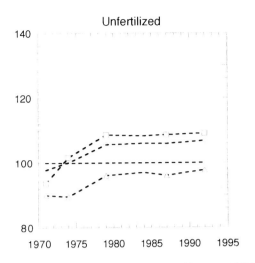

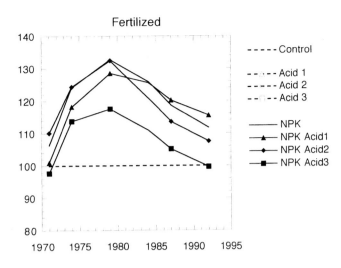

Fig. 7. Relative stem volume 1971–1992 in expt E57 Norrliden, expressed as in Fig. 6. (From Tamm and Popovic 1989, updated).

stand, volume growth during a subsequent period is linearly related to the expression $\sqrt{n} \cdot h^2$, where n stands for the number or trees per plot and h for their arithmetic mean height at the beginning of the experiment. The usefulness of Bjørgung's index (BI) has been well confirmed in our experience, but the correlation between BI and volume or volume growth decreases after canopy closure (Tamm 1985). In experiment E57 Norrliden, 54% of the total between-plot variation in volume growth 1972–1984 of unfertilized plots was accounted for by the BI; the corresponding figure for fertilized plots was 32%, (Tamm and Popovic 1989, Table 3.4). In experiments E57 Norrliden and E42 Lisselbo we have also made independent growth measurements on stem cores from trees cut in thinnings.

Table 6. Average stem growth 1974–1987 (A) and treatment effects (B) on stem growth in expt E63 Åseda. Stem growth is expressed as dm^3 $tree^{-1}$ during the whole period, as means for four replicate plots, and is adjusted by analysis of covariance for differences in tree volume 1973 (prior to treatments). The number of trees measured in each plot is variable (mean 75). Some plots (9 out of the 16 plots receiving nitrogen) had suffered from drought and insect damage and only apparently healthy trees were included here. Treatment and interaction effects are also given as dm^3 $tree^{-1}$ and with symbols for statistical significance: *** = $p<0.001$, ** $p<0.01$, * $p<0.05$, (*) $p<0.1$. Unpublished data kindly supplied by Hallbäcken.

A. Average stem growth

Acidity treatment	Fertilizer treatment			
	None	N	N+PK	PK
None	28.5	45.7	62.7	31.8
S	32.7	49.4	56.9	27.9

B. Treatments and interactions

Treatment main effect	Volume growth differences
	1974–1987
S	−0.45
N	23.5***
PK	5.8(*)
Interactions	
S*N	−0.61
N*PK	6.5*
S*PK	−4.4
S*N*PK	−0.32

Results

Growth results from the acidification experiments have been presented in several publications. Tamm and Wiklander (1980) showed that the acid application gave a positive effect on volume growth in expt E57 Norrliden, if given alone, but a negative effect when given together with NPK. Later volume measurements confirmed the positive effect of acid without NPK, but the negative effect of acid+NPK was not statistically significant when calculated for a longer time period. However, the slope of the curve was still similar to that for the period 1972–1979 (Tamm and Popovic 1989). As shown in Table 5, the basal area measurements showed statistically significant effects of the acid treatment for both fertilized and unfertilized plots. The only difference compared to the volume measurements was that the larger amounts of acid (without NPK) resulted in poorer growth than acid 1, leading to a curvilinear relationship (Table 5).

Curves for relative tree volumes and relative basal areas are shown in Figs 6 and 7, respectively, both adjusted by analysis of covariance for the initial differences as described, and expressed as percentages of controls. It seems clear that the positive effect of acid (without NPK) started early during the period of acid application, and then levelled off. The negative effect of acid to fertilized

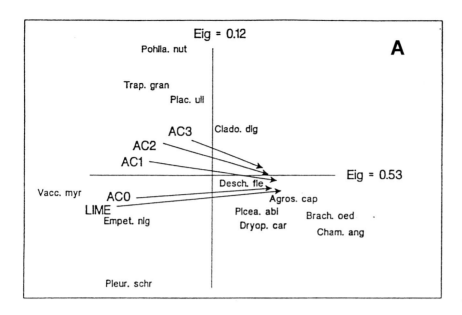

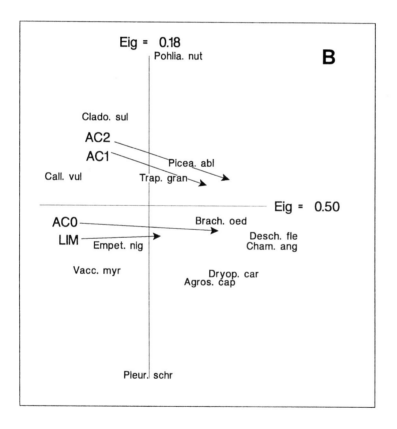

Fig. 8 A and B. Ordination diagrams (distance biplots) of species scores and treatment scores, based on RDA with model Acid*NPK and blocks as covariate. Species scores (for a selection of common species) are located in the centre of the abbreviated name. The treatment scores are given as centroids for treatment scores without NPK (base of arrow), connected to treatment scores with NPK (head of arrow). The eigenvalues (Eig) of the first two axes are shown. Species: Agros. cap = *Agrostis capillaris*, Brach. oed = *Brachythecium oedipodium*[3], Call. vulg = *Calluna vulgaris*, Chame. ang = *Chamerion angustifolium*, Clado. dig = *Cladonia digitata*[1], Clado. sulc = *Cladonia sulcata*, Desch. fle = *Deschampsia flexuosa*, Dryop. car = *Dryopteris carthusiana*, Empet. nig = *Empetrum nigrum*, Picea. abi = *Picea abies*, Placy. uli = *Placynthella uliginosa*[1], Pleur. sch = *Pleurozium schreberi*[2], Pohlia. nut = *Pohlia nutans*[2], Trap. gran = *Trapeliopsis granulosa*[1], Vacc. myr = *Vaccinium myrtillus*. [1]Lichen species usually occurring on decaying wood, [2]moss species usually occurring in poor sites, [3]moss species usually occurring in rich sites. A. Experiment E57 Norrliden, 30 plots, total number of species 105. The biplot explains 87% of the variance in the fitted abundance values. B. Experiment E42 Lisselbo, 20 plots, total number of species 85. The biplot explains 89% of the variance in the fitted abundance values. (Redrawn from van Dobben 1993).

plots started somewhat later (first in the basal area diagram) and then persisted even after 1976, when no more acid was applied. The curves for acid 1, with or without NPK, deviate to some extent from those with larger amounts of acid, which may be explained at least partly by differences in initial values. In any case the effects of acid are small compared with the NPK effect (at most one third), even when they are statistically significant (Table 5, Table 3.5 in Tamm and Popovic 1989). Considering the moderate effects of treatments acid 2 and 3, we find it unlikely that the deviations of the acid 1 curves from control are caused by the treatment and believe that they reflect the natural variation always present in field experiments.

Data on relative stem volume development in expt E42 Lisselbo have been published by Tamm and Popovic (1989). As at Norrliden, any acidification effect is much smaller than the effect of fertilization, and the small size of the experiment makes it impossible to obtain statistical significance except for large differences. We now have access to basal area measurements, which largely confirm the curves in the relative stem volume diagram, with small or no effects of acid 1, and a weak positive tendency for acid 2. The only discrepancy compared with the Norrliden results seems to be a lack of clear negative interaction acid2*NPK at Lisselbo, but considering the small number of plots and the occurrence of a between-plot-variation, the importance of this is doubtful.

From expt E63 Åseda, Tamm (1989) reported a weak positive effect of sulphur application on tree volume in 1983 together with a strong nitrogen effect and significant interactions S*PK (negative) and N*PK (positive). However, while the volume data given in Tamm (1989) are correct, the degrees of freedom used in the statistical tests were incorrectly calculated, and only the nitrogen effect and the interaction N*PK were significant. Table 6 shows a new calculation, this time for stem volume growth 1974–1987, showing that for this variable none of the effects of sulphur or sulphur interactions is statistically significant. The nitrogen effect is still highly significant and the N*PK interaction significant at the 5% level, a result supported by the near-significant figure for the PK effect. The measurements at Åseda may contain a slight bias with respect to the nitrogen effect, as only apparently healthy trees were measured in 1987. Nine of the 16 plots receiving nitrogen had been damaged by drought and insect outbreaks in 1982–1983, while damage was rare on unfertilized plots. There seems to be no reason to suspect similar bias with regard to the sulphur and PK effects.

In conclusion it seems unlikely that the sulphur addition has affected forest growth at Åseda, in spite of the established (but small) decrease in soil pH. The result that nitrogen and PK additions increase spruce growth is well in line with other results from south Sweden, even from areas with a higher atmospheric nitrogen deposition than Åseda (see e.g. Nilsson and Wiklund 1992).

Conventional measurements on standing volumes have been made in expt E63 Fexboda 1980, 1986, 1989 and 1992. All treatments showed negative trends in stem volume growth in comparison with the control plots, particularly the acid plots, but the differences were not large and were not statistically significant. It is hoped that increment cores for more accurate growth measurements will be available later.

From the Farabol experiment the data available are volume and basal area data in condensed form (Andersson et al. 1994). Both types of measurements show positive nitrogen effects (differences from control statistically significant for volume growth), negative but not sig-

nificant differences for sulphur application, and no effect of liming.

Vegetation changes

Detailed studies of ground vegetation were made at Lisselbo, August-September 1987 and at Norrliden, August 1988 (van Dobben et al. 1992, van Dobben 1993).

Figure 8 A and B are ordination diagrams, redrawn from van Dobben (1993). In an ordination diagram ("bi-plot") each species is represented by an axis, and the abundance of a species in a plot can be inferred by projecting the point representing that plot on the axis representing that species. The representation in Fig. 8 A and B deviates from the standard biplot in that a) the species axes have not been drawn (the position of the point in the middle of each species name represents the endpoint of an axis), and b) the plots are represented by centroids (means) for each treatment, and those representing the NPK levels are connected with arrows (base = unfertilized, head = fertilized). In this way, the shift in the plot's position in the space defined by the species gives an impression of the effect of a treatment on the species' abundances. The drawn axes represent the most important directions of variation, their relative importance being represented by their "eigenvalue" (in our scale equal to the fraction of explained variance).

It is apparent from the diagrams that all arrows are pointing in a direction left to right, i.e. more or less parallel to axis 1. Evidently axis 1 expresses much of the fertilizer effects. This axis also accounts for a much larger part of the total variation in species composition than axis 2, as shown by the higher eigenvalue. Axis 2 seems to be related to the acidity factor (or something related to acid application and liming). The tendency to convergence of the arrows, particularly visible in Fig. 8 A, may have to do with the disturbance caused by all treatments and especially by treatment combinations, disfavouring several of the species characteristic of the least disturbed plots, e.g. *Vaccinium myrtillus* and *V. vitis-idea*, *Empetrum nigrum*, *Cladonia rangiferina* and *Cetraria islandica*. The species favoured by fertilization and thus found to the right in the diagram, are fewer, but include *Chamaeneion angustifolium* and *Deschampsia flexuosa*, together with tree seedlings.

It must be remembered that the quantitative vegetation data available represent one point in time only, 11 and 12 yr after the last acid treatment, but still with annual additions of nitrogen (PK given every third year). The acid application, especially in combination with NPK, had strong shock effects on the existing vegetation, and provided space for a later colonization by mainly opportunistic species from outside or from the seedbank in the soil. When the acid applications ended, the recovery of survivors of the pretreatment vegetation began, particularly at lower acid regimes. The effects of the acid application are still visible > 12 yr after the last application, while the effect of liming was much less conspicuous, in spite of the far stronger influence on soil acidity (Tamm and Popovic 1989). It should also be mentioned that the moss and lichen species in acidified plots at the relevées were mostly those that usually grow on decaying wood, e.g. *Trapeliopsis granulosa*, *Cladonia cenotea* and *C. digitata*.

Systematic vegetation studies have been carried out in expt E67 Fexboda, 1985 and in more detailed form 1986 (Svanberg unpubl.). Sulphuric acid plots had lower frequencies of most species, mostly significantly lower than in the control treatment. (The statistical tests were made on 16 species fulfilling certain requirements for frequency and distribution evenness within the same treatment). Bryophytes (five frequent and eight less frequent species) were completely absent from acid plots. There were also differences in species, some of them statistically significant between controls and the two other treatments, sulphur powder and dolomite, but the differences went in both directions and were seldom large.

Vegetation studies have also been made on four occasions at Farabol, 1976 before the treatments, 1977, 1981, and 1991. The results are briefly reported by Andersson et al. (1994). There were effects on the vegetation of all treatments. The sulphur treatment decreased the number of species but favoured *Calluna vulgaris* and *Pohlia nutans*. Spots without vegetation were common on sulphur plots. Nitrogen and lime favoured the development of grasses; in addition, liming increased the number of species, and certain species occurred on limed plots only.

Biomass of ground vegetation

As ground vegetation competes with the trees for nutrients, it would be of interest to know the biomass of this vegetation and its nutrient content. The data available from our acidification plots concern some of the treatments in expt E57 Norrliden, which were sampled in the late summer of 1980 (Tamm et al. 1995). However, these data concern above-ground parts of the field vegetation, and are complete only for weight and nutrient concentrations of current growth of dwarf-shrubs and standing crops of herbs and grasses. Only very rough estimates of vegetation biomass can thus be made, for the most common dwarf-shrub *Vaccinium myrtillus* using data from Flower-Ellis (1971), and for the moss cover, data from Tamm (1953). The calculations made and the assumption made are described by Tamm et al. (1995). The field layer of the control plots in expt E57 Norrliden had in 1980 a biomass production around 1250 kg ha^{-1}, (range 1010–1470), 93% of which was dwarf shrubs. This biomass contained around 15 kg N ha^{-1}. Using rough estimates of the ratio current growth/total above-ground biomass and above/below ground ratio, both taken from Flower-Ellis (1971), we find it is likely that field layer nitrogen store above and below ground would be between 50 and 100 kg N ha^{-1}.

To this should be added the amounts in the bottom layer. The dominating moss species was *Pleurozium schreberi*, but covered only 41% of the control plots in 1988 (van Dobben 1993) with two additional per cent for other mosses and 19% for lichens. According to Tamm (1953) a well developed carpet of the closely related moss *Hylocomium splendens* may sequester up to 10 kg N ha^{-1} yr^{-1} under favourable conditions. At a site not far from Norrliden, Kulbäcksliden experimental forest, the nitrogen uptake by *Hylocomium splendens* was estimated to about half as much (Tamm 1953). Of course a moss and lichen cover contains more than the last year's uptake at a given point in time, perhaps 4–5 times as much. On the other hand, mosses and lichens covered less than two thirds of the ground. We consider it likely the amounts in the field layer estimated earlier, should be increased with about 10 kg N ha^{-1} to include the bottom layer, giving a total of 60–110 kg ha^{-1} on control plots. With acid treatments, the biomass decreased regularly to rather low amounts on acid 3 plots.

Discussion

Soil chemical changes induced by sulphur compounds

It has already been concluded that the experimental addition of sulphur compounds in amounts comparable with those in ambient acid deposition acidified the rooting zone, or at least part of it.

For data on soil chemical changes in our experiments, other than those in pH values, e.g. in magnesium availability and aluminium solubility, the reader is referred to Tamm and Popovic (1989) and other quoted references.

The statement that additions of sulphur compounds acidify soils may appear as a truism not worthy of further comment. However, the degree to which different soil horizons are affected by a given amount of acidifier is of interest, as well as the recovery rate. As the humus layer in northern coniferous forests often has a pH$_{H_2O}$ near 4 (Lundmark 1974), the acidification caused by acid precipitation with similar or only slightly lower pH may not have large biological effects. However, the protons in the unpolluted humus are balanced by often high-molecular and relatively immobile organic acids ("acidoids"), not by sulphate and nitrate. These latter "mobile anions" (Seip 1980) may act as carriers of acidity down to the mineral soil, where acidity is normally lower (up to one pH unit higher in the B horizon, on our sites often at a depth around 10–20 cm). It is known that sulphate ions can be absorbed to iron and aluminium compounds in the accumulation horizon, but this mechanism does not prevent leaching, merely retards it (Singh 1980).

The cation exchange capacity (per unit soil weight) is much lower in the mineral soil than in the humus layer and so is the "base saturation" (Fig. 2). Within the "ion exchange range" (pH 4.2–5.0, Ulrich et al. 1979) buffering capacity against acidification is thus limited. When the acidification process approaches the boundary between the cation exchange range and the aluminium buffer range at pH 4.2, dissolution of aluminium will start. There is also a change in aluminium speciation in the soil solution around pH 4.2, from chelation with dissolved organic substances and polymerized hydroxides to monomeric Al^{+++} (Ulrich 1983).

Several authors have discussed the release of inorganic monomeric aluminium in the mineral soil as a possible cause of root damage, and in any case this process means a radical change from the normal podzolisation process, where aluminium is precipitated rather than dissolved in the B horizon. Judging from the soil data presented here, our acid treatments have seldom depressed (bulk) soil $pH_{H_2O} < 4.5$ in the horizon 10–20 cm, as has happened in horizons above 10 cm at expt E67 Fexboda and, when combined with NPK also at expt E57 Norrliden. The published pH data from expt E42 Lisselbo do not include deeper horizons than 5 cm, but for expt E63 Åseda we may state that all treatments except NPKS had pH > 4.4 at 10–20 cm depth in 1986 (Table 4). According to Andersson et al. (1994) acidified plots at Farabol had pH values 4.2–4.3 at the same depth, even if control plots also had low pH values (4.4) on this site.

However, soil is a very heterogeneous material, often deviating from chemical equilibrium. Persson et al. (1991) have shown, with an example from Fexboda control plots, that rhizosphere pH may be 0.4 units lower than bulk soil pH. Andersson et al. (1994) also found larger amounts of total acidity at the root/soil interface than in the bulk soil at Farabol. Other small-scale spatial variation in soil pH has also been described (Nykvist and Skyllberg 1989). As some dissolution of aluminium may occur even at higher pH values than 4.2, increased and perhaps harmful aluminium concentrations may occur at microsites when bulk soil pH values are c. 4.5, even if this aluminium does not leave the soil profile to the same extent as at lower pH.

The results from the SNSF project in Norway (Stuanes and Abrahamsen 1994) agree reasonably well with our results, if we compare their strongest treatment (irrigation with water adjusted to pH 2 or 2.5). This is equivalent to additions of SO_4-S amounts similar and up to twice those at level S2 at Farabol. Only in very few cases do their experiments show pH values at or slightly below pH 4.4 in the Bs2-horizon (lower accumulation horizon) after the treatment has ended, while there are slightly lower values in the Bs1-horizon above (in our samples the 10–20 cm stratum would normally contain parts of these two horizons). There are also some occurrences of low pH values earlier in the sampling series, before the end of the acid irrigation. Abrahamsen et al. (1994b) conclude, from their lysimeter experiments, that treatments with strong acid (pH = 2) lead to overestimates of aluminium dissolution, compared with "normal" acid rain. This further emphasizes the difficulties in drawing conclusions from what happens during the active treatment phase. It may

still be of interest to study conditions some years later, as well as the rate of recovery.

The very ambitious Höglwald project in Bavaria (Kreutzer and Göttlein 1991) has used irrigation with acidified water (pH 2.7–2.8) during a period of six years, i.e. a "mild" acidification treatment. The total amount of sulphur applied (768 kg S ha^{-1}) is more than in our experiments, except at level S2 at Farabol. Some of the effects on soil chemistry are in the same direction as we have found, but detailed comparisons are difficult, because of large differences in climate and initial conditions (high productivity of spruce in spite of a profile already in the aluminium buffering range down to 50 cm depth).

Soil acidification by nitrogen compounds

Normally the changes in pH after nitrogen fertilization go in the opposite direction in the humus layer and the in mineral soil. The humus layer (O) becomes less acid: the mineral soil is acidified by nitrogen fertilizers, to a lesser extent by PK.

Acidification of mineral soil caused by application of nitrogen or NPK is documented in earlier papers by us and others (Tamm and Popovic 1974, 1989, Hallbäcken and Popovic 1985, Nilsson et al. 1988). Nohrstedt (1988, 1990) maintains that ammonium nitrate does not acidify the soil, but admits that the mineral soil in fertilized plots had somewhat lower pH values in one of his experiments and that in another experiment with three applications of fertilizers, 1967–1981, the upper B horison had lower pH values in samples taken 1987. The differences from controls were not significant in Nohrstedt's experiments, except in a case where heavily fertilized plots had been clear-felled (Nohrstedt et al. 1994) with acidification of lysimeter water as consequence. Berdén (1994) has recently published similar results from expt E26A Stråsan. The lysimeter results are well in line with our earlier results (Tamm and Popovic 1974) and vigorous nitrification seems to explain at least much of the acidification after clearfelling.

Nitrogen additions will also influence the soil acidity by stimulated uptake of "basic cations" (K, Ca, Mg, Na) from the rooting zone. Most of the cations taken up are allocated to the foliage and thus to a large extent are returned to the top of the soil profile with the litterfall and throughfall. The humus layer is thus compensated for the plant uptake of base cations, while the rest of the rooting zone is losing base cations both by plant uptake and by leaching caused by the acid soil solution percolating downwards (Berdén et al. 1987). The result is that illustrated in Fig. 4, viz. a pH increase in the humus layer and a pH decrease in the mineral soil, with the 0–5 cm horizon in an intermediate position.

It is clearly seen in Figs 2 and 4 that the effect of nitrogen fertilizer on mineral soil pH is much stronger than that of acid application, particularly at levels N2 and N3. It should then be remembered that level N3 means far larger numbers of moles of charge, if recalculated to hydrogen ions, than acid 3. Still, level N3 is comparable to amounts deposited in, e.g., forest borders in the most exposed parts of Europe. Figure 4 thus clearly shows that excessive nitrogen deposition is able to bring the accumulation horizon in a typical north Swedish podzol profile very close to the aluminium buffering range. The low pH values in the mineral soil at level N3 are also associated with extremely low calcium concentrations (Tamm et al. 1995), so the Ca/Al ratio in the soil solution would be very low.

Deposited nitrate ions may be taken up by roots and soil organisms in exchange for hydroxyl ions or ions of weak acids, and ammonium ions in exchange for hydrogen ions, both processes affecting soil acidity. Ammonium sulphate deposition is very effective as a soil acidifier, as the ammonium ions are eagerly taken up, while much of the sulphate ions remain in the soil solution and percolate down, together with equivalent amounts of cations, at least some of which are "basic cations" (Ca, K, Mg, Na).

After fertilization with ammonium nitrate, any excess of nitrate ions will also percolate down in the same way as the sulphate ions. Ammonium ions not taken up by biota may to a large extent be absorbed by colloids, but a permanent high concentration of ammonium ions in the soil is likely to lead to nitrification, which is a strongly acidifying process, also providing nitrate ions which increase the leaching of base cations.

In the case of urea addition, this is for most purposes equal to adding ammonia. This means a local increase in pH around the hydrolyzing urea pellets, but as the ammonium ions formed are rapidly absorbed by the acid forest humus, high ionic concentrations in the soil solution are unlikely, except temporarily at microsites. Certain losses by volatilization of ammonia may occur. However, urea is known to stimulate nitrification (Popovic 1977, 1985) and this is a likely explanation of the mineral soil acidification with higher additions of urea in expt E55 Norrliden.

A treatment with ammonium nitrate adds two of the main components of the acid deposition, so the crucial question is whether the difference in dosage between the treatment (once annually) and ambient deposition (distributed over the year) is important. The answer may be yes, particularly with the higher nitrogen regimes (including the first years of N1, when the annual additions were larger). Aber et al. (1989) warn against comparisons of increased deposition with fertilizer experiments, but the experiments they refer to have used far larger applications of nitrogen than our annual additions. In natural ecosystems the concentration of available nitrogen fluctuates within rather wide limits, depending on season and weather conditions and neither nitrate nor ammonium ions are very toxic in the soil. Our conclusion is that even if we may have changed biologically important chemical conditions in space or time (e.g. ionic strength in the soil solution), these changes are not likely to have affected the directions of processes in the soil, although possibly their rates, as compared with addition of nitrogen compounds by atmospheric deposition.

The third type of fertilizer, PK addition, should not affect soil acidity much, according to agricultural experience. Its effect is also limited, compared with that of nitrogen and sulphur compounds. However, PK fertilizer contains both "basic cations" (K, Ca) and mobile anions (sulphate, some chloride), and when applied to an acid forest soil with low base saturation there are evident effects: pH increases in the humus layer and decreases in the mineral soil (see further Tamm et al. 1995).

In the boreal forest, some plants clearly prefer ammonium ions as a source of nitrogen (e.g. ericaceous species, Lee and Stewart 1978) and there are also several reports indicating that many conifer species prefer ammonium nitrogen to nitrate, or at least absorb it faster (e.g. Gosz 1981, Schulze 1989). An excess of nitrogen, either originating directly from deposition or fertilizers, or formed by nitrification, will thus leave the ecosystem as nitrate ions, accompanied by different cations. The exception would be gaseous losses by denitrification, but this process requires a special environment which appears to be uncommon on well-drained forest soils (Nohrstedt et al. 1994).

Biological effects of acidification with sulphuric acid or elementary sulphur

Because of the relatively limited changes in soil acidity properties caused by the acid treatments we should not expect drastic changes in biota, if, for the moment we disregard poisoning effects by the high acid concentrations used. Trees, other vegetation, and soil organisms in a typical northern coniferous forest are with few exceptions acid-tolerant. On the other hand, some changes in a negative direction in production of trees and other primary producers would not be unexpected, as well as species changes in communities of organisms. We have found clear changes in the frequency of several species of both higher plants and soil organisms. For trees, in some cases, we have found a moderate and temporary growth increase upon acidification, in one case statistically significant. There are also cases of a lower tree growth after acidification. At expt E57 Norrliden, the basal area growth (but not the less accurately measured volume growth) eventually reacted negatively to at least the highest level of acid, and the combination of NPK and acid 2 or acid 3 gave lower production of both volume and basal area than NPK alone. At Farabol, the tree growth rates were consistently lower for treatments with sulphur than without, even if the differences were not statistically significant. At Åseda, any first order effect of sulphur was lacking, only the negative S*PK interaction had a magnitude approaching (but not attaining) that for significant effects.

Besides acidifying the mineral soil, nitrogen addition

at the highest rates (with or without PK) has, in all our dosage experiments, reduced tree growth below that at lower nitrogen addition rates rates. However, because of the various ways in which excessive nitrogen affects plant metabolism, it would be premature to ascribe this growth decline to the observed soil acidification.

The question now arises, how much of the observed effects can be ascribed to the "unnatural" treatments and therefore be regarded as artifacts, not produced by "normal" acid deposition. The dilute sulphuric acid is undoubtedly a poison for most organisms, until it becomes still more diluted by subsequent irrigation, by rain, or by reaction with buffering substances in litter or soil. We were at first surprised that the vegetation appeared relatively little affected by the first applications at Lisselbo, but later on the damage became much more marked, although normally not completely lethal with acid alone. Vegetation damage similar to that at Norrliden and Lisselbo has also been observed in SNSF plots irrigated with water adjusted to pH 2.5 (Nygaard 1994). Acidification with elementary sulphur avoids the strong shock caused by the liquid acid, except possibly in microsites. However, selective effects on soil organisms cannot be excluded, judging from the use of sulphur powder in horticulture both as acidifier and for preparation of a traditional spray in orchards.

It is relevant to ask whether the amount of nitrogen released by killing the ground vegetation (estimated as 60–110 kg ha^{-1}) is large enough to explain the observed positive growth effect on the trees in expt E57 Norrliden. The fertilizer effect demonstrated in Figs 6 and 7 was obtained with an addition of 900 kg N ha^{-1}, but the response/dose curve is not linear in our experiments; it rises more steeply at low additions. Moreover, the additions in recent years had not yet had their full effect when the measurements were taken. In expt E55, adjacent to expt E57, and with similar stand and ground vegetation, volume growth in regime N1 (510 kg N ha^{-1} added 1971–1983) had increased by c. 30% for the period 1972–1984 (Tamm et al. 1995). The increase in volume growth for regime acid 3 in expt E57 for the same period can be calculated as 13% from the regression presented by Tamm and Popovic (1989, Table 3.4), although this may be an overestimate, especially as the basal area measurements indicate less growth at the highest acid regime. Because of the range of uncertainty in our estimates of the field layer biomass and store of nitrogen, it is difficult to make firm conclusions, but the amounts of nitrogen released from killed vegetatation are likely to cause a positive growth reaction. The trees may also take advantage of the temporarily reduced competition for nitrogen.

As mentioned earlier, there are indications that our treatments with strong acid had shock effects on other organisms than the ground vegetation. The rapid decrease in soil respiration is perhaps the strongest evidence. The increased release of ammonium nitrogen in incubation experiments has been confirmed by several authors

(Lohm et al. 1984, Persson et al. 1987, 1989, and Persson and Wirén 1993). It has been suggested that this ammonia nitrogen comes from killed soil organisms (Persson et al. 1987, 1989, Persson and Wirén 1993). However, these effects also occur with elementary sulphur, where the pH changes are smaller, and the ammonium release is slower and lower in magnitude (Fig. 5). It is interesting that some of the effects on soil organisms were similar to those found in the more sophisticated Norwegian acidification experiments (Hågvar 1994b, Høiland and Jenssen 1994, Olsen 1994). Thus, it appears entirely possible that the disturbance of microbiota by acidification could explain a part of the tree growth increase.

Many of the results concerning biological effects of experimental acidification reviewed here were obtained during the period of active treatment and are thus not well suited for separating acute and chronic effects. However, Persson and Wirén (1993) reported that carbon mineralization was still retarded at E67 Fexboda four years after the last treatment and at E57 Norrliden 11 yr after the last treatment. At Norrliden, this had led to an increase in the carbon pool. The net mineralization of nitrogen at both cases had been increased in short-term tests by acidification in flasks, but not in incubation of soil samples from acidified plots collected four or eleven years after the last treatments. In material from acid 3 plots at Norrliden, however, nitrogen mineralization was more rapid than expected from the carbon mineralization, but not significantly different from the controls.

Some of the organism studies mentioned earlier also indicate long-term changes. Bååth et al. (1984) found FDA-active fungal biomass to be lower than in controls in all three acid treatments at expt E57 Norrliden four years after the last treatment. There were also some changes in fungal species isolated. Counts of fungal fruiting-bodies at Lisselbo and Norrliden (Gahne pers. comm.) also seem to indicate persistent changes in the fungal community. However, Olsen (1994) concludes from the Norwegian experiments that ambient acid rain is likely to have only minor effects on decomposition rate, in spite of the observed changes in mycoflora and mycorrhiza in more extreme treatments. On the other hand, the "mild" acid treatment at Höglwald (Kreutzer and Göttlein 1991) affected decomposition rate and also nitrogen turnover in the same direction as in our experiments. Hågvar (1994a) found reduced decomposition rate in acidified incubation samples and also in one of the SNSF field experiments.

Changes in species distribution and activity of soil organisms after nitrogen application deserve a discussion of their own, but as it is impossible in our field experiments to separate the effects of acidification by nitrogen application from other effects of a changed supply of the nitrogen, we shall only discuss some aspects related to the process of nitrification. Nitrification is a strongly acidifying process, which explains some of the soil acidification found with larger nitrogen additions in our experiments. On the other hand, there are many indications that mor humus layers in northern coniferous forests have an ability to resist nitrification, until a threshold is crossed (Popovic 1977, 1985). The threshold value for undisturbed sites should be related to the "critical load" proposed by Nilsson and Grennfelt (1988) among others.

Concerning the threshold for nitrification in a fertilized stand, it is interesting that Högberg et al. (1986) found no difference in nitrate availability between regime N1 urea and controls, at expt E55 Norrliden in 1985. Their method was an bioassay of nitrate reductase activity (NRA) in leaves of the grass *Deschampsia flexuosa*. Higher rates of urea application (as well as of ammonium nitrate) resulted in 2–3 times higher NRA values, indicating presence of nitrate in the soil. This can be taken as an indication that regime N1 had not yet attained the "critical load" for nitrogen application in 1985, if this is set at the level where nitrogen availability exceeds the needs of vegetation and microorganisms (Aber et al. 1989). Regimes N2 and N3 had reached or exceeded this threshold.

The vegetation studies (Fig. 8, van Dobben 1993) indicate persistent changes in species composition as well as in abundance. However, most of the changes ascribed to the acid treatment are decreases in frequency of previously established species or increases in certain species with preferences for specific substrates, e.g. lichens growing on decaying wood, or which are otherwise opportunistic (e.g. *Pohlia nutans*).

The fertilizer effect is far stronger than the acid effect, although the species favoured or disfavoured are not the same. A detailed comparison between the treatments is difficult, as the fertilizer treatment was still going on when the observations were made, while the acid additions ended 11–12 yr earlier. We have already concluded that the fertilizer effect of nitrogen application was not quite as "unnatural" as the effect of sulphuric acid. The species favoured or disfavoured by fertilization were the same as those reacting to other disturbances, anthropogenic or natural, which increased nitrogen availability (clear-felling, fire, insect outbreaks, etc.).

In our observations of treatment effects, the best time resolution is of course given by the tree growth observations. The relative volume development (Fig. 6) and the basal area development (Fig. 7) do not always present exactly the same picture, probably because of differences in initial level not accounted for by the adjustment by the analysis of covariance. However, the diagrams are in agreement in that the positive effects of acid 2 and acid 3 occurred during the period of active treatment and then levelled off. Acid 1 deviates from this pattern by showing increases during the last part of the treatment period, or afterwards, but since the effect of this treatment was weak in all other respects, we do not pay too much attention to this deviation. In expt E42 Lisselbo, there was also a slight volume growth increase during the active treatment period at the highest level of acidification (acid 2), while the curve afterwards levelled off, and the acid 1 curve was never very different from the control (Tamm and Popovic 1989).

When acid was given together with NPK and compared with NPK alone, in expt E57 Norrliden (Figs 6 and 7), there was no positive initial effect either in volume or in basal area for the treatments acid 2 and acid 3, only a pronounced negative effect after some time. The acid 1 curves again behaved somewhat differently, but at least in Fig. 7 this seems to be caused by differences in initial levels: acid 1 + NPK and NPK curves run parallel during the entire period.

Conclusions

1) Soil chemical changes were observed after addition of sulphuric acid or sulphur and affected temporarily all the studied horizons, at least at higher levels of acidification. There is no evidence here that the treatments increased mobilization of base cations from less available soil fractions, but field experiments are not the best way to study this. Acidification treatments decreased the soil stores of exchangeable cations, particularly calcium, magnesium and manganese. High acidity increased mobility of aluminium. All these changes were most consistent in the mineral soil. At one site (Farabol) a large amount of acidifier reduced the pH value of the accumulation horizon (bulk soil) to the aluminium buffering range. Values at or below pH 4.5 at the level of 10–20 cm occurred in treatments where sulphuric acid or sulphur was combined with fertilization in expts E57 Norrliden and E63 Åseda.

2) Annual additions of nitrogen as ammonium nitrate or urea (in amounts corresponding to 30, 60, or 90 kg ha^{-1} a^{-1} nitrogen during most of the studied period) also acidified the mineral soil (but not the humus layer). When the source of nitrogen was urea in the smallest amounts used, the acidification was insignificant. Combined addition of N and PK fertilizers gave more acidification than nitrogen alone, also when urea was the source of nitrogen. Addition once per year (for PK once every third year) might possibly lead to larger leaching losses (by uneven percolation), but the effects of our treatments, at least for lower nitrogen applications, should be comparable with those of other natural or anthropogenic impacts causing increased nitrogen availability. Large rates of nitrogen additions, especially when combined with PK (Norrliden, Åseda) or PKS (Åseda) brought the mineral soil pH$_{H_2O}$ close to the aluminium buffering range.

3) Soil biological changes have occurred, as a consequence of treatments with acid or sulphur. Many of these changes may be considered "artifacts" caused by shock effects induced by the strong acid. But decreased organism activity (soil respiration, amounts of active fungal hyphae and fruiting-body formation, and behaviour of nitrogen mineralizers) persisted for several years after treatment. Considering i) that the trees have been remarkably little affected by the acidification treatments and ii) that microorganisms and also soil microfauna normally have good dispersal ability, it seems more likely that the

changed chemical environment is the cause of the observed long-term effects. Persistent changes in ground vegetation were also recorded.

4) The acidification experiments described here can be criticized in several respects: the amounts applied have been too low in comparison with deposition in exposed areas; the treatments with sulphuric acid have been "unnatural"; the follow up with analytical work on deeper soil horizons or analysis of certain elements, e.g. aluminium, might have been more systematic.

On the other hand, like the SNSF experiments (Abrahamsen et al. 1994a), our results have shown that dose/response relations exist between added acidic substances and the effects on soil properties, that soil acidity increases as a result of such additions in spite of the buffering capacity of the soil profile, something not generally accepted when the experiments were designed. Both series of experiments have stimulated a number of studies of biological effects of acidification, providing a better insight into the mechanisms operating. The tree growth studies, first showing an unexpected positive growth effect, may have helped to sober the debate on "forest death" without denying the seriousness of the problem. "Forest death" was not an immediate threat to Scandinavian forests in general in the late 1970's and the early 1980's. The expected negative changes in tree growth after acidification, well documented in some of the Norwegian experiments (Tveite et al. 1994), have not been demonstrated in a statistically convincing way in the Swedish experiments, but the trend in the longer series of observations is negative. Positive tree growth effects of acidification have also been reported from some of the Norwegian experiments, but only for short periods (1–3 yr) at the beginning of the measuring period.

5) An important result concerning tree growth is the clear negative interaction found in expt E57 Norrliden between sulphuric acid addition and NPK fertilization. The combined treatments lead to poorer growth than expected from the results of either treatment alone. This conclusion has a clear bearing on the pollution situation in most industrialized countries, where both sulphur and nitrogen compounds are emitted and deposited.

6) Compared with our additions of sulphur compounds, the experiments with addition of nitrogen alone or in combination with other nutrients have been more satisfactory as real world simulations of increased deposition. There seems to be little reason to suspect that annual additions of 20–40 kg N ha^{-1}, with ammonium nitrate or urea, should have an impact on the forest ecosystem which is basically different from that of corresponding amounts received by atmospheric deposition. However, our additions were larger during the first years of our experiments, and we also used regimes with larger additions. Our application methods might then aggravate some conditions (e.g. nitrate leaching) more than a more even supply, but in our opinion the differences would be quantitative rather than qualitative. Soil changes, such as acidification, dissolution of precipitated aluminium and losses of magnesium and calcium, might happen either somewhat earlier with annual applications than with continuous supply, or later (if much of the added fertilizer is rapidly leached by channeled flow).

7) It is thus the firm belief of the authors that dosage experiments of the type described here constitute a valuable method of studying effects of increasing nitrogen deposition, and to check results obtained by site comparisons, laboratory tests, and simulation models. When dealing with the element nitrogen, which participates in a number of complicated processes in an ecosystem, full-scale field experiments represent "ground truth" in a way difficult to obtain otherwise. Of course this does not mean that all experiments are equally valuable, or that the field results should not be checked using other methods. For example Högberg (1991) and Högberg et al. (1992), using differences in isotope ratio ($^{15}N/^{14}N$), have confirmed our main results about large losses of nitrogen in regimes N2 and N3 in expts E55 and E26A. On the other hand, they have shown with the same method that there is contamination between plots, so the plot size was not sufficient for experiments extended over long periods.

Acknowledgements – Since the late 1960's the experiments described here, belonging to the projects "Optimum nutrition experiments in forest stands" and "Accelerated acidification", have been central activities for what was then the Section for Forest Ecology at the College of Forestry, Stockholm, (later at the Forest Faculty of the Swedish University of Agricultural Sciences, Uppsala). Practically all the scientific and technical staff of the Section has in some way been involved in the projects. It is not possible to thank all persons whithin and outside the section who have contributed to the results reported here, but an exception must be made for two collaborators, whose outstanding efforts are not documented in the list of references: H. Burgtorf, who has been responsible for the field work, and B. Hultin, who in a similar way has guaranteed the quality and continuity of laboratory work from sample preparation to final data.

The generosity of the Forest Owners' Association, of the Forest Research Institute, and of the Section of Systems Ecology, SLU, in connection with our use of Farabol results here, is gratefully acknowledged.

The project "Optimum nutrition experiments in forest stands" has been generously supported by the Swedish Research Council for Forestry and Agriculture. Continuous support has also been received from the Swedish Foundation of Plant Nutrient Research. The acidification experiments have been supported by the Environmental Protection Board from 1977 until 1989. We gratefully acknowledge the funding received and the understanding from the funding agencies that field studies of the kind described here need a longer duration of funding than is normally granted for a standard project.

References

Aber, J. D., Nadelhoffer, K. J., Steudler, P., and Melillo, J. M. 1989. Nitrogen saturation in northern forest ecosystems. – BioScience 39: 378–386.

Abrahamsen, G., Stuanes, A. O. and Tveite, B. (eds) 1994a. Long-term experiments with acid rain in Norwegian forest ecosystems. – Ecol. Stud. 104.

– , Stuanes, A. O. and Sogn, T. A. 1994b. Monolith lysimeters. – In: Abrahamsen, G., Stuanes, A. O. and Tveite, B. (eds),

Long-term experiments with acid rain in Norwegian forest ecosystems. – Ecol. Stud. 104: 239–286.

Andersson, F., Bergholm, J., Hallbäcken, L., Möller, G., Pettersson, F. and Popovic, B. 1994. Farabolförsöket – effekter av svavel, kväve och kalk i en sydöstsvensk granskog. – Swedish Univ. Agricult. Sci. Uppsala, Dept Ecol. and Environ. Res., Report No. 70, (in Swedish).

Bååth, E., Berg, B., Lohm, U., Lundgren, B., Lundkvist, H., Rosswall, T., Söderström, B. and Wirén, A. 1980. Effects of experimental acidification and liming on soil organisms and decomposition in a Scots pine forest. – Pedobiologia 20: 85–100.

– , Lundgren, B. and Söderström, B. 1984. Fungal populations in podzolic soil experimentally acidified to simulate acid rain. – Microb. Ecol. 10: 197–203.

Berdén, M. 1994. Ion leaching and soil acidification in a forest Haplic podzol: effects of nitrogen application and clear-cutting. – Swedish Univ. Agricult. Sci., Uppsala, Dept Ecol. and Environ. Res., Report No. 73.

– , Nilsson, S. I., Rosén, K. and Tyler, G. 1987. Soil acidification – extent, causes and consequences. – Swedish Environ. Protect. Board, Solna, Report 3292.

Berg, B. 1986a. The influence of experimental acidification on needle litter decomposition in a Picea abies L. forest. – Scand. J. For. Res. 1: 317–322.

– 1986b. Nutrient release from litter and humus in coniferous forest soils – a minireview. – Scand. J. For. Res. 1: 359–369.

– 1986c. The influence of experimental acidification on nutrient release and decomposition rates of needle and root litter in the forest floor. – For. Ecol. Manage. 15: 195–213.

Bjørgung, E. 1968. Om kovariansanalyse av volum i gjødslingsforsøk. – Norsk Skogbruk 9: 229–231, (in Norwegian).

FAO-Unesco 1989. Soil map of the world, revised legend. – Techn. Paper No. 20, ISRIC, Wageningen.

Farrell, E. P., Wiklander, G., Nilsson, S. I. and Tamm, C. O. 1984. Distribution of nitrogen in lysimeters previously treated with sulphuric acid and a combination of acid and fertiliser. – For. Ecol. Manage. 8: 265–279.

Flower-Ellis, J. G. K. 1971. Age structure and dynamics in stands of bilberry (Vaccinium myrtillus L.). – Dept Forest Ecol. Forest Soils, Royal Coll. Forestry, Stockholm, Research Note No. 9.

Gosz, J. R. 1981. Nitrogen cycling in coniferous ecosystems. – In: Clark, F. E. and Rosswall, T. (eds), Terrestrial nitrogen cycles. – Ecol. Bull. (Stockholm) 33: 405–426.

Hågvar, S. 1994a. Soil Biology: Decomposition and soil acidity. – In: Abrahamsen, G., Stuanes, A. O. and Tveite, B. (eds), Long-term experiments with acid rain in Norwegian forest ecosystems. – Ecol. Stud. 104: 136–139.

– 1994b. Soil biology: Soil animals and soil acidity. – In: Abrahamsen, G., Stuanes, A. O. and Tveite, B. (eds), Long-term experiments with acid rain in Norwegian forest ecosystems. – Ecol. Stud. 104: 101–121.

Hallbäcken, L. and Popovic, B. 1985. Effects of forest liming on soil chemistry. Investigation of Swedish liming experiments. – Swedish Environ. Protect. Board, Solna, Report 1880.

Helmisaari, H. and Helmisaari, H.-S. 1992. Long-term forest fertilization experiments in Finland and Sweden – their use for vitality and nutrient balance studies. – Swedish Environ. Protect. Agency, Solna, Report 4099.

Högberg, P. 1991. Development of ^{15}N enrichment in a nitrogen-fertilized forest soil-plant system. – Soil Biol. Biochem. 23: 335–338.

– , Granström, A., Johansson, T., Lundmark-Thelin, A. and Näsholm, T. 1986. Plant nitrate reductase activity as an indicator of availability of nitrate in forest soils. – Can. J. For. Res. 16: 1165–1169.

– , Tamm, C. O. and Högberg, M. 1992. Variations in ^{15}N abundance in a forest fertilization trial: Critical loads of N, N saturation, contamination and effects of revitalization. – Plant Soil 142: 211–219.

Høiland, K. and Jenssen, H. B. 1994. Ground vegetation: Myco-

flora. – In: Abrahamsen, G., Stuanes, A. O. and Tveite, B. (eds), Long-term experiments with acid rain in Norwegian forest ecosystems. – Ecol. Stud. 104: 230–238.

Holmen, H., Nilsson, Å., Popovic, B. and Wiklander, G. 1976. The optimum nutrition experiment Norrliden. A brief description of an experiment in a young stand of Scots pine (Pinus silvestris L.). – Dept Forest Ecol. and Forest Soils, Royal College of Forestry, Stockholm, Research Note 26.

Hultberg, H., Lee, Y. H., Nyström, U. and Nilsson, S. I. 1990. Chemical effects on surface-, ground- and soil-water of adding acid and neutral sulphate to catchments in south-west Sweden. – In: Mason, B. J. (ed.), Surface waters acidification programme. Cambridge Univ. Press, Cambridge, pp. 167–182.

Kreutzer, K. and Göttlein, A. (eds) 1991. Ökosystemforschung Höglwald. Beiträge zur Auswirkung von saurer Beregnung und Kalkung in einem Fichtenaltbestand. – P. Parey, Hamburg.

Lee, J. A. and Stewart, G. R. 1978. Ecological aspects of nitrogen assimilation. – Adv. Bot. Res. 6: 1–43.

Lohm, U. 1989. Increased weathering rate in experimental field acidification. – In: Barto-Kyriakidis, A. (ed.), Weathering; its products and deposits. Vol. I. Processes. Theophrastus Publ., Athens, Greece.

– , Larsson, K. and Nömmik, H. 1984. Acidification and liming of coniferous forest soil: long-term effects on turn-over rates of carbon and nitrogen during an incubation experiment. – Soil Biol. Biochem. 16: 343–364.

Lundkvist, H. 1977. Effects of artificial acidification on the abundance of Enchytraeidae in a Scots pine forest in northern Sweden. – In: Lohm, U. and Persson, T. (eds), Soil organisms as components of ecosystems. Ecol. Bull. (Stockholm) 25: 570–573.

Lundmark, J.-E. 1974. Use of site properties for assessing site index in stands of Scots pine and Norway spruce. – Dept of Forest Ecol. and Forest Soils, Royal College of Forestry, Stockholm, Research Note No. 16.

Nilsson, J. and Grennfelt, P. (eds) 1988. Critical loads for sulphur and nitrogen. – Miljörapport 1988, No. 15, Nordic Council of Ministers, Copenhagen.

Nilsson, L. O. and Wiklund, K. 1992. Influence of nutrient and water stress on Norway spruce production in south Sweden – the role of air pollutants. – Plant Soil 147: 251–265.

Nilsson, S. I., Berdén, M. and Popovic, B. 1988. Experimental work related to nitrogen deposition, nitrification and soil acidification – a case study. – Environ. Pollut. 54: 233–248.

Nohrstedt, H.-Ö. 1988. Studies of soil chemistry in two fertilization experiments. – Inst. Forest Improvement, Report No. 2, Uppsala.

– 1990. Effects of repeated nitrogen fertilization with different doses on soil properties in a Pinus sylvestris stand. – Scand. J. For. Res. 5: 3–15.

– , Ring, E., Klemedtsson, L. and Nilsson, Å. 1994. Nitrogen losses and soil water acidity after clear-felling of fertilized experimental plots in a Pinus sylvestris stand. – For. Ecol. Manage. 66: 69–86.

Nygaard, P. H. 1994. Ground vegetation: The B-2 experiment. – In: Abrahamsen, G., Stuanes, A. O. and Tveite, B. (eds), Long-term experiments with acid rain in Norwegian forest ecosystems. – Ecol. Stud. 104: 221–229.

Nykvist, N. and Skyllberg, U. 1989. The spatial variation of pH in the mor layer of some coniferous forest stands in northern Sweden. – Scand. J. For. Res. 4: 3–11.

Odén S. 1968. The acidification of air and precipitation and its consequences for the natural environment. – Bull. Ecol. Comm. Stockholm 1: 1–86, (in Swedish, English translation, published by Translation Consultants Ltd. Arlington, VA).

Olsen, R. A. 1994. Soil biology: Soil microflora and soil acidity. – In: Abrahamsen, G., Stuanes, A. O. and Tveite, B. (eds), Long-term experiments with acid rain in Norwegian forest ecosystems. – Ecol. Stud. 104: 122–135.

Persson, H., Ahlström, K., Clemensson-Lindell, A. and Majdi,

H. 1991. Experimental approaches to the study of the effects of air pollution on trees – root studies. – In: Persson, H. (ed.), Above and below-ground interactions in forest trees in acidified soils. CEC Air pollution research report 32: 8–16.

Persson, T. and Wirén, A. 1993. Effects of experimental acidification on C and N mineralization in forest soils. – Agricult. Ecosyst. Environ. 47: 159–174.

– , Hyvönen, R. and Lundkvist, H. 1987. Influence of acidification and liming on nematodes and oligochaetes in two coniferous forests. – In: Striganova, B. R. (ed.), Soil fauna and soil fertility. Proc. IX Int. Coll. Soil Zool. Moscow 'Nauka' pp. 191.

– , Lundkvist, H., Wirén, A., Hyvönen, R. and Wessén, B. 1989. Effects of acidification and liming on carbon and nitrogen mineralization and soil organisms in mor humus. – Water Air Soil Pollut. 45: 77–96.

Popovic, B. 1977. Effect of ammonium nitrate and urea fertilizers on nitrogen mineralization, especially nitrification, in a forest soil. – Dept Forest Ecol. and Forest Soils, Royal College of Forestry, Stockholm, Research Notes 30.

– 1984. Mineralization of carbon and nitrogen in humus from field acidification studies. – For. Ecol. Manage. 8: 81–93.

– 1985. The effect of nitrogenous fertilizers on the nitrification of forest soils. – Fert. Res. 6: 139–147.

Schulze, E.-D. 1989. Air pollution and forest decline in a spruce (Picea abies) forest. – Science 244: 774–783.

Seip, H. M. 1980. Acidification of freshwater – sources and mechanisms. – In: Drabløs, D. and Tollan, A. (eds), Ecological impact of acid precipitation. Oslo-Ås, pp. 358–366.

Singh, B. R. 1980. Effects of acid precipitation on soil and forest. 3. Sulfate sorption by acid forest soils. – In: Drabløs, D. and Tollan, A. (eds), Ecological impact of acid precipitation. Oslo-Ås, pp. 194–195.

Stuanes, A. O. and Abrahamsen, G. 1994. Soil chemistry. – In: Abrahamsen, G., Stuanes, A. O. and Tveite, B. (eds), Long-term experiments with acid rain in Norwegian forest ecosystems. – Ecol. Stud. 104: 37–100.

Tamm, C. O. 1953. Growth, yield and nutrition in carpets of a forest moss (Hylocomium splendens). – Medd. Stat. Skogsforskn. Inst. Stockholm, 43(1).

– 1980. Responses of spruce forest ecosystems to controlled changes in nutrient regime, maintained over periods up to 13 years. – In: Klimo, E. (ed.), Stability of spruce forest ecosystems. Int. Symp., Brno, pp. 423–433.

– 1985. The Swedish optimum nutrition experiments in forest stands – aims, methods, yield results. – K. Skogs- Lantbr. Akad. Tidskr. Suppl. 17: 9–29.

– 1989 Comparative and experimental approaches to the study of acid deposition effects on soils as substrate for forest growth. – Ambio 18: 184–191.

– and Popovic, B. 1974. Intensive fertilization with nitrogen as

a stressing factor in a spruce ecosystem. I. Soil effects. – Stud. For. Suec. 121: 1–32.

– and Wiklander, G. 1980. Effects of artificial acidification with sulphuric acid on tree growth in Scots pine forest. – In: Drabløs, D. and Tollan, A. (eds), Ecological impact of acid precipitation. Oslo-Ås, pp. 188–189.

– and Popovic, B. 1989. Acidification experiments in pine forests. – Swedish Environ. Protect. Board, Solna, Report 3589.

– , Aronsson, A. and Burgtorf, H. 1974a. The optimum nutrition experiment Stråsan. A brief description of an experiment in a young stand of Norway spruce (Picea abies Karst.). – Dept of Forest Ecology and Forest Soils, Royal College of Forestry, Stockholm, Research Note 17.

– , Nilsson, Å. and Wiklander, G. 1974b. The optimum nutrition experiment Lisselbo. A brief description of an experiment in a young stand of Scots pine (Pinus silvestris L.). – Dept of Forest Ecology and Forest Soils, Royal College of Forestry, Stockholm, Research Note 18.

– , Wiklander, G. and Popovic, B. 1977. Effects of application of sulphuric acid to poor pine forests. – Water Air Soil Pollut. 8: 75–87.

– , Aronsson, A. and Popovic, B. 1995. Nutrient optimisation and nitrogen saturation experiments in pine forest. – Stud. For. Suec.

Tveite, B., Abrahamsen, G. and Huse, M. 1994. Trees: Growth. – In: Abrahamsen, G., Stuanes, A. O. and Tveite, B. (eds), Long-term experiments with acid rain in Norwegian forest ecosystems. Ecol. Stud. 104: 180–203.

Ulrich, B. 1983. An ecosystem oriented hypothesis on the effect of air pollution on forest ecosystems. – Swedish Environ. Protect. Board, Solna, Report 1636, pp. 221–231.

– , Mayer, R. and Khanna, P. K. 1979. Deposition von Luftverunreinigungen und ihre Auswirkungen in Waldökosystemen im Solling. – Schriften Forstl. Fak. Univ. Göttingen Bd 58.

van Dobben, H. F. 1993. Vegetation as a monitor for deposition of nitrogen and acidity. – Ph. D. thesis Univ. Utrecht.

– , Dirkse, G. M., ter Braak, C. J. F. and Tamm, C. O. 1992. Effects of acidification, liming and fertilization on the undergrowth of a pine forest stand in central Sweden. – DLO Inst. Forestry and Nature Research, RIN report 92/21, Wageningen.

Wästerlund, I. 1982. Do pine mycorrhizae disappear following fertilization? – Svensk Bot. Tidskr. 76: 411–417, (in Swedish with English Summary).

Westling, O., Hallgren Larsson, E., Sjöblad, K. and Lövblad, G. 1992. Deposition and effects of air pollutants in south and middle Sweden. – Swedish Environ. Res. Inst. IVL Report B 1079, Aneboda/Stockholm, (in Swedish).

Ecological Bulletins 44: 322–334. Copenhagen 1995

NITREX project: ecosystem response to chronic additions of nitrogen to a spruce-forested catchment at Gårdsjön, Sweden

Richard F. Wright, Tor-Erik Brandrud, Anna Clemensson-Lindell, Hans Hultberg, O. Janne Kjønaas, Filip Moldan, Hans Persson and Arne O. Stuanes

Wright, R. F., Brandrud, T.-E., Clemensson-Lindell, A., Hultberg, H., Kjønaas, O. J., Moldan, F., Persson, H. and Stuanes, A. O. 1995. NITREX project: ecosystem response to chronic additions of nitrogen to a spruce-forested catchment at Gårdsjön, Sweden. – Ecol. Bull. (Copenhagen) 44: 322–334.

We are conducting a whole-catchment experimental addition of nitrogen at Gårdsjön, SW Sweden, to investigate the risk and consequences of nitrogen saturation in coniferous forests typical of southern Scandinavia. The Gårdsjön experiment is part of the European NITREX project (Nitrogen Saturation Experiments). Beginning April 1991 we add in weekly portions c. 35 kg NH_4NO_3-N ha^{-1} yr^{-1} in 5% extra water to the ambient 12 kg N ha^{-1} yr^{-1} in throughfall. Elevated concentrations of nitrate in runoff occurred in April 1991 during the first two weeks of treatment, but then were very low during the growing season. The signal of nitrate leakage appeared again in mid-November 1991 and continued during the winter and throughout the second and third years of treatment. The inorganic nitrogen lost during the first, second and third years was c. 0.6%, 1.1% and 5.2%, respectively, of the total inorganic nitrogen input. The volume-weighted average nitrate concentrations in the soil solution increased from the pre-treatment year through the treatment years. There was no treatment effect on ammonium or organic N concentrations. Mineralization measurements from the first 2 yr of treatment, 1991 and 1992, show no significant differences between the 3 catchments and no systematic change from 1991 to 1992. There are indications of increased nitrification of ammonium to nitrate and reduced immobilization of nitrate in the soil. There was a tendency towards decreased fine-root biomass in the FH layer in the moss-dominated plot at G2 NITREX catchment. In contrast, at G1 ROOF catchment there was an increase of fine-root biomass. Nitrogen-sensitive species of mycorrhiza became less abundant with lower fruit body production. Together the results from G2 NITREX catchment indicate that the three years of nitrogen addition have resulted in changes across the forest ecosystem boundary as well as changes within the ecosystem. Most apparent and rapid is the increase in nitrate concentrations in runoff, whereas the changes in components and processes within the ecosystem are apparently slower and more difficult to detect. As yet there are insufficient data available to assess the effects of nitrogen addition (and nitrogen removal at G1 ROOF) on the tree growth and vitality. Over the long term, the nitrogen addition can be expected to affect the productivity of trees and other vegetation, alter the relative abundance of lower vegetation, and change species composition. The nitrogen input-output data indicate that the forested catchment ecosystem G2 NITREX is proceeding rapidly through several stages of "nitrogen saturation". Both the frequency and magnitude of nitrate peaks in runoff increased during the non-growing season in the first year (stage 1) and throughout the second year (stage 2). The results from NITREX Gårdsjön demonstrate that nitrogen saturation can be induced over a relatively short time by increasing atmospheric deposition of nitrogen. The rate of response suggests that at ambient nitrogen deposition of 12 kg N ha^{-1} yr^{-1} the ecosystem is near the threshold at which additional N inputs cause significant nitrate leaching.

R. F. Wright and T.-E. Brandrud, Norwegian Inst. for Water Research (NIVA), Box 173, Kjelsås, N-0411 Oslo, Norway. – A. Clemensson-Lindell and H. Persson, Dept of Ecology and Environmental Res., Swedish Univ. of Agricultural Sciences, Box 7072, S-750 07 Uppsala, Sweden. – H. Hultberg and F. Moldan, Swedish Environmental Res. Inst. (IVL), Box 47086, S-402 58 Göteborg, Sweden. – O. J. Kjønaas, Norwegian Forest Res. Inst. (NISK), Høgskoleveien 12, N-1432 Ås, Norway. – A. O. Stuanes, Dept of Soil and Water Sciences, Agricultural Univ. of Norway, Box 5028, N-1432 Ås, Norway.

Introduction

Nitrogen is one of the most important macro-nutrients in terrestrial ecosystems. Low nitrogen supply often limits tree growth. High supply stimulates growth and may lead to deficiency of other nutrients resulting in ecosystem malfunction such as decreased frost hardiness, disturbed mycorrhiza development and function, reduced root growth and increased forest damage by animals, fungi, bacteria and viruses (Nihlgård 1985).

In regions where the natural cycle of nitrogen is relatively little affected by human activities, the primary source of available nitrogen is biological fixation. In large areas of Europe and North America, however, deposition of nitrogen from the atmosphere comprises a major supplementary source.

In nitrogen-limited forest ecosystems typical of Scandinavia and northern North America, the internal nitrogen cycle is very efficient. Little or no nitrate is lost to the water below the rooting zone. In areas where nitrogen deposition exceeds c. 10–15 kg N ha^{-1}, however, forested catchments show increased leaching of nitrate (Grennfelt and Hultberg 1986, Hauhs et al. 1989, Dise and Wright 1995). Increasing concentration of nitrate in surface waters is commonly used as a criterion for "nitrogen saturation" (Aber et. al. 1989). Nitrogen saturation has potentially serious impacts on soil chemistry, ground water and surface water quality, forest health and emissions of "greenhouse gasses".

To investigate the risk and consequences of nitrogen saturation in coniferous forests typical of southern Scandinavia, we are conducting a large-scale experimental addition of nitrogen to precipitation at a whole forested headwater catchment, located near Lake Gårdsjön, SW Sweden (Fig. 1). The Gårdsjön G2 catchment manipulation is part of the European NITREX project (Nitrogen Saturation Experiments) (Dise and Wright 1992). We measure the fluxes of water and chemicals across the ecosystem boundary to obtain a measure of the whole-ecosystem response to the nitrogen addition. Further we measure the responses in key components and processes within the ecosystem to identify mechanisms for whole-ecosystem changes as well as to evaluate the risk of deleterious effects on ecosystem components. Here we summarise results from input-output measurements (Moldan et al. 1995), soil solution (Stuanes et al. 1995), soil microbial processes (Kjønaas unpubl.), fine-roots (Clemensson-Lindell and Persson 1995) and mycorrhiza studies (Brandrud 1995) for the first 2–3 yr of treatment.

Site description

The experimental site at Gårdsjön is located at 135–145 m elevation c. 10 km from the Swedish west coast (58°04'N, 12°01'E), 50 km north of Gothenburg. The Gårdsjön region has a humid climate, with 1100 mm mean annual precipitation, 586 mm mean annual runoff, and mean annual temperature of 6.4°C. The area is characterised by an acid lake whose terrestrial catchment is dominated by shallow podzolic soils with inclusions of barren rock and peat soils. The bedrock consists mainly of granites and granodiorites (Olsson et al. 1985).

The G2 NITREX catchment has an area of 0.52 ha. The forest is mainly a naturally-regenerated mixture of 66–90 yr old Norway spruce (*Picea abies* L. Karst) and some Scots pine (*Pinus sylvestris* L.). Ground vegetation is dominated by grasses and mosses (*Dicranum majus* Sm., *Leucobryum glaucum* (Hedw.) Fries, *Deschampsia flexuosa* (L.) Tin) in the upper catchment, *Vaccinium myrtillus* L./*Vaccinium vitis-idea* L. in drier outcrops, *Sphagnum* (predominantly *Sphagnum girgensohnii* Russ.) in the wetter lower parts, and *Calluna vulgaris* (L.) Hill among the most exposed ridges (Stuanes et al. 1992).

Soils are predominantly acidic silty and sandy loams, drier in the upper catchment and more peaty in the lower parts. They are classified as orthic humic podzols, orthic

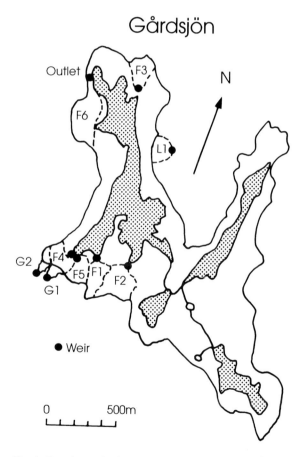

Fig. 1. Experimental sub-catchments within the Lake Gårdsjön basin used for large-scale manipulations. F1 is control (since 1979). The G2 NITREX catchment and G1 ROOF catchment lie just outside Gårdsjön's drainage basin. Grennfelt and Hultberg (1986) give details on manipulations in remaining sub-catchments.

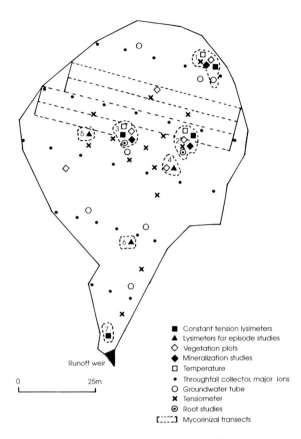

Experimental design

We simulate increased atmospheric deposition of nitrogen by weekly experimental additions of ammonium nitrate solution at a rate of c. 35 kg N ha^{-1} yr^{-1} to the G2 NITREX catchment. NH$_4$NO$_3$ is dissolved in de-ionised water and applied by means of 270 ground-level sprinklers installed in a 5×5 m grid over the whole catchment. Additions are done weekly in amounts proportional to the volume of ambient throughfall which occurred the previous week. The additional water comprises c. 5% of total throughfall volume. The average ammonium nitrate addition is in 1.4 mm of water and added during 30 minutes. This amount and intensity was chosen such that discharge is not affected on an event or an annual basis.

Methods
Input-output budgets

Atmospheric inputs are measured by two bulk precipitation collectors and 28 and 15 throughfall collectors in G2 NITREX and F1 CONTROL, respectively. Samples are bulked to one sample per catchment. During the growing season (April – September) pairs of light-proof throughfall sampling vessels are used with iodine added to one of each pair to inhibit microbial activity. There was no significant NH$_4^+$ or NO$_3^-$ concentration difference between preserved and unpreserved samples; concentrations of organic nitrogen in unpreserved samples, however, were consistently lower by c. 45% relative to preserved samples. The preserved samples were used for organic N input calculations. Samples for chemical analysis are collected biweekly or monthly.

Runoff from the G2 NITREX, G1 ROOF and F1 CONTROL catchments is gauged continuously. Water samples for chemical analysis are collected weekly or biweekly as flow-proportional samples (grab samples prior to November 1991). Routine chemical analysis of precipitation, throughfall and runoff includes pH, SO$_4$, Cl, NO$_3$, NH$_4$, Kjeldahl-N, Ca, Mg, Mn, K, Na, total P, TOC, colour and conductivity. Anions are determined by ion chromatography, NH$_4$ colorimetrically by flow injection analysis, and base cations by atomic adsorption spectroscopy. Estimated uncertainty in flux calculations is < 10%. Details are given by Moldan et al. (1995).

Soil solution

Soil solutions are collected using polytetrafluorethene (PTFE) cups (21-mm outer diameter, 50-mm porous length, pore size 7~5 μm) (Prenart Equipment, Denmark) placed at 5, 10, 20, 40, and 70 cm depth or down to the lithic contact where the soils are shallow. Two to three lysimeters are placed at each depth and connected to the same sampling bottle; a constant suction of 20 to 30 kPa

Fig. 2. Map of the G2 NITREX catchment at Gårdsjön showing location of vegetation plots, lysimeters, tensiometers, throughfall collectors, mineralization studies, mycorrhiza transects, and groundwater wells.

Legend:
- ■ Constant tension lysimeters
- ▲ Lysimeters for episode studies
- ◇ Vegetation plots
- ◆ Mineralization studies
- □ Temperature
- • Throughfall collector, major ions
- ○ Groundwater tube
- ✕ Tensiometer
- ◉ Root studies
- [::::] Mycorinizal transects

Runoff weir

0   25m

ferro-humic podzols, gleyed humo-ferric podzols, and, at the shallow outcrops, typic folisols. The C/N ratio is higher than 30 in the organic and A horizons. Soil depth ranges from 0 cm to > 100 cm with an average of 38 cm (Kjønaas et al. 1992, Stuanes et al. 1992).

A separate manipulation experiment run in parallel with G2 NITREX is conducted at the adjacent catchment G1 ROOF (0.63 ha) (Hultberg et al. 1993). At G1 ROOF ambient inputs of N and S are removed by means of roof beneath the canopy and clean deionized water with the addition of natural sea salts spread underneath the roof by means of a sprinkling system. Water is sprinkled weekly beneath the roof.

A nearby 3.7-ha catchment (F1 CONTROL) with comparable soils and vegetation serves as a control for both G2 NITREX and G1 ROOF. Here inputs and outputs of major ions have been measured continuously since 1979.

The Gårdsjön area receives moderately high deposition of sulphate, nitrate and ammonium; 14-yr mean (minimum and maximum) throughfall inputs at F1 CONTROL for 1979–1992 were 25.0 (18.9 to 31.6) kg SO$_4$-S ha^{-1} yr^{-1}, 7.3 (4.6 to 11.7) kg NO$_3$-N ha^{-1} yr^{-1} and 4.8 (2.4 to 7.7) kg NH$_4$-N ha^{-1} yr^{-1}, respectively.

Table 1. Fluxes of nitrogen species in bulk precipitation, throughfall (tf) and runoff during the 2-yr pre-treatment period April 1989–March 1991, and the 3 yr of treatment April 1991–March 1994 at catchments F1 CONTROL, G1 ROOF and G2 NITREX. Data from Hultberg et al. (1993, 1994) and Moldan et al. (1994). Units: mmol m^{-2} yr^{-1} (1 mmol = 1 meq; 1 mmol m^{-2} yr^{-1} = 0.14 kg ha^{-1} yr^{-1}).

| | F1 CONTROL | | | | G1 ROOF | | | | G2 NITREX | | | | | |
| | NO3 | | NH4 | | NO3 | | NH4 | | NO3 | | | NH4 | | |
	in tf	out runoff	in tf	out runoff	in tf	out runoff	in tf	out runoff	in tf	in added	out runoff	in tf	in added	out runoff
pre-treatment														
Apr 89–Mar 90	66	1	37	0	80	1	40	0	63	0	0	33	0	0
Apr 90–Mar 91	51	0	26	0	46	0	22	0	42	0	0	21	0	0
treatment														
Apr 91–Mar 92	56	1	32	1	0	0	0	1	54	115	2	36	115	1
Apr 92–Mar 93	51	0	22	0	0	0	0	2	52	133	4	26	133	1
Apr 93–Mar 94	36	1	17	2	8	0	0	1	40	146	17	30	146	2

is applied to all lysimeters by use of electric vacuum pumps. The collection bottles are placed in insulated containers to keep them dark and cool during summer. To prevent freezing during winter the containers are heated with lamps. Catchment F1 CONTROL soil solutions are collected at two locations, while in the catchment G2 NITREX soil solutions are collected from locations covering a moisture gradient from the upper drier to the wetter lower parts. Locations G2LY1, G2LY2, and F1LY1 are in drier parts while G2LY3, G2LY7, and F1LY2 are in wetter parts (Stuanes et al. 1992). Subsamples for chemical analysis are taken biweekly and analyzed by methods described by Ogner et al. (1991). Details are given by Stuanes et al. (1995).

Soil nitrogen transformation processes

A soil-core incubation technique is used to measure the rate and amount of microbiological conversion of organic-N to ammonium (mineralization), ammonium to nitrate (nitrification), and nitrate to gaseous N_2O/N_2 (denitrification). Rates are compared between plots under similar moisture regimes at G2 NITREX, G1 ROOF, and F1 CONTROL catchments.

The incubations are made in situ with ion-exchange resin bags placed at the bottom of soil cores. Resin bags are also placed on the surface adjacent to the cores to estimate the relative contribution of incoming nitrogen. Soil cores are collected before each mineralization study to measure pre-incubation levels of NH_4, NO_3 and total N. Ten replicates are used at each plot. Details are given by Kjønaas (unpubl.).

Incubations are conducted over 2-month intervals during the growing season (May–June, July–August, September–October) and over 5 months during the winter (November–May). The cores are placed in a 1 × 1 m grid within each vegetation type, with sample collection proceeding in an upslope direction. The vegetation in the grid is recorded in detail. The holes from the soil cores

are filled with soil from similar vegetation types outside the catchment to minimize mineralization in the catchments due to disturbance. Soil temperature is measured adjacent to the mineralization plots at 5, 10, 20, 40, and 70 cm depth by thermistors and datalogger. The mineralization studies are conducted close to the lysimeter installations to facilitate comparison with NH_4 and NO_3 concentrations in soil solution.

Fine roots

Root sampling is performed by two methods; i) soil cores taken with a 4.5-cm diameter steel corer and ii) ingrowth cores installed in the holes left by soil coring (Vogt and Persson 1991). Sets of 7–10 soil cores (depth c. 30 cm) are taken early October each year at two vegetation types (heather- and moss-dominated, respectively) within each catchment. Investigation of the distribution pattern of fine roots indicates no dependency in the distance to the nearest tree. The soil cores are divided into humus layer and mineral soil and further sub-divided into 5-cm layers. The samples are stored at −4°C until sorting could take place.

Immediately after thawing, roots are picked out from each soil layer and sorted into 3 diameter classes: 0–1, 1–2 and 2–5 mm. The fine roots 0–1 mm are separated into 3 classes, based on the following morphological characteristics: Class 1: the roots are more or less suberized and well-branched, with the main part of the root tips being light colored and turgid. The stele is white and elastic. Class 2: the roots are darkened. White root tips are few or lacking. The stele is still elastic and light to slightly brown. Class 3: the roots in this class are normally referred to as dead. The stele is brownish and easily broken off. No elasticity remains.

Small diameter roots (1–2 mm and 2–5 mm in diameter respectively) are sorted into a living and dead root fraction based on the colour and strength of the stele.

Ingrowth cores of a nylon net filled with perlite are

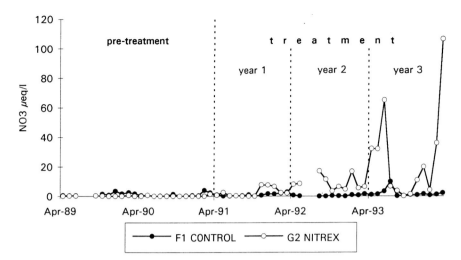

Fig. 3. Monthly mean volume-weighted nitrate concentrations in runoff from catchment G2 NITREX (NH₄NO₃ addition) and F1 CONTROL (untreated control) at Gårdsjön during the 2 yr prior to and 3 yr of N addition.

placed into the holes produced by the core sampling each year (Vogt and Persson 1991). After two growing seasons they are resampled and stored at −4°C. After thawing, each core is divided into the upper 10 cm, and the lower 20 cm. The roots are separated into diameter classes 0–1 mm, 1–2 mm and 2–5 mm respectively, without any further classification. Finally, the roots are dried at 70°C

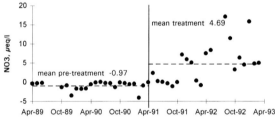

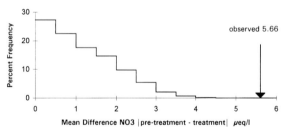

Fig. 4. Differences in nitrate concentrations (upper panel) in runoff at G2 NITREX less F1 CONTROL for the pre-treatment period through March 1991 and the first 2 yr of treatment April 1991–March 1993. Results of randomised intervention analysis (RIA) (lower panel) of nitrate concentrations in runoff at G2 NITREX and F1 CONTROL catchments at Gårdsjön. Shown is the frequency distribution of mean nitrate concentration differences for 1000 random permutations of pre-treatment period and treatment period. The actual observed mean difference of 5.66 µeq/l falls well beyond the tail and thus RIA indicates that the observed change is statistically significant at the p < 0.001 level (from Moldan et al. 1995).

for 48 hours and weighed. Details are given by Clemensson-Lindell and Persson (1995).

Mycorrhiza

At each catchment the above-ground production of fruit bodies of ectomycorrhizal fungi is recorded 3–4 times during the fungal season (August–October) in three transects per catchment. The three transects lie adjacent to each other, and represent a cross-section of the catchment. These are divided into 52 5 × 5 m permanent plots per catchment. Catchment G2 NITREX was studied 1990–93, while G1 ROOF and F1 CONTROL were first studied in 1992. Details are given by Brandrud (1995).

Results

Runoff

Nitrogen addition resulted in increased concentrations of nitrate in runoff at catchment G2 NITREX (Table 1). Elevated concentrations of nitrate occurred in runoff during the first two weeks of treatment in April 1991, but then concentrations were very low during the rest of the spring, summer and early autumn (Fig. 3). The signal of nitrate leakage appeared again in mid-November 1991 and continued much of the winter (Hultberg et al. 1994).

Losses of nitrate continued and increased during the second and third years of treatment. Concentrations of nitrate were higher also during the growing season (Fig. 3). The inorganic nitrogen (mostly as nitrate) lost during the first year of treatment was c. 0.6% ± 0.06%, the second year was 1.1% ± 0.1% and the third year was 5.2% ± 0.5% of the total inorganic nitrogen input, respectively. Output of nitrate from the G2 NITREX catchment during pre-treatment years and similarly from the F1

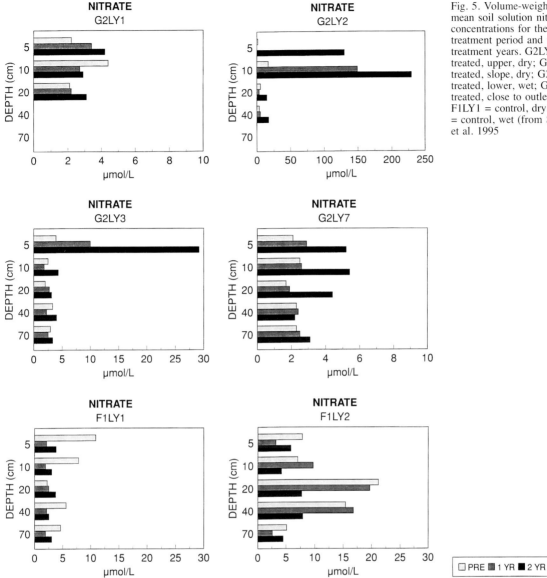

Fig. 5. Volume-weighted mean soil solution nitrate concentrations for the pre-treatment period and two treatment years. G2LY1 = treated, upper, dry; G2LY2 = treated, slope, dry; G2LY3 = treated, lower, wet; G2LY7 = treated, close to outlet, wet; F1LY1 = control, dry; F1LY2 = control, wet (from Stuanes et al. 1995

CONTROL catchment during all years was <0.1% of the nitrogen input (Moldan et al. 1995) (Table 1).

The statistical significance of the increase in nitrate concentrations in runoff can be tested by means of Randomised Intervention Analysis (RIA) (Carpenter et. al. 1989). This technique uses paired samples from the treated and control catchments to evaluate possible changes in the mean difference between the pairs prior to and following onset of treatment. Comparison of the difference in nitrate concentrations in runoff from G2 NITREX and F1 CONTROL shows that G2 NITREX had significantly higher (p < 0.01) nitrate concentrations during the first 2 yr of treatment relative to the 2 yr pre-treatment period (Fig. 4).

Soil-solution

With few exceptions, the volume-weighted average nitrate concentrations in the soil solution increased from the pre-treatment year through the treatment years in G2 NITREX (Fig. 5). No such time trend occurred at F1 CONTROL. On the contrary, in F1 CONTROL nitrate concentrations were generally highest in the pre-treatment year. The response was most pronounced in the upper soil solutions. Nitrate concentrations at or below 20 cm soil depth were generally <5 μmol/l. The highest nitrate concentrations were found at 5 and 10 cm depth at location G2LY2 (in excess of 120 μmol/l). Lysimeter location F1LY2 in the untreated control catchment exhib-

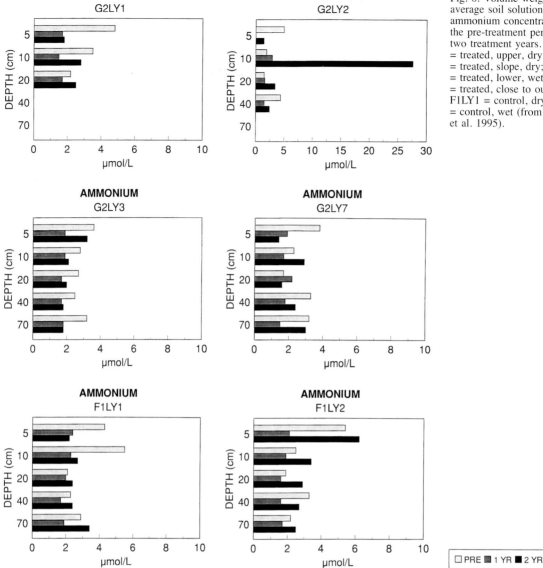

Fig. 6. Volume weighted average soil solution ammonium concentrations for the pre-treatment period and two treatment years. G2LY1 = treated, upper, dry; G2LY2 = treated, slope, dry; G2LY3 = treated, lower, wet; G2LY7 = treated, close to outlet, wet; F1LY1 = control, dry; F1LY2 = control, wet (from Stuanes et al. 1995).

ited consistently high nitrate concentrations throughout the soil profile. This is perhaps due to soil heterogeneity, differences in microclimate or hydrology. The solutions in the treated catchment had generally lower baseline values for nitrate, especially at greater soil depth, but showed higher post-treatment peaks (Stuanes et al. 1995). The G2LY2 location is characterized by a relatively thick, mostly decomposed mor mat and dry conditions. Despite a low water flux, the nitrate flux from the 10-cm depth at G2LY2 was higher than that at the other lysimeter sites.

Ammonium in solution followed a different trend: concentrations were generally higher in the pre-treatment year and the second treatment year compared to those from the first treatment year for both the treated and the control catchment (Fig. 6). The addition of nitrogen thus caused a shift in the dissolved inorganic nitrogen pool towards a greater contribution from nitrate, especially in the upper part of the soil profile. Concentrations of organic nitrogen in the soil solution remained unaffected by treatment and were generally highest in the upper 10 cm of the soil in both catchments (Stuanes et al. 1995).

Soil nitrogen transformation processes

Mineralization measurements from the first 2 yr of treatment, 1991 and 1992, show no significant differences

NET MINERALIZATION
Immoblization of NO3

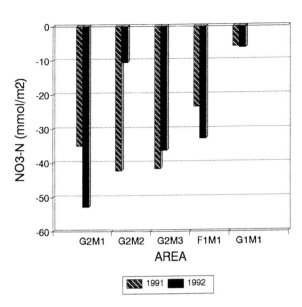

Fig. 7. Immobilization of nitrate in soil at G2 NITREX, F1 CONTROL and G1 ROOF catchments in 1991 and 1992 as measured by soil core incubation technique (from Kjønaas unpubl.).

NET NITRIFICATION

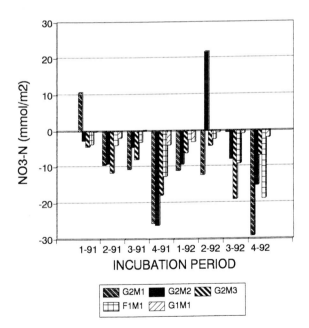

Fig. 8. Net nitrification in soil at G2 NITREX, F1 CONTROL and G1 ROOF catchments in 1991 and 1992 as measured by soil core incubation technique (from Kjønaas unpubl.).

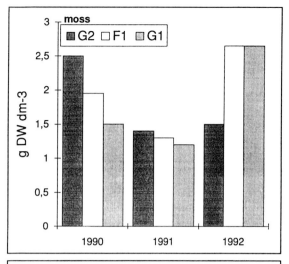

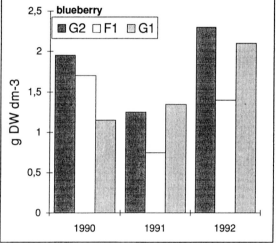

Fig. 9. Living fine roots (classes 1 and 2) in soil core samples from the organic soil layer collected from catchments G2 NITREX, F1 CONTROL, and G1 ROOF in 1990 (pre-treatment), 1991 and 1992 (data from Clemensson-Lindell and Persson 1995).

between the 3 catchments and no systematic change from 1991 to 1992. The fairly high C/N ratio of the soil, with a range between 35–52 in the organic layer suggests that immobilisation of added nitrogen is an important process during the seasons of microbial activity. The immobilisation of nitrate, however, appears to be reduced at site M2 in catchment G2 NITREX during the second year of treatment (Fig. 7). This is reflected in increased levels of nitrate in the ion exchange resins under the soil cores. The same trend is found in the corresponding lysimeters. This is probably due to an increase in nitrification during the summer 1992 at this site (Fig. 8) (Kjønaas unpubl.). Additional years data are necessary to reveal trends over time.

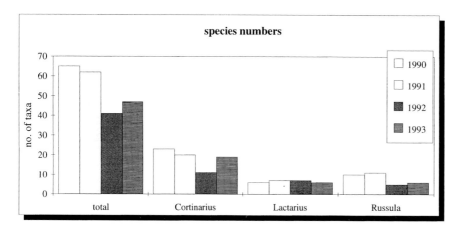

Fig. 10. Species numbers of mycorrhizal fungi in the G2 NITREX catchment 1990–1993. Total number of taxa (species and varieties) and taxa of dominant genera recorded in the 1300-m² transect (from Brandrud 1995).

Fine-roots

There were large variations in the different root fractions in the core samples from all catchment areas between different years. Few significant changes were observed in diameter classes 1–2 and 2–5. The majority of changes in fine roots were observed in the humus layer (FH 0–5 and 5–10 cm, respectively) (Clemensson-Lindell and Persson 1995).

In the G2 NITREX catchment there was a tendency towards a decrease in the biomass of living fine roots in the FH layer in the moss-dominated plots (Fig. 9). In contrast, at G1 ROOF catchment there was a significant increase in the amount of living fine roots in the FH layer. In F1 CONTROL there were no significant trends over time (Clemensson-Lindell and Persson 1995).

Mycorrhiza

During the four-year study period 82 taxa of mycorrhizal fungi were recorded in the 1300-m² transect area at G2

NITREX. The dominant *Vaccinium myrtillus* spruce forest type (Eu-Piceetum myrtilletosum) included 77 species, while 5 species were recorded exclusively in plots of dry pine forest (8 plots out of 52) and mesic (swampy) spruce forest with dominance of *Sphagnum* (3 plots) (Brandrud 1995).

The species numbers decreased after N-treatment, from >60 species yr⁻¹ in 1990 and 1991 (half a year before and after the start of treatment, respectively), to <50 in 1992 and 1993 (Fig. 10). A decrease occurred also in the two dominant genera, *Cortinarius* and *Russula*, while the genus *Lactarius* exhibited little change.

The total mycorrhizal fungus production was 2300–2500 fruit bodies yr⁻¹ in 1990–1991, while in 1992–1993 the production decreased to 1700–1800 fruit bodies yr⁻¹. The decrease was most pronounced for the dominant genus *Cortinarius* for which the production went from >1000 fruit bodies in 1990 to <300 in 1992 and 1993.

A few species such as the dominant *Cantharellus tubaeformis*, as well as *Lactarius rufus*, *Lactarius theiogalus* and *Paxillus involutus*, showed increases in fruit body production after treatment. Typical for all these species

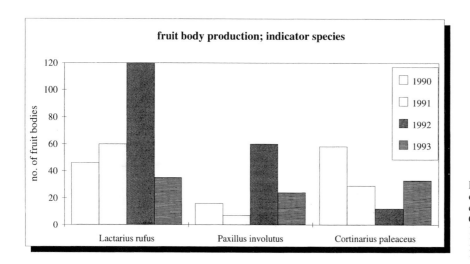

Fig. 11. Fruitbody production of selected indicator species of mycorrhizal fungi in the G2 NITREX catchment 1990–1993. Transect area: 1300 m² (from Brandrud 1995).

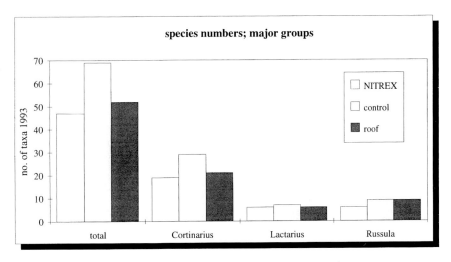

Fig. 12. Species numbers of mycorrhizal fungi in the G2 NITREX catchment, the F1 CONTROL catchment and G1 ROOF catchment in 1993. Total number of taxa (species and varieties) and taxa of dominant genera recorded in the 1300-m² transects in each catchment (from Brandrud 1995).

was a peak production in 1992, with a decrease in 1993, even though 1993 was a generally more productive season in the area (Fig. 11).

Results from 1993 at all 3 catchments indicate that species diversity of the G2 NITREX catchment was similar to that of the G1 ROOF catchment, but lower than in the F1 CONTROL catchment (Fig. 12). The diversity of the F1 CONTROL catchment was more comparable to that of the G2 NITREX catchment before, and just after start of treatment (Fig. 10). The genus *Lactarius* showed a different pattern with similar diversity in 1993 at all the three catchments (Fig. 12).

In spite of the fairly low diversity, the fruit body production of the G1 ROOF catchment in 1993 was approximately twice as high as that of the other two catchments. The stress-tolerant taxa *Lactarius* spp. (species such as *L. rufus*, *L. theiogalus*), *Cantharellus tubaeformis* and *Paxillus involutus* exhibited a comparatively low production at the F1 CONTROL catchment, and a high production at the G1 ROOF and G2 NITREX catch-

ments (Fig. 13). The sensitive groups *Cortinarius* and *Russula* showed the highest production at the G1 ROOF and F1 CONTROL catchments and the lowest at the G2 NITREX catchment.

Discussion

Together the results from G2 NITREX catchment indicate that the three years of nitrogen addition have resulted in changes across the forest ecosystem boundary as well as changes within the ecosystem. Most apparent and rapid is the increase in nitrate concentrations in runoff, whereas the changes in components and processes within the ecosystem are apparently slower and more difficult to detect.

The increased nitrate concentrations in runoff are related to the increased nitrate concentrations in soil solution in several of the lysimeters. Spatial differences

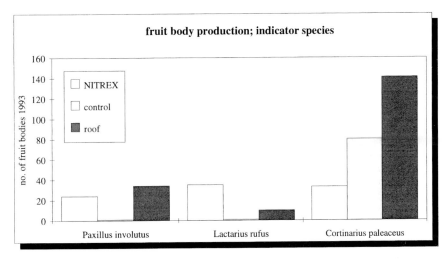

Fig. 13. Fruitbody production of some selected indicator species of mycorrhizal fungi in the G2 NITREX catchment, the F1 CONTROL catchment and G1 ROOF catchment in 1993. Each catchment includes a 1300-m² transect (from Brandrud 1995).

in nitrogen demand by plant uptake and microbial immobilization and nitrogen supply through mineralization probably account for the large variation in nitrate concentration observed between the lysimeters at various points within the catchment. The soil nitrogen transformation results suggest that the increased nitrate concentrations observed in soil solution (and hence runoff) are due to increased nitrification of ammonium to nitrate and reduced immobilization of nitrate in the soil. This is especially apparent in the uppermost soil layers where the major sources of ammonium are throughfall and the added ammonium nitrate.

Over the long term the nitrogen addition can be expected to affect productivity of trees and other vegetation, alter the relative abundance of annual and perennial lower vegetation, and change species composition. Thus far these types of changes are most apparent in the mycorrhizal fungus flora. The results after three years of treatment indicate that nitrogen sensitive species have become less abundant with lower fruit body production and nitrogen tolerant species have increased their fruit body production.

The fine-root data also point to changes due to the nitrogen addition. At G2 NITREX the amount of fine roots in the forest floor layer decreased while at G1 ROOF the amount increased. These changes are consistent with the hypothesis that nitrogen suppresses fine-root production.

As yet there are insufficient data available to assess the effects of nitrogen addition (and nitrogen removal at G1 ROOF) on the tree growth and vitality. The trees in both catchments are growing slowly and measurement of changes in growth parameters such as height and diameter require several years of data. Of particular interest are observations on crown density, needle loss and nutrient concentrations in foliage. Chronic nitrogen deposition at the levels applied at G2 NITREX (c. 50 kg N ha^{-1} yr^{-1} total deposition) cause major changes including discoloration of needles, needle loss and reduced growth in coniferous forests in areas such as Germany (Ulrich 1989, Hauchs 1990) and the Netherlands (Van Breemen and Van Dijk 1988). Further years of nitrogen addition at G2 NITREX might ultimately result in such changes.

A delayed response in vegetation to changes in nitrogen deposition is apparently typical in coniferous forests. At the NITREX site in a Scots pine stand at Ysselsteyn, the Netherlands, changes in nutrient concentrations in needles, for example, were first manifest after four years of treatment, while at the NITREX site in a Douglas fir stand at Speuld, the Netherlands, no changes have occurred after four years (Boxman et al. 1995). When nitrogen is added in massive doses in fertilization experiments, however, vegetation changes occur more rapidly (Kellner 1993).

Results from a similar nitrogen addition study underway at a 10-ha forested catchment in Maine, USA (Bear Brook Watershed Manipulation), indicate that most of the added nitrogen (in this case monthly additions of granular

ammonium sulfate) is retained in the organic soil with only a minor fraction going to the vegetation (Nadelhofer et al. 1993). Here ^{15}N is enriched in the added N and used as a tracer. Most of the added nitrogen at Gårdsjön G2 NITREX catchment probably also is retained in the upper soil horizons, as is indicated by the soil microbial process studies (Kjønaas unpubl.). Forthcoming data on stable N isotope abundances will provide direct information on the fate of the added N at Gårdsjön. The ammonium nitrate added during the second year of treatment April 1992–April 1993 was enriched to 1500 per mil ^{15}NH$_4$^{15}NO$_3$, and measurements of isotopic ratios and N pool sizes are in progress following a standard protocol developed for the European NITREX project (Kjønaas et al. 1993).

The results from the NITREX experiment at Gårdsjön indicate that the forested catchment ecosystem G2 NITREX is moving towards "nitrogen saturation". A widely-used definition of nitrogen saturation is the condition under which "the availability of ammonium and nitrate is in excess of total combined plant and microbial demand" and that this is manifest by "an increased leaching of NO$_3$$^-$ (or NH$_4$$^+$) below the rooting zone" (Aber et al. 1989). The situation at Gårdsjön G2 NITREX is unclear; whereas the nitrate concentrations in runoff clearly show increases, the lysimeter data do not indicate increased leaching below the rooting zone.

Stoddard (1994) suggested that ecosystems pass through several stages of nitrogen saturation. In stage 0 (pre-treatment conditions) nearly all nitrogen is retained with the possible exception of specific short-term hydrologic events such as snow melt. This is the case for G2 NITREX prior to treatment and F1 CONTROL during all years (Fig. 3). These peaks (maximum F1 CONTROL nitrate concentration in runoff point sample runoff recorded since 1988 was 17 μeq l^{-1}) are mainly coincident with snow melt or heavy autumn rains. At all other times the concentration of nitrate remains close to or below the analytical detection limit of 0.4 μeq l^{-1}. At high flow situations ground water approaches or even rises above the soil surface in discharge areas where the hydraulic conductivity of the uppermost organic soil horizon is often high (10^{-3} to 10^{-4} m s^{-1}) (Johansson and Nilsson 1985). Therefore, the water near the soil surface moves rapidly towards the catchment outlet. As a result of short retention time of water, nitrate entering the catchment or melting from the snowpack may not be immobilised or taken up. A. Henriksen (pers. comm.) has termed this "hydrological nitrate". Furthermore during the late autumn, winter and spring biological immobilisation and uptake of nitrate by vegetation is reduced by low soil temperature and physiological status of the vegetation. These nitrate peaks thus reflect hydrological conditions rather than a "saturation" of the physiological demand of the plants and microbes.

In stage 1 of nitrogen saturation, significant nitrate leaching occurs during the non-growing season (late autumn, winter and early spring). This is the case for G2 NITREX during the first year of treatment (April 1991–

March 1992). In stage 2 the ecosystem begins to loose nitrate during all seasons as is the case for G2 NITREX during the second and third years of treatment (Fig. 3).

After the nitrogen addition began in April 1991 we observed an increase of both the frequency and magnitude of nitrate peaks in runoff during the non-growing season the first year (stage 1) and all through the second and third years (stage 2). These peaks are, however, always followed by very low nitrate concentrations.

The nitrogen retention capacity at catchment G2 NITREX is exceeded only during short periods related to experimental addition or by large episodic atmospheric inputs. Nitrate pulses found in runoff from the G2 NITREX catchment after NH_4NO_3 additions are in part "hydrological nitrate". The catchment is, however, retaining less and less of the incoming nitrogen, as the total loss increased from 0.6% of the inorganic N input the first treatment year to 1.1% the second and 5.2% the third treatment year. These losses, however, represents a > 10-fold increase in nitrate concentrations in runoff. Soil solution samples collected from the G2 NITREX catchment also point to a temporal change in nitrate concentrations as a result of the experimental nitrogen additions (Stuanes et al. 1995). The G2 NITREX catchment continues to accumulate most of the nitrogen input but at an increasingly slower rate.

These results are consistent with those from the experiment at the Bear Brook Watershed in Maine, USA (Kahl et al. 1993). Here the catchment was apparently at Stage 1 with significant concentrations of nitrate in streamwater during the winter already prior to treatment, despite the relatively low ambient N deposition of 8 kg N ha^{-1} yr^{-1}. The experimental addition of c. 25 kg N ha^{-1} yr^{-1} caused immediate increase in nitrate concentrations both during the winter but also during the growing season; the catchment has apparently moved to stage 2 of nitrogen saturation.

Indeed the rapid response of nitrate concentrations in runoff and soil solution is characteristic for many such large-scale whole ecosystem manipulation experiments with nitrogen deposition. In the European NITREX project, treatment of increased or decreased N deposition resulted in significant change in nitrate concentrations at all 7 sites across the gradient of N deposition from < 3 to > 50 kg N ha^{-1} yr^{-1} (Wright et al. 1995).

With additional years of treatment we expect that catchment G2 NITREX will loose an increasing proportion of the nitrogen input. Nitrate concentrations in runoff will increase and be accompanied by increasing concentrations of acid cations as H^+ and Al^{3+}. The results from NITREX Gårdsjön and the Bear Brook watershed clearly demonstrate that nitrogen saturation can be induced over a relatively short time by increasing atmospheric deposition of nitrogen. The response of runoff chemistry was manifest already during the first year of treatment in both cases.

At Gårdsjön as well as most of the other NITREX sites (Wright et al. 1995), the runoff responds immediately to changed nitrogen deposition, but the vegetation and soils lag behind by one or more years. Generally, biological responses are difficult to detect during the first stage of nitrogen saturation (Aber et al. 1989). The expected response of parameters such as nutrient content of foliage and growth of trees occur only several years after the onset of treatment. Apparently, this is the case for both increased as well as decreased (by roof) nitrogen deposition. The ecosystem as a whole shows a response to treatment (changed nitrate concentration in runoff), but only modest responses in individual components and internal processes are evident during the first two years of treatment.

The Gårdsjön area receives intermediate deposition of nitrogen relative to pristine areas in northern Scandinavia, on the one hand, and heavily-impacted areas in central Europe on the other hand (Wright et al. 1995). The rate at which catchment G2 NITREX responded to nitrogen addition suggests that at ambient nitrogen deposition of 12 kg inorganic N ha^{-1} yr^{-1} the ecosystem is near the threshold at which additional N inputs cause significant nitrate leaching.

Acknowledgements – This research was funded in part by the Swedish Environmental Protection Agency (SNV), the Norwegian Research Council (NFR Programme Committee on Transport and Effects of Airborne Pollutants TVLF), the Commission of European Communities (STEP-CV90-0056, ENVIRONMENT-CT93-0264), the Swedish Environmental Research Institute (IVL), the Norwegian Institute for Water Research (NIVA) and the Norwegian Forest Research Institute. We thank I. Andersson, M. Huse, and V. Timmermann for assistance and helpful discussions.

References

Aber, J. D., Nadelhoffer, K. J., Steudler, P. and Melillo, J. 1989. Nitrogen saturation in northern forest ecosystems. – Bioscience 39: 378–386.

Boxman, A. W., Van Dam, D., Van Dijk, H. F. G., Hogervorst, R. F. and Koopmans, C. 1995. Ecosystem responses to reduced nitrogen and sulphur inputs into two coniferous forest stands in the Netherlands. – Forest Ecol. Manage. 71: 7–29.

Brandrud, T. E. 1995. The effects of experimental nitrogen addition on the ectomycorrhizal fungi in an oligotrophic spruce forest at Gårdsjön, Sweden. – Forest Ecol. Manage. 71: 111–122.

Carpenter, S. R., Frost, T. M., Heisey, D. and Kratz, T. K. 1989. Randomized intervention analysis and the interpretation of whole-ecosystem experiments. – Ecology 70: 1142–1152.

Clemensson-Lindell, A. and Persson, H. 1995. The effects of nitrogen addition and removal on Norway spruce vitality and distribution in three catchment areas at Gårdsjön. – Forest Ecol. Manage. 71: 123–131.

Dise, N. B. and Wright, R. F. (eds) 1992. The NITREX project (Nitrogen Saturation experiments). – Ecosystems Res. Rep. 2, Comm. European Commun., Brussels.

– and Wright, R. F. 1995. Nitrogen in relation to nitrogen deposition leaching from European forests. – Forest Ecol. Manage. 71: 153–161.

Grennfelt, P. and Hultberg, H. 1986. Effects of nitrogen deposition on the acidification of terrestrial and aquatic ecosystems. – Water Air Soil Pollut. 30: 945–963.

Hauhs, M. 1990. Lange Bramke – an ecosystem study of a

forested catchment. – In: Adriano, D.C. and Havas, M. (eds), Acidic precipitation, case studies, Vol. 5. Springer, New York, pp. 275–305.

– , Rost-Siebert, K., Raben, G., Paces, T. and Vigerust, B., 1989. Summary of European data. – In: Malanchuk, J.L. and Nilsson, J. (eds) The role of nitrogen in the acidification of soils and surface waters. Miljørapport 1989:10, Nordic Council of Ministers, Copenhagen.

Hultberg, H., Andersson, B.I. and Moldan, F. 1993. The covered catchment – an experimental approach to reversal of acidification in a forest ecosystem. – In: Rasmussen, L., Brydges, T. and Mathy, P. (eds), Experimental manipulations of biota and biogeochemical cycling in ecosystems. Ecosystems Res. Rep. 4, Comm. European Commun., Brussels, pp. 46–55.

– , Dise, N.B., Wright, R.F., Andersson, I. and Nyström, U. 1994. Nitrogen saturation induced during winter by experimental NH_4NO_3 addition to a forested catchment. – Environ. Pollut. 84: 145–147.

Johansson, S. and Nilsson, T., 1985. Hydrology of the Lake Gårdsjön area. – In: Andersson, F. and Olsson, B. (eds), Lake Gårdsjön – an acid lake and its catchment. Ecol. Bull. (Stockholm) 37: 86–100.

Kahl, J.S., Norton, S.A., Fernandez, I.J., Nadelhofer, K.J., Driscoll, C.T. and Aber, J.D. 1993. Experimental inducement of nitrogen saturation by $(NH_4)_2 SO_4$ at the watershed scale. – Environ. Sci. Tech. 27: 565–568.

Kellner, O. 1993. Effects of fertilization on forest flora and vegetation. – Acta Univ. Uppsala, Comprehensive summaries of Uppsala dissertations from the Faculty of Science 464, Uppsala, Sweden.

Kjønaas, O.J., Stuanes, A.O. and Huse, M. 1992. Soils and soil solution. – In: Dise, N.B. and Wright, R.F. (eds), NITREX Gårdsjön. Status report for 1990–91. Norwegian Inst. for Water Research, NIVA, Oslo, pp. 23–29.

– , Emmett, B., Gundersen, P., Koopmans, C. and Tietema, A. 1993. ^{15}N approach within NITREX: 1. Natural abundance along a pollution gradient. – In: Rasmussen, L., Brydges, T. and Mathy, P. (eds), Experimental manipulations of biota and biogeochemical cycling in ecosystems. Ecosystems Res. Rep. 4, Comm. European Commun., Brussels, pp. 232–234.

Moldan, F., Hultberg, H., Nyström, U. and Wright, R.F. 1995. Nitrogen saturation at Gårdsjön, southwest Sweden, induced by experimental addition of nitrogen. – Forest Ecol. Manage. 71: 89–97.

Nadelhofer, K.J., Aber, J.D., Downs, M.R., Fry, B. and Melillo, J.M. 1993. Biological sinks for nitrogen additions to a forested catchment. – In: Rasmussen, L., Brydges, T. and Mathy, P. (eds), Experimental manipulations of biota and biogeochemical cycling in ecosystems. Ecosystems Res. Rep. 4, Comm. European Commun., Brussels, pp. 64–70.

Nihlgård, B. 1985. The ammonium hypothesis – an additional explanation to the forest dieback in Europe. – Ambio 14: 2–8.

Ogner, G., Opem, M., Remedios, G., Sjøtveit, G. and Sørlie, B. 1991. The chemical analysis program of the Norwegian Forest Res. Inst. 1991. – Norwegian Forest Res. Inst., Ås-NLH, Norway.

Olsson, B., Hallbäcken, L., Johansson, S., Melkerud, P.-A., Nilsson, S.I. and Nilsson, T. 1985. The Lake Gårdsjön area – physiographical and biological features. – Ecol. Bull. (Stockholm) 37: 10–28.

Stoddard, J.L. 1994. Long-term changes in watershed retention of nitrogen: its causes and aquatic consequences. – In: Baker, L.A. (ed.), Environmental chemistry of lakes and reservoirs. Adv. Chemistry Ser. No. 237, Am. Chemical Soc., Washington, pp. 223–284.

Stuanes, A.O., Andersson, I., Dise, N., Hultberg, H., Kjønaas, J., Nygaard, P.H. and Nyström, U. 1992. Gårdsjön, Sweden. – In: Dise, N.B. and Wright, R.F. (eds), The NITREX project (Nitrogen saturation experiments). Ecosystems Res. Rep. 2, Comm. European Comm., Brussels, pp. 24–34.

– , Kjønaas, O.J. and van Miegroet, H. 1995. Soil solution response to experimental addition of nitrogen to a forested catchment at Gårdsjön, Sweden. – Forest. Ecol. Manage. 71: 99–100.

Ulrich, B. 1989. Effects of acid precipitation on forest ecosystems in Europe. – In: Adriano, D.C. and Johansson, H.H. (eds), Advances in environmental sciences, acidic precipitation Vol. 2. Springer, New York, pp. 189–272.

Van Breemen, N. and Van Dijk, H.F.G. 1988. Ecosystem effects of atmospheric deposition of nitrogen in the Netherlands. – Environ. Pollut. 54: 249–274.

Vogt, K.A. and Persson, H. 1991. Measuring growth and development of roots. – In: Lassoie, J.P. and Hincley, T.M. (eds), Techniques and approaches in forest tree ecophysiology. CRS Press, Boca Raton, pp. 477–501.

Wright, R.F., Roelofs, J.G.M., Bredemeier, M., Blanck, K., Boxman, A.W., Emmett, B.A., Gundersen, P., Hultberg, H., Kjønaas, O.J., Moldan, F., Tietema, A., van Breemen, N. and van Dijk, H.F.G. 1995. NITREX: responses of coniferous forest ecosystems to experimentally-changed deposition of nitrogen. – Forest Ecol. Manage. 71:163–169.

Ecological Bulletins 44: 335–351. Copenhagen 1995

Past and future changes in soil acidity and implications for forest growth under different deposition scenarios

Harald Sverdrup and Per Warfvinge

Sverdrup, H. and Warfvinge, P. 1995. Past and future changes in soil acidity and implications for forest growth under different deposition scenarios. – Ecol. Bull. (Copenhagen) 44: 335–351.

Calculations with the dynamic model SAFE were used to assess the effect on soil chemistry and implications for forest growth under different future deposition scenarios. The calculations show that there is consistency between observed and calculated chemical changes in soils exposed to high acidity input loads, and that the acidification models are capable of reconstructing the recorded chemical history. Studies of the individual terms in a mass balance of acidity for the soil reveal that acid deposition is the most significant acidity input to the soil and the greatest cause of chemical change. The model calculations indicate that a large part of these chemical changes can be reversed within 10–30 yr if the acidity input is significantly reduced. If laboratory sensitivity of trees to Al is transferable to field conditions, then the present rate of deposition should cause concern about soil acidification effects on forest growth. Regional assessments based on the Swedish Forest Inventory suggest that deposition reductions of 80% for S and 50% for N as compared to the 1980 level will result in soil chemistry at steady state that will not be harmful to trees.

H. Sverdrup and P. Warfvinge, Dept of Chemical Engineering II, Lund Inst. of Technology, Box 124, Chemical Center of Lund Univ., S-221 00 Lund, Sweden.

Introduction

Dynamic models for soil acidification are being used to assess the effects of acid deposition on forest soils and aquatic ecosystems. Such assessments, together with mapping of critical loads of acidity, are being used as input to European negotiations on sulphur and nitrogen emission reductions (Cosby et al. 1985, Alcamo et al. 1985, Kämäri 1986, Mulder et al. 1987, de Vries 1987, Wright et al. 1988, de Vries and Kros 1989, de Vries et al. 1989a,b, Sverdrup et al. 1990, Warfvinge et al. 1992, 1993). In Sweden, the SAFE and PROFILE models were developed as a tool for assessing the temporal impact of acid deposition on soil chemistry and forest vitality (Warfvinge and Sverdrup 1992a,b,c, Sverdrup et al. 1993, Sverdrup et al. 1995a,b). These models have been operated at individual research sites, and in a regional context, for development, testing, verification and assessment. The Swedish Forest Inventory is the basis for input data acquisition used in the regional assessments (Kempe et al. 1992).

Scope and objective

The objective of this study is to report from some of the assessments that have been made with respect to effects of acid deposition to Swedish forest ecosystems. Two different tasks in the assessment work is described here: 1) Illustrate the effect of acid deposition decrease on soil chemical recovery at a specific site, using the Swedish acidification model SAFE. 2) Discuss potential regional effects on soil chemistry and possible forest growth effects from chemical changes in the soil under different deposition scenarios, as explored with the Swedish acidification model PROFILE.

The models have been tested and described in detail earlier (Warfvinge and Sverdrup 1992a,b, Sverdrup and Warfvinge 1993a), and the objective here is more focussed towards their use in practical assessments.

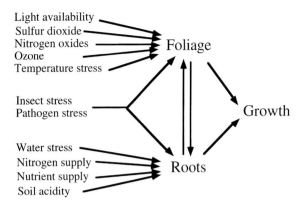

Fig. 1. Several causes may affect defoliation, vitality and growth of trees in forests, and soil acidification is only one of these.

Effects of acid deposition

Acidity input to soils

Acidity input to soils in excess of the weathering neutralization rate will cause leaching of base cations from the ion exchange complex of the soil. The soil solution will become acidic when the base saturation is exhausted and can no longer neutralize incoming acidity. In the acidic solution, inorganic Al will stay in solution and the pH value will be low. During the leaching phase, when the base saturation is decreasing, base cation concentrations may temporarily increase significantly, but then decrease again as the storage is gradually emptied. The final result is that both the solid soil phase and the soil solution have been acidified and base cations are limited by weathering and deposition.

Causes of forest decline

It is not only soil acidification which may affect vitality, growth and the state of defoliation of trees in forests (Ulrich 1985b, Pitelka and Raynal 1989, Roberts et al. 1989, Barnard and Lucier 1991, Becker 1991, Schulze and Freer-Smith 1991). Soil acidification is only one of several causes acting simultaneously; as illustrated in Fig. 1. Changes in defoliation and forest vitality are caused by changes in any of the factors listed in the Fig. 1. Several of the tree response mechanisms may be strongly interrelated. Soil acidification can cause changes in root distribution, making the tree more vulnerable to drought or wind.

It can be seen in Fig. 1 that if system analysis begins from a particular stress, it will be possible to find connections to root damage, defoliation and growth change, since those responses will be stratified with respect to the cause selected. However, working the other way, from growth changes back, any of the listed causes may have caused the damage, making the likelihood of finding

good correlations between one specific cause and growth change extremely unlikely.

Better growth than ever and soil acidification?

Despite all the data available, it is still sometimes disputed whether effects of soil acidification on trees exist at all; it has been said that Sweden's forests are currently producing more biomass than ever. Swedish forest inventory statistics show this to be true for a large part of Sweden (SOU 1992:76).

During the last 60 yr, nitrogen deposition in southern Sweden has increased from c. 1–2 kg N ha^{-1} yr^{-1} to 20–30 kg ha^{-1} yr^{-1}; in the north the change has been much less, from 0.5–2 kg N ha^{-1} yr^{-1} to 7–12 kg ha^{-1} yr^{-1} (Johnston et al. 1986, Sverdrup et al. 1995b). Trees in the boreal zone are evolutionarily adapted to a deficiency situation for N, and hence increased N will enhance tree growth. Efficient forestry practices have made possible the exploitation of the increased amount of N. Acidification has made base cations in the ion exchange reservoir available for the nitrogen-induced increased growth. If the weathering rate is insufficient to supply this increased demand for base cations, the accelerated base cation uptake will rapidly deplete the forest reservoir of base cations.

The additional presence of high concentrations of soluble inorganic Al and low pH may reduce uptake efficiency and plant vitality significantly (Ulrich 1983, 1985a, Schulze 1987, Cronan et al. 1987, 1989, Arovaara and Ilvesniemi 1990, Cronan 1991, Sverdrup and Warfvinge 1993b).

Measurements have shown that base saturation has been reduced to very low levels in southern parts of Sweden and that this has occurred during the last 60 yr. The change can be explained by the increase in acid deposition and to a lesser degree by increased growth (Tamm and Hallbäcken 1985, Berdén et al. 1987, Falkengren-Grerup 1987, Falkengren-Grerup et al. 1987, Falkengren-Grerup and Eriksson 1990, Warfvinge et al. 1993, Sverdrup et al. 1995b). During the same time-period, increases in defoliation and tree vitality has been observed over large parts of Europe (Moseholm et al. 1988). These have been seen as partly caused by acid deposition.

Model description

Model requirements

Good growth in the past cannot be extrapolated into the future, especially if new factors affecting growth are involved. Traditional yield models all have in common that they assume soil chemistry to stay constant over time. Such models cannot predict sudden changes in the system caused by soil acidification. For such events, explanatory prediction tools based on actual mechanisms are needed.

Soil chemistry

The soil chemistry model SAFE has been developed with the objective of studying the effects of acid deposition on soils and groundwater. Detailed descriptions of the model can be found in Warfvinge and Sverdrup (1992a,b,c), Sverdrup and Warfvinge (1993a). It calculates the values of different chemical state variables as a function of time. The basic principle in both models are mass balances for alkalinity, nitrogen and base cations. PROFILE has mass balances for individual ions; in SAFE, base cations and nitrogen are both lumped parameters. These models can therefore be used to study the process of acidification and recovery, as affected by deposition rates, soil parameters and hydrological variations. PROFILE is a steady state model closely related to SAFE. It bypasses the changes in soil state over time, and calculates the final steady state directly. In PROFILE, the differentials for change in the soil solution and ion exchange reservoirs are set to zero.

SAFE and PROFILE are based on the same basic principles for describing individual processes. SAFE, which takes the gradual changes in soil state over time into account, contains the following chemical subsystems: 1) Deposition, leaching and accumulation of dissolved chemical components. 2) Chemical weathering reactions of soil minerals with the soil solution, depending on mineralogy, texture, soil moisture, pH, Al, organic acids and base cation concentration. 3) Cation exchange reactions between Al and base cations. 4) The reactions of N–compounds; nitrification and denitrification. 5) Cycling of N and base cations in the canopy, using input data for as canopy exchange, litterfall, uptake, mineralization and immobilization. 6) Solution equilibrium reactions involving CO_2, Al and organic acids. 7) A gibbsite type forcing function for Al control.

The SAFE model is structured in different compartments to represent the natural vertical differences in soils, which are reflected in the large difference in physical and chemical properties between soil layers. To assemble input data to the model, the soil was divided into four layers, based on the natural soil stratification, as the soil horizons are the largest chemically isotropic elements in the system. In general, these corresponds to the O-, E-, B- and C-layers of a podsol.

Basic assumptions in the model are several. The model accounts for vertical flow only. Horizontal flow can only be included if the model is coupled with a hydrological model with such features. In the model, sulphate absorption is considered insignificant. Cation exchange capacity (CEC) is assumed to remain constant throughout the simulation, and the same is assumed for the selectivity coefficient. The amount of organic material in the soil is assumed to be at or near steady state.

The outstanding difference between SAFE and other comparable soil chemistry models is that the weathering rate is calculated from independent geophysical properties of the soil system (Sverdrup 1990, Sverdrup and Warfvinge 1993a). This reduces the degrees of freedom

in the model and reduces the need for calibration. PRO-FILE also considers chemical feedback on the rate of denitrification and nitrification.

Uptake in biomass

In SAFE and PROFILE, cycling of nutrients between the canopy and the soil is considered. Nutrients are taken from the soil layers in proportion to root distribution, and a part of this is returned as litterfall to the top layer. Of the litterfall, 90% is assumed to decompose, the residual is N immobilization. Decomposition is considered to be at steady state with respect to litterfall.

Trees satisfy their need for nutrients from the soil, in proportion to availability in the soil. The transport process from the bulk of the solution to the root surface is governed by solute transport mechanisms such as solute flow, mass diffusion and ion exchange (Nye and Tinker 1977, Cronan 1991).

$$G = G_{max} \cdot f(BC/Al) \cdot f(Nutr) \qquad (1)$$

where f(Nutr) expresses the action of Liebig's law and f(BC/Al) the action of soil acidity. The two effects may act independently, but by decreasing the general uptake from a soil layer, the effect of f(BC/Al) is to inhibit uptake of necessary base cation nutrients. G_{max} is the maximum growth obtainable, not considering any base cation nutrient limitations or any limiting effects caused by soil acidification. The reference level for growth in our calculations is the growth calculated using standard yield tables for growth modified for site characteristics as implemented in the model HUGIN (Kempe et al. 1992).

In the model, growth is also subject to nutrient restrictions. Growth is assumed to be proportional to uptake of the limiting nutrient, according to "Liebig's law": In the model, nitrogen and base cation uptake is coupled, and either nitrogen or base cation deficit will eventually reduce total growth. The long term sustainable growth based on available nutrients.

Limits for uptake is determined by the availability of nutrient either by deposition (N) or deposition and weathering (BC, P). Thus, the "Liebig's law function is expressed as the ratio between nutrient demand to sustain growth according to the yield model, and what is available in the soil:

$$f(Nutr) = \min_{i=nutrients} \sum_{j=1}^{4 \text{ layers}} \frac{W_{j,i} + BCdep_{j,i}}{G_{max} \cdot x_i} \qquad (2)$$

x_i is the content of the nutrient i in the tree.

If the amount required by the plant is less than the amount available, then the function has the value 1, and this nutrient will not limit growth.

We allow both soil chemistry and supply to restrict growth, but net growth is maximized to present net uptake (Warfvinge et al. 1992) in the steady state PROFILE

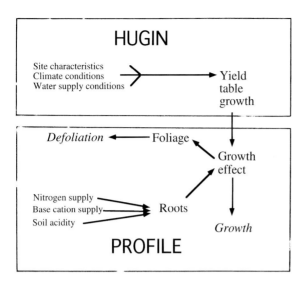

Fig. 2. The diagram show how soil acidification interaction with tree growth and defoliation was implemented in PROFILE.

calculations. In the PROFILE and SAFE models, base cation uptake occurs from each soil layer, but uptake is stopped if the base cation soil solution concentration falls below 7 µeq l^{-1} for Mg and Ca and below 1 µeq l^{-1} for K in PROFILE. In SAFE when the base cation concentration falls <15 µeq l^{-1} (Arovaara and Ilvesniemi 1990, Sverdrup and Warfvinge 1993b). The minimum leaching is introduced because experiments show that there is a lower limit for uptake. It can also be interpreted as taking into account that the roots do not perfectly penetrate the soil, and some nutrients will always escape uptake.

Plant-soil chemistry interactions

Some of the basic conditions for nutrient uptake change when the soil is acidified. High concentrations of H and soluble inorganic Al may cause less nutrients and more Al to be available at the root-soil interface. The tree will have to overcome a "counterpressure" of Al at the root-soil interface to get the proper nutrient flow into the plant. The root surface will be occupied by an abundance of H and Al in addition to Ca, Mg or K, and the root surface occupied by H and Al will therefore not be effective in uptake of other useful ions (Cronan 1991, Sverdrup and Warfvinge 1993b). Further, if Al and heavy metals released in the soil are taken up, they may disturb physiological processes in the plant.

It is known from laboratory studies that Al in the soil solution will be able to disturb the growth of trees, and that this can be expressed in terms of the BC/Al ratio (Schulze 1987, Cronan et al. 1987, 1989, Arovaara and Ilvesniemi 1990, Sverdrup and Warfvinge 1993b). The response function uses bulk concentrations of base cations and inorganic Al. Concentrations may well be different at the root surface because of local conditions

there, but the experimental design recorded the connection between bulk soil solution chemistry and plant growth, integrating surface effects in the empirical relation obtained.

In the model, growth reduction is calculated as relative growth in percent of a normal level defined as the level in absence of limitations from soil acidity. The reference growth level is set to 100%. Relative growth effect of soil acidity is calculated by using the following response function (Sverdrup and Warfvinge 1993b):

$$f(BC/Al) = \frac{[BC^{2+}]^m}{[BC^{2+}]^m + k \cdot ([Al] + m \cdot [H^+])^n} \quad (3)$$

where n = m = 1 for spruce and m = 3 and n = 2 for pine and beech. For Norway spruce, a value of k = 0.35±0.05 was obtained through a statistical fit to the laboratory data. The data had a standard deviation of ±15% (Sverdrup and Warfvinge 1993b). Al is the molar concentration of dissolved inorganic charged Al. The expression takes the adverse effect of both inorganic Al and low pH into account, but also the ameliorating effect of base cations.

The relation for Norway spruce (n = m = 1) can be derived from the data on a purely empirical basis, but an uptake model based on an ion exchange analogy according to Nye and Tinker (1977), Cronan (1991) and Sverdrup and Warfvinge (1993b), will also yield Eq. 1.

The response expression derived for single plants in the laboratory has been scaled up, and applied to the whole stand. But it differs from the laboratory response expression in one very important way. If the uptake specified for a certain layer cannot be satisfied, either due to deficiency or low BC/Al-ratio, then the residual uptake is taken from another layer. If this reallocation cannot secure supply, then uptake, and hence growth is reduced. This allows the tree to optimize its uptake over the soil profile, and partly counteract the effect of Al by reallocating uptake to a horizon with less Al. This algorithm predicts that rooting depth should become shallower in very acid soils. Local variation in soil quality as well as interactions between individual trees in a stand has been ignored. Figure 2 shows how nutrient limitations and soil acidity were used to modify the growth calculated using a yield model.

Calibration and initial conditions

All dynamic models used for calculating changes in soil chemistry must be triggered from an equilibrium situation. They converge towards a steady state under constant inputs. These initial values are calculated with PROFILE, a steady state version of SAFE (Warfvinge and Sverdrup 1992a,b,c, Sverdrup and Warfvinge 1993a).

Parameter	Unit	1	2	3	4
Morphological characterization		O	A	B	C
Soil layer thickness	m	0.08	0.15	0.5	0.5
Soil depth	m	0–0.08	0.08–0.25	0.25–0.75	0.75–1.25
Moisture content	$m^3\ m^{-3}$	0.25	0.2	0.2	0.25
Soil bulk density	$kg\ m^{-3}$	800	1100	1400	1500
Specific surface area	$m^2\ m^{-3} \cdot 10^6$	0.47	0.71	0.88	0.9
Cation exchange capacity (CEC)	$keq\ kg^{-1} \cdot 10^{-6}$	12.2	3.8	0.96	0.33
CO_2 pressure	times ambient	2	5	20	30
Dissolved organic carbon	$mg\ l^{-1}$	30	15	2	0.5
log gibbsite eq. constant	$kmol^2\ m^{-3}$	6.5	7.5	8.5	9.2
Inflow	% of precipitation	100	64	50	50
Percolation	% of precipitation	64	50	50	50
Mg+Ca+K uptake	% of total max	50	40	10	0
N uptake	% of total max	50	40	10	0
1800 steady state soil H		5.3	5.6	6.3	6.5
1800 steady state [BC]	$\mu mol(+)\ m^{-3}$	200	200	160	180
1990 measured base saturation	%	45	4	4	8
Mineral	% of total				
K-Feldspar		31	31	30	29
Oligoclase		29	29	30	28
Hornblende		0	2	4	4
Biotite		3	3	3	4
Vermiculite		0	2	5	3
Apatite		0.1	0.2	0.3	0.3

The PROFILE model cannot be calibrated; all inputs are determined by measurement. SAFE is only calibrated on initial base saturation.

Model validation

For the models used in soil acidification assessments, credibility is of great importance. If the model is to be used for making predictions, it is a requirement that it is also capable of reconstructing the past. Both SAFE and PROFILE have been tested in many countries and seem to be applicable to a wide range of soils and climates (Sverdrup and Warfvinge 1993a). The SAFE model has been tested on historical data at three sites from Skåne, southern Sweden (Warfvinge et al. 1993), and at the Rothamsted site in England (Sverdrup et al. 1995b).

The soil samples from the site at Rothamsted are unique. For the Rothamsted site, soil properties and complete soil chemistry data from 1883, 1904, 1964 and 1991 were available, and the original soil samples are still available. The deposition measurements at the site began in 1844. All input variables to the model are known through measurements, including initial conditions. No parameter was available for calibration, and a true test of the model was performed. The model performed well in the test. Both pH and base saturation were reproduced well, layer by layer. Base cations showed less good fit with the available data, but the mass balance for base cations was correct.

Fårahall case study
Site description

The Fårahall catchment is located on the northern side of the Hallandsåsen Hill in southern Sweden, between 140 and 180 m a.s.l., 30 km from the coast. Annual rainfall is 1050 mm, annual runoff is 45 mm. The site has a 60–80 yr old Norway spruce stand, on a former *Calluna* heathland used for sheep grazing. The site is drained by a small stream, too small to have fish (0.09 km^2). The soil appears to be acidified down to c. 2 m depth (Jacks 1994, Knutsson et al. 1995). According to estimates, deposition was 1.5–1.8 kEq S $ha^{-1}\ yr^{-1}$, 0.6–0.8 kEq NO_3 $ha^{-1}\ yr^{-1}$, 0.5–0.7 kEq NH_4 $ha^{-1}\ yr^{-1}$, and 1.04 kEq BC $ha^{-1}\ yr^{-1}$ (Lövblad et al. 1992). The historical development of deposition was partly derived from a study by EMEP (Mylona 1993) and from the deposition monitoring from Rothamsted going back to 1844 (Johnston et al. 1986, Sverdrup et al. 1995b). Detailed information, including everything needed for modelling the soil at Fårahall, can be found in Knutsson et al. (1995). The site was selected to represent a sensitive, exposed and high deposition Norway spruce site in southern Sweden. The data required by SAFE, with exception of inputs required as

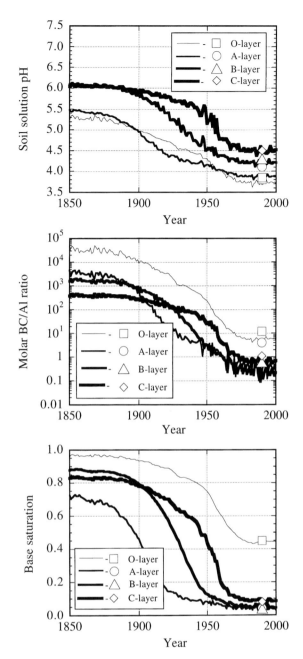

Fig. 3. Calculated development during the period 1850–2000 of soil solution pH, base saturation and soil BC/Al ratio in the forest soil at Fårahall, a Norway spruce stand on the Hallands-åsen Hill in southern Sweden. The dots represent measurements.

time series, such as deposition and uptake, have been listed in Table 1.

Soil chemistry results

Figure 3 shows soil pH, base saturation and BC/Al ratio for Fårahall catchment during 1850–2000. The model

calculation passes through the current observed soil chemistry. The measured soil chemistry in 1991 is included in the diagram, generally this is reproduced by the model calculations. At a nearby power-line clear-cut, soil pH varied from pH 5.2 in the top layer to pH 6 at 1.0 m depth as measured in samples taken in 1951. For 1951, pH 4.6 is calculated for the forested site, i.e. 0.6 pH units lower than for the clear-cut site under the power line. This appears reasonable, considering the clear-cutting effects on soils.

The lower diagram in Fig. 3 shows the development of base saturation with time. The model run was calibrated by varying the base saturation at the starting point in 1800, the 1991 base saturation as target values. The model indicates that base saturation of the soil was significantly higher in the past. This result has been criticized, but it it strongly supported by the historical data available from different sites in Sweden. Studies going back to the 1940's and 1950's, show a consistent pattern of base saturation values in the range from 25 to 65%, that can hardly be ignored (10 sites in southern Sweden, Falkengren-Grerup 1987, Falkengren-Grerup et al. 1987, and 40 sites in southern Sweden, Falkengren-Grerup and Eriksson 1990).

Figure 3 shows that in 1991, soil solution BC/Al values measured in the O-, A-, B- and C-layers were 12, 4, 0.6 and 1.1. This compares favorably with the calculated BC/Al values for these layers, which were 8, 0.8, 0.5, 1. It can be seen that the soil is close to steady state with respect to the current deposition rate.

The measured base cation concentrations showed very much scatter down through the soil profile, whereas the model predicts much less valiation.

Reconstructing the weathering rate

At Fårahall, the weathering rate was estimated in the field, using two different methods (Jacks et al. 1989). First, the historical post-glacial weathering rate was estimated using titanium as a conservative tracer. This method estimates the amount of material lost in the soil profile since it was formed. The average weathering rate since the last glaciation, 12 000 yr ago, is c. 60 meq m^{-2} yr^{-1}.

Secondly, the weathering rate was estimated using the strontium isotope ratio method (Jacks et al. 1989). This estimates the present weathering rate for calcium. The weathering rate is calculated assuming Sr weathering to be proportional to Ca weathering. If the ratio between the weathering rate for Ca and other base cations are currently the same as they have been historically then the present weathering rate is in the range of 67–73 meq m^{-2} yr^{-1}.

The calculations using the SAFE model reproduce these two values excellently. The SAFE estimate was based on soil mineralogy and texture, and by considering

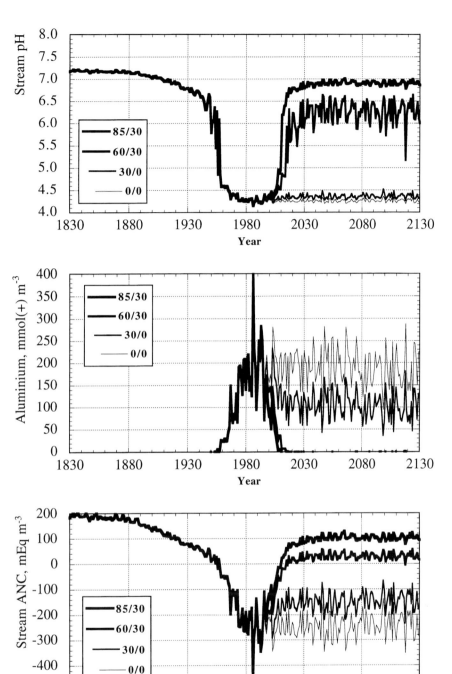

Fig. 4. Calculated past and future development of stream pH, stream inorganic Al and ANC at Fårahall. The diagrams show the response to continued deposition at the present rate and three scenario future deposition rates. In 1991 the average stream chemistry was pH 4.1, inorganic Al varied in the range 30–120 meq m^{-3} and ANC was c. −170 meq m^{-3}.

soil conditions like pH, cation concentrations, organic acids and temperature. The model reconstructs the historic rate as c. 59 meq m^{-2} yr^{-1} (measured was 60 meq m^{-2} yr^{-1}) and the current as 71 meq m^{-2} yr^{-1} (measured was 67–73 meq m^{-2} yr^{-1}) for the upper 1 m of the soil.

Stream chemistry results

Figure 4 shows calculated water chemistry for the stream draining the Fårahall site. The upper diagram shows changes in stream water pH with time. The middle dia-

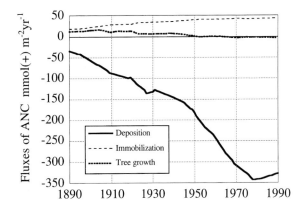

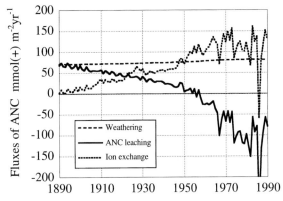

Fig. 5. Mass balance for acidity at Fårahall using output from the model. The upper diagram shows the inputs of acidity to the soil. The lower diagram shows how the input acidity is handled by the system.

Mass balance for acidity

Figure 5 shows a mass balance for acidity at Fårahall calculated from the model outputs. The upper diagram shows the inputs of acidity to the soil. Deposition of acidity is the largest input. The effect of tree growth is calculated as the difference between N uptake plus N immobilization in organic matter and base cation uptake.

The lower diagram shows how the acidity input is handled by the system. The acidity is partly neutralized by weathering of minerals and rocks in the soil. This term does not vary much over time, but it permanently removes acidity from the system. A large part of the incoming acidity is initially neutralized by cation exchange, but as the base cation exchange reservoir is depleted, the amount of acidity that is neutralized becomes less with time. It is evident that the base saturation is consumed by incoming acids and the soil is leached of base cations in the process.

gram shows the calculated concentration of total inorganic Al and the bottom diagram shows stream ANC. In 1991, the measured average pH in the stream was pH 4.1, while the calculated pH value varied between 4.1 and 4.3. The observed inorganic Al concentration varied in the range 30–120 meq m^{-3}, the calculated concentration varied from 70–270 meq m^{-3}, a slight overprediction. Observed annual average ANC was c. −170 meq m^{-3} during 1988–1992, and it varied between −150 to −350 meq m^{-3}, which is slightly more acid than the observed.

The calculations indicate that the stream had already started to acidify in 1880, in 1955 ANC had been lowered to 0 meq m^{-3} and pH had reached pH 6.0. The inorganic Al concentration was >50 meq m^{-3} in 1968, according to the calculations. Conditions for fish reproduction were very unfavourable after 1968; according to this reconstruction, the stream would have been empty of fish by 1970. The natural pH of this stream was pH 6.5 to 7.5 and with an ANC of 200 meq m^{-3}, assuming that today's concentration of DOC is the same as 1840. This is the typical surface water acidification story.

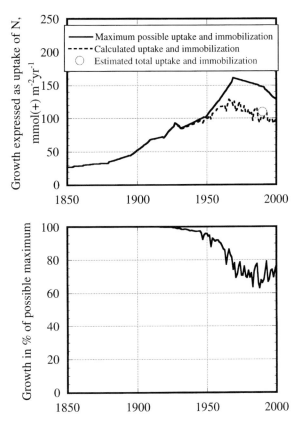

Fig. 6. The diagram shows the calculated effect of soil acidity on growth as expressed by N uptake. Maximum possible N uptake as limited by deposition is reduced by the value of the BC/Al function calculated by the model, to give the actual uptake. This uptake is close to estimated uptake and immobilization.

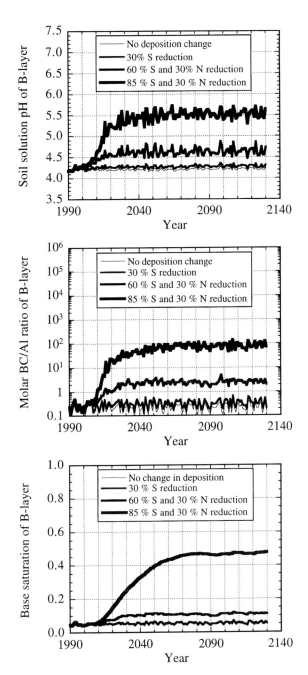

Fig. 7. Calculated development in soil pH, BC/Al ratio and base saturation at Fårahall under different scenarios.

possible growth as set by N availability, and the dotted line the actual calculated growth. For Fårahall no record of earlier tree growth exist, but the dot represents an estimate of present N uptake. The difference between the two lines represents the reduced uptake and hence growth lost due to acid deposition.

Effect of deposition reductions

The different deposition scenarios studied in both the dynamic calculations and the regional scenarios later, were: 1) No deposition change after 1988. 2) 30% S deposition reduction. 3) 60% N and 30% N deposition reduction. 4) 85% S and 30% N deposition reduction.

Figure 7 show the effect of the scenarios from 1990 to 2130. The middle diagram shows the recovery of the BC/Al ratio and the bottom diagram the recovery of the base saturation in the B-layer, during the period 2000–2130.

Development of soil solution pH and BC/Al ratio in the B-layer, corresponding to 0.23–0.73 m soil depth, at the Fårahall site is shown in Fig. 7. The model calculations show that the recovery time of the soil BC/Al ratio to return to values >1.0 anywhere in the soil profile, is c. 25 yr under the 60% N and 30% N deposition reduction scenario. The recovery time for soil pH is of the same magnitude as for the BC/Al ratio. The recovery time for base saturation is much longer, c. 80 yr. When evaluating the 80 yr recovery time, it must be kept in mind that the site has below average CEC for Swedish forest sites, and that the recovery time will be much longer for other Swedish soils.

The right sides of the diagrams in Fig. 4 show the results of different deposition reductions for the stream at Fårahall. It is evident that at least 60% S and 30% N deposition reduction is required for the stream to have conditions favorable to salmonid fish. The recovery time, starting with deposition reductions in 1994, will be 16 yr before the stream can be recolonized by fish again, if the reduction in acidic deposition corresponds to a 60% S reduction and 30% N deposition reduction. The recovery time for the stream is significantly shorter than the recovery time of the deeper soil layers.

Effect on tree growth

Figure 6 shows the calculated effect of soil acidification on tree growth. The model calculates the reduction in growth as a result of the BC/Al response expression and any nutrient limitations. This must be multiplied by the reference growth level to give the actual growth at the site. In Fig. 6 the upper line represents the maximum

Regional assessment

The PROFILE model was used to make assessment of the final soil chemical states under different deposition rates. With the dynamic model we can estimate the time required for certain changes. The PROFILE model is used to explore how some of these effects will be geographically distributed in Sweden. PROFILE was applied to 1804 sites evenly distributed over the forested area of Sweden. The results of the calculations were also used to assess potential effects on growth.

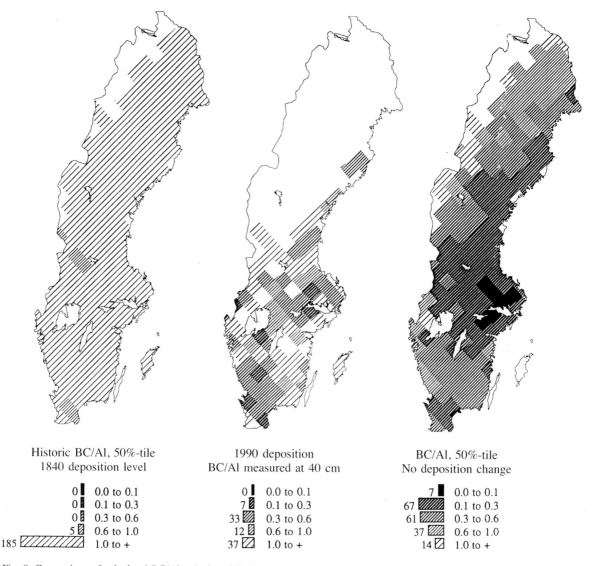

Historic BC/Al, 50%-tile
1840 deposition level

0	■	0.0 to 0.1
0	▧	0.1 to 0.3
0	▨	0.3 to 0.6
5	▨	0.6 to 1.0
185	▨▨▨▨	1.0 to +

1990 deposition
BC/Al measured at 40 cm

0	■	0.0 to 0.1
7	▧	0.1 to 0.3
33	▨	0.3 to 0.6
12	▨	0.6 to 1.0
37	▨	1.0 to +

BC/Al, 50%-tile
No deposition change

7	■	0.0 to 0.1
67	▨	0.1 to 0.3
61	▨	0.3 to 0.6
37	▨	0.6 to 1.0
14	▨	1.0 to +

Fig. 8. Comparison of calculated BC/Al ratio for 1840 (left map), observed BC/Al ratio in the 1985–87 Forest Inventory (middle map) and predicted BC/Al at steady state (right map), assuming no further change in acid deposition in relation to 1990. The number associated with the bars in the legend indicate the numbers of 50 × 60 km squares in each category, shown values apply to the upper B-horizon.

All data in this work are based on the framework of the Swedish Forest Inventory. They are the same as for the calculation of critical loads described in Sverdrup et al. (1992, 1993) and Sverdrup and Warfvinge (1995a). The assessments are all based on data derived ultimately from measurements at each site and analysis of soil samples from 1804 sites. All key parameters derive from actual measurement.

Deposition data files were constructed by the Swedish Environmental Res. Inst. (IVL) in Göteborg. The deposition pattern was calculated using EMEP model calculations calibrated against field deposition monitoring stations.

Regional deposition scenarios explored with the model, are with respect to our regional deposition data base; −30%S, −60%S/−30%N and −85%S/−30%N scenarios in relation to the 1988 deposition level. 1840 deposition level was taken to be 0.05 keq ha^{-1} yr^{-1} of S, 0.01 keq ha^{-1} yr^{-1} of nitrate and 0.15 keq ha^{-1} yr^{-1} of ammonium.

Results

BC/Al ratios

Steady state chemical conditions were calculated for forest soils, streams and groundwater, in order to be able to

344

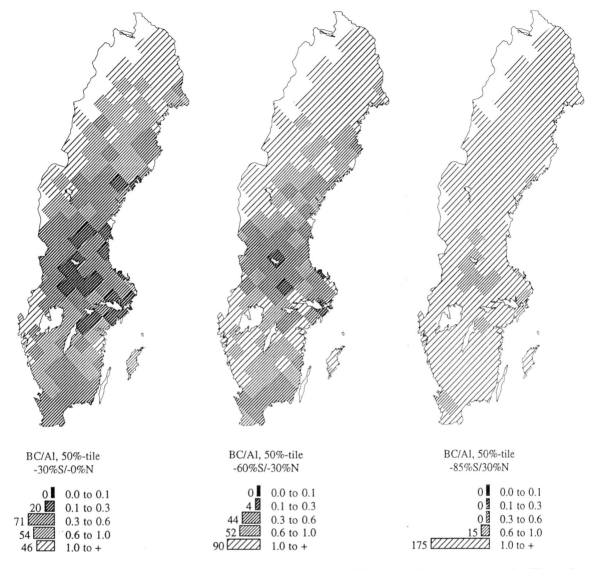

Fig. 9. Predicted future steady state BC/Al ratios in Swedish forests under different deposition reduction scenarios. The numbers associated with the bars in the legend indicate the numbers of 50 × 50 km squares in each category, shown values apply to the upper B-horizon.

judge effects of different levels of critical loads exceedances. For forest soils, pH, soil solution ANC, Al, base cation concentration and BC/Al was calculated at steady state under the different deposition scenarios.

A pre-industrial deposition rate (1840) was chosen as a deposition scenario in order to reconstruct soil chemistry for the time before the advent of acidic deposition and intensive silviculture.

The regional model was used to estimate regional values of BC/Al in 1840, and at steady state with the 1988 deposition rate, these values are compared to measurements from the samples taken in the forest inventory of 1983–1985 (Fig. 8). The comparison show that the presently observed soil BC/Al values are significantly lower than those reconstructed for 1840, but that the observed values are approaching the predicted future values in the southern part of Sweden. It can be seen that the sites in southernmost Sweden are approaching steady state, and according to the dynamic calculations, this has in many cases occurred fairly recently.

Predicted changes in the geographical distribution of BC/Al values and how they will change when the deposition is changed according to the −30%S/−0%N, −60%S/ −30%N and −85%S/−30%N scenarios, are shown in Fig. 9. It can be seen that both the scenarios −30%S/−0%N and −60%S/−30%N are insufficient to fully protect Swedish forest soils from acidification. Reducing acid deposition according to −85%S/−30%N will provide ade-

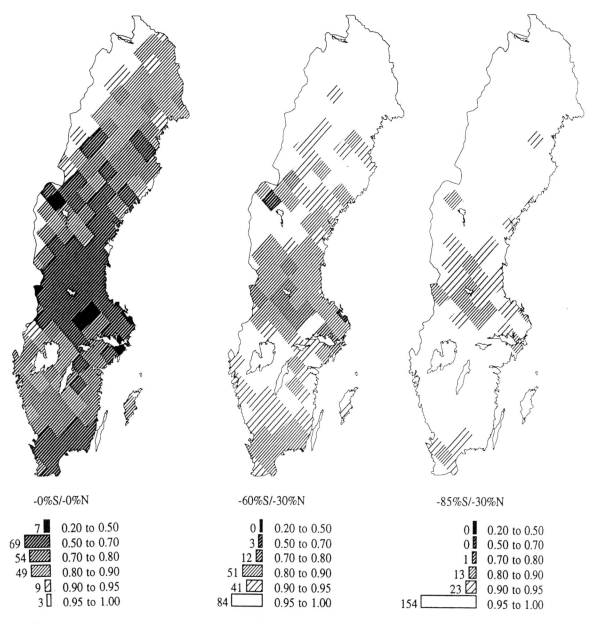

-0%S/-0%N

7 ■	0.20 to 0.50
69 ▨	0.50 to 0.70
54 ▨	0.70 to 0.80
49 ▨	0.80 to 0.90
9 ▨	0.90 to 0.95
3 ▯	0.95 to 1.00

-60%S/-30%N

0 ▮	0.20 to 0.50
3 ▮	0.50 to 0.70
12 ▨	0.70 to 0.80
51 ▨	0.80 to 0.90
41 ▨	0.90 to 0.95
84 ▭	0.95 to 1.00

-85%S/-30%N

0 ■	0.20 to 0.50
0 ▮	0.50 to 0.70
1 ▨	0.70 to 0.80
13 ▨	0.80 to 0.90
23 ▨	0.90 to 0.95
154 ▭	0.95 to 1.00

Fig. 10. Predicted future impact on forest growth under different deposition scenarios. The maps show relative remaining fraction of the growth potential. 1.0 corresponds to the growth potential with no limitations due to soil acidification. The numbers associated with the bars in the legend indicate the number of 50×50 km squares in each category.

quate protection for 95% of the Swedish forest area. It is also evident that even the −85%S/−30%N reduction scenario results in a pattern different from that calculated for 1840.

Future effects on forest growth

Calculated BC/Al was used to estimate future reductions in forest growth potential in Sweden, utilizing the BC/Al response function described in Eq. (1). The results obtained for the 50%-tile is shown in Fig. 10 for different deposition rates. The results have been summarized in Table 2. The calculations suggest that drastic reductions in acidic deposition are required, to avoid risks for adverse effects on the forest production of Sweden. 5% exceedance is achieved at an acidity reduction corresponding to −85%S/−30%N reduction, implying that 95% of the area would get less acid deposition than the critical load.

Table 2. The effect of acidification on forest growth, expressed as fraction of reference level growth. The last column show the area weighted average value for Sweden. The results were calculated with the PROFILE model. The final effect in the field is a result of this acidification effect together with promoting effects of temperature and nitrogen deposition.

| Region | BC/Al-ratio range | | | | Growth reduction |
| | ≥1.0 | 1.0–0.3 | 0.3–0.1 | ≤0.1 | function value in % |
	% of area weighted growth				
1990 level	19	34	46	1	20
−30% S	44	34	22	1	14
−60% S	54	28	18	–	8
−60% S/−30% N	83	14	3	–	3
−85% S/−30% N	96	3	1	–	0
1840 level	100	–	–	–	0

The development of soil acidification over time varies quite significantly between sites. The time aspect of these regional dynamic calculations have been summarized in Table 3. The time of changes in soil chemistry in many cases coincides with recent changes in needle loss in southern Sweden.

Discussion

Uncertainties

The uncertainty of the results of model application can be divided into different elements. Uncertainty may arise from uncertainty in model formulation and basic principles, from uncertainty in input data estimation and from uncertainties in the representativeness of the input data used.

Effect of parameter uncertainties

For the PROFILE and SAFE model, some of these aspects have been systematically investigated. Jönsson et al. (1994) carried out a systematical Monte Carlo analysis of error propagation and uncertainties in the output from the PROFILE model, caused by variations in the input data. A sensitivity analysis of the SAFE model was done by Warfvinge and Sandén (1992) with much emphasis on the effect of uncertainties in soil capacity properties and hydrological variation. Sverdrup and Warfvinge (1993a) tested the accuracy of the calculation of weathering rates in the field, by testing the PROFILE model against field observations.

Uncertainties are to a certain degree controllable, be-

cause of the heavy reliance on measured parameters for the model, and the lack of calibration for PROFILE.

The SAFE sensitivity analysis showed the model to be sensitive to hydrology with respect to short term chemical changes, but very robust with respect to long term trends. The tests of the PROFILE model's capability to calculate field weathering rates were excellent, in general the weathering rate is predicted with ±10% of the observed value, but often significantly better. The weathering rate estimate is totally dependent on the accuracy of the important input data such as mineralogy, soil moisture saturation and soil texture. The trend in the model is that very high accuracy in inputs yield the same very high accuracy in weathering rate, supporting the idea that the model formulation is basically sound. The quality of the weathering calculation is a major strength of the model.

At Rothamsted, CEC changes significantly over the period 1884–1990, but the ion exchange selectivity coefficients stayed constant. Though both are assumed to stay constant in the model, the model was still able to reproduce soil pH and base saturation with good accuracy over the whole period.

Representativeness of data

The Monte Carlo analysis of the PROFILE model indicates that the uncertainty in the calculated critical load will be in the order of ±15% if input data with the standard deviations reported in the literature are used (Jönsson et al. 1995), the weathering rate can be calculated with an accuracy better than ±8% (Sverdrup and Warfvinge 1993a). For regional studies, the accuracy of the predictions becomes dependent on the number of calculation points in each grid. Jönsson et al. (1995) were

Table 3. Time when the BC/Al ratio will pass below the critical limit 1, continuing 1990 deposition rate.

Region	Latitude	Spruce high yield	Spruce low yield	Pine high yield	Pine low yield
North	62–70°N	2010–2030	2010–2050	2010–2030	2030–2060
Middle	58–62°N	1990–2010	2000–2030	1990–2020	2020–2030
South	55–58°N	1985–2000	2000–2030	1990–2010	2000–2010

able to show that if the error in input data is stochastically randomly distributed, the cumulative distributions of critical loads, BC/Al ratios etc, will be stable if there are >7 points per 50 × 50 km grid, implying >1370 points for Sweden. This implies that there will be equally as many errors making the values too small as there are making the values too large. In terms of mapping, this could be stated that we are able to calculate the right values, but they are not always placed exactly in the correct position on the map. For Sweden the number of points was 1804.

Model structure

Large uncertainties can arise from the basic structure given to a model, and the way processes are formulated. In the model, one of the uncertainties arise from the scaling up of process descriptions from laboratory bench-scale to ecosystem units up to 12 km² in size. For some processes (weathering) this scaling up has been tested and verified as valid, for others (tree response to Al), this type of validation has has failed for a number of reasons. For a few sites in central Europe with much stronger soil acidification and larger variation in tree damage, a correlation between soil acidity, weathering rate and tree damage can be demonstrated (Bonneau 1991, Sverdrup and Warfvinge 1993b, Nelleman and Frogner 1994), but for areas with less obvious damage, this has not been demonstrated. Thus when we scale up laboratory experiments to ecosystem scale, we are estimating an effect with a certain probaility of occuring, but which by no means is to be considered as certain.

An other element of uncertainty propagation in the calculations is related to how the physical structure of a soil is represented. The number of layers used in the model can be shown to be important; a four-layer model gives significantly better correlation between field measurement and calculated soil or stream chemistry than a one-layer model (Warfvinge et al. 1992, Sverdrup et al. 1995b). The importance of the multi-layer structure for stream chemistry predictions arises from the fact that runoff from different soil depths dominates under different hydrological regimes. A multi-layer model can also be tested against lysimeter data, for one-layer models this will be less relevant.

Model strength and deficiencies

The foremost strength of the model is the good performance of the weathering rate sub-model. It has repeatedly, in both deliberate and blind tests against field data, been able to predict the weathering rate within ±15%. The model takes into account the effect of H^+, hydrolysis with water, CO_2, the dissolving effect of different organic acids, as well as product inhibition caused by base cations and Al, in good consistency with both laboratory and field data. The inclusion of the weathering rate sub-model the model, implies that it is possible to restrict calibration to one single parameter, initial base saturation, making the application procedure for the model transparent.

There are some significant weaknesses in the model

formulation which will be subject to future revision. At present, the models utilize the gibbsite forcing function which has the appearance of an equilibrium equation. It almost goes without saying that the gibbsite forcing function has several serious shortcomings. The coefficient is a standardized calibrated value, specific for each horizon; it cannot be determined from laboratory experiments. The expression is not valid in soils with pH <3.9. Probably, several non-equilibrium processes are responsible for the results the gibbsite forcing function is meant to simulate.

In the models, organic matter decomposition or accumulation is only taken into account as input numbers for each year. No feedback between decomposition and soil chemistry has been implemented in the model. Obviously, the organic pool is not at steady state at all times, and this is a source of error.

In summary, the performance of the dynamic model performance is "good" for the following parameters; soil solution pH, alkalinity, base saturation and weathering. It is "acceptable" for long term average of base cation concentrations and BC/Al ratio, but "poor" for Al. The performance for nitrate and ammonium cannot be judged since the model is incomplete in this respect. The performance of the model on the effect of soil acidity on growth cannot be tested at present because of lack of suitable data for the purpose.

Soil chemistry development

The calculations indicate that the soil chemistry of Sweden will continue to change unless deposition reductions can be achieved. The implementation of the Oslo Protocol will be of importance for avoiding severe soil and water acidification. The soils of southern Sweden will continue to acidify under the full implementation of the Oslo Protocol, still 45% of the forest area will approach BC/Al values <1. Through the effect of the protocol, the rate of soil and water acidification will become significantly slower, but the recovery will in many areas be modest. Soil with high weathering rate and at present low excedance, have best possibilities for showing a recovery. Areas with high present exceedance will probably show no recovery, but only slower acidification.

Implications for forest growth

An explanation of how better growth and significant adverse effects on vitality can occur at the same time is illustrated in Fig. 6.

The diagram shows calculated tree growth at Fårahall, expressed as uptake of N to trees. The upper line represents calculated potential growth, expressed as uptake of N at the site. If uptake is limited by N availability alone, it is equal to total N deposition. The lower line is the same growth adjusted for the effect of soil acidity. The adjustment is done using the response function based on the

BC/Al ratio, and applied in PROFILE. It is apparent that even if growth as expressed by N uptake increased from 1930 to 1990, there is a significant impact of soil acidification. Growth is almost 30% less than it would have been if N was the only limiting factor. Thus it is not a contradiction that the forest grows better than ever, at the same time as the effect of soil acidification on growth is significant. The calculated "loss in growth potential" is the difference between the two lines, growth that would have occurred without the effect of soil acidification. Any limitations to growth set by management methods, water stress, climate, ozone etc, are not considered in Fig. 6.

In Sweden there is 267 000 km^2 of forest, of which 235 000 km^2 is considered productive silviculture. The Swedish Forest Inventory estimated total stem growth in 1990–1993 to 90–95 million m^3 yr^{-1}, and c. 70 million m^3 yr^{-1} of this is presently harvested.

We can try to estimate present growth losses using an empirical model connecting observed needle loss to growth (Söderberg unpubl.). If presently observed increase in needle loss since 1984 is entered into Söderbergs regressions, it can be calculated that this needle loss will cause a loss of growth of c. 5–10%. Our model predicts that growth losses due to soil acidity will rise to c. 15–25% by 2040, if the deposition continues at the 1988 rate, under the Oslo Protocol this will be only 7–10% in 2040. Converting the above calculated growth potential loss to needle loss by using the Söderberg regressions for southern Sweden, give as result less than the currently observed needle loss. This suggests that other factors in addition to soil acidification may influence needle loss in these regions. Using the present growth loss estimated by the Söderberg regression, we can estimate that present growth unaffected by soil acidity factors may be 100 million m^3 yr^{-1} (5% less growth than possible at present).

Nutrient limitations with respect to base cations are estimated to limit forest growth in the future to c. 80–85 million m^3 yr^{-1}, quite independently of any soil acidification. This number is obtained by estimating how much stemwood the available base cations can produce, using todays nutrient composition of stemwood. The limitation is set by weathering plus base cation deposition, assuming weathering in the upper 0.5 m of the soil to be available for trees. For Norway spruce, average rooting depth is sometimes significantly <0.5 m, making this a conservative estimate. This is the long term sustainable reference level for forest growth. Any reduction caused by other factors (acidification, gas effects, pathogens) will in the long-term start from that reference level.

Thus, in the short run, observable growth losses will be small, since there is at present an excess growth which is supplied by deposited N and base cations leached by acidity.

The growth reductions predicted under different scenarios are shown in Table 2. The last column in the table show the area weighted value for Sweden of the integrated soil profile value of the BC/Al response function

as applied in the model. This value should be applied to the long term sustainable growth to estimate longterm growth loss from soil acidification. A growth reduction of 15–20% will imply that c. 63 million m^3 yr^{-1} will be available for harvest. In a better case, where the upper limit of the weathering estimate is used, and assuming a high tolerance to Al (BC/Al = 0.5), this may increase to 73 million m^3 yr^{-1} available for harvesting. Assuming a low case for weathering and more sensitive trees (BC/Al = 2), the amount available may shrink to 55 million m^3 yr^{-1}.

Taking the optimistic case for weathering and tree sensitivity, assuming that the Oslo Protocol will be implemented, we can estimate that only 80–85 million m^3 yr^{-1} of today's growth of 100 million m^3 yr^{-1} can be harvested in the longterm as a result of base cation limitations. Loss of growth potential corresponding to 5–7 million m^3 yr^{-1} as a result of soil acidification effects by 2040 may persist. With the implementation of the Oslo Protocol, unsustainability with respect to base cations appear as the greater problem.

In areas where base cations can still be taken from the base saturation, no immediate significant growth reductions will be seen. At present, growth may be accellerating because of the extra N available and the steady inprovement of stand through management. At such sites, growth will be limited by temperature and water. N promotion and high base cation leaching will maintain growth and keep BC/Al ratios >1 for some time yet. This corresponds to the present situation in all northern Sweden and many sites in middle Sweden.

At sites where no more base cations can leached, a stage may be reached where adverse effects appear quickly, and the effects on growth may be significant. Nutrient limitations will become evident through deficiency symptoms. Further Al effects will become visible, and growth and uptake will be restricted. This corresponds to the situation in a majority of the sites in southernmost Sweden, and a minority of the sites in middle Sweden.

Necessary deposition reductions

The assessment indicates that significant reduction in deposition rates will result in a fairly rapid chemical recovery of the soils, and soil acidification seems to be reversible. Whether ecosystem recovery in terms of biology is fully reversible, is unknown.

In order to reach a situation where 95% of the forested area will have no exceedance, acid deposition will have to be reduced by at least 75% of the 1985 rate. This implies 80% reduction in S and 30% reduction in N, but, for the sake of economic optimization, the reductions may be distributed differently between S and N. It should be made quite clear that abatement of S alone will not be sufficient. S accounts only for c. 65% of the deposited acidity in southern Sweden. Interesting of course, is the future effect of the protocol on S reductions signed in 1994 in Oslo. After implementation of this protocol, S deposition in southern Sweden will be reduced by up to 75%. In northern Sweden, approximate estimates indicate a reduction of 55–60% by 2010 compared to the situation in 1980. Assuming that S deposition accounts for 65% of the total acidity deposition, this implies that the new S protocol will lead to a reduction in deposited acidity of 50% in the south and 35% in the north. The 60/30 scenario used as an illustration above is in relation to 1988, and imply a 65% reduction in deposited acidity as compared to 1980, the protocol reference. Thus the 60/30 scenario gives an approximate picture of the result to be expected from the present S protocol and current N reduction plans.

The sulphur protocol signed in Oslo in 1994, will probably imply an average of 60% S deposition reduction in Sweden as compared to 1980, but more in the south and significantly less in the far north.

Conclusions

Acid deposition has the capacity to cause significant soil chemical changes in >75% of the forest area in Sweden. It is obvious that this will change some of the important factors that influence growth and health of trees and ground vegetation. If the sensitivity of tree growth measured in the laboratory can be transferred to field conditions as described here, then there is cause for concern that productive growth may be affected. No massive forest die-back is predicted. Instead a slowly progressing decline in growth potential is the most likely result unless the rate of deposition changes. The growth potential could drop to 80–90% of maximum sustainable growth.

Reducing acid deposition does reverse the major soil chemical effects of soil acidification. Deposition of acidifying compounds like sulphuric acid, nitric acid and ammonium is the dominating cause of soil acidification in all parts of Sweden. Fluxes related to natural podsolization and increased uptake of nutrients by trees are significant but smaller sources of acidity.

The results stresses the necessity to make risk assessments, and to make decisions on the amount of reductions to be implemented. Reductions cannot be achieved overnight; the response time for a reduction to take effect may be 10–30 yr. For the recovery of the base saturation to historical levels, 50–300 yr are required. This makes it important that decisions are made now on the necessity for and the eventual implementation of preliminary mitigative actions, to protect against adverse effects until the long term reductions have taken effect.

Both models are available free of cost from the authors for Macintosh and PC computers. PROFILE exists in a scientific version with a user-friendly interface and a users manual. Regional variants of the models are available.

References

Alcamo, J., Hordijk, J., Kämäri, J., Kauppi, P., Posch, M. and Runca, E. 1985. Integrated analysis of acidification in Europe. – J. Environ. Manage. 21: 47–61.

Arovaara, H. and Ilvesniemi, H. 1990. The effect of soluble inorganic aluminium and nutrient imbalances on Pinus sylvestris and Picea abies seedlings. – In: Kauppi, P. (ed.), Acidification in Finland. Springer, Berlin, pp. 715–733.

Barnard, J. and Lucier, A. 1991. Changes in forest health and productivity in the United States and Canada. – In: Irving, P. (ed.), Acidic deposition; State of science and technology. Summary report of the U.S. national acid precipitation assessment program. U.S. Governm. Printing Office, Washington, pp. 135–138.

Becker, M. 1991. Impacts of climate, soil and silviculture on forest growth and health. – In: Landmann, G. (ed.), French research into forest decline – DEFORPA programme, 2nd report. Ecole Nationale du Génie Rural, des Euax et des Forest, Nancy, pp. 23–38.

Berdén, M., Nilsson, S., Rosen, K. and Tyler, G. 1987. Soil acidification, extent, causes and consequences. – An evaluation of literature information and current research. – National Swedish Environ. Protect. Board, Solna, Report 3292.

Bonneau, M. 1991. Effects of atmospheric pollution via the soil. – In: Landmann, G. (ed.), French research into forest decline – DEFORPA programme, 2nd report. Ecole Nationale du Genie Rural, des Euax et des Forest, Nancy, pp. 87–100.

Cosby, J., Hornberger, G. and Wright, R. 1985. Estimating time delays and extent of regional de-acidification in southern Norway in response to several deposition scenarios. – In: Kämäri, J., Jenkins, A., Brakke, D.F., Wright, R.F. and Norton, S.A. (eds), Regional acidification models. Springer, Berlin, pp. 151–166.

Cronan, C.S. 1989. Aluminium toxicity in forests exposed to acidic deposition; The Albios results. – Water Air Soil Pollut. 48: 181–192.

– , Kelly, J.M., Schofield, C. and Goldstein, R.A. 1987. Aluminium geochemistry and tree toxicity in forests exposed to acidic deposition. – Selper, London, pp. 649–656.

– , April, R., Barlett, R.J., Bloom, P.M., Driscoll, C.T., Gherihrin, S.A., Henderson, G.S., Joslin, J.D., Kelly, J.M., Newton, R.M., Parnell, R.A., Patterson, H.H., Raynar, D.J., Schaedele, M., Schoefield, C.L., Sucoff, E.I., Tepper, H.B. and Thornton, F.C. 1991. Differential adsorbtion of Al, Ca and Mg by roots of red spruce (Picea rubens sarg.). – Tree Physiol. 8: 227–237.

Falkengren-Grerup, U. 1987. Long-term changes in pH of forest soils in southern Sweden. – Environ. Pollut. 43: 79–90.

– and Eriksson, H. 1990. Changes in soil, vegetation and forest yield between 1947 and 1988 in beech and oak sites of southern Sweden. – Forest Ecol. Manage. 38: 37–53.

– , Linnermark, N. and Tyler, G. 1987. Changes in soil acidity and cation pools of south Swedish soils between 1949 and 1985. – Chemosphere 16: 2239–2248.

Jacks, G. 1994. The Fårahall hillslope. – In: Maxe, L. (ed.), Effects of acidification on groundwater in Sweden. – Hydrological and hydrochemical processes. Dept Environ. Engineering, Royal Inst. of Technology, Stockholm, pp. 15–30.

– , Åberg, G. and Hamilton, P.J. 1989. Calcium budgets for catchments as interpreted by strontium isotopes. – Nordic Hydrol. 20: 85–96.

Johnston, A., Goulding, K. and Poulton, P. 1986. Soil acidification during more than 100 years under permanent grassland and woodland at Rothamsted. – Soil Use Manage. 2: 3–10.

Jönsson, C., Warfvinge, P. and Sverdrup, H. 1995. Uncertainty in prediction of weathering rate and environmental stress factors with the PROFILE model. – Water Air Soil Pollut. 77: in press.

Kämäri, J. 1986. Critical deposition limits for surface water assessed by a process-oriented model. – In: Nilsson, J. (ed.),

Critical loads for nitrogen and sulphur. Nordic Council of Ministers, Copenhagen, 11: 121–142.

Kempe, G., Toet, H., Magnusson, P.H. and Bergstedt, N.-J. 1992. The Swedish National Forest Inventory 1983–1987. – Dept Forest Survey, Swedish Univ. of Agricult. Sci., Umeå, Vol. 51.

Knutsson, G., Bergström, S., Danielsson, L.-G., Jacks, G., Lundin, L., Maxe, L., Sanden, P., Sverdrup, H. and Warvinge, P. 1995. Acidification of groundwater on forested till areas. – Ecol. Bull. (Copenhagen) 44: 271–300.

Lövblad, G., Amann, M., Andersen, B., Hovmand, M., Joffre, S. and Pedersen, U. 1992. Deposition of sulfur and nitrogen in the Nordic countries: Present and future. – Ambio 21: 339–347.

Moseholm, L., Andersen, B. and Johnsen, I. 1988. Acid deposition and novel forest decline in central and northern Europe. – Miljørapport 1988: 9, Nordic Council of Ministers, Copenhagen.

Mulder, J., van Grinsven, J. and van Breemen, N. 1987. Impacts of acid atmospheric deposition on woodland soils in the Netherlands. III Aluminium chemistry. – Soil Sci. Soc. Am. 51: 1640–1646.

Mylona, S. 1993. Trends of sulphur dioxide emissions, air concentrations and depositions of sulphur in europe since 1880. – Tech. report EMEP-MSC-W, Norwegian Meteorolog. Inst., Oslo, Norway.

Nelleman, C. and Frogner, T. 1994. Spatial patterns of spruce defoliation: Relation to acid deposition, critical loads, and natural growth conditions in Norway. – Ambio 23: 255–259.

Nye, P.H. and Tinker, P.B. 1977. Solute movement in the soil-root system. – Stud. in Ecol. 4, Blackwell Scientific Publ.

Pitelka, L. and Raynal, D. 1989. Forest decline and acidic deposition. – Ecology 70: 2–10.

Roberts, R.M., Skeffington, R. and Blank, L.W. 1989. Cause of type 1 spruce decline in Europe. – Forestry 62: 255–259.

Schulze, E. and Freer-Smith, P. 1991. An avaluation of forest decline based on field observations focussed on Norway spruce, *Picea abies*. – Proc. Royal Soc. Edinbourgh 97B: 155–191.

Szhulze, E.-D. 1987. Tree response to acid deposition into the soil, a summary of the COST workshop at Juelich 1985. – In: Mathy, P. (ed.), Air pollution and ecosystems. D. Reidel Publ. Co., pp. 225–241.

SOU 1992. Skogspolitiken inför 2000-talet. – Statens Offentliga Utredningar, 1992: 76. Almänna Förlaget, Stockholm.

Sverdrup, H. 1990. The kinetics of chemical weathering. – Lund Univ. Press, Lund.

– and Warfvinge, P. 1988. Weathering of primary silicate minerals in the natural soil environment in relation to a chemical weathering model. – Water Air Soil Pollut. 38: 387–408.

– and Warfvinge, P. 1993a. Calculating field weathering rates using a mechanistic geochemical model – PROFILE. – J. Appl. Geochem. 8: 273–283.

– and Warfvinge, P. 1993b. Soil acidification effect on growth of trees, grasses and herbs, expressed by the (Ca+Mg)/Al ratio. – Lund Univ. Rep. Environ. Engineering Ecol. 1: 1–1348.

– , de Vries, W. and Henriksen, A. 1990. Mapping critical loads. – Miljørapport 1990: 15, Nord 1990: 98, Nordic Council of Ministers, Copenhagen.

– , Warfvinge, P., Frogner, T., Håöya, A.O., Johansson, M. and Andersen, B. 1992. Critical loads for forest soils in the Nordic countries. – Ambio 21: 348–355.

– , Warfvinge, P. and Jönsson, C. 1993. Critical loads of acidity and nitrogen for forest soils, groundwater and first-order streams in Sweden. – In: Hornung, M. and Skeffington, R. (eds), Critical loads, concepts and applications. NMSO, pp. 54–67.

– , Warfvinge, P. and Nihlgård, B. 1995a. A risk assessment on ecological effects and economic impacts of acidification on forestry in Sweden. – Water Air Soil Pollut. 78: 1–36.

– , Warfvinge, P., Blake, L. and Goulding, K. 1995b. Modelling recent and historic soil data from the Rothamsted experimental station, England, using SAFE. – Agricult. Ecosyst. Pollut. 53: 161–177.

Tamm, C. and Hallbäcken, L. 1985. Changes in soil pH over a 50-year period under different canopies in SW Sweden. – In: Martin, H.C. (ed.), International conference on acid precipitation, Muskoka, Canada, Part 2. – Riedel Publ. Co., pp. 337–341.

Ulrich, B. 1983. An ecosystem oriented hypothesis on the effect of air pollution on forest ecosystems. – In: Persson, G. and Jernelöv, A. (eds), Ecological effects of acid deposition. Swedish Environ. Protect. Board, Solna, PM 1636, pp. 221–231.

– 1985a. Interaction of indirect and direct effects of air pollutants in forests. – In: Troyanowsky, C. (ed.), Air pollution and plants. Gesellsch. Deut. Chem., VCH Verlagsgesellschaft, Weinheim, pp. 149–181.

– 1985b. Stability and destabilization of central European forest ecosystems, a theoretical data based approach. – In: Cooley, J. and Golley, F. (eds), Trends in ecological research for the 1980's. NATO Conference Series, pp. 217–237.

Vries, W. de 1987. The role of soil data in assessing the large scale impact of atmospheric pollutants on the quality of soil water. – In: Duivenboden, W. van and Waege, J.H. (eds), Proc. int. conf. vulnerability of soil and groundwater to pollutants, pp. 897–910.

– and Kros, J. 1989. Modelling time patterns of forest soil acidification for various scenarios. – In: Kämäri, J., Bracke, D., Jenkins, A., Norton, S. and Wright, R. (eds), Regional acidification models. Geographic extent and time development. Kluwer Academic Publ., pp. 113–128.

– , Posch, M. and Kämäri, J. 1989a. Simulation of the long-term response to acid deposition in various buffer ranges. – Water Air Soil Pollut 48: 349–390.

– , Schoumans, O.F., Kragt, J.F. and Vreeuwsma, A. 1989b. Use of models and soil survey information in assessing regional water quality. – In: Jousma, G., Haimes, Y.Y. and Walter, F. (eds), Groundwater contamination, use of models in decision making. Springer, Berlin, pp. 419–432.

Warfvinge, P. and Sandén, P. 1992. Sensitivity analysis. – In: Warfvinge, P. and Sandén, P. (eds), Modelling groundwater response to acidification. SMHI, Norrköping, pp. 119–146.

– and Sverdrup, H. 1992a. Calculating critical loads of acid deposition with PROFILE – a steady-state soil chemistry model. – Water Air Soil Pollut. 63: 119–143.

– and Sverdrup, H. 1992b. Hydrochemical modelling. – In: Warfvinge, P. and Sandén, P. (eds), Modelling groundwater response to acidification. SMHI, Norrköping, pp. 79–117.

– and Sverdrup, H. 1992c. Scenarios for acidification of groundwater. – In: Warfvinge, P. and Sandén, P. (eds), Modelling groundwater response to acidification. SMHI, Norrköping, pp. 147–156.

– , Sverdrup, H., Ågren, G. and Rosen, K. 1992. Effekter av luftföroreningar på framtida skogstillväxt. Skogspolitiken inför 2000-talet – 1900 års skogspolitiska kommittee. – Statens Offentliga Utredningar; 1992: 76, Allmänna Förlaget, Stockholm, pp. 377–412, (in Swedish).

– , Falkengren-Grerup, U., Sverdrup, H. and Andersen, B. 1993. Modelling long-term cation supply to acidified forest stands. – Environ. Pollut. 80: 209–221.

Wright, R.F., Kämäri, J. and Forsius, M. 1988. Critical loads for sulfur, modelling time response of water chemistry to changes in loading. Miljørapport 15, Nordic Council of Ministers, Copenhagen, pp. 201–224.

Ecological Bulletins 44: 352–362. Copenhagen 1995

Complexity versus simplicity in modelling acid deposition effects on forest growth

Harmke van Oene and Göran I. Ågren

Van Oene, H. and Ågren, G.I. 1995. Complexity versus simplicity in modelling acid deposition effects on forest growth. – Ecol. Bull. (Copenhagen) 44: 352–362.

Models originating from different research fields such as soil science or tree physiology are used to analyse effects of deposition of anthropogenic air pollutants on forests. These different origins have resulted in models with different complexities for similar processes. In this paper, we compare the consequences of process formulations for modelling effects on forest growth and nutrient cycling. The differences in complexities between models lead to different views on the buffer capacity of the ecosystem. We discuss how much complexity is justified and required to be able to address effects of acid deposition.

H. van Oene and G.I. Ågren, Dept of Ecology and Environmental Research, Swedish Univ. of Agricultural Sciences, P.O. Box 7072, S-750 07 Uppsala, Sweden.

Introduction

A variety of models has been developed to analyse effects of air pollutants. Since the adverse effects were first noticed as acidification of bodies of freshwater, the emphasis was initially on lake and associated soil chemistry. Models specialized on soil (chemical) processes, sometimes coupled with detailed hydrology, were used for this purpose, e.g. ILWAS (Gherini et al. 1985), MAGIC (Cosby et al. 1985), and ETD (Nikolaidis et al. 1988). The development of this type of model has continued with NUCSAM (Groenenberg et al. 1995), SAFE (Warfvinge et al. 1993), and SOILN (Johnsson et al. 1987). When it later became apparent that there were both direct and indirect impacts of air pollution on the vegetation, some existing specialized forest growth models were used independently, or coupled together with, or used input from, soil acidification models. Examples of these are FORGRO (Mohren and Van Veen 1995), PNET-CN (Aber and Federer 1992), and TREEDYN (Bossel and Schäfer 1989). The need to link effects of anthropogenic deposition and soil acidification on forest growth has also called for an integrated approach in which the models have been designed from the beginning to include the whole atmosphere-vegetation-soil complex (FIWALD, Schall 1991; FORMS, Oja et al. 1995; FOR-

SOL, Van Minnen et al. 1995; NAP, Van Oene 1992; NIICE, Van Dam 1995, and SOILVEG, Berdowski et al. 1991). A representation of these models in the Plant Physiology Complexity – Soil Processes Complexity phase space is given in Fig. 1. We will use the amount of information (qualitatively estimated) required to use a model as a measure of the complexity of the model. There is some arbitrariness in assigning values to the complexities of the models in Fig. 1, but it is apparent that soil process models, forest growth models, and integrated models form three distinct classes, each with a large variability in complexity.

The different origins (soil-based, forest growth, or integrated) have led to models with different basic assumptions and process formulations. Thus a large variation exists, from very complex models, containing almost all involved processes described in great detail, to simpler models, containing only a limited number of processes and with highly aggregated descriptions. There are also differences in how the complexity of the models is distributed over different parts of the model. Differences in predictions from the models can therefore be attributable to different process descriptions, or to different parameter values and initial state variables. The uncertainty of some models and their sensitivity to parameter values has been analysed (Schecher and Driscoll 1988, Kros et al. 1993,

352

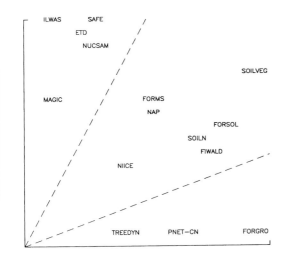

Fig. 1. Representation of models in the Plant Physiology Complexity – Soil Processes Complexity phase space. The assignment of coordinates to the models is based on a qualitative estimation of the amount of required input information needed to use the models.

Van Oene and De Vries 1994). The consequences of process formulations are, however, hardly touched upon. Rose et al. (1991a,b) and Cook et al. (1992) made a systematic comparison of three models with an input mapping procedure for parameters, thereby providing consistent inputs. The aggregation over temporal and spatial scales implicit in process formulation is, however, a far from trivial problem, Rastetter et al. (1992).

In this paper we want to compare the consequences of process descriptions in models used for predicting effects of air pollutants on forest growth. We want to address such questions as: How much do the processes or assumptions used affect the output of the models? How much complexity is required to address the effects of air pollution on forest ecosystems? How much complexity is justified in view of the knowledge we have? Does an increase in complexity actually increase our understanding or does it obscure our view of the system?

We are interested in descriptions of the impacts of air pollutants at the ecosystem scale. We also restrict ourselves to time scales of a whole rotation period. At this level of resolution we can distinguish three major components – vegetation, soil, and atmosphere – and two interacting linkages – nutrient and water cycling – that we need to scrutinize. We will concentrate on soil-mediated effects acting through the nutrient supply on forest growth and not pay any attention to the atmospheric part of the environment, because direct effects attributable to SO_2 and ozone can be considered as special cases. Normally the impact is regarded as small (Roberts et al. 1989). We do not intend to go into the details of particular models but rather to select a number of processes the description of which varies between models.

Selected processes and approach

We will restrict our interest to the way in which models describe processes affecting nutrient availability, uptake by trees, nutrient-induced growth limitations, and the feedback mechanisms between soil and plants. We do this because we believe these processes to be among the most important but also because the approach we will choose is most suitable for studying these processes, which will be analysed by modelling alternative formulations. The complexity with which soil profile, soil processes and hydrology are described is discussed in relation to their importance and to their influence on nutrient availability.

To analyse the effects of different process descriptions we will use our own model, NAP (Nutrient Availability and Productivity), as a pivotal point where we will increase or decrease the complexity with which certain processes are described. NAP is an integrated ecosystem model that was built for analysing the effects of decreasing soil nutrient stores on growth of forests affected by acid deposition (Van Oene 1992, Van Oene and Ågren 1995). NAP may be characterized as a small to medium sized model that focuses on nutrient interactions between the plants and the soil. The model describes a forest, with three compartments for trees: fine roots, wood (stem, branches and coarse roots), and needles. Tree growth is limited by N or Mg. Nitrogen is the growth-limiting nutrient under normal conditions and Mg is taken as a potential limiting nutrient in forests affected by acid deposition. Magnesium has been reported as deficient in some German forests (Zöttl 1987). Variable nutrient concentrations of N and Mg are simulated in the needle biomass, whereas the other compartments have fixed nutrient concentrations. NAP has one soil compartment for inorganic chemical reactions and includes cation exchange reactions, Al-hydroxide equilibrium, CO_2 dissolution, and sulphate adsorption. Additionally, there is one pool of organic matter with a fixed yearly mineralisation rate. All N not taken up by the trees is nitrified and leached. Input data are the initial biomass, nutrient and organic matter data, soil chemistry parameters, deposition, volumetric water content and the outflow of water. The model runs with a time step of one year.

We changed the description of processes in NAP to represent those used in other models and compared the outputs. Since the models are very different in construction it is not possible to copy a process description from one model into another. We have instead worked with generic process formulations when representing other types of model.

For the comparison, we have used characteristics of two ecosystems, Solling (Germany) and Skogaby (Sweden), which are affected by acidification to different extents. We emphasize that our interest is in comparing process descriptions. The choice of the sites is based on data accessibility and by using two data-sets with different site conditions the effects of process descriptions can emerge more clearly. The assumptions behind one pro-

Table 1. Input data, initial amounts, parameters and simulated variables that characterize the Skogaby and the Solling* site (soil depth 0.50 m).

	Skogaby	Solling[a]
Total deposition (kg ha^{-1} yr^{-1})		
N/S/Mg/Ca	16/16/3.6/4.0[b]	40/58/2.2/12.3
Initial amounts (kg ha^{-1})		
wood dry weight	76390[c]	81719
needle dry weight	14520[c]	18187
organic N	917	565
organic Mg	31	33
exchangeable Mg	42[d]	34
exchangeable Ca	148[d]	99
Initial base saturation	0.10	0.02
(Mg and Ca)		
Mg weathering (kg ha^{-1} yr^{-1})	1.7[e]	3
Mineralization rate (yr^{-1})	0.05[f]	0.02
Simulated variables		
Needle growth rate	2500	3200
(averaged; kg ha^{-1} yr^{-1})		
Limiting nutrient	N	first 12 yr N, thereafter Mg

*adjusted Solling site (see text), [a]all Solling data from Van Oene and Ågren (1995), [b]pers. comm. O. Westling, [c]Nilsson and Wiklung (1992), [d]pers. comm. J. Bergholm, [e]Van Oene (1992), [f]derived data T. Persson.

cess formulation might hold for one site, but not for the other. The Solling site is an old spruce forest characterized by a very high acid load and almost depleted nutrient stores (very low base saturation) (Ellenberg et al. 1986). To avoid problems in the regeneration phase, or caused by senescence, we adjusted this data-set to correspond to a 20-year-old stand with a closed canopy as the initial condition (Van Oene and Ågren 1995). The Skogaby site is a 30-year-old spruce forest, situated near the southwest coast of Sweden with low to moderate deposition (Nilsson and Wiklund 1992). Characteristic data for these sites are given in Table 1. Consequences of different model formulations were evaluated at year 55 (Skogaby) and 47 (Solling) of the simulation, when wood biomasses of two alternatives (see below) are equal to each other, and at year 80.

Growth and nutrient limitations

Most soil process models typically use highly aggregated plant growth models and lump all plant biomass into one compartment, see Fig. 1. In this approach "growth" is an input variable required to calculate uptake of nutrients and normally based on static vegetation data. The integrated models and the forest growth models separate plant biomass into morpho-physiological parts and assume physiologically based growth processes. Within these groups, some models use a more empirical approach and describe growth with equations based on regional yield tables or other empirical relations.

These different approaches provide an ability and flexibility of the models to simulate forest growth in response to changing nutrient availability. The soil process models either do not change growth, i.e. uptake of nutrients, or reduce it by applying soil solution based criteria (Al concentration, Ca/Al ratio etc.). Physiologically based models can reduce the maximum (light determined) photosynthesis with nutrient limitations by empirically derived equations (FORGRO). Another approach is to reduce growth on the basis of the concentration of the most limiting nutrient according to Liebig's law of the minimum (SOILVEG). The approach we use in our model NAP is based on the nutrient productivity concept (Ågren 1983, 1985, 1988) and relates growth to the amounts of nutrients in the needle biomass.

We will compare four alternative descriptions.

Alternative G1

Nutrient-determined growth, which is the description used in NAP:

$$dW_l/dt = \min_i(P_iZ_i) \text{ with } P_i = a_i - b_iW_l \qquad (1)$$

$$dW_w/dt = k_{all}dW_l/dt \qquad (2)$$

where W is biomass (kg ha^{-1}), P_i is the productivity of nutrient i [(kg biomass) (kg nutrient)$^{-1}$ yr^{-1}], Z_i is the amount of nutrient i in needles (kg ha^{-1}), a_i and b_i are nutrient- and species-specific parameters, k_{all} is an allocation parameter [g dw wood (g dw needle)$^{-1}$]. Subscript l stands for leaf, w for wood (i.e. stem, branches, coarse roots), and i stands for nutrient N or Mg.

Alternative G2

Growth unlimited by nutrients. We model this alternative by setting a constant N concentration of 25 mg g^{-1} in eq. 1. Normally, this approach would require a model in which temperature and light control growth. Such a model often has a detailed canopy (e.g. FORGRO) requiring a large number of additional parameters.

$$dW_l/dt = P_N0.025W_l \text{ with } P_N = a_N - b_NW_l \qquad (3)$$

$$dW_w/dt = k_{all}dW_l/dt \qquad (4)$$

Alternative G3

Constant growth and uptake rate of nutrients. The uptake data represent net growth uptake, i.e. accumulation of nutrients in wood biomass. We set this rate equal to the one calculated during the first year of growth in the model. The needle biomass and needle nutrient concentrations remain at their initial level.

Table 2. Simulated biomass (ton ha⁻¹) and needle concentrations (mg g⁻¹) for Skogaby and Solling after 55/47 and 80 yr of simulation. (Note that alternatives G3 and G4 have fixed concentrations during the whole simulation period and therefore represent the concentrations at year 1 for all alternatives.) (–: The simulation period is too long, there is not enough N and Mg and the trees would already have died.)

	Skogaby				Solling			
	G1	G2	G3	G4	G1	G2	G3	G4
Year 55/47								
Wood biomass	382	610	358	381	480	714	523	478
Needle biomass	18	30	15	15	18	27	18	18
Needle N concentration	14.0	1.5	13.5	13.5	21.7	10.0	17.7	17.7
Needle Mg concentration	1.00	0.11	0.94	0.94	0.67	0.15	0.78	0.78
N:Mg ratio	14.0	13.6	14.4	14.4	32.4	66.7	22.7	22.7
Year 80								
Wood biomass	582	848	485	436	791	1154	832	536
Needle biomass	18	30	15	15	20	27	18	18
Needle N concentration	14.2	–	13.5	13.5	25.7	13.2	17.7	17.7
Needle Mg concentration	1.00	–	0.94	0.94	0.71	0.19	0.78	0.78
N:Mg ratio	14.2	–	14.4	14.4	36.2	69.5	22.7	22.7

Alternative G4

Logistic growth. A variation on alternative G3, but instead of a constant growth and uptake rate, net growth rate is modelled with a logistic function:

$$dW_w/dt = r \cdot W_w \cdot (1 - W_w/W_{w,max})$$ (5)

where r is the annual growth rate (yr⁻¹), 0.06 and 0.09 for Skogaby and Solling, respectively. $W_{w,max}$ is the maximum amount of wood biomass and is 450 and 540 ton ha⁻¹ for Skogaby and Solling, respectively. The maximum wood production in Skogaby is based on yield tables for site class G32 in the Swedish classification system (the height of the dominant spruce trees is 32 m at age 100 yr, Eriksson 1976). The maximum at Solling is derived from the Skogaby data, based on a 20% higher needle growth rate in alternative G1. These maximum

levels are estimated growth of these stands under "unpolluted" conditions. The needle biomass and needle nutrient concentrations remain at their initial level.

The biomass production and the nutritional status of the simulated alternatives are given in Table 2. The differences between the sites are obvious. The Skogaby site has a good nutritional status (N concentration < 2%, N:Mg < 25), whereas the Solling site has a rather unbalanced nutritional status. Note though, that the production at Solling is higher than at Skogaby in spite of this imbalance and even with the Mg limited growth of alternative G1.

Alternative G2 is unrealistic. Growth controlled by light and temperature implies a high enough availability of nutrients to maintain normal nutrient concentrations. In the cases here, the amounts of nutrients available are far too small to sustain the growth obtained under only light and temperature limitations. However, physiologically based models which do not consider nutrient limi-

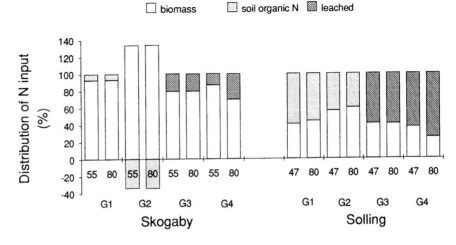

Fig. 2. Distribution of the N input over the system as modelled with different alternative process formulations for growth. The accumulated N input is 880 (kg ha⁻¹) (year 55) and 1280 (year 80) at Skogaby and 1898 (year 47) and 3230 (year 80) at Solling.

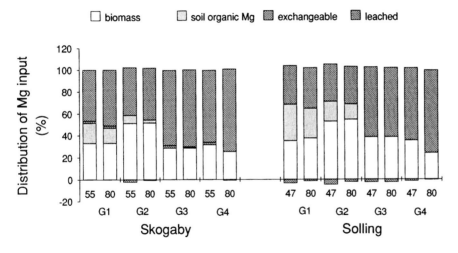

Fig. 3. Distribution of Mg input over the system as modelled with different alternative process formulations for growth. Negative values mean decreases of the compartment. The accumulated Mg input is 292 (kg ha⁻¹) (year 55) and 424 (year 80) at Skogaby and 246 (year 47) and 418 (year 80) at Solling.

tations on growth may handle limitations in nutrient uptake indirectly.

The effects of an approach in which growth is related to nutrients as compared to a predefined growth, can be seen by comparing alternatives G1 and G3. These alternatives work in opposite directions at the two sites. For the Skogaby site G3 underestimates production compared to G1 whereas for the Solling site the reverse occurs. The two sites represent different stages of acidification: N-enhanced growth at the Skogaby site, where acidification does not yet show negative effects, and Mg-limited growth at the Solling site, where acidification (external and internal) has almost depleted the nutrient stores. The occurrence of both N-enhanced growth and Mg-limited growth during the simulation period cannot be handled by alternative G3. The Mg limitation on growth at Solling in alternative G1 emerges in the reinforced nutrient imbalance during the simulation, cf. in Table 2 the N:Mg-ratios at years 47 and 80, respectively.

The logistic uptake function of alternative G4 indicates that comparisons are dependent on the year chosen for evaluation. The differences between G1, G3, and G4 are small at the first evaluation but have become much larger at 80 yr.

The different biomass productions imply differences in nutrient usage. Figs 2 and 3 give the budgets for N and Mg in which the input to the system (deposition for N, and deposition plus weathering for Mg) is distributed to different parts of the ecosystem. The alternatives differ much in the amounts of N leached. Differences in accumulation of N in the biomass explain part of these differences. At the Skogaby site, the N-stimulated growth in G1 accumulates more N in the wood biomass than in the predefined growth alternative G3 (Table 3). In addition, extra N is stored in the needle biomass, resulting in a somewhat higher N concentration (Table 2). At the Solling site, the situation is more complex. After 47 yr, the greater growth in G3, with a higher N accumulation in wood than in G1, is offset by the extra N stored in the needle biomass in G1 because of its flexible nutrient concentration. Therefore, the differences in N leaching between the alternatives caused by the accumulation in biomass, are negligible at this point. After 80 yr however, a larger amount of N is stored in the needle biomass in

Table 3. Accumulation of N and Mg (kg ha⁻¹) in wood and needle biomass in alternatives G1 and G3.

	N						Mg					
	Skogaby			Solling			Skogaby			Solling		
	G1	G3	G1–G3	G1	G3	G1–G3	G1	G3	G1–G3	G1	G3	G1–G3
Year 55/47												
Wood	961	895	66	846	920	−74	115	106	9	106	112	−6
Needle	252	196	56	391	318	73	18	14	4	12	14	−2
Sum	1213	1091	122	1237	1238	−1	133	120	13	118	126	−8
Year 80												
Wood	1325	1213	112	1393	1464	−71	159	143	15	174	178	−4
Needle	261	196	65	506	318	188	18	14	5	14	14	0
Sum	1586	1409	177	1899	1782	117	177	157	20	188	192	−4

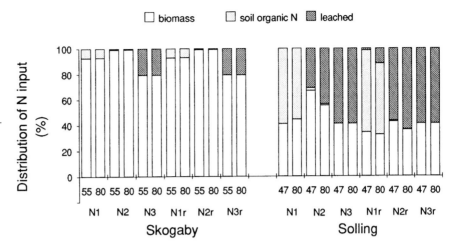

Fig. 4. Distribution of N input over the system as modelled with different alternative process formulations for nutrient cyling. The accumulated N input is 880 (kg ha⁻¹) (year 55) and 1280 (year 80) at Skogaby and 1898 (year 47) and 3230 (year 80) at Solling.

G1. This results in a larger total N accumulation in the biomass in G1, in spite of the higher production in G3. In G4, leaching increases with time as a result of the decline in growth rate and thus less N uptake.

Nutrient cycling

The annual circulation through uptake, litterfall and mineralization normally provides the trees with the larger part of their uptake. Nitrogen in particular is tightly circulated with very low losses. Anthropogenic N deposition can drastically change this situation. In combination with leaching of cationic nutrients induced by acid deposition, N deposition may cause a change in the nutritional balance of the system.

The integrated models and forest growth models all include the processes of uptake, litterfall and mineralization in an explicit way. Differences between models may be in the detail of these processes, e.g. litterfall divided over needle age classes and organic material

divided into several pools with different mineralization rates. The other models use uptake as an input variable but some include litterfall and mineralization; these last two processes are, however, modelled in a simple way assuming that mineralization equals litterfall, i.e. a steady state situation. This assumption also lies behind the exclusion of these processes in some models.

We will consider three different alternatives to describe nutrient cycling. Other processes are described as in G1.

Alternative N1

Non-steady state nutrient cycling with growth response to nutrient availability. This is the NAP model with specific death rates for the total needle biomass and the roots and a single pool of organic matter with a constant mineralization rate.

$$dW_{org}/dt = -k_{org}W_{org} + k_{d,l}W_l + k_{d,r}W_r \qquad (6)$$

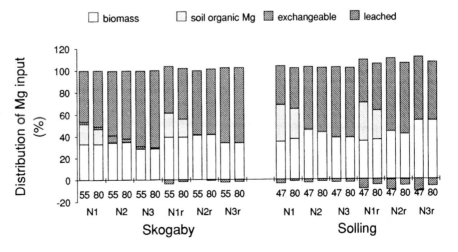

Fig. 5. Distribution of Mg input over the system as modelled with different alternative process formulations for nutrient cycling. Negative values mean decreases of the compartment. For N1, N2 and N3, the accumulated Mg input is 292 (kg ha⁻¹) (year 55) and 424 (year 80) at Skogaby and 246 (year 47) and 418 (year 80) at Solling. For N1r, N2r and N3r, the accumulated Mg input is 245 (kg ha⁻¹) (year 55) and 356 (year 80) at Skogaby and 175 (year 47) and 298 (year 80) at Solling.

Table 4. Differences between N1 and N3 in the amounts stored and partitioning of this difference over different compartments.

Year 80	N		Mg	
	Skogaby	Solling	Skogaby	Solling
Difference in stored amount N1–N3 (kg ha^{-1})	264	1910	82	110
Caused by				
growth	41%	–4%	19%	–4%
variable concentrations	25%	10%	6%	0%
accumulation as soil organic	34%	94%	71%	104%
accumulation as exchangeable			5%	–1%

$$dZ_{org,i}/dt = -k_{org}Z_{org,i} + (1 - f_{l,ret,i})k_{d,l}Z_{l,i} + (1 - f_{r,ret,i})k_{d,r}Z_{r,i} \quad (7)$$

where k_d is death rate (yr^{-1}). Subscript r stands for root, ret for retained, and org for organic. For N, the uptake is determined by either the amount of N mineralized plus deposition or the uptake capacity of the plants. For Mg, the uptake is dependent on both root biomass and soil solution concentration, up to the point where Mg is in excess and the uptake capacity of the plants is saturated.

Alternative N2

Steady state assumption with growth response to nutrient availability. Soil organic matter is in steady state, i.e. mineralization equals litterfall. Uptake is modelled as in N1. The nutrient concentration in the needle biomass is variable.

$$k_{org}Z_{org,i} = (1 - f_{l,ret,i})k_{d,l}Z_{l,i} + (1 - f_{r,ret,i})k_{d,r}Z_{r,i} \quad (8)$$

Alternative N3

Steady state assumption without growth response to nutrient availability. The same steady state assumption as in N2 for soil organic nutrients, but uptake is litterfall plus uptake for wood growth. The nutrient concentrations in the needle biomass are fixed and wood growth is an input variable. There is therefore, no growth response to nutrient availability in this alternative, unlike N1 and N2.

These different approaches result in large differences in the way in which the inputs of N and Mg are divided between different parts of the ecosystem with subsequent effects on losses, Figs 4 and 5. The consequences of the assumption of a steady state nutrient cycling are clearly seen by comparing alternative N1 (no steady state) with N2 (steady state) for the two sites. At the N-limited Skogaby, over 90% of the input accumulates in the trees with only a small accumulation as soil organic N. Thus the differences between N1 and N2 are small and both have a tight N cycling without N losses. At Solling, a large amount of N is accumulated as soil organic N in the

N1 alternative, whereas the steady state alternative (N2) is incapable of accumulating more N and subsequently N losses occur. The differences between the two steady state alternatives N2 and N3 result from the larger accumulation of N in the trees in N2 by N-stimulated growth or higher N needle concentrations (see section above). In Table 4 the differences in storage of N between N1 and N3 is divided between effects attributable to growth, to variable nutrient concentration in the needle biomass and to accumulation in soil organic matter. At Skogaby, the smaller leaching losses of N in alternative N1, which are a consequence of the more efficient storage mechanisms, are 66% explained by the accumulation in the biomass, of which extra growth contributes somewhat more than a higher N content. The remaining 34% is attributable to accumulation in soil organic N. At Solling the picture is quite different and is only 6% explained by accumulation in the biomass, whereas as much as 94% is attributable to accumulation of soil organic N.

The differences in Mg losses between N1 and N3 are accounted for in an almost similar way (Table 4). Compared to N, the accumulation of soil organic Mg is more important at both sites. Additionally, a small part of the difference in Mg leaching is caused by accumulation (Skogaby) or depletion (Solling) of exchangeable Mg.

After 80 yr, the differences in exchangeable amounts of Mg between the alternatives are at most 8 kg ha^{-1} (Skogaby) and 1 kg ha^{-1} (Solling), in spite of the smaller amounts of N leached in N1 (the corresponding values for Ca are 4 kg ha^{-1} (Skogaby) and 13 kg ha^{-1} (Solling). Thus, the different model approaches result in an almost identical depletion of exchangeable nutrients. However, there are major differences in Mg cycling between the model formulations. In alternative N1, the depletion of exchangeable nutrients is to a large extent attributable to internal acidification (uptake by forest), whereas leaching of external inputs is the major cause of the soil acidification in N3. Alternatives N2 and N3 exhibit similar behaviour at both sites with almost similar fractions of Mg losses. On the other hand, alternative N1 differentiates between the sites by showing a more conservative behaviour of Mg at the Solling site where this nutrient is scarce. We analysed this behaviour further by running the alternatives N1, N2 and N3 with a reduced Mg input; weathering rates were halved at both sites. The results (Figs 4 and 5, N1r, N2r and N3r) confirm the differentiation between sites in N1r and now also in N2r. Alternative N3r decreases leaching at both sites as a result of depletion of exchangeable Mg by uptake and thus lower concentrations in the soil solution. Similarly, N1r and N2r reduce leaching at the Skogaby site, but increase it at the Solling site. The reduced availability of Mg affects growth negatively at Solling and therefore decreases the uptake capacity for N, which subsequently results in higher N and Mg leaching. This emphasizes once more the importance of the coupling between growth and nutrient limitations that is present in N1r and N2r, but not in N3r. These simulations may be seen as examples of model flexibility required to address situations such as those reported by Hedin et al. (1994), in which reductions in sulphate deposition were offset by even stronger declining base cation depositions, thereby contributing to an increased sensitivity of poorly buffered ecosystems.

Discussion

The complexity with which processes are included in a model is strongly dependent on the origin of the model. Models originating as pure soil models invariably include many soil chemical and biochemical processes, often with the soil profile divided into several layers, but have lumped plant biomass compartments and simple carbon and nutrient cycling. These models are found in the upper left-hand part of Fig. 1. Physiologically based models of forest (tree) growth models which did not initially include soil chemical and biochemical processes, except for nutrient cycling, typically try to connect to existing soil acidification models. These models are found along the x-axis of Fig. 1, but over a broad range. Each of these two groups of models has enlarged its field of application by including the "missing" parts of the ecosystem. The integrated models, on the other hand, which are intermediate in both physiological and soil chemical respects, originate from a problem-oriented approach, i.e. the model is built for a specific problem. These models form a long band around the y=x-line; in this class there are both very simple and very complex models.

Nutrient cycling, nutrient limitations and ecosystem buffer capacity

The different levels of complexity of the models, or rather how the complexity is distributed over the model, imply different relative importance attributed to different processes. The models considered in this paper all deal with nutrient cycling and availability. The discrepancy in the view on acidification lies in the explanation of the causes of acidification. In some models, acidification is to a great extent externally driven and the buffer capacity of the ecosystem lies in the soil chemical processes which consume protons. In other models, internal acidification through uptake by trees is a stronger factor and an additional buffer capacity is added with the biotic component (Figs 4, 5). The accumulation of N in the stand biomass and soil organic matter increases the resilience of the ecosystem to N losses to a considerable extent. The differences in views, particularly of the buffer capacity of the system, may have consequences for the forest and nutrient dynamics in the long run. In models where acidification is more externally driven, a large part of the N input is rapidly lost from the system, whereas in the models with a larger internal acidification, a large part of the N input remains in the system.

These differences in view arise mainly because of the inclusion or exclusion of the feedback mechanisms between the plants and the soil. The mechanistic approach in which tree growth is dynamically responding to soil nutrient availability, enables a dynamic simulation of nutrient accumulation in the tree biomass. In addition, the implementation of nutrient cycling explicitly without any a priori assumptions, emphasizes the importance of the biological buffer capacity. The greater complexity involved with this mechanistic approach results in a greater generality of the models and an increased insight into underlying mechanisms. These feedback mechanisms can sometimes fairly easily be included in a model.

Nutrient availability, soil profile and soil processes

Soil processes and heterogeneity of soil profiles are also of importance for nutrient availability to plants. The complexity of soil profile description and soil processes varies between models. The knowledge of many below-ground processes, however, is still limited. Several soil processes are included in a very simplified way, e.g. organic acid chemistry, and (de)nitrification. In addition, processes that are described more mechanistically such as cation exchange reactions and Al chemistry, suffer from large uncertainty in the parameter values (Schecher and Driscoll 1987, Kros et al. 1993). A multi-layered description of a soil profile to address the heterogeneity in nutrient availability requires detailed information for all layers. At this moment, lack of knowledge about these processes makes a meaningful parameterization of these models impossible. Furthermore, the top soil layers (with the highest nutrient availability) are particularly sensitive to parameter changes in soil processes that are unknown, such as nitrification (Van Oene and De Vries 1994).

Nutrient uptake and hydrological processes

The complexity with which the models describe hydrology varies. Several models describe a complete water balance with precipitation, interception, evaporation, transpiration, soil water retention characteristics, outflow and snow melt. This complexity of description of hydrology does not correlate simply with either the complexity of soil processes or of plant growth. Instead, the greatest complexity is found in the more physiological models as well as in the soil process models. Only the physiological models include a complete water balance for simulation of transpiration coupled with photosynthesis, whereas the soil process models incorporate a water balance for simulating the leaching fluxes. The mechanisms of nutrient uptake with high time resolution are still rather poorly understood. However, uptake is dependent on the supply of nutrients to the roots by diffusion and mass flow. Transpiration by plants is a major component of the water balance, as well as of major importance for the mass flow of nutrients to plant roots. To estimate mass flow of nutrients to plant roots accurately, the flow of water to the roots has to be apportioned between the soil horizons. Thus a good model of transpiration by the plants is necessary (Grigal 1990). The simulation of transpiration, however, can show a large variation between models, even at a single site (Bouten and Jansson 1995). Additionally, the importance of mass flow for supply of nutrients is dependent on the soil solution concentrations and thus might vary for the same nutrient between sites (Grigal 1990). The modelling of nutrient uptake coupled to water uptake in a multi-layered soil profile requires an adequate knowledge of the root distribution, the transpiration and the nutrient concentrations throughout the soil profile.

How much complexity is required?

The complexity of models with regard to the description of soil profile, soil chemistry, and hydrology should be viewed together with the aim of the model. Simulating effects of acid deposition on forest growth implies a time scale from years to centuries. The limitations in modelling uptake of nutrients affect the complexity required to describe soil profile, soil processes and hydrology. The lack of knowledge about the mechanisms of nutrient uptake probably justifies the aggregation of this process to only one soil compartment. Spatial distribution through the soil profile implies increased uncertainty through nutrient concentrations, transpiration and uptake parameters. The choice of which time step to use is also dependent on the aim and spatial resolution of the model. It seems unreasonable to expect that in the immediate future nutrient uptake in a forest stand can be estimated with a higher time resolution than one year. This makes the use of soil modules running with a high time step or high spatial resolution superfluous for nutrient availability purposes, while for the hydrology input and output, data can be given at a resolution of one day.

How much complexity will increase our understanding?

In addition to these processes of nutrient limitations on growth and the implementation of nutrient cycling, there are other processes which need a stronger emphasis. These relate to the complexity of the ecosystem in its responses to changed environmental conditions.

Uptake

One of the key processes for modelling the effects of anthropogenic deposition on forest growth is uptake of nutrients. The soil-mediated effects of acid deposition on uptake occur through changes in the chemical composition of the soil solution around the roots. Some qualitative effects of this change in soil solution composition on nutrient supply such as the inhibiting effect of Al on uptake of base cations (e.g. Göransson and Eldhuset 1991), are well known. High ratios of NH_4^+/K^+ and NH_4^+/Mg^{2+} also affect uptake (Roelofs et al. 1985). Other effects, such as the antagonism between Ca and Mg (Van Oene 1994), are less known. However, the quantitative effects of these interactions, their possible contribution to nutrient imbalances, and their feedback effect on soil solution chemistry, are not understood. In addition, trees have a large flexibility to adjust their root uptake by increasing the activity of roots in nutrient-rich parts of the soil. These responses and interactions are hard to model without increased understanding of uptake processes.

Soil organic matter – chemistry

The effects of acid deposition on soil solution composition also requires understanding of how soil chemistry affects soil organic matter. Increased Al and SO_4^{2-} concentrations in the soil solution enhance the binding of these ions to organic compounds, with effects on the amounts and speciation of dissolved organic carbon (DOC). Since DOC is an important component in the pH buffering of soils and surface waters (David et al. 1989, Driscoll et al. 1994), changes can affect pH buffering. Weathering induced by organic acids may also be affected. The consequences of the (ir)reversible binding of SO_4^{2-} to organic matter for buffer capacity and long-term soil chemical conditions are not yet known.

Soil organic matter – biology

The chemical reactions of soil solution compounds with soil organic matter also result in qualitative changes of soil organic matter for micro-organisms. The efficiency of microbial decomposition of organic matter (Ulrich 1989), and thus also mineralization of nutrients, may decrease. The effects on forest growth and buffer capacity of the ecosystem can be large. The consequences of the resulting accumulation of organic matter for the long-term nutrient availability need more attention. In addition, there is the question of how long the quality changes in organic matter will affect the system, whether these changes are reversible, and whether improvement will occur with decreased anthropogenic deposition?

Allocation

The growth-stimulating effects of N deposition and the growth limitations through other nutrients result in behavioural responses of trees by changes in allocation pattern (Ericsson 1994). These changes in allocation pattern have both direct and indirect effects on forests. The direct effects, i.e. decreased root biomass and increased above-ground biomass, are well recognised. Still, the mechanistic understanding is insufficient to model these changes. The indirect effects are less recognised but may have substantial effects on nutrient cycling through the effect on litterfall composition. The composition of the litter is affected by changes in the relative amounts of root necromass and needle biomass in the litter input. In addition, the chemical composition of the litter is changed with higher N and lower base cation concentrations. Both these changes affect litter quality, with subsequent effects on decomposition.

Conclusions

Different types of models, with different complexity, emphasize different aspects of the changes in a forest affected by acid deposition. Models that consider the effects of nutrient limitations on forest growth and describe nutrient cycling mechanistically are required to handle the differences in nutrient availability between forest stands. However, it is clear that our models are not yet capable of addressing all responses of the system satisfactorily, because of gaps in knowledge. Extending models to be able to handle these responses requires an increased fundamental knowledge of the processes involved. Modelling can increase and accelerate understanding through a strong interaction between models and experiments. This requires models that address specific problems but are still simple enough to elucidate the mechanisms involved. In the balance between simplicity, such that we can understand what the model is doing, and complexity, such that all relevant processes are included, we believe that increasing the complexity of a model is only warranted when it entails new strong feed-back mechanisms. Otherwise, increased complexity can obscure what we are doing, add to the uncertainty of the model, or both. Further, we believe that model comparisons such as we have made in this paper, improve our insight into conceptual views and basic relationships. Models should be used more frequently as instruments to increase our understanding and not merely as predictive tools.

Acknowledgements – This study was financed by the Swedish Environmental Protection Agency.

References

Aber, J.D. and Federer, C.A. 1992. A generalized, lumped-parameter model of photosynthesis, evatranspiration and net primary production in temperate and boreal forest ecosystems. – Oecologia 92: 463–474.

Ågren, G.I. 1983. Nitrogen productivity of some conifers. – Can. J. For. Res. 13: 494–500.

– 1985. Theory for growth of plants derived from the nitrogen productivity concept. – Physiol. Plant 64: 17–28.

– 1988. Ideal nutrient productivities and nutrient proportions in plant growth. – Plant, Cell and Environ. 11: 613–620.

Berdowski, J.J.M., Van Heerden, C., Van Grinsven, J.J.M., Van Minnen, J.G. and De Vries, W. 1991. SoilVeg: A model to evaluate effects of acid atmospheric deposition on soil and forest. Vol 1: Model principles and application procedures. – Dutch Priority Programme Acidific., Report no. 114.1–02, RIVM, Bilthoven.

Bossel, H. and Schäfer, H. 1989. Generic simulation model of forest growth, carbon and nitrogen dynamics, and application to tropical acacia and European spruce. – Ecol. Model. 48: 221–265.

Bouten, W. and Jansson, P.-E. 1995. Water balance of the Solling spruce stand as simulated with various Forest-Soil-Atmosphere models. – Ecol. Model.

Cook, R.B., Rose, K.A., Brenkert, A.L. and Ryan, P.F. 1992. Systematic comparison of ILWAS, MAGIC, and EDT watershed acidification models: III. Mass balance budgets for acid neutralizing capacity. – Environ. Pollut. 77: 235–242.

Cosby, B.J., Hornberger, G.M., Galloway, J.N. and Wright, R.F. 1985. Modeling the effects of acid deposition: assessment of a lumped parameter model of soil water and streamwater chemistry. – Water Resour. Res. 21: 51–63.

David, M.B., Vance, G.F., Rissing, J.M. and Stevenson, F.J. 1989. Organic carbon fractions in extracts of O and B horizons from a New England spodosol: effect of acid treatment. – J. Environ. Qual. 18: 212–217.

Driscoll, C.T., Lehtinen, M.D. and Sullivan, T.J. 1994. Modeling the acid-base chemistry of organic solutes in Adirondack, New York, lakes. – Water Resour. Res. 30: 297–306.

Ellenberg, H., Mayer, R. and Schauerman, J. 1986. Ökosystemforschung Ergebnisse des Sollingprojekts 1966–1986. – Eugen Ulmer Verlag, Stuttgart.

Ericsson, T. in press. Growth and shoot: root ratio of seedlings in relation to nutrient availability. – Plant Soil.

Eriksson, H. 1976. Yield of Norway spruce in Sweden. – Research notes no. 41, Dept Forest Yield Res., Royal Coll. Forestry, Stockholm, (in Swedish with English summary).

Gherini, S.A., Mok, L., Hudson, R.J.M., Davis, G.F., Chen, C.W. and Goldstein, R.A. 1985. The ILWAS model: formulation and application. – Water Air Soil Pollut. 26: 425–459.

Grigal, D.F. 1990. Mechanistic modeling of nutrient acquisition by trees. – In: Dixon, R.K., Mehldahl, R.S., Ruark, G.A. and Warren, W.G. (eds), Process modeling of forest growth responses to environmental stress. Timber Press, Portland Oregon, pp. 113–123.

Göransson, A. and Eldhuset, T.D. 1991. Effects of aluminium on growth and nutrient uptake of small *Picea abies* and *Pinus sylvestris* plants. – Trees 5: 136–142.

Groenenberg, J.E., Kros, J., Van der Salm, C. and De Vries, W. 1995. Application of the model NUCSAM to the Solling spruce site. – Ecol. Model.

Hedin L.O., Granat, L., Likens, G.E., Buishand, T.A., Galloway, J.N., Butler, T.J. and Rodhe, H. 1994. Steep declines in atmospheric base cations in regions of Europe and North America. – Nature 367: 351–354.

Johnsson, H., Bergström, L., Jansson, P.-E. and Paustian, K.

1987. Simulated nitrogen dynamics and losses in a layered agricultural soil. – Agric. Ecosyst. Environ. 18: 333–356.

Kros, J., De Vries, W., Janssen, P. H. M. and Bak, C. I. 1993. The uncertainty in forecasting trends of forest soil acidification. – Water Air Soil Pollut. 66: 29–58.

Mohren, G. M. J. and Van Veen, J. R. 1995. Forest growth in relation to site conditions: Application of the model FOR-GRO to the Solling spruce site. – Ecol. Model.

Nikolaidis, N. P., Rajaram, H., Schnoor, J. L. and Georgakakos, K. P. 1988. A generalized soft water acidification model. – Water Resour. Res. 24: 1983–1996.

Nilsson L.-O. and Wiklund, K. 1992. Influence of nutrient and water stress on Norway spruce production in south Sweden – the role of air pollutants. – Plant Soil 147: 251–265.

Oja, T., Yin, X. and Arp, P. A. 1995. The forest model series FORM-S: applications to the Solling spruce site. – Ecol. Model.

Rastetter, E. B., King, A. W., Cosby, B. J., Hornberger, G. M., O'Neill, R. V. and Hobbie, J. E. 1992. Aggregating fine-scale ecological knowledge to model coarser-scale attributes of ecosystems. – Ecol. Model. 2: 55–70.

Roberts, T. M., Skeffington, R. A. and Blank, L. W. 1989. Causes of type I spruce decline in Europe. – Forestry 62: 179–223.

Roelofs, J. G. M., Kempers, A. J., Houdijk, A. L. F. M. and Jansen, J. 1985. The effect of air-borne sulphate on *Pinus nigra* var. *maritima* in the Netherlands. – Plant Soil 84: 45–56.

Rose, K. A., Brenkert, A. L., Cook, R. B. and Gardner, R. H. 1991a. Systematic comparison of ILWAS, MAGIC, and ETD watershed acidification models. II. Monte Carlo analysis under regional variability. – Water Resour. Res. 27: 2591–2603.

– , Cook, R. B., Brenkert, A. L. and Gardner, R. H. 1991b. Systematic comparison of ILWAS, MAGIC, and ETD watershed acidification models. I. Mapping among model input and deterministic results. – Water Resour. Res. 27: 2577–2589.

Schall P. 1991. Productivity and vitality of spruce stands: dynamic feedback simulation for responses to different annual and seasonal levels of magnesium supply from soil. – Vegetatio 92: 111–118.

Schecher, W. D. and Driscoll, C. T. 1987. An evaluation of uncertainty associated with aluminum equilibrium calculations. – Water Resour. Res. 24: 525–534.

– and Driscoll, C. T. 1988. An evaluation of the equilibrium calculations within acidification models: the effect of uncertainty in measured chemical components. – Water Resour. Res. 24: 533–540.

Ulrich, B. 1989. Effects of acidic precipitation on forest ecosystems in Europe. – In: Adriano, D. C. and Johnson, A. H. (eds), Acidic precipitation Vol. 2. Biological and ecological effects. Springer, New York, pp. 189–272.

Van Dam, D. 1995. Application of the model NIICE to the Solling spruce site. – Ecol. Model.

Van Minnen, J. G., Meijers, R. and Braat, L. C. 1995. Application of the FORSOL model to the Spruce site at Solling. – Ecol. Model.

Van Oene, H. 1992. Acid deposition and forest nutrient imbalances: a modeling approach. – Water Air Soil Pollut. 63: 33–50.

– 1994. A mechanistic model for the inhibiting effects of aluminium on the uptake of cations. – In: Van Oene, H. (ed.), Modelling the impacts of sulphur and nitrogen deposition on forests. Diss., Rapport 68, Dept Ecol. Environ. Res., Swedish Univ. Agricul. Sci., Uppsala, Sweden.

– and De Vries, W. 1994. Comparison of measured and simulated changes in base cation amounts using a one-layer and a multi-layer soil acidification model. – Water Air Soil Pollut. 72: 41–65.

– and Ågren, G. I. 1995. The application of the model NAP to the Solling spruce site. – Ecol. Model.

Warfvinge, P., Falkengren-Grerup, U., Sverdrup, H. and Andersen, B. 1993. Modelling long-term cation supply in acidified forest stands. – Environ. Pollut. 80: 209–221.

Zöttl, H. W. 1987. Responses of forests in decline to experimental fertilization. – In: Hutchinson, T. C. and Meema, K. M. (eds), Effects of atmospheric pollutants on forests, wetlands, and agricultural ecosystems. Springer, pp. 255–265.

Ecological Bulletins 44: 363–369. Copenhagen 1995

Effects of Air Pollutants and Acidification – List of publications 1989–1994

This is a list of publications from projects funded within the research programme Effects of Air Pollutants and Acidification, phase II. This programme was administered by the Swedish Environmental Protection Agency during the period 1988/89–1992/93. In order to keep the liste reasonably short only open literature publications and major conference proceedings have been included, leaving out the numerous reports published by university departments and research institutes.

The bibliography only covers studies in the terrestrial environment. A review of acid water research from Sweden can be found in an Ambio volume, 22(5) from 1993, as well in the book Liming of acidified surface waters edited by L. Henriksson and Y. W. Brodin (SpringerVerlag 1995).

References

Alavi, G. and Jansson, P.-E. 1994. Transpiration and soil moisture dynamics for spruce stands of different canopy densities and water availability. – In: Nilsson, L.-O., Hüttl, R. F. and Mathy, P. (eds), Nutrient uptake and cycling in forest ecosystems. C E C, Brussells, Ecosystems Research Report 13.

Andersson, F. 1989. Air pollution impact on Swedish forests – present evidence and future development. – Environ. Monitor. Assess. 12: 29–38.

– 1990. A field experimental approach to understanding of forest damage. – In: Klimo, E. and Materna, J. (eds), Verification of hypotheses on the mechanism of damage and possibilities of recovery of forest ecosystems. Int. workshop, Brno 4–8 September 1989, pp. 157–166.

Andersson, M. 1992. Effects of pH and aluminum on growth of *Galium odoratum* L. Scop. in flowing solution culture. – Environ. Exp. Bot. 32: 497–504.

– , Balsberg-Påhlsson, A.-M., Berlin, G., Falkengren-Grerup, U. and Tyler, G. 1989. Environment and mineral nutrients of beech *Fagus sylvatica* L. in south Sweden. – Flora (Jena) 183: 405–415.

Andersson, M. E. 1993. Aluminium toxicity as a factor limiting the distribution of *Allium ursinum* L. – Ann. Bot. 72: 607–611.

– and Brunet, J. 1993. Sensitivity to H and Al ions limiting growth and distribution of the woodland grass *Bromus benekenii*. – Plant Soil 15: 243–254.

Andersson, S., Nilsson, S. I. and Valeur, I. 1993. Influence of lime on soil respiration, leaching of DOC and C/S relationships in the mor humus of a Haplic Podsol. – Environ. Int. 20: 81–88.

Andrén, C., Mårdén, M. and Nilson, G. 1989. Tolerance to low pH in a population of moor frogs, *Rana arvalis* Nilsson, from an acid and a neutral environment – a possible case of rapid evolutionary response to acidification. – Oikos 56: 215–223.

Arnebrant, K., Ek, H., Finlay, R. D. and Söderström, B. 1993. Translocation of nitrogen between *Alnus glutinosa* seedlings inoculated with *Frankia* sp. and *Pinus contorta* seedlings connected by a common ectomycorrhizal fungus. – New Phytol. 124: 231–242.

Asp, H. and Berggren, D. 1990. Phosphate and calcium uptake in beech *Fagus sylvatica* in the presence of aluminium and natural fulvic acids. – Physiol. Plant. 80: 307–314.

– , Bengtsson, B. and Jensén, P. 1991. Influence of aluminium on phosphorus and calcium localization in roots on beech *Fagus sylvatica*. – Physiol. Plant. 83: 41–46.

Balsberg-Påhlsson, A.-M. 1989. Toxicity of heavy metals (Zn, Cu, Cd, Pb) to vascular plants. A literature review. – Water Air Soil Poll. 47: 287–319.

– 1989. Effects of heavy metal and SO_2 pollution on the concentrations of carbohydrates and nitrogen in tree leaves. – Can. J. Bot. 67: 2106–2113.

– 1989. Mineral nutrients, carbohydrates and phenolic compounds in leaves of beech, *Fagus sylvatica* L. in southern Sweden as related to environmental factors. – Tree Physiol 5: 485–495.

– 1990. Influence of aluminium on biomass, nutrients, soluble, carbohydrates and phenols in beech *Fagus sylvatica*. – Physiol. Plant. 78: 79–84.

– 1992. Influence of nitrogen fertilization on minerals, carbohydrates, amino acids and phenolic compounds in beech *Fagus sylvatica* L. leaves. – Tree Physiol. 10: 93–100.

Bengtsson, B. 1992. Influence of aluminium and nitrogen on uptake and distribution of minerals in beech roots (*Fagus sylvatica*). – Vegetatio 101: 35–41.

– , Asp, H., Jensén, P. and Berggren, D. 1988. Influence of aluminium on phosphate and calcium uptake, in beech *Fagus sylvatica* grown in nutrient solution and soil solution. – Physiol. Plant. 74: 299–305.

– , Asp, H. and Jensén, P. 1993. Uptake and distribution of calcium and phosphorus in beech *Fagus sylvatica* as influenced by aluminium and nitrogen. – Tree Physiol.: 14: 63–73.

Berdén, M. and Berggren, D. 1990. Gel filtration chromatography of humic substances in soil solutions using HPLC-determination of the molecular weight distribution. – J. Soil Sci. 41: 61–72.

– , Clarke, N., Danielsson, L.-G. and Sparén, A. 1994. Aluminium speciation: variations caused by the choice of analytical method and by sample storage. – Water Air Soil Poll. 72: 213–239.

Berggren, D. 1989. Speciation of aluminium, cadmium, copper and lead in humic soil solutions. A comparison of the ion exchange column procedure and equilibrium dialysis. – Int. J. Environ. Anal. Chem. 35: 1–15.

– 1990. Specification of cadmium (II) using Donnan dialysis and differential pulse anodic stripping voltammetry in a flow-injection system. – Int. J. Environ. Anal. Chem. 41: 133–148.

– 1992. Speciation and mobilization of aluminium and cadmium in podzols and cambisols of S. Sweden. – Water Air Soil Poll. 62: 125–156.

– , Bergkvist, B., Falkengren-Grerup, U., Folkeson, L. and Tyler, G. 1990. Metal solubility and pathways in acidified forest ecosystems of south Sweden. – Science of the Total Environ. 96: 103–114.

– , Bergkvist, B., Falkengren-Grerup, U. and Tyler, G. 1992. Influence of acid deposition on metal solubility and fluxes in natural soils. – In: Låg, J. (eds), Chemical climatology and geomedical problems. Norw. Acad. Sci. Lett., Oslo, pp. 141–154.

Bergholm, J., Jansson, P.-E., Johansson, U., Majdi, H., Nilsson, L. O., Persson, H., Rosengren, U. and Wiklund, K. 1994. The Skogaby project – A general description of soil, soil water, nutritional and biomass status in crown and roots, and climate at start of the experiment. – In: Nilsson, L.-O., Hüttl, R. F. and Mathy, P. (eds), Nutrient uptake and cycling in forest ecosystems. Comm. European Commun. Ecosyst. Res. Report 13: (in press).

Bergkvist, B. and Folkeson, L. 1992. Soil acidification and element fluxes of a *Fagus sylvatica* forest as influenced by simulated nitrogen deposition. – Water Air Soil Poll. 65: 111–133.

– , Folkeson, L. and Berggren, D. 1989. Fluxes of Cu, Zn, Pb, Cd, Cr and Ni in temperate forest ecosystems. A literature review. – Water Air Soil Poll. 47: 217– 286.

Bergström, S. 1991. Principles and confidence in hydrological modelling. – Nord. Hydrol. 22: 123–136.

– and Lindström, G. 1992. Recharge and discharge areas in hydrological modelling – a new approach. – Vannet i Norden, No. 3.

Bishop, K. H. 1991. Is there more to acidity in organic-rich surface waters than air pollution. – Vatten 47: 342–347.

– and Richards, K. S. 1993. Shallow flow pathways vs. pre-event water in the storm hydrograph: How great a conflict? – In: Howells, G. and Dalziel, T. R. K. (eds), Restoring acid waters, Loch Fleet 1984–1990. Elsevier Science Publ., pp. 121–133.

– , Grip, H. and O'Neill, A. 1990. The origins of acid run off in a hillslope during storm events. – J. Hydrol. 116: 35–61.

– , Grip, H. and Piggott, E. 1990. Spate-specific flow pathways in an episodically acid stream. – In: Mason, B. J. (ed.), The surface water acidification programme. Cambridge Univ. Press, pp. 107–120.

– , Lundström, U. S. and Giesler, R. 1993. The transfer of organic carbon from forest soils to surface waters: example from northern Sweden. – Appl. Geochem. Spec. Iss. 2: 11–15.

Björkdahl, G. and Eriksson, H. 1989. Effects of crown decline on increment in Norway spruce, *Picea abies* (L.) Karst, in southern Sweden. – Medd. Norsk Inst. Skogforsk. 42: 19–55.

Bricker, O., Pačes, T., Johnson, C. and Sverdrup, H. 1994. Weathering and erosion aspects of small catchment research. – In: Moldan, B. and Cherny, J. (eds), Biogeochemistry of small catchments, SCOPE No. 51. John Wiley and Sons, pp. 85–106.

Brunet, J. and Neymark, M. 1992. Importance of soil acidity to the distribution of rare forest grasses in south Sweden. – Flora (Jena) 187: 317–326.

Čermák, J., Cienciala, E., Kučera, J. and Hällgren, J.-E. 1992. Radial velocity profiles of water flow in trunks of Norway spruce and oak and response to serving. – Tree Physiol. 10: 367–380.

Chalot, M., Brun, A., Finlay, R. D. and Söderström, B. 1994. Metabolism of [^{14}C]glutamate and [^{14}C]glutamine by the ectomycorrhizal fungus *Paxillus involutus*. – Microbiology 140: 1641–1649.

Cienciala, E., Lindroth, A., Čermák, J., Hällgren, J.-E. and Kučera, J. 1992. Assessment of transpiration estimates for *Picea abies* trees during a growing season. – Trees Struct. Funct. 6: 121–127.

– , Lindroth, A., Čermák, J., Hällgren, J.-E. and Kučera, J. 1994. The effects of water availability on transpiration, water potential and growth of *Picea abies* during a growing season. – J. Hydrol. 155: 57–71.

– , Eckersten, H., Lindroth, A. and Hällgren, J.-E. 1994. Simulated and measured water uptake by *Picea abies* under non-limiting soil water conditions. – J. Agricult. For. Meteorol. 71: 147–164.

Clarke, N., Danielsson, L.-G. and Sparén, A. 1991. Organic bound aluminium and its interaction with a new method for the determination of aluminium in natural waters – Fin. Humus News 3: 253–258.

– , Danielsson, L.-G. and Sparén, A. 1992. The determination of quickly reacting aluminium in natural waters by kinetic discrimination in a flow system. – Int. J. Environ. Anal. Chem. 48: 77–100.

Clegg, S., Clarholm, M. and Gobran, G. R. 1993. Can phosphorus availability be manipulated in forest soils? – In: Rasmussen, L., Brydges, T. and Mathy, P. (eds), Experimental manipulations of biota and biogeochemical cycling in ecosystems. CEC. Ecosyst Res. Rep. 4: 202–205.

Dise, N. B. and Wright, R. F. (eds) 1992. The NITREX project (Nitrogen saturation experiments). – Comm. European Commun., Brussels. Ecosyst. Res. Rep. 2.

Edfast, A.-B., Näsholm, T. and Ericsson, A. 1990. Free amino acids in needles of Norway spruce and Scots pine trees on different sites in areas with two levels of nitrogen deposition. – Can. J. For. Res. 20: 1132–1136.

Ek, H., Finlay, R. D., Odham, G. and Söderström, B. 1990. Capillary gas chromatographic determination of free ^{15}N-labelled ammonium and ^{15}N-labelled total nitrogen in plant and fungal tissues using flame ionization and mass spectrometric determination. – J. Microbiol. Meth. 11: 169–176.

Eklund, L., Cienciala, E. and Hällgren, J.-E. 1992. No relation between drought stress and ethylene production in stems of Norway spruce trees. – Physiol. Plant. 86: 297–300.

Ericsson, A., Nordén, L.-G., Näsholm, T. and Walheim, M. 1993. Mineral nutrient imbalances and arginine concentrations in needles of *Picea abies* (L.) Karst. from two areas with different levels of airborne deposition. – Trees 8: 67–74.

Eriksson, E., Karltun, E. and Lundmark, J.-E. 1992. Acidification of forest soils in Sweden. – Ambio 21: 150–153.

Eriksson, H. M. and Rosén, K. 1993. Nutrient distribution in a tree species experiment in south-western Sweden – Plant Soil 164: 51–59.

Eriksson, M. O. G. 1994. Susceptibility to freshwater acidification by two species of loon: red-throated loon *Gavia stellata* and arctic loon *Gavia arctica* in south-west Sweden. – Hydrobiologia 279/280: 439–444.

– and Sundberg, P. 1991. The choice of fishing lakes in red-throated diver *Gavia stellata* and black-throated diver *G. arctica* during the breeding season in south-west Sweden. – Bird Study 38: 135–144.

– , Blomqvist, D., Hake, M. and Johansson, O. C. 1990. Parental feeding in the red-throated diver *Gavia stellata*. – Ibis 132: 1–13.

– , Johansson, I. and Ahlgren, C.-G. 1992. Levels of mercury in eggs of red-throated diver *Gavia stellata* and black-throated diver *Gavia arctica* in south-west Sweden. – Ornis Svecia 2: 29–36.

Erland, S. E. and Finlay, R. D. 1992. Effects of temperature and incubation time on the ability of three ectomycorrhizal fungi to colonize *Pinus sylvestris* roots. – Mycol. Res. 96: 270–272.

Erland, S., Finlay, R. D. and Söderström, B. 1989. The effects of liming on mycelial colonization and carbon allocation on ectomycorrhizal mycelia attached to *Pinus sylvestris* plants. – In: Mejstrik, V. (ed.), Proceedings of the 2nd European symposium on mycorrhizae, ecology and practical applications of mycorrhizae. Agriculture, Ecosystems and Environment 28: 111–113.

– , Finlay, R. D. and Söderström, B. 1991. The influence of substrate pH on carbon translocation in ectomycorrhizal and non-mycorrhizal pine seedlings. – New Phytol. 119: 235–242.

Falkengren-Grerup, U. 1989. Effect of stemflow on beech forest soils and vegetation in southern Sweden. – J. Appl. Ecol. 26: 341–352.

- 1989. Soil acidification and its impact on ground vegetation. – Ambio 18: 179–183.
- 1990. Distribution of field layer species in Swedish deciduous forests in 1929–54 and 1979–88 as a related to soil pH. – Vegetatio 86: 143–150.
- 1990. Biometric and chemical analysis of five herbs in a regional acid-base gradient in Swedish beech forest soils. – Acta Oecol. 11: 755–766.
- 1993. Effects on beech forest species of experimentally enhanced nitrogen deposition. – Flora 188: 85–91.
- 1994. Importance of soil solution chemistry to field performance of *Galium odoratum* and *Stellaria nemorum*. – J. Appl. Ecol. 31: 182–192.
- and Eriksson, H. 1990. Changes in soil, vegetation and forest yield between 1947 and 1988 in beech and oak sites of southern Sweden. – For. Ecol. Manage. 38: 37–53.
- and Björk, L. 1991. Reversibility of stemflow-induced soil acidification in Swedish beech forest. – Environ. Pollut. 74: 31–37.
- and Tyler, G. 1991. Changes of cation pools of the topsoil in south Swedish beech forests between 1979 and 1989. – Scand. J. For. Res. 6: 145–152.
- and Tyler, G. 1991. Dynamic floristic changes of Swedish beech forest in relation to soil acidity and stand management. – Vegetatio 95: 149–158.
- and Tyler, G. 1992. Chemical conditions limiting survival and growth of *Galium odoratum* (L.) Scop. in acid forest soil. – Acta Oecol. 13: 169–180.
- and Tyler, G. 1992. Changes since 1950 of mineral pools in the upper C-horizon of Swedish deciduous forest soils. – Water Air Soil Poll. 64: 495–501.
- and Tyler, G. 1993. Soil chemical properties excluding field-layer species from beech forest mor. – Plant Soil 148: 185–191, (errata in 150: 323).
- and Tyler, G. 1993. The importance of soil acidity, moisture, exchangeable cation pools and organic matter solubility to the cationic composition of beech forest *Fagus sylvatica* (L.) soil solution. – Z. Pflanzenernähr. Bodenkd. 156: 365–370.
- and Lakkenborg-Kristensen, H. 1994. Importance of ammonium and nitrate to the performance of forest herbs and grasses. – Environ. Exp. Bot. 34: 31–38.
- and Tyler, G. 1994. Experimental evidence for the relative sensitivity of deciduous forest plants to high soil acidity. – For. Ecol. Manage. 60: 311–326.
- , Rühling, Å. and Tyler, G. 1994. Effects of phosphorus application on vascular plants and macrofungi in an acid beech forest soil. – Sci. Total Environ. 151: 125–130.
- , Quist, M. E., and Tyler, G. in press. Relative importance of exchangeable and soil solution cation concentrations to the distribution of vascular plants. – Environ. Exp. Bot.
Finlay, R. D. 1989. Functional aspects of incompatible ectomycorrhizal associations. – In: Mejstrik, V. (ed.), Proceedings of 2nd European symposium on mycorrhizae, ecology and practical applications of mycorrhizae. Agriculture, Ecosystems and Environment 28: 127–132.
- 1990. Functional aspects of phosphorus uptake and carbon translocation in incompatible ectomycorrhizal associations between *Pinus sylvestris* and *Suillus grevillei* and *Boletinus cavipes*. – New Phytol. 112: 185–192.
- 1993. Uptake and mycelial translocation of nutrients by ectomycorrhizal fungi. – In: Read, D. J., Lewis, D. H., Fitter, A. H. and Alexander, I. J. (eds), Mycorrhiza in ecosystems. Proc. of 3rd European Symp. Mycorrhizas, CAB Int., pp. 91–97.
- and Söderström, B. 1989. Mycorrhiza and their role in soil and plant communities. – In: Clarholm, M. and Bergström, L. (eds), Developments in plant and soil sciences Vol. 39. The ecology of arable land – Perspectives and challengers. Kluwer Academic Publ., Dordrecht, pp. 139–148.
- , Ek, H., Odham, G. and Söderström, B. 1989. Uptake, translocation and assimilation of nitrogen from ^{15}N-labelled ammonium and nitrate sources by intact ectomycorrhizal systems of *Fagus sylvatica* infected with *Paxillus involutus*. – New Phytol. 113: 47–55.
- , Ek, H., Odham, G. and Söderström, B. 1990. Mycelial uptake, translocation and assimilation of inorganic and organic ^{15}N-labelled compounds by ectomycorrhizal *Pinus sylvestris* plants. – In: Mejstrik, V. (ed.), Proceedings of 2nd European symposium of mycorrhizae, ecology and practical applications of mycorrhizae. Agriculture, Ecosystem and Environment 28: 133–137.
- , Frostegård, Å. and Sonnerfeldt, A.-M. 1992. Utilization of organic and inorganic nitrogen sources by ectomycorrhizal fungi of different successional stages grown in pure culture and symbiosis with *Pinus contorta* (Dougl. ex Loud). – New Phytol. 120: 105–115.
Fiskesjö, G. 1989. Aluminium toxicity in root tips of *Picea abies* (L.) Karst., *Fagus sylvatica* (L.) and *Quercus robur* (L.). – Hereditas 111: 149–157.
Flower-Ellis, J. G. K. 1993. Dry-matter allocation in Norway spruce branches: a demographic approach. – Stud. Forest. Suec. 191: 51–73.
Folkeson, L., Nyholm, N. E. I. and Tyler, G. 1990. Influence of acidity and other soil properties on metal concentrations in forest plants and animals. – Sci. Total Environ. 96: 211–233.
Gärdenfors, U. 1992. Effects of artificial liming on land snail populations. – J. Appl. Ecol. 29: 50–54.
Giesler, R. and Lundström, U. S. 1993. Soil solution chemistry : The effects of bulking soil samples. – Soil Sci. Soc. Am., J. 57: 1283–1288.
Gobran, G. R. and Clegg, S. 1992. Relationship between total dissolved organic carbon and SO_4^{-2} in soil and waters. – Sci. Total Environ. 117/118: 449–461.
- and Tipping, E. 1993. Modelling the chemistry of humic-rich soil leachates. – Appl. Geochem. 2: 121–124.
- , Fenn, L. B., Persson, H. and Al Windi, I. 1993. Nutrition response of Norway spruce and willow to varying levels of calcium and aluminium. – Fert. Res. 34: 181–189.
Göransson, A. and Eldhuset, T. D. 1991. Effects of aluminium on growth and nutrient uptake of *Picea abies* and *Pinus sylvestris* seedlings. – Trees 5: 136–142.
Granat, L. and Hällgren, J.-E. 1992. Relation between estimated dry deposition and throughfall in a coniferous forest exposed to controlled levels of SO_2 and NO_2. – Environ. Pollut. 75: 237–242.
Grennfelt, P. 1989. The deposition of acids to forest ecosystems. – In: Tomlinson, G. (ed.), Effects of acid deposition on the forests of Europe and North America. CRC Press.
Gunnarsson, B. 1990. Vegetation structure and the abundance and size distribution of spruce-living spiders. – J. Anim. Ecol. 59: 242–253.
- and Johnsson, J. 1989. Effects of simulated acid rain on growth rate in a spruce-living spider. – Environ. Pollut. 56: 311–317.
Gustafsson, J. P. and Johnsson, L. 1992. Selenium retention in organic matter of forest soils. – J. Soil Sci. 43: 461–472.
- and Jacks, G. 1993. Sulphur status in some Swedish podzols as influenced by acidic deposition and extractable carbon. – Environ. Pollut. 81: 185–191.
- and Johnsson, L. 1994. The association between selenium and humic substance in in forested ecosystems – laboratory evidence. – Appl. Organomet. Chem. 8: 141–147.
- , Jacks, G., Stegmann, B. and Ross, H. B. 1993. Soil acidity and adsorbed anions in Swedish forest soils – Long-term changes. – Agric. Ecosyst. Environ. 47: 103–115.
Hake, M. 1991. The effects of needle loss in coniferous forests in south-west Sweden on the winter foraging behaviour of willow tits *Passer montanus*. – Biol. Conserv. 58: 357–366.
- and Eriksson, M. O. G. 1990. Passerine birds and their prey affected by forest dieback: a tentative work model. – Fauna Norvegica Ser. C, Cinclus, Suppl. 1: 13–16.
Hallbäcken, L. 1992. Long term changes of base cation pools in soil and biomass in a beech and a spruce forest of southern Sweden. – Z. Pflanzenernähr. Bodenkd. 155: 51–60.

Hallingbäck, T. 1992. The effect of air pollution on mosses in southern Sweden. – Biol. Conserv. 59: 163–170.
– and Kellner, O. 1992. Effects of simulated nitrogen rich and acid rain on the nitrogen-fixing lichen *Peltigera aphthosa* (L.) Willd. – New Phytol. 120: 99–103.
Hansen, P. A. 1991. Improved interpretation of soil – macrofungal relations in south Swedish beech forests using two complementary regression analyses. – Vegetatio 93: 47–55.
– and Tyler, G. 1992. Statistical evaluation of tree species affinity and soil preference of the macrofungal flora in south Swedish beech, oak and hornbeam forests. – Cryptogam. Bot. 2: 355–361.
Hendershot, W., Warfvinge, P., Courchesne, F. and Sverdrup, H. 1991. The mobile anion concept – Time for a reappraisal? – J. Environ. Qual. 20: 505–509.
Högberg, P., Johannisson, C., Nicklasson, H. and Högbom, L. 1990. Shoot nitrate reductase activities of field-layer species in different forest types. I. Preliminary surveys in northern Sweden. – Scand. J. For. Res. 5: 449–456.
– , Högbom, L., Johannisson, C. and Näsholm, T. 1993. ^{15}N abundance of forests is correlated with losses of nitrogen. – Plant Soil 157: 147–150.
Högbom, L. 1994. Shoot nitrate reductase activities of field-layer species in different forest types. III. A preliminary survey in beech forests in southern Sweden. – Scand. J. For. Res. 9: 124–128.
– and Högberg, P. 1991. Nitrate nutrition of *Deschampsia flexuosa* (L.) i relation to nitrogen deposition in Sweden. – Oecologia 87: 488–494.
Hult, M., Bengtsson, B. T., Larsson, N. P.-O. and Yang, C. 1992. Particle induced X-ray emission microanalysis of root samples from beech *Fagus sylvatica*. – Scanning Microscopy 6: 581–592.
Hultberg, H. and Likens, E. G. 1991. Sulphur deposition to forested catchments in northern Europe and North America – large-scale variations and long-term dynamics. – In: Schwarts, S. E. and Slinn, W. G. N. (eds), Proc. of 5th Int. Conf. on: precipitation scavenging and atmosphere-surface exchange. Hemisphere Publ. Corp., Vol. 3: 1343–1365.
– and Grennfelt, P. 1992. Sulphur and seasalt deposition as reflected by throughfall and runoff chemistry in forested catchments. – Environ. Pollut. 75: 215–222.
– , Lee, Y.-H., Nyström, U. and Nilsson, S. I. 1990. Chemical effects on surface-, ground-and soil-water of adding acid and neutral sulphate to catchments in southwest Sweden. – In: Mason, B. J. (ed.), SWAP, The surface water acidification programme conference. Cambridge Univ. Press, pp. 167–182.
– , Andersson, B. I. and Moldan, F. 1992. The covered catchment – An experimental approach to reversal of acidification in a forest ecosystem. – In: The CEC Int. Symp. on experimental manipulation of biota and biogeochemical cycling in ecosystems. Copenhagen 18–20 May, 1992. CEC, Brussels, Ecosystems Research Report 4.
– , Apsimon, H., Church, R. M., Grennfelt, P., Mitchell, M. J., Moldan, F. and Ross, H. B. 1994. Sulphur. – In: Moldan, B. and Cherny, J. (eds), Biogeochemistry of small catchments: A tool for environmental research. John Wiley and Sons, SCOPE 51, pp. 229–254.
– , Dise, N. B., Wright, R. F., Andersson, I. and Nyström, U. 1994. Nitrogen saturation induced during winter by experimental NH_4NO_3 addition to a forested catchment. – Environ. Pollut. 84: 145–147.
Jacks, G. 1990. Acidification of soils – processes and present state. – In: Kucera, V. (ed.), Effects of water and soil acidification on corrosion. Nordic Council of Ministers, Copenhagen, Miljørapport 1990:9: 124–133.
– 1992. Acidification of soil and water below the highest holocene shoreline. – In: Robertsson, A. M., Ringberg, B., Miller, U. and Brunnberg, L. (eds), Quarternary stratigraphy, glacial morphology and environmental changes. Swedish Geol. Surv. Ser. Ca 81: 145–148.

– 1993. Groundwater acidification. – In: Alley, W. and Van Norstrand, R. (eds), Regional groundwater quality. New York, pp. 405–421.
Jensén, P., Pettersson, S., Drakenberg, T. and Asp, H. 1989. Aluminium effects on vacuolar phosphorus in roots of beech, *Fagus sylvatica* L. – J. Plant Physiol. 134: 37–42.
Jönsson, C., Sverdrup, H. and Warfvinge, P. 1994. Uncertainty in predicting weathering rate and environmental stress factors with the PROFILE model. – Water Air Soil Pollut. 78: 37–55.
Kämäri, J., Amann, M., Brodin, Y.-W., Chadwick, M., Henriksen, A., Hettelingh, J.-P., Kuylenstierna, J., Posch, M. and Sverdrup, H. 1992. The use of critical loads for the assessment of future alternatives to acidification. – Ambio 21: 377–386.
Karpinski, S., Wingsle, G., Karpinska, B. and Hällgren, J.-E. 1992. Differential expression of CuZn-superoxide dismutases in *Pinus sylvestris* (L.) needles exposed to SO_2 and NO_2. – Physiol. Plant. 85: 689–696.
– , Wingsle, G., Olsson, O. and Hällgren, J.-E. 1992. Characterization of cDNAs encoding CuZn-superoxide dismutases in Scots pine. – Plant Molecul. Biol. 18: 545–555.
– , Karpinska, B., Wingsle, G. and Hällgren, J.-E. 1993. Molecular responses to photooxidative stress in *Pinus sylvestris* (L.): II. Differential expression of CuZn-superoxide dismutases and glutathione reductase. – Plant Physiol. 103: 1385–1391.
Karltun, E. and Gustafsson, J. P. 1993. Interference by organic complexation of Fe and Al on the sulphate adsorption in spodic B-horizons in Sweden. – J. Soil Sci. 44: 625–632.
Knutsson, G. 1992. Studies of some acid springs in till, Lofsdalen, Sweden. – Nordisk Hydrologisk Program NHP, Oslo, Report 30: 73–84.
– 1994. Trends in the acidification of groundwater. Groundwater quality management. – International Association of Hydrological Sciences (IAHS) 20: 107–118.
– 1994. Acidification effects on groundwater prognosis of the risks for the future. – In: Future groundwater resources and risk. International Association of Hydrological Sciences (IAHS) 222: 3–17.
Kucera, V. 1992. How do acidifying air pollutants affect buildings and monuments – an international exposure programme within UN/ECE. – In: Proceedings from NKM 12, 12th Scandinavian corrosion congress and Eurocorr '92. Vol. 1. The Corrosion Society of Finland, Helsinki, pp. 9–24.
– , Tidblad, J. and Leygraf, C. 1992. Acid deposition effects on materials: evaluation of silver after 4 years of exposure. – In: Proceedings from NKM 12, 12th Scandinavian corrosion congress and Eurocorr '92, The Corrosion Society of Finland, Helsingfors, Vol 1: 179–188.
Kutschera, L., Hübl, E., Lichtenegger, E., Persson, H. and Sobotnik, M. (eds) 1992. Root ecology and its practical application 2. – Verein für Wurzelforschung, Klagenfurt.
Larsson, S. and Björkman, C. 1993. Performance of chewing and phloem-feeding insects on stressed trees. – Scand. J. For. Res. 8: 550–559.
Lee, Y.-H. and Hultberg, H. 1990. Studies of strong and weak acid speciation in fresh waters in experimental forested catchments. – In: Mason, B. J. (ed.), SWAP, The surface water acidification programme conference. Cambridge Univ. Press, pp 183–192.
Levlin, E. 1991. Corrosion of underground structures due to acidification: laboratory investigation. – Brit. Corr. J. 26: 36–66.
Lind, B. B. and Lundin, L. 1990. Saturated hydraulic conductivity of Scandinavian tills. – Nordic Hydrol. 21: 107–118.
Linder, S. and Flower-Ellis, J. G. K. 1992. Environmental and physiological constraints to forest yield. – In: Teller, A., Mathy, P. and Jeffers, J. N. R. (eds), Responses of forest ecosystems to environmental changes. Elsevier Applied Science, pp. 149–164.
– and McDonald, A. J. S. 1993. Plant nutrition and the in-

terpretation of growth response to elevated concentrations of atmospheric carbon dioxide. – In: Schulze, E. D. and Mooney, H. (eds.), Design and execution of experiments on CO_2-enrichment. C E C, Brussels, Ecosyst. Res. Rep. 6: 73–82.

Lindström, G. and Rodhe, A. 1992. Transit times of water in soil lysimeters from modelling of oxygen-18. – Water Air Soil Pollut. 65: 83–100.

Livsey, S. and Barklund, P. 1992. *Lophodermium piceae* and *Rhizosphaera kalkhoffii* in fallen needles of Norway spruce *Picea abies*. – Eur. J. For. Pathol. 22: 204–216.

Ljungström, M. and Stjernquist, I. 1993. Factors toxic to beech *Fagus sylvatica* L. seedlings in acid soils. – Plant and Soil 157: 19–29.

– Gyllin, M. and Nihlgård, B. 1990. Effects of liming on soil acidity and beech *Fagus sylvatica* L. regenertion on acid soils in south Swedish beech forests. – Scand. J. For. Res. 5: 243–254.

Lohm, U. 1989. Increased weathering rate in experimental field acidification. – In: Augustithis, S. S. (ed.), Weathering; Its products and deposits. Vol. 1. Processes – Theophrastus Publ., Athens, pp. 363–366.

Lundin, L. 1991. Influence of silviculture on content of organic matter and metals in water. – The third international nordic symposium on humic substances, 21–23 August 1991, Turku/Åbo, Finland. Finnish Humus News, Vol. 3: 21–26.

Lundström, U. 1990. Laboratory and lysimeter studies of chemical weathering. – In: Mason, B. J. (ed.), The surface acidification programme. Cambridge Univ. Press, Cambridge, pp. 267–274.

– 1993. The role of organic acids in soil solution chemistry in a podzolized soil. – J. Soil Sci. 44: 121–133.

– 1994. The significance of organic acids for weathering and the podzolisation process. – Environ. Int. 20: 21–30.

– and Öman, L.-O. 1990. Dissolution of feldspars in the presence of natural, organic solutes. – J. Soil Sci. 41: 359–369.

Majdi, H. and Persson, H. 1993. Spatial distribution of fine roots, rhizoshere and bulk-soil chemistry in an acidified *Picea abies* stand. – Scand. J. For. Res. 8: 147–155.

– and Rosengren-Brinck, U. 1994. Effects of ammonium sulphate application on the fine root and needle chemistry in a *Picea abies* stand – Plant Soil 162: 71–80.

– , Smucker, A. J. M. and Persson, H. 1992. A comparison between minirhizotron and monolith sampling methods for measuring root growth of maize *Zea mays* L. – Plant Soil 147: 127–134.

Malmer, N. 1990. Constant or increasing nitrogen concentrations in *Sphagnum* mosses on mires in southern Sweden during the last few decades? – Aquilo Ser. Bot. 28: 57–65.

Maxe, L. (ed.) 1994. Effects of acidification on groundwater in Sweden – Hydrological and hydrochemical processes. – Swedish Environ. Protect. Agency, Solna, Report 4388.

Näsholm, T. 1994. Removal of nitrogen during needle senescence in Scots pine *(Pinus sylvestris L.)*. – Oecologia 99: 290–296.

– , Högberg, P. and Edfast, A.-B. 1991. Uptake of NO_x by mycorrhizal and non-mycorrhizal Scots pine seedlings: quantities and effects on amino acid and protein concentrations. – New Phytol. 119: 83–92.

– , Edfast, A.-B., Ericsson, A. and Nordén, L.-G. 1994. Accumulation of amino acids in boreal coniferous forest plants in response to increased nitrogen availability. – New Phytol. 126: 137–143.

Nihlgård, B. 1990. Relationship of forest damage to air pollution in the Nordic countries. – Agric. For. Meteorol. 50: 87–98.

– , Ljungström, M. and Gyllin, M. 1990. Effects of liming on soil acidity and *Fagus sylvatica* (L.) regeneration on acid soils in south Swedish beech forests. – Scand. J. For. Res. 2: 243–254.

Nilsson, L. O. 1992. The Skogaby Project – Air pollution, tree vitality, forest damage and production. Growth performance after 4 years of treatment. – In: Rasmussen, L., Brydges, T.

and Mathy, P. (eds), Experimental manipulation of the biota and biogeochemical cycling in ecosystems. C E C, Brussels, Ecosyst. Res. Rep. 4: 155–164.

– 1993. Carbon sequestration in Norway spruce in south Sweden as influenced by air pollution, water availability and fertilization. – Water Air Soil Pollut. 70: 177–186.

– and Wiklund, K. 1992. Influence of nutrient and water stress on Norway spruce production in south Sweden – the role of air pollutants. – Plant Soil 147: 251–265.

– and Wiklund, K. 1994. Nitrogen uptake in Norway spruce stand following ammonium sulphate application, fertigation, irrigation, drought and nitrogen-free-fertilisation. – Plant Soil 164: 221–229.

– , Hüttl, R. and Mathy, P. (eds) 1994. Nutrient uptake and cycling in forest ecosystems. – CEC, Brussels, Ecosyst. Res. Rep. 13: (in press).

Nordén, U. 1991. Acid deposition and throughfall fluxes of elements as related to tree species in deciduous forests of south Sweden. – Water Air Soil Pollut. 60: 209–230.

– 1991. Acid deposition and throughfall fluxes as related to tree species. – In: Teller, A., Mathy, P. and Jeffers, J. N. R. (eds), Responses of forest ecosystems to environmental changes. Elsevier Applied Science, London, pp. 693–694.

– 1994. Influence of tree species on acidification and mineral pools in deciduous forest soils of south Sweden. – Water Air Soil Pollut. 76: 363–381.

Norrström, A.-C. 1993. Retention and chemistry of aluminium in groundwater discharge areas. – J. Soil Sci. 43: 461–472.

Nyberg, L., Bishop, K. H. and Rodhe, A. 1993. Importance of hydrology in the reversal of acidification in the soils, Gårdsjön, Sweden. – Appl. Geochem., Special issue 2: 61–66.

Ojanperä, K., Sutinen, S., Pleijel, H. and Selldén, G. 1991. Exposure of spring wheat, *Triticum aestivum* (L.), cv. Drabant, to different concentrations of ozone in open-top chambers: effects on the ultrastructure of flag leaf cells. – New Phytol. 120: 39–48.

Pernestål, K., Jonsson, B., Hällgren, J.-E. and Li, H.-K. 1991. Analysis of tree samples from acidic and lime environment by means of PIXE. – Int. J. PIXE 1: 281–296.

Persson, H. 1992. Factors affecting root dynamics of tree. – Suoseura – Finnish Peatland Society (SOU) 43: 163–172.

– and Majdi, H. 1993. The use of minirhizotron system to investigate fine-root dynamics. – In: Rasmussen, L., Brydges, T. and Mathy, P. (eds), Experimental manipulations of biota and biogeochemical cycling in ecosystems. CEC, Brussels, Ecosyst. Res. Rep. 4: 260–262.

– and Yehna, S. 1993. Fine-root distribution in a Norway spruce stand subjected to drought and ammonium sulphate application. – In: Rasmussen, L., Brydges, T. and Mathy, P. (eds), Experimental manipulations of biota and biogeochemical cycling in ecosystems. CEC, Brussels, Ecosyst. Res. Rep. 4: 262–265.

– , Ahlström, K., Clemensson-Lindell, A. and Majdi, H. 1991. Experimental approaches to the study of the effects of air pollution in trees – root studies. – In: Persson, H. (ed.), Above and below ground interactions in forest trees in acidified soils. Proc. of a COST-workshop May 21–23 1990, Simlångsdalen, Sweden. Air Pollut. Res. Rep. 32: 8–16.

Persson, T., Wirén, A. and Andersson, S. 1990/91. Effects of liming on carbon and nitrogen mineralization in coniferous forests. – Water Air Soil Pollut. 54: 351–364.

Pleijel, H., Skärby, L., Wallin, G. and Selldén, G. 1991. Yield and grain quality of spring wheat, *Triticum Aestivum* (L.), cv. drabant exposed to different concentrations of ozone in open-top chambers. – Environ. Pollut. 69: 151–168.

– , Skärby, L., Ojanperä, K. and Selldén, G. 1992. Yield and quality of spring barley, *Hordeum Vulgare* (L.) exposed to different concentrations of ozone in open-top chambers. – Agric. Ecosyst. and Environ. 38: 21–29.

Posch, M., Falkengren-Grerup, U. and Kauppi, P. 1989. RAINS soil models: Application of two soil acidification models to historical soil chemistry data from Sweden. – In: Kämäri, J.,

Brakke, D. F., Jenkins, A., Norton, S. A. and Wright, R. F. (eds), Regional acidification models. Geographic extent and time development. Springer, Berlin, pp. 241–251.

Redbo-Torstensson, P. 1993. The demographic consequences of nitrogen fertilization of a population of sundew, *Drosera rotundifolia*. – Acta Bot. Neerl. 43: 175–188.

Rühling, Å. and Tyler, G. 1990. Soil factors influencing the distribution of macrofungi in oak forests of southern Sweden. – Holarct. Ecol. 13: 11–18.

– and Tyler, G. 1991. Effects of simulated nitrogen deposition to the forest floor on the macrofungal flora of a beech forest. Ambio 20: 261–263.

Sandén, P. 1992. A simple soil temperature model. – Contribution to the Nordic Hydrological Conference, Alta, Norway, NHP-rapport, No. 30.

– , Gardelin, M. and Espeby, B. 1992. Vertically distributed soil moisture simulations. – Contribution to the Nordic Hydrological Conference, Alta, Norway, NHP-rapport, No. 30.

Sjöström, J. and Qvarfort, U. 1992. Long-term changes of soil chemistry in central Sweden. – Soil Sci. 154: 450–457.

Söderström, B. 1992. The ecological potential of the mycorrhizal mycelium. – In: Read, D., Lewis, D., Fitter, A. and Alexander, I. J. (eds), Mycorrhiza in ecosystems. Proc. 3rd European Symp. on Mycorrhizas, CAB Int., pp. 77–83.

Ström, L., Olsson, T. and Tyler, G. in press. Differences between calcifuge and acidifuge plants in root exudation of low-molecular organic acids. – Plant Soil.

Sutinen, S., Skärby, L., Wallin, G. and Selldén, G. 1990. Long-term exposure of Norway spruce, *Picea abies* (L.) Karst., to ozone in opentop chambers. II. Effects on the ultrastructure of needles. – New Phytol. 115: 345–355.

Sverdrup, H. 1990. The kinetics of chemical weathering. – Lund Univ. Press.

– and Warfvinge, P. 1990. The role of forest growth and weathering in soil acidification. – Water Air Soil Pollut. 52: 71–78.

– and Warfvinge, P. 1992. Modelling regional soil mineralogy and weathering rates. – Proc. 7th int. symp. on water-rock interaction, Utah, 13–19 July, 1992, Balkemaa Publ., pp. 603–606.

– and Warfvinge, P. 1993. Calculation field weathering rates using a mechanistic geochemical model-PROFILE. – J. Appl. Geochem. 8: 1–11.

– and Warfvinge, P. 1994. Soil acidification effect on growth of coniferous trees as expressed by the (Ca+Mg)/Al ratio. – Comm. European Commun. Brussels, Ecosyst. Res. Rep. 13.

– and de Vries, W. 1994. Calculating critical loads for acidity with the simple mass balanced method. – Water Air Soil Pollut. 72: 143–162.

Sverdrup, H., Warfvinge, P. and Rosén, K. 1991. A model for the impact of soil solution Ca:Al ratio soil moisture and temperature on tree base cation uptake. – Water Air Soil Pollut. 61: 365–383.

– , Warfvinge, P., Janicki, A., Morgan, R., Rabenhorst, M. and Bowman, M. 1992. Mapping critical loads and steady state stream chemistry in the state of Maryland. – Environ. Pollut. 77: 195–203.

– , Warfvinge, P., Frogner, T., Håøya, A. O., Johansson, M. and Andersen, B. 1992. Critical loads for forest soils in the Nordic countries. – Ambio 21: 348–355.

– , Warfvinge, P. and Jönsson, C. 1993. Critical loads of acidity for forest soils, groundwater and first order streams in Sweden. – In: Hornung, M., Delve, J. and Skeffington, R. (eds), Critical loads: Concepts and applications. Her Majesty's Stationary Office Publ., pp. 54–67.

– , Warfvinge, P. and Nihlgård, B. 1994. Assessment of soil acidification effects on forest growth in Sweden. – Water Air Soil Pollut. 78: 1–36.

– , Warfvinge, P. and Rosén, K. 1994. Critical loads of acidity and nitrogen for Swedish forest ecosystems, and the relationship to soil weathering. – In: Raitio, H. and Kilponen, T.

(eds), Critical loads and critical limit values. The Finnish Forest Research Inst., Res. Papers 513: 109–138.

Tamm, C. O. 1989. Comparative and experimental approaches to the study of acid deposition effects on soils as substrate for forest growth. – Ambio 18: 184–191.

– 1989. The role of field experiments in the study of soil mediated effects of air pollution on forest trees. – Medd. Norsk Inst. Skogsforskn. Meddelande 42 (1): 179–196.

Taugbøl, G., Seip, H. M., Bishop, K. H. and Grip, H. 1994. Hydrochemical modelling of a stream dominated by organic acids and organically bound aluminium. – Water Air Soil Pollut. 77: 1–36.

Tidblad, J., Leygraf, C. and Kucera, V. 1993. Acid deposition effect on materials: evaluation of nickel after four years of exposure. – J. Electrochem. Soc. 140, 7: 1912–1917.

Townsend, G. S., Bishop, K. H. and Bache, B. W. 1990. Aluminium speciation during episodes. – In: Mason, B. J. (ed.), The surface water acidification programme conference. Cambridge Univ. Press, Cambridge, pp. 275–278.

Tyler, G. 1989. Edaphical distribution and sporophore dynamics of macrofungi in hornbeam *Carpinus betulus* (L.) stands of south Sweden. – Nova Hedwigia 49: 239–253.

– 1989. Edaphical distribution patterns of macrofungal species in deciduous forest of south Sweden. – Acta Oecol./Oecol. Gen. 10: 309–326.

– 1989. The interacting effects of soil acidity and canopy cover on the species composition of field layer vegetation in oak-hornbeam forests. – For. Ecol. Manage. 28: 101–114.

– 1991. Effects of litter treatments on the sporophore production of beech forest macrofungi. – Mycol. Res. 95: 1137–1139.

– 1991. Ecology of the genus Mycena in beech *Fagus sylvatica*, oak *Quercus robur* and hornbeam *Carpinus betulus* forest of S. Sweden. – Nord. J. Bot. 11: 111–121.

– 1992. Tree species affinity of decomposer and ectomycorrhizal macrofungi in beech *Fagus sylvatica* (L.). oak *Quercus robur* (L.) and hornbeam *Carpinus betulus* (L.) forests. – For. Ecol. Manage. 47: 269–284.

– 1992. Inability to solubilize phosphate in limestone soils – key factor controlling calcifuge habit of plants. – Plant and Soil 145: 65–70.

– 1993. Soil solution chemistry controlling the field distribution of *Melica ciliata* (L.). – Ann. Bot. 71: 295–301.

– 1994. A new approach to understanding the calcifuge habit of plants. – Ann. Bot. 71: 295–301.

– 1994. Plant uptake of aluminium from calcareous soils. – Experientia 50: 701–703.

– and Olsson, P. A. 1993. The calcifuge behaviour of *Viscaria vulgaris*. – J. Veg. Sci. 4: 29–36.

– , Balsberg-Påhlsson, A.-M., Bengtsson, G., Bååth, E. and Tranvik, L. 1989. Heavy-metal ecology of terrestrial plants, microorganisms and invertebrates. – Water Air Soil Pollut. 47: 189–215.

– , Balsberg-Påhlsson, A.-M., Bergkvist, B., Falkengren-Grerup, U., Folkeson, L., Nihlgård, B., Rühling, Å. and Stjernquist, I. 1992. Chemical and biological effects of simulated nitrogen deposition to the ground in a Swedish beech forest. – Scand. J. For. Res. 7: 515–532.

Valeur, I. and Nilsson, S. I. 1993. Effects of lime and two incubation techniques on sulphur mineralization in a forest soil. – Soil Biol. Biochem. 25: 1343–1350.

Van Oene, H. 1992. Acid deposition and forest nutrient imbalances: a modelling approach. – Water Air Soil Pollut. 63: 33–50.

– and De Vries, W. 1994. Comparison of measured and simulated changes in base cation amounts using an one-layer soil acidification model. – Water Air Soil Pollut. 72: 41–65.

Vinka, T.-G. 1990. The influence of acidification on the corrosion on structures in contact with soil and water – a compilation of knowledge. – In: Kucera, V. (ed.), Effects of water and soil acidification on corrosion. Nordic Council of Ministers, Copenhagen, Miljørapport 1990:9: 26–76.

– and Sederholm, B. 1990. Corrosion of metals in acid soil. Laboratory exposure. Progress report. – In: Kucera, V. (ed.), Effects of water and soil acidification on corrosion. Nordic Council of Ministers, Copenhagen, Miljørapport 1990:9: 150–169.

Waldén, H. W., Gärdenfors, U. and Wäreborn, I. 1992. The impact of acid rain and heavy metals on the terrestrial mollusc fauna. – Proc. 10th Int. Malacological Congress, Tübingen 2: 425–435.

Wallander, H., Persson, H. and Ahlström, K. 1991. Effects of nitrogen fertilization on fungal biomass in ectomycorrhizal roots and surrounding soils. – In: Persson, H. (ed.), Above and below-ground interaction in forest trees in acidified soils. Proc. of a Cost-workshop, May 21–23 1990, Simlångsdalen, Sweden. Air Pollut. Res. Rep. 32: 99–102.

Wallin, G. and Skärby, L. 1992. The influence of ozone on the stomatal and non-stomatal limitation of photosynthesis in Norway spruce, *Picea abies* (L.) Karst., exposed to soil moisture deficit. – Trees 6: 128–136.

– , Skärby, L. and Selldén, G. 1990. Long-term exposure of Norway spruce, *Picea abies* (L.) Karst., to ozone in opentop chambers. I. Effects on the capacity of net photosynthesis, dark respiration and leaf conductance of shoots of different ages. – New Phytol. 115: 335–344.

– , Skärby, L. and Selldén, G. 1992. Long-term exposure of Norway spruce, *Picea abies* (L.) Karst., to ozone in opentop chambers. III. Effects on the light response of net photosynthesis in shoots of different ages. – New Phytol. 121: 387–394.

– , Ottosson, S. and Selldén, G. 1992. Long-term exposure of Norway spruce, *Picea abies* (L.) Karst., to ozone in open-top chambers. IV. Effects on the stomatal and non-stomatal limitation of photosynthesis and on the carboxylation efficiency. – New Phytol. 121: 395–401.

Wäreborn, I. 1992. Changes in the land mollusc fauna and soil chemistry in an inland district in southern Sweden. – Ecography 15: 62–69.

Warfvinge, P. and Sverdrup, H. 1992. Calculating critical loads of acid deposition with PROFILE – a steady-state soil chemistry model. – Water Air Soil Pollut. 63: 119–143.

– , Falkengren-Grerup, U., Sverdrup, H. and Andersen, B. 1993. Modelling long-term cation supply in acidified forest stands. – Environ. Pollut. 80: 209–221.

Widell, S., Asp, H. and Jensen, P. 1994. Ativities of plasma membrane-bound enzymes isolated from roots of spruce (*Picea abies*) grown in the presence of aluminium. – Physiologia Plantarum 92: 459–466.

Wickman, T. and Jacks, G. 1992. Strontioum isotopes i weathering budgeting. – In: Kharaka, Y. K. and Maest, A. N. (ed.), Water-rock interaction. Balkerna, Rotterdam, pp. 611–614.

– and Jacks, G. 1993. Base cation nutrition for pine stands on lithic soils in the vicinty of Stockholm, Sweden. – Appl. Geochem. Suppl. Issue 2: 199–202.

Wiklander, G., Nordlander, G. and Andersson, R. 1991. Leaching of nitrogen from a forest catchment at Söderåsen in southern Sweden. – Water Air Soil Pollut. 55: 263–282.

Wiman, B. L. B., Unsworth, M. H., Lindberg, S. E., Bergkvist, B., Jaenicke, R. and Hansson, H.-C. 1990. Perspectives on aerosol deposition to natural surfaces: interactions between aerosol residence times, removal processes, the biosphere and global environmental change. – J. Aerosol Sci. 21: 313–338.

Wingsle, G. and Hällgren, J.-E. 1993. Influence of SO_2 and NO_2 exposure on superoxide dismutase and glutathione reductase activities in *Pinus sylvestris* (L.). – J. Exp. Bot. 44: 463–470.

– , Sandberg, G. and Hällgren, J.-E. 1989. Determination of gluthathione in Scots pine needles by high-performance liquid chromatography as its monobromobimane derivative. – J. Chromatogr. 479: 335–344.

– , Mattson, A., Ekblad, A., Hällgren, J.-E. and Selstam, E. 1992. Activities of glutathione reductase and superoxide dismutase in relation to changes of lipids and pigments due to ozone in seedlings of *Pinus sylvestris* (L.). – Plant Sci. 82: 167–178.

– , Strand, M., Karpinski, S. and Hällgren, J.-E. 1992. Influence of SO_2 and NO_2 exposure on antioxidants and photosynthesis in *Pinus sylvestris* (L.). – Phyton 32: 145–151.

Wright, R. F. 1989. RAIN Project: Role of organic acids in moderating pH change following reduction in acid deposition. – Water Air Soil Pollut. 46: 251–259.

– R. F., Cosby, B. J., Flaten, M. B. and Reuss, J. O. 1990. Evaluation of an acidification model with data from manipulated catchments in Norway. – Nature 343: 53–55.

– , Lotse, E. and Semb, A. 1993. RAIN project: results after 8 years of experimentally reduced acid deposition to a whole catchment. – Can. J. Fish. Aquat. Sci. 50: 258–268.

– , Lotse, E. and Semb, A. 1994. Experimental acidification of alpine catchments at Sogndal, Norway: results after 8 years. – Water Air Soil Pollut. 72: 297–315.

Xu, H., Allard, B. and Grimvall, A. 1991. Effects of acidification and natural organic materials on the mobility of arsenic in the environment. – Water Air Soil Pollut. 57/58: 269–278.

– , Grimvall, A. and Allard, B. 1994. Estimation of human exposure to and uptake of arsenic found in drinking water. – In: Nriagu, J. O. (ed.), Arsenic in the Environment. Part II. Human health and ecosystem effects. John Wiley and Sons, New York, pp. 173–183.